Lectures on the Geometry of Manifolds

Second Edition

Lectures on the Geometry of Manifolds

Second Edition

Liviu I Nicolaescu

University of Notre Dame, USA

World Scientific

NEW JERSEY · LONDON · SINGAPORE · BEIJING · SHANGHAI · HONG KONG · TAIPEI · CHENNAI

Published by

World Scientific Publishing Co. Pte. Ltd.

5 Toh Tuck Link, Singapore 596224

USA office: 27 Warren Street, Suite 401-402, Hackensack, NJ 07601

UK office: 57 Shelton Street, Covent Garden, London WC2H 9HE

Library of Congress Cataloging-in-Publication Data
Nicolaescu, Liviu I.
        Lectures on the geometry of manifolds / by L.I. Nicolaescu. -- 2nd ed.
            p. cm.
        Includes bibliographical references and index.
        ISBN-13 978-981-270-853-3 (hardcover : alk. paper)
        ISBN-10 981-270-853-7 (hardcover : alk. paper)
        ISBN-13 978-981-277-862-8 (paperback : alk. paper)
        ISBN-10 981-277-862-4 (paperback : alk. paper)
        1. Geometry, Differential. 2. Manifolds (Mathematics). I. Title.
QA649.N53 2007
516.3'62--dc22

                                                                    2007025469

cover image: courtesy of Vanessa Hill
copyright of the designs of the cover belongs to Vanessa Hill

British Library Cataloguing-in-Publication Data
A catalogue record for this book is available from the British Library.

Printed in Singapore.

To the memory of my friend, Gheorghe 'Prinţu' ' Ionesei

Preface

Shape is a fascinating and intriguing subject which has stimulated the imagination of many people. It suffices to look around to become curious. Euclid did just that and came up with the first pure creation. Relying on the common experience, he created an abstract world that had a life of its own. As the human knowledge progressed so did the ability of formulating and answering penetrating questions. In particular, mathematicians started wondering whether Euclid's "obvious" absolute postulates were indeed obvious and/or absolute. Scientists realized that Shape and Space are two closely related concepts and asked whether they really look the way our senses tell us. As Felix Klein pointed out in his Erlangen Program, there are many ways of looking at Shape and Space so that various points of view may produce different images. In particular, the most basic issue of "measuring the Shape" cannot have a clear cut answer. This is a book about Shape, Space and some particular ways of studying them.

Since its inception, the differential and integral calculus proved to be a very versatile tool in dealing with previously untouchable problems. It did not take long until it found uses in geometry in the hands of the Great Masters. This is the path we want to follow in the present book.

In the early days of geometry nobody worried about the natural context in which the methods of calculus "feel at home". There was no need to address this aspect since for the particular problems studied this was a non-issue. As mathematics progressed as a whole the "natural context" mentioned above crystallized in the minds of mathematicians and it was a notion so important that it had to be given a name. The geometric objects which can be studied using the methods of calculus were called smooth manifolds. Special cases of manifolds are the curves and the surfaces and these were quite well understood. B. Riemann was the first to note that the low dimensional ideas of his time were particular aspects of a higher dimensional world.

The first chapter of this book introduces the reader to the concept of smooth manifold through abstract definitions and, more importantly, through many we believe relevant examples. In particular, we introduce at this early stage the notion of Lie group. The main geometric and algebraic properties of these objects will be gradually described as we progress with our study of the geometry of manifolds. Besides their obvious usefulness in geometry, the Lie groups are academically very friendly. They provide a marvelous testing ground for abstract results. We have consistently taken advantage of this feature throughout this book. As a bonus, by the end of these lectures the reader will feel comfortable manipulating basic Lie theoretic concepts.

To apply the techniques of calculus we need things to derivate and integrate. These "things" are introduced in Chapter 2. The reason why smooth manifolds have many differentiable objects attached to them is that they can be locally very well approximated by linear spaces called tangent spaces . Locally, everything looks like traditional calculus. Each point has a tangent space attached to it so that we obtain a "bunch of tangent spaces" called the tangent bundle. We found it appropriate to introduce at this early point the notion of vector bundle. It helps in structuring both the language and the thinking.

Once we have "things to derivate and integrate" we need to know how to explicitly perform these operations. We devote the Chapter 3 to this purpose. This is perhaps one of the most unattractive aspects of differential geometry but is crucial for all further developments. To spice up the presentation, we have included many examples which will found applications in later chapters. In particular, we have included a whole section devoted to the representation theory of compact Lie groups essentially describing the equivalence between representations and their characters.

The study of Shape begins in earnest in Chapter 4 which deals with Riemann manifolds. We approach these objects gradually. The first section introduces the reader to the notion of geodesics which are defined using the Levi-Civita connection. Locally, the geodesics play the same role as the straight lines in an Euclidian space but globally new phenomena arise. We illustrate these aspects with many concrete examples. In the final part of this section we show how the Euclidian vector calculus generalizes to Riemann manifolds.

The second section of this chapter initiates the local study of Riemann manifolds. Up to first order these manifolds look like Euclidean spaces. The novelty arises when we study "second order approximations " of these spaces. The Riemann tensor provides the complete measure of how far is a Riemann manifold from being flat. This is a very involved object and, to enhance its understanding, we compute it in several instances: on surfaces (which can be easily visualized) and on Lie groups (which can be easily formalized). We have also included Cartan's moving frame technique which is extremely useful in concrete computations. As an application of this technique we prove the celebrated Theorema Egregium of Gauss. This section concludes with the first global result of the book, namely the Gauss-Bonnet theorem. We present a proof inspired from [25] relying on the fact that all Riemann surfaces are Einstein manifolds. The Gauss-Bonnet theorem will be a recurring theme in this book and we will provide several other proofs and generalizations.

One of the most fascinating aspects of Riemann geometry is the intimate correlation "local-global". The Riemann tensor is a local object with global effects. There are currently many techniques of capturing this correlation. We have already described one in the proof of Gauss-Bonnet theorem. In Chapter 5 we describe another such technique which relies on the study of the global behavior of geodesics. We felt we had the moral obligation to present the natural setting of this technique and we briefly introduce the reader to the wonderful world of the calculus of variations. The ideas of the calculus of variations produce remarkable results when applied to Riemann manifolds. For example, we explain in rigorous terms why "very curved manifolds" cannot be "too long" .

In Chapter 6 we leave for a while the "differentiable realm" and we briefly discuss the fundamental group and covering spaces. These notions shed a new light on the results of Chapter 5. As a simple application we prove Weyl's theorem that the semisimple Lie groups with definite Killing form are compact and have finite fundamental group.

Chapter 7 is the topological core of the book. We discuss in detail the cohomology of smooth manifolds relying entirely on the methods of calculus. In writing this chapter we could not, and would not escape the influence of the beautiful monograph [17], and this explains the frequent overlaps. In the first section we introduce the DeRham cohomology and the Mayer-Vietoris technique. Section 2 is devoted to the Poincaré duality, a feature which sets the manifolds apart from many other types of topological spaces. The third section offers a glimpse at homology theory. We introduce the notion of (smooth) cycle and then present some applications: intersection theory, degree theory, Thom isomorphism and we prove a higher dimensional version of the Gauss-Bonnet theorem at the cohomological level. The fourth section analyzes the role of symmetry in restricting the topological type of a manifold. We prove Élie Cartan's old result that the cohomology of a symmetric space is given by the linear space of its bi-invariant forms. We use this technique to compute the lower degree cohomology of compact semisimple Lie groups. We conclude this section by computing the cohomology of complex Grassmannians relying on Weyl's integration formula and Schur polynomials. The chapter ends with a fifth section containing a concentrated description of Čech cohomology.

Chapter 8 is a natural extension of the previous one. We describe the Chern-Weil construction for arbitrary principal bundles and then we concretely describe the most important examples: Chern classes, Pontryagin classes and the Euler class. In the process, we compute the ring of invariant polynomials of many classical groups. Usually, the connections in principal bundles are defined in a global manner, as horizontal distributions. This approach is geometrically very intuitive but, at a first contact, it may look a bit unfriendly in concrete computations. We chose a local approach build on the reader's experience with connections on vector bundles which we hope will attenuate the formalism shock. In proving the various identities involving characteristic classes we adopt an invariant theoretic point of view. The chapter concludes with the general Gauss-Bonnet-Chern theorem. Our proof is a variation of Chern's proof.

Chapter 9 is the analytical core of the book.[1] Many objects in differential geometry are defined by differential equations and, among these, the elliptic ones play an important role. This chapter represents a minimal introduction to this subject. After presenting some basic notions concerning arbitrary partial differential operators we introduce the Sobolev spaces and describe their main functional analytic features. We then go straight to the core of elliptic theory. We provide an almost complete proof of the elliptic a priori estimates (we left out only the proof of the Calderon-Zygmund inequality). The regularity results are then deduced from the a priori estimates via a simple approximation technique. As a first application of these results we consider a Kazhdan-Warner type equation which recently found applications in solving the Seiberg-Witten equations on a Kähler manifold. We adopt a variational approach. The uniformization theorem for compact Riemann surfaces is then a nice bonus. This may not be the most direct proof but it has an academic advantage. It builds a circle of ideas with a wide range of applications. The last section of this chapter is devoted to Fredholm theory. We prove that the elliptic operators on compact manifolds are Fredholm and establish the homotopy invariance of the index. These are very general Hodge type theorems. The classical one follows immediately from these results. We conclude with a few facts about the spectral properties of elliptic operators.

The last chapter is entirely devoted to a very important class of elliptic operators namely the Dirac operators. The important role played by these operators was singled out in the works of Atiyah and Singer and, since then, they continue to be involved in the most dramatic advances of modern geometry. We begin by first describing a general notion of Dirac operators and their natural geometric environment, much like in [11]. We then isolate a special subclass we called *geometric Dirac operators*. Associated to each such operator is a very concrete Weitzenböck formula which can be viewed as a bridge between geometry and analysis, and which is often the source of many interesting applications. The abstract considerations are backed by a full section describing many important concrete examples.

In writing this book we had in mind the beginning graduate student who wants to specialize in global geometric analysis in general and gauge theory in particular. The second half of the book is an extended version of a graduate course in differential geometry we taught at the University of Michigan during the winter semester of 1996.

The minimal background needed to successfully go through this book is a good knowledge of vector calculus and real analysis, some basic elements of point set topology and linear algebra. A familiarity with some basic facts about the differential geometry of curves of surfaces would ease the understanding of the general theory, but this is not a must. Some parts of the chapter on elliptic equations may require a more advanced background in functional analysis.

---

[1] In the new edition, this chapter has become Chapter 10.

The theory is complemented by a large list of exercises. Quite a few of them contain technical results we did not prove so we would not obscure the main arguments. There are however many non-technical results which contain additional information about the subjects discussed in a particular section. We left hints whenever we believed the solution is not straightforward.

Personal note It has been a great personal experience writing this book, and I sincerely hope I could convey some of the magic of the subject. Having access to the remarkable science library of the University of Michigan and its computer facilities certainly made my job a lot easier and improved the quality of the final product.

I learned differential equations from Professor Viorel Barbu, a very generous and enthusiastic person who guided my first steps in this field of research. He stimulated my curiosity by his remarkable ability of unveiling the hidden beauty of this highly technical subject. My thesis advisor, Professor Tom Parker, introduced me to more than the fundamentals of modern geometry. He played a key role in shaping the manner in which I regard mathematics. In particular, he convinced me that behind each formalism there must be a picture, and uncovering it, is a very important part of the creation process. Although I did not directly acknowledge it, their influence is present throughout this book. I only hope the filter of my mind captured the full richness of the ideas they so generously shared with me.

My friends Louis Funar and *Gheorghe Ionesei*[2] read parts of the manuscript. I am grateful to them for their effort, their suggestions and for their friendship. I want to thank Arthur Greenspoon for his advice, enthusiasm and relentless curiosity which boosted my spirits when I most needed it. Also, I appreciate very much the input I received from the graduate students of my "Special topics in differential geometry" course at the University of Michigan which had a beneficial impact on the style and content of this book.

At last, but not the least, I want to thank my family who supported me from the beginning to the completion of this project.

Ann Arbor, 1996.

Preface to the second edition

Rarely in life is a man given the chance to revisit his "youthful indiscretions". With this second edition I have been given this opportunity, and I have tried to make the best of it.

The first edition was generously sprinkled with many typos, which I can only attribute to the impatience of youth. In spite of this problem, I have received very

---

[2] He passed away while I was preparing this new edition. He was the ultimate poet of mathematics.

good feedback from a very indulgent and helpful audience, from all over the world.

In preparing the new edition, I have been engaged on a massive typo hunting, supported by the wisdom of time, and the useful comments that I have received over the years from many readers. I can only say that the number of typos is substantially reduced. However, experience tells me that Murphy's Law is still at work, and there are still typos out there which will become obvious only in the printed version.

The passage of time has only strengthened my conviction that, in the words of Isaac Newton, "in learning the sciences examples are of more use than precepts". The new edition continues to be guided by this principle. I have not changed the old examples, but I have polished many of my old arguments, and I have added quite a large number of new examples and exercises.

The only major addition to the contents is a new chapter (Chapter 9) on classical integral geometry. This is a subject that captured my imagination over the last few years, and since the first edition developed all the tools needed to understand some of the juiciest results in this area of geometry, I could not pass the chance to share with a curious reader my excitement about this line of thought.

One novel feature in our presentation of the classical results of integral geometry is the use of tame geometry. This is a recent extension of the better know area of real algebraic geometry which allowed us to avoid many heavy analytical arguments, and present the geometric ideas in as clear a light as possible.

Notre Dame, 2007.

Contents

Chapter 1

Manifolds

1.1 Preliminaries

1.1.1 *Space and Coordinatization*

Mathematics is a natural science with a special modus operandi. It replaces concrete natural objects with mental abstractions which serve as intermediaries. One studies the properties of these abstractions in the hope they reflect facts of life. So far, this approach proved to be very productive.

The most visible natural object is the Space, the place where all things happen. The first and most important mathematical abstraction is the notion of number. Loosely speaking, the aim of this book is to illustrate how these two concepts, Space and Number, fit together.

It is safe to say that geometry as a rigorous science is a creation of ancient Greeks. Euclid proposed a method of research that was later adopted by the entire mathematics. We refer of course to the axiomatic method. He viewed the Space as a collection of points, and he distinguished some basic objects in the space such as lines, planes etc. He then postulated certain (natural) relations between them. All the other properties were derived from these simple axioms.

Euclid's work is a masterpiece of mathematics, and it has produced many interesting results, but it has its own limitations. For example, the most complicated shapes one could reasonably study using this method are the conics and/or quadrics, and the Greeks certainly did this. A major breakthrough in geometry was the discovery of *coordinates* by René Descartes in the 17th century. Numbers were put to work in the study of Space.

Descartes' idea of producing what is now commonly referred to as Cartesian coordinates is familiar to any undergraduate. These coordinates are obtained using a very special method (in this case using three concurrent, pairwise perpendicular lines, each one endowed with an orientation and a unit length standard. What is important here is that they produced a one-to-one mapping

$$\text{Euclidian Space} \rightarrow \mathbb{R}^3, \quad P \longmapsto (x(P), y(P), z(P)).$$

We call such a process *coordinatization*. The corresponding map is called (in this case) *Cartesian system of coordinates.* A line or a plane becomes via coordinatization an algebraic object, more precisely, an equation.

In general, any coordinatization replaces geometry by algebra and we get a two-way correspondence

Study of Space ⟷ Study of Equations.

The shift from geometry to numbers is beneficial to geometry as long as one has efficient tools do deal with numbers and equations. Fortunately, about the same time with the introduction of coordinates, Isaac Newton created the differential and integral calculus and opened new horizons in the study of equations.

The Cartesian system of coordinates is by no means the unique, or the most useful coordinatization. Concrete problems dictate other choices. For example, the polar coordinates represent another coordinatization of (a piece of the plane) (see Figure 1.1).

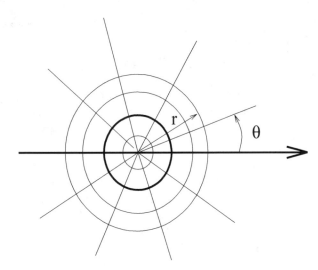

Fig. 1.1   *Polar coordinates*

$$P \mapsto (r(P), \theta(P)) \in (0, \infty) \times (-\pi, \pi).$$

This choice is related to the Cartesian choice by the well known formulae

$$x = r \cos \theta \quad y = r \sin \theta. \tag{1.1.1}$$

A remarkable feature of (1.1.1) is that $x(P)$ and $y(P)$ depend smoothly upon $r(P)$ and $\theta(P)$.

As science progressed, so did the notion of Space. One can think of Space as a *configuration set*, i.e., the collection of all possible states of a certain phenomenon. For example, we know from the principles of Newtonian mechanics that the motion of a particle in the ambient space can be completely described if we know the position and the velocity of the particle at a given moment. The space associated with this problem consists of all pairs *(position, velocity)* a particle can possibly have. We can coordinatize this space using six functions: three of them will describe the position, and the other three of them will describe the velocity. We say the configuration space is 6-dimensional. We cannot visualize this space, but it helps to think of it as an Euclidian space, only "roomier".

There are many ways to coordinatize the configuration space of a motion of a particle, and for each choice of coordinates we get a different description of the motion. Clearly, all these descriptions must "agree" in some sense, since they all reflect the same phenomenon. In other words, these descriptions should be *independent of coordinates*. Differential geometry studies the objects which are independent of coordinates.

The coordinatization process had been used by people centuries before mathematicians accepted it as a method. For example, sailors used it to travel from one point to another on Earth. Each point has a latitude and a longitude that completely determines its position on Earth. This coordinatization is not a global one. There exist four domains delimited by the Equator and the Greenwich meridian, and each of them is then naturally coordinatized. Note that the points on the Equator or the Greenwich meridian admit two different coordinatizations which are smoothly related.

The manifolds are precisely those spaces which can be piecewise coordinatized, with smooth correspondence on overlaps, and the intention of this book is to introduce the reader to the problems and the methods which arise in the study of manifolds. The next section is a technical interlude. We will review the implicit function theorem which will be one of the basic tools for detecting manifolds.

1.1.2 *The implicit function theorem*

We gather here, with only sketchy proofs, a collection of classical analytical facts. For more details one can consult [26].

Let X and Y be two Banach spaces and denote by $L(X,Y)$ the space of bounded linear operators $X \to Y$. For example, if $X = \mathbb{R}^n$, $Y = \mathbb{R}^m$, then $L(X,Y)$ can be identified with the space of $m \times n$ matrices with real entries.

Definition 1.1.1. Let $F : U \subset X \to Y$ be a continuous function (U is an open subset of X). The map F is said to be (Fréchet) differentiable at $u \in U$ if there exists $T \in L(X,Y)$ such that

$$\|F(u_0 + h) - F(u_0) - Th\|_Y = o(\|h\|_X) \ \text{ as } h \to 0. \qquad \square$$

Loosely speaking, a continuous function is differentiable at a point if, near that point, it admits a " best approximation " by a linear map.

When F is differentiable at $u_0 \in U$, the operator T in the above definition is uniquely determined by

$$Th = \frac{d}{dt}\big|_{t=0} F(u_0 + th) = \lim_{t \to 0} \frac{1}{t}\left(F(u_0 + th) - F(u_0)\right).$$

We will use the notation $T = D_{u_0}F$ and we will call T the *Fréchet derivative* of F at u_0.

Assume that the map $F : U \to Y$ is differentiable at each point $u \in U$. Then F is said to be of class C^1, if the map $u \mapsto D_u F \in L(X, Y)$ is continuous. F is said to be of class C^2 if $u \mapsto D_u F$ is of class C^1. One can define inductively C^k and C^∞ (or *smooth*) maps.

Example 1.1.2. Consider $F : U \subset \mathbb{R}^n \to \mathbb{R}^m$. Using Cartesian coordinates $x = (x^1, \ldots, x^n)$ in $\mathbb{R}^n$ and $u = (u^1, \ldots, u^m)$ in $\mathbb{R}^m$ we can think of F as a collection of m functions on U

$$u^1 = u^1(x^1, \ldots, x^n), \ldots, u^m = u^m(x^1, \ldots, x^n).$$

The map F is differentiable at a point $p = (p^1, \ldots, p^n) \in U$ if and only if the functions u^i are differentiable at p in the usual sense of calculus. The Fréchet derivative of F at p is the linear operator $D_p F : \mathbb{R}^n \to \mathbb{R}^m$ given by the *Jacobian matrix*

$$D_p F = \frac{\partial(u^1, \ldots, u^m)}{\partial(x^1, \ldots, x^n)} = \begin{bmatrix} \frac{\partial u^1}{\partial x^1}(p) & \frac{\partial u^1}{\partial x^2}(p) & \cdots & \frac{\partial u^1}{\partial x^n}(p) \\ \frac{\partial u^2}{\partial x^1}(p) & \frac{\partial u^2}{\partial x^2}(p) & \cdots & \frac{\partial u^2}{\partial x^n}(p) \\ \vdots & \vdots & \vdots & \vdots \\ \frac{\partial u^m}{\partial x^1}(p) & \frac{\partial u^m}{\partial x^2}(p) & \cdots & \frac{\partial u^m}{\partial x^n}(p) \end{bmatrix}.$$

The map F is smooth if and only if the functions $u^i(x)$ are smooth.   □

Exercise 1.1.3. (a) Let $\mathcal{U} \subset L(\mathbb{R}^n, \mathbb{R}^n)$ denote the set of invertible $n \times n$ matrices. Show that $\mathcal{U}$ is an open set.

(b) Let $F : \mathcal{U} \to \mathcal{U}$ be defined as $A \to A^{-1}$. Show that $D_A F(H) = -A^{-1}HA^{-1}$ for any $n \times n$ matrix H.

(c) Show that the Fréchet derivative of the map $\det : L(\mathbb{R}^n, \mathbb{R}^n) \to \mathbb{R}$, $A \mapsto \det A$, at $A = \mathbb{1}_{\mathbb{R}^n} \in L(\mathbb{R}^n, \mathbb{R}^n)$ is given by $\operatorname{tr} H$, i.e.,

$$\frac{d}{dt}\big|_{t=0} \det(\mathbb{1}_{\mathbb{R}^n} + tH) = \operatorname{tr} H, \quad \forall H \in L(\mathbb{R}^n, \mathbb{R}^n).$$ □

Theorem 1.1.4 (Inverse function theorem). *Let X, Y be two Banach spaces, and $F : U \subset X \to Y$ a smooth function. If at a point $u_0 \in U$ the derivative $D_{u_0}F \in L(X, Y)$ is invertible, then there exits a neighborhood U_1 of u_0 in U such that $F(U_1)$ is an open neighborhood of $v_0 = F(u_0)$ in Y and $F : U_1 \to F(U_1)$ is bijective, with smooth inverse.*   □

The spirit of the theorem is very clear: the invertibility of the derivative $D_{u_0}F$ "propagates" locally to F because $D_{u_0}F$ is a very good local approximation for F.

More formally, if we set $T = D_{u_0}F$, then

$$F(u_0 + h) = F(u_0) + Th + r(h),$$

where $r(h) = o(\|h\|)$ as $h \to 0$. The theorem states that, for every v sufficiently close to v_0, the equation $F(u) = v$ has a unique solution $u = u_0 + h$, with h very small. To prove the theorem one has to show that, for $\|v - v_0\|_Y$ sufficiently small, the equation below

$$v_0 + Th + r(h) = v$$

has a unique solution. We can rewrite the above equation as $Th = v - v_0 - r(h)$ or, equivalently, as $h = T^{-1}(v - v_0 - r(h))$. This last equation is a fixed point problem that can be approached successfully via the Banach fixed point theorem.

Theorem 1.1.5 (Implicit function theorem). *Let X, Y, Z be Banach spaces, and $F : X \times Y \to Z$ a smooth map. Let $(x_0, y_0) \in X \times Y$, and set $z_0 = F(x_0, y_0)$. Set $F_2 : Y \to Z$, $F_2(y) = F(x_0, y)$. Assume that $D_{y_0}F_2 \in L(Y, Z)$ is invertible. Then there exist neighborhoods U of $x_0 \in X$, V of $y_0 \in Y$, and a smooth map $G : U \to V$ such that the set S of solution (x, y) of the equation $F(x, y) = z_0$ which lie inside $U \times V$ can be identified with the graph of G, i.e.,*

$$\left\{ (x, y) \in U \times V \; ; \; F(x, y) = z_0 \right\} = \left\{ (x, G(x)) \in U \times V \; ; \; x \in U \right\}.$$

In pre-Bourbaki times, the classics regarded the coordinate y as a function of x defined implicitly by the equality $F(x, y) = z_0$.

Proof. Consider the map

$$H : X \times Y \to X \times Z, \;\; \xi = (x, y) \mapsto (x, F(x, y)).$$

The map H is a smooth map, and at $\xi_0 = (x_0, y_0)$ its derivative $D_{\xi_0}H : X \times Y \to X \times Z$ has the block decomposition

$$D_{\xi_0}H = \begin{bmatrix} \mathbb{1}_X & 0 \\ D_{\xi_0}F_1 & D_{\xi_0}F_2 \end{bmatrix}.$$

Above, DF_1 (respectively DF_2) denotes the derivative of $x \mapsto F(x, y_0)$ (respectively the derivative of $y \mapsto F(x_0, y)$). The linear operator $D_{\xi_0}H$ is invertible, and its inverse has the block decomposition

$$(D_{\xi_0}H)^{-1} = \begin{bmatrix} \mathbb{1}_X & 0 \\ -(D_{\xi_0}F_2)^{-1} \circ (D_{\xi_0}F_1) & (D_{\xi_0}F_2)^{-1} \end{bmatrix}.$$

Thus, by the inverse function theorem, the equation $(x, F(x, y)) = (x, z_0)$ has a unique solution $(\tilde{x}, \tilde{y}) = H^{-1}(x, z_0)$ in a neighborhood of (x_0, y_0). It obviously satisfies $\tilde{x} = x$ and $F(\tilde{x}, \tilde{y}) = z_0$. Hence, the set $\{(x, y) \; ; \; F(x, y) = z_0\}$ is locally the graph of $x \mapsto H^{-1}(x, z_0)$. $\qquad\square$

1.2   Smooth manifolds

1.2.1   *Basic definitions*

We now introduce the object which will be the main focus of this book, namely, the concept of (smooth) manifold. It formalizes the general principles outlined in Subsection 1.1.1.

Definition 1.2.1. A *smooth manifold* of dimension m is a locally compact, paracompact Hausdorff space M together with the following collection of data (henceforth called *atlas* or *smooth structure*) consisting of the following.

(a) An open cover $\{U_i\}_{i \in I}$ of M;

(b) A collection of continuous, injective maps $\{ \Psi_i : U_i \to \mathbb{R}^m; \quad i \in I \}$ (called *charts* or *local coordinates*) such that, $\Psi_i(U_i)$ is open in $\mathbb{R}^m$, and if $U_i \cap U_j \neq \emptyset$, then the transition map

$$\Psi_j \circ \Psi_i^{-1} : \Psi_i(U_i \cap U_j) \subset \mathbb{R}^m \to \Psi_j(U_i \cap U_j) \subset \mathbb{R}^m$$

is smooth. (We say the various charts are *smoothly compatible*; see Figure 1.2).□

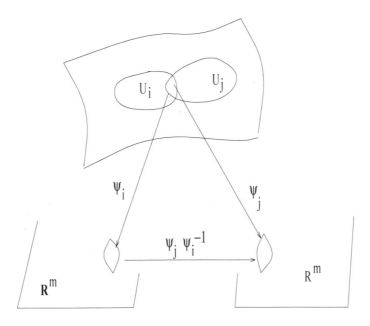

Fig. 1.2   *Transition maps*

Each chart Ψ_i can be viewed as a collection of m functions $(x^1, \ldots, x^m)$ on U_i. Similarly, we can view another chart Ψ_j as another collection of functions $(y^1, \ldots, y^m)$. The transition map $\Psi_j \circ \Psi_i^{-1}$ can then be interpreted as a collection of maps

$$(x^1, \ldots, x^m) \mapsto \big(y^1(x^1, \ldots, x^m), \ldots, y^m(x^1, \ldots, x^m) \big).$$

The first and the most important example of manifold is $\mathbb{R}^n$ itself. The natural smooth structure consists of an atlas with a single chart, $\mathbb{1}_{\mathbb{R}^n} : \mathbb{R}^n \to \mathbb{R}^n$. To construct more examples we will use the implicit function theorem .

Definition 1.2.2. (a) Let M, N be two smooth manifolds of dimensions m and respectively n. A continuous map $f : M \to N$ is said to be *smooth* if, for any local charts ϕ on M and ψ on N, the composition $\psi \circ f \circ \phi^{-1}$ (whenever this makes sense) is a smooth map $\mathbb{R}^m \to \mathbb{R}^n$.
(b) A smooth map $f : M \to N$ is called a *diffeomorphism* if it is invertible and its inverse is also a smooth map. □

Example 1.2.3. The map $t \mapsto e^t$ is a diffeomorphism $(-\infty, \infty) \to (0, \infty)$. The map $t \mapsto t^3$ is a homeomorphism $\mathbb{R} \to \mathbb{R}$ but it is not a diffeomorphism! □

If M is a smooth manifold we will denote by $C^\infty(M)$ the linear space of all smooth functions $M \to \mathbb{R}$.

Remark 1.2.4. Let U be an open subset of the smooth manifold M (dim $M = m$) and $\psi : U \to \mathbb{R}^m$ a smooth, one-to one map with open image and smooth inverse. Then ψ defines local coordinates over U compatible with the existing atlas of M. Thus (U, ψ) can be added to the original atlas and the new smooth structure is diffeomorphic with the initial one. Using Zermelo's Axiom we can produce a *maximal atlas* (no more compatible local chart can be added to it). □

Our next result is a general recipe for producing manifolds. Historically, this is how manifolds entered mathematics.

Proposition 1.2.5. *Let M be a smooth manifold of dimension m and $f_1, \ldots, f_k \in C^\infty(M)$. Define*

$$\mathcal{Z} = \mathcal{Z}(f_1, \ldots, f_k) = \{ p \in M ; f_1(p) = \cdots = f_k(p) = 0 \}.$$

Assume that the functions $f_1, \ldots, f_k$ are functionally independent along $\mathcal{Z}$, i.e., for each $p \in \mathcal{Z}$, there exist local coordinates $(x^1, \ldots, x^m)$ defined in a neighborhood of p in M such that $x^i(p) = 0$, $i = 1, \ldots, m$, and the matrix

$$\frac{\partial \vec{f}}{\partial \vec{x}}|_p := \begin{bmatrix} \frac{\partial f_1}{\partial x^1} & \frac{\partial f_1}{\partial x^2} & \cdots & \frac{\partial f_1}{\partial x^m} \\ \vdots & \vdots & \vdots & \vdots \\ \frac{\partial f_k}{\partial x^1} & \frac{\partial f_k}{\partial x^2} & \cdots & \frac{\partial f_k}{\partial x^m} \end{bmatrix}_{x^1 = \cdots = x^m = 0}$$

has rank k. Then $\mathcal{Z}$ has a natural structure of smooth manifold of dimension $m - k$.

Proof. **Step 1: Constructing the charts.** Let $p_0 \in \mathcal{Z}$, and denote by $(x^1, \ldots, x^m)$ local coordinates near p_0 such that $x^i(p_0) = 0$. One of the $k \times k$ minors of the matrix

$$\frac{\partial \vec{f}}{\partial \vec{x}}|_p := \begin{bmatrix} \frac{\partial f_1}{\partial x^1} & \frac{\partial f_1}{\partial x^2} & \cdots & \frac{\partial f_1}{\partial x^m} \\ \vdots & \vdots & \vdots & \vdots \\ \frac{\partial f_k}{\partial x^1} & \frac{\partial f_k}{\partial x^2} & \cdots & \frac{\partial f_k}{\partial x^m} \end{bmatrix}_{x^1 = \cdots = x^m = 0}$$

is nonzero. Assume this minor is determined by the last k columns (and all the k lines).

We can think of the functions $f_1, \ldots, f_k$ as defined on an open subset U of $\mathbb{R}^m$. Split $\mathbb{R}^m$ as $\mathbb{R}^{m-k} \times \mathbb{R}^k$, and set

$$x' := (x^1, \ldots, x^{m-k}), \quad x'' := (x^{m-k+1}, \ldots, x^m).$$

We are now in the setting of the implicit function theorem with

$$X = \mathbb{R}^{m-k}, \quad Y = \mathbb{R}^k, \quad Z = \mathbb{R}^k,$$

and $F : X \times Y \to Z$ given by

$$x \mapsto \begin{bmatrix} f_1(x) \\ \vdots \\ f_k(x) \end{bmatrix} \in \mathbb{R}^k.$$

In this case, $DF_2 = \left(\frac{\partial F}{\partial x''} \right) : \mathbb{R}^k \to \mathbb{R}^k$ is invertible since its determinant corresponds to our nonzero minor. Thus, in a product neighborhood $U_{p_0} = U'_{p_0} \times U''_{p_0}$ of p_0, the set $\mathcal{Z}$ is the graph of some function

$$g : U'_{p_0} \subset \mathbb{R}^{m-k} \longrightarrow U''_{p_0} \subset \mathbb{R}^k,$$

i.e.,

$$\mathcal{Z} \cap U_{p_0} = \left\{ (x', g(x')) \in \mathbb{R}^{m-k} \times \mathbb{R}^k; \ x' \in U'_{p_0}, \ |x'| \text{ small} \right\}.$$

We now define $\psi_{p_0} : \mathcal{Z} \cap U_{p_0} \to \mathbb{R}^{m-k}$ by

$$(x', g(x')) \overset{\psi_{p_0}}{\longmapsto} x' \in \mathbb{R}^{m-k}.$$

The map ψ_{p_0} is a local chart of $\mathcal{Z}$ near p_0.

Step 2. The transition maps for the charts constructed above are smooth. The details are left to the reader. $\square$

Exercise 1.2.6. Complete Step 2 in the proof of Proposition1.2.5. $\square$

Definition 1.2.7. Let M be a m-dimensional manifold. A *codimension k submanifold* of M is a subset $N \subset M$ locally defined as the common zero locus of k functionally independent functions $f_1, \ldots, f_k \in C^\infty(M)$. $\square$

Proposition1.2.5 shows that any submanifold $N \subset M$ has a natural smooth structure so it becomes a manifold *per se*. Moreover, the inclusion map $i : N \hookrightarrow M$ is smooth.

1.2.2 *Partitions of unity*

This is a very brief technical subsection describing a trick we will extensively use in this book.

Definition 1.2.8. Let M be a smooth manifold and $(U_\alpha)_{\alpha \in \mathcal{A}}$ an open cover of M. A (smooth) *partition of unity* subordinated to this cover is a family $(f_\beta)_{\beta \in \mathcal{B}} \subset C^\infty(M)$ satisfying the following conditions.

(i) $0 \leq f_\beta \leq 1$.
(ii) $\exists \phi : \mathcal{B} \to \mathcal{A}$ such that $\operatorname{supp} f_\beta \subset U_{\phi(\beta)}$.
(iii) The family $(\operatorname{supp} f_\beta)$ is locally finite, i.e., any point $x \in M$ admits an open neighborhood intersecting only finitely many of the supports $\operatorname{supp} f_\beta$.
(iv) $\sum_\beta f_\beta(x) = 1$ for all $x \in M$. □

We include here for the reader's convenience the basic existence result concerning partitions of unity. For a proof we refer to [95].

Proposition 1.2.9. *(a) For any open cover $\mathcal{U} = (U_\alpha)_{\alpha \in \mathcal{A}}$ of a smooth manifold M there exists at least one smooth partition of unity $(f_\beta)_{\beta \in \mathcal{B}}$ subordinated to $\mathcal{U}$ such that $\operatorname{supp} f_\beta$ is compact for any β.*
(b) If we do not require compact supports, then we can find a partition of unity in which $\mathcal{B} = \mathcal{A}$ and $\phi = \mathbb{1}_\mathcal{A}$. □

Exercise 1.2.10. Let M be a smooth manifold and $S \subset M$ a *closed* submanifold. Prove that the restriction map

$$r : C^\infty(M) \to C^\infty(S) \quad f \mapsto f|_S$$

is surjective. □

1.2.3 *Examples*

Manifolds are everywhere, and in fact, to many physical phenomena which can be modelled mathematically one can naturally associate a manifold. On the other hand, many problems in mathematics find their most natural presentation using the language of manifolds. To give the reader an idea of the scope and extent of modern geometry, we present here a short list of examples of manifolds. This list will be enlarged as we enter deeper into the study of manifolds.

Example 1.2.11. (The n-dimensional sphere). This is the codimension 1 submanifold of $\mathbb{R}^{n+1}$ given by the equation

$$|x|^2 = \sum_{i=0}^{n} (x^i)^2 = r^2, \quad x = (x^0, \ldots, x^n) \in \mathbb{R}^{n+1}.$$

One checks that, along the sphere, the differential of $|x|^2$ is nowhere zero, so by Proposition 1.2.5, S^n is indeed a smooth manifold. In this case one can explicitly construct an atlas (consisting of two charts) which is useful in many applications. The construction relies on *stereographic projections*.

Let N and S denote the North and resp. South pole of S^n ($N = (0, \ldots, 0, 1) \in \mathbb{R}^{n+1}$, $S = (0, \ldots, 0, -1) \in \mathbb{R}^{n+1}$). Consider the open sets $U_N = S^n \setminus \{N\}$ and $U_S = S^n \setminus \{S\}$. They form an open cover of S^n. The stereographic projection from the North pole is the map $\sigma_N : U_N \to \mathbb{R}^n$ such that, for any $P \in U_N$, the point $\sigma_N(P)$ is the intersection of the line NP with the hyperplane $\{x^n = 0\} \cong \mathbb{R}^n$.

The stereographic projection from the South pole is defined similarly. For $P \in U_N$ we denote by $(y^1(P), \ldots, y^n(P))$ the coordinates of $\sigma_N(P)$, and for $Q \in U_S$, we denote by $(z^1(Q), \ldots, z^n(Q))$ the coordinates of $\sigma_S(Q)$. A simple argument shows the map

$$\big(y^1(P), \ldots, y^n(P) \big) \mapsto \big(z^1(P), \ldots, z^n(P) \big), \quad P \in U_N \cap U_S,$$

is smooth (see the exercise below). Hence $\{(U_N, \sigma_N), (U_S, \sigma_S)\}$ defines a smooth structure on S^n. $\qquad\square$

Exercise 1.2.12. Show that the functions y^i, z^j constructed in the above example satisfy

$$z^i = \frac{y^i}{\left(\sum_{j=1}^n (y^j)^2 \right)}, \quad \forall i = 1, \ldots, n. \qquad\qquad\square$$

Example 1.2.13. (The n-dimensional torus). This is the codimension n submanifold of $\mathbb{R}^{2n}(x_1, y_1; \ldots; x_n, y_n)$ defined as the zero locus

$$x_1^2 + y_1^2 = \cdots = x_n^2 + y_n^2 = 1.$$

Note that T^1 is diffeomorphic with the 1-dimensional sphere S^1 (unit circle). As a set T^n is a direct product of n circles $T^n = S^1 \times \cdots \times S^1$ (see Figure 1.3). $\qquad\square$

Fig. 1.3   *The 2-dimensional torus*

The above example suggests the following general construction.

Example 1.2.14. Let M and N be smooth manifolds of dimension m and respectively n. Then their topological direct product has a natural structure of smooth manifold of dimension $m + n$. $\qquad\square$

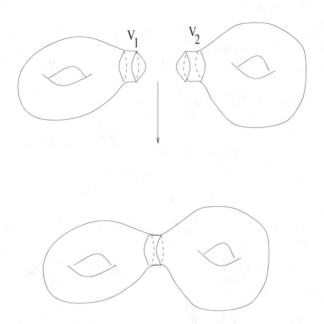

Fig. 1.4 *Connected sum of tori*

Example 1.2.15. (The connected sum of two manifolds). Let M_1 and M_2 be two manifolds of the same dimension m. Pick $p_i \in M_i$ $(i = 1, 2)$, choose small open neighborhoods U_i of p_i in M_i and then local charts ψ_i identifying each of these neighborhoods with $B_2(0)$, the ball of radius 2 in $\mathbb{R}^m$.

Let $V_i \subset U_i$ correspond (via ψ_i) to the annulus $\{1/2 < |x| < 2\} \subset \mathbb{R}^m$. Consider

$$\phi : \left\{\, 1/2 < |x| < 2 \,\right\} \to \left\{\, 1/2 < |x| < 2 \,\right\}, \quad \phi(x) = \frac{x}{|x|^2}.$$

The action of ϕ is clear: it switches the two boundary components of $\{1/2 < |x| < 2\}$, and reverses the orientation of the radial directions.

Now "glue" V_1 to V_2 using the "prescription" given by $\psi_2^{-1} \circ \phi \circ \psi_1 : V_1 \to V_2$. In this way we obtain a new topological space with a natural smooth structure induced by the smooth structures on M_i. Up to a diffeomorphism, the new manifold thus obtained is independent of the choices of local coordinates ([19]), and it is called *the connected sum of M_1 and M_2* and is denoted by $M_1 \# M_2$ (see Figure 1.4). $\square$

Example 1.2.16. (The real projective space $\mathbb{RP}^n$). As a topological space $\mathbb{RP}^n$ is the quotient of $\mathbb{R}^{n+1}$ modulo the equivalence relation

$$x \sim y \overset{\text{def}}{\Longleftrightarrow} \exists \lambda \in \mathbb{R}^* : \quad x = \lambda y.$$

The equivalence class of $x = (x^0, \dots, x^n) \in \mathbb{R}^{n+1} \setminus \{0\}$ is usually denoted by $[x^0, \dots, x^n]$. Alternatively, $\mathbb{RP}^n$ is the set of all lines (directions) in $\mathbb{R}^{n+1}$. Traditionally, one attaches a point to each direction in $\mathbb{R}^{n+1}$, the so-called "point at infinity" along that direction, so that $\mathbb{RP}^n$ can be thought as the collection of all points at infinity along all directions.

The space $\mathbb{RP}^{n+1}$ has a natural structure of smooth manifold. To describe it consider the sets

$$U_k = \left\{ [x^0, \ldots, x^n] \in \mathbb{RP}^n \; ; \; x^k \neq 0 \right\}, \quad k = 0, \ldots, n.$$

Now define

$$\psi_k : U_k \to \mathbb{R}^n \quad [x^0, \ldots, x^n] \mapsto (x^0/x^k, \ldots, x^{k-1}/x^k, x^{k+1}/x^k, \ldots x^n).$$

The maps ψ_k define local coordinates on the projective space. The transition map on the overlap region $U_k \cap U_m = \{[x^0, \ldots, x^n] \; ; \; x^k x^m \neq 0\}$ can be easily described. Set

$$\psi_k([x^0, \ldots, x^n]) = (\xi_1, \ldots, \xi_n), \;\; \psi_m([x^0, \ldots, x^n]) = (\eta_1, \ldots, \eta_n).$$

The equality

$$[x^0, \ldots, x^n] = [\xi_1, \ldots, \xi_{k-1}, 1, \xi_k, \ldots, \xi_n] = [\eta_1, \ldots, \eta_{m-1}, 1, \eta_m, \ldots, \eta_n]$$

immediately implies (assume $k < m$)

$$\begin{cases} \xi_1 = \eta_1/\eta_k, & \ldots, \quad \xi_{k-1} = \eta_{k-1}/\eta_k, \quad \xi_{k+1} = \eta_k \\ \xi_k = \eta_{k+1}/\eta_k, & \ldots, \quad \xi_{m-2} = \eta_{m-1}/\eta_k, \xi_{m-1} = 1/\eta_k \\ \xi_m = \eta_m \eta_k, & \ldots, \qquad \xi_n = \eta_n/\eta_k \end{cases} \qquad (1.2.1)$$

This shows the map $\psi_k \circ \psi_m^{-1}$ is smooth and proves that $\mathbb{RP}^n$ is a smooth manifold. Note that when $n = 1$, $\mathbb{RP}^1$ is diffeomorphic with S^1. One way to see this is to observe that the projective space can be alternatively described as the quotient space of S^n modulo the equivalence relation which identifies antipodal points. □

Example 1.2.17. (The complex projective space $\mathbb{CP}^n$). The definition is formally identical to that of $\mathbb{RP}^n$. $\mathbb{CP}^n$ is the quotient space of $\mathbb{C}^{n+1} \setminus \{0\}$ modulo the equivalence relation

$$x \sim y \overset{\text{def}}{\Longleftrightarrow} \exists \lambda \in \mathbb{C}^* : \; x = \lambda y.$$

The open sets U_k are defined similarly and so are the local charts $\psi_k : U_k \to \mathbb{C}^n$. They satisfy transition rules similar to (1.2.1) so that $\mathbb{CP}^n$ is a smooth manifold of dimension $2n$. □

Exercise 1.2.18. Prove that $\mathbb{CP}^1$ is diffeomorphic to S^2. □

In the above example we encountered a special (and very pleasant) situation: the gluing maps not only are smooth, they are also *holomorphic* as maps $\psi_k \circ \psi_m^{-1} : U \to V$ where U and V are open sets in $\mathbb{C}^n$. This type of gluing induces a "rigidity" in the underlying manifold and it is worth distinguishing this situation.

Definition 1.2.19. (Complex manifolds). A complex manifold is a smooth, $2n$-dimensional manifold M which admits an atlas $\{(U_i, \psi_i) : U_i \to \mathbb{C}^n\}$ such that all transition maps are holomorphic. □

The complex projective space is a complex manifold. Our next example naturally generalizes the projective spaces described above.

Example 1.2.20. (The real and complex Grassmannians $\mathbf{Gr}_k(\mathbb{R}^n)$, $\mathbf{Gr}_k(\mathbb{C}^n)$).

Suppose V is a real vector space of dimension n. For every $0 \leq k \leq n$ we denote by $\mathbf{Gr}_k(V)$ the set of k-dimensional vector subspaces of V. We will say that $\mathbf{Gr}_k(V)$ is the *linear Grassmannian of k-planes in E*. When $V = \mathbb{R}^n$ we will write $\mathbf{Gr}_{k,n}(\mathbb{R})$ instead of $\mathbf{Gr}_k(\mathbb{R}^n)$.

We would like to give several equivalent descriptions of the natural structure of smooth manifold on $\mathbf{Gr}_k(V)$. To do this it is very convenient to fix a Euclidean metric on V.

Any k-dimensional subspace $L \subset V$ is uniquely determined by the orthogonal projection onto L which we will denote by P_L. Thus we can identify $\mathbf{Gr}_k(V)$ with the set of rank k projectors

$$\mathbf{Proj}_k(V) := \{ P : V \to V; \ P^* = P = P^2, \ \operatorname{rank} P = k \}.$$

We have a natural map

$$P : \mathbf{Gr}_k(V) \to \mathbf{Proj}_k(V), \ L \mapsto P_L$$

with inverse $P \mapsto \operatorname{Range}(P)$.

The set $\mathbf{Proj}_k(V)$ is a subset of the vector space of symmetric endomorphisms

$$\operatorname{End}^+(V) := \{ A \in \operatorname{End}(V), \ A^* = A \}.$$

The sapce $\operatorname{End}^+(V)$ is equipped with a natural inner product

$$(A, B) := \frac{1}{2} \operatorname{tr}(AB), \ \ \forall A, B \in \operatorname{End}^+(V). \tag{1.2.2}$$

The set $\mathbf{Proj}_k(V)$ is a closed and bounded subset of $\operatorname{End}^+(V)$. The bijection

$$P : \mathbf{Gr}_k(V) \to \mathbf{Proj}_k(V)$$

induces a topology on $\mathbf{Gr}_k(V)$. We want to show that $\mathbf{Gr}_k(V)$ has a natural structure of smooth manifold compatible with this topology. To see this, we define for every $L \subset \mathbf{Gr}_k(V)$ the set

$$\mathbf{Gr}_k(V, L) := \{ U \in \mathbf{Gr}_k(V); \ U \cap L^\perp = 0 \}.$$

Lemma 1.2.21. *(a) Let $L \in \mathbf{Gr}_k(V)$. Then*

$$U \cap L^\perp = 0 \Longleftrightarrow \mathbb{1} - P_L + P_U : V \to V \ \text{is an isomorphism.} \tag{1.2.3}$$

(b) The set $\mathbf{Gr}_k(V, L)$ is an open subset of $\mathbf{Gr}_k(V)$.

Proof. (a) Note first that a dimension count implies that

$$U \cap L^\perp = 0 \Longleftrightarrow U + L^\perp = V \Longleftrightarrow U^\perp \cap L = 0.$$

Let us show that $U \cap L^\perp = 0$ implies that $\mathbb{1} - P_L + P_L$ is an isomorphism. It suffices to show that

$$\ker(\mathbb{1} - P_L + P_U) = 0.$$

Suppose $v \in \ker(\mathbb{1} - P_L + P_U)$. Then

$$0 = P_L(\mathbb{1} - P_L + P_U)v = P_L P_U v = 0 \implies P_U v \in U \cap \ker P_L = U \cap L^\perp = 0.$$

Hence $P_U v = 0$, so that $v \in U^\perp$. From the equality $(\mathbb{1} - P_L - P_U)v = 0$ we also deduce $(\mathbb{1} - P_L)v = 0$ so that $v \in L$. Hence

$$v \in U^\perp \cap L = 0.$$

Conversely, we will show that if $\mathbb{1} - P_L + P_U = P_{L^\perp} + P_U$ is onto, then $U + L^\perp = V$.
Indeed, let $v \in V$. Then there exists $x \in V$ such that

$$v = P_{L^\perp} x + P_U x \in L^\perp + U.$$

(b) We have to show that, for every $K \in \mathbf{Gr}_k(V, L)$, there exists $\varepsilon > 0$ such that any U satisfying

$$\|P_U - P_K\| < \varepsilon$$

intersects $L^\perp$ trivially. Since $K \in \mathbf{Gr}_k(V, L)$ we deduce from (a) that the map $\mathbb{1} - P_L - P_K : V \to V$ is an isomorphism. Note that

$$\|(\mathbb{1} - P_L - P_K) - (\mathbb{1} - P_L - P_U)\| = \|P_K - P_U\|.$$

The space of isomorphisms of V is an open subset of $\mathrm{End}(V)$. Hence there exists $\varepsilon > 0$ such that, for any subspace U satisfying $\|P_U - P_K\| < \varepsilon$, the endomorphism $(\mathbb{1} - P_L - P_U)$ is an isomorphism. We now conclude using part (a).   □

Since $L \in \mathbf{Gr}_k(V, L)$, $\forall L \in \mathbf{Gr}_k(V)$, we have an open cover of $\mathbf{Gr}_k(V)$

$$\mathbf{Gr}_k(V) = \bigcup_{L \in \mathbf{Gr}_k(V)} \mathbf{Gr}_k(V, L).$$

Note that for every $L \in \mathbf{Gr}_k(V)$ we have a natural map

$$\Gamma : \mathrm{Hom}(L, L^\perp) \to \mathbf{Gr}_k(V, L),$$

which associates to each linear map $S : L \to L^\perp$ its graph (see Figure 1.5)

$$\Gamma_S = \{x + Sx \in L + L^\perp = V; \ x \in L\}.$$

We will show that this is a homeomorphism by providing an explicit description of the orthogonal projection P_{Γ_S}
Observe first that the orthogonal complement of Γ_S is the graph of $-S^* : L^\perp \to L$. More precisely,

$$\Gamma_S^\perp = \Gamma_{-S^*} = \{y - S^*y \in L^\perp + L = V; \ y \in L^\perp \}.$$

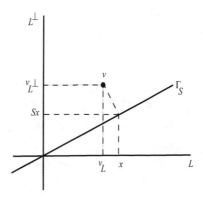

Fig. 1.5 *Subspaces as graphs of linear operators.*

Let $v = P_L v + P_{L^\perp} v = v_L + v_{L+} \in V$ (see Figure 1.5). Then

$$P_{\Gamma_S} v = x + Sx, \quad x \in L \iff v - (x + Sx) \in \Gamma_S^\perp$$

$$\iff \exists x \in L, \ y \in L^\perp \text{ such that } \begin{cases} x + S^* y = v_L \\ Sx - y = v_{L^\perp} \end{cases}.$$

Consider the operator $\mathcal{S} : L \oplus L^\perp \to L \oplus L^\perp$ which has the block decomposition

$$\mathcal{S} = \begin{bmatrix} \mathbb{1}_L & S^* \\ S & -\mathbb{1}_L^\perp \end{bmatrix}.$$

Then the above linear system can be rewritten as

$$\mathcal{S} \cdot \begin{bmatrix} x \\ y \end{bmatrix} = \begin{bmatrix} v_L \\ v_{L^\perp} \end{bmatrix}.$$

Now observe that

$$\mathcal{S}^2 = \begin{bmatrix} \mathbb{1}_L + S^* S & 0 \\ 0 & \mathbb{1}_{L^\perp} + SS^* \end{bmatrix}.$$

Hence $\mathcal{S}$ is invertible, and

$$\mathcal{S}^{-1} = \begin{bmatrix} (\mathbb{1}_L + S^* S)^{-1} & 0 \\ 0 & (\mathbb{1}_{L^\perp} + SS^*)^{-1} \end{bmatrix} \cdot \mathcal{S}$$

$$= \begin{bmatrix} (\mathbb{1}_L + S^* S)^{-1} & (\mathbb{1}_L + S^* S)^{-1} S^* \\ (\mathbb{1}_{L^\perp} + SS^*)^{-1} S & -(\mathbb{1}_{L^\perp} + SS^*)^{-1} \end{bmatrix}.$$

We deduce

$$x = (\mathbb{1}_L + S^* S)^{-1} v_L + (\mathbb{1}_L + S^* S)^{-1} S^* v_{L^\perp}$$

and

$$P_{\Gamma_S} v = \begin{bmatrix} x \\ Sx \end{bmatrix}.$$

Hence P_{Γ_S} has the block decomposition

$$P_{\Gamma_S} = \begin{bmatrix} \mathbb{1}_L \\ S \end{bmatrix} \cdot [(\mathbb{1}_L + S^*S)^{-1} \quad (\mathbb{1}_L + S^*S)^{-1}S^*]$$

$$= \begin{bmatrix} (\mathbb{1}_L + S^*S)^{-1} & (\mathbb{1}_L + S^*S)^{-1}S^* \\ S(\mathbb{1}_L + S^*S)^{-1} & S(\mathbb{1}_L + S^*S)^{-1}S^* \end{bmatrix}. \qquad (1.2.4)$$

Note that if $U \in \mathbf{Gr}_k(V, L)$, then with respect to the decomposition $V = L + L^\perp$ the projector P_U has the block form

$$P_U = \begin{bmatrix} A & B \\ C & D \end{bmatrix} = \begin{bmatrix} P_L P_U I_L & P_L P_U I_{L^\perp} \\ P_{L^\perp} P_U I_L & P_L L^\perp P_U I_{L^\perp} \end{bmatrix},$$

where for every subspace $K \hookrightarrow V$ we denoted by $I_K : K \to V$ the canonical inclusion, then $U = \Gamma_S$, where $S = CA^{-1}$. This shows that the graph map

$$\mathrm{Hom}(L, L^\perp) \ni S \mapsto \Gamma_S \in \mathbf{Gr}_k(V)$$

is a homeomorphism. Moreover, the above formulæ show that if $U \in \mathbf{Gr}_k(V, L_0) \cap \mathbf{Gr}_k(V, L_1)$, then we can represent U in two ways,

$$U = \Gamma_{S_0} = \Gamma_{S_1}, \quad S_i \in \mathrm{Hom}(L_i, L_i^\perp), \quad i = 0, 1,$$

and the correspondence $S_0 \to S_1$ is smooth. This shows that $\mathbf{Gr}_k(V)$ has a natural structure of smooth manifold of dimension

$$\dim \mathbf{Gr}_k(V) = \dim \mathrm{Hom}(L, L^\perp) = k(n - k).$$

$\mathbf{Gr}_k(\mathbb{C}^n)$ is defined as the space of complex k-dimensional subspaces of $\mathbb{C}^n$. It can be structured as above as a smooth manifold of dimension $2k(n - k)$. Note that $\mathbf{Gr}_1(\mathbb{R}^n) \cong \mathbb{R}\mathbb{P}^{n-1}$, and $\mathbf{Gr}_1(\mathbb{C}^n) \cong \mathbb{C}\mathbb{P}^{n-1}$. The Grassmannians have important applications in many classification problems. □

Exercise 1.2.22. Show that $\mathbf{Gr}_k(\mathbb{C}^n)$ is a *complex* manifold of complex dimension $k(n - k)$. □

Example 1.2.23. (Lie groups). A *Lie group* is a smooth manifold G together with a group structure on it such that the map

$$G \times G \to G \quad (g, h) \mapsto g \cdot h^{-1}$$

is smooth. These structures provide an excellent way to formalize the notion of symmetry.

(a) $(\mathbb{R}^n, +)$ is a commutative Lie group.

(b) The unit circle S^1 can be alternatively described as the set of complex numbers of norm one and the complex multiplication defines a Lie group structure on it. This is a commutative group. More generally, the torus T^n is a Lie group as a direct product of n circles[1].

---

[1] One can show that any connected commutative Lie group has the form $T^n \times \mathbb{R}^m$.

(c) The general linear group $GL(n, \mathbb{K})$ defined as the group of invertible $n \times n$ matrices with entries in the field $\mathbb{K} = \mathbb{R}$, $\mathbb{C}$ is a Lie group. Indeed, $GL(n, \mathbb{K})$ is an open subset (see Exercise 1.1.3) in the linear space of $n \times n$ matrices with entries in $\mathbb{K}$. It has dimension $d_{\mathbb{K}} n^2$, where $d_{\mathbb{K}}$ is the dimension of $\mathbb{K}$ as a linear space over $\mathbb{R}$. We will often use the alternate notation $GL(\mathbb{K}^n)$ when referring to $GL(n, \mathbb{K})$.

(d) The orthogonal group $O(n)$ is the group of real $n \times n$ matrices satisfying

$$T \cdot T^t = \mathbb{1}.$$

To describe its smooth structure we will use the Cayley transform trick as in [84] (see also the classical [100]). Set

$$M_n(\mathbb{R})^\# := \big\{ T \in M_n(\mathbb{R}) \; ; \; \det(\mathbb{1} + T) \neq 0 \big\}.$$

The matrices in $M_n(\mathbb{R})^\#$ are called non exceptional. Clearly $\mathbb{1} \in O(n)^\# = O(n) \cap M_n(\mathbb{R})^\#$ so that $O(n)^\#$ is a *nonempty* open subset of $O(n)$. The *Cayley transform* is the map $\# : M_n(\mathbb{R})^\# \to M_n(\mathbb{R})$ defined by

$$A \mapsto A^\# = (\mathbb{1} - A)(\mathbb{1} + A)^{-1}.$$

The Cayley transform has some very nice properties.

(i) $A^\# \in M_n(\mathbb{R})^\#$ for every $A \in M_n(\mathbb{R})^\#$.

(ii) $\#$ is involutive, i.e., $(A^\#)^\# = A$ for any $A \in M_n(\mathbb{R})^\#$.

(iii) For every $T \in O(n)^\#$ the matrix $T^\#$ is skew-symmetric, and conversely, if $A \in M_n(\mathbb{R})^\#$ is skew-symmetric then $A^\# \in O(n)$.

Thus the Cayley transform is a homeomorphism from $O(n)^\#$ to the space of non-exceptional, skew-symmetric, matrices. The latter space is an open subset in the linear space of real $n \times n$ skew-symmetric matrices, $\underline{o}(n)$.

Any $T \in O(n)$ defines a self-homeomorphism of $O(n)$ by *left translation in the group*

$$L_T : O(n) \to O(n) \quad S \mapsto L_T(S) = T \cdot S.$$

We obtain an open cover of $O(n)$:

$$O(n) = \bigcup_{T \in O(n)} T \cdot O(n)^\#.$$

Define $\Psi_T : T \cdot O(n)^\# \to \underline{o}(n)$ by $S \mapsto (T^{-1} \cdot S)^\#$. One can show that the collection

$$\Big(T \cdot O(n)^\#, \Psi_T \Big)_{T \in O(n)}$$

defines a smooth structure on $O(n)$. In particular, we deduce

$$\dim O(n) = n(n-1)/2.$$

Inside $O(n)$ lies a normal subgroup (the *special orthogonal group*)

$$SO(n) = \{T \in O(n) \; ; \; \det T = 1\}.$$

The group $SO(n)$ is a Lie group as well and $\dim SO(n) = \dim O(n)$.

(e) The unitary group $U(n)$ is defined as

$$U(n) = \{T \in \mathrm{GL}(n, \mathbb{C}) \; ; \; T \cdot T^* = \mathbb{1}\},$$

where T^* denotes the conjugate transpose (adjoint) of T. To prove that $U(n)$ is a manifold one uses again the Cayley transform trick. This time, we coordinatize the group using the space $\underline{u}(n)$ of skew-adjoint (skew-Hermitian) $n \times n$ complex matrices $(A = -A^*)$. Thus $U(n)$ is a smooth manifold of dimension

$$\dim U(n) = \dim \underline{u}(n) = n^2.$$

Inside $U(n)$ sits the normal subgroup $SU(n)$, the kernel of the group homomorphism $\det : U(n) \to S^1$. $SU(n)$ is also called the *special unitary group*. This a smooth manifold of dimension $n^2 - 1$. In fact the Cayley transform trick allows one to coordinatize $SU(n)$ using the space

$$\underline{su}(n) = \{A \in \underline{u}(n) \; ; \; \mathrm{tr}\, A = 0\}. \qquad \square$$

Exercise 1.2.24. (a) Prove the properties (i)-(iii) of the Cayley transform, and then show that $\left(T \cdot O(n)^{\#}, \Psi_T\right)_{T \in O(n)}$ defines a smooth structure on $O(n)$.
(b) Prove that $U(n)$ and $SU(n)$ are manifolds.
(c) Show that $O(n)$, $SO(n)$, $U(n)$, $SU(n)$ are compact spaces.
(d) Prove that $SU(2)$ is diffeomorphic with S^3 (Hint: think of S^3 as the group of unit quaternions.) $\qquad \square$

Exercise 1.2.25. Let $SL(n; \mathbb{K})$ denote the group of $n \times n$ matrices of determinant 1 with entries in the field $\mathbb{K} = \mathbb{R}, \mathbb{C}$. Using the Cayley trick show that $SL(n; \mathbb{K})$ is a smooth manifold modeled on the linear space

$$\underline{sl}(n, \mathbb{K}) = \{A \in M_{n \times n}(\mathbb{K}) \; ; \; \mathrm{tr}\, A = 0\}.$$

In particular, it has dimension $d_{\mathbb{K}}(n^2 - 1)$, where $d_{\mathbb{K}} = \dim_{\mathbb{R}} \mathbb{K}$. $\qquad \square$

Exercise 1.2.26. (Quillen). Suppose V_0, V_1 are two real, finite dimensional Euclidean space, and $T : V_0 \to V_1$ is a linear map. We denote by T^* is adjoint, $T^* : V_1 \to V_0$, and by Γ_T the graph of T,

$$\Gamma_T = \left\{(v_0, v_1) \in V_0 \oplus V_1; \; v_1 = Tv_0\, l\right\}.$$

We form the skew-symmetric operator

$$X : V_0 \oplus V_1 \to V_0 \oplus V_1, \quad X \cdot \begin{bmatrix} v_0 \\ v_1 \end{bmatrix} = \begin{bmatrix} 0 & T^* \\ -T & 0 \end{bmatrix} \cdot \begin{bmatrix} v_0 \\ v_1 \end{bmatrix}.$$

We denote by C_T the Cayley transform of X,

$$C_T = (\mathbb{1} - X)(\mathbb{1} + X)^{-1},$$

and by $R_0 : V_0 \oplus V_1 \to V_0 \oplus V_1$ the reflection

$$R_0 = \begin{bmatrix} \mathbb{1}_{V_0} & 0 \\ 0 & -\mathbb{1}_{V_1} \end{bmatrix}.$$

Show that $R_T = C_T R_0$ is an orthogonal involution, i.e.,

$$R_T^2 = \mathbb{1}, \quad R_T^* = R_T,$$

and $\ker(\mathbb{1} - R_T) = \Gamma_T$. In other words, R_T is the orthogonal reflection in the subspace Γ_T,

$$R_T = 2P_{\Gamma_T} - \mathbb{1},$$

where P_{Γ_T} denotes the orthogonal projection onto Γ_T. $\qquad\qquad\square$

Exercise 1.2.27. Suppose G is a Lie group, and H is an abstract subgroup of G. Prove that the closure of H is also a subgroup of G.

Exercise 1.2.28. (a) Let G be a *connected* Lie group and denote by U a neighborhood of $1 \in G$. If H is the subgroup algebraically generated by U show that H is dense in G.
(b) Let G be a *compact* Lie group and $g \in G$. Show that $1 \in G$ lies in the closure of $\{g^n; \ n \in \mathbb{Z} \setminus \{0\}\}$. $\qquad\qquad\square$

Remark 1.2.29. If G is a Lie group, and H is a *closed* subgroup of G, then H is in fact a smooth submanifold of G, and with respect to this smooth structure H is a Lie group. For a proof we refer to [44, 95]. In view of Exercise 1.2.27, this fact allows us to produce many examples of Lie groups. $\qquad\qquad\square$

1.2.4 *How many manifolds are there?*

The list of examples in the previous subsection can go on forever, so one may ask whether there is any coherent way to organize the collection of all possible manifolds. This is too general a question to expect a clear cut answer. We have to be more specific. For example, we can ask

Question 1: *Which are the compact, connected manifolds of a given dimension d?*

For $d = 1$ the answer is very simple: the only compact connected 1-dimensional manifold is the circle S^1. (Can you prove this?)

We can raise the stakes and try the same problem for $d = 2$. Already the situation is more elaborate. We know at least two surfaces: the sphere S^2 and the torus T^2. They clearly look different but we have not yet proved rigorously that they are indeed not diffeomorphic. This is not the end of the story. We can connect sum two tori, three tori or any number g of tori. We obtain doughnut-shaped surface as in Figure 1.6.

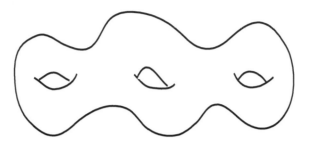

Fig. 1.6　*Connected sum of 3 tori*

Again we face the same question: do we get non-diffeomorphic surfaces for different choices of g? Figure 1.6 suggests that this may be the case but this is no rigorous argument.

We know another example of compact surface, the projective plane $\mathbb{RP}^2$, and we naturally ask whether it looks like one of the surfaces constructed above. Unfortunately, we cannot visualize the real projective plane (one can prove rigorously it does not have enough room to exist inside our 3-dimensional Universe). We have to decide this question using a little more than the raw geometric intuition provided by a picture. To kill the suspense, we mention that $\mathbb{RP}^2$ does not belong to the family of donuts. One reason is that, for example, a torus has two faces: an inside face and an outside face (think of a car rubber tube). $\mathbb{RP}^2$ has a weird behavior: it has "no inside" and "no outside". It has only one side! One says the torus is *orientable* while the projective plane is not.

We can now connect sum any numbers of $\mathbb{RP}^2$'s to any donut an thus obtain more and more surfaces, which we cannot visualize and we have yet no idea if they are pairwise distinct. A classical result in topology says that all compact surfaces can be obtained in this way (see [68]), but in the above list some manifolds are diffeomorphic, and we have to describe which. In dimension 3 things are not yet settled[2] and, to make things look hopeless, in dimension ≥ 4 Question 1 is algorithmically undecidable.

We can reconsider our goals, and look for all the manifolds with a given property X. In many instances one can give fairly accurate answers. Property X may refer to more than the (differential) topology of a manifold. Real life situations suggest the study of manifolds with additional structure. The following problem may give the reader a taste of the types of problems we will be concerned with in this book.

Question 2 *Can we wrap a planar piece of canvas around a metal sphere in a one-to-one fashion? (The canvas is flexible but not elastic.)*

A simple do-it-yourself experiment is enough to convince anyone that this is not possible. Naturally, one asks for a rigorous explanation of what goes wrong.

---

[2]Things are still not settled in 2007, but there has been considerable progress due to G. Perelman's proof of the Poincaré conjecture.

The best explanation of this phenomenon is contained in the celebrated Theorema Egregium (Golden Theorem) of Gauss. Canvas surfaces have additional structure (they are made of a special material), and for such objects there is a rigorous way to measure "how curved" are they. One then realizes that the problem in Question 2 is impossible, since a (canvas) sphere is curved in a different way than a plane canvas.

There are many other structures Nature forced us into studying them, but they may not be so easily described in elementary terms.

A word to the reader. The next two chapters are probably the most arid in geometry but, keep in mind that, behind each construction lies a natural motivation and, even if we do not always have the time to show it to the reader, it is there, and it may take a while to reveal itself. Most of the constructions the reader will have to "endure" in the next two chapters constitute not just some difficult to "swallow" formalism, but the basic language of geometry. It might comfort the reader during this less than glamorous journey to carry in the back of his mind Hermann Weyl's elegantly phrased advise.

> "It is certainly regrettable that we have to enter into the purely formal aspect in such detail and to give it so much space but, nevertheless, it cannot be avoided. Just as anyone who wishes to give expressions to his thoughts with ease must spend laborious hours learning language and writing, so here too the only way we can lessen the burden of formulæ is to master the technique of tensor analysis to such a degree that we can turn to real problems that concern us without feeling any encumbrance, our object being to get an insight into the nature of space [...]. Whoever sets out in quest of these goals must possess a perfect mathematical equipment from the outset."
> *H. Weyl: Space, Time, Matter.*

Chapter 2

Natural Constructions on Manifolds

The goal of this chapter is to introduce the basic terminology used in differential geometry. The key concept is that of tangent space at a point which is a first order approximation of the manifold near that point. We will be able to transport many notions in linear analysis to manifolds via the tangent space.

2.1 The tangent bundle

2.1.1 *Tangent spaces*

We begin with a simple example which will serve as a motivation for the abstract definitions.

Example 2.1.1. Consider the sphere

$$(S^2) : x^2 + y^2 + z^2 = 1 \quad \text{in } \mathbb{R}^3.$$

We want to find the plane passing through the North pole $N(0, 0, 1)$ that is "closest" to the sphere. The classics would refer to such a plane as an *osculator plane*.

The natural candidate for this osculator plane would be a plane given by a linear equation that best approximates the defining equation $x^2 + y^2 + z^2 = 1$ in a neighborhood of the North pole. The linear approximation of $x^2 + y^2 + z^2$ near N seems like the best candidate. We have

$$x^2 + y^2 + z^2 - 1 = 2(z - 1) + O(2),$$

where $O(2)$ denotes a quadratic error. Hence, the osculator plane is $z = 1$, Geometrically, it is the horizontal *affine* plane through the North pole. The *linear* subspace $\{z = 0\} \subset \mathbb{R}^3$ is called the *tangent space* to S^2 at N.

The above construction has one deficiency: it is *not intrinsic*, i.e., it relies on objects "outside" the manifold S^2. There is one natural way to fix this problem. Look at a smooth path $\gamma(t)$ on S^2 passing through N at $t = 0$. Hence, $t \mapsto \gamma(t) \in \mathbb{R}^3$, and

$$|\gamma(t)|^2 = 1. \tag{2.1.1}$$

If we differentiate (2.1.1) at $t = 0$ we get $(\dot{\gamma}(0), \gamma(0)) = 0$, i.e., $\dot{\gamma}(0) \perp \gamma(0)$, so that $\dot{\gamma}(0)$ lies in the linear subspace $z = 0$. We deduce that the tangent space consists of the tangents to the curves on S^2 passing through N.

This is apparently no major conceptual gain since we still regard the tangent space as a subspace of $\mathbb{R}^3$, and this is still an extrinsic description. However, if we use the stereographic projection from the South pole we get local coordinates (u, v) near N, and any curve $\gamma(t)$ as above can be viewed as a curve $t \mapsto (u(t), v(t))$ in the (u, v) plane. If $\phi(t)$ is another curve through N given in local coordinates by $t \mapsto (\underline{u}(t), \underline{v}(t))$, then

$$\dot{\gamma}(0) = \dot{\phi}(0) \iff \big(\dot{u}(0), \dot{v}(0)\big) = \big(\underline{\dot{u}}(0), \underline{\dot{v}}(0)\big).$$

The right hand side of the above equality defines an equivalence relation $\sim$ on the set of smooth curves passing trough $(0,0)$. Thus, there is a bijective correspondence between the tangents to the curves through N, and the equivalence classes of "$\sim$". This equivalence relation is now intrinsic modulo one problem: "$\sim$" may depend on the choice of the local coordinates. Fortunately, as we are going to see, this is a non-issue.   □

Definition 2.1.2. Let M^m be a smooth manifold and p_0 a point in M. Two smooth paths α, $\beta : (-\varepsilon, \varepsilon) \to M$ such that $\alpha(0) = \beta(0) = p_0$ are said to have a *first order contact* at p_0 if there exist local coordinates $(x) = (x^1, \ldots, x^m)$ near p_0 such that

$$\dot{x}_\alpha(0) = \dot{x}_\beta(0),$$

where $\alpha(t) = (x_\alpha(t)) = (x_\alpha^1(t), \ldots, x_\alpha^m(t))$, and $\beta(t) = (x_\beta(t)) = (x_\beta^1(t), \ldots, x_\beta^m(t))$. We write this $\alpha \sim_1 \beta$.   □

Lemma 2.1.3. $\sim_1$ *is an equivalence relation.*

Sketch of proof. The binary relation $\sim_1$ is obviously reflexive and symmetric, so we only have to check the transitivity. Let $\alpha \sim_1 \beta$ and $\beta \sim_1 \gamma$. Thus there exist local coordinates $(x) = (x^1, \ldots, x^m)$ and $(y) = (y^1, \ldots, y^m)$ near p_0 such that $(\dot{x}_\alpha(0)) = (\dot{x}_\beta(0))$ and $(\dot{y}_\beta(0)) = (\dot{y}_\gamma(0))$. The transitivity follows from the equality

$$\dot{y}_\gamma^i(0) = \dot{y}_\beta^i(0) = \sum_j \frac{\partial y^i}{\partial x^j} \dot{x}_\beta^j(0) = \sum_j \frac{\partial y^i}{\partial x^j} \dot{x}_\alpha^j(0) = \dot{y}_\alpha^j(0). \qquad \square$$

Definition 2.1.4. A *tangent vector* to M at p is a first-order-contact equivalence class of curves through p. The equivalence class of a curve $\alpha(t)$ such that $\alpha(0) = p$ will be temporarily denoted by $[\dot{\alpha}(0)]$. The set of these equivalence classes is denoted by T_pM, and is called the *tangent space* to M at p.   □

Lemma 2.1.5. T_pM *has a natural structure of vector space.*

Proof. Choose local coordinates $(x^1, \ldots, x^m)$ near p such that $x^i(p) = 0$, $\forall i$, and let α and β be two smooth curves through p. In the above local coordinates the curves α, β become $\left(x^i_\alpha(t)\right)$, $\left(x^i_\beta(t)\right)$. Construct a new curve γ through p given by

$$\left(x^i_\gamma(t)\right) = \left(x^i_\alpha(t) + x^i_\beta(t)\right).$$

Set $[\dot\alpha(0)] + [\dot\beta(0)] := [\dot\gamma(0)]$. For this operation to be well defined one has to check two things.

(a) The equivalence class $[\dot\gamma(0)]$ is independent of coordinates.
(b) If $[\dot\alpha_1(0)] = [\dot\alpha_2(0)]$ and $[\dot\beta_1(0)] = [\dot\beta_2(0)]$ then

$$[\dot\alpha_1(0)] + [\dot\beta_1(0)] = [\dot\alpha_2(0)] + [\dot\beta_2(0)].$$

We let the reader supply the routine details. $\qquad\square$

Exercise 2.1.6. Finish the proof of the Lemma 2.1.5. $\qquad\square$

From this point on we will omit the brackets $[\,-\,]$ in the notation of a tangent vector. Thus, $[\dot\alpha(0)]$ will be written simply as $\dot\alpha(0)$.

As one expects, all the above notions admit a nice description using local coordinates. Let $(x^1, \ldots, x^m)$ be coordinates near $p \in M$ such that $x^i(p) = 0$, $\forall i$. Consider the curves

$$e_k(t) = (t\delta^1_k, \ldots, t\delta^m_k), \quad k = 1, \ldots, m,$$

where δ^i_j denotes Kronecker's delta symbol. We set

$$\cdot\ \frac{\partial}{\partial x^k}(p) := \dot e_k(0). \qquad (2.1.2)$$

Note that these vectors depend on the local coordinates $(x^1, \ldots, x^m)$. Often, when the point p is clear from the context, we will omit it in the above notation.

Lemma 2.1.7. $\left(\frac{\partial}{\partial x^k}(p)\right)_{1 \leq k \leq m}$ *is a basis of* T_pM.

Proof. It follows from the obvious fact that any path through the origin in $\mathbb{R}^m$ has first order contact with a linear one $t \mapsto (a_1 t, \ldots, a_m t)$. $\qquad\square$

Exercise 2.1.8. Let $F : \mathbb{R}^N \to \mathbb{R}^k$ be a smooth map. Assume that
(a) $M = F^{-1}(0) \neq \emptyset$;
(b) $\operatorname{rank} D_x F = k$, for all $x \in M$.
Then M is a smooth manifold of dimension $N - k$ and

$$T_x M = \ker D_x F, \quad \forall x \in M. \qquad\square$$

Example 2.1.9. We want to describe $T_1 G$, where G is one of the Lie groups discussed in Section 1.2.2.
(a) $G = O(n)$. Let $(-\varepsilon, \varepsilon) \ni s \mapsto T(s)$ be a smooth path of orthogonal matrices such that $T(0) = 1$. Then $T^t(s) \cdot T(s) = 1$. Differentiating this equality at $s = 0$ we get

$$\dot T^t(0) + \dot T(0) = 0.$$

The matrix $\dot{T}(0)$ defines a vector in $T_1O(n)$, so the above equality states that this tangent space lies inside the space of skew-symmetric matrices, i.e., $T_1O(n) \subset \underline{o}(n)$. On the other hand, we proved in Section 1.2.2 that $\dim G = \dim \underline{o}(n)$ so that

$$T_1O(n) = \underline{o}(n).$$

(b) $G = SL(n; \mathbb{R})$. Let $(-\varepsilon, \varepsilon) \ni s \mapsto T(s)$ be a smooth path in $SL(n; \mathbb{R})$ such that $T(0) = 1$. Then $\det T(s) = 1$ and differentiating this equality at $s = 0$ we get (see Exercise 1.1.3)

$$\operatorname{tr} \dot{T}(0) = 0.$$

Thus, the tangent space at 1 lies inside the space of traceless matrices, i.e. $T_1SL(n; \mathbb{R}) \subset \underline{sl}(n; \mathbb{R})$. Since (according to Exercise 1.2.25) $\dim SL(n; \mathbb{R}) = \dim \underline{sl}(n; \mathbb{R})$ we deduce

$$T_1SL(n; \mathbb{R}) = \underline{sl}(n; \mathbb{R}). \qquad \square$$

Exercise 2.1.10. Show that $T_1U(n) = \underline{u}(n)$ and $T_1SU(n) = \underline{su}(n)$. $\qquad \square$

2.1.2 *The tangent bundle*

In the previous subsection we have naturally associated to an arbitrary point p on a manifold M a vector space T_pM. It is the goal of the present subsection to coherently organize the family of tangent spaces $(T_pM)_{p \in M}$. In particular, we want to give a rigorous meaning to the intuitive fact that T_pM depends smoothly upon p.

We will organize the disjoint union of all tangent spaces as a smooth manifold TM. There is a natural surjection

$$\pi : TM = \bigsqcup_{p \in M} T_pM \to M, \quad \pi(v) = p \iff v \in T_pM.$$

Any local coordinate system $x = (x^i)$ defined over an open set $U \subset M$ produces a natural basis $\left(\frac{\partial}{\partial x^i}(p) \right)$ of T_pM, for any $p \in U$. Thus, an element $v \in TU = \bigsqcup_{p \in U} T_pM$ is completely determined if we know

- which tangent space does it belong to, i.e., we know $p = \pi(v)$,
- the coordinates of v in the basis $\left(\frac{\partial}{\partial x^i}(p) \right)$,

$$v = \sum_i X^i(v) \left(\frac{\partial}{\partial x^i}(p) \right).$$

We thus have a bijection

$$\Psi_x : TU \to U^x \times \mathbb{R}^m \subset \mathbb{R}^m \times \mathbb{R}^m,$$

where U^x is the image of U in $\mathbb{R}^m$ via the coordinates (x^i). We can now use the map Ψ_x to transfer the topology on $\mathbb{R}^m \times \mathbb{R}^m$ to TU. Again, we have to make sure this topology is independent of local coordinates.

To see this, pick a different coordinate system $y = (y^i)$ on U. The coordinate independence referred to above is equivalent to the statement that the transition map

$$\Psi_y \circ \Psi_x^{-1} : U^x \times \mathbb{R}^m \longrightarrow TU \longrightarrow U^y \times \mathbb{R}^m$$

is a homeomorphism.

Let $A := (\overline{x}, X) \in U^x \times \mathbb{R}^m$. Then $\Psi_x^{-1}(A) = (p, \dot{\alpha}(0))$, where $x(p) = \overline{x}$, and $\alpha(t) \subset U$ is a path through p given in the x coordinates as

$$\alpha(t) = \overline{x} + tX.$$

Denote by $F : U^x \to U^y$ the transition map $x \mapsto y$. Then

$$\Psi_y \circ \Psi_x^{-1}(A) = \big(y(\overline{x}); Y^1, \dots, Y^m\big),$$

where $\dot{\alpha}(0) = (\dot{y}_\alpha^j(0)) = \sum Y^j \frac{\partial}{\partial y_j}(p)$, and $\big(y_\alpha(t)\big)$ is the description of the path $\alpha(t)$ in the coordinates y^j. Applying the chain rule we deduce

$$Y^j = \dot{y}_\alpha^j(0) = \sum_i \frac{\partial y^j}{\partial x^i} \dot{x}^i(0) = \sum_i \frac{\partial y^j}{\partial x^i} X^i. \tag{2.1.3}$$

This proves that $\Psi_y \circ \Psi_x^{-1}$ is actually smooth.

The natural topology of TM is obtained by patching together the topologies of TU_γ, where $(U_\gamma, \phi_\gamma)_\gamma$ is an atlas of M. A set $D \subset TM$ is open if its intersection with any TU_γ is open in TU_γ. The above argument shows that TM is a smooth manifold with (TU_γ, Ψ_γ) a defining atlas. Moreover, the natural projection $\pi : TM \to M$ is a smooth map.

Definition 2.1.11. The smooth manifold TM described above is called the *tangent bundle* of M. $\qquad\square$

Proposition 2.1.12. *A smooth map $f : M \to N$ induces a smooth map $Df : TM \to TN$ such that*
(a) $Df(T_pM) \subset T_{f(p)}N$, $\forall p \in M$
(b) The restriction to each tangent space $D_pF : T_pM \to T_{f(p)}N$ is linear. The map Df is called the differential of f, and one often uses the alternate notation $f_* = Df$.

Proof. Recall that T_pM is the space of tangent vectors to curves through p. If $\alpha(t)$ is such a curve $(\alpha(0) = p)$, then $\beta(t) = f(\alpha(t))$ is a smooth curve through $q = f(p)$, and we define

$$Df(\dot{\alpha}(0)) := \dot{\beta}(0).$$

One checks easily that if $\alpha_1 \sim_1 \alpha_2$, then $f(\alpha_1) \sim_1 f(\alpha_2)$, so that Df is well defined. To prove that the map $Df : T_pM \to T_qN$ is linear it suffices to verify this in any particular local coordinates $(x^1, \dots, x^m)$ near p, and $(y^1, \dots, y^n)$ near q, such that

$x^i(p) = 0$, $y^j(q) = 0$, $\forall i, j$, since any two choices differ (infinitesimally) by a linear substitution. Hence, we can regard f as a collection of maps

$$(x^1, \ldots, x^m) \mapsto (y^1(x^1, \ldots, x^m), \ldots, y^n(x^1, \ldots, x^m)).$$

A basis in $T_p M$ is given by $\left\{ \frac{\partial}{\partial x_i} \right\}$, while a basis of $T_q N$ is given by $\left\{ \frac{\partial}{\partial y_j} \right\}$.

If $\alpha, \beta : (-\varepsilon, \varepsilon) \to M$ are two smooth paths such that $\alpha(0) = \beta(0) = p$, then in local coordinates they have the description

$$\alpha(t) = (x_\alpha^1(t), \ldots, x_\alpha^m(t)), \quad \beta(t) = (x_\beta^1(t), \ldots, x_\beta^m(t)),$$

$$\dot{\alpha}(0) = (\dot{x}_\alpha^1(0), \ldots, \dot{x}_\alpha^m(0)), \quad \dot{\beta}(0) = (\dot{x}_\beta^1(0), \ldots, \dot{x}_\beta^m(0))$$

Then

$$F(\alpha(t)) = \left(y^1(x_\alpha^i(t)), \ldots, y^n(x_\alpha^i(t)) \right), \quad F(\beta(t)) = \left(y^1(x_\beta^i(t)), \ldots, y^n(x_\beta^i(t)) \right),$$

$$F(\alpha(t) + \beta(t)) = \left(y^1(x_\alpha^i(t) + x_\beta^i(t)), \ldots, y^n(x_\alpha^i(t) + x_\beta^i(t)) \right),$$

$$DF(\dot{\alpha}(0)) = \sum_{j=1}^n \left(\sum_{i=1}^m \frac{\partial y^j}{\partial x^i} \dot{x}_\alpha^i \right) \frac{\partial}{\partial y_j}, \quad DF(\dot{\beta}(0)) = \sum_{j=1}^n \left(\sum_{i=1}^m \frac{\partial y^j}{\partial x^i} \dot{x}_\beta^i \right) \frac{\partial}{\partial y_j},$$

$$DF(\dot{\alpha}(0) + \dot{\beta}(0)) = \frac{d}{dt}\Big|_{t=0} F(\alpha(t) + \beta(t)) = \sum_{j=1}^n \left(\sum_{i=1}^m \frac{\partial y^j}{\partial x^i} (\dot{x}_\alpha^i + \dot{x}_\beta^i) \right) \frac{\partial}{\partial y_j}$$

$$= DF(\dot{\alpha}(0)) + DF(\dot{\beta}(0)).$$

This shows that $Df : T_p M \to T_q N$ is the linear operator given in these bases by the matrix $\left(\frac{\partial y^j}{\partial x^i} \right)_{1 \le j \le n, \, 1 \le i \le m}$. In particular, this implies that Df is also smooth.

□

2.1.3　Sard's Theorem

In this subsection we want to explain rigorously a phenomenon with which the reader may already be intuitively acquainted. We describe it first in a special case.

Suppose M is a submanifold of dimension 2 in $\mathbb{R}^3$. Then, a simple thought experiment suggests that *most* horizontal planes will not be tangent to M. Equivalently, Iif we denote by f the restriction of the function z to M, then for most real numbers h the level set $f^{-1}(h)$ does not contain a point where the differential of f is zero, so that most level sets $f^{-1}(h)$ are smooth submanifolds of M of codimension 1, i.e., smooth curves on M.

We can ask a more general question. Given two smooth manifolds X, Y, a smooth map $f : X \to Y$, is it true that for "most" $y \in Y$ the level set $f^{-1}(y)$ is a *smooth* submanifold of X of codimension $\dim Y$? This question has a positive answer, known as Sard's theorem.

Definition 2.1.13. Suppose that Y is a smooth, connected manifold of dimension m.

(a) We say that a subset $S \subset Y$ is *negligible* if, for any coordinate chart of Y, $\Psi : U \to \mathbb{R}^m$, the set $\Psi(S \cap U) \subset \mathbb{R}^m$ has Lebesgue measure zero in $\mathbb{R}^m$.

(b) Suppose $F : X \to Y$ is a smooth map, where X is a smooth manifold. A point $x \in X$ is called a *critical* point of F, if the differential $D_x F : T_x X \to T_{F(x)} Y$ is not surjective.

We denote by $\boldsymbol{Cr}_F$ the set of critical points of F, and by $\Delta_F \subset Y$ its image via F. We will refer to Δ_F as the discriminant set of F. The points in Δ_F are called the *critical values* of F. □

Exercise 2.1.14. Define

$$\mathcal{Z} := \left\{ (x, a, b, c) \in \mathbb{R}^4; \ ax^2 + bx + c = 0; \ a \neq 0 \right\}.$$

(a) Prove that $\mathcal{Z}$ is a smooth submanifold of $\mathbb{R}^4$.

(b) Define $\pi : \mathcal{Z} \to \mathbb{R}^3$ by $(x, a, b, c) \overset{\pi}{\longmapsto} (a, b, c)$. Compute the discriminant set of π. □

Suppose U, V are finite dimensional real Euclidean vector spaces, $\mathcal{O} \subset U$ is an open subset, and $F : \mathcal{O} \to V$ is a smooth map. Then a point $u \in \mathcal{O}$ is a *critical point* of F if and only if

$$\operatorname{rank}(D_u F : U \to V) < \dim V.$$

Exercise 2.1.15. Show that for every $q \in N \setminus \Delta_F$ the fiber $f^{-1}(q)$ is either empty, or a submanifold of M of codimension $\dim N$. □

Theorem 2.1.16 (Sard). *Let $f : \mathcal{O} \to V$ be a smooth map as above. Then the discriminant set Δ_F is negligible.*

Proof. We follow the elegant approach of J. Milnor [74] and L. Pontryagin [82]. Set $n = \dim U$, and $m = \dim V$. We will argue inductively on the dimension n.

For every positive integer k we denote by $\boldsymbol{Cr}_F^k \subset \boldsymbol{Cr}_F$ the set of points $u \in \mathcal{O}$ such that all the partial derivatives of F up to order k vanish at u. We obtain a decreasing filtration of closed sets

$$\boldsymbol{Cr}_F \supset \boldsymbol{Cr}_F^1 \supset \boldsymbol{Cr}_F^2 \supset \cdots.$$

The case $n = 0$ is trivially true so we may assume $n > 0$, and the statement is true for any $n' < n$, and any m. The inductive step is divided into three intermediary steps.

Step 1. The set $F(\boldsymbol{Cr}_F \setminus \boldsymbol{Cr}_F^1)$ is negligible.

Step 2. The set $F(\boldsymbol{Cr}_F^k \setminus \boldsymbol{Cr}_F^{k+1})$ is negligible for all $k \geq 1$.

Step 3. The set $F(\boldsymbol{Cr}_F^k)$ is negligible for some sufficiently large k.

Step 1. Set $\boldsymbol{Cr}_F' := \boldsymbol{Cr}_F \setminus \boldsymbol{Cr}_F^1$. We will show that there exists a countable open cover $\{\mathcal{O}_j\}_{j \geq 1}$ of $\boldsymbol{Cr}_F'$ such that $F(\mathcal{O}_j \cap \boldsymbol{Cr}_F')$ is negligible for all $j \geq 1$. Since

Cr'_F is separable, it suffices to prove that every point $u \in Cr'_F$ admits an open neighborhood $\mathcal{N}$ such that $F(\mathcal{N} \cap Cr'_F)$ is negligible.

Suppose $u_0 \in Cr'_F$. Assume first that there exist local coordinates $(x^1, \ldots, x^n)$ defined in a neighborhood $\mathcal{N}$ of u_0, and local coordinates $(y^1, \ldots, y^m)$ near $v_0 = F(u_0)$ such that,

$$x^i(u_0) = 0, \quad \forall i = 1, \ldots n, \quad y^j(v_0) = 0, \quad \forall j = 1, \ldots m,$$

and the restriction of F to $\mathcal{N}$ is described by functions $y^j = y^j(x^1, \ldots, x^m)$ such that $y^1 = x^1$.

For every $t \in \mathbb{R}$ we set

$$\mathcal{N}_t := \{(x^1, \ldots, x^n) \in \mathcal{N}; \ x^1 = t\},$$

and we define

$$G_t : \mathcal{N}_t \to \mathbb{R}^{m-1}, \quad (t, x^2, \ldots, x^n) \mapsto \left(y^2(t, x^2, \ldots, x^n), \ldots, y^m(t, x^2, \ldots, x^n)\right).$$

Observe that

$$\mathcal{N} \cap Cr'_F = \bigcup_t Cr_{G_t}.$$

The inductive assumption implies that the sets Cr_{G_t} have trivial $(n-1)$-dimensional Lebesgue measure. Using Fubini's theorem we deduce that $\mathcal{N} \cap Cr'_F$ has trivial n-dimensional Lebesgue measure.

To conclude Step 1 is suffices to prove that the above simplifying assumption concerning the existence of nice coordinates is always fulfilled. To see this, choose local coordinates $(s^1, \ldots, s^n)$ near u_0 and coordinates $(y^1, \ldots, y^m)$ near v_0 such that

$$s^i(u_0) = 0, \quad \forall i = 1, \ldots, n, \quad y^j(v_0) = 0, \quad \forall j = 1, \ldots, m,$$

The map F is then locally described by a collection of functions $y^j(s^1, \ldots, s^n)$, $j = 1, \ldots, n$. Since $u \in Cr'_F$, we can assume, after an eventual re-labelling of coordinates, that $\frac{\partial y^1}{\partial s^1}(u_0) \neq 0$. Now define

$$x^1 = y^1(s^1, \ldots, s^n), \quad x^i = s^i, \quad \forall i = 2, \ldots, n.$$

The implicit function theorem shows that the collection of functions $(x^1, \ldots, x^n)$ defines a coordinate system in a neighborhood of u_0. We regard y^j as functions of x^i. From the definition we deduce $y^1 = x^1$.

Step 2. Set $Cr_F^{(k)} := Cr_F^k \setminus Cr_F^{k+1}$. Since $u_0 \in Cr_F^{(k)}$, we can find local coordinates $(s^1, \ldots, s^n)$ near u_0 and coordinates $(y^1, \ldots, y^m)$ near v_0 such that

$$s^i(u_0) = 0, \quad \forall i = 1, \ldots, n, \quad y^j(v_0) = 0, \quad \forall j = 1, \ldots, m,$$

$$\frac{\partial^j y^1}{\partial (s^1)^j}(u_0) = 0, \quad \forall j = 1, \ldots, k,$$

and

$$\frac{\partial^{k+1} y^1}{\partial (s^1)^{k+1}}(u_0) \neq 0.$$

Define

$$x^1(s) = \frac{\partial^k y^1}{\partial (s^1)^k},$$

and set $x^i := s^i$, $\forall i = 2, \ldots, n$.

Then the collection (x^i) defines smooth local coordinates on an open neighborhood $\mathcal{N}$ of u_0, and $\boldsymbol{Cr}^k_F \cap \mathcal{N}$ is contained in the hyperplane $\{x^1 = 0\}$. Define

$$G : \mathcal{N} \cap \{x^1 = 0\} \to V; \quad G(x^2, \ldots, x^m) = F(0, x^2, \ldots, x^n).$$

Then

$$\boldsymbol{Cr}^k_G \cap \mathcal{N} = \boldsymbol{Cr}^k_G, \quad F(\boldsymbol{Cr}^k_F \cap \mathcal{N}) = G(\boldsymbol{Cr}^k_G),$$

and the induction assumption implies that $G(\boldsymbol{Cr}^k_G)$ is negligible. By covering $\boldsymbol{Cr}^{(k)}_F$ with a countably many open neighborhood $\{\mathcal{N}_\ell\}_{\ell > 1}$ such that $F(\boldsymbol{Cr}^k_F \cap \mathcal{N}_\ell)$ is negligible we conclude that $F(\boldsymbol{Cr}^{(k)}_F)$ is negligible.

Step 3. Suppose $k > \frac{n}{m}$. We will prove that $F(\boldsymbol{Cr}^k_F)$ is negligible. More precisely, we will show that, for every compact subset $S \subset \mathcal{O}$, the set $F(S \cap \boldsymbol{Cr}^k_F)$ is negligible.

From the Taylor expansion around points in $\boldsymbol{Cr}^k_F \cap S$ we deduce that there exist numbers $r_0 \in (0, 1)$ and $\lambda_0 > 0$, depending only on S, such that, if C is a cube with edge $r < r_0$ which intersects $\boldsymbol{Cr}^k_F \cap S$, then

$$\operatorname{diam} F(C) < \lambda_0 r^k,$$

where for every set $A \subset V$ we define

$$\operatorname{diam}(A) := \sup\{ |a_1 - a_2|; \ a_1, a_2 \in A \}.$$

In particular, if μ_m denotes the m-dimensional Lebesgue measure on V, and μ_n denotes the n-dimensional Lebesgue measure on U, we deduce that there exists a constant $\lambda_1 > 0$ such that

$$\mu_m(F(C)) \leq \lambda_1 r^{mk} = \lambda_1 \mu_n(C)^{mk/n}.$$

Cover $\boldsymbol{Cr}^k_F \cap S$ by finitely many cubes $\{C_\ell\}_{1 \leq \ell \leq L}$, of edges $< r_0$, such that their interiors are disjoint. For every positive integer P, subdivide each of the cubes C_ℓ into P^n sub-cubes C^i_ℓ of equal sizes. For every sub-cube C^σ_ℓ which intersects $\boldsymbol{Cr}^k_F$ we have

$$\mu_m(C^\sigma_\ell) \leq \lambda_1 \mu_n(C^\sigma_\ell)^{mk/n} = \frac{\lambda_1}{P^{mk}} \mu_n(C_\ell).$$

We deduce that

$$\mu_m(C_\ell \cap \boldsymbol{Cr}^k_F) \leq \sum_\sigma \mu_m(C^\sigma_\ell \cap \boldsymbol{Cr}^k_F) \leq P^{n-mk} \mu_n(C).$$

If we let $P \to \infty$ in the above inequality, we deduce that when $k > \frac{n}{m}$ we have

$$\mu_m(C_\ell \cap \boldsymbol{Cr}^k_F) = 0, \quad \forall \ell = 1, \ldots, L. \qquad \square$$

Theorem 2.1.16 admits the following immediate generalization.

Corollary 2.1.17 (Sard). *Suppose $F : X \to Y$ is a smooth map between two smooth manifolds. Then its discriminant set is negligible.* □

Definition 2.1.18. A smooth map $f : M \to N$ is called *immersion* (resp. *submersion*) if for every $p \in M$ the differential $D_p f : T_p M \to T_{f(p)} N$ is injective (resp. surjective). A smooth map $f : M \to N$ is called an *embedding* if it is an injective immersion. □

Suppose $F : M \to N$ is a smooth map, and $\dim M \geq \dim N$. Then F is a submersion if and only if the discriminant set Δ_F is empty.

Exercise 2.1.19. Suppose $F : M \to N$ is a smooth map, and $S \subset N$ is a smooth submanifold of N. We say that F is *transversal* to S if for every $x \in F^{-1}(S)$ we have

$$T_{F(x)} N = T_{F(x)} S + D_x F(T_x M).$$

Prove that if F is transversal to S, then $F^{-1}(S)$ is a submanifold of M whose codimension is equal to the codimension of S in N. □

Exercise 2.1.20. Suppose Λ, X, Y are smooth, connected manifolds, and $F : \Lambda \times X \to Y$ is a smooth map

$$\Lambda \times X \ni (\lambda, x) \mapsto F_\lambda(x) \in Y.$$

Suppose S is a submanifold of Y such that F is transversal to S. Define

$$\mathcal{Z} = F^{-1}(S) \subset \Lambda \times X,$$

$$\Lambda_0 = \left\{ \lambda \in \Lambda; \ \ F_\lambda : X \to Y \text{ is not transversal to } S \right\}.$$

Prove that Λ_0 is contained in the discriminant set of the natural projection $\mathcal{Z} \to \Lambda$. In particular, Λ_0 must be negligible. □

2.1.4   Vector bundles

The tangent bundle TM of a manifold M has some special features which makes it a very particular type of manifold. We list now the special ingredients which enter into this special structure of TM since they will occur in many instances. Set for brevity $E := TM$, and $F := \mathbb{R}^m$ ($m = \dim M$). We denote by $\mathrm{Aut}(F)$ the Lie group $\mathrm{GL}(n, \mathbb{R})$ of linear automorphisms of F. Then

(a) E is a smooth manifold, and there exists a surjective submersion $\pi : E \to M$. For every $U \subset M$ we set $E \mid_U := \pi^{-1}(U)$.

(b) From (2.1.3) we deduce that there exists a *trivializing cover*, i.e., an open cover $\mathcal{U}$ of M, and for every $U \in \mathcal{U}$ a diffeomorphism

$$\Psi_U : E \mid_U \to U \times F, \quad v \mapsto (p = \pi(v), \Phi_p^U(v))$$

(b1) Φ_p is a diffeomorphism $E_p \to F$ for any $p \in U$.

(b2) If U, $V \in \mathcal{U}$ are two trivializing neighborhoods with non empty overlap $U \cap V$ then, for any $p \in U \cap V$, the map $\Phi_{VU}(p) = \Phi_p^V \circ (\Phi_p^U)^{-1} : F \to F$ is a linear isomorphism, and moreover, the map

$$p \mapsto \Phi_{VU}(p) \in \mathrm{Aut}(F)$$

is smooth.

In our special case, the map $\Phi_{VU}(p)$ is explicitly defined by the matrix (2.1.3)

$$A(p) = \left(\frac{\partial y^j}{\partial x^i}(p) \right)_{1 \le i, j \le m}.$$

In the above formula, the functions (x^i) are local coordinates on U, and the functions (y^j) are local coordinates on V.

The properties (a) and (b) make no mention of the special relationship between $E = TM$ and M. There are many other quadruples (E, π, M, F) with these properties and they deserve a special name.

Definition 2.1.21. A *vector bundle* is a quadruple (E, π, M, F) such that

- E, M are smooth manifolds,
- $\pi : E \to M$ is a surjective submersion,
- F is a vector space over the field $\mathbb{K} = \mathbb{R}, \mathbb{C}$, and
- the conditions (a) and (b) above are satisfied.

The manifold E is called the *total space*, and M is called the *base space*. The vector space F is called the *standard fiber*, and its dimension (over the field of scalars $\mathbb{K}$) is called the *rank* of the bundle. A *line bundle* is a vector bundle of rank one.☐

Roughly speaking, a vector bundle is a smooth family of vector spaces. Note that the properties (b1) and (b2) imply that the fibers $\pi^{-1}(p)$ of a vector bundle have a natural structure of linear space. In particular, one can add elements in the same fiber. Moreover, the addition and scalar multiplication operations on $\pi^{-1}(p)$ depend smoothly on p. The smoothness of the addition operation this means that the addition is a smooth map

$$+ : E \times_M E = \big\{ (u, v) \in E \times E; \ \pi(u) = \pi(v) \big\} \to E.$$

The smoothness of the scalar multiplication means that it is smooth map

$$\mathbb{R} \times E \to E.$$

There is an equivalent way of defining vector bundles. To describe it, let us introduce a notation. For any vector space F over the field $\mathbb{K} = \mathbb{R}, \mathbb{C}$ we denote by $\mathrm{GL}_{\mathbb{K}}(F)$, (or simply $\mathrm{GL}(F)$ if there is no ambiguity concerning the field of scalars $\mathbb{K}$) the Lie group of linear automorphisms $F \to F$.

According to Definition 2.1.21, we can find an open cover (U_α) of M such that each of the restrictions $E_\alpha = E|_{U_\alpha}$ is isomorphic to a product $\Psi_\alpha : E_\alpha \cong F \times U_\alpha$. Moreover, on the overlaps $U_\alpha \cap U_\beta$, *the transition maps* $g_{\alpha\beta} = \Psi_\alpha \Psi_\beta^{-1}$ can be viewed as smooth maps

$$g_{\alpha\beta} : U_\alpha \cap U_\beta \to \mathrm{GL}(F).$$

They satisfy the *cocycle condition*

(a) $g_{\alpha\alpha} = \mathbb{1}_F$
(b) $g_{\alpha\beta}g_{\beta\gamma}g_{\gamma\alpha} = \mathbb{1}_F$ over $U_\alpha \cap U_\beta \cap U_\gamma$.

Conversely, given an open cover (U_α) of M, and a collection of smooth maps

$$g_{\alpha\beta} : U_\alpha \cap U_\beta \to \mathrm{GL}(F)$$

satisfying the cocycle condition, we can reconstruct a vector bundle by gluing the product bundles $E_\alpha = F \times U_\alpha$ on the overlaps $U_\alpha \cap U_\beta$ according to the gluing rules

the point $(v, x) \in E_\alpha$ is identified with the point $\big(g_{\beta\alpha}(x)v, x\big) \in E_\beta \ \forall x \in U_\alpha \cap U_\beta$.

The details are carried out in the exercise below.

We will say that the map $g_{\beta\alpha}$ *is the transition from* the α-trivialization to the β-trivialization, and we will refer to the collection of maps $(g_{\beta\alpha})$ satisfying the cocycle condition as a *gluing cocycle*. We will refer to the cover (U_α) as above as a *trivializing cover*.

Exercise 2.1.22. Consider a smooth manifold M, a vector space V, an open cover (U_α), and smooth maps

$$g_{\alpha\beta} : U_\alpha \cap U_\beta \to \mathrm{GL}(V)$$

satisfying the cocycle condition. Set

$$X := \bigcup_\alpha V \times U_\alpha \times \{\alpha\}.$$

We topologize X as the disjoint union of the topological spaces $U_\alpha \times V$, and we define a relation $\sim \subset X \times X$ by

$$V \times U_\alpha \times \{\alpha\} \ni (u, x, \alpha) \sim (v, x, \beta) \in V \times U_\beta \times \{\beta\} \stackrel{def}{\Longleftrightarrow} x = y, \ v = g_{\beta\alpha}(x)u.$$

(a) Show that $\sim$ is an equivalence relation, and $E = X/\sim$ equipped with the quotient topology has a natural structure of smooth manifold.
(b) Show that the projection $\pi : X \to M$, $(u, x, \alpha) \mapsto x$ descends to a submersion $E \to M$.
(c) Prove that (E, π, M, V) is naturally a smooth vector bundle. $\square$

Definition 2.1.23. A description of a vector bundle in terms of a trivializing cover, and a gluing cocycle is called a *gluing cocycle description* of that vector bundle. $\square$

Exercise 2.1.24. Find a gluing cocycle description of the tangent bundle of the 2-sphere. □

In the sequel, we will prefer to think of vector bundles in terms of gluing cocycles.

Definition 2.1.25. (a) A *section* in a vector bundle $E \xrightarrow{\pi} M$ defined over the open subset $u \in M$ is a smooth map $s : U \to E$ such that

$$s(p) \in E_p = \pi^{-1}(p), \ \ \forall p \in U \iff \pi \circ s = \mathbb{1}_U.$$

The space of smooth sections of E over U will be denoted by $\Gamma(U, E)$ or $C^\infty(U, E)$. Note that $\Gamma(U, E)$ is naturally a vector space.

(b) A section of the tangent bundle of a smooth manifold is called a *vector field* on that manifold. The space of vector fields over on open subset U of a smooth manifold is denoted by Vect (U). □

Proposition 2.1.26. *Suppose $E \to M$ is a smooth vector bundle with standard fiber F, defined by an open cover $(U_\alpha)\alpha \in \mathcal{A}$, and gluing cocycle*

$$g_{\beta\alpha} : U_{\alpha\beta} \to \mathrm{GL}(F).$$

Then there exists a natural bijection between the vector space of smooth sections of E, and the set of families of smooth maps $\{ s_\alpha : U_\alpha \to F; \ \alpha \in \mathcal{A} \}$, satisfying the following gluing condition on the overlaps

$$s_\alpha(x) = g_{\alpha\beta}(x)s_\beta(x), \ \ \forall x \in U_\alpha \cap U_\beta.$$ □

Exercise 2.1.27. Prove the above proposition. □

Definition 2.1.28. (a) Let $E^i \xrightarrow{\pi_i} M_i$ be two smooth vector bundles. A *vector bundle map* consists of a pair of smooth maps $f : M_1 \to M_2$ and $F : E^1 \to E^2$ satisfying the following properties.

- The map F covers f, i.e., $F(E_p^1) \subset E_{f(p)}^2$, $\forall p \in M_1$. Equivalently, this means that the diagram below is commutative

$$
\begin{array}{ccc}
E^1 & \xrightarrow{\ F\ } & E^2 \\
\pi_1 \downarrow & & \downarrow \pi_2 \\
M_1 & \xrightarrow{\ f\ } & M_2
\end{array}
$$

- The induced map $F : E_p^1 \to E_{f(p)}^2$ is linear.

The composition of bundle maps is defined in the obvious manner and so is the identity morphism so that one can define the notion of bundle isomorphism in the standard way.

(b) If E and F are two vector bundles over the same manifold, then we denote by $\mathrm{Hom}(E, F)$ the space of bundle maps $E \to F$ which *cover the identity* $\mathbb{1}_M$. Such bundle maps are called *bundle morphisms*. □

For example, the differential Df of a smooth map $f : M \to N$ is a bundle map $Df : TM \to TN$ covering f.

Definition 2.1.29. Let $E \xrightarrow{\pi} M$ be a smooth vector bundle. A *bundle endo-morphism* of E is a bundle morphism $F : E \to E$. An automorphism (or *gauge transformation*) is an invertible endomorphism.   □

Example 2.1.30. Consider the trivial vector bundle $\mathbb{R}^n_M \to M$ over the smooth manifold M. A section of this vector bundle is a smooth map $u : M \to \mathbb{R}^n$. We can think of u as a smooth family of vectors $(u(x) \in \mathbb{R}^n)_{x \in M}$.

An endomorphism of this vector bundle is a smooth map $A : M \to \operatorname{End}_{\mathbb{R}}(\mathbb{R}^n)$. We can think of A as a smooth family of $n \times n$ matrices

$$A_x = \begin{bmatrix} a^1_1(x) & a^1_2(x) & \cdots & a^1_n(x) \\ a^2_1(x) & a^2_2(x) & \cdots & a^2_n(x) \\ \vdots & \vdots & \vdots & \vdots \\ a^n_1(x) & a^n_2(x) & \cdots & a^n_n(x) \end{bmatrix}, \quad a^i_j(x) \in C^\infty(M).$$

The map A is a gauge transformation if and only if $\det A_x \neq 0$, $\forall x \in M$.   □

Exercise 2.1.31. Suppose $E_1, E_2 \to M$ are two smooth vector bundles over the smooth manifold with standard fibers F_1, and respectively F_2. Assume that both bundle are defined by a common trivializing cover $(U_\alpha)_{\alpha \in A}$ and gluing cocycles

$$g_{\beta\alpha} : U_{\alpha\beta} \to \operatorname{GL}(F_1), \quad h_{\beta\alpha} : U_{\alpha\beta} \to \operatorname{GL}(F_2).$$

Prove that there exists a bijection between the vector space of bundle morphisms $\operatorname{Hom}(E, F)$, and the set of families of smooth maps

$$\{ T_\alpha : U_\alpha \to \operatorname{Hom}(F_1, F_2); \ \alpha \in A \},$$

satisfying the gluing conditions

$$T_\beta(x) = h_{\beta\alpha}(x) T_\alpha(x)(x) g_{\beta\alpha}^{-1}, \quad \forall x \in U_{\alpha\beta}.$$ □

Exercise 2.1.32. Let V be a vector space, M a smooth manifold, $\{U_\alpha\}$ an open cover of M, and $g_{\alpha\beta}$, $h_{\alpha\beta} : U_\alpha \cap U_\beta \to \operatorname{GL}(V)$ two collections of smooth maps satisfying the cocycle conditions. Prove the two collections define isomorphic vector bundles if and only they are *cohomologous*, i.e., there exist smooth maps $\phi_\alpha : U_\alpha \to \operatorname{GL}(V)$ such that

$$h_{\alpha\beta} = \phi_\alpha g_{\alpha\beta} \phi_\beta^{-1}.$$ □

2.1.5 *Some examples of vector bundles*

In this section we would like to present some important examples of vector bundles and then formulate some questions concerning the global structure of a bundle.

Example 2.1.33. (The tautological line bundle over $\mathbb{RP}^n$ and $\mathbb{CP}^n$). First, let us recall that a rank one vector bundle is usually called a *line bundle*. We consider only the complex case. The total space of the *tautological* or *universal* line bundle over $\mathbb{CP}^n$ is the space

$$\mathcal{U}_n = \mathcal{U}_n^{\mathbb{C}} = \left\{ (z, L) \in \mathbb{C}^{n+1} \times \mathbb{CP}^n; \ z \text{ belongs to the line } L \subset \mathbb{C}^{n+1} \right\}.$$

Let $\pi : \mathcal{U}_n^{\mathbb{C}} \to \mathbb{CP}^n$ denote the projection onto the second component. Note that for every $L \in \mathbb{CP}^n$, the fiber through $\pi^{-1}(L) = \mathcal{U}_{n,L}^{\mathbb{C}}$ coincides with the one-dimensional subspace in $\mathbb{C}^{n+1}$ defined by L. $\square$

Example 2.1.34. (The tautological vector bundle over a Grassmannian). We consider here for brevity only complex Grassmannian $\mathbf{Gr}_k(\mathbb{C}^n)$. The real case is completely similar. The total space of this bundle is

$$\mathcal{U}_{k,n} = \mathcal{U}_{k,n}^{\mathbb{C}} = \left\{ (z, L) \in \mathbb{C}^n \times \mathbf{Gr}_k(\mathbb{C}^n) \ ; \ z \text{ belongs to the subspace } L \subset \mathbb{C}^n \right\}.$$

If π denotes the natural projection $\pi : \mathcal{U}_{k,n} \to \mathbf{Gr}_k(\mathbb{C}^n)$, then for each $L \in \mathbf{Gr}_k(\mathbb{C}^n)$ the fiber over L coincides with the subspace in $\mathbb{C}^n$ defined by L. Note that $\mathcal{U}_{n-1}^{\mathbb{C}} = \mathcal{U}_{1,n}^{\mathbb{C}}$. $\square$

Exercise 2.1.35. Prove that $\mathcal{U}_n^{\mathbb{C}}$ and $\mathcal{U}_{k,n}^{\mathbb{C}}$ are indeed smooth vector bundles. Describe a gluing cocycle for $\mathcal{U}_n^{\mathbb{C}}$. $\square$

Example 2.1.36. A complex line bundle over a smooth manifold M is described by an open cover $(U_\alpha)_{\alpha \in \mathcal{A}}$, and smooth maps

$$g_{\beta\alpha} : U_\alpha \cap U_\beta \to \mathrm{GL}(1, \mathbb{C}) \cong \mathbb{C}^*,$$

satisfying the cocycle condition

$$g_{\gamma\alpha}(x) = g_{\gamma\beta}(x) \cdot g_{\beta\alpha}(x), \ \forall x \in U_\alpha \cap U_\beta \cap U_\gamma.$$

Consider for example the manifold $M = S^2 \subset \mathbb{R}^3$. Denote as usual by N and S the North and respectively South pole. We have an open cover

$$S^2 = U_0 \cup U_\infty, \ U_0 = S^2 \setminus \{S\}, \ U_1 = S^2 \setminus \{N\}.$$

In this case, we have only a single nontrivial overlap, $U_N \cap U_S$. Identify U_0 with the complex line $\mathbb{C}$, so that the North pole becomes the origin $z = 0$.

For every $n \in \mathbb{Z}$ we obtain a complex line bundle $L_n \to S^2$, defined by the open cover $\{U_0, U_1\}$ and gluing cocycle

$$g_{10}(z) = z^{-n}, \ \forall z \in \mathbb{C}^* = U_0 \setminus \{0\}.$$

A smooth section of this line bundle is described by a pair of smooth functions

$$u_0 : U_0 \to \mathbb{C}, \quad u_1 : U_1 \to \mathbb{C},$$

which along the overlap $U_0 \cap U_1$ satisfy the equality $u_1(z) = z^{-n} u_0(z)$. For example, if $n \geq 0$, the pair of functions

$$u_0(z) = z^n, \quad u_1(p) = 1, \quad \forall p \in U_1$$

defines a smooth section of L_n.   □

Exercise 2.1.37. We know that $\mathbb{CP}^1$ is diffeomorphic to S^2. Prove that the universal line bundle $\mathcal{U}_n \to \mathbb{CP}^1$ is isomorphic with the line bundle L_{-1} constructed in the above example.   □

Exercise 2.1.38. Consider the incidence set

$$\mathfrak{I} := \left\{ (x, L) \in (\mathbb{C}^{n+1} \setminus \{0\}) \times \mathbb{CP}^n; \ z \in L \right\}.$$

Prove that the closure of $\mathfrak{I}$ in $\mathbb{C}^{n+1} \times \mathbb{CP}^n$ is a smooth manifold diffeomorphic to the total space of the universal line bundle $\mathcal{U}_n \to \mathbb{CP}^n$. This manifold is called the *complex blowup* of $\mathbb{C}^{n+1}$ at the origin.   □

   The family of vector bundles is very large. The following construction provides a very powerful method of producing vector bundles.

Definition 2.1.39. Let $f : X \to M$ be a smooth map, and E a vector bundle over M defined by an open cover (U_α) and gluing cocycle $(g_{\alpha\beta})$. The *pullback of E by f* is the vector bundle f^*E over X defined by the open cover $f^{-1}(U_\alpha)$, and the gluing cocycle $(g_{\alpha\beta} \circ f)$.   □

   One can check easily that the isomorphism class of the pullback of a vector bundle E is independent of the choice of gluing cocycle describing E. The pullback operation defines a linear map between the space of sections of E and the space of sections of f^*E.

   More precisely, if $s \in \Gamma(E)$ is defined by the open cover (U_α), and the collection of smooth maps (s_α), then the pullback f^*s is defined by the open cover $f^{-1}(U_\alpha)$, and the smooth maps $(s_\alpha \circ f)$. Again, there is no difficulty to check the above definition is independent of the various choices.

Exercise 2.1.40. For every positive integer k consider the map

$$p_k : \mathbb{CP}^1 \to \mathbb{CP}^1, \quad p_k([z_0, z_1]) = [z_0^k, z_1^k].$$

Show that $p_k^* L_n \cong L_{kn}$, where L_n is the complex line bundle $L_n \to \mathbb{CP}^1$ defined in Example 2.1.36.   □

Exercise 2.1.41. Let $E \to X$ be a rank k (complex) smooth vector bundle over the manifold X. Assume E is *ample*, i.e. there exists a finite family $s_1, \ldots, s_N$ of smooth sections of E such that, for any $x \in X$, the collection $\{s_1(x), \ldots, s_N(x)\}$ spans E_x. For each $x \in X$ we set

$$S_x := \Big\{ v \in \mathbb{C}^N \; ; \; \sum_i v^i s_i(x) = 0 \Big\}.$$

Note that $\dim S_x = N - k$. We have a map $F : X \to \mathbf{Gr}_k(\mathbb{C}^N)$ defined by $x \mapsto S_x^\perp$.
(a) Prove that F is smooth.
(b) Prove that E is isomorphic with the pullback $F^* \mathcal{U}_{k,N}$. $\qquad\square$

Exercise 2.1.42. Show that any vector bundle over a smooth *compact* manifold is ample. Thus any vector bundle over a compact manifold is a pullback of some tautological bundle! $\qquad\square$

The notion of vector bundle is trickier than it may look. Its definition may suggest that a vector bundle looks like a direct product *manifold* × vector space since this happens at least locally. We will denote by $\mathbb{K}_M^n$ the bundle $\mathbb{K}^r \times M \to M$.

Definition 2.1.43. A rank r vector bundle $E \xrightarrow{\pi} M$ (over the field $\mathbb{K} = \mathbb{R}, \mathbb{C}$) is called *trivial* or *trivializable* if there exists a bundle isomorphism $E \cong \mathbb{K}_M^r$. A bundle isomorphism $E \to \mathbb{K}_M^r$ is called a *trivialization* of E, while an isomorphism $\mathbb{K}^r \to E$ is called a *framing* of E.

A pair (trivial vector bundle, trivialization) is called a *trivialized*, or *framed* bundle. $\qquad\square$

Remark 2.1.44. Let us explain why we refer to a bundle isomorphism $\varphi : \mathbb{K}_M^r \to E$ as a framing.

Denote by $(e_1, \ldots, e_r)$ the canonical basis of $\mathbb{K}^r$. We can also regard the vectors e_i as constant maps $M \to \mathbb{K}^r$, i.e., as (special) sections of $\mathbb{K}_M^r$. The isomorphism φ determines sections $f_i = \varphi(e_i)$ of E with the property that for every $x \in M$ the collection $(f_1(x), \ldots, f_r(x))$ is a *frame* of the fiber E_x.

This observation shows that we can regard any framing of a bundle $E \to M$ of rank r as a collection of r sections $u_1, \ldots, u_r$ which are *pointwise* linearly independent. $\qquad\square$

One can naively ask the following question. Is every vector bundle trivial? We can even limit our search to tangent bundles. Thus we ask the following question.

Is it true that for every smooth manifold M the tangent bundle TM is trivial (as a vector bundle)?

Let us look at some positive examples.

Example 2.1.45. $TS^1 \cong \mathbb{R}_{S^1}$ Let θ denote the angular coordinate on the circle. Then $\frac{\partial}{\partial \theta}$ is a globally defined, nowhere vanishing vector field on S^1. We thus get a map

$$\mathbb{R}_{S^1} \to TS^1, \quad (s,\theta) \mapsto (s\frac{\partial}{\partial\theta}, \theta) \in T_\theta S^1$$

which is easily seen to be a bundle isomorphism.

Let us carefully analyze this example. Think of S^1 as a Lie group (the group of complex numbers of norm 1). The tangent space at $z = 1$, i.e., $\theta = 0$, coincides with the subspace $\textbf{Re}\, z = 0$, and $\frac{\partial}{\partial\theta}|_1$ is the unit vertical vector $\textbf{\textit{j}}$.

Denote by R_θ the counterclockwise rotation by an angle θ. Clearly R_θ is a diffeomorphism, and for each θ we have a linear isomorphism

$$D_\theta|_{\theta=0}\, R_\theta : T_1 S^1 \to T_\theta S^1.$$

Moreover,

$$\frac{\partial}{\partial\theta} = D_\theta|_{\theta=0}\, R_\theta \textbf{\textit{j}}.$$

The existence of the trivializing vector field $\frac{\partial}{\partial\theta}$ is due to to our ability to "move freely and coherently" inside S^1. One has a similar freedom inside a Lie group as we are going to see in the next example.   □

Example 2.1.46. For any Lie group G the tangent bundle TG is trivial.

To see this let $n = \dim G$, and consider $e_1, \ldots, e_n$ a basis of the tangent space at the origin, $T_1 G$. We denote by R_g the right translation (by g) in the group defined by

$$R_g : \; x \mapsto x \cdot g, \quad \forall x \in G.$$

R_g is a diffeomorphism with inverse $R_{g^{-1}}$ so that the differential DR_g defines a linear isomorphism $DR_g : T_1 G \to T_g G$. Set

$$E_i(g) = DR_g(e_i) \in T_g G, \quad i = 1, \cdots, n.$$

Since the multiplication $G \times G \to G$, $(g, h) \mapsto g \cdot h$ is a smooth map we deduce that the vectors $E_i(g)$ define *smooth* vector fields over G. Moreover, for every $g \in G$, the collection $\{E_1(g), \ldots, E_n(g)\}$ is a basis of $T_g G$ so we can define without ambiguity a map

$$\Phi : \mathbb{R}^n_G \to TG, \quad (g; X^1, \ldots X^n) \mapsto (g; \sum X^i E_i(g)).$$

One checks immediately that Φ is a vector bundle isomorphism and this proves the claim. In particular TS^3 is trivial since the sphere S^3 is a Lie group (unit quaternions). (Using the Cayley numbers one can show that TS^7 is also trivial; see [83] for details.)   □

We see that the tangent bundle TM of a manifold M is trivial if and only if there exist vector fields $X_1, \ldots, X_m$ ($m = \dim M$) such that for each $p \in M$, $X_1(p), \ldots, X_m(p)$ span $T_p M$. This suggests the following more refined question.

Problem *Given a manifold M, compute $v(M)$, the maximum number of pointwise linearly independent vector fields over M. Obviously $0 \le v(M) \le \dim M$ and TM is trivial if and only if $v(M) = \dim M$. A special instance of this problem is the celebrated vector field problem: compute $v(S^n)$ for any $n \ge 1$.*

We have seen that $v(S^n) = n$ for $n = 1, 3$ and 7. Amazingly, these are the only cases when the above equality holds. This is a highly nontrivial result, first proved by J.F.Adams in [2] using very sophisticated algebraic tools. This fact is related to many other natural questions in algebra. For a nice presentation we refer to [69].

The methods we will develop in this book will not suffice to compute $v(S^n)$ for any n, but we will be able to solve "half" of this problem. More precisely we will show that

$$v(S^n) = 0 \text{ if and only if } n \text{ is an even number.}$$

In particular, this shows that TS^{2n} is not trivial. In odd dimensions the situation is far more elaborate (a complete answer can be found in [2]).

Exercise 2.1.47. $v(S^{2k-1}) \ge 1$ for any $k \ge 1$. $\qquad\qquad\qquad\qquad\qquad$ $\square$

The quantity $v(M)$ can be viewed as a measure of nontriviality of a tangent bundle. Unfortunately, its computation is highly nontrivial. In the second part of this book we will describe more efficient ways of measuring the extent of nontriviality of a vector bundle.

2.2 A linear algebra interlude

We collect in this section some classical notions of linear algebra. Most of them might be familiar to the reader, but we will present them in a form suitable for applications in differential geometry. This is perhaps the least glamorous part of geometry, and unfortunately cannot be avoided.

☞ **Convention.** *All the vector spaces in this section will tacitly be assumed finite dimensional, unless otherwise stated.*

2.2.1 Tensor products

Let E, F be two vector spaces over the field $\mathbb{K}$ ($\mathbb{K} = \mathbb{R}, \mathbb{C}$). Consider the (infinite) direct sum

$$\mathcal{T}(E,F) = \bigoplus\nolimits_{(e,f)\in E\times F} \mathbb{K}.$$

Equivalently, the vector space $\mathcal{T}(E, F)$ can be identified with the space of functions $c : E \times F \to \mathbb{K}$ with finite support. The space $\mathcal{T}(E, F)$ has a natural basis consisting of "Dirac functions"

$$\delta_{e,f} : E \times F \to \mathbb{K}, \quad (x,y) \mapsto \begin{cases} 1 \text{ if } (x,y) = (e,f) \\ 0 \text{ if } (x,y) \neq (e,f). \end{cases}$$

In particular, we have an injection[1]

$$\delta : E \times F \to \mathcal{T}(E,F), \quad (e,f) \mapsto \delta_{e,f}.$$

Inside $\mathcal{T}(E,F)$ sits the linear subspace $\mathcal{R}(E,F)$ spanned by

$$\lambda\delta_{e,f} - \delta_{\lambda e,f}, \; \lambda\delta_{e,f} - \delta_{e,\lambda f}, \; \delta_{e+e',f} - \delta_{e,f} - \delta_{e',f}, \; \delta_{e,f+f'} - \delta_{e,f} - \delta_{e,f'},$$

where $e, e' \in E$, $f, f' \in F$, and $\lambda \in \mathbb{K}$. Now define

$$E \otimes_{\mathbb{K}} F := \mathcal{T}(E,F)/\mathcal{R}(E,F),$$

and denote by π the canonical projection $\pi : \mathcal{T}(E,F) \to E \otimes F$. Set

$$e \otimes f := \pi(\delta_{e,f}).$$

We get a natural map

$$\iota : E \times F \to E \otimes F, \; e \times f \mapsto e \otimes f.$$

Obviously ι is bilinear. The vector space $E \otimes_{\mathbb{K}} F$ is called the *tensor product* of E and F over $\mathbb{K}$. Often, when the field of scalars is clear from the context, we will use the simpler notation $E \otimes F$. The tensor product has the following universality property.

Proposition 2.2.1. *For any bilinear map* $\phi : E \times F \to G$ *there exists a* unique *linear map* $\Phi : E \otimes F \to G$ *such that the diagram below is commutative.*

$$
\begin{array}{ccc}
E \times F & \xrightarrow{\;\iota\;} & E \otimes F \\
& \searrow{\phi} & \Big\downarrow{\Phi} \\
& & G
\end{array}
$$
□

The proof of this result is left to the reader as an exercise. Note that if (e_i) is a basis of E, and (f_j) is a basis of F, then $(e_i \otimes f_j)$ is a basis of $E \otimes F$, and therefore

$$\dim_{\mathbb{K}} E \otimes_{\mathbb{K}} F = (\dim_{\mathbb{K}} E) \cdot (\dim_{\mathbb{K}} F).$$

Exercise 2.2.2. Using the universality property of the tensor product prove that there exists a natural isomorphism $E \otimes F \cong F \otimes E$ uniquely defined by $e \otimes f \mapsto f \otimes e$.
□

The above construction can be iterated. Given three vector spaces E_1, E_2, E_3 over the same field of scalars $\mathbb{K}$ we can construct two triple tensor products:

$$(E_1 \otimes E_2) \otimes E_3 \; \text{ and } \; E_1 \otimes (E_2 \otimes E_3).$$

Exercise 2.2.3. Prove there exists a natural isomorphism of $\mathbb{K}$-vector spaces

$$(E_1 \otimes E_2) \otimes E_3 \cong E_1 \otimes (E_2 \otimes E_3).$$
□

---

[1] *A word of caution:* δ *is not linear!*

The above exercise implies that there exists a unique (up to isomorphism) triple tensor product which we denote by $E_1 \otimes E_2 \otimes E_3$. Clearly, we can now define multiple tensor products: $E_1 \otimes \cdots \otimes E_n$.

Definition 2.2.4. (a) For any two vector spaces U, V over the field $\mathbb{K}$ we denote by $\mathrm{Hom}(U, V)$, or $\mathrm{Hom}_{\mathbb{K}}(U, V)$ the space of $\mathbb{K}$-linear maps $U \to V$.
(b) The dual of a $\mathbb{K}$-linear space E is the linear space E^* defined as the space $\mathrm{Hom}_{\mathbb{K}}(E, \mathbb{K})$ of $\mathbb{K}$-linear maps $E \to \mathbb{K}$. For any $e^* \in E^*$ and $e \in E$ we set

$$\langle e^*, e \rangle := e^*(e). \qquad \square$$

The above constructions are functorial. More precisely, we have the following result.

Proposition 2.2.5. *Suppose E_i, F_i, G_i, $i = 1, 2$ are $\mathbb{K}$-vector spaces. Let $T_i \in \mathrm{Hom}(E_i, F_i)$, $S_i \in \mathrm{Hom}(F_i, G_i)$, $i = 1, 2$, be two linear operators. Then they naturally induce a linear operator*

$$T = T_1 \otimes T_2 : E_1 \otimes E_2 \to F_1 \otimes F_2, \quad S_1 \otimes S_2 : F_1 \otimes F_2$$

uniquely defined by

$$T_1 \otimes T_2(e_1 \otimes e_2) = (T_1 e_1) \otimes (T_2 e_2), \quad \forall e_i \in E_i,$$

and satisfying

$$(S_1 \otimes S_2) \circ (T_1 \otimes T_2) = (S_1 \circ T_1) \otimes (S_2 \circ T_2).$$

(b) Any linear operator $A : E \to F$ induces a linear operator $A^\dagger : F^* \to E^*$ uniquely defined by

$$\langle A^\dagger f^*, e \rangle = \langle f^*, Ae \rangle, \quad \forall e \in E, \; f^* \in F^*.$$

The operator $A^\dagger$ is called the transpose *or* adjoint *of A. Moreover,*

$$(A \circ B)^\dagger = B^\dagger \circ A^\dagger, \quad \forall A \in \mathrm{Hom}(F, G), \; B \in \mathrm{Hom}(E, F). \qquad \square$$

Exercise 2.2.6. Prove the above proposition. $\qquad \square$

Remark 2.2.7. Any basis $(e_i)_{1 \le i \le n}$ of the n-dimensional $\mathbb{K}$-vector space determines a basis $(e^i)_{1 \le i \le n}$ of the dual vector space V^* uniquely defined by the conditions

$$\langle e^i, e_j \rangle = \delta^i_j = \begin{cases} 1 & i = j \\ 0 & i \ne j. \end{cases}$$

We say that the basis (e^i) is *dual* to the basis (e_j). The quantity (δ^i_j) is called the *Kronecker symbol*.

A vector $\boldsymbol{v} \in V$ admits a decomposition

$$\boldsymbol{v} = \sum_{i=1}^{n} v^i \boldsymbol{e}_i,$$

while a vector $\boldsymbol{v}^* \in V^*$ admits a decomposition

$$\boldsymbol{v}^* = \sum_{i=1}^{n} v_i^* \boldsymbol{e}^i.$$

Moreover,

$$\langle \boldsymbol{v}^*, \boldsymbol{v} \rangle = \sum_{i=1}^{n} v_i^* v^i.$$

Classically, a vector $\boldsymbol{v}$ in V is represented by a one-column matrix

$$\boldsymbol{v} = \begin{bmatrix} v^1 \\ \vdots \\ v^n \end{bmatrix},$$

while a vector $\boldsymbol{v}^*$ is represented by a one-row matrix

$$\boldsymbol{v}^* = [v_1^* \ \ \cdots \ \ v_n^*].$$

Then

$$\langle \boldsymbol{v}^*, \boldsymbol{v} \rangle = [v_1^* \ \ \cdots \ \ v_n^*] \cdot \begin{bmatrix} v^1 \\ \vdots \\ v^n \end{bmatrix},$$

where the $\cdot$ denotes the multiplication of matrices. $\qquad\square$

Using the functoriality of the tensor product and of the dualization construction one proves easily the following result.

Proposition 2.2.8. *(a) There exists a natural isomorphism*

$$E^* \otimes F^* \cong (E \otimes F)^*,$$

uniquely defined by

$$E^* \otimes F^* \ni e^* \otimes f^* \longmapsto L_{e^* \otimes f^*} \in (E \otimes F)^*,$$

where

$$\langle L_{e^* \otimes f^*}, x \otimes y \rangle = \langle e^*, x \rangle \langle f^*, y \rangle, \ \ \forall x \in E, \ \ y \in F.$$

In particular, this shows $E^* \otimes F^*$ can be naturally identified with the space of bilinear maps $E \times F \to \mathbb{K}$.

(b) The adjunction morphism $E^* \otimes F \to \operatorname{Hom}(E, F)$, given by

$$E^* \otimes F \ni e^* \otimes f \longmapsto T_{e^* \otimes f} \in \operatorname{Hom}(E, F),$$

where

$$T_{e^* \otimes f}(x) := \langle e^*, x \rangle f, \ \ \forall x \in E,$$

is an isomorphism.[2] $\qquad\square$

---

[2]The finite dimensionality of E is absolutely necessary. This adjunction formula is known in Bourbaki circles as *"formule d'adjonction chèr à Cartan"*.

Exercise 2.2.9. Prove the above proposition. □

Let V be a vector space. For $r, s \geq 0$ set

$$\mathcal{T}_s^r(V) := V^{\otimes r} \otimes (V^*)^{\otimes s},$$

where by definition $V^{\otimes 0} = (V^*)^{\otimes 0} = \mathbb{K}$. An element of $\mathcal{T}_s^r$ is called *tensor of type* (r,s).

Example 2.2.10. According to Proposition 2.2.8 a tensor of type $(1,1)$ can be identified with a linear endomorphism of V, i.e.,

$$\mathcal{T}_1^1(V) \cong \operatorname{End}(V),$$

while a tensor of type $(0, k)$ can be identified with a k-linear map

$$\underbrace{V \times \cdots \times V}_{k} \to \mathbb{K}. \qquad \square$$

A tensor of type $(r, 0)$ is called *contravariant*, while a tensor of type $(0, s)$ is called *covariant*. The *tensor algebra* of V is defined to be

$$\mathcal{T}(V) := \bigoplus_{r,s} \mathcal{T}_s^r(V).$$

We use the term algebra since the tensor product induces bilinear maps

$$\otimes : \mathcal{T}_s^r \times \mathcal{T}_{s'}^{r'} \to \mathcal{T}_{s+s'}^{r+r'}.$$

The elements of $\mathcal{T}(V)$ are called *tensors*.

Exercise 2.2.11. Show that $\big(\mathcal{T}(V), +, \otimes\big)$ is an associative algebra. □

Example 2.2.12. It is often useful to represent tensors using coordinates. To achieve this pick a basis (e_i) of V, and let (e^i) denote the dual basis in V^* uniquely defined by

$$\langle e^i, e_j \rangle = \delta_j^i = \begin{cases} 1 \text{ if } i = j \\ 0 \text{ if } i \neq j \end{cases}.$$

We then obtain a basis of $\mathcal{T}_s^r(V)$

$$\{e_{i_1} \otimes \cdots \otimes e_{i_r} \otimes e^{j_1} \otimes \cdots \otimes e^{j_s} \,/\, 1 \leq i_\alpha, j_\beta \leq \dim V\}.$$

Any element $T \in \mathcal{T}_s^r(V)$ has a decomposition

$$T = T_{j_1 \ldots j_s}^{i_1 \ldots i_r} e_{i_1} \otimes \cdots \otimes e_{i_r} \otimes e^{j_1} \otimes \cdots, \otimes e^{j_s},$$

where we use Einstein convention to sum over indices which appear twice, once as a *superscript*, and the second time as *subscript*.

Using the adjunction morphism in Proposition 2.2.8, we can identify the space $\mathcal{T}_1^1(V)$ with the space $\operatorname{End}(V)$ a linear isomorphisms. Using the bases (e_i) and (e^j), and Einstein's convention, the adjunction identification can be described as the correspondence which associates to the tensor $A = a_j^i e_i \otimes e^j \in \mathcal{T}_1^1(V)$, the linear operator $L_A : V \to V$ which maps the vector $v = v^j e_j$ to the vector $L_A v = a_j^i v^j e_i$. □

On the tensor algebra there is a natural *contraction* (or *trace*) operation

$$\mathrm{tr} : \mathcal{T}^r_s \to \mathcal{T}^{r-1}_{s-1}$$

uniquely defined by

$$\mathrm{tr}\,(v_1 \otimes \cdots \otimes v_r \otimes u^1 \otimes \cdots \otimes u^s) := \langle u^1, v_1 \rangle v_2 \otimes \cdots v_r \otimes u^2 \otimes \cdots \otimes u^s,$$

$\forall v_i \in V$, $u^j \in V^*$.

In the coordinates determined by a basis (e_i) of V, the contraction can be described as

$$(\mathrm{tr}\, T)^{i_2 \ldots i_r}_{j_2 \ldots j_s} = \left(T^{i i_2 \ldots i_r}_{i j_2 \ldots j_s} \right),$$

where again we use Einstein's convention. In particular, we see that the contraction coincides with the usual trace on $\mathcal{T}^1_1(V) \cong \mathrm{End}\,(V)$.

2.2.2   *Symmetric and skew-symmetric tensors*

Let V be a vector space over $\mathbb{K} = \mathbb{R}, \mathbb{C}$. We set $\mathcal{T}^r(V) := \mathcal{T}^r_0(V)$, and we denote by $\mathcal{S}_r$ the group of permutations of r objects. When $r = 0$ we set $\mathcal{S}_0 := \{1\}$.

Every permutation $\sigma \in \mathcal{S}_r$ determines a linear map $\mathcal{T}^r(V) \to \mathcal{T}^r(V)$, uniquely determined by the correspondences

$$v_1 \otimes \cdots \otimes v_r \longmapsto v_{\sigma(1)} \otimes \cdots \otimes v_{\sigma(r)}, \quad \forall v_1, \ldots, v_r \in V.$$

We denote this action of $\sigma \in \mathcal{S}_r$ on an arbitrary element $t \in \mathcal{T}^r(V)$ by σt.

In this subsection we will describe two subspaces invariant under this action. These are special instances of the so called *Schur functors*. (We refer to [34] for more general constructions.) Define

$$\boldsymbol{S}_r : \mathcal{T}^r(V) \to \mathcal{T}^r(V), \quad \boldsymbol{S}_r(t) := \frac{1}{r!} \sum_{\sigma \in \mathcal{S}_r} \sigma t,$$

and

$$\boldsymbol{A}_r : \mathcal{T}^r(V) \to \mathcal{T}^r(V), \quad \boldsymbol{A}_r(t) := \begin{cases} \frac{1}{r!} \sum_{\sigma \in \mathcal{S}_r} \epsilon(\sigma) \sigma t & \text{if } r \leq \dim V \\ 0 & \text{if } r > \dim V \end{cases}.$$

Above, we denoted by $\epsilon(\sigma)$ the signature of the permutation σ. Note that

$$\boldsymbol{A}_0 = \boldsymbol{S}_0 = \mathbb{1}_{\mathbb{K}}.$$

The following results are immediate. Their proofs are left to the reader as exercises.

Lemma 2.2.13. *The operators $\boldsymbol{A}_r$ and $\boldsymbol{S}_r$ are projectors of $\mathcal{T}^r(V)$, i.e.,*

$$\boldsymbol{S}_r^2 = \boldsymbol{S}_r, \quad \boldsymbol{A}_r^2 = \boldsymbol{A}_r.$$

Moreover,

$$\sigma \boldsymbol{S}_r(t) = \boldsymbol{S}_r(\sigma t) = \boldsymbol{S}_r(t), \ \sigma \boldsymbol{A}_r(t) = \boldsymbol{A}_r(\sigma t) = \epsilon(\sigma) \boldsymbol{A}_r(t), \ \forall t \in \mathcal{T}^r(V). \qquad \square$$

Definition 2.2.14. A tensor $T \in \mathcal{T}^r(V)$ is called *symmetric* (respectively *skew-symmetric*) if

$$\boldsymbol{S}_r(T) = T \quad (\text{respectively } \boldsymbol{A}_r(T) = T).$$

The nonnegative integer r is called the *degree* of the (skew-)symmetric tensor.

The space of symmetric tensors (respectively skew-symmetric ones) of degree r will be denoted by $\boldsymbol{S}^r V$ (and respectively $\Lambda^r V$). □

Set

$$\boldsymbol{S}^\bullet V := \bigoplus_{r \geq 0} \boldsymbol{S}^r V, \text{ and } \Lambda^\bullet V := \bigoplus_{r \geq 0} \Lambda^r V.$$

Definition 2.2.15. The *exterior product* is the bilinear map

$$\wedge : \Lambda^r V \times \Lambda^s V \to \Lambda^{r+s} V,$$

defined by

$$\omega^r \wedge \eta^s := \frac{(r+s)!}{r!s!} \boldsymbol{A}_{r+s}(\omega \otimes \eta), \quad \forall \omega^r \in \Lambda^r V, \ \eta^s \in \Lambda^s V. \qquad □$$

Proposition 2.2.16. *The exterior product has the following properties.*
(a) (Associativity)

$$(\alpha \wedge \beta) \wedge \gamma = \alpha \wedge (\beta \wedge \gamma), \quad \forall \alpha, \beta \gamma \in \Lambda^\bullet V.$$

In particular,

$$v_1 \wedge \cdots \wedge v_k = k! \boldsymbol{A}_k(v_1 \otimes \ldots \otimes v_k) = \sum_{\sigma \in \mathcal{S}_k} \epsilon(\sigma) v_{\sigma(1)} \otimes \ldots \otimes v_{\sigma(k)}, \quad \forall v_i \in V.$$

(b) (Super-commutativity)

$$\omega^r \wedge \eta^s = (-1)^{rs} \eta^s \wedge \omega^r, \quad \forall \omega^r \in \Lambda^r V, \ \omega^s \in \Lambda^s V.$$

Proof. We first define a new product "$\wedge_1$" by

$$\omega^r \wedge_1 \eta^s := \boldsymbol{A}_{r+s}(\omega \otimes \eta),$$

which will turn out to be associative and will force $\wedge$ to be associative as well.

To prove the associativity of $\wedge_1$ consider the quotient algebra $\mathcal{Q}^* = \mathcal{T}^*/\mathcal{J}^*$, where $\mathcal{T}^*$ is the *associative* algebra $(\bigoplus_{r \geq 0} \mathcal{T}^r(V), +, \otimes)$, and $\mathcal{J}^*$ is the bilateral ideal generated by the set of squares $\{v \otimes v / v \in V\}$. Denote the (obviously associative) multiplication in $\mathcal{Q}^*$ by $\cup$. The natural projection $\pi : \mathcal{T}^* \to \mathcal{Q}^*$ induces a linear map $\pi : \Lambda^\bullet V \to \mathcal{Q}^*$. We will complete the proof of the proposition in two steps.

Step 1. We will prove that the map $\pi : \Lambda^\bullet V \to \mathcal{Q}^*$ is a linear isomorphism, and moreover

$$\pi(\alpha \wedge_1 \beta) = \pi(\alpha) \cup \pi(\beta). \tag{2.2.1}$$

In particular, $\wedge_1$ is an associative product.

The crucial observation is

$$\pi(T) = \pi(\boldsymbol{A}_r(T)), \quad \forall t \in \mathcal{T}^r(V). \tag{2.2.2}$$

It suffices to check (2.2.2) on monomials $T = e_1 \otimes \cdots \otimes e_r$, $e_i \in V$. Since

$$(u + v)^{\otimes 2} \in \mathcal{J}^*, \quad \forall u, v \in V$$

we deduce $u \otimes v = -v \otimes u \pmod{\mathcal{J}^*}$. Hence, for any $\sigma \in \mathcal{S}_r$

$$\pi(e_1 \otimes \cdots \otimes e_r) = \epsilon(\sigma)\pi(e_{\sigma(1)} \otimes \cdots \otimes e_{\sigma(r)}). \tag{2.2.3}$$

When we sum over $\sigma \in \mathcal{S}_r$ in (2.2.3) we obtain (2.2.2).

To prove the injectivity of π note first that $\boldsymbol{A}_*(\mathcal{J}^*) = 0$. If $\pi(\omega) = 0$ for some $\omega \in \Lambda^\bullet V$, then $\omega \in \ker \pi = \mathcal{J}^* \cap \Lambda^\bullet V$ so that

$$\omega = \boldsymbol{A}_*(\omega) = 0.$$

The surjectivity of π follows immediately from (2.2.2). Indeed, any $\pi(T)$ can be alternatively described as $\pi(\omega)$ for some $\omega \in \Lambda^\bullet V$. It suffices to take $\omega = \boldsymbol{A}_*(T)$.

To prove (2.2.1) it suffices to consider only the special cases when α and β are monomials:

$$\alpha = \boldsymbol{A}_r(e_1 \otimes \cdots \otimes e_r), \quad \beta = \boldsymbol{A}_s(f_1 \otimes \cdots \otimes f_s).$$

We have

$$\pi(\alpha \wedge_1 \beta) = \pi\left(\boldsymbol{A}_{r+s}(\boldsymbol{A}_r(e_1 \otimes \cdots \otimes e_r) \otimes \boldsymbol{A}_s(f_1 \otimes \cdots \otimes f_s))\right)$$

$$\stackrel{(2.2.2)}{=} \pi\left(\boldsymbol{A}_r(e_1 \otimes \cdots \otimes e_r) \otimes \boldsymbol{A}_s(f_1 \otimes \cdots \otimes f_s)\right)$$

$$\stackrel{def}{=} \pi(\boldsymbol{A}_r(e_1 \otimes \cdots \otimes e_r)) \cup \pi(\boldsymbol{A}_s(f_1 \otimes \cdots \otimes f_s)) = \pi(\alpha) \cup \pi(\beta).$$

Thus $\wedge_1$ is associative.

Step 2. The product $\wedge$ is associative. Consider $\alpha \in \Lambda^r V$, $\beta \in \Lambda^s V$ and $\gamma \in \Lambda^t V$. We have

$$(\alpha \wedge \beta) \wedge \gamma = \left(\frac{(r+s)!}{r!s!}\alpha \wedge_1 \beta\right) \wedge \gamma = \frac{(r+s)!}{r!s!}\frac{(r+s+t)!}{(r+s)!t!}(\alpha \wedge_1 \beta) \wedge_1 \gamma$$

$$= \frac{(r+s+t)!}{r!s!t!}(\alpha \wedge_1 \beta) \wedge_1 \gamma = \frac{(r+s+t)!}{r!s!t!}\alpha \wedge_1 (\beta \wedge_1 \gamma) = \alpha \wedge (\beta \wedge \gamma).$$

The associativity of $\wedge$ is proved. The computation above shows that

$$e_1 \wedge \cdots \wedge e_k = k!\boldsymbol{A}_k(e_1 \otimes \cdots \otimes e_k).$$

(b) The supercommutativity of $\wedge$ follows from the supercommutativity of $\wedge_1$ (or $\cup$). To prove the latter one uses (2.2.2). The details are left to the reader. $\square$

Exercise 2.2.17. Finish the proof of part (b) in the above proposition. $\square$

The space $\Lambda^\bullet V$ is called the *exterior algebra* of V. $\wedge$ is called the *exterior product* . The exterior algebra is a $\mathbb{Z}$-graded algebra, i.e.,

$$(\Lambda^r V) \wedge (\Lambda^s V) \subset \Lambda^{r+s} V, \ \ \forall r, s.$$

Note that $\Lambda^r V = 0$ for $r > \dim V$ (pigeonhole principle).

Definition 2.2.18. Let V be an n-dimensional $\mathbb{K}$-vector space. The one dimensional vector space $\Lambda^n V$ is called the *determinant line* of V, and it is denoted by $\det V$. $\qquad\qquad\square$

There exists a natural injection $\iota_V : V \hookrightarrow \Lambda^\bullet V$, $\iota_V(v) = v$, such that

$$\iota_V(v) \wedge \iota_V(v) = 0, \ \ \forall v \in V.$$

This map enters crucially into the formulation of the following universality property.

Proposition 2.2.19. *Let V be a vector space over $\mathbb{K}$. For any $\mathbb{K}$-algebra A, and any linear map $\phi : V \to A$ such that $(\phi(x))^2 = 0$, there exists an unique morphism of $\mathbb{K}$-algebras $\Phi : \Lambda^\bullet V \to A$ such that the diagram below is commutative*

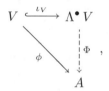

i.e., $\Phi \circ \iota_V = \phi$.

Exercise 2.2.20. Prove Proposition 2.2.19. $\qquad\qquad\square$

The space of symmetric tensors $\boldsymbol{S}^\bullet V$ can be similarly given a structure of associative algebra with respect to the product

$$\alpha \cdot \beta := \boldsymbol{S}_{r+s}(\alpha \otimes \beta), \ \ \forall \alpha \in \boldsymbol{S}^r V, \ \beta \in \boldsymbol{S}^s V.$$

The symmetric product "$\cdot$" is also commutative.

Exercise 2.2.21. Formulate and prove the analogue of Proposition 2.2.19 for the algebra $\boldsymbol{S}^\bullet V$. $\qquad\qquad\square$

It is often convenient to represent (skew-)symmetric tensors in coordinates. If $e_1, \ldots, e_n$ is a basis of the vector space V then, for any $1 \leq r \leq n$, the family

$$\{e_{i_1} \wedge \cdots \wedge e_{i_r} / 1 \leq i_1 < \cdots < i_r \leq n\}$$

is a basis for $\Lambda^r V$ so that any degree r skew-symmetric tensor ω can be uniquely represented as

$$\omega = \sum_{1 \leq i_1 < \cdots < i_r \leq n} \omega^{i_1 \cdots i_r} e_{i_1} \wedge \cdots \wedge e_{i_r}.$$

Symmetric tensors can be represented in a similar way.

The $\Lambda^\bullet$ and $S^\bullet$ constructions are functorial in the following sense.

Proposition 2.2.22. *Any linear map $L : V \to W$ induces a natural morphisms of algebras*

$$\Lambda^\bullet L : \Lambda^\bullet V \to \Lambda^\bullet W, \quad S^\bullet L : S^\bullet V \to S^\bullet W$$

uniquely defined by their actions on monomials

$$\Lambda^\bullet L(v_1 \wedge \cdots \wedge v_r) = (Lv_1) \wedge \cdots \wedge (Lv_r),$$

and

$$S^\bullet L(v_1, \ldots, v_r) = (Lv_1) \cdots (Lv_r).$$

Moreover, if $U \xrightarrow{A} V \xrightarrow{B} W$ are two linear maps, then

$$\Lambda^r(BA) = (\Lambda^r B)(\Lambda^r A), \quad S^r(BA) = (S^r B)(S^r A). \qquad \square$$

Exercise 2.2.23. Prove the above proposition. $\qquad \square$

In particular, if $n = \dim_{\mathbb{K}} V$, then any linear endomorphism $L : V \to V$ defines an endomorphism

$$\Lambda^n L : \det V = \Lambda^n V \to \det V.$$

Since the vector space $\det V$ is 1-dimensional, the endomorphism $\Lambda^n L$ can be identified with a scalar $\det L$, the determinant of the endomorphism L.

Definition 2.2.24. Suppose V is a finite dimensional vector space and $A : V \to V$ is an endomorphism of V. For every positive integer r we denote by $\sigma_r(A)$ the trace of the induced endomorphism

$$\Lambda^r A : \Lambda^r V \to \Lambda^r V,$$

and by $\psi_r(A)$ the trace of the endomorphism

$$S^r(A) : S^r V \to S^r V.$$

We define

$$\sigma_0(A) = 1, \quad \psi_0(A) = \dim V. \qquad \square$$

Exercise 2.2.25. Suppose V is a complex n-dimensional vector space, and A is an endomorphism of V.
(a) Prove that if A is diagonalizable and its eigenvalues are $a_1, \ldots, a_n$, then

$$\sigma_r(A) = \sum_{1 \le i_1 < \cdots < i_r \le n} a_{i_1} \cdots a_{i_r}, \quad \psi_r(A) = a_1^r + \cdots + a_n^r.$$

(b) Prove that for every sufficiently small $z \in \mathbb{C}$ we have the equalities

$$\det(\mathbb{1}_V + zA) = \sum_{j \ge 0} \sigma_r(A) z^r, \quad -\frac{d}{dz} \log \det(\mathbb{1} - zA) = \sum_{r \ge 1} \psi_r(A) z^{r-1}. \qquad \square$$

The functors $\Lambda^\bullet$ and $S^\bullet$ have an *exponential like* behavior, i.e., there exists a *natural* isomorphism

$$\Lambda^\bullet (V \oplus W) \cong \Lambda^\bullet V \otimes \Lambda^\bullet W. \tag{2.2.4}$$

$$S^\bullet (V \oplus W) \cong S^\bullet V \otimes S^\bullet W. \tag{2.2.5}$$

To define the isomorphism in (2.2.4) consider the bilinear map

$$\phi : \Lambda^\bullet V \times \Lambda^\bullet W \to \Lambda^\bullet (V \oplus W),$$

uniquely determined by

$$\phi(v_1 \wedge \cdots \wedge v_r , \; w_1 \wedge \cdots \wedge w_s) = v_1 \wedge \cdots \wedge v_r \wedge w_1 \wedge \cdots \wedge w_s.$$

The universality property of the tensor product implies the existence of a linear map

$$\Phi : \Lambda^\bullet V \otimes \Lambda^\bullet W \to \Lambda^\bullet (V \oplus W),$$

such that $\Phi \circ \iota = \phi$, where ι is the inclusion of $\Lambda^\bullet V \times \Lambda^\bullet W$ in $\Lambda^\bullet V \otimes \Lambda^\bullet W$. To construct the inverse of Φ, note that $\Lambda^\bullet V \otimes \Lambda^\bullet W$ is naturally a $\mathbb{K}$-algebra by

$$(\omega \otimes \eta) * (\omega' \otimes \eta') = (-1)^{\deg \eta \cdot \deg \omega'} (\omega \wedge \omega') \otimes (\eta \wedge \eta').$$

The vector space $V \oplus W$ is naturally embedded in $\Lambda^\bullet V \otimes \Lambda^\bullet W$ via the map given by

$$(v, w) \mapsto \psi(v, w) = v \otimes 1 + 1 \otimes w \in \Lambda^\bullet V \otimes \Lambda^\bullet W.$$

Moreover, for any $x \in V \oplus W$ we have $\psi(x) * \psi(x) = 0$. The universality property of the exterior algebra implies the existence of a unique morphism of $\mathbb{K}$-algebras

$$\Psi : \Lambda^\bullet (V \oplus W) \to \Lambda^\bullet V \otimes \Lambda^\bullet W,$$

such that $\Psi \circ \iota_{V \oplus W} = \psi$. Note that Φ is also a morphism of $\mathbb{K}$-algebras, and one verifies easily that

$$(\Phi \circ \Psi) \circ \iota_{V \oplus W} = \iota_{V \oplus W}.$$

The uniqueness part in the universality property of the exterior algebra implies $\Phi \circ \Psi = identity$. One proves similarly that $\Psi \circ \Phi = identity$, and this concludes the proof of (2.2.4).

We want to mention a few general facts about $\mathbb{Z}$-graded vector spaces, i.e., vector spaces equipped with a direct sum decomposition

$$V = \bigoplus_{n \in \mathbb{Z}} V_n.$$

(We will always assume that each V_n is finite dimensional.) The vectors in V_n are said to be *homogeneous, of degree n*. For example, the ring of polynomials $\mathbb{K}[x]$ is a $\mathbb{K}$-graded vector space. The spaces $\Lambda^\bullet V$ and $S^\bullet V$ are $\mathbb{Z}$-graded vector spaces.

The direct sum of two $\mathbb{Z}$-graded vector spaces V and W is a $\mathbb{Z}$-graded vector space with

$$(V \oplus W)_n := V_n \oplus W_n.$$

The tensor product of two $\mathbb{Z}$-graded vector spaces V and W is a $\mathbb{Z}$-graded vector space with

$$(V \otimes W)_n := \bigoplus_{r+s=n} V_r \otimes W_s.$$

To any $\mathbb{Z}$-graded vector space V one can naturally associate a formal series $P_V(t) \in \mathbb{Z}[[t, t^{-1}]]$ by

$$P_V(t) := \sum_{n \in \mathbb{Z}} (\dim_{\mathbb{K}} V_n) t^n.$$

The series $P_V(t)$ is called the *Poincaré series* of V.

Example 2.2.26. The Poincaré series of $\mathbb{K}[x]$ is

$$P_{\mathbb{K}[x]}(t) = 1 + t + t^2 + \cdots + t^{n-1} + \cdots = \frac{1}{1-t}. \qquad \square$$

Exercise 2.2.27. Let V and W be two $\mathbb{Z}$-graded vector spaces. Prove the following statements are true (whenever they make sense).
(a) $P_{V \oplus W}(t) = P_V(t) + P_W(t)$.
(b) $P_{V \otimes W}(t) = P_V(t) \cdot P_W(t)$.
(c) $\dim V = P_V(1)$. $\qquad \square$

Definition 2.2.28. Let V be a $\mathbb{Z}$-graded vector space. The *Euler characteristic* of V, denoted by $\chi(V)$, is defined by

$$\chi(V) := P_V(-1) = \sum_{n \in \mathbb{Z}} (-1)^n \dim V_n,$$

whenever the sum on the right-hand side makes sense. $\qquad \square$

Remark 2.2.29. If we try to compute $\chi(\mathbb{K}[x])$ using the first formula in Definition 2.2.28 we get $\chi(\mathbb{K}[x]) = 1/2$, while the second formula makes no sense (divergent series). $\qquad \square$

Proposition 2.2.30. *Let V be a $\mathbb{K}$-vector space of dimension n. Then*

$$P_{\Lambda^\bullet V}(t) = (1+t)^n \text{ and } P_{S^\bullet V}(t) = \left(\frac{1}{1-t} \right)^n = \frac{1}{(n-1)!} \left(\frac{d}{dt} \right)^{n-1} \left(\frac{1}{1-t} \right).$$

In particular, $\dim \Lambda^\bullet V = 2^n$ and $\chi(\Lambda^\bullet V) = 0$.

Proof. From (2.2.4) and (2.2.5) we deduce using Exercise 2.2.27 that for any vector spaces V and W we have

$$P_{\Lambda^\bullet(V\oplus W)}(t) = P_{\Lambda^\bullet V}(t) \cdot P_{\Lambda^\bullet W}(t) \text{ and } P_{S^\bullet(V\oplus W)}(t) = P_{S^\bullet V}(t) \cdot P_{S^\bullet W}(t).$$

In particular, if V has dimension n, then $V \cong \mathbb{K}^n$ so that

$$P_{\Lambda^\bullet V}(t) = (P_{\Lambda^\bullet \mathbb{K}}(t))^n \text{ and } P_{S^\bullet V}(t) = (P_{S^\bullet \mathbb{K}}(t))^n.$$

The proposition follows using the equalities

$$P_{\Lambda^\bullet \mathbb{K}}(t) = 1 + t, \text{ and } P_{S^\bullet \mathbb{K}}(t) = P_{\mathbb{K}[x]}(t) = \frac{1}{1-t}. \qquad \square$$

2.2.3 The "super" slang

The aim of this very brief section is to introduce the reader to the "super" terminology. We owe the "super" slang to the physicists. In the quantum world many objects have a special feature not present in the Newtonian world. They have parity (or chirality), and objects with different chiralities had to be treated differently.

The "super" terminology provides an algebraic formalism which allows one to deal with the different parities on an equal basis. From a strictly syntactic point of view, the "super" slang adds the attribute super to most of the commonly used algebraic objects. In this book, the prefix "*s-*" will abbreviate the word "super".

Definition 2.2.31. (a) A *s-space* is a $\mathbb{Z}_2$-graded vector space, i.e., a vector space V equipped with a direct sum decomposition $V = V_0 \oplus V_1$.

(b) A *s-algebra* over $\mathbb{K}$ is a $\mathbb{Z}_2$-graded $\mathbb{K}$-algebra, i.e., a $\mathbb{K}$-algebra $\mathcal{A}$ together with a direct sum decomposition $\mathcal{A} = \mathcal{A}^0 \oplus \mathcal{A}^1$ such that $\mathcal{A}^i \cdot \mathcal{A}^j \subset \mathcal{A}^{i+j \,(mod\,2)}$. The elements in $\mathcal{A}^i$ are called *homogeneous of degree* i. For any $a \in \mathcal{A}^i$ we denote its degree (mod 2) by $|a|$. The elements in $\mathcal{A}^0$ are said to be *even* while the elements in $\mathcal{A}^1$ are said to be *odd*.

(c) The *supercommutator* in a s-algebra $\mathcal{A} = \mathcal{A}^0 \oplus \mathcal{A}^1$ is the bilinear map

$$[\bullet, \bullet]_s : \mathcal{A} \times \mathcal{A} \to \mathcal{A},$$

defined on homogeneous elements $\omega^i \in \mathcal{A}^i$, $\eta^j \in \mathcal{A}^j$ by

$$[\omega^i, \eta^j]_s := \omega^i \eta^j - (-1)^{ij} \eta^j \omega^j.$$

An *s*-algebra is called *s-commutative*, if the suppercommutator is trivial, $[\bullet, \bullet]_s \equiv 0$. $\qquad \square$

Example 2.2.32. Let $E = E^0 \oplus E^1$ be a *s*-space. Any linear endomorphism $T \in \operatorname{End}(E)$ has a block decomposition

$$T = \begin{bmatrix} T_{00} & T_{01} \\ T_{10} & T_{11} \end{bmatrix},$$

where $T_{ji} \in \mathrm{End}\,(E^i, E^j)$. We can use this block decomposition to describe a structure of s-algebra on $\mathrm{End}\,(E)$. The even endomorphisms have the form

$$\begin{bmatrix} T_{00} & 0 \\ 0 & T_{11} \end{bmatrix},$$

while the odd endomorphisms have the form

$$\begin{bmatrix} 0 & T_{01} \\ T_{10} & 0 \end{bmatrix}. \qquad \square$$

Example 2.2.33. Let V be a finite dimensional space. The exterior algebra $\Lambda^\bullet V$ is naturally a s-algebra. The even elements are gathered in

$$\Lambda^{even} V = \bigoplus_{r\ even} \Lambda^r V,$$

while the odd elements are gathered in

$$\Lambda^{odd} V = \bigoplus_{r\ odd} \Lambda^r V.$$

The s-algebra $\Lambda^\bullet V$ is s-commutative. $\qquad \square$

Definition 2.2.34. Let $\mathcal{A} = \mathcal{A}^0 \oplus \mathcal{A}^1$ be a s-algebra. An *s-derivation* on $\mathcal{A}$ is a linear operator on $D \in \mathrm{End}\,(\mathcal{A})$ such that, for any $x \in \mathcal{A}$,

$$[D, L_x]_s^{\mathrm{End}(\mathcal{A})} = L_{Dx}, \qquad (2.2.6)$$

where $[\ ,\ ]_s^{\mathrm{End}(\mathcal{A})}$ denotes the supercommutator in $\mathrm{End}\,(\mathcal{A})$ (with the s-structure defined in Example 2.2.32), while for any $z \in \mathcal{A}$ we denoted by L_z the left multiplication operator $a \mapsto z \cdot a$.

An s-derivation is called *even* (respectively *odd*), if it is even (respectively *odd*) as an element of the s-algebra $\mathrm{End}\,(\mathcal{A})$. $\qquad \square$

Remark 2.2.35. The relation (2.2.6) is a super version of the usual Leibniz formula. Indeed, assuming D is homogeneous (as an element of the s-algebra $\mathrm{End}\,(\mathcal{A})$) then equality (2.2.6) becomes

$$D(xy) = (Dx)y + (-1)^{|x||D|} x(Dy),$$

for any homogeneous elements $x, y \in \mathcal{A}$. $\qquad \square$

Example 2.2.36. Let V be a vector space. Any $u^* \in V^*$ defines an odd s-derivation of $\Lambda^\bullet V$ denoted by i_{u^*} uniquely determined by its action on monomials.

$$i_{u^*}(v_0 \wedge v_1 \wedge \cdots \wedge v_r) = \sum_{i=0}^{r} (-1)^i \langle u^*, v_i \rangle v_0 \wedge v_1 \wedge \ldots \wedge \hat{v}_i \wedge \cdots \wedge v_r.$$

As usual, a hat indicates a missing entry. The derivation i_{u^*} is called the *interior derivation* by u^* or the *contraction* by u^*. Often, one uses the alternate notation $u^*\lrcorner$ to denote this derivation. $\qquad \square$

Exercise 2.2.37. Prove the statement in the above example. □

Definition 2.2.38. Let $\mathcal{A} = (\mathcal{A}^0 \oplus \mathcal{A}^1, +, [\ ,\ ])$ be an s-algebra over $\mathbb{K}$, not necessarily associative. For any $x \in \mathcal{A}$ we denote by R_x the right multiplication operator, $a \mapsto [a, x]$. $\mathcal{A}$ is called an s-*Lie algebra* if it is s-anticommutative, i.e.,

$$[x, y] + (-1)^{|x||y|}[y, x] = 0, \quad \text{for all homogeneous elements } x, y \in \mathcal{A},$$

and $\forall x \in \mathcal{A}$, R_x is a s-derivation.

When $\mathcal{A}$ is purely even, i.e., $\mathcal{A}^1 = \{0\}$, then $\mathcal{A}$ is called simply a *Lie algebra*. The multiplication in a $(s\text{-})$ Lie algebra is called the $(s\text{-})$bracket. □

The above definition is highly condensed. In down-to-earth terms, the fact that R_x is a s-derivation for all $x \in \mathcal{A}$ is equivalent with the *super Jacobi identity*

$$[[y, z], x] = [[y, x], z] + (-1)^{|x||y|}[y, [z, x]], \tag{2.2.7}$$

for all homogeneous elements $x, y, z \in \mathcal{A}$. When $\mathcal{A}$ is a purely even $\mathbb{K}$-algebra, then $\mathcal{A}$ is a *Lie algebra* over $\mathbb{K}$ if $[\ ,\ ]$ is anticommutative and satisfies (2.2.7), which in this case is equivalent with the classical *Jacobi identity*,

$$[[x, y], z] + [[y, z], x] + [[z, x], y] = 0, \quad \forall x, y, z \in \mathcal{A}. \tag{2.2.8}$$

Example 2.2.39. Let E be a vector space (purely even). Then $\mathcal{A} = \text{End}(E)$ is a Lie algebra with bracket given by the usual commutator: $[a, b] = ab - ba$. □

Proposition 2.2.40. *Let $\mathcal{A} = \mathcal{A}^0 \oplus \mathcal{A}^1$ be a s-algebra, and denote by $\text{Der}_s(\mathcal{A})$ the vector space of s-derivations of $\mathcal{A}$.*
(a) For any $D \in \text{Der}_s(\mathcal{A})$, its homogeneous components D^0, $D^1 \in \text{End}(\mathcal{A})$ are also s-derivations.
(b) For any D, $D' \in \text{Der}_s(\mathcal{A})$, the s-commutator $[D, D']_s^{\text{End}(\mathcal{A})}$ is again an s-derivation.
(c) $\forall x \in \mathcal{A}$ the bracket $B^x : a \mapsto [a, x]_s$ is a s-derivation called the bracket derivation determined by x. Moreover

$$[B^x, B^y]_s^{\text{End}(\mathcal{A})} = B^{[x,y]_s}, \quad \forall x, y \in \mathcal{A}. \qquad \square$$

Exercise 2.2.41. Prove Proposition 2.2.40. □

Definition 2.2.42. Let $E = E^0 \oplus E^1$ and $F = F^0 \oplus F^1$ be two s-spaces. Their s-tensor product is the s-space $E \otimes F$ with the $\mathbb{Z}_2$-grading,

$$(E \otimes F)^\epsilon := \bigoplus_{i+j \equiv \epsilon\,(2)} E^i \otimes F^j, \quad \epsilon = 0, 1.$$

To emphasize the super-nature of the tensor product we will use the symbol "$\widehat{\otimes}$" instead of the usual "$\otimes$". □

Exercise 2.2.43. Show that there exists a natural isomorphism of s-spaces

$$V^* \widehat{\otimes} \Lambda^\bullet V \cong \mathrm{Der}_s(\Lambda^\bullet V),$$

uniquely determined by $v^* \times \omega \mapsto D^{v^* \otimes \omega}$, where $D^{v^* \otimes \omega}$ is s-derivation defined by

$$D^{v^* \otimes \omega}(v) = \langle v^*, v \rangle \omega, \quad \forall v \in V.$$

Notice in particular that any s-derivation of $\Lambda^\bullet V$ is uniquely determined by its action on $\Lambda^1 V$. (When $\omega = 1$, $D^{v^* \otimes 1}$ coincides with the internal derivation discussed in Example 2.2.36.) $\square$

Let $\mathcal{A} = \mathcal{A}^0 \oplus \mathcal{A}^1$ be an s-algebra over $\mathbb{K} = \mathbb{R}, \mathbb{C}$. A *supertrace* on $\mathcal{A}$ is a $\mathbb{K}$-linear map $\tau : \mathcal{A} \to \mathbb{K}$ such that,

$$\tau([x, y]_s) = 0 \quad \forall x, y \in \mathcal{A}.$$

If we denote by $[\mathcal{A}, \mathcal{A}]_s$ the linear subspace of $\mathcal{A}$ spanned by the supercommutators

$$\big\{ [x, y]_s \; ; \; x, y \in \mathcal{A} \big\},$$

then we see that the space of s-traces is isomorphic with the dual of the quotient space $\mathcal{A}/[\mathcal{A}, \mathcal{A}]_s$.

Proposition 2.2.44. *Let $E = E_0 \oplus E_1$ be a finite dimensional s-space, and denote by $\mathcal{A}$ the s-algebra of endomorphisms of E. Then there exists a canonical s-trace tr_s on $\mathcal{A}$ uniquely defined by*

$$\mathrm{tr}_s \mathbb{1}_E = \dim E_0 - \dim E_1.$$

In fact, if $T \in \mathcal{A}$ has the block decomposition

$$T = \begin{bmatrix} T_{00} & T_{01} \\ T_{10} & T_{11} \end{bmatrix},$$

then

$$\mathrm{tr}_s T = \mathrm{tr}\, T_{00} - \mathrm{tr}\, T_{11}. \qquad \square$$

Exercise 2.2.45. Prove the above proposition. $\square$

2.2.4 Duality

Duality is a subtle and fundamental concept which permeates all branches of mathematics. This section is devoted to those aspects of the atmosphere called duality which are relevant to differential geometry. In the sequel, *all vector spaces will be tacitly assumed finite dimensional,* and we will use Einstein's convention without mentioning it. $\mathbb{K}$ will denote one of the fields $\mathbb{R}$ or $\mathbb{C}$.

Definition 2.2.46. A *pairing* between two $\mathbb{K}$-vector spaces V and W is a bilinear map $B : V \times W \to \mathbb{K}$. $\square$

Any pairing $B : V \times W \to \mathbb{K}$ defines a linear map

$$\mathbb{I}_B : V \to W^*, \quad v \mapsto B(v, \bullet) \in W^*,$$

called the *adjunction morphism* associated to the pairing.

Conversely, any linear map $L : V \to W^*$ defines a pairing

$$B_L : V \times W \to \mathbb{K}, \quad B(v, w) = (Lv)(w), \quad \forall v \in V, \ w \in W.$$

Observe that $\mathbb{I}_{B_L} = L$. A pairing B is called a *duality* if the adjunction map $\mathbb{I}_B$ is an isomorphisms.

Example 2.2.47. The natural pairing $\langle \bullet, \bullet \rangle : V^* \times V \to \mathbb{K}$ is a duality. One sees that $\mathbb{I}_{\langle \bullet, \bullet \rangle} = \mathbb{1}_{V^*} : V^* \to V^*$. This pairing is called the *natural duality* between a vector space and its dual. □

Example 2.2.48. Let V be a finite dimensional real vector space. Any symmetric nondegenerate quadratic form $(\bullet, \bullet) : V \times V \to \mathbb{R}$ defines a (self)duality, and in particular a natural isomorphism

$$\mathcal{L} := \mathbb{I}_{(\bullet, \bullet)} : V \to V^*.$$

When $(\bullet, \bullet)$ is positive definite, then the operator $\mathcal{L}$ is called *metric duality*. This operator can be nicely described in coordinates as follows. Pick a basis (e_i) of V, and set

$$g_{ij} := (e_i, e_j).$$

Let (e^j) denote the dual basis of V^* defined by

$$\langle e^j, e_i \rangle = \delta_i^j, \quad \forall i, j.$$

The action of $\mathcal{L}$ is then

$$\mathcal{L} e_i = g_{ij} e^j.$$ □

Example 2.2.49. Consider V a real vector space and $\omega : V \times V \to \mathbb{R}$ a skew-symmetric bilinear form on V. The form ω is said to be *symplectic* if this pairing is a duality. In this case, the induced operator $\mathbb{I}_\omega : V \to V^*$ is called *symplectic duality*. □

Exercise 2.2.50. Suppose that V is a real vector space, and $\omega : V \times V \to \mathbb{R}$ is a symplectic duality. Prove the following.
(a) The V has even dimension.
(b) If (e_i) is a basis of V, and $\omega_{ij} := \omega(e_i, e_j)$, then $\det(\omega_{ij})_{1 \le i, j \le \dim V} \ne 0$. □

The notion of duality is compatible with the functorial constructions introduced so far.

Proposition 2.2.51. *Let* $B_i : V_i \times W_i \to \mathbb{R}$ *($i = 1, 2$) be two pairs of spaces in duality. Then there exists a natural duality*

$$B = B_1 \otimes B_2 : (V_1 \otimes V_2) \times (W_1 \otimes W_2) \to \mathbb{R},$$

uniquely determined by

$$\mathbb{I}_{B_1 \otimes B_2} = \mathbb{I}_{B_1} \otimes \mathbb{I}_{B_2} \Longleftrightarrow B(v_1 \otimes v_2, w_1 \otimes w_2) = B_1(v_1, w_1) \cdot B_2(v_2, w_2). \qquad \square$$

Exercise 2.2.52. Prove Proposition 2.2.51. $\qquad \square$

Proposition 2.2.51 implies that given two spaces in duality $B : V \times W \to \mathbb{K}$ there is a naturally induced duality

$$B^{\otimes n} : V^{\otimes r} \times W^{\otimes r} \to \mathbb{K}.$$

This defines by restriction a pairing

$$\Lambda^r B : \Lambda^r V \times \Lambda^r W \to \mathbb{K}$$

uniquely determined by

$$\Lambda^r B \left(v_1 \wedge \cdots \wedge v_r, w_1 \wedge \cdots \wedge w_r \right) := \det \left(B \left(v_i, w_j \right) \right)_{1 \le i, j \le r}.$$

Exercise 2.2.53. Prove the above pairing is a duality. $\qquad \square$

In particular, the natural duality $\langle \bullet, \bullet \rangle : V^* \times V \to \mathbb{K}$ induces a duality

$$\langle \bullet, \bullet \rangle : \Lambda^r V^* \times \Lambda^r V \to \mathbb{R},$$

and thus defines a natural isomorphism

$$\Lambda^r V^* \cong (\Lambda^r V)^*.$$

This shows that we can regard the elements of $\Lambda^r V^*$ as skew-symmetric r-linear forms $V^r \to \mathbb{K}$.

A duality $B : V \times W \to \mathbb{K}$ naturally induces a duality $B^\dagger : V^* \times W^* \to \mathbb{K}$ by

$$B^\dagger(v^*, w^*) := \langle v^*, \mathbb{I}_B^{-1} w^* \rangle,$$

where $\mathbb{I}_B : V \to W^*$ is the adjunction isomorphism induced by the duality B.

Now consider a (*real*) Euclidean vector space V. Denote its inner product by $(\bullet, \bullet)$. The self-duality defined by $(\bullet, \bullet)$ induces a self-duality

$$(\bullet, \bullet) : \Lambda^r V \times \Lambda^r V \to \mathbb{R},$$

determined by

$$(v_1 \wedge \cdots \wedge v_r, w_1 \wedge \cdots \wedge w_r) := \det \left((v_i, w_j) \right)_{1 \le i, j \le r}. \qquad (2.2.9)$$

The right hand side of (2.2.9) is a Gramm determinant, and in particular, the bilinear form in (2.2.9) is symmetric and positive definite. Thus, we have proved the following result.

Corollary 2.2.54. *An inner product on a real vector space V naturally induces an inner product on the tensor algebra $\mathfrak{T}(V)$, and in the exterior algebra $\Lambda^\bullet V$.* $\qquad \square$

In a Euclidean vector space V the inner product induces the metric duality $\mathcal{L} : V \to V^*$. This induces an operator $\mathcal{L} : T^r_s(V) \to T^{r-1}_{s+1}(V)$ defined by

$$\mathcal{L}(v_1 \otimes \ldots \otimes v_r \otimes u^1 \otimes \cdots \otimes u^s) = (v_2 \otimes \cdots \otimes v_r) \otimes ((\mathcal{L}v_1 \otimes u^1 \otimes \cdots \otimes u^s). \quad (2.2.10)$$

The operation defined in (2.2.10) is classically referred to as *lowering the indices*.

The reason for this nomenclature comes from the coordinate description of this operation. If $T \in \mathcal{T}^r_s(V)$ is given by

$$T = T^{i_1 \ldots i_r}_{j_1 \ldots j_s} e_{i_1} \otimes \cdots \otimes e_{i_r} \otimes e^{j_1} \otimes \cdots \otimes e^{j_s},$$

then

$$(\mathcal{L}T)^{i_2 \ldots i_r}_{j j_1 \ldots j_r} = g_{ij} T^{i i_2 \ldots i_r}_{j_1 \ldots j_s},$$

where $g_{ij} = (e_i, e_j)$. The inverse of the metric duality $\mathcal{L}^{-1} : V^* \to V$ induces a linear operation $\mathcal{T}^r_s(V) \to \mathcal{T}^{r+1}_{s-1}(V)$ called *raising the indices*.

Exercise 2.2.55 (Cartan). *Let V be an Euclidean vector space. For any $v \in V$ denote by e_v (resp. i_v) the linear endomorphism of $\Lambda^* V$ defined by $e_v \omega = v \wedge \omega$ (resp. $i_v = \imath_{v^*}$ where $\imath_{v^*}$ denotes the interior derivation defined by $v^* \in V^*$-the metric dual of v; see Example 2.2.36). Show that for any $u, v \in V$*

$$[e_v, i_u]_s = e_v i_u + i_u e_v = (u, v) \mathbb{1}_{\Lambda^\bullet V}. \qquad \square$$

Definition 2.2.56. Let V be a *real* vector space. A *volume form* on V is a non-trivial linear form on the determinant line of V, $\mu : \det V \to \mathbb{R}$. $\qquad \square$

Equivalently, a volume form on V is a nontrivial element of $\det V^*$ ($n = \dim V$). Since $\det V$ is 1-dimensional, a choice of a volume form corresponds to a choice of a basis of $\det V$.

Definition 2.2.57. (a) An *orientation* on a vector space V is a *continuous, surjective* map

$$\boldsymbol{or} : \det V \setminus \{0\} \to \{\pm 1\}.$$

We denote by $\boldsymbol{Or}(V)$ the set of orientations of V. Observe that $\boldsymbol{Or}(V)$ consists of precisely two elements.

(b) A pair $(V, \boldsymbol{or})$, where V is a vector space, and $\boldsymbol{or}$ is an orientation on V is called an *oriented vector space*.

(c) Suppose $\boldsymbol{or} \in \boldsymbol{Or}(V)$. A basis ω of $\det V$ is said to be *positively oriented* if $\boldsymbol{or}(\omega) > 0$. Otherwise, the basis is said to be *negatively oriented.* $\qquad \square$

There is an equivalent way of looking at orientations. To describe it, note that any nontrivial volume form μ on V uniquely specifies an orientation $\boldsymbol{or}_\mu$ given by

$$\boldsymbol{or}_\mu(\omega) := \operatorname{sign} \mu(\omega), \quad \forall \omega \in \det V \setminus \{0\}.$$

We define an equivalence relation on the space of nontrivial volume forms by declaring

$$\mu_1 \sim \mu_2 \iff \mu_1(\omega)\mu_2(\omega) > 0, \quad \forall \omega \in \det V \setminus \{0\}.$$

Then

$$\mu_1 \sim \mu_2 \iff \boldsymbol{or}_{\mu_1} = \boldsymbol{or}_{\mu_2}.$$

To every orientation $\boldsymbol{or}$ we can associate an equivalence class $[\mu]_{\boldsymbol{or}}$ of volume forms such that

$$\mu(\omega)\boldsymbol{or}(\omega) > 0, \quad \forall \omega \in \det V \setminus \{0\}.$$

Thus, we can identify the set of orientations with the set of equivalence classes of nontrivial volume forms.

Equivalently, to specify an orientation on V it suffices to specify a basis ω of $\det V$. The associated orientation $\boldsymbol{or}_\omega$ is uniquely characterized by the condition

$$\boldsymbol{or}_\omega(\omega) = 1.$$

To any basis $\{e_1, ..., e_n\}$ of V one can associate a basis $e_1 \wedge \cdots \wedge e_n$ of $\det V$. Note that a permutation of the indices $1, \ldots, n$ changes the associated basis of $\det V$ by a factor equal to the signature of the permutation. Thus, to define an orientation on a vector space, it suffices to specify a total ordering of a given basis of the space.

An *ordered* basis of an *oriented* vector space $(V, \boldsymbol{or})$ is said to be *positively oriented* if so is the associated basis of $\det V$.

Definition 2.2.58. Given two orientations $\boldsymbol{or}_1, \boldsymbol{or}_2$ on the vector space V we define

$$\boldsymbol{or}_1/\boldsymbol{or}_2 \in \{\pm 1\}$$

to be

$$\boldsymbol{or}_1/\boldsymbol{or}_2 := \boldsymbol{or}_1(\omega)\boldsymbol{or}_2(\omega), \quad \forall \omega \in \det V \setminus \{0\}.$$

We will say that $\boldsymbol{or}_1/\boldsymbol{or}_2$ is the *relative signature* of the pair of orientations $\boldsymbol{or}_1, \boldsymbol{or}_2$. □

Assume now that V is an Euclidean space. Denote the Euclidean inner product by $g(\bullet, \bullet)$. The vector space $\det V$ has an induced Euclidean structure, and in particular, there exist *exactly* two length-one-vectors in $\det V$. If we fix one of them, call it ω, and we think of it as a basis of $\det V$, then we achieve two things.

- First, it determines a volume form μ_g defined by

$$\mu_g(\lambda\omega) = \lambda.$$

- Second, it determines an orientation on V.

Conversely, an orientation $\boldsymbol{or} \in \boldsymbol{Or}(V)$ uniquely selects a length-one-vector $\omega = \omega_{\boldsymbol{or}}$ in $\det V$, which determines a volume form $\mu_g = \mu_g^{\boldsymbol{or}}$. Thus, we have proved the following result.

Proposition 2.2.59. *An orientation $\boldsymbol{or}$ on an Euclidean vector space (V, g) canonically selects a volume form on V, henceforth denoted by $\mathrm{Det}_g = \mathrm{Det}_g^{\boldsymbol{or}}$.* $\square$

Exercise 2.2.60. Let (V, g) be an n-dimensional Euclidean vector space, and $\boldsymbol{or}$ an orientation on V. Show that, for any basis $v_1, \dots v_n$ of V, we have

$$\mathrm{Det}_g^{\boldsymbol{or}}(v_1 \wedge \cdots \wedge v_n) = \boldsymbol{or}(v_1 \wedge \cdots \wedge v_n)\sqrt{(\det g(v_i, v_j))}.$$

If $V = \mathbb{R}^2$ with its standard metric, and the orientation given by $e_1 \wedge e_2$, prove that

$$|\mathrm{Det}_g^{\boldsymbol{or}}(v_1 \wedge v_2)|$$

is the area of the parallelogram spanned by v_1 and v_2. $\square$

Definition 2.2.61. Let $(V, g, \boldsymbol{or})$ be an oriented, Euclidean space and denote by $\mathrm{Det}_g^{\boldsymbol{or}}$ the associated volume form. The *Berezin integral* or *(berezinian)* is the linear form

$$\overrightarrow{\int_g} : \Lambda^\bullet V \to \mathbb{R},$$

defined on homogeneous elements by

$$\overrightarrow{\int_g} \omega = \begin{cases} 0 & \text{if } \deg \omega < \dim V \\ \mathrm{Det}_g^{\boldsymbol{or}} \omega & \text{if } \deg \omega = \dim V \end{cases}.$$
$\square$

Definition 2.2.62. Let $\omega \in \Lambda^2 V$, where $(V, g, \boldsymbol{or})$ is an oriented, Euclidean space. We define its *pfaffian* as

$$\boldsymbol{Pf}(\omega) = \boldsymbol{Pf}_g^{\boldsymbol{or}}(\omega) := \overrightarrow{\int_g} \exp \omega = \begin{cases} 0 & \text{if } \dim V \text{ is odd} \\ \frac{1}{n!}\mathrm{Det}_g^{\boldsymbol{or}}(\omega^{\wedge n}) & \text{if } \dim V = 2n \end{cases},$$

where $\exp \omega$ denotes the exponential in the (nilpotent) algebra $\Lambda^\bullet V$,

$$\exp \omega := \sum_{k \geq 0} \frac{\omega^k}{k!}.$$
$\square$

If (V, g) is as in the above definition, $\dim V = N$, and $A : V \to V$ is a skew-symmetric endomorphism of V, then we can define $\omega_A \in \Lambda^2 V$ by

$$\omega_A = \sum_{i<j} g(Ae_i, e_j)e_i \wedge e_j = \frac{1}{2} \sum_{i,j} g(Ae_i, e_j)e_i \wedge e_j,$$

where $(e_1, \dots, e_N)$ is a *positively oriented orthonormal basis* of V. The reader can check that ω_A is independent of the choice of basis as above. Notice that

$$\omega_A(u, v) = g(Au, v), \quad \forall u, v \in V.$$

We define the *pfaffian* of A by

$$\boldsymbol{Pf}(A) := \boldsymbol{Pf}(\omega_A).$$

Example 2.2.63. Let $V = \mathbb{R}^2$ denote the standard Euclidean space oriented by $e_1 \wedge e_2$, where $e_1 e_2$ denotes the standard basis. If

$$A = \begin{bmatrix} 0 & -\theta \\ \theta & 0 \end{bmatrix},$$

then $\omega_A = \theta e_1 \wedge e_2$ so that $\boldsymbol{Pf}(A) = \theta$. □

Exercise 2.2.64. Let $A : V \to V$ be a skew-symmetric endomorphism of an oriented Euclidean space V. Prove that $\boldsymbol{Pf}(A)^2 = \det A$. □

Exercise 2.2.65. Let $(V, g, \boldsymbol{or})$ be an oriented Euclidean space of dimension $2n$. Consider $A : V \to V$ a skewsymmetric endomorphism and a positively oriented orthonormal frame $e_1, \ldots, e_{2n}$. Prove that

$$\boldsymbol{Pf}_g^{\boldsymbol{or}}(A) = \frac{(-1)^n}{2^n n!} \sum_{\sigma \in \mathcal{S}_{2n}} \epsilon(\sigma) a_{\sigma(1)\sigma(2)} \cdots a_{\sigma(2n-1)\sigma(2n)}$$

$$= (-1)^n \sum_{\sigma \in \mathcal{S}'_{2n}} \epsilon(\sigma) a_{\sigma(1)\sigma(2)} \cdots a_{\sigma(2n-1)\sigma(2n)},$$

where $a_{ij} = g(e_i, A e_j)$ is the (i, j)-th entry in the matrix representing A in the basis (e_i), and $\mathcal{S}'_{2n}$ denotes the set of permutations $\sigma \in \mathcal{S}_{2n}$ satisfying

$$\sigma(2k-1) < \min\{\sigma(2k), \sigma(2k+1)\}, \quad \forall k. \qquad \square$$

Let $(V, g, \boldsymbol{or})$ be an n-dimensional, oriented, real Euclidean vector space. The metric duality $\mathcal{L}_g : V \to V^*$ induces both a metric, and an orientation on V^*. In the sequel we will continue to use the same notation $\mathcal{L}_g$ to denote the metric duality $\mathcal{T}_s^r(V) \to \mathcal{T}_s^r(V^*) \cong \mathcal{T}_s^r(V)$.

Definition 2.2.66. Suppose $(V, g, \boldsymbol{or})$ oriented Euclidean space, and r is a nonnegative integer, $r \leq \dim V$. The r-th *Hodge pairing* is the pairing

$$\Xi = \Xi_{g,\boldsymbol{or}}^r : \Lambda^r V^* \times \Lambda^{n-r} V \to \mathbb{R},$$

defined by

$$\Xi(\omega^r, \eta^{n-r}) := \text{Det}_g^{\boldsymbol{or}}\left(\mathcal{L}_g^{-1}\omega^r \wedge \eta^{n-r}\right), \quad \omega^r \in \Lambda^r V^*, \quad \eta^{n-r} \in \Lambda^{n-r} V. \qquad \square$$

Exercise 2.2.67. Prove that the Hodge pairing is a duality. □

Definition 2.2.68. The Hodge $*$-operator is the adjunction isomorphism

$$* = \mathbb{I}_\Xi : \Lambda^r V^* \to \Lambda^{n-r} V^*$$

induced by the Hodge duality. □

The above definition obscures the meaning of the $*$-operator. We want to spend some time clarifying its significance.

Let $\alpha \in \Lambda^r V^*$ so that $*\alpha \in \Lambda^{n-r} V^*$. Denote by $\langle \bullet, \bullet \rangle$ the standard pairing

$$\Lambda^{n-r} V^* \times \Lambda^{n-r} V \to \mathbb{R},$$

and by $(\bullet, \bullet)$ the induced metric on $\Lambda^{n-r} V^*$. Then, by definition, for every $\beta \in \Lambda^{n-s} V^*$ the operator $*$ satisfies

$$\mathrm{Det}_g(\mathcal{L}_g^{-1}\alpha \wedge \mathcal{L}_g^{-1}\beta) = \langle *\alpha, \mathcal{L}_g^{-1}\beta \rangle = (*\alpha, \beta), \quad \forall \beta \in \Lambda^{n-r} V^*. \tag{2.2.11}$$

Let ω denote the unit vector in $\det V$ defining the orientation. Then (2.2.11) can be rewritten as

$$\langle \alpha \wedge \beta, \omega \rangle = (*\alpha, \beta), \quad \forall \beta \in \Lambda^{n-r} V^*.$$

Thus

$$\alpha \wedge \beta = (*\alpha, \beta)\,\mathrm{Det}_g^{or}, \quad \forall \alpha \in \Lambda^r V^*, \quad \forall \beta \in \Lambda^{n-r} V^*. \tag{2.2.12}$$

Equality (2.2.12) uniquely determines the action of $*$.

Example 2.2.69. Let V be the standard Euclidean space $\mathbb{R}^3$ with standard basis e_1, e_2, e_3, and orientation determined by $e_1 \wedge e_2 \wedge e_3$. Then

$$*e_1 = e_2 \wedge e_3, \quad *e_2 = e_3 \wedge e_1, \quad *e_3 = e_1 \wedge e_2,$$

$$*1 = e_1 \wedge e_2 \wedge e_3, \quad *(e_1 \wedge e_2 \wedge e_3) = 1,$$

$$*(e_2 \wedge e_3) = e_1, \quad *(e_3 \wedge e_1) = e_2, \quad *(e_1 \wedge e_2) = e_3. \qquad \square$$

The following result is left to the reader as an exercise.

Proposition 2.2.70. *Suppose (V, g, or) is an oriented, real Euclidean space of dimension n. Then the associated Hodge $*$-operator satisfies*

$$*(*\omega) = (-1)^{p(n-p)}\omega \quad \forall \omega \in \Lambda^p V^*,$$

$$\mathrm{Det}_g(*1) = 1,$$

and

$$\alpha \wedge *\beta = (\alpha, \beta) * 1, \quad \forall \alpha \in \Lambda^k V^*, \ \forall \beta \in \Lambda^{n-k} V^*. \qquad \square$$

Exercise 2.2.71. Let (V, g, ε) be an n-dimensional, oriented, Euclidean space. For every $t > 0$ Denote by g_t the rescaled metric $g_t = t^2 g$. If $*$ is the Hodge operator corresponding to the metric g and orientation or, and $*_t$ is the Hodge operator corresponding to the metric g_t and the same orientation, show that

$$\mathrm{Det}_{tg}^{or} = t^n \, \mathrm{Det}_g^{or},$$

and

$$*_t\omega = t^{n-2p} * \omega \quad \forall \omega \in \Lambda^p V^*. \qquad \square$$

We conclude this subsection with a brief discussion of densities.

Definition 2.2.72. Let V be a real vector space. For any $r \geq 0$ we define an *r-density* to be a function $f : \det V \to \mathbb{R}$ such that

$$f(\lambda u) = |\lambda|^r f(u), \quad \forall u \in \det V \setminus \{0\}, \; \forall \lambda \neq 0. \qquad \square$$

The linear space of r-densities on V will be denoted by $|\Lambda|_V^r$. When $r = 1$ we set $|\Lambda|_V := |\Lambda|_V^1$, and we will refer to 1-densities simply as densities.

Example 2.2.73. Any Euclidean metric g on V defines a canonical 1-density $|\operatorname{Det}_g| \in |\Lambda|_V^1$ which associated to each $\omega \in \det V$ its length, $|\omega|_g$. $\qquad \square$

Observe that an orientation $\boldsymbol{or} \in \boldsymbol{Or}(V)$ defines a natural *linear* isomorphism

$$\imath_{\boldsymbol{or}} : \det V^* \to |\Lambda|_V \quad \det V^* \ni \mu \mapsto \imath_{\boldsymbol{or}}\mu \in |\Lambda|_V, \qquad (2.2.13)$$

where

$$\imath_{\boldsymbol{or}}\mu(\omega) = \boldsymbol{or}(\omega)\mu(\omega), \quad \forall \omega \in \det V \setminus \{0\}.$$

In particular, an orientation in a Euclidean vector space canonically identifies $|\Lambda|_V$ with $\mathbb{R}$.

2.2.5 *Some complex linear algebra*

In this subsection we want to briefly discuss some aspects specific to linear algebra over the field of complex numbers.

Let V be a complex vector space. Its *conjugate* is the complex vector space $\overline{V}$ which coincides with V as a *real* vector space, but in which the multiplication by a scalar $\lambda \in \mathbb{C}$ is defined by

$$\lambda \cdot v := \overline{\lambda}v, \quad \forall v \in V.$$

The vector space V has a complex dual V_c^* that can be identified with the space of complex linear maps $V \to \mathbb{C}$. If we forget the complex structure we obtain a *real* dual V_r^* consisting of all real-linear maps $V \to \mathbb{R}$.

Definition 2.2.74. A *Hermitian metric* is a complex bilinear map

$$(\bullet, \bullet) : V \times \overline{V} \to \mathbb{C}$$

satisfying the following properties.

- The bilinear from $(\bullet, \bullet)$ is positive definite, i.e.

$$(v, v) > 0, \quad \forall v \in V \setminus \{0\}.$$

- For any $u, v \in V$ we have $(u, v) = \overline{(v, u)}$. $\qquad \square$

A Hermitian metric defines a duality $V \times \overline{V} \to \mathbb{C}$, and hence it induces a *complex linear* isomorphism

$$\mathcal{L} : \overline{V} \to V_c^*, \quad v \mapsto (\cdot, v) \in V_c^*.$$

If V and W are complex Hermitian vector spaces, then any complex linear map $A : V \to W$ induces a complex linear map

$$A^* : \overline{W} \to V_c^* \quad A^* w := \left(v \mapsto \langle Av, w \rangle \right) \in V_c^*,$$

where $\langle \bullet, \bullet \rangle$ denotes the natural duality between a vector space and its dual. We can rewrite the above fact as

$$\langle Av, w \rangle = \langle v, A^* w \rangle.$$

A complex linear map $\overline{W} \to V_c^*$ is the same as a complex linear map $W \to \overline{V_c}^*$. The metric duality defines a complex linear isomorphism $\overline{V_c}^* \cong V$ so we can view the *adjoint* A^* as a complex linear map $W \to V$.

Let $h = (\bullet, \bullet)$ be a Hermitian metric on the complex vector space V. If we view $(\bullet, \bullet)$ as an object over $\mathbb{R}$, i.e., as an $\mathbb{R}$-bilinear map $V \times V \to \mathbb{C}$, then the Hermitian metric decomposes as

$$h = \boldsymbol{Re}\, h - \boldsymbol{i}\omega, \quad \boldsymbol{i} := \sqrt{-1}, \quad \omega = -\boldsymbol{Im}\, h.$$

The real part is an inner product on the *real* space V, while ω is a *real*, skew-symmetric bilinear form on V, and thus can be identified with an element of $\Lambda_{\mathbb{R}}^2 V^*$. ω is called the *real 2-form associated to the Hermitian metric.*

It is convenient to have a coordinate description of the abstract objects introduced above. Let V be an n-dimensional complex vector space and h a Hermitian metric on it. Pick an unitary basis $e_1, ..., e_n$ of V, i.e., $n = \dim_{\mathbb{C}} V$, and $h(e_i, e_j) = \delta_{ij}$. For each j, we denote by f_j the vector $\boldsymbol{i}e_j$. Then the collection $e_1, f_1, \ldots, e_n, f_n$ is an $\mathbb{R}$-basis of V. Denote by $e^1, f^1, \ldots, e^n, f^n$ the dual $\mathbb{R}$-basis in V^*. Then

$$\boldsymbol{Re}\, h\, (e_i, e_j) = \delta_{ij} = \boldsymbol{Re}\, h(f_i, f_j) \text{ and } \boldsymbol{Re}\, h\, (e_i, f_j) = 0,$$

i.e.,

$$\boldsymbol{Re}\, h = \sum_i (e^i \otimes e^i + f^i \otimes f^i).$$

Also

$$\omega(e_i, f_j) = -\boldsymbol{Im}\, h(e_i, \boldsymbol{i}e_j) = \delta_{ij}, \quad \omega(e_i, e_j) = \omega(f_i, f_j) = 0 \ \forall i, j,$$

which shows that

$$\omega = -\boldsymbol{Im}\, h = \sum_i e^i \wedge f^i.$$

Any complex space V can be also thought of as a real vector space. The multiplication by $i = \sqrt{-1}$ defines a *real* linear operator which we denote by J. Obviously J satisfies $J^2 = -\mathbb{1}_V$.

Conversely, if V is a real vector space then any real operator $J : V \to V$ as above defines a complex structure on V by

$$(a + bi)v = av + bJv, \quad \forall v \in V, \ a + bi \in \mathbb{C}.$$

We will call an operator J as above a *complex structure*.

Let V be a real vector space with a complex structure J on it. The operator J has no eigenvectors on V. The natural extension of J to the complexification of V, $V_{\mathbb{C}} = V \otimes \mathbb{C}$, has two eigenvalues $\pm i$, and we have a splitting of $V_{\mathbb{C}}$ as a direct sum of *complex* vector spaces (eigenspaces)

$$V_{\mathbb{C}} = \ker (J - i) \oplus (\ker J + i).$$

Exercise 2.2.75. Prove that we have the following isomorphisms of *complex* vector spaces

$$V \cong \ker (J - i) \qquad \overline{V} \cong (\ker J + i). \qquad \qquad \square$$

Set

$$V^{1,0} := \ker(J - i) \cong_{\mathbb{C}} V \qquad V^{0,1} := \ker(J + i) \cong_{\mathbb{C}} \overline{V}.$$

Thus $V_{\mathbb{C}} \cong V^{1,0} \oplus V^{0,1} \cong V \oplus \overline{V}$. We obtain an isomorphism of $\mathbb{Z}$-**graded complex** vector spaces

$$\Lambda^{\bullet} V_{\mathbb{C}} \cong \Lambda^{\bullet} V^{1,0} \otimes \Lambda^{\bullet} V^{0,1}.$$

If we set

$$\Lambda^{p,q} V := \Lambda^p V^{1,0} \otimes_{\mathbb{C}} \Lambda^q V^{0,1},$$

then the above isomorphism can be reformulated as

$$\Lambda^k V_{\mathbb{C}} \cong \bigoplus_{p+q=k} \Lambda^{p,q} V. \qquad (2.2.14)$$

Note that the complex structure J on V induces by duality a complex structure J^* on V_r^*, and we have an isomorphism of *complex vector spaces*

$$V_c^* = (V, J)_c^* \cong (V_r^*, J^*).$$

We can define similarly $\Lambda^{p,q} V^*$ as the $\Lambda^{p,q}$-construction applied to the real vector space V_r^* equipped with the complex structure J^*. Note that

$$\Lambda^{1,0} V^* \cong (\Lambda^{1,0} V)_c^*, \ {}^{\backprime}$$

$$\Lambda^{0,1} V^* \cong (\Lambda^{0,1} V)_c^*,$$

and, more generally

$$\Lambda^{p,q}V^* \cong (\Lambda^{p,q}V)_c^*.$$

If h is a Hermitian metric on the complex vector space (V, J), then we have a natural isomorphism of complex vector spaces

$$V_c^* \cong (V_r^*, J^*) \cong_{\mathbb{C}} (V, -J) \cong_{\mathbb{C}} \overline{V},$$

so that

$$\Lambda^{p,q}V^* \cong_{\mathbb{C}} \Lambda^{q,p}V.$$

The Euclidean metric $g = \boldsymbol{Re}\, h$, and the associated 2-form $\omega = -\boldsymbol{Im}\, h$ are related by

$$g(u, v) = \omega(u, Jv), \quad \omega(u, v) = g(Ju, v), \quad \forall u, v \in V. \tag{2.2.15}$$

Moreover, ω is a $(1,1)$-form. To see this it suffices to pick a unitary basis (e_i) of V, and construct as usual the associated real orthonormal basis $\{e_1, f_1, \cdots, e_n, f_n\}$ $(f_i = Je_i)$. Denote by $\{e^i, f^i \; ; \; i = 1, \ldots, n\}$ the dual orthonormal basis in V_r^*. Then $J^*e^i = -f^i$, and if we set

$$\varepsilon^i := \frac{1}{\sqrt{2}}(e^i + \boldsymbol{i}f^i), \quad \bar{\varepsilon}^j := \frac{1}{\sqrt{2}}(e^j - \boldsymbol{i}f^j),$$

then

$$\Lambda^{1,0}V^* = \operatorname{span}_{\mathbb{C}}\{\varepsilon^i\} \quad \Lambda^{0,1} = \operatorname{span}_{\mathbb{C}}\{\bar{\varepsilon}^j\},$$

and

$$\omega = \boldsymbol{i} \sum \varepsilon^i \wedge \bar{\varepsilon}^i.$$

Let V be a complex vector space, and $e_1, \ldots, e_n$ be a basis of V over $\mathbb{C}$. This is not a *real* basis of V since $\dim_{\mathbb{R}} V = 2\dim_{\mathbb{C}} V$. We can however complete this to a real basis. More precisely, the vectors $e_1, \boldsymbol{i}e_1, \ldots, e_n, \boldsymbol{i}e_n$ form a real basis of V.

Proposition 2.2.76. *Suppose $(e_1, \ldots, e_n)$ and $(f_1, \ldots, f_n)$ are two complex bases of V, and $Z = (z_k^j)_{1 \le j,k}$ is the complex matrix describing the transition from the basis e to the basis f, i.e.,*

$$f_k = \sum_j z_k^j e_j, \quad \forall 1 \le k \le n.$$

Then

$$f_1 \wedge \boldsymbol{i}f_1 \wedge \cdots \wedge f_n \wedge \boldsymbol{i}f_n = |\det Z|^2 e_1 \wedge \boldsymbol{i}e_1 \wedge \cdots \wedge e_n \wedge \boldsymbol{i}e_n.$$

Proof.   We write

$$z_k^j = x_k^j + iy_k^j, \quad x_k^j, y_k^j \in \mathbb{R}, \quad \hat{e}_j = ie_j, \quad \hat{f}_k = if_k.$$

Then

$$f_k = \sum_j (x_k^j + iy_k^j)e_j = \sum_j x_k^j e_j + \sum_j y_k^j \hat{e}_j,$$

and

$$\hat{f}_k = -\sum_j y_k^j e_j + \sum_j x_k^j \hat{e}_j.$$

Then, if we set $\epsilon_n = (-1)^{n(n-1)/2}$, we deduce

$$f_1 \wedge if_1 \wedge \cdots \wedge f_n \wedge if_n = \epsilon_n f_1 \wedge \cdots \wedge f_n \wedge \hat{f}_1 \wedge \cdots \wedge \hat{f}_n$$

$$= \epsilon_n(\det \hat{Z})e_1 \wedge \cdots \wedge e_n \wedge \hat{e}_1 \wedge \cdots \wedge \hat{e}_n = (\det \hat{Z})e_1 \wedge ie_1 \wedge \ldots \wedge e_n \wedge ie_n,$$

where $\hat{Z}$ is the $2n \times 2n$ real matrix

$$\hat{Z} = \begin{bmatrix} \boldsymbol{Re}\, Z & -\boldsymbol{Im}\, Z \\ \boldsymbol{Im}\, Z & \boldsymbol{Re}\, Z \end{bmatrix} = \begin{bmatrix} X & -Y \\ Y & X \end{bmatrix},$$

and X, Y denote the $n \times n$ real matrices with entries (x_k^j) and respectively (y_k^j).

We want to show that

$$\det \hat{Z} = |\det Z|^2, \quad \forall Z \in \text{End}_{\mathbb{C}}(\mathbb{C}^n). \tag{2.2.16}$$

Let

$$\mathcal{A} := \left\{ Z \in \text{End}_{\mathbb{C}}(\mathbb{C}^n); \ |\det Z|^2 = \det \hat{Z} \right\}.$$

We will prove that $\mathcal{A} = \text{End}_{\mathbb{C}}(\mathbb{C}^n)$.

The set $\mathcal{A}$ is nonempty since it contains all the diagonal matrices. Clearly $\mathcal{A}$ is a closed subset of $\text{End}_{\mathbb{C}}(\mathbb{C}^n)$, so it suffices to show that $\mathcal{A}$ is dense in $\text{End}_{\mathbb{C}}(\mathbb{C}^n)$.

Observe that the correspondence

$$\text{End}_{\mathbb{C}}(\mathbb{C}^n) \ni Z \to \hat{Z} \in \text{End}_{\mathbb{R}}(\mathbb{R})$$

is an endomorphism of $\mathbb{R}$-algebras. Then, for every $Z \in \text{End}_{\mathbb{C}}(\mathbb{C}^n)$, and any complex linear automorphism $T \in \text{Aut}_{\mathbb{C}}(\mathbb{C}^n)$, we have

$$\widehat{TZT^{-1}} = \hat{T}\hat{Z}\hat{T}^{-1}.$$

Hence

$$Z \in \mathcal{A} \iff TZT^{-1} \in \mathcal{A}, \quad \forall T \in \text{Aut}_{\mathbb{C}}(\mathbb{C}^n).$$

In other words, if a complex matrix Z satisfies (2.2.16), then so will any of its conjugates. In particular, $\mathcal{A}$ contains *all the diagonalizable $n \times n$ complex matrices*, and these form a dense subset of $\text{End}_{\mathbb{C}}(\mathbb{C}^n)$ (see Exercise 2.2.77). $\square$

Exercise 2.2.77. Prove that the set of diagonalizable $n \times n$ complex matrices form a dense subset of the vector space $\text{End}_{\mathbb{C}}(\mathbb{C}^n)$. $\square$

Definition 2.2.78. The *canonical orientation* of a complex vector space V, $\dim_{\mathbb{C}} V = n$, is the orientation defined by $e_1 \wedge \boldsymbol{i}e_1 \wedge \ldots \wedge e_n \wedge \boldsymbol{i}e_n \in \Lambda_{\mathbb{R}}^{2n} V$, where $\{e_1, \ldots, e_n\}$ is any complex basis of V. $\qquad\square$

Suppose h is a Hermitian metric on the complex vector space V. Then $g = \boldsymbol{Re}\, h$ defines is real, Euclidean metric on V regarded as a real vector space. The canonical orientation $\boldsymbol{or}_c$ on V, and the metric g define a volume form $\mathrm{Det}_g^{\boldsymbol{or}_c} \in \Lambda_{\mathbb{R}}^{2n} V_r^*$, $n = \dim_{\mathbb{C}} V$, and a pffafian

$$\boldsymbol{Pf}_h = \boldsymbol{Pf}_h^{\boldsymbol{or}_c} : \Lambda_{\mathbb{R}}^2 V^* \to \mathbb{R}, \quad \Lambda^2 V_r^* \ni \eta \mapsto \frac{1}{n!} g(\eta^n, \mathrm{Det}_g).$$

If $\omega = -\boldsymbol{Im}\, h$ is real 2-form associated with the Hermitian metric h, then

$$\mathrm{Det}_g = \mathrm{Det}_g^{\boldsymbol{or}_c} = \frac{1}{n!} \omega^n,$$

and we conclude

$$\boldsymbol{Pf}_h \omega = 1.$$

2.3 Tensor fields

2.3.1 *Operations with vector bundles*

We now return to geometry, and more specifically, to vector bundles.

Let $\mathbb{K}$ denote one of the fields $\mathbb{R}$ or $\mathbb{C}$, and let $E \to M$ be a rank r $\mathbb{K}$-vector bundle over the smooth manifold M. According to the definition of a vector bundle, we can find an open cover (U_α) of M such that each restriction $E\,|_{U_\alpha}$ is trivial: $E\,|_{U_\alpha} \cong V \times U_\alpha$, where V is an r-dimensional vector space over the field $\mathbb{K}$. The bundle E is obtained by gluing these trivial pieces on the overlaps $U_\alpha \cap U_\beta$ using a collection of transition maps $g_{\alpha\beta} : U_\alpha \cap U_\beta \to \mathrm{GL}(V)$ satisfying the cocycle condition.

Conversely, a collection of gluing maps as above satisfying the cocycle condition uniquely defines a vector bundle. In the sequel, we will exclusively think of vector bundles in terms of gluing cocycles.

Let E, F be two vector bundles over the smooth manifold M with standard fibers V_E and respectively V_F, given by a (common) open cover (U_α), and gluing cocycles

$$g_{\alpha\beta} : U_{\alpha\beta} \to \mathrm{GL}(V_E), \quad \text{and respectively} \quad h_{\alpha\beta} : U_{\alpha\beta} \to \mathrm{GL}(V_F).$$

Then the collections

$$g_{\alpha\beta} \oplus h_{\alpha\beta} : U_{\alpha\beta} \to \mathrm{GL}(V_E \oplus V_F), \quad g_{\alpha\beta} \otimes h_{\alpha\beta} : U_{\alpha\beta} \to \mathrm{GL}(V_E \otimes V_F),$$

$$(g_{\alpha\beta}^\dagger)^{-1} : U_{\alpha\beta} \to \mathrm{GL}(V_E^*), \quad \Lambda^r g_{\alpha\beta} : U_{\alpha\beta} \to \mathrm{GL}(\Lambda^r V_E),$$

where † denotes the transpose of a linear map, satisfy the cocycle condition, and therefore define vector bundles which we denote by $E \oplus F$, $E \otimes F$, E^*, and respectively $\Lambda^r E$. In particular, if $r = \mathrm{rank}_{\mathbb{K}} E$, the bundle $\Lambda^r E$ has rank 1. It is called the *determinant line bundle* of E, and it is denoted by $\det E$.

The reader can check easily that these vector bundles are independent of the choices of transition maps used to characterize E and F (use Exercise 2.1.32). The bundle E^* is called the *dual* of the vector bundle E. The direct sum $E \oplus F$ is also called the *Whitney sum* of vector bundles. All the functorial constructions on vector spaces discussed in the previous section have a vector bundle correspondent. (Observe that a vector space can be thought of as a vector bundle over a point.)

These above constructions are natural in the following sense. Let E' and F' be vector bundles over the same smooth manifold M'. Any bundle maps $S : E \to E'$ and $T : F \to F'$, both covering the *same diffeomorphism* $\phi : M \to M'$, induce bundle morphisms

$$S \oplus T : E \oplus F \to E' \oplus T', \quad S \otimes T : E \otimes F \to E' \otimes F',$$

covering ϕ, a morphism

$$S^\dagger : (E')^* \to E^*,$$

covering ϕ^{-1} etc.

Exercise 2.3.1. Prove the assertion above. $\square$

Example 2.3.2. Let E, F, E' and F' be vector bundles over a smooth manifold M. Consider bundle isomorphisms $S : E \to E'$ and $T : F \to F'$ covering the same diffeomorphism of the base, $\phi : M \to M$. Then $(S^{-1})^\dagger : E^* \to (E')^*$ is a bundle isomorphism covering ϕ, so that we get an induced map $(S^{-1})^\dagger \otimes T : E^* \otimes F \to (E')^* \otimes F'$. Note that we have a natural identification

$$E^* \otimes F \cong \mathrm{Hom}\,(E, F).$$ $\square$

Definition 2.3.3. Let $E \to M$ be a $\mathbb{K}$-vector bundle over M. A *metric* on E is a section h of $E^* \otimes_{\mathbb{K}} \overline{E^*}$ ($\overline{E} = E$ if $\mathbb{K} = \mathbb{R}$) such that, for any $m \in M$, $h(m)$ defines a metric on E_m (Euclidean if $\mathbb{K} = \mathbb{R}$ or Hermitian if $\mathbb{K} = \mathbb{C}$). $\square$

2.3.2   *Tensor fields*

We now specialize the previous considerations to the special situation when E is the tangent bundle of M, $E \cong TM$. The *cotangent bundle* is then

$$T^*M := (TM)^*.$$

We define the tensor bundles of M

$$\mathcal{T}^r_s(M) := \mathcal{T}^r_s(TM) = (TM)^{\otimes r} \otimes (T^*M)^{\otimes s}.$$

Definition 2.3.4. (a) A *tensor field* of type (r, s) over the open set $U \subset M$ is a section of $\mathcal{T}_s^r(M)$ over U.

(b) A degree r *differential form* (r-form for brevity) is a section of $\Lambda^r(T^*M)$. The space of (smooth) r-forms over M is denoted by $\Omega^r(M)$. We set

$$\Omega^\bullet(M) := \bigoplus_{r \geq 0} \Omega^r(M).$$

(c) A *Riemannian metric* on a manifold M is a metric on the tangent bundle. More precisely, it is a symmetric $(0, 2)$-tensor g, such that for every $x \in M$, the bilinear map

$$g_x : T_x M \times T_x M \to \mathbb{R}$$

defines a Euclidean metric on $T_x M$. □

If we view the tangent bundle as a smooth family of vector spaces, then a tensor field can be viewed as a smooth selection of a tensor in each of the tangent spaces. In particular, a Riemann metric defines a smoothly varying procedure of measuring lengths of vectors in tangent spaces.

Example 2.3.5. It is often very useful to have a local description of these objects. If $(x^1, \ldots, x^n)$ are local coordinates on an open set $U \subset M$, then the vector fields $(\frac{\partial}{\partial x^1}, \ldots, \frac{\partial}{\partial x^n})$ trivialize $TM|_U$, i.e., they define a framing of the restriction of TM to U. We can form a dual framing of $T^*M|_U$, using the 1-forms dx^i, $i = 1, \ldots, n$. They satisfy the duality conditions

$$\langle dx^i, \frac{\partial}{\partial x_j} \rangle = \delta_j^i, \quad \forall i, j.$$

A basis in $\mathcal{T}_s^r(T_x M)$ is given by

$$\left\{ \frac{\partial}{\partial x^{i_1}} \otimes \ldots \otimes \frac{\partial}{\partial x^{i_r}} \otimes dx^{j_1} \otimes \ldots \otimes dx^{j_s}; \ 1 \leq i_1, \ldots, i_r \leq n, \ 1 \leq j_1, \ldots, j_s \leq n \right\}.$$

Hence, any tensor $T \in \mathcal{T}_s^r(M)$ has a local description

$$T = T_{j_1 \ldots j_s}^{i_1 \ldots i_r} \frac{\partial}{\partial x^{i_1}} \otimes \ldots \otimes \frac{\partial}{\partial x^{i_r}} \otimes dx^{j_1} \otimes \ldots \otimes dx^{j_s}.$$

In the above equality we have used Einstein's convention. In particular, an r-form ω has the local description

$$\omega = \sum_{1 \leq i_1 < \cdots < i_r \leq n} \omega_{i_1 \ldots i_r} dx^{i_1} \wedge \cdots \wedge dx^{i_r},$$

while a Riemann metric g has the local description

$$g = \sum_{i,j} g_{ij} dx^i \otimes dx^j, \quad g_{ij} = g_{ji}. \qquad \Box$$

Remark 2.3.6. (a) A covariant tensor field, i.e., a $(0, s)$-tensor field S, naturally defines a $C^\infty(M)$-multilinear map

$$S : \bigoplus_1^s \mathrm{Vect}\,(M) \to C^\infty(M),$$

$$(X_1, \ldots, X_s) \mapsto \big(p \mapsto S_p\big(X_1(p), \ldots, X_s(p) \big) \big) \in C^\infty(M).$$

Conversely, any such map uniquely defines a $(0, s)$-tensor field. In particular, an r-form η can be identified with a skew-symmetric $C^\infty(M)$-multilinear map

$$\eta : \bigoplus_1^r \mathrm{Vect}\,(M) \to C^\infty(M).$$

Notice that the wedge product in the exterior algebras induces an associative product in $\Omega^\bullet(M)$ which we continue to denote by $\wedge$.

(b) Let $f \in C^\infty(M)$. Its Fréchet derivative $Df : TM \to T\mathbb{R} \cong \mathbb{R} \times \mathbb{R}$ is naturally a 1-form. Indeed, we get a smooth $C^\infty(M)$-linear map $df : \mathrm{Vect}\,(M) \to C^\infty(M)$ defined by

$$df(X)_m := Df(X)_{f(m)} \in T_{f(m)}\mathbb{R} \cong \mathbb{R}.$$

In the sequel we will always regard the differential of a smooth function f as a 1-form and to indicate this we will use the notation df (instead of the usual Df). □

Any diffeomorphism $f : M \to N$ induces bundle isomorphisms $Df : TM \to TN$ and $(Df^{-1})^\dagger : T^*M \to T^*N$ covering f. Thus, a diffeomorphism f induces a linear map

$$f_* : \mathcal{T}_s^r(M) \to \mathcal{T}_s^r(N), \tag{2.3.1}$$

called the *push-forward* map. In particular, the group of diffeomorphisms of M acts naturally (and linearly) on the space of tensor fields on M.

Example 2.3.7. Suppose $f : M \to N$ is a diffeomorphism, and S is a $(0, k)$-tensor field on M, which we regard as a $C^\infty(M)$-multilinear map

$$S : \underbrace{\mathrm{Vect}\,(M) \times \cdots \times \mathrm{Vect}\,(M)}_{k} \to C^\infty(M).$$

Then f_*S is a $(0, k)$ tensor field on N. Let $q \in N$, and set $p := f^{-1}(q)$. Then, for any $Y_1, \ldots, Y_k \in T_qN$, we have

$$(f_*S)_q(Y_1, \ldots, Y_k) = S_p\big(f_*^{-1}Y_1, \ldots, (f_*)^{-1}Y_k \big)$$

$$= S_p\big((D_pf)^{-1}Y_1, \ldots, (D_pf)^{-1}Y_k \big). \qquad □.$$

For covariant tensor fields a more general result is true. More precisely, any smooth map $f : M \to N$ defines a linear map

$$f^* : \mathcal{T}_s^0(N) \to \mathcal{T}_s^0(M),$$

called the *pullback* by f. Explicitly, if S is such a tensor defined by a $C^\infty(M)$-multilinear map

$$S : (\text{Vect}\,(N))^s \to C^\infty(N),$$

then f^*S is the covariant tensor field defined by

$$(f^*S)_p(X_1(p), \ldots, X_s(p)) := S_{f(p)}\big(D_p f(X_1), \ldots, D_p f(X_s)\big),$$

$\forall X_1, \ldots, X_s \in \text{Vect}\,(M),\ p \in M$. Note that when f is a *diffeomorphism* we have

$$f^* = (f_*^{-1})^\dagger,$$

where f_* is the push-forward map defined in (2.3.1).

Example 2.3.8. Consider the map

$$F_* : (0, \infty) \times (0, 2\pi) \to \mathbb{R}^2, \quad (r, \theta) \mapsto (x = r \cos\theta, y = r \sin\theta).$$

The map F defines the usual polar coordinates. It is a diffeomorphism onto the open subset

$$U := \mathbb{R}^2 \setminus \{(x, 0);\ x \geq 0\}.$$

For simplicity, we will write ∂_r, ∂_x instead of $\frac{\partial}{\partial r}, \frac{\partial}{\partial x}$ etc. We have

$$F^*dx = d(r \cos\theta) = \cos\theta dr - r\sin\theta d\theta, \quad F^*dy = d(r \sin\theta) = \sin\theta dr + r\cos\theta d\theta,$$

$$F^*(dx \wedge dy) = d(r \cos\theta) \wedge d(r\sin\theta) = (\cos\theta dr - r\sin\theta d\theta) \wedge (\sin\theta dr + r\cos\theta d\theta)$$

$$= r(\cos^2\theta + \sin^2\theta)dr \wedge d\theta = rdr \wedge d\theta.$$

To compute $F_*\partial_r$ and $F_*\partial_\theta$ we use the chain rule which implies

$$F_*\partial_r = \frac{\partial x}{\partial r}\partial_x + \frac{\partial y}{\partial r}\partial_y = \cos\theta\partial_x + \sin\theta\partial_y = \frac{1}{r}(x\partial_x + y\partial_y)$$

$$= \frac{1}{(x^2 + y^2)^{1/2}}(x\partial_x + y\partial_y).$$

The Euclidean metric is described over U by the symmetric $(0, 2)$–tensor $g = dx^2 + dy^2$. The pullback of g by F is the symmetric $(0, 2)$–tensor

$$F^*(dx^2 + dy^2) = \big(d(r \cos\theta) \big)^2 + \big(d(r \sin\theta) \big)^2$$

$$= (\cos\theta dr - r\sin\theta d\theta)^2 + (\sin\theta dr + r\cos\theta d\theta)^2 = dr^2 + r^2 d\theta^2.$$

To compute F_*dr we need to express r as a function of x and y, $r = (x^2 + y^2)^{1/2}$, and then we have

$$F_*dr = (F^{-1})^*dr = d(x^2 + y^2)^{1/2} = x(x^2 + y^2)^{-1/2}dx + y(x^2 + y^2)^{-1/2}dy. \quad \square$$

All the operations discussed in the previous section have natural extensions to tensor fields. There exists a tensor multiplication, a Riemann metric defines a duality $\mathcal{L} : \mathrm{Vect}\,(M) \to \Omega^1(M)$ etc. In particular, there exists a contraction operator

$$\mathrm{tr} : \mathfrak{T}^{r+1}_{s+1}(M) \to \mathfrak{T}^r_s(M)$$

defined by

$$\mathrm{tr}\,(X_0 \otimes \cdots \otimes X_r) \otimes (\omega_0 \otimes \cdots \otimes \omega_s) = \omega_0(X_0)(X_1 \otimes \cdots X_r \otimes \omega_1 \otimes \cdots \otimes \omega_s),$$

$\forall X_i \in \mathrm{Vect}\,(M)$, $\forall \omega_j \in \Omega^1(M)$. In local coordinates the contraction has the form

$$\{\mathrm{tr}\,(T^{i_0\ldots i_r}_{j_0\ldots j_s})\} = \{T^{ii_1\ldots i_r}_{ij_1\ldots j_s}\}.$$

Let us observe that a Riemann metric g on a manifold M induces metrics in all the associated tensor bundles $\mathfrak{T}^r_s(M)$. If we choose local coordinates (x^i) on an open set U then, as explained above, the metric g can be described as

$$g = \sum_{i,j} g_{ij} dx^i \otimes dx^j,$$

while a tensor field T of type (r, s) can be described as

$$T = T^{i_1\ldots i_r}_{j_1\ldots j_s} \frac{\partial}{\partial x^{i_1}} \otimes \ldots \otimes \frac{\partial}{\partial x^{i_r}} \otimes dx^{j_1} \otimes \ldots \otimes dx^{j_s}.$$

If we denote by (g^{ij}) the inverse of the matrix (g_{ij}), then, for every point $p \in U$, the length of $T(p) \in \mathfrak{T}^r_s(M)_p$ is the number $|T(p)|_g$ defined by

$$|T(p)|_g = g_{i_1 k_1} \cdots g_{i_r k_r} g^{j_1 \ell_1} \cdots g^{j_s \ell_s} T^{i_1\ldots i_r}_{j_1\ldots j_s} T^{k_1\ldots k_r}_{\ell_1\ldots \ell_s},$$

where in the above equalities we have used Einstein's convention.

The exterior product defines an exterior product on the space of smooth differential forms

$$\wedge : \Omega^\bullet(M) \times \Omega^\bullet(M) \to \Omega^\bullet(M).$$

The space $\Omega^\bullet(M)$ is then an associative algebra with respect to the operations $+$ and $\wedge$.

Proposition 2.3.9. *Let $f : M \to N$ be a smooth map. The pullback by f defines a morphism of associative algebras $f^* : \Omega^\bullet(N) \to \Omega^\bullet(M)$.* □

Exercise 2.3.10. Prove the above proposition. □

2.3.3   *Fiber bundles*

We consider it useful at this point to bring up the notion of fiber bundle. There are several reasons to do this.

On one hand, they arise naturally in geometry, and they impose themselves as worth studying. On the other hand, they provide a very elegant and concise language to describe many phenomena in geometry.

We have already met examples of fiber bundles when we discussed vector bundles. These were "smooth families of vector spaces". A fiber bundle wants to be a smooth family of copies of the same manifold. This is a very loose description, but it offers a first glimpse at the notion about to be discussed.

The model situation is that of direct product $X = F \times B$, where B and F are smooth manifolds. It is convenient to regard this as a family of manifolds $(F_b)_{b \in B}$. The manifold B is called the base, F is called the standard (model) fiber, and X is called the total space. This is an example of *trivial* fiber bundle.

In general, a fiber bundle is obtained by gluing a bunch of trivial ones according to a prescribed rule. The gluing may encode a symmetry of the fiber, and we would like to spend some time explaining what do we mean by symmetry.

Definition 2.3.11. (a) Let M be a smooth manifold, and G a Lie group. We say the group G acts on M from the left (respectively right), if there exists a smooth map

$$\Phi : G \times M \to M, \quad (g, m) \mapsto T_g m,$$

such that $T_1 \equiv \mathbb{1}_M$ and

$$T_g(T_h m) = T_{gh} m \quad (\text{respectively } T_g(T_h m) = T_{hg} m) \ \forall g, h \in G, \ m \in M.$$

In particular, we deduce that $\forall g \in G$ the map T_g is a diffeomorphism of M. For any $m \in M$ the set

$$G \cdot m = \{T_g m; \ g \in G\}$$

is called the *orbit* of the action through m.

(b) Let G act on M. The action is called *free* if $\forall g \in G$, and $\forall m \in M$ $T_g m \neq m$. The action is called *effective* if, $\forall g \in G$, $T_g \neq \mathbb{1}_M$. $\qquad \square$

It is useful to think of a Lie group action on a manifold as encoding a symmetry of that manifold.

Example 2.3.12. Consider the unit sphere

$$S^1 = \left\{ (x, yz) \in \mathbb{R}^3; \ x^2 + y^2 + z^2 = 1 \right\}.$$

Then the counterclockwise rotations about the z-axis define a smooth left action of S^1 on S^2. More formally, if we use cylindrical coordinates (r, θ, z),

$$x = r \cos \theta, \quad y = r \sin \theta, \quad z = 0,$$

then for every $\varphi \in \mathbb{R}$ mod $2\pi \cong S^1$ we define $T_\varphi : S^2 \to S^2$ by

$$T_\varphi(r, \theta, z) = (r, (\theta + \varphi) \bmod 2\pi, z).$$

The resulting map $T : S^1 \times S^2 \to S^2$, $(\varphi, p) \mapsto T_\varphi(p)$ defines a left action of S^1 on S^2 encoding the rotational symmetry of S^2 about the z-axis. $\qquad \square$

Example 2.3.13. Let G be a Lie group. A *linear representation* of G on a vector space V is a left action of G on V such that each T_g is a linear map. One says V is a *G-module*. For example, the tautological linear action of $SO(n)$ on $\mathbb{R}^n$ defines a linear representation of $SO(n)$.   □

Example 2.3.14. Let G be a Lie group. For any $g \in G$ denote by L_g (resp. R_g) the left (resp right) translation by g. In this way we get the tautological left (resp. right) action of G on itself.   □

Definition 2.3.15. A smooth *fiber bundle* is an object composed of the following:

(a) a smooth manifold E called the *total space*;
(b) a smooth manifold F called the *standard fiber*;
(c) a smooth manifold B called the *base*;
(d) a surjective submersion $\pi : E \to B$ called the *natural projection*;
(e) a collection of *local trivializations*, i.e., an open cover (U_α) of the base B, and diffeomorphisms $\Psi_\alpha : F \times U_\alpha \to \pi^{-1}(U_\alpha)$ such that

$$\pi \circ \Psi_\alpha(f, b) = b, \quad \forall (f, b) \in F \times U_\alpha,$$

i.e., the diagram below is commutative.

$$F \times U_\alpha \xrightarrow{\quad \Psi_\alpha \quad} \pi^{-1}(U_\alpha)$$
$$\searrow \qquad \swarrow \pi$$
$$U_\alpha$$

We can form the transition (gluing) maps $\Psi_{\alpha\beta} : F \times U_{\alpha\beta} \to F \times U_{\alpha\beta}$, where $U_{\alpha\beta} = U_\alpha \cap U_\beta$, defined by $\Psi_{\alpha\beta} = \psi_\alpha^{-1} \circ \psi_\beta$. According to (e), these maps can be written as

$$\psi_{\alpha\beta}(f, b) = (T_{\alpha\beta}(b)f, b),$$

where $T_{\alpha\beta}(b)$ is a diffeomorphism of F depending smoothly upon $b \in U_{\alpha\beta}$.

We will denote this fiber bundle by (E, π, F, B).

If G is a Lie group, then the bundle (E, π, F, B) is called a *G-fiber bundle* if it satisfies the following additional conditions.

(f) There exists an effective left action of the Lie group G on F,

$$G \times F \ni (g, x) \mapsto g \cdot x = T_g x \in F.$$

The group G is called the *symmetry group of the bundle*.

(g) There exist smooth maps $g_{\alpha\beta} : U_{\alpha\beta} \to G$ satisfying the cocycle condition

$$g_{\alpha\alpha} = 1 \in G, \quad g_{\gamma\alpha} = g_{\gamma\beta} \cdot g_{\beta\alpha}, \quad \forall \alpha, \beta, \gamma,$$

and such that

$$T_{\alpha\beta}(b) = T_{g_{\alpha\beta}(b)}.$$

We will denote a G-fiber bundle by (E, π, F, B, G). $\qquad\qquad\qquad\qquad\square$

The choice of an open cover (U_α) in the above definition is a source of arbitrariness since there is no natural prescription on how to perform this choice. We need to describe when two such choices are equivalent.

Two open covers (U_α) and (V_i), together with the collections of local trivializations

$$\Phi_\alpha : F \times U_\alpha \to \pi^{-1}(U_\alpha) \ \text{ and } \ \Psi_i : F \times V_i \to \pi^{-1}(V_i)$$

are said to be equivalent if, for all α, i, there exists a smooth map

$$T_{\alpha i} : U_\alpha \cap V_i \to G,$$

such that, for any $x \in U_\alpha \cap V_i$, and any $f \in F$, we have

$$\Phi_\alpha^{-1} \Psi_i(f, x) = (T_{\alpha i}(x)f, x).$$

A G-bundle structure is defined by an equivalence class of trivializing covers.

As in the case of vector bundles, a collection of gluing data determines a G-fiber bundle. Indeed, if we are given a cover $(U_\alpha)_{\alpha \in A}$ of the base B, and a collection of transition maps $g_{\alpha\beta} : U_\alpha \cap U_\beta \to G$ satisfying the cocycle condition, then we can get a bundle by gluing the trivial pieces $U_\alpha \times F$ along the overlaps.

More precisely, if $b \in U_\alpha \cap U_\beta$, then the element $(f, b) \in F \times U_\alpha$ is identified with the element $(g_{\beta\alpha}(b) \cdot f, b) \in F \times U_\beta$.

Definition 2.3.16. Let $E \xrightarrow{\pi} B$ be a G-fiber bundle. A G-automorphism of this bundle is a diffeomorphism $T : E \to E$ such that $\pi \circ T = \pi$, i.e., T maps fibers to fibers, and for any trivializing cover (U_α) (as in Definition 2.3.15) there exists a smooth map $g_\alpha : U_\alpha \to G$ such that

$$\Psi_\alpha^{-1} T \Psi_\alpha(f, b) = (g_\alpha(b)f, b), \ \ \forall b, f. \qquad\qquad\qquad \square$$

Definition 2.3.17. (a) A *fiber bundle* is an object defined by conditions (a)-(d) and (f) in the above definition. (One can think the structure group is the group of diffeomorphisms of the standard fiber).
(b) A *section* of a fiber bundle $E \xrightarrow{\pi} B$ is a smooth map $s : B \to E$ such that $\pi \circ s = \mathbb{1}_B$, i.e., $s(b) \in \pi^{-1}(b), \forall b \in B$. $\qquad\qquad\qquad \square$

Example 2.3.18. A rank r vector bundle (over $\mathbb{K} = \mathbb{R}, \mathbb{C}$) is a $\mathrm{GL}(r, \mathbb{K})$-fiber bundle with standard fiber $\mathbb{K}^r$, and where the group $\mathrm{GL}(r, \mathbb{K})$ acts on $\mathbb{K}^r$ in the natural way. $\qquad\qquad\qquad \square$

Example 2.3.19. Let G be a Lie group. A *principal G-bundle* is a G-fiber bundle with fiber G, where G acts on itself by left translations. Equivalently, a principal G-bundle over a smooth manifold M can be described by an open cover $\mathcal{U}$ of M and a G-cocycle, i.e., a collection of smooth maps

$$g_{UV} : U \cap V \to G \ \ U, V \in \mathcal{U},$$

such that $\forall x \in U \cap V \cap W$ $(U, V, W \in \mathcal{U})$

$$g_{UV}(x)g_{VW}(x)g_{WU}(x) = 1 \in G. \qquad \square$$

Exercise 2.3.20. (Alternative definition of a principal bundle). Let P be a fiber bundle with fiber a Lie group G. Prove the following are equivalent.
(a) P is a principal G-bundle.
(b) There exists a free, right action of G on G,

$$P \times G \to P, \quad (p, g) \mapsto p \cdot g,$$

such that its orbits coincide with the fibers of the bundle P, and there exists a trivializing cover

$$\left\{ \, \Psi_\alpha : G \times U_\alpha \to \pi^{-1}(U_\alpha) \, \right\},$$

such that

$$\Psi_\alpha(hg, u) = \Psi_\alpha(h, u) \cdot g, \quad \forall g, h \in G, \ u \in U_\alpha. \qquad \square$$

Exercise 2.3.21. (The frame bundle of a manifold). Let M^n be a smooth manifold. Denote by $F(M)$ the set of frames on M, i.e.,

$$F(M) = \{(m; X_1, \dots, X_n); \ m \in M, \ X_i \in T_m M \text{ and } \operatorname{span}(X_1, \dots, X_n) = T_m M\} \,.$$

(a) Prove that $F(M)$ can be naturally organized as a smooth manifold such that the natural projection $p : F(M) \to M$, $(m; X_1, \dots, X_n) \mapsto m$ is a submersion.
(b) Show $F(M)$ is a principal $\operatorname{GL}(n, \mathbb{R})$-bundle. The bundle $F(M)$ is called the *frame bundle* of the manifold M.
Hint: A matrix $T = (T_j^i) \in \operatorname{GL}(n, \mathbb{K})$ acts on the right on $F(M)$ by

$$(m; X_1, ..., X_n) \mapsto (m; (T^{-1})_1^i X_i, ..., (T^{-1})_n^i X_i). \qquad \square$$

Example 2.3.22. (Associated fiber bundles). Let $\pi : P \to G$ be a principal G-bundle. Consider a trivializing cover $(U_\alpha)_{\alpha \in A}$, and denote by $g_{\alpha\beta} : U_\alpha \cap U_\beta \to G$ a collection of gluing maps determined by this cover. Assume G acts (on the left) on a smooth manifold F

$$\tau : G \times F \to F, \quad (g, f) \mapsto \tau(g)f.$$

The collection $\tau_{\alpha\beta} = \tau(g_{\alpha\beta}) : U_{\alpha\beta} \to \operatorname{Diffeo}(F)$ satisfies the cocycle condition and can be used (exactly as we did for vector bundles) to define a G-fiber bundle with fiber F. This new bundle is independent of the various choices made (cover (U_α) and transition maps $g_{\alpha\beta}$). (Prove this!) It is called the bundle associated to P via τ and is denoted by $P \times_\tau F$. $\qquad \square$

Exercise 2.3.23. Prove that the tangent bundle of a manifold M^n is associated to $F(M)$ via the natural action of $\operatorname{GL}(n, \mathbb{R})$ on $\mathbb{R}^n$. $\qquad \square$

Exercise 2.3.24. (The Hopf bundle) If we identify the unit odd dimensional sphere S^{2n-1} with the submanifold

$$\left\{ (z_1, \ldots, z_n) \in \mathbb{C}^n; \; |z_0|^2 + \cdots + |z_n|^2 = 1 \right\}$$

then we detect an S^1-action on S^{2n-1} given by

$$e^{i\theta} \cdot (z_1, \ldots, z_n) = (e^{i\theta} z_1, \ldots, e^{i\theta} z_n).$$

The space of orbits of this action is naturally identified with the complex projective space $\mathbb{CP}^{n-1}$.

(a) Prove that $p : S^{2n-1} \to \mathbb{CP}^{n-1}$ is a principal S^1 bundle called Hopf bundle. (p is the obvious projection). Describe one collection of transition maps.

(b) Prove that the tautological line bundle over $\mathbb{CP}^{n-1}$ is associated to the Hopf bundle via the natural action of S^1 on $\mathbb{C}^1$. □

Exercise 2.3.25. Let E be a vector bundle over the smooth manifold M. Any metric h on E (euclidian or Hermitian) defines a submanifold $S(E) \subset E$ by

$$S(E) = \{v \in E; \; |v|_h = 1\}.$$

Prove that $S(E)$ is a fibration over M with standard fiber a sphere of dimension rank $E - 1$. The bundle $S(E)$ is usually called the *sphere bundle* of E. □

Chapter 3

Calculus on Manifolds

This chapter describes the "kitchen" of differential geometry. We will discuss how one can operate with the various objects we have introduced so far. In particular, we will introduce several derivations of the various algebras of tensor fields, and we will also present the inverse operation of integration.

3.1 The lie derivative

3.1.1 *Flows on manifolds*

The notion of flow should be familiar to anyone who has had a course in ordinary differential equations. In this section we only want to describe some classical analytic facts in a geometric light. We strongly recommend [4] for more details, and excellent examples.

A neighborhood $\mathcal{N}$ of $\{0\} \times M$ in $\mathbb{R} \times M$ is called *balanced* if, $\forall m \in M$, there exists $r \in (0, \infty]$ such that

$$(\mathbb{R} \times \{m\}) \cap \mathcal{N} = (-r, r) \times \{m\}.$$

Note that any continuous function $f : M \to (0, \infty)$ defines a balanced open

$$\mathcal{N}_f := \{(t, m) \in \mathbb{R} \times M; \ |t| < f(m)\}.$$

Definition 3.1.1. A *local flow* is a smooth map $\Phi : \mathcal{N} \to M$, $(t, m) \mapsto \Phi^t(m)$, where $\mathcal{N}$ is a balanced neighborhood of $\{0\} \times M$ in $\mathbb{R} \times M$, such that

(a) $\Phi^0(m) = m$, $\forall m \in M$.
(b) $\Phi^t(\Phi^s(m)) = \Phi^{t+s}(m)$ for all $s, t \in \mathbb{R}$, $m \in M$ such that

$$(s, m), (s + t, m), \ (t, \Phi^s(m)) \in \mathcal{N}.$$

When $\mathcal{N} = \mathbb{R} \times M$, Φ is called a *flow*. □

The conditions (a) and (b) above show that a flow is nothing but a left action of the additive (Lie) group $(\mathbb{R}, +)$ on M.

Example 3.1.2. Let A be an $n \times n$ real matrix. It generates a flow Φ_A^t on $\mathbb{R}^n$ by

$$\Phi_A^t x = e^{tA} x = \left(\sum_{k=0}^{\infty} \frac{t^k}{k!} A^k \right) x. \qquad \square$$

Definition 3.1.3. Let $\Phi : \mathbb{N} \to M$ be a local flow on M. The *infinitesimal generator* of Φ is the vector field X on M defined by

$$X(p) = X_\Phi(p) := \frac{d}{dt} \big|_{t=0} \Phi^t(p), \quad \forall p \in M,$$

i.e., $X(p)$ is the tangent vector to the smooth path $t \mapsto \Phi^t(p)$ at $t = 0$. This path is called the *flow line* through p. $\qquad \square$

Exercise 3.1.4. Show that X_Φ is a *smooth* vector field. $\qquad \square$

Example 3.1.5. Consider the flow e^{tA} on $\mathbb{R}^n$ generated by an $n \times n$ matrix A. Its generator is the vector field X_A on $\mathbb{R}^n$ defined by

$$X_A(u) = \frac{d}{dt} \big|_{t=0} e^{tA} u = Au. \qquad \square$$

Proposition 3.1.6. *Let M be a smooth n-dimensional manifold. The map*

$$X : \{\text{Local flows on } M\} \to \mathrm{Vect}\,(M), \quad \Phi \mapsto X_\Phi,$$

is a surjection. Moreover, if $\Phi_i : \mathbb{N}_i \to M$ (i=1,2) are two local flows such that $X_{\Phi_1} = X_{\Phi_2}$, then $\Phi_1 = \Phi_2$ on $\mathbb{N}_1 \cap \mathbb{N}_2$.

Proof. *Surjectivity.* Let X be a vector field on M. An *integral curve* for X is a smooth curve $\gamma : (a, b) \to M$ such that

$$\dot{\gamma}(t) = X(\gamma(t)).$$

In local coordinates (x^i) over on open subset $U \subset M$ this condition can be rewritten as

$$\dot{x}^i(t) = X^i\big(x^1(t), ..., x^n(t)\big), \quad \forall i = 1, \ldots, n, \qquad (3.1.1)$$

where $\gamma(t) = (x^1(t), ..., x^n(t))$, and $X = X^i \frac{\partial}{\partial x^i}$. The above equality is a system of ordinary differential equations. Classical existence results (see e.g. [4, 43]) show that, for any precompact open subset $K \subset U$, there exists $\varepsilon > 0$ such that, for all $x \in K$, there exists a unique integral curve for X, $\gamma_x : (-\varepsilon, \varepsilon) \to M$ satisfying

$$\gamma_x(0) = x. \qquad (3.1.2)$$

Moreover, as a consequence of the smooth dependence upon initial data we deduce that the map

$$\Phi_K : \mathbb{N}_K = (-\varepsilon, \varepsilon) \times K \to M, \quad (x, t) \mapsto \gamma_x(t),$$

is smooth.

Now we can cover M by open, precompact, local coordinate neighborhoods $(K_\alpha)_{\alpha \in \mathcal{A}}$, and as above, we get smooth maps $\Phi_\alpha : \mathcal{N}_\alpha = (-\varepsilon_\alpha, \varepsilon_\alpha) \times K_\alpha \to M$ solving the initial value problem (3.1.1-2). Moreover, by uniqueness, we deduce

$$\Phi_\alpha = \Phi_\alpha \text{ on } \mathcal{N}_\alpha \cap \mathcal{N}_\beta.$$

Define

$$\mathcal{N} := \bigcup_{\alpha \in \mathcal{A}} \mathcal{N}_\alpha,$$

and set $\Phi : \mathcal{N} \to M$, $\Phi = \Phi_\alpha$ on $\mathcal{N}_\alpha$.

The uniqueness of solutions of initial value problems for ordinary differential equations implies that Φ satisfies all the conditions in the definition of a local flow. Tautologically, X is the infinitesimal generator of Φ. The second part of the proposition follows from the uniqueness in initial value problems. $\square$

The family of local flows on M with the same infinitesimal generator $X \in \text{Vect}(M)$ is naturally ordered according to their domains,

$$(\Phi_1 : \mathcal{N}_1 \to M) \prec (\Phi_2 : \mathcal{N}_2 \to M)$$

if and only if $\mathcal{N}_1 \subset \mathcal{N}_2$. This family has a *unique* maximal element which is called the *local flow generated by* X, and it is denoted by Φ_X.

Exercise 3.1.7. Consider the unit sphere

$$S^2 = \left\{(x, y, z) \in \mathbb{R}^3; \ x^2 + y^2 + z^2 = 1\right\}.$$

For every point $p \in S^2$ we denote by $X(p) \in T_p\mathbb{R}^3$, the orthogonal projection of the vector $\boldsymbol{k} = (0, 0, 1)$ onto $T_p S^2$.
(a) Prove that $p \mapsto X(p)$ is a smooth vector field on S^2, and then describe it in cylindrical coordinates (z, θ), where

$$x = r \cos \theta, \ y = r \sin \theta, \ r = (x^2 + y^2)^{1/2}.$$

(b) Describe explicitly the flow generated by X. $\square$

3.1.2 *The Lie derivative*

Let X be a vector field on the smooth n-dimensional manifold M and denote by $\Phi = \Phi_X$ the local flow it generates. For simplicity, we assume Φ is actually a flow so its domain is $\mathbb{R} \times M$. The local flow situation is conceptually identical, but notationally more complicated.

For each $t \in \mathbb{R}$, the map Φ^t is a diffeomorphism of M and so it induces a push-forward map on the space of tensor fields. If S is a tensor field on M we define its *Lie derivative* along the direction given by X as

$$L_X S_m := -\lim_{t \to 0} \frac{1}{t} \left((\Phi_*^t S)_m - S_m \right) \ \forall m \in M. \tag{3.1.3}$$

Intuitively, $L_X S$ measures how fast is the flow Φ changing[1] the tensor S.

If the limit in (3.1.3) exists, then one sees that $L_X S$ is a tensor of the same type as S. To show that the limit exists, we will provide more explicit descriptions of this operation.

Lemma 3.1.8. *For any $X \in \text{Vect}\,(M)$ and $f \in C^\infty(M)$ we have*

$$Xf := L_X f = \langle df, X \rangle = df(X).$$

Above, $\langle \bullet, \bullet \rangle$ denotes the natural duality between T^*M and TM,

$$\langle \bullet, \bullet \rangle : C^\infty(T^*M) \times C^\infty(TM) \to C^\infty(M),$$

$$C^\infty(T^*M) \times C^\infty(TM) \ni (\alpha, X) \mapsto \alpha(X) \in C^\infty(M).$$

In particular, L_X is a derivation of $C^\infty(M)$.

Proof.   Let $\Phi^t = \Phi_X^t$ be the local flow generated by X. Assume for simplicity that it is defined for all t. The map Φ^t acts on $C^\infty(M)$ by the pullback of its inverse, i.e.

$$\Phi_*^t = (\Phi^{-t})^*.$$

Hence, for point $p \in M$ we have

$$L_X f(p) = \lim_{t \to 0} \frac{1}{t}(f(p) - f(\Phi^{-t}p)) = -\frac{d}{dt}\mid_{t=0} f(\Phi^{-t}p) = \langle df, X \rangle_p. \qquad \square$$

Exercise 3.1.9. Prove that any derivation of the algebra $C^\infty(M)$ is of the form L_X for some $X \in \text{Vect}(M)$, i.e.

$$\text{Der}\,(C^\infty(M)) \cong \text{Vect}\,(M). \qquad \square$$

Lemma 3.1.10. *Let $X, Y \in \text{Vect}\,(M)$. Then the Lie derivative of Y along X is a new vector field $L_X Y$ which, viewed as a derivation of $C^\infty(M)$, coincides with the commutator of the two derivations of $C^\infty(M)$ defined by X and Y i.e.*

$$L_X Y f = [X, Y]f, \ \ \forall f \in C^\infty(M).$$

The vector field $[X, Y] = L_X Y$ is called the **Lie bracket** *of X and Y. In particular the Lie bracket induces a Lie algebra structure on $\text{Vect}\,(M)$.*

---

[1] Arnold refers to the Lie derivative L_X as the *"fisherman's derivative"*. Here is the intuition behind this very suggestive terminology. We place an observer (fisherman) at a fixed point $p \in M$, and we let him keep track of the the sizes of the tensor S carried by the flow at the point p. The Lie derivatives measures the rate of change in these sizes.

Proof. We will work in local coordinates (x^i) near a point $m \in M$ so that

$$X = X^i \frac{\partial}{\partial x_i} \quad \text{and} \quad Y = Y^j \frac{\partial}{\partial x_j}.$$

We first describe the commutator $[X, Y]$. If $f \in C^\infty(M)$, then

$$[X, Y]f = (X^i \frac{\partial}{\partial x_i})(Y^j \frac{\partial f}{\partial x^j}) - (Y^j \frac{\partial}{\partial x_j})(X^i \frac{\partial f}{\partial x^i})$$

$$= \left(X^i Y^j \frac{\partial^2 f}{\partial x^i \partial x^j} + X^i \frac{\partial Y^j}{\partial x^i} \frac{\partial f}{\partial x^j} \right) - \left(X^i Y^j \frac{\partial^2 f}{\partial x^i \partial x^j} + Y^j \frac{\partial X^i}{\partial x^j} \frac{\partial f}{\partial x^i} \right),$$

so that the commutator of the two derivations is the derivation defined by the vector field

$$[X, Y] = \left(X^i \frac{\partial Y^k}{\partial x^i} - Y^j \frac{\partial X^k}{\partial x^j} \right) \frac{\partial}{\partial x_k}. \tag{3.1.4}$$

Note in particular that $[\frac{\partial}{\partial x_i}, \frac{\partial}{\partial x_j}] = 0$, i.e., the basic vectors $\frac{\partial}{\partial x_i}$ commute as derivations.

So far we have not proved the vector field in (3.1.4) is independent of coordinates. We will achieve this by identifying it with the intrinsically defined vector field $L_X Y$.

Set $\gamma(t) = \Phi^t m$ so that we have a parametrization $\gamma(t) = (x^i(t))$ with $\dot{x}^i = X^i$. Then

$$\Phi^{-t} m = \gamma(-t) = \gamma(0) - \dot{\gamma}(0)t + O(t^2) = \left(x_0^i - tX^i + O(t^2) \right),$$

and

$$Y^j_{\gamma(-t)} = Y^j_m - tX^i \frac{\partial Y^j}{\partial x^i} + O(t^2). \tag{3.1.5}$$

Note that $\Phi_*^{-t} : T_{\gamma(0)}M \to T_{\gamma(-t)}M$ is the linearization of the map

$$(x^i) \mapsto \left(x_0^i - tX^i + O(t^2) \right),$$

so it has a matrix representation

$$\Phi_*^{-t} = \mathbb{1} - t \left(\frac{\partial X^i}{\partial x^j} \right)_{i,j} + O(t^2). \tag{3.1.6}$$

In particular, using the geometric series

$$(\mathbb{1} - A)^{-1} = \mathbb{1} + A + A^2 + \cdots,$$

where A is a matrix of operator norm strictly less than 1, we deduce that the differential

$$\Phi_*^t = (\Phi_*^{-t})^{-1} : T_{\gamma(-t)}M \to T_{\gamma(0)}M,$$

has the matrix form

$$\Phi_*^t = \mathbb{1} + t \left(\frac{\partial X^i}{\partial x^j} \right)_{i,j} + O(t^2). \tag{3.1.7}$$

Using (3.1.7) in (3.1.5) we deduce

$$Y_m^k - \left(\Phi_*^t Y_{\Phi^{-t}m}\right)^k = t\left(X^i \frac{\partial Y^k}{\partial x^i} - Y^j \frac{\partial X^k}{\partial x^j}\right) + O(t^2).$$

This concludes the proof of the lemma.   □

Lemma 3.1.11. *For any differential form* $\omega \in \Omega^1(M)$, *and any vector fields* $X, Y \in Vect(M)$ *we have*

$$(L_X\omega)(Y) = L_X\left(\omega(Y)\right) - \omega([X, Y]),   \tag{3.1.8}$$

where $X \cdot \omega(Y)$ *denotes the (Lie) derivative of the function* $\omega(Y)$ *along the vector field* X.

Proof.   Denote by Φ^t the local flow generated by X. We have $\Phi_*^t\omega = (\Phi^{-t})^*\omega$, i.e., for any $p \in M$, and any $Y \in \text{Vect}\,(M)$, we have

$$(\Phi_*^t\omega)_p(Y_p) = \omega_{\Phi^{-t}p}\left(\Phi_*^{-t}Y_p\right).$$

Fix a point $p \in M$, and choose local coordinates (x^i) near p. Then

$$\omega = \sum_i \omega^i dx^i, \quad X = \sum_i X^i \frac{\partial}{\partial x^i}, \quad Y = \sum_i Y^i \frac{\partial}{\partial x^i}.$$

Denote by $\gamma(t)$ the path $t \mapsto \Phi^t(p)$. We set $\omega_i(t) = \omega_i(\gamma(t))$, $X_0^i = X^i(p)$, and $Y_0^i = Y^i(p)$. Using (3.1.6) we deduce

$$(\Phi_*^t\omega)_p(Y_p) = \sum_i \omega_i(-t) \cdot \left(Y_0^i - t\sum_j \frac{\partial X^i}{\partial x^j} Y_0^j + O(t^2)\right).$$

Hence

$$-(L_X\omega)Y = \frac{d}{dt}\Big|_{t=0}(\Phi_*^t\omega)_p(Y_p) = -\sum_i \dot\omega_i(0)Y_0^i - \sum_{i,j} \omega_i(0)\frac{\partial X^i}{\partial x^j}Y_0^j.$$

On the other hand, we have

$$X \cdot \omega(Y) = \frac{d}{dt}\Big|_{t=0} \sum_i \omega_i(t)Y^i(t) = \sum_i \dot\omega_i(0)Y_0^i + \sum_{i,j} \omega_i(0)X_0^j \frac{\partial Y^i}{\partial x^j}.$$

We deduce that

$$X \cdot \omega(Y) - (L_X\omega)Y = \sum_{i,j} \omega_i(0)\left(X_0^j \frac{\partial Y^i}{\partial x^j} - \frac{\partial X^i}{\partial x^j}Y_0^j\right) = \omega_p([X, Y]_p).   □$$

Observe that if S, T are two tensor fields on M such that both $L_X S$ and $L_X T$ exist, then using (3.1.3) we deduce that $L_X(S \otimes T)$ exists, and

$$L_X(S \otimes T) = L_X S \otimes T + S \otimes L_X T.   \tag{3.1.9}$$

Since any tensor field is locally a linear combination of tensor monomials of the form

$$X_1 \otimes \cdots \otimes X_r \otimes \omega_1 \otimes \cdots \otimes \omega_s, \quad X_i \in \text{Vect}\,(M), \quad \omega_j \in \Omega^1(M),$$

we deduce that the Lie derivative $L_X S$ exists for every $X \in \text{Vect}\,(M)$, and any smooth tensor field S. We can now completely describe the Lie derivative on the algebra of tensor fields.

Proposition 3.1.12. *Let X be a vector field on the smooth manifold M. Then the Lie derivative L_X is the unique derivation of $\mathcal{T}_*^*(M)$ with the following properties.*
(a) $L_X f = \langle df, X \rangle = Xf$, $\forall f \in C^\infty(M)$.
(b) $L_X Y = [X, Y]$, $\forall X, Y \in \text{Vect}\,(M)$.
(c) L_X commutes with the contraction $\text{tr} : \mathcal{T}_{s+1}^{r+1}(M) \to \mathcal{T}_s^r(M)$.
    Moreover, L_X is a natural operation, i.e., for any diffeomorphism $\phi : M \to N$ we have $\phi_* \circ L_X = L_{\phi_* X} \circ \phi_*$, $\forall X \in \text{Vect}\,(M)$, i.e., $\phi_*(L_X) = L_{\phi_* X}$.

Proof.   The fact that L_X is a derivation follows from (3.1.9). Properties (a) and (b) were proved above. As for part (c), in its simplest form, when $T = Y \otimes \omega$, where $Y \in \text{Vect}\,(M)$, and $\omega \in \Omega^1(M)$, the equality

$$L_X \, \text{tr}\, T = \text{tr}\, L_X T$$

is equivalent to

$$L_X \, (\omega(Y)) = (L_X \omega)(Y) + \omega(L_X(Y)), \qquad (3.1.10)$$

which is precisely (3.1.8).

    Since L_X is a derivation of the algebra of tensor fields, its restriction to $C^\infty(M) \oplus \text{Vect}\,(M) \oplus \Omega^1(M)$ uniquely determines the action on the entire algebra of tensor fields which is generated by the above subspace. The reader can check easily that the general case of property (c) follow from this observation coupled with the product rule (3.1.9).

    The naturality of L_X is another way of phrasing the coordinate independence of this operation. We leave the reader to fill in the routine details.   □

Corollary 3.1.13. *For any $X, Y \in \text{Vect}\,(M)$ we have*

$$[L_X, L_Y] = L_{[X,Y]},$$

as derivations of the algebra of tensor fields on M. In particular, this says that $\text{Vect}\,(M)$ as a space of derivations of $\mathcal{T}_*^*(M)$ is a Lie subalgebra of the Lie algebra of derivations.

Proof.   The commutator $[L_X, L_Y]$ is a derivation (as a commutator of derivations). By Lemma 3.1.10, $[L_X, L_Y] = L_{[X,Y]}$ on $C^\infty(M)$. Also, a simple computation shows that

$$[L_X, L_Y] Z = L_{[X,Y]} Z, \quad \forall Z \in \text{Vect}\,(M),$$

so that $[L_X, L_Y] = L_{[X,Y]}$ on $\text{Vect}\,(M)$. Finally, since the contraction commutes with both L_X and L_Y it obviously commutes with $L_X L_Y - L_Y L_X$. The corollary is proved.   □

Exercise 3.1.14. Prove that the map
$$\mathcal{D} : \operatorname{Vect}(M) \oplus \operatorname{End}(TM) \to Der(\mathcal{T}^*_*(M))$$
given by $\mathcal{D}(X, S) = L_X + S$ is well defined and is a linear isomorphism. Moreover,
$$[\mathcal{D}(X_1, S_1), \mathcal{D}(X_2, S_2)] = \mathcal{D}([X_1, X_2], [S_1, S_2]). \qquad \square$$

L_X is a derivation of $\mathcal{T}^*_*$ with the remarkable property
$$L_X(\Omega^*(M)) \subset \Omega^*(M).$$
The wedge product makes $\Omega^*(M)$ an s-algebra, and it is natural to ask whether L_X is an s-derivation with respect to this product.

Proposition 3.1.15. *The Lie derivative along a vector field X is an even s-derivation of $\Omega^*(M)$, i.e.*
$$L_X(\omega \wedge \eta) = (L_X\omega) \wedge \eta + \omega \wedge (L_X\eta), \quad \forall \omega, \, \eta \in \Omega^*(M).$$

Proof. As in Subsection 2.2.2, denote by $\mathcal{A}$ the anti-symmetrization operator $\mathcal{A} : (T^*M)^{\otimes k} \to \Omega^k(M)$. The statement in the proposition follows immediately from the straightforward observation that the Lie derivative commutes with this operator (which is a projector). We leave the reader to fill in the details. $\qquad \square$

Exercise 3.1.16. Let M be a smooth manifold, and suppose that $\Phi, \Psi : \mathbb{R} \times M \to M$ are two smooth flows on M with infinitesimal generators X and respectively Y. We say that the two flows commute if
$$\Phi^t \circ \Psi^s = \Psi^s \circ \Psi^t, \quad \forall s, t \in \mathbb{R}.$$
Prove that if
$$\{\, p \in M; \;\; X(p) = 0 \,\} = \{\, p \in M; \;\; Y(p) = 0 \,\},$$
the two flows commute if and only if $[X, Y] = 0$. $\qquad \square$

3.1.3   *Examples*

Example 3.1.17. Let $\omega = \omega_i dx^i$ be a 1-form on $\mathbb{R}^n$. If $X = X^j \frac{\partial}{\partial x^j}$ is a vector field on $\mathbb{R}^n$ then $L_X\omega = (L_X\omega)_k dx^k$ is defined by
$$(L_X\omega)_k = (L_X\omega)(\frac{\partial}{\partial x_k}) = X\omega(\frac{\partial}{\partial x_k}) - \omega(L_X \frac{\partial}{\partial x_k}) = X \cdot \omega_k + \omega\left(\frac{\partial X^i}{\partial x^k} \frac{\partial}{\partial x_i}\right).$$
Hence
$$L_X\omega = \left(X^j \frac{\partial \omega_k}{\partial x^j} + \omega_j \frac{\partial X^j}{\partial x^k}\right) dx^k.$$
In particular, if $X = \partial_{x^i} = \sum_j \delta^{ij} \partial_{x^j}$, then
$$L_X\omega = L_{\partial_{x^i}}\omega = \sum_{k=1}^{n} \frac{\partial \omega_k}{\partial x^i} dx^k.$$
If X is the radial vector field $X = \sum_i x^i \partial_{x^i}$, then
$$L_X\omega = \sum_k (X \cdot \omega_k + \omega_k) dx^k. \qquad \square$$

Example 3.1.18. Consider a smooth vector field $X = F\frac{\partial}{\partial x} + G\frac{\partial}{\partial y} + H\frac{\partial}{\partial z}$ on $\mathbb{R}^3$. We want to compute $L_X dv$, where dv is the volume form on $\mathbb{R}^3$, $dv = dx \wedge dy \wedge dz$. Since L_X is an even s-derivation of $\Omega^*(M)$, we deduce

$$L_X(dx \wedge dy \wedge dz) = (L_X dx) \wedge dy \wedge dz + dx \wedge (L_X dy) \wedge dz + dx \wedge dy \wedge (L_X dz).$$

Using the computation in the previous example we get

$$L_X(dx) = dF := \frac{\partial F}{\partial x}dx + \frac{\partial F}{\partial y}dy + \frac{\partial F}{\partial z}dz, \ \ L_X(dy) = dG, \ \ L_X(dz) = dH$$

so that

$$L_X(dv) = \left(\frac{\partial F}{\partial x} + \frac{\partial G}{\partial y} + \frac{\partial H}{\partial z} \right) dv = (\boldsymbol{div}\, X)dv.$$

In particular, we deduce that if $\boldsymbol{div}\, X = 0$, the local flow generated by X preserves the form dv. We will get a better understanding of this statement once we learn integration on manifolds, later in this chapter. □

Example 3.1.19. (The exponential map of a Lie group). Consider a Lie group G. Any element $g \in G$ defines two diffeomorphisms of G: the left (L_g), and the right translation (R_g) on G,

$$L_g(h) = g \cdot hy, \ \ R_g(h) = h \cdot g, \ \ \forall h \in G.$$

A tensor field T on G is called *left* (respectively *right*) *invariant* if for any $g \in G$ $(L_g)_* T = T$ (respectively $(R_g)_* T = T$). The set of left invariant vector fields on G is denoted by $\mathcal{L}_G$. The naturality of the Lie bracket implies

$$(L_g)_*[X, Y] = [(L_g)_* X, (L_g)_* Y],$$

so that $\forall X, Y \in \mathcal{L}_G$, $[X, Y] \in \mathcal{L}_G$. Hence $\mathcal{L}_G$ is a Lie subalgebra of $\text{Vect}\,(G)$. It is called called the *Lie algebra* of the group G.

Fact 1. $\dim \mathcal{L}_G = \dim G$. Indeed, the left invariance implies that the restriction map $\mathcal{L}_G \to T_1 G$, $X \mapsto X_1$ is an isomorphism (*Exercise*). We will often find it convenient to identify the Lie algebra of G with the tangent space at 1.

Fact 2. Any $X \in \mathcal{L}_G$ defines a local flow Φ_X^t on G that is is defined for all $t \in \mathbb{R}$. In other wors, Φ_X^t is a flow. (*Exercise*) Set

$$\exp(tX) : = \Phi_X^t(1).$$

We thus get a map

$$\exp : T_1 G \cong \mathcal{L}_G \to G, \ \ X \mapsto \exp(X)$$

called the *exponential map of the group* G.

Fact 3. $\Phi_X^t(g) = g \cdot \exp(tX)$, i.e.,

$$\Phi_X^t = R_{\exp(tX)}.$$

Indeed, it suffices to check that

$$\frac{d}{dt}\Big|_{t=0} (g \cdot \exp{(tX)}) = X_g.$$

We can write $(g \cdot \exp(tX)) = L_g \exp(tX)$ so that

$$\frac{d}{dt}\Big|_{t=0} (L_g \exp(tX)) = (L_g)_* \left(\frac{d}{dt}\Big|_{t=0} \exp(tX) \right) = (L_g)_* X = X_g, \quad \text{(left invariance)}.$$

The reason for the notation $\exp{(tX)}$ is that when $G = \mathrm{GL}(n, \mathbb{K})$, then the Lie algebra of G is the Lie algebra $\underline{\mathrm{gl}}(n, \mathbb{K})$ of all $n \times n$ matrices with the bracket given by the commutator of two matrices, and for any $X \in \mathcal{L}_G$ we have (*Exercise*)

$$\exp{(X)} = e^X = \sum_{k \geq 0} \frac{1}{k!} X^k. \qquad \square$$

Exercise 3.1.20. Prove the statements left as exercises in the example above. $\square$

Exercise 3.1.21. Let G be a matrix Lie group, i.e., a Lie subgroup of some general linear group $\mathrm{GL}(N, \mathbb{K})$. This means the tangent space $T_1 G$ can be identified with a linear space of matrices. Let $X, Y \in T_1 G$, and denote by $\exp(tX)$ and $\exp(tY)$ the 1-parameter groups with they generate, and set

$$g(s, t) = \exp(sX) \exp(tY) \exp(-sX) \exp(-tY).$$

(a) Show that

$$g_{s,t} = 1 + [X, Y]_{alg} st + O((s^2 + t^2)^{3/2}) \quad \text{as} \quad s, t \to 0,$$

where the bracket $[X, Y]_{alg}$ (temporarily) denotes the commutator of the two matrices X and Y.
(b) Denote (temporarily) by $[X, Y]_{geom}$ the Lie bracket of X and Y viewed as left invariant vector fields on G. Show that at $1 \in G$

$$[X, Y]_{alg} = [X, Y]_{geom}.$$

(c) Show that $\underline{o}(n) \subset \underline{\mathrm{gl}}(n, \mathbb{R})$ (defined in Section 1.2.2) is a Lie subalgebra with respect to the commutator $[\cdot, \cdot]$. Similarly, show that $\underline{u}(n)$, $\underline{su}(n) \subset \underline{\mathrm{gl}}(n, \mathbb{C})$ are *real* Lie subalgebras of $\underline{\mathrm{gl}}(n, \mathbb{C})$, while $\underline{su}(n, \mathbb{C})$ is even a *complex* Lie subalgebra of $\underline{\mathrm{gl}}(n, \mathbb{C})$.
(d) Prove that we have the following isomorphisms of *real* Lie algebras. $\mathcal{L}_{O(n)} \cong \underline{o}(n)$, $\mathcal{L}_{U(n)} \cong \underline{u}(n)$, $\mathcal{L}_{SU(n)} \cong \underline{su}(n)$ and $\mathcal{L}_{SL(n,\mathbb{C})} \cong \underline{sl}(n, \mathbb{C})$. $\square$

Remark 3.1.22. In general, in a non-commutative matrix Lie group G, the traditional equality

$$\exp(tX) \exp(tY) = \exp(t(X + Y))$$

no longer holds. Instead, one has the *Campbell-Hausdorff formula*

$$\exp(tX) \cdot \exp(tY) = \exp\left(td_1(X, Y) + t^2 d_2(X, Y) + t^3 d_3(X, Y) + \cdots \right),$$

where d_k are homogeneous polynomials of degree k in X, and Y with respect to the multiplication between X and Y *given by their bracket.* The d_k's are usually known as Dynkin polynomials. For example,

$$d_1(X,Y) = X + Y, \quad d_2(X,Y) = \frac{1}{2}[X,Y],$$

$$d_3(X,Y) = \frac{1}{12}([X,[X,Y]] + [Y,[Y,X]]) \ \text{ etc.}$$

For more details we refer to [41, 84]. □

3.2 Derivations of $\Omega^\bullet(M)$

3.2.1 *The exterior derivative*

The super-algebra of exterior forms on a smooth manifold M has additional structure, and in particular, its space of derivations has special features. This section is devoted precisely to these new features.

The Lie derivative along a vector field X defines an even derivation in $\Omega^\bullet(M)$. The vector field X also defines, via the contraction map, an odd derivation i_X, called the *interior derivation along X*, or the *contraction by X*,

$$i_X\omega : = \text{tr}\,(X \otimes \omega), \ \ \forall \omega \in \Omega^r(M).$$

More precisely, $i_X\omega$ is the $(r-1)$-form determined by

$$(i_X\omega)(X_1,\ldots,X_{r-1}) = \omega(X,X_1,\ldots,X_{r-1}), \ \ \forall X_1,\ldots,X_{r-1} \in \text{Vect}\,(M).$$

The fact that i_X is an odd s-derivation is equivalent to

$$i_X(\omega \wedge \eta) = (i_X\omega) \wedge \eta + (-1)^{\deg\omega}\omega \wedge (i_X\eta), \ \ \forall \omega, \eta \in \Omega^*(M).$$

Often the contraction by X is denoted by $X\lrcorner$.

Exercise 3.2.1. Prove that the interior derivation along a vector field is a s-derivation. □

Proposition 3.2.2. *(a)* $[i_X, i_Y]_s = i_X i_Y + i_Y i_X = 0$.
(b) The super-commutator of L_X and i_Y as s-derivations of $\Omega^*(M)$ is given by

$$[L_X, i_Y]_s = L_X i_Y - i_Y L_X = i_{[X,Y]}.$$ □

The proof uses the fact that the Lie derivative commutes with the contraction operator, and it is left to the reader as an exercise.

The above s-derivations by no means exhaust the space of s-derivations of $\Omega^\bullet(M)$. In fact we have the following fundamental result.

Proposition 3.2.3. *There exists an odd s-derivation d on the s-algebra of differential forms $\Omega^\bullet(\,\cdot\,)$ uniquely characterized by the following conditions.*

(a) *For any smooth function $f \in \Omega^0(M)$, df coincides with the differential of f.*

(b) *$d^2 = 0$.*

(c) *d is natural, i.e., for any smooth function $\phi : N \to M$, and for any form ω on M, we have*

$$d\phi^*\omega = \phi^* d\omega (\Longleftrightarrow [\phi^*, d] = 0).$$

The derivation d is called the exterior derivative.

Proof.   *Uniqueness.* Let U be a local coordinate chart on M^n with local coordinates $(x^1, \dots, x^n)$. Then, over U, any r-form ω can be described as

$$\omega = \sum_{1 \le i_1 < \cdots < i_r \le n} \omega_{i_1 \dots i_r} dx^{i_1} \wedge \dots \wedge dx^{i_r}.$$

Since d is an s-derivation, and $d(dx^i) = 0$, we deduce that, over U

$$d\omega = \sum_{1 \le i_1 < \cdots < i_r \le n} (d\omega_{i_1 \dots i_r}) \wedge (dx^{i_1} \wedge \cdots \wedge dx^{i_r})$$

$$= \sum_{1 \le i_1 < \cdots < i_r \le n} \left(\frac{\partial \omega_{i_1 \dots i_r}}{\partial x^i} dx^i \right) \wedge (dx^{i_1} \wedge \cdots \wedge dx^{i_r}). \tag{3.2.1}$$

Thus, the form $d\omega$ is uniquely determined on any coordinate neighborhood, and this completes the proof of the uniqueness of d.

Existence. Consider an r-form ω. For each coordinate neighborhood U we define $d\omega |_U$ as in (3.2.1). To prove that this is a well defined operation we must show that, if U, V are two coordinate neighborhoods, then

$$d\omega |_U = d\omega |_V \quad \text{on } U \cap V.$$

Denote by $(x^1, \dots, x^n)$ the local coordinates on U, and by $(y^1, \dots, y^n)$ the local coordinates along V, so that on the overlap $U \cap V$ we can describe the y's as functions of the x's. Over U we have

$$\omega = \sum_{1 \le i_1 < \cdots < i_r \le n} \omega_{i_1 \dots i_r} dx^{i_1} \wedge \dots \wedge dx^{i_r}$$

$$d\omega = \sum_{1 \le i_1 < \cdots < i_r \le n} \left(\frac{\partial \omega_{i_1 \dots i_r}}{\partial x^i} dx^i \right) \wedge (dx^{i_1} \wedge \dots \wedge dx^{i_r}),$$

while over V we have

$$\omega = \sum_{1 \le j_1 < \cdots < j_r \le n} \widehat{\omega}_{j_1 \dots j_r} dy^{j_1} \wedge \cdots \wedge dy^{j_r}$$

$$d\omega = \sum_{1 \le j_1 < \cdots < j_r \le n} \left(\frac{\partial \widehat{\omega}_{j_1 \dots j_r}}{\partial y^j} dy^j \right) (dy^{j_1} \wedge \cdots \wedge dy^{j_r}).$$

The components $\omega_{i_1 \dots i_r}$ and $\widehat{\omega}_{j_1 \dots j_r}$ are skew-symmetric, i.e., $\forall \sigma \in S_r$,

$$\omega_{i_{\sigma(1)} \dots i_{\sigma(r)}} = \epsilon(\sigma) \omega_{i_1 \dots i_r},$$

and similarly for the $\widehat{\omega}$'s. Since $\omega|_U = \omega|_V$ over $U \cap V$ we deduce

$$\omega_{i_1 \dots i_r} = \frac{\partial y^{j_1}}{\partial x^{i_1}} \cdots \frac{\partial y^{j_r}}{\partial x^{i_r}} \widehat{\omega}_{j_1 \dots j_r}.$$

Hence

$$\frac{\partial \omega_{i_1 \dots i_r}}{\partial x^i} = \sum_{k=1}^{r} \left(\frac{\partial y^{j_1}}{\partial x^{i_1}} \cdots \frac{\partial^2 y^{j_k}}{\partial x^i \partial x^{i_k}} \cdots \frac{\partial y^{j_r}}{\partial x^{i_r}} \widehat{\omega}_{j_1 \dots j_r} + \frac{\partial y^{j_1}}{\partial x^{i_1}} \cdots \frac{\partial y^{j_r}}{\partial x^{i_r}} \frac{\partial \widehat{\omega}_{j_1 \dots j_r}}{\partial x^i} \right),$$

where in the above equality we also sum over the indices $j_1, \dots, j_r$ according to Einstein's convention. We deduce

$$\sum_{1 \leq i_1 < \dots < i_r \leq n} \frac{\partial \omega_{i_1 \dots i_r}}{\partial x^i} dx^i \wedge dx^{i_1} \wedge \dots \wedge dx^{i_r}$$

$$= \sum_i \sum_{k=1}^{r} \frac{\partial y^{j_1}}{\partial x^{i_1}} \cdots \frac{\partial^2 x^{j_k}}{\partial x^i \partial x^{i_k}} \cdots \frac{\partial y^{j_r}}{\partial x^{i_r}} \widehat{\omega}_{j_1 \dots j_r} dx^i \wedge dx^{i_1} \wedge \dots \wedge dx^{i_r}$$

$$+ \sum_i \sum_{k=1}^{r} \frac{\partial y^{j_1}}{\partial x^{i_1}} \cdots \frac{\partial y^{j_r}}{\partial x^{i_r}} \frac{\partial \widehat{\omega}_{j_1 \dots j_r}}{\partial x^i} dx^i \wedge dx^{i_1} \wedge \dots \wedge dx^{i_r}. \qquad (3.2.2)$$

Notice that

$$\frac{\partial^2}{\partial x^i \partial x^{i_k}} = \frac{\partial^2}{\partial x^{i_k} \partial x^i},$$

while $dx^i \wedge dx^{i_k} = -dx^{i_k} \wedge dx^i$ so that the first term in the right hand side of (3.2.2) vanishes. Consequently on $U \cap V$

$$\frac{\partial \omega_{i_1 \dots i_r}}{\partial x^i} dx^i \wedge dx^{i_1} \cdots \wedge dx^{i_r} = \frac{\partial y^{j_1}}{\partial x^{i_1}} \cdots \frac{\partial y^{j_r}}{\partial x^{i_r}} \frac{\partial \widehat{\omega}_{j_1 \dots j_r}}{\partial x^i} \wedge dx^i \wedge dx^{i_1} \cdots dx^{i_r}$$

$$= \left(\frac{\partial \widehat{\omega}_{j_1 \dots j_r}}{\partial x^i} dx^i \right) \wedge \left(\frac{\partial y^{j_1}}{\partial x^{i_1}} dx^{i_1} \right) \wedge \cdots \wedge \left(\frac{\partial y^{j_r}}{\partial x^{i_r}} dx^{i_r} \right)$$

$$= (d\widehat{\omega}_{j_1 \dots j_r}) \wedge dy^{j_1} \wedge \cdots \wedge dy^{j_r} = \frac{\partial \widehat{\omega}_{j_1 \dots j_r}}{\partial y^j} dy^j \wedge dy^{j_1} \wedge \cdots \wedge dy^{j_r}.$$

This proves $d\omega|_U = d\omega|_V$ over $U \cap V$. We have thus constructed a well defined linear map

$$d : \Omega^\bullet(M) \to \Omega^{\bullet+1}(M).$$

To prove that d is an odd s-derivation it suffices to work in local coordinates and show that the (super)product rule on monomials.

Let $\theta = f dx^{i_1} \wedge \cdots \wedge dx^{i_r}$ and $\omega = g dx^{j_1} \wedge \cdots \wedge dx^{j_s}$. We set for simplicity

$$dx^I := dx^{i_1} \wedge \cdots \wedge dx^{i_r} \quad \text{and} \quad dx^J := dx^{j_1} \wedge \cdots \wedge dx^{j_s}.$$

Then

$$d(\theta \wedge \omega) = d(fgdx^I \wedge dx^J) = d(fg) \wedge dx^I \wedge dx^J$$

$$= (df \cdot g + f \cdot dg) \wedge dx^I \wedge dx^J$$

$$= df \wedge dx^I \wedge dx^J + (-1)^r (f \wedge dx^I) \wedge (dg \wedge dx^J)$$

$$= d\theta \wedge \omega + (-1)^{\deg \theta} \theta \wedge d\omega.$$

We now prove $d^2 = 0$. We check this on monomials fdx^I as above.

$$d^2(fdx^I) = d(df \wedge dx^I) = (d^2 f) \wedge dx^I.$$

Thus, it suffices to show $d^2 f = 0$ for all smooth functions f. We have

$$d^2 f = \frac{\partial f^2}{\partial x^i \partial x^j} dx^i \wedge dx^j.$$

The desired conclusion follows from the identities

$$\frac{\partial f^2}{\partial x^i \partial x^j} = \frac{\partial f^2}{\partial x^j \partial x^i} \quad \text{and} \quad dx^i \wedge dx^j = -dx^j \wedge dx^i.$$

Finally, let ϕ be a smooth map $N \to M$ and $\omega = \sum_I \omega_I dx^I$ be an r-form on M. Here I runs through all ordered multi-indices $1 \le i_1 < \cdots < i_r \le \dim M$. We have

$$d_N(\phi^* \omega) = \sum_I \left(d_N(\phi^* \omega_I) \wedge \phi^*(dx^I) + \phi^* \omega^I \wedge d(\phi^* dx^I) \right).$$

For functions, the usual chain rule gives $d_N(\phi^* \omega_I) = \phi^*(d_M \omega_I)$. In terms of local coordinates (x^i) the map ϕ looks like a collection of n functions $\phi^i \in C^\infty(N)$ and we get

$$\phi^*(dx^I) = d\phi^I = d_N \phi^{i_1} \wedge \cdots \wedge d_N \phi^{i_r}.$$

In particular, $d_N(d\phi^I) = 0$. We put all the above together and we deduce

$$d_N(\phi^* \omega) = \phi^*(d_M \omega^I) \wedge d\phi^I = \phi^*(d_M \omega^I) \wedge \phi^* dx^I = \phi^*(d_M \omega).$$

The proposition is proved.   □

Proposition 3.2.4. *The exterior derivative satisfies the following relations.*
(a) $[d, d]_s = 2d^2 = 0$.
(b) **(Cartan's homotopy formula)** $[d, i_X]_s = di_X + i_X d = L_X \ \forall X \in \text{Vect}(M)$.
(c) $[d, L_X]_s = dL_X - L_X d = 0, \ \forall X \in \text{Vect}(M)$.

An immediate consequence of the homotopy formula is the following invariant description of the exterior derivative:

$$(d\omega)(X_0, X_1, ..., X_r) = \sum_{i=0}^r (-1)^i X_i(\omega(X_0, ..., \hat{X}_i, ..., X_r))$$

$$+ \sum_{0 \le i < j \le r} (-1)^{i+j} \omega([X_i, X_j], X_0, ..., \hat{X}_i, ..., \hat{X}_j, ..., X_r). \tag{3.2.3}$$

Above, the hat indicates that the corresponding entry is missing, and $[\, , \,]_s$ denotes the super-commutator in the s-algebra of real endomorphisms of $\Omega^\bullet(M)$.

Proof. To prove the homotopy formula set
$$\mathcal{D} := [d, i_X]_s = d i_X + i_X d.$$
$\mathcal{D}$ is an even s-derivation of $\Omega^*(M)$. It is a local s-derivation, i.e., if $\omega \in \Omega^*(M)$ vanishes on some open set U then $\mathcal{D}\omega$ vanishes on that open set as well. The reader can check easily by direct computation that $\mathcal{D}\omega = L_X\omega$, $\forall \omega \in \Omega^0(M) \oplus \Omega^1(M)$. The homotopy formula is now a consequence of the following technical result left to the reader as an exercise.

Lemma 3.2.5. *Let $\mathcal{D}$, $\mathcal{D}'$ be two local s-derivations of $\Omega^\bullet(M)$ which have the same parity, i.e., they are either both even or both odd. If $\mathcal{D} = \mathcal{D}'$ on $\Omega^0(M) \oplus \Omega^1(M)$, then $\mathcal{D} = \mathcal{D}'$ on $\Omega^\bullet(M)$.* $\square$

Part (c) of the proposition is proved in a similar way. Equality (3.2.3) is a simple consequence of the homotopy formula. We prove it in two special case $r = 1$ and $r = 2$.

The case $r = 1$. Let ω be an 1-form and let $X, Y \in \mathrm{Vect}\,(M)$. We deduce from the homotopy formula
$$d\omega(X, Y) = (i_X d\omega)(Y) = (L_X\omega)(Y) - (d\omega(X))(Y).$$
On the other hand, since L_X commutes with the contraction operator, we deduce
$$X\omega(Y) = L_X(\omega(Y)) = (L_X\omega)(Y) + \omega([X, Y]).$$
Hence
$$d\omega(X, Y) = X\omega(Y) - \omega([X, Y]) - (d\omega(X))(Y) = X\omega(Y) - Y\omega(X) - \omega([X, Y]).$$
This proves (3.2.3) in the case $r = 1$.

The case $r = 2$. Consider a 2-form ω and three vector fields X, Y and Z. We deduce from the homotopy formula
$$(d\omega)(X, Y, Z) = (i_X d\omega)(Y, Z) = (L_X - d i_X)\omega(Y, Z). \tag{3.2.4}$$
Since L_X commutes with contractions we deduce
$$(L_X\omega)(Y, X) = X(\omega(Y, Z)) - \omega([X, Y], Z) - \omega(Y, [X, Z]). \tag{3.2.5}$$
We substitute (3.2.5) into (3.2.4) and we get
$$(d\omega)(X, Y, Z) = X(\omega(Y, Z)) - \omega([X, Y], Z) - \omega(Y, [X, Z]) - d(i_X\omega)(Y, X). \tag{3.2.6}$$
We apply now (3.2.3) for $r = 1$ to the 1-form $i_X\omega$. We get
$$d(i_X\omega)(Y, X) = Y(i_X\omega(Z)) - Z(i_X\omega(Y)) - (i_X\omega)([Y, Z])$$
$$= Y\omega(X, Z) - Z\omega(X, Y) - \omega(X, [Y, Z]). \tag{3.2.7}$$
If we use (3.2.7) in (3.2.6) we deduce
$$(d\omega)(X, Y, Z) = X\omega(Y, Z) - Y\omega(X, Z) + Z\omega(X, Y)$$
$$- \omega([X, Y], Z) + \omega([X, Z], Y) - \omega([Y, Z], X). \tag{3.2.8}$$
The general case in (3.2.3) can be proved by induction. The proof of the proposition is complete. $\square$

Exercise 3.2.6. Prove Lemma 3.2.5. $\square$

Exercise 3.2.7. Finish the proof of (3.2.3) in the general case. $\square$

3.2.2   *Examples*

Example 3.2.8. (The exterior derivative in $\mathbb{R}^3$).
(a) Let $f \in C^\infty(\mathbb{R}^3)$. Then

$$df = \frac{\partial f}{\partial x}dx + \frac{\partial f}{\partial y}dy + \frac{\partial f}{\partial z}dz.$$

The differential df looks like the gradient of f.
(b) Let $\omega \in \Omega^1(\mathbb{R}^3)$, $\omega = Pdx + Qdy + Rdz$. Then

$$d\omega = dP \wedge dx + dQ \wedge dy + dR \wedge dz$$

$$= \left(\frac{\partial Q}{\partial x} - \frac{\partial P}{\partial y}\right) dx \wedge dy + \left(\frac{\partial R}{\partial y} - \frac{\partial Q}{\partial z}\right) dy \wedge dz + \left(\frac{\partial P}{\partial z} - \frac{\partial R}{\partial x}\right) dz \wedge dx,$$

so that $d\omega$ looks very much like a curl.
(c) Let $\omega = Pdy \wedge dz + Qdz \wedge dx + Rdx \wedge dy \in \Omega^2(\mathbb{R}^3)$. Then

$$d\omega = \left(\frac{\partial P}{\partial x} + \frac{\partial Q}{\partial y} + \frac{\partial R}{\partial z}\right) dx \wedge dy \wedge dz.$$

This looks very much like a divergence.   □

Example 3.2.9. Let G be a connected Lie group. In Example 3.1.11 we defined the Lie algebra $\mathcal{L}_G$ of G as the space of left invariant vector fields on G. Set

$$\Omega^r_{left}(G) = \text{left invariant } r\text{-forms on } G.$$

In particular, $\mathcal{L}_G^* \cong \Omega^1_{left}(G)$. If we identify $\mathcal{L}_G^* \cong T_1^*G$, then we get a natural isomorphism

$$\Omega^r_{left}(G) \cong \Lambda^r \mathcal{L}_G^*.$$

The exterior derivative of a form in Ω^*_{left} can be described *only in terms of the algebraic structure of $\mathcal{L}_G$*.

Indeed, let $\omega \in \mathcal{L}_G^* = \Omega^1_{left}(G)$. For $X, Y \in \mathcal{L}_G^*$ we have (see (3.2.3))

$$d\omega(X, Y) = X\omega(Y) - Y\omega(X) - \omega([X, Y]).$$

Since ω, X and Y are left invariant, the scalars $\omega(X)$ and $\omega(Y)$ are constants. Thus, the first two terms in the above equality vanish so that

$$d\omega(X, Y) = -\omega([X, Y]).$$

More generally, if $\omega \in \Omega^r_{left}$, then the same arguments applied to (3.2.3) imply that for all $X_0, ..., X_r \in \mathcal{L}_G$ we have

$$d\omega(X_0, X_1, ..., X_r) = \sum_{0 \leq i < j \leq r} (-1)^{i+j}\omega([X_i, X_j], X_1, ..., \hat{X}_i, ..., \hat{X}_j, ..., X_r). \quad (3.2.9)$$

□

3.3 Connections on vector bundles

3.3.1 *Covariant derivatives*

We learned several methods of differentiating tensor objects on manifolds. However, the tensor bundles are not the only vector bundles arising in geometry, and very often one is interested in measuring the "oscillations" of sections of vector bundles.

Let E be a $\mathbb{K}$-vector bundle over the smooth manifold M ($\mathbb{K} = \mathbb{R}, \mathbb{C}$). Technically, one would like to have a procedure of measuring the rate of change of a section u of E along a direction described by a vector field X. For such an arbitrary E, we encounter a problem which was not present in the case of tensor bundles. Namely, the local flow generated by the vector field X on M no longer induces bundle homomorphisms.

For tensor fields, the transport along a flow was a method of comparing objects in different fibers which otherwise are abstract linear spaces with no natural relationship between them.

To obtain something that looks like a derivation we need to formulate clearly what properties we should expect from such an operation.

(a) It should measure how fast is a given section changing along a direction given by a vector field X. Hence it has to be an operator

$$\nabla : \text{Vect}\,(M) \times C^\infty(E) \to C^\infty(E), \quad (X, u) \mapsto \nabla_X u.$$

(b) If we think of the usual directional derivative, we expect that after "rescaling" the direction X the derivative along X should only rescale by the same factor, i.e.,

$$\forall f \in C^\infty(M): \ \nabla_{fX} u = f \nabla_X u.$$

(c) Since ∇ is to be a derivation, it has to satisfy a sort of (Leibniz) product rule. The only product that exists on an abstract vector bundle is the multiplication of a section with a smooth function. Hence we require

$$\nabla_X(fu) = (Xf)u + f\nabla_X u, \ \forall f \in C^\infty(M), \ u \in C^\infty(E).$$

The conditions (a) and (b) can be rephrased as follows: for any $u \in C^\infty(E)$, the map

$$\nabla u : \text{Vect}\,(M) \to C^\infty(E), \ X \mapsto \nabla_X u,$$

is $C^\infty(M)$-linear so that it defines a bundle map

$$\nabla u \in \text{Hom}\,(TM, E) \cong C^\infty(T^*M \otimes E).$$

Summarizing, we can formulate the following definition.

Definition 3.3.1. A *covariant derivative* (or *linear connection*) on E is a $\mathbb{K}$-linear map

$$\nabla : C^\infty(E) \to C^\infty(T^*M \otimes E),$$

such that, $\forall f \in C^\infty(M)$, and $\forall u \in C^\infty(E)$, we have

$$\nabla(fu) = df \otimes u + f\nabla u. \qquad \square$$

Example 3.3.2. Let $\underline{\mathbb{K}}^r_M \cong \mathbb{K}^r \times M$ be the rank r trivial vector bundle over M. The space $C^\infty(\underline{\mathbb{K}}_M)$ of smooth sections coincides with the space $C^\infty(M, \mathbb{K}^r)$ of $\mathbb{K}^r$-valued smooth functions on M. We can define

$$\nabla^0 : C^\infty(M, \mathbb{K}^r) \to C^\infty(M, T^*M \otimes \mathbb{K}^r)$$

$$\nabla^0(f_1, ..., f_r) = (df_1, ..., df_r).$$

One checks easily that ∇ is a connection. This is called the *trivial connection*.   □

Remark 3.3.3. Let ∇^0, ∇^1 be two connections on a vector bundle $E \to M$. Then for any $\alpha \in C^\infty(M)$ the map

$$\nabla = \alpha\nabla^1 + (1 - \alpha)\nabla^0 : C^\infty(E) \to C^\infty(T^* \otimes E)$$

is again a connection.   □

✏**Notation.**   For any vector bundle F over M we set

$$\Omega^k(F) : = C^\infty(\Lambda^k T^*M \otimes F).$$

We will refer to these sections as *differential k-forms with coefficients in the vector bundle F*.   □

Proposition 3.3.4. *Let E be a vector bundle. The space $\mathcal{A}(E)$ of linear connections on E is an affine space modeled on $\Omega^1(\mathrm{End}\,(E))$.*

Proof.   We first prove that $\mathcal{A}(E)$ is not empty. To see this, choose an open cover $\{U_\alpha\}$ of M such that $E\,|_{U_\alpha}$ is trivial $\forall\alpha$. Next, pick a smooth partition of unity (μ_β) subordinated to this cover.

Since $E\,|_{U_\alpha}$ is trivial, it admits at least one connection, the trivial one, as in the above example. Denote such a connection by ∇^α. Now define

$$\nabla := \sum_{\alpha,\beta} \mu_\beta\nabla^\alpha.$$

One checks easily that ∇ is a connection so that $\mathcal{A}(E)$ is nonempty. To check that $\mathcal{A}(E)$ is an affine space, consider two connections ∇^0 and ∇^1. Their difference $A = \nabla^1 - \nabla^0$ is an operator

$$A : C^\infty(E) \to C^\infty(T^*M \otimes E),$$

satisfying $A(fu) = fA(u)$, $\forall u \in C^\infty(E)$. Thus,

$$A \in C^\infty(\mathrm{Hom}\,(E, T^*M \otimes E)) \cong C^\infty(T^*M \otimes E^* \otimes E) \cong \Omega^1(E^* \otimes E) \cong \Omega^1(\mathrm{End}E).$$

Conversely, given $\nabla^0 \in \mathcal{A}(E)$ and $A \in \Omega^1(\mathrm{End}E)$ one can verify that the operator

$$\nabla^A = \nabla^0 + A : C^\infty(E) \to \Omega^1(E).$$

is a linear connection. This concludes the proof of the proposition.   □

The tensorial operations on vector bundles extend naturally to *vector bundles with connections*. The guiding principle behind this fact is the product formula. More precisely, if E_i $(i = 1, 2)$ are two bundles with connections ∇^i, then $E_1 \otimes E_2$ has a naturally induced connection $\nabla^{E_1 \otimes E_2}$ uniquely determined by the product rule,

$$\nabla^{E_1 \otimes E_2}(u_1 \otimes u_2) = (\nabla^1 u_1) \otimes u_2 + u_1 \otimes \nabla^2 u_2.$$

The dual bundle E_1^* has a natural connection ∇^* defined by the identity

$$X\langle v, u \rangle = \langle \nabla_X^* v, u \rangle + \langle v, \nabla_X^1 u \rangle, \quad \forall u \in C^\infty(E_1), \ v \in C^\infty(E_1^*), \ X \in \text{Vect}(M),$$

where

$$\langle \bullet, \bullet \rangle : C^\infty(E_1^*) \times C^\infty(E_1) \to C^\infty(M)$$

is the pairing induced by the natural duality between the fibers of E_1^* and E_1. In particular, any connection ∇^E on a vector bundle E induces a connection $\nabla^{\text{End}(E)}$ on $\text{End}(E) \cong E^* \otimes E$ by

$$(\nabla^{\text{End}(E)} T)(u) = \nabla^E(Tu) - T(\nabla^E u) = [\nabla^E, T]u, \tag{3.3.1}$$

$\forall T \in \text{End}(E), \ u \in C^\infty(E)$.

It is often useful to have a local description of a covariant derivative. This can be obtained using Cartan's moving frame method.

Let $E \to M$ be a $\mathbb{K}$-vector bundle of rank r over the smooth manifold M. Pick a coordinate neighborhood U such $E|_U$ is trivial. A *moving frame*[2] is a bundle isomorphism $\phi : \underline{\mathbb{K}}_U^r = \mathbb{K}^r \times M \to E|_U$.

Consider the sections $e_\alpha = \phi(\delta_\alpha)$, $\alpha = 1, ..., r$, where δ_α are the natural basic sections of the trivial bundle $\underline{\mathbb{K}}_U^r$. As x moves in U, the collection $(e_1(x), ..., e_r(x))$ describes a basis of the moving fiber E_x, whence the terminology moving frame. A section $u \in C^\infty(E|_U)$ can be written as a linear combination

$$u = u^\alpha e_\alpha \quad u^\alpha \in C^\infty(U, \mathbb{K}).$$

Hence, if ∇ is a covariant derivative in E, we have

$$\nabla u = du^\alpha \otimes e_\alpha + u^\alpha \nabla e_\alpha.$$

Thus, the covariant derivative is completely described by its action on a moving frame.

To get a more concrete description, pick local coordinates (x^i) over U. Then $\nabla e_\alpha \in \Omega^1(E|_U)$ so that we can write

$$\nabla e_\alpha = \Gamma_{i\alpha}^\beta dx^i \otimes e_\beta, \quad \Gamma_{i\alpha}^\beta \in C^\infty(U, \mathbb{K}).$$

Thus, for any section $u^\alpha e_\alpha$ of $E|_U$ we have

$$\nabla u = du^\alpha \otimes e_\alpha + \Gamma_{i\alpha}^\beta u^\alpha dx^i \otimes e_\beta. \tag{3.3.2}$$

---

[2] A moving frame is what physicists call a choice of *local gauge*.

It is convenient to view $\left(\Gamma_{i\alpha}^{\beta}\right)$ as an $r \times r$-matrix valued 1-form, and we write this as

$$\left(\Gamma_{i\alpha}^{\beta}\right) = dx^i \otimes \Gamma_i.$$

The form $\Gamma = dx^i \otimes \Gamma_i$ is called the *connection 1-form* associated to the choice of local gauge. A moving frame allows us to identify sections of $E|_U$ with $\mathbb{K}^r$-valued functions on U, and we can rewrite (3.3.2) as

$$\nabla u = du + \Gamma u. \tag{3.3.3}$$

A natural question arises: how does the connection 1-form changes with the change of the local gauge?

Let $\boldsymbol{f} = (f_\alpha)$ be another moving frame of $E|_U$. The correspondence $e_\alpha \mapsto f_\alpha$ defines an automorphism of $E|_U$. Using the local frame $\boldsymbol{e}$ we can identify this correspondence with a smooth map $g : U \to \mathrm{GL}(r; \mathbb{K})$. The map g is called *the local gauge transformation* relating $\boldsymbol{e}$ to $\boldsymbol{f}$.

Let $\hat{\Gamma}$ denote the connection 1-form corresponding to the new moving frame, i.e.,

$$\nabla f_\alpha = \hat{\Gamma}_\alpha^\beta f_\beta.$$

Consider a section σ of $E|_U$. With respect to the local frame (e_α) the section σ has a decomposition

$$\sigma = u^\alpha e_\alpha,$$

while with respect to (f_β) it has a decomposition

$$\sigma = \hat{u}^\beta f_\beta.$$

The two decompositions are related by

$$u = g\hat{u}. \tag{3.3.4}$$

Now, we can identify the E-valued 1-form $\nabla\sigma$ with a $\mathbb{K}^r$-valued 1-form in two ways: either using the frame $\boldsymbol{e}$, or using the frame $\boldsymbol{f}$. In the first case, the derivative $\nabla\sigma$ is identified with the $\mathbb{K}^r$-valued 1-form

$$du + \Gamma u,$$

while in the second case it is identified with

$$d\hat{u} + \hat{\Gamma}\hat{u}.$$

These two identifications are related by the same rule as in (3.3.4):

$$du + \Gamma u = g(d\hat{u} + \hat{\Gamma}\hat{u}).$$

Using (3.3.4) in the above equality we get

$$(dg)\hat{u} + g d\hat{u} + \Gamma g\hat{u} = g d\hat{u} + g\hat{\Gamma}\hat{u}.$$

Hence

$$\hat{\Gamma} = g^{-1}dg + g^{-1}\Gamma g.$$

The above relation is the transition rule relating two local gauge descriptions of the same connection.

☞**A word of warning.** The identification

$$\{moving\ frames\} \cong \{local\ trivialization\}$$

should be treated carefully. These are like an object and its image in a mirror, and there is a great chance of confusing the right hand with the left hand.

More concretely, if $t_\alpha : E_\alpha \xrightarrow{\cong} \mathbb{K}^r \times U_\alpha$ (respectively $t_\beta : E_\beta \xrightarrow{\cong} \mathbb{K}^r \times U_\beta$) is a trivialization of a bundle E over an open set U_α (respectively U_β), then the transition map "from α to β" over $U_\alpha \cap U_\beta$ is $g_{\beta\alpha} = t_\beta \circ t_\alpha^{-1}$. The standard basis in $\mathbb{K}^r$, denoted by (δ_i), induces two local moving frames on E:

$$\boldsymbol{e}_{\alpha,i} = t_\alpha^{-1}(\delta_i) \text{ and } \boldsymbol{e}_{\beta,i} = t_\beta^{-1}(\delta_i).$$

On the overlap $U_\alpha \cap U_\beta$ these two frames are related by the local gauge transformation

$$\boldsymbol{e}_{\beta,i} = g_{\beta\alpha}^{-1}\boldsymbol{e}_{\alpha,i} = g_{\alpha\beta}\boldsymbol{e}_{\alpha,i}.$$

This is precisely the opposite way the two trivializations are identified. □

The above arguments can be reversed producing the following global result.

Proposition 3.3.5. *Let $E \to M$ be a rank r smooth vector bundle, and (U_α) a trivializing cover with transition maps $g_{\alpha\beta} : U_\alpha \cap U_\beta \to \mathrm{GL}(r; \mathbb{K})$. Then any collection of matrix valued 1-forms $\Gamma_\alpha \in \Omega^1(\mathrm{End}\,\underline{\mathbb{K}}_{U_\alpha}^r)$ satisfying*

$$\Gamma_\beta = (g_{\alpha\beta}^{-1}dg_{\alpha\beta}) + g_{\alpha\beta}^{-1}\Gamma_\alpha g_{\alpha\beta} = -(dg_{\beta\alpha})g_{\beta\alpha}^{-1} + g_{\beta\alpha}\Gamma_\alpha g_{\beta\alpha}^{-1} \text{ over } U_\alpha \cap U_\beta, \quad (3.3.5)$$

uniquely defines a covariant derivative on E. □

Exercise 3.3.6. Prove the above proposition. □

We can use the local description in Proposition 3.3.5 to define the notion of pullback of a connection. Suppose we are given the following data.

- A smooth map $f : N \to M$.
- A rank r $\mathbb{K}$-vector bundle $E \to M$ defined by the open cover (U_α), and transition maps $g_{\beta\alpha} : U_\alpha \cap U_\beta \to \mathrm{GL}(\mathbb{K}^r)$.
- A connection ∇ on E defined by the 1-forms $\Gamma_\alpha \in \Omega^1(\mathrm{End}\,\underline{\mathbb{K}}_{U_\alpha}^r)$ satisfying the gluing conditions (3.3.5).

Then, these data define a connection $f^*\nabla$ on f^*E described by the open cover $f^{-1}(U_\alpha)$, transition maps $g_{\beta\alpha} \circ f$ and 1-forms $f^*\Gamma_\alpha$. This connection is independent of the various choices and it is called the *pullback of ∇ by f*.

Example 3.3.7. (Complex line bundles). Let $L \to M$ be a complex line bundle over the smooth manifold M. Let $\{U_\alpha\}$ be a trivializing cover with transition maps $z_{\alpha\beta} : U_\alpha \cap U_\beta \to \mathbb{C}^* = \mathrm{GL}(1, \mathbb{C})$. The bundle of endomorphisms of L, $\mathrm{End}\,(L) \cong L^* \otimes L$ is trivial since it can be defined by transition maps $(z_{\alpha\beta})^{-1} \otimes z_{\alpha\beta} = 1$. Thus, the space of connections on L, $\mathcal{A}\,(L)$ is an affine space modelled by the linear space of complex valued 1-forms. A connection on L is simply a collection of $\mathbb{C}$-valued 1-forms ω^α on U_α related on overlaps by

$$\omega^\beta = \frac{dz_{\alpha\beta}}{z_{\alpha\beta}} + \omega^\alpha = d\log z_{\alpha\beta} + \omega^\alpha. \qquad\qquad \square$$

3.3.2   *Parallel transport*

As we have already pointed out, the main reason we could not construct natural derivations on the space of sections of a vector bundle was the lack of a canonical procedure of identifying fibers at different points. We will see in this subsection that such a procedure is all we need to define covariant derivatives. More precisely, we will show that once a covariant derivative is chosen, it offers a simple way of identifying different fibers.

Let $E \to M$ be a rank r $\mathbb{K}$-vector bundle and ∇ a covariant derivative on E. For any smooth path $\gamma : [0, 1] \to M$ we will define a *linear* isomorphism $T_\gamma : E_{\gamma(0)} \to E_{\gamma(1)}$ called the *parallel transport* along γ. More exactly, we will construct an entire family of linear isomorphisms

$$T_t : E_{\gamma(0)} \to E_{\gamma(t)}.$$

One should think of this T_t as identifying different fibers. In particular, if $u_0 \in E_{\gamma(0)}$ then the path $t \mapsto u_t = T_t u_0 \in E_{\gamma(t)}$ should be thought of as a "constant" path. The rigorous way of stating this "constancy" is via derivations: a quantity is "constant" if its derivatives are identically 0. Now, the only way we know how to derivate sections is via ∇, i.e., u_t should satisfy

$$\nabla_{\frac{d}{dt}} u_t = 0, \quad \text{where} \quad \frac{d}{dt} = \dot\gamma.$$

The above equation suggests a way of defining T_t. For any $u_0 \in E_{\gamma(0)}$, and any $t \in [0, 1]$, define $T_t u_0$ as the value at t of the solution of the initial value problem

$$\begin{cases} \nabla_{\frac{d}{dt}} u(t) = 0 \\ u(0) = u_0 \end{cases}. \qquad\qquad (3.3.6)$$

The equation (3.3.6) is a system of linear ordinary differential equations in disguise.

To see this, let us make the simplifying assumption that $\gamma(t)$ lies entirely in some coordinate neighborhood U with coordinates $(x^1, ..., x^n)$, such that $E\,|_U$ is trivial. This is always happening, at least on every small portion of γ. Denote by $(e_\alpha)_{1 \leq \alpha \leq r}$ a local moving frame trivializing $E\,|_U$ so that $u = u^\alpha e_\alpha$. The connection 1-form corresponding to this moving frame will be denoted by $\Gamma \in \Omega^1(\mathrm{End}\,(\mathbb{K}^r))$. Equation (3.3.6) becomes (using Einstein's convention)

$$\begin{cases} \frac{du^\alpha}{dt} + \Gamma^\alpha_{t\beta} u^\beta = 0 \\ u^\alpha(0) = u^\alpha_0 \end{cases}, \tag{3.3.7}$$

where

$$\Gamma_t = \frac{d}{dt} \lrcorner \Gamma = \dot\gamma \lrcorner \Gamma \in \Omega^0(\mathrm{End}\,(\mathbb{K}^r)) = \mathrm{End}\,(\mathbb{K}^r).$$

More explicitly, if the path $\gamma(t)$ is given by the smooth map

$$t \mapsto \gamma(t) = \big(x^1(t), \ldots, x^n(t) \big),$$

then Γ_t is the endomorphism given by

$$\Gamma_t e_\beta = \dot x^i \Gamma^\alpha_{i\beta} e_\alpha.$$

The system (3.3.7) can be rewritten as

$$\begin{cases} \frac{du^\alpha}{dt} + \Gamma^\alpha_{i\beta} \dot x^i u^\beta = 0 \\ \\ u^\alpha(0) = u^\alpha_0. \end{cases} \tag{3.3.8}$$

This is obviously a system of linear ordinary differential equations whose solutions exist for any t. We deduce

$$\dot u(0) = -\Gamma_t u_0. \tag{3.3.9}$$

This gives a geometric interpretation for the connection 1-form Γ: for any vector field X, the contraction $-i_X \Gamma = -\Gamma(X) \in \mathrm{End}\,(E)$ describes the infinitesimal parallel transport along the direction prescribed by the vector field X, in the non-canonical identification of nearby fibers via a local moving frame.

In more intuitive terms, if $\gamma(t)$ is an integral curve for X, and T_t denotes the parallel transport along γ from $E_{\gamma(0)}$ to $E_{\gamma(t)}$, then, given a local moving frame for E in a neighborhood of $\gamma(0)$, T_t is identified with a t-dependent matrix which has a Taylor expansion of the form

$$T_t = \mathbb{1} - \Gamma_0 t + O(t^2), \quad t \text{ very small}, \tag{3.3.10}$$

with $\Gamma_0 = (i_X \Gamma)|_{\gamma(0)}$.

3.3.3   *The curvature of a connection*

Consider a rank r smooth $\mathbb{K}$-vector bundle $E \to M$ over the smooth manifold M, and let $\nabla : \Omega^0(E) \to \Omega^1(E)$ be a covariant derivative on E.

Proposition 3.3.8. *The connection ∇ has a natural extension to an operator*

$$d^\nabla : \Omega^r(E) \to \Omega^{r+1}(E)$$

uniquely defined by the requirements,

(a) $d^\nabla \mid_{\Omega^0(E)} = \nabla$,
(b) $\forall \omega \in \Omega^r(M)$, $\eta \in \Omega^s(E)$

$$d^\nabla(\omega \wedge \eta) = d\omega \wedge \eta + (-1)^r \omega \wedge d^\nabla \eta.$$

Outline of the proof   *Existence.* For $\omega \in \Omega^r(M)$, $u \in \Omega^0(E)$ set

$$d^\nabla(\omega \otimes u) = d\omega \otimes u + (-1)^r \omega \nabla u. \tag{3.3.11}$$

Using a partition of unity one shows that any $\eta \in \Omega^r(E)$ is a locally finite combination of monomials as above so the above definition induces an operator $\Omega^r(E) \to \Omega^{r+1}(E)$. We let the reader check that this extension satisfies conditions (a) and (b) above.

Uniqueness. Any operator with the properties (a) and (b) acts on monomials as in (3.3.11) so it has to coincide with the operator described above using a given partition of unity.   □

Example 3.3.9.   The trivial bundle $\mathbb{K}_M$ has a natural connection ∇^0- the trivial connection. This coincides with the usual differential $d : \Omega^0(M) \otimes \mathbb{K} \to \Omega^1(M) \otimes \mathbb{K}$. The extension d^{∇^0} is the usual exterior derivative.   □

There is a major difference between the usual exterior derivative d, and an arbitrary d^∇. In the former case we have $d^2 = 0$, which is a consequence of the commutativity $[\frac{\partial}{\partial x_i}, \frac{\partial}{\partial x_j}] = 0$, where (x^i) are local coordinates on M. In the second case, the equality $(d^\nabla)^2 = 0$ does not hold in general. Still, something very interesting happens.

Lemma 3.3.10.   *For any smooth function $f \in C^\infty(M)$, and any $\omega \in \Omega^r(E)$ we have*

$$(d^\nabla)^2(f\omega) = f\{(d^\nabla)^2\omega\}.$$

Hence $(d^\nabla)^2$ is a bundle morphism $\Lambda^r T^*M \otimes E \to \Lambda^{r+2}T^*M \otimes E$.

Proof.   We compute

$$(d^\nabla)^2(f\omega) = d^\nabla(df \wedge \omega + f d^\nabla \omega)$$

$$= -df \wedge d^\nabla \omega + df \wedge d^\nabla \omega + f(d^\nabla)^2 \omega = f(d^\nabla)^2 \omega. \qquad \square$$

As a map $\Omega^0(E) \to \Omega^2(E)$, the operator $(d^\nabla)^2$ can be identified with a section of

$$\mathrm{Hom}_{\mathbb{K}}\left(E, \Lambda^2 T^*M \otimes_{\mathbb{R}} E\right) \cong E^* \otimes \Lambda^2 T^*M \otimes_{\mathbb{R}} E \cong \Lambda^2 T^*M \otimes_{\mathbb{R}} \mathrm{End}_{\mathbb{K}}(E).$$

Thus, $(d^\nabla)^2$ is an $\mathrm{End}_{\mathbb{K}}(E)$-valued 2-form.

Definition 3.3.11. For any connection ∇ on a smooth vector bundle $E \to M$, the object $(d^\nabla)^2 \in \Omega^2(\mathrm{End}_{\mathbb{K}}(E))$ is called the *curvature* of ∇, and it is usually denoted by $F(\nabla)$. □

Example 3.3.12. Consider the trivial bundle $\underline{\mathbb{K}}_M^r$. The sections of this bundle are smooth $\mathbb{K}^r$-valued functions on M. The exterior derivative d defines the trivial connection on $\underline{\mathbb{K}}_M^r$, and any other connection differs from d by a $M_r(\mathbb{K})$-valued 1-form on M. If A is such a form, then the curvature of the connection $d + A$ is the 2-form $F(A)$ defined by

$$F(A)u = (d + A)^2 u = (dA + A \wedge A)u, \quad \forall u \in C^\infty(M, \mathbb{K}^r).$$

The $\wedge$-operation above is defined for any vector bundle E as the bilinear map

$$\Omega^r(\mathrm{End}(E)) \times \Omega^s(\mathrm{End}(E)) \to \Omega^{r+s}(\mathrm{End}(E)),$$

uniquely determined by

$$(\omega^r \otimes A) \wedge (\eta^s \otimes B) = \omega^r \wedge \eta^s \otimes AB, \quad A, B \in \mathrm{End}(E). \qquad \square$$

We conclude this subsection with an alternate description of the curvature which hopefully will shed some light on its analytical significance.

Let $E \to M$ be a smooth vector bundle on M and ∇ a connection on it. Denote its curvature by $F = F(\nabla) \in \Omega^2(\mathrm{End}(E))$. For any $X, Y \in \mathrm{Vect}(M)$ the quantity $F(X, Y)$ is an endomorphism of E. In the remaining part of this section we will give a different description of this endomorphism.

For any vector field Z, we denote by $i_Z : \Omega^r(E) \to \Omega^{r-1}(E)$ the $C^\infty(M)$-*linear* operator defined by

$$i_Z(\omega \otimes u) = (i_Z \omega) \otimes u, \quad \forall \omega \in \Omega^r(M), \ u \in \Omega^0(E).$$

The covariant derivative ∇_Z extends naturally to elements of $\Omega^r(E)$ by

$$\nabla_Z(\omega \otimes u) := (L_Z \omega) \otimes u + \omega \otimes \nabla_Z u.$$

The operators d^∇, i_Z, ∇_Z satisfy the usual super-commutation identities.

$$i_Z d^\nabla + d^\nabla i_Z = \nabla_Z. \tag{3.3.12}$$

$$i_X i_Y + i_Y i_X = 0. \tag{3.3.13}$$

$$\nabla_X i_Y - i_Y \nabla_X = i_{[X,Y]}. \tag{3.3.14}$$

For any $u \in \Omega^0(E)$ we compute using (3.3.12)-(3.3.14)

$$F(X,Y)u = i_Y i_X (d^\nabla)^2 u = i_Y (i_X d^\nabla) \nabla u$$

$$= i_Y (\nabla_X - d^\nabla i_X) \nabla u = (i_Y \nabla_X) \nabla u - (i_Y d^\nabla) \nabla_X u$$

$$= (\nabla_X i_Y - i_{[X,Y]}) \nabla u - \nabla_Y \nabla_X u = (\nabla_X \nabla_Y - \nabla_Y \nabla_X - \nabla_{[X,Y]}) u.$$

Hence

$$F(X,Y) = [\nabla_X, \nabla_Y] - \nabla_{[X,Y]}. \tag{3.3.15}$$

If in the above formula we take $X = \frac{\partial}{\partial x_i}$ and $Y = \frac{\partial}{\partial x_j}$, where (x^i) are local coordinates on M, and we set $\nabla_i := \nabla_{\frac{\partial}{\partial x_i}}$, $\nabla_j = \nabla_{\frac{\partial}{\partial x_j}}$, then we deduce

$$F_{ij} = -F_{ji} := F\left(\frac{\partial}{\partial x_i}, \frac{\partial}{\partial x_j}\right) = [\nabla_i, \nabla_j]. \tag{3.3.16}$$

Thus, the endomorphism F_{ij} measures the extent to which the partial derivatives ∇_i, ∇_j fail to commute. This is in sharp contrast with the classical calculus and an analytically oriented reader may object to this by saying we were careless when we picked the connection. Maybe an intelligent choice will restore the classical commutativity of partial derivatives so we should concentrate from the very beginning to covariant derivatives ∇ such that $F(\nabla) = 0$.

Definition 3.3.13. A connection ∇ such that $F(\nabla) = 0$ is called *flat*.   □

A natural question arises: given an arbitrary vector bundle $E \to M$ do there exist flat connections on E? If E is trivial then the answer is obviously positive. In general, the answer is negative, and this has to do with the *global structure* of the bundle. In the second half of this book we will discuss in more detail this fact.

3.3.4 *Holonomy*

The reader may ask a very legitimate question: why have we chosen to name curvature, the deviation from commutativity of a given connection. In this subsection we describe the geometric meaning of curvature, and maybe this will explain this choice of terminology. Throughout this subsection we will use Einstein's convention.

Let $E \to M$ be a rank r smooth $\mathbb{K}$-vector bundle, and ∇ a connection on it. Consider local coordinates $(x^1, ..., x^n)$ on an open subset $U \subset M$ such that $E|_U$ is trivial. Pick a moving frame $(e_1, ..., e_r)$ of E over U. The connection 1-form associated to this moving frame is

$$\Gamma = \Gamma_i dx^i = (\Gamma_{i\beta}^\alpha) dx^i, \quad 1 \le \alpha, \beta \le r.$$

It is defined by the equalities $(\nabla_i := \nabla_{\frac{\partial}{\partial x_i}})$

$$\nabla_i e_\beta = \Gamma_{i\beta}^\alpha e_\alpha. \tag{3.3.17}$$

Using (3.3.16) we compute

$$F_{ij}e_\beta = (\nabla_i\nabla_j - \nabla_j\nabla_i)e_\beta = \nabla_i(\Gamma_j e_\beta) - \nabla_j(\Gamma_i e_\beta)$$

$$= \left(\frac{\partial\Gamma^\alpha_{j\beta}}{\partial x^i} - \frac{\partial\Gamma^\alpha_{j\beta}}{\partial x^j}\right)e_\alpha + \left(\Gamma^\gamma_{j\beta}\Gamma^\alpha_{i\gamma} - \Gamma^\gamma_{i\beta}\Gamma^\alpha_{j\gamma}\right)e_\alpha,$$

so that

$$F_{ij} = \left(\frac{\partial\Gamma_j}{\partial x^i} - \frac{\partial\Gamma_i}{\partial x^j} + \Gamma_i\Gamma_j - \Gamma_j\Gamma_i\right). \tag{3.3.18}$$

Though the above equation looks very complicated it will be the clue to understanding the geometric significance of curvature.

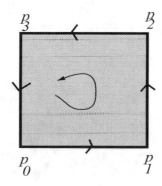

Fig. 3.1   *Parallel transport along a coordinate parallelogram.*

Assume for simplicity that the point of coordinates $(0, ..., 0)$ lies in U. Denote by T^s_1 the parallel transport (using the connection ∇) from $(x^1, ..., x^n)$ to $(x^1 + s, x^2, ..., x^n)$ along the curve $\tau \mapsto (x^1 + \tau, x^2, ..., x^n)$. Define T^t_2 in a similar way using the coordinate x^2 instead of x^1.

Look at the parallelogram $P_{s,t}$ in the "plane" (x^1, x^2) described in Figure 3.1, where

$$p_0 = (0, \ldots, 0), \quad p_1 = (s, 0, \ldots, 0), \quad p_2 = (s, t, 0, \ldots, 0), \quad p_3 = (0, t, 0, \ldots, 0).$$

We now perform the counterclockwise parallel transport along the boundary of $P_{s,t}$. The outcome is a linear map $\mathfrak{T}_{s,t} : E_0 \to E_0$, where E_0 is the fiber of E over p_0. Set $F_{12} := F(\frac{\partial}{\partial x^1}\frac{\partial}{\partial x^2})\,|_{(0,\ldots,0)}$. F_{12} is an endomorphism of E_0.

Proposition 3.3.14.   *For any $u \in E_0$ we have*

$$F_{12}u = -\frac{\partial^2}{\partial s \partial t}\mathfrak{T}_{s,t}u.$$

We see that the parallel transport of an element $u \in E_0$ along a closed path may not return it to itself. The curvature is an infinitesimal measure of this deviation.

Proof.　　The parallel transport along $\partial P_{s,t}$ can be described as

$$\mathcal{T}_{s,t} = T_2^{-t} T_1^{-s} T_2^t T_1^s.$$

The parallel transport $T_1^s : E_0 \to E_{p_1}$ can be approximated using (3.3.9)

$$u_1 = u_1(s,t) = T_1^s u_0 = u_0 - s\Gamma_1(p_0)u_0 + C_1 s^2 + O(s^3). \tag{3.3.19}$$

C_1 is a constant vector in E_0 whose exact form is not relevant to our computations. In the sequel the letter C (eventually indexed) will denote constants.

$$u_2 = u_2(s,t) = T_2^t T_1^s u = T_2^t u_1 = u_1 - t\Gamma_2(p_1)u_1 + C_2 t^2 + O(t^3)$$

$$= u_0 - s\Gamma_1(p_0)u_0 - t\Gamma_2(p_1)(u_0 - s\Gamma_1(p_0)u_0) + C_1 s^2 + C_2 t^2 + O(3)$$

$$= \left\{ 1 - s\Gamma_1(p_0) - t\Gamma_2(p_1) + ts\Gamma_2(p_1)\Gamma_1(p_0) \right\} u_0 + C_1 s^2 + C_2 t^2 + O(3).$$

$O(k)$ denotes an error $\leq C(s^2 + t^2)^{k/2}$ as $s, t \to 0$. Now use the approximation

$$\Gamma_2(p_1) = \Gamma_2(p_0) + s\frac{\partial \Gamma_2}{\partial x^1}(p_0) + O(2),$$

to deduce

$$u_2 = \left\{ 1 - s\Gamma_1 - t\Gamma_2 - st\left(\frac{\partial \Gamma_2}{\partial x^1} - \Gamma_2\Gamma_1 \right) \right\} |_{p_0} u_0$$

$$+ C_1 s^2 + C_2 t^2 + O(3). \tag{3.3.20}$$

Similarly, we have

$$u_3 = u_3(s,t) = T_1^{-s} T_2^t T_1^s u_0 = T_1^{-s} u_2 = u_2 + s\Gamma_1(p_2)u_2 + C_3 s^2 + O(3).$$

The Γ-term in the right-hand side of the above equality can be approximated as

$$\Gamma_1(p_2) = \Gamma_1(p_0) + s\frac{\partial \Gamma_1}{\partial x^1}(p_0) + t\frac{\partial \Gamma_1}{\partial x^2}(p_0) + O(2).$$

Using u_2 described as in (3.3.20) we get after an elementary computation

$$u_3 = u_3(s,t) = \left\{ 1 - t\Gamma_2 + st\left(\frac{\partial \Gamma_1}{\partial x^2} - \frac{\partial \Gamma_2}{\partial x^1} + \Gamma_2\Gamma_1 - \Gamma_1\Gamma_2 \right) \right\} |_{p_0} u_0$$

$$+ C_4 s^2 + C_5 t^2 + O(3). \tag{3.3.21}$$

Finally, we have

$$u_4 = u_4(s,t) = T_2^{-t} = u_3 + t\Gamma_2(p_3)u_3 + C_6 t^2 + O(3),$$

with

$$\Gamma_2(p_3) = \Gamma_2(p_0) + t\frac{\partial \Gamma_2}{\partial x^2}(p_0) + C_7 t^2 + O(3).$$

Using (3.3.21) we get

$$u_4(s,t) = u_0 + st\left(\frac{\partial\Gamma_1}{\partial x^2} - \frac{\partial\Gamma_2}{\partial x^1} + \Gamma_2\Gamma_1 - \Gamma_1\Gamma_2\right)\Big|_{p_0} u_0$$

$$+C_8 s^2 + C_9 t^2 + O(3)$$

$$= u_0 - st F_{12}(p_0)u_0 + C_8 s^2 + C_9 t^2 + O(3).$$

Clearly $\frac{\partial^2 u_4}{\partial s \partial t} = -F_{12}(p_0)u_0$ as claimed. □

Remark 3.3.15. If we had kept track of the various constants in the above computation we would have arrived at the conclusion that $C_8 = C_9 = 0$ i.e.

$$\mathcal{T}_{s,t} = \mathbb{1} - st F_{12} + O(3).$$

Alternatively, the constant C_8 is the second order correction in the Taylor expansion of $s \mapsto \mathcal{T}_{s,0} \equiv$, so it has to be 0. The same goes for C_9. Thus we have

$$-F_{12} = \frac{d\mathcal{T}_{s,t}}{d\,\mathrm{area}P_{s,t}} = \frac{d\mathcal{T}_{\sqrt{s},\sqrt{s}}}{ds}.$$

Loosely speaking, the last equality states that the curvature is the the "amount of holonomy per unit of area". □

The result in the above proposition is usually formulated in terms of holonomy.

Definition 3.3.16. Let $E \to M$ be a vector bundle with a connection ∇. The *holonomy* of ∇ along a closed path γ is the parallel transport along γ. □

We see that the curvature measures the holonomy along infinitesimal parallelograms. A connection can be viewed as an analytic way of trivializing a bundle. We can do so along paths starting at a fixed point, using the parallel transport, but using different paths ending at the same point we may wind up with trivializations which differ by a twist. The curvature provides an infinitesimal measure of that twist.

Exercise 3.3.17. Prove that any vector bundle E over the Euclidean space $\mathbb{R}^n$ is trivializable.
Hint: Use the parallel transport defined by a connection on the vector bundle E to produce a bundle isomorphism $E \to E_0 \times \mathbb{R}^n$, where E_0 is the fiber of E over the origin. □

3.3.5   The Bianchi identities

Consider a smooth $\mathbb{K}$-vector bundle $E \to M$ equipped with a connection ∇. We have seen that the associated exterior derivative $d^{\nabla} : \Omega^p(E) \to \Omega^{p+1}(E)$ does not satisfy the usual $(d^{\nabla})^2 = 0$, and the curvature is to blame for this. The Bianchi identity describes one remarkable algebraic feature of the curvature.

Recall that ∇ induces a connection in any tensor bundle constructed from E. In particular, it induces a connection in $E^* \otimes E \cong \mathrm{End}\,(E)$ which we continue to denote by ∇. This extends to an "exterior derivative" $D : \Omega^p(\mathrm{End}\,(E)) \to \Omega^{p+1}(\mathrm{End}\,(E))$.

Proposition 3.3.18. (The Bianchi identity) *Let $E \to M$ and ∇ as above. Then*

$$DF(\nabla) = 0.$$

Roughly speaking, the Bianchi identity states that $(d^{\nabla})^3$ is 0.

Proof.   We will use the identities (3.3.12)–(3.3.14). For any vector fields X, Y, Z we have

$$i_X D = \nabla_X - D i_X.$$

Hence,

$$(DF)(X, Y, Z) = i_Z i_Y i_X DF = i_Z i_Y (\nabla_X - D i_X) F$$

$$= i_Z (\nabla_X i_Y - i_{[X,Y]}) F - i_Z (\nabla_Y - D i_Y) i_X F$$

$$= (\nabla_X i_Z i_Y - i_{[X,Z]} i_Y - i_Z i_{[X,Y]}) F - (\nabla_Y i_Z i_X - i_{[Y,Z]} i_X - \nabla_Z i_Y i_X) F$$

$$= (i_{[X,Y]} i_Z + i_{[Y,Z]} i_X + i_{[Z,X]} i_Y) F - (\nabla_X i_Y i_Z + \nabla_Y i_Z i_X + \nabla_Z i_X i_Y) F.$$

We compute immediately

$$i_{[X,Y]} i_Z F = F(Z, [X, Y]) = \big[\nabla_Z, \nabla_{[X,Y]}\big] - \nabla_{[Z,[X,Y]]}.$$

Also for any $u \in \Omega^0(E)$ we have

$$(\nabla_X i_Y i_Z F) u = \nabla_X (F(Z, Y) u) - F(Z, Y) \nabla_X u = \big[\nabla_X, F(Z, Y)\big] u$$

$$= \big[\nabla_X, \nabla_{[Y,Z]}\big] u - \big[\nabla_X, [\nabla_Y, \nabla_Z]\big] u.$$

The Bianchi identity now follows from the classical Jacobi identity for commutators.   □

Example 3.3.19.   Let $\underline{\mathbb{K}}$ be the trivial line bundle over a smooth manifold M. Any connection on $\underline{\mathbb{K}}$ has the form $\nabla^{\omega} = d + \omega$, where d is the trivial connection, and ω is a $\mathbb{K}$-valued 1-form on M. The curvature of this connection is

$$F(\omega) = d\omega.$$

The Bianchi identity is in this case precisely the equality $d^2 \omega = 0$.   □

3.3.6 Connections on tangent bundles

The tangent bundles are very special cases of vector bundles so the general theory of connections and parallel transport is applicable in this situation as well. However, the tangent bundles have some peculiar features which enrich the structure of a connection.

Recall that, when looking for a local description for a connection on a vector bundle, we have to first choose local coordinates on the manifolds, and then a local moving frame for the vector bundle. For an arbitrary vector bundle there is no correlation between these two choices.

For tangent bundles it happens that, once local coordinates (x^i) are chosen, they automatically define a moving frame of the tangent bundle, $\left(\partial_i = \frac{\partial}{\partial x_i} \right)$, and it is thus very natural to work with this frame. Hence, let ∇ be a connection on TM. With the above notations we set

$$\nabla_i \partial_j = \Gamma^k_{ij} \partial_k \quad (\nabla_i = \nabla_{\partial_i}).$$

The coefficients Γ^k_{ij} are usually known as the *Christoffel symbols* of the connection. As usual we construct the curvature tensor

$$F(X,Y) = [\nabla_X, \nabla_Y] - \nabla_{[X,Y]} \in \text{End}\,(TM).$$

Still, this is not the only tensor naturally associated to ∇.

Lemma 3.3.20. *For $X, Y \in \text{Vect}\,(M)$ consider*

$$T(X,Y) = \nabla_X Y - \nabla_Y X - [X,Y] \in \text{Vect}\,(M).$$

Then $\forall f \in C^\infty(M)$

$$T(fX,Y) = T(X, fY) = fT(X,Y),$$

so that $T(\bullet, \bullet)$ is a tensor $T \in \Omega^2(TM)$, i.e., a 2-form whose coefficients are vector fields on M. The tensor T is called the torsion *of the connection ∇.* □

The proof of this lemma is left to the reader as an exercise. In terms of Christoffel symbols, the torsion has the description

$$T(\partial_i, \partial_j) = (\Gamma^k_{ij} - \Gamma^k_{ji}) \partial_k.$$

Definition 3.3.21. A connection on TM is said to be *symmetric* if $T = 0$. □

We guess by now the reader is wondering how the mathematicians came up with this object called torsion. In the remaining of this subsection we will try to sketch the geometrical meaning of torsion.

To find such an interpretation, we have to look at the finer structure of the tangent space at a point $x \in M$. It will be convenient to regard $T_x M$ as an affine space modeled by $\mathbb{R}^n$, $n = \dim M$. Thus, we will no longer think of the elements of T_xM as vectors, but instead we will treat them as points. The tangent space T_xM can be coordinatized using *affine frames*. These are pairs $(p; e)$, where p is a point in T_xM, and e is a basis of the underlying vector space. A frame allows one to identify T_xM with $\mathbb{R}^n$, where p is thought of as the origin.

Suppose that A, B are two affine spaces, both modelled by $\mathbb{R}^n$, and $(p; e)$, $(p; f)$ are affine frames of A and respectively B. Denote by (x^i) the coordinates in A induced by the frame $(p; e)$, and by (y^j) the coordinates in B induced by the frame $(q; f)$. An affine isomorphism $T : A \to B$ can then be described using these coordinates as

$$T : \mathbb{R}^n_x \to \mathbb{R}^n_y \quad x \mapsto y = Sx + v,$$

where v is a vector in $\mathbb{R}^n$, and S is an invertible $n \times n$ real matrix. Thus, an affine map is described by a "rotation" S, followed by a translation v. This vector measures the "drift" of the origin. We write $T = S \hat{+} x$

If now (x^i) are local coordinates on M, then they define an affine frame $\mathcal{A}_x$ at each $x \in M$: $(\mathcal{A}_x = (0; (\partial_i)))$. Given a connection ∇ on TM, and a smooth path $\gamma : I \to M$, we will construct a family of affine isomorphisms $T_t : T_{\gamma(0)} \to T_{\gamma(t)}$ called the *affine transport of ∇ along γ*. In fact, we will determine T_t by imposing the initial condition $T_0 = \mathbb{1}$, and then describing $\dot{T}_t$.

This is equivalent to describing the infinitesimal affine transport at a given point $x_0 \in M$ along a direction given by a vector $X = X^i \partial_i \in T_{x_0}M$. The affine frame of $T_{x_0}M$ is $\mathcal{A}_{x_0} = (0; (\partial_i))$.

If x_t is a point along the integral curve of X, close to x_0 then its coordinates satisfy

$$x_t^i = x_0^i + tX^i + O(t^2).$$

This shows the origin x_0 of $\mathcal{A}_{x_0}$ "drifts" by $tX + O(t^2)$. The frame (∂_i) suffers a parallel transport measured as usual by $\mathbb{1} - ti_X\Gamma + O(t^2)$. The total affine transport will be

$$T_t = (\mathbb{1} - ti_X\Gamma) \hat{+} tX + O(t^2).$$

The holonomy of ∇ along a closed path will be an affine transformation and as such it has two components: a "rotation"" and a translation. As in Proposition 3.3.14 one can show the torsion measures the translation component of the holonomy along an infinitesimal parallelogram, i.e., the "amount of drift per unit of area". Since we will not need this fact we will not include a proof of it.

Exercise 3.3.22. Consider the vector valued 1-form $\omega \in \Omega^1(TM)$ defined by

$$\omega(X) = X \quad \forall X \in \text{Vect}(M).$$

Show that if ∇ is a linear connection on TM, then $d^\nabla \omega = T^\nabla$, where T^∇ denotes the torsion of ∇. $\square$

Exercise 3.3.23. Consider a smooth vector bundle $E \to M$ over the smooth manifold M. We assume that both E and TM are equipped with connections and moreover the connection on TM is torsionless. Denote by $\hat{\nabla}$ the induced connection on $\Lambda^2 T^* M \otimes \mathrm{End}\,(E)$. Prove that $\forall X, Y, Z \in \mathrm{Vect}\,(M)$

$$\hat{\nabla}_X F(Y, Z) + \hat{\nabla}_Y F(Z, X) + \hat{\nabla}_Z F(X, Y) = 0. \qquad \square$$

3.4 Integration on manifolds

3.4.1 *Integration of 1-densities*

We spent a lot of time learning to differentiate geometrical objects but, just as in classical calculus, the story is only half complete without the reverse operation, integration.

Classically, integration requires a background measure, and in this subsection we will describe the differential geometric analogue of a measure, namely the notion of 1-density on a manifold.

Let $E \to M$ be a rank k, smooth real vector bundle over a manifold M defined by an open cover (U_α) and transition maps $g_{\beta\alpha} : U_{\alpha\beta} \to \mathrm{GL}(k, \mathbb{R})$ satisfying the cocycle condition. For any $r \in \mathbb{R}$ we can form the real line bundle $|\Lambda|^r(E)$ defined by the same open cover and transition maps

$$t_{\beta\alpha} := |\det g_{\beta\alpha}|^{-r} = |\det g_{\alpha\beta}|^r : U_{\alpha\beta} \to \mathbb{R}_{>0} \hookrightarrow \mathrm{GL}(1, \mathbb{R}).$$

The fiber at $p \in M$ of this bundle consists of r-densities on E_p (see Subsection 2.2.4).

Definition 3.4.1. Let M be a smooth manifold and $r \geq 0$. The bundle of r-densities on M is

$$|\Lambda|_M^r := |\Lambda|^r(TM).$$

When $r = 1$ we will use the notation $|\Lambda|_M = |\Lambda|_M^1$. We call $|\Lambda|_M$ the *density bundle* of M. $\qquad \square$

Denote by $C^\infty(|\Lambda|_M)$ the space of smooth sections of $|\Lambda|_M$, and by $C_0^\infty(|\Lambda|_M)$ its subspace consisting of compactly supported densities.

It helps to have local descriptions of densities. To this aim, pick an open cover of M consisting of coordinate neighborhoods (U_α). Denote the local coordinates on U_α by (x_α^i). This choice of a cover produces a trivializing cover of TM with transition maps

$$T_{\alpha\beta} = \left(\frac{\partial x_\alpha^i}{\partial x_\beta^j} \right)_{1 \leq i, j \leq n},$$

where n is the dimension of M. Set $\delta_{\alpha\beta} = |\det T_{\alpha\beta}|$. A 1-density on M is then a collection of functions $\mu_\alpha \in C^\infty(U_\alpha)$ related by

$$\mu_\alpha = \delta_{\alpha\beta}^{-1}\mu_\beta.$$

It may help to think that for each point $p \in U_\alpha$ the basis $\frac{\partial}{\partial x_\alpha^1}, ..., \frac{\partial}{\partial x_\alpha^n}$ of T_pM spans an infinitesimal parallelepiped and $\mu_\alpha(p)$ is its "volume". A change in coordinates should be thought of as a change in the measuring units. The gluing rules describe how the numerical value of the volume changes from one choice of units to another.

The densities on a manifold resemble in many respects the differential forms of maximal degree. Denote by $\det TM = \Lambda^{\dim M}TM$ the determinant line bundle of TM. A density is a map

$$\mu : C^\infty(\det TM) \to C^\infty(M),$$

such that $\mu(fe) = |f|\mu(e)$, for all smooth functions $f : M \to \mathbb{R}$, and all $e \in C^\infty(\det TM)$. In particular, any smooth map $\phi : M \to N$ between manifolds of the same dimension induces a pullback transformation

$$\phi^* : C^\infty(|\Lambda|_N) \to C^\infty(|\Lambda|_M),$$

described by

$$(\phi^*\mu)(e) = \mu\big(\,(\det \phi_*) \cdot e\,\big) = |\det \phi_*|\,\mu(e), \ \ \forall e \in C^\infty(\det TM).$$

Example 3.4.2. (a) Consider the special case $M = \mathbb{R}^n$. Denote by $e_1, ..., e_n$ the canonical basis. This extends to a trivialization of $T\mathbb{R}^n$ and, in particular, the bundle of densities comes with a natural trivialization. It has a nowhere vanishing section $|dv_n|$ defined by

$$|dv_n|(e_1 \wedge ... \wedge e_n) = 1.$$

In this case, any smooth density on $\mathbb{R}^n$ takes the form $\mu = f|dv_n|$, where f is some smooth function on $\mathbb{R}^n$. The reader should think of $|dv_n|$ as the standard Lebesgue measure on $\mathbb{R}^n$.

If $\phi : \mathbb{R}^n \to \mathbb{R}^n$ is a smooth map, viewed as a collection of n smooth functions

$$\phi_1 = \phi_1(x^1, ..., x^n), \ldots, \phi_n = \phi_n(x^1, ..., x^n),$$

then,

$$\phi^*(|dv_n|) = \left|\det\left(\frac{\partial \phi_i}{\partial x^j}\right)\right| \cdot |dv_n|.$$

(b) Suppose M is a smooth manifold of dimension m. Then any top degree form $\omega \in \Omega^m(M)$ defines a density $|\omega|$ on M which associates to each section $e \in C^\infty(\det TM)$ the smooth function

$$x \mapsto |\omega_x(\,e(x)\,)|$$

Observe that $|\omega| = |-\omega|$, so this map $\Omega^m(M) \to C^\infty(|\Lambda|_M)$ *is not linear*.

(c) Suppose g is a Riemann metric on the smooth manifold M. The *volume density* defined by g is the density denoted by $|dV_g|$ which associates to each $e \in C^\infty(\det TM)$ the pointwise length

$$x \mapsto |e(x)|_g.$$

If $(U_\alpha, (x_\alpha^i))$ is an atlas of M, then on each U_α we have top degree forms

$$dx_\alpha := dx_\alpha^1 \wedge \cdots \wedge dx_\alpha^m,$$

to which we associate the density $|dx_\alpha|$. In the coordinates (x_α^i) the metric g can be described as

$$g = \sum_{i,j} g_{\alpha;ij} dx_\alpha^i \otimes dx_\alpha^j.$$

We denote by $|g_\alpha|$ the determinant of the symmetric matrix $g_\alpha = (g_{\alpha;ij})_{1 \le i,j \le m}$. Then the restriction of $|dV_g|$ to U_α has the description

$$|dV_g| = \sqrt{|g_\alpha|}\, |dx_\alpha|. \qquad \square$$

The importance of densities comes from the fact that they are exactly the objects that can be integrated. More precisely, we have the following abstract result.

Proposition 3.4.3. *There exists a natural way to associate to each smooth manifold M a linear map*

$$\int_M : C_0^\infty(|\Lambda|_M) \to \mathbb{R}$$

uniquely defined by the following conditions.

(a) $\int_M$ is invariant under diffeomorphisms, i.e., for any smooth manifolds M, N of the same dimension n, any diffeomorphism $\phi : M \to N$, and any $\mu \in C_0^\infty(|\Lambda|_M)$, we have

$$\int_M \phi^* \mu = \int_N \mu.$$

(b) $\int_M$ is a local operation, i.e., for any open set $U \subset M$, and any $\mu \in C_0^\infty(|\Lambda|_M)$ with $\operatorname{supp} \mu \subset U$, we have

$$\int_M \mu = \int_U \mu.$$

(c) For any $\rho \in C_0^\infty(\mathbb{R}^n)$ we have

$$\int_{\mathbb{R}^n} \rho |dv_n| = \int_{\mathbb{R}^n} \rho(x)\, dx,$$

where in the right-hand side stands the Lebesgue integral of the compactly supported function ρ. $\int_M$ is called the integral *on M.*

Proof. To establish the existence of an integral we associate to each manifold M a collection of data as follows.

(i) A smooth partition of unity $\mathcal{A} \subset C_0^\infty(M)$ such that $\forall \alpha \in \mathcal{A}$ the support $\operatorname{supp} \alpha$ lies entirely in some precompact coordinate neighborhood U_α, and such that the cover (U_α) is locally finite.

(ii) For each U_α we pick a collection of local coordinates (x_α^i), and we denote by $|dx_\alpha|$ $(n = \dim M)$ the density on U_α defined by

$$|dx_\alpha| \left(\frac{\partial}{\partial x_\alpha^1} \wedge \dots \wedge \frac{\partial}{\partial x_\alpha^n} \right) = 1.$$

For any $\mu \in C^\infty(|\Lambda|)$, the product $\alpha\mu$ is a density supported in U_α, and can be written as

$$\alpha\mu = \mu_\alpha |dx_\alpha|,$$

where μ_α is some smooth function compactly supported on U_α. The local coordinates allow us to interpret μ_α as a function on $\mathbb{R}^n$. Under this identification $|dx_\alpha^i|$ corresponds to the Lebesgue measure $|dv_n|$ on $\mathbb{R}^n$, and μ_α is a compactly supported, smooth function. We set

$$\int_{U_\alpha} \alpha\mu := \int_{\mathbb{R}^n} \mu_\alpha |dx_\alpha|.$$

Finally, define

$$\int_M^{\mathcal{A}} \mu = \int_M \mu \overset{def}{=} \sum_{\alpha \in \mathcal{A}} \int_{U_\alpha} \alpha\mu.$$

The above sum contains only finitely many nonzero terms since $\operatorname{supp} \mu$ is compact, and thus it intersects only finitely many of the $U'_\alpha s$ which form a locally finite cover.

To prove property (a) we will first prove that the integral defined as above is independent of the various choices: the partition of unity $\mathcal{A} \subset C_0^\infty(M)$, and the local coordinates $(x_\alpha^i)_{\alpha \in \mathcal{A}}$.

• **Independence of coordinates.** Fix the partition of unity $\mathcal{A}$, and consider a new collection of local coordinates (y_α^i) on each U_α. These determine two densities $|dx_\alpha^j|$ and respectively $|dy_\alpha^j|$. For each $\mu \in C_0^\infty(|\Lambda|_M)$ we have

$$\alpha\mu = \alpha \mu_\alpha^x |dx_\alpha| = \alpha \mu_\alpha^y |dy_\alpha|,$$

where $\mu_\alpha^x, \mu_\alpha^y \in C_0^\infty(U_\alpha)$ are related by

$$\mu_\alpha^y = \left| \det \left(\frac{\partial x_\alpha^i}{\partial y_\alpha^j} \right) \right| \mu_\alpha^x.$$

The equality

$$\int_{\mathbb{R}^n} \mu_\alpha^x |dx_\alpha| = \int_{\mathbb{R}^n} \mu_\alpha^y |dy_\alpha|$$

is the classical change in variables formula for the Lebesgue integral.

• **Independence of the partition of unity.** Let $\mathcal{A}, \mathcal{B} \subset C_0^\infty(M)$ two partitions of unity on M. We will show that

$$\int_M^{\mathcal{A}} = \int_M^{\mathcal{B}}.$$

Form the partition of unity

$$\mathcal{A} * \mathcal{B} := \left\{ \alpha\beta \, ; \, (\alpha, \beta) \in \mathcal{A} \times \mathcal{B} \right\} \subset C_0^\infty(M).$$

Note that $\operatorname{supp} \alpha\beta \subset U_{\alpha\beta} = U_\alpha \cap U_\beta$. We will prove

$$\int_M^\mathcal{A} = \int_M^{\mathcal{A}*\mathcal{B}} = \int_M^\mathcal{B}.$$

Let $\mu \in C_0^\infty(|\Lambda|_M)$. We can view $\alpha\mu$ as a compactly supported function on $\mathbb{R}^n$. We have

$$\int_{U_\alpha} \alpha\mu = \sum_\beta \int_{U_\alpha \subset \mathbb{R}^n} \beta\alpha\mu = \sum_\beta \int_{U_{\alpha\beta}} \alpha\beta\mu. \tag{3.4.1}$$

Similarly

$$\int_{U_\beta} \beta\mu = \sum_\alpha \int_{U_{\alpha\beta}} \alpha\beta\mu. \tag{3.4.2}$$

Summing (3.4.1) over α and (3.4.2) over β we get the desired conclusion.

To prove property (a) for a diffeomorphism $\phi : M \to N$, consider a partition of unity $\mathcal{A} \subset C_0^\infty(N)$. From the classical change in variables formula we deduce that, for any coordinate neighborhood U_α containing the support of $\alpha \in \mathcal{A}$, and any $\mu \in C_0^\infty(|\Lambda|_N)$ we have

$$\int_{\phi^{-1}(U_\alpha)} (\phi^* \alpha)\phi^* \mu = \int_{U_\alpha} \alpha\mu.$$

The collection $\left(\phi^* \alpha = \alpha \circ \phi\right)_{\alpha \in \mathcal{A}}$ forms a partition of unity on M. Property (a) now follows by summing over α the above equality, and using the independence of the integral on partitions of unity.

To prove property (b) on the local character of the integral, pick $U \subset M$, and then choose a partition of unity $\mathcal{B} \subset C_0^\infty(U)$ subordinated to the open cover $(V_\beta)_{\beta \in \mathcal{B}}$. For any partition of unity $\mathcal{A} \subset C_0^\infty(M)$ with associated cover $(V_\alpha)_{\alpha \in \mathcal{A}}$ we can form a new partition of unity $\mathcal{A} * \mathcal{B}$ of U with associated cover $V_{\alpha\beta} = V_\alpha \cap V_\beta$. We use this partition of unity to compute integrals over U. For any density μ on M supported on U we have

$$\int_M \mu = \sum_\alpha \int_{V_\alpha} \alpha\mu = \sum_{\alpha \in \mathcal{A}} \sum_{\beta \in \mathcal{B}} \int_{V_{\alpha\beta}} \alpha\beta\mu = \sum_{\alpha\beta \in \mathcal{A}*\mathcal{B}} \int_{V_{\alpha\beta}} \alpha\beta\mu = \int_U \mu.$$

Property (c) is clear since, for $M = \mathbb{R}^n$, we can assume that all the local coordinates chosen are Cartesian. The uniqueness of the integral is immediate, and we leave the reader to fill in the details. $\qquad\square$

3.4.2   *Orientability and integration of differential forms*

Under some mild restrictions on the manifold, the calculus with densities can be replaced with the richer calculus with differential forms. The mild restrictions referred to above have a *global nature*. More precisely, we have to require that the background manifold is *oriented*.

Roughly speaking, the oriented manifolds are the "2-sided manifolds", i.e., one can distinguish between an "inside face" and an "outside face" of the manifold. (Think of a 2-sphere in $\mathbb{R}^3$ (a soccer ball) which is naturally a "2-faced" surface.)

The 2-sidedness feature is such a frequent occurrence in the real world that for many years it was taken for granted. This explains the "big surprise" produced by the famous counter-example due to Möbius in the first half of the 19th century. He produced a 1-sided surface nowadays known as the Möbius band using paper and glue. More precisely, he glued the opposite sides of a paper rectangle attaching arrow to arrow as in Figure 3.2. The 2-sidedness can be formulated rigorously as

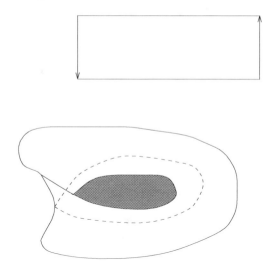

Fig. 3.2   *The Mobius band.*

follows.

Definition 3.4.4. A smooth manifold M is said to be *orientable* if the determinant line bundle $\det TM$ (or equivalently $\det T^*M$) is trivializable.   □

We see that $\det T^*M$ is trivializable if and only if it admits a nowhere vanishing section. Such a section is called a *volume form* on M. We say that two volume forms ω_1 and ω_2 are equivalent if there exists $f \in C^\infty(M)$ such that

$$\omega_2 = e^f \omega_1.$$

This is indeed an equivalence relation, and an equivalence class of volume forms will be called an *orientation* of the manifold. We denote by $\boldsymbol{Or}(M)$ the set of orientations on the smooth manifold M. A pair (*orientable manifold, orientation*) is called an *oriented* manifold.

Let us observe that if M is orientable, and thus $\boldsymbol{Or}(M) \neq \emptyset$, then for every point $p \in M$ we have a natural map

$$\boldsymbol{Or}(M) \to \boldsymbol{Or}(T_pM), \quad \boldsymbol{Or}(M) \ni \boldsymbol{or} \mapsto \boldsymbol{or}_p \in \boldsymbol{Or}(T_pM),$$

defined as follows. If the orientation $\boldsymbol{or}$ on M is defined by a volume form ω, then $\omega_p \in \det T_p^*M$ is a nontrivial volume form on T_pM, which canonically defines an orientation $\boldsymbol{or}_{\omega,p}$ on T_pM. It is clear that if ω_1 and ω_2 are equivalent volume form then $\boldsymbol{or}_{\omega_1,p} = \boldsymbol{or}_{\omega_2,p}$.

This map is clearly a surjection because $\boldsymbol{or}_{-\omega,p} = -\boldsymbol{or}_{\omega,p}$, for any volume form ω.

Proposition 3.4.5. *If M is a connected, orientable smooth manifold M, then for every $p \in M$ the map*

$$\boldsymbol{Or}(M) \ni \boldsymbol{or} \mapsto \boldsymbol{or}_p \in \boldsymbol{Or}(T_pM)$$

is a bijection.

Proof. Suppose $\boldsymbol{or}$ and $\boldsymbol{or}'$ are two orientations on M such that $\boldsymbol{or}_p = \boldsymbol{or}'_p$. The function

$$M \ni q \mapsto \epsilon(q) = \boldsymbol{or}'_q / \boldsymbol{or}_q \in \{\pm 1\}$$

is *continuous*, and thus *constant*. In particular, $\epsilon(q) = \epsilon(p) = 1, \forall q \in M$.

If $\boldsymbol{or}$ is given by the volume form ω and $\boldsymbol{or}'$ is given by the volume form ω', then there exists a *nowhere vanishing smooth function* $\rho : M \to \mathbb{R}$ such that $\omega' = \rho\omega$. We deduce

$$\operatorname{sign} \rho(q) = \epsilon(q), \quad \forall q \in M.$$

This shows that the two forms ω' and ω are equivalent and thus $\boldsymbol{or} = \boldsymbol{or}'$. □

The last proposition shows that on a *connected, orientable* manifold, a choice of an orientation of one of its tangent spaces uniquely determines an orientation of the manifold. A natural question arises.

How can one decide whether a given manifold is orientable or not.

We see this is just a special instance of the more general question we addressed in Chapter 2: how can one decide whether a given vector bundle is trivial or not. The orientability question can be given a very satisfactory answer using topological techniques. However, it is often convenient to decide the orientability issue using ad-hoc arguments. In the remaining part of this section we will describe several simple ways to detect orientability.

Example 3.4.6. If the tangent bundle of a manifold M is trivial, then clearly TM is orientable. In particular, all Lie groups are orientable. □

Example 3.4.7. Suppose M is a manifold such that the Whitney sum $\underline{\mathbb{R}}^k_M \oplus TM$ is trivial. Then M is orientable. Indeed, we have

$$\det(\underline{\mathbb{R}}^k \oplus TM) = \det\underline{\mathbb{R}}^k \otimes \det TM.$$

Both $\det\underline{\mathbb{R}}^k$ and $\det(\underline{\mathbb{R}}^k \oplus TM)$ are trivial. We deduce $\det TM$ is trivial since

$$\det TM \cong \det(\underline{\mathbb{R}}^k \oplus TM) \otimes (\det\underline{\mathbb{R}}^k)^*.$$

This trick works for example when $M \cong S^n$. Indeed, let ν denote the normal line

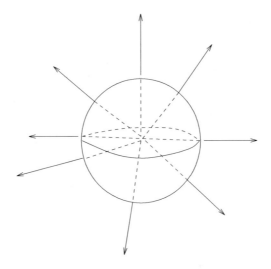

Fig. 3.3   *The normal line bundle to the round sphere.*

bundle. The fiber of ν at a point $p \in S^n$ is the 1-dimensional space spanned by the position vector of p as a point in $\mathbb{R}^n$; (see Figure 3.3). This is clearly a trivial line bundle since it has a tautological nowhere vanishing section $p \mapsto p \in \nu_p$. The line bundle ν has a remarkable feature:

$$\nu \oplus TS^n = \underline{\mathbb{R}}^{n+1}.$$

Hence all spheres are orientable. $\square$

☞**Important convention.**   The *canonical orientation* on $\mathbb{R}^n$ is the orientation defined by the volume form $dx^1 \wedge \cdots \wedge dx^n$, where $x^1, ..., x^n$ are the canonical Cartesian coordinates.

The unit sphere $S^n \subset \mathbb{R}^{n+1}$ is orientable. In the sequel we will exclusively deal with its *canonical orientation*. To describe this orientation it suffices to describe a positively oriented basis of $\det T_p M$ for some $p \in S^n$. To this aim we will use the relation

$$\mathbb{R}^{n+1} \cong \nu_p \oplus T_p S^n.$$

An element $\omega \in \det T_p S^n$ defines the canonical orientation if $\vec{p} \wedge \omega \in \det \mathbb{R}^{n+1}$ defines the canonical orientation of $\mathbb{R}^{n+1}$. Above, by $\vec{p}$ we denoted the position vector of p as a point inside the Euclidean space $\mathbb{R}^{n+1}$. We can think of $\vec{p}$ as the "outer" normal to the round sphere. We call this orientation *outer normal first*. When $n = 1$ it coincides with the counterclockwise orientation of the unit circle S^1. $\square$

Lemma 3.4.8. *A smooth manifold M is orientable if and only if there exists an open cover $(U_\alpha)_{\alpha \in \mathcal{A}}$, and local coordinates $(x_\alpha^1, ..., x_\alpha^n)$ on U_α such that*

$$\det \left(\frac{\partial x_\alpha^i}{\partial x_\beta^j} \right) > 0 \quad \text{on } U_\alpha \cap U_\beta. \tag{3.4.3}$$

Proof. 1. We assume that there exists an open cover with the properties in the lemma, and we will prove that $\det T^* M$ is trivial by proving that there exists a volume form.

Consider a partition of unity $\mathcal{B} \subset C_0^\infty(M)$ subordinated to the cover $(U_\alpha)_{\alpha \in \mathcal{A}}$, i.e., there exists a map $\varphi : \mathcal{B} \to \mathcal{A}$ such that

$$\operatorname{supp} \beta \subset U_{\varphi(\beta)} \quad \forall \beta \in \mathcal{B}.$$

Define

$$\omega := \sum_\beta \beta \omega_{\varphi(\beta)},$$

where for all $\alpha \in \mathcal{A}$ we define $\omega_\alpha := dx_\alpha^1 \wedge \cdots \wedge dx_\alpha^n$. The form ω is nowhere vanishing since condition (3.4.3) implies that on an overlap $U_{\alpha_1} \cap \cdots \cap U_{\alpha_m}$ the forms $\omega_{\alpha_1}, ..., \omega_{\alpha_m}$ differ by a *positive* multiplicative factor.

2. Conversely, let ω be a volume form on M and consider an atlas $(U_\alpha; (x_\alpha^i))$. Then

$$\omega |_{U_\alpha} = \mu_\alpha dx_\alpha^1 \wedge \cdots \wedge dx_\alpha^n,$$

where the smooth functions μ_α are nowhere vanishing, and on the overlaps they satisfy the gluing condition

$$\Delta_{\alpha\beta} = \det \left(\frac{\partial x_\alpha^i}{\partial x_\beta^j} \right) = \frac{\mu_\beta}{\mu_\alpha}.$$

A permutation φ of the variables $x_\alpha^1, ..., x_\alpha^n$ will change $dx_\alpha^1 \wedge \cdots \wedge dx_\alpha^n$ by a factor $\epsilon(\varphi)$ so we can always arrange these variables in such an order so that $\mu_\alpha > 0$. This will insure the positivity condition

$$\Delta_{\alpha\beta} > 0.$$

The lemma is proved. $\square$

We can rephrase the result in the above lemma in a more conceptual way using the notion of *orientation bundle*. Suppose $E \to M$ is a *real* vector bundle of rank r on the smooth manifold M described by the open cover $(U_\alpha)_{\alpha \in \mathcal{A}}$, and gluing cocycle

$$g_{\beta\alpha} : U_{\alpha\beta} \to \mathrm{GL}(r, \mathbb{R}).$$

The *orientation bundle associated to E* is the real line bundle $\Theta(E) \to M$ described by the open cover $(U_\alpha)_{\alpha \in \mathcal{A}}$, and gluing cocycle

$$\epsilon_{\beta\alpha} := \mathrm{sign}\, \det g_{\beta\alpha} : U_{\alpha\beta} \to \mathbb{R}^* = \mathrm{GL}(1, \mathbb{R}).$$

We define *orientation bundle* Θ_M of a smooth manifold M as the orientation bundle associated to the tangent bundle of M, $\Theta_M := \Theta(TM)$.

The statement in Lemma 3.4.8 can now be rephrased as follows.

Corollary 3.4.9. *A smooth manifold M is orientable if and only if the orientation bundle Θ_M is trivializable.* $\square$

From Lemma 3.4.8 we deduce immediately the following consequence.

Proposition 3.4.10. *The connected sum of two orientable manifolds is an orientable manifold.* $\square$

Exercise 3.4.11. Prove the above result. $\square$

Using Lemma 3.4.8 and Proposition 2.2.76 we deduce the following result.

Proposition 3.4.12. *Any complex manifold is orientable. In particular, the complex Grassmannians $\mathbf{Gr}_k(\mathbb{C}^n)$ are orientable.* $\square$

Exercise 3.4.13. Supply the details of the proof of Proposition 3.4.12. $\square$

The reader can check immediately that the product of two orientable manifolds is again an orientable manifold. Using connected sums and products we can now produce many examples of manifolds. In particular, the connected sums of g tori is an orientable manifold.

By now the reader may ask where does orientability interact with integration. The answer lies in Subsection 2.2.4 where we showed that an orientation $\boldsymbol{or}$ on a vector space V induces a *canonical, linear isomorphism* $\imath_{\boldsymbol{or}} : \det V^* \to |\Lambda|_V$; see (2.2.13).

Similarly, an orientation $\boldsymbol{or}$ on a smooth manifold M defines an isomorphism

$$\imath_{\boldsymbol{or}} : C^\infty(\det T^*M) \to C^\infty(|\Lambda|_M).$$

For any compactly supported differential form ω on M of maximal degree we define its integral by

$$\int_M \omega := \int_M \imath_{or}\omega.$$

We want to emphasize that this definition *depends on the choice of orientation*.

We ought to pause and explain in more detail the isomorphism $\imath_{or} : C^\infty(\det T^*M) \to C^\infty(|\Lambda|_M)$. Since M is oriented we can choose a coordinate atlas $(U_\alpha,\ (x_\alpha^i))$ such that

$$\det\left[\frac{\partial x_\beta^i}{\partial x_\alpha^j}\right]_{1\leq i,j\leq M} > 0, \quad n = \dim M, \tag{3.4.4}$$

and on each coordinate patch U_α the orientation is given by the top degree from $dx_\alpha = dx_\alpha^1 \wedge \cdots \wedge dx_\alpha^n$.

The differential form is described by a collection of forms

$$\omega_\alpha = \rho_\alpha dx_\alpha, \quad \rho_\alpha \in C^\infty(U_\alpha),$$

and due to the condition (3.4.4) the collection of desities

$$\mu_\alpha = \rho_\alpha |dx_\alpha| \in C^\infty(U_\alpha, |\,|\Lambda|\,|_M)$$

satisfy $\mu_\alpha = \mu_\beta$ on the overlap $U_{\alpha\beta}$. Thus they glue together to a density on M, which is precisely $\imath_{or}\omega$.

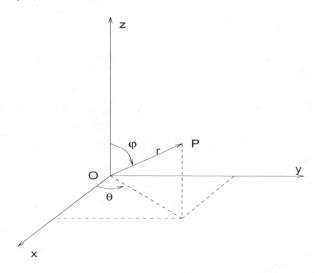

Fig. 3.4 *Spherical coordinates.*

Example 3.4.14. Consider the 2-form on $\mathbb{R}^3$, $\omega = xdy \wedge dz$, and let S^2 denote the unit sphere. We want to compute $\int_{S^2} \omega|_{S^2}$, where S^2 has the canonical orientation.

To compute this integral we will use spherical coordinates (r, φ, θ). These are defined by (see Figure 3.4)

$$\begin{cases} x = r \sin \varphi \cos \theta \\ y = r \sin \varphi \sin \theta \\ z = r \cos \varphi \end{cases}.$$

At the point $p = (1, 0, 0)$ we have

$$\partial_r = \partial_x = \vec{p}, \ \ \partial_\theta = \partial_y \ \ \partial_\varphi = -\partial_z,$$

so that the standard orientation on S^2 is given by $d\varphi \wedge d\theta$. On S^2 we have $r \equiv 1$ and $dr \equiv 0$ so that

$$xdy \wedge dz \mid_{S^2} = \sin \varphi \cos \theta \left(\cos \theta \sin \varphi d\theta + \sin \theta \cos \varphi d\varphi) \right) \wedge (-\sin \varphi) d\varphi$$

$$= \sin^3 \varphi \cos^2 \theta d\varphi \wedge d\theta.$$

The standard orientation associates to this form de density $\sin^3 \varphi \cos^2 \theta |d\varphi d\theta|$, and we deduce

$$\int_{S^2} \omega = \int_{[0,\pi] \times [0,2\pi]} \sin^3 \varphi \cos^2 \theta |d\varphi d\theta| = \int_0^\pi \sin^3 \varphi d\varphi \cdot \int_0^{2\pi} \cos^2 \theta d\theta$$

$$= \frac{4\pi}{3} = \text{volume of the unit ball } B^3 \subset \mathbb{R}^3.$$

As we will see in the next subsection the above equality is no accident. $\square$

Example 3.4.15. (Invariant integration on compact Lie groups). Let G be a compact, connected Lie group. Fix once and for all an orientation on the Lie algebra $\mathcal{L}_G$. Consider a positively oriented volume element $\omega \in \det \mathcal{L}_G^*$. We can extend ω by left translations to a left-invariant volume form on G which we continue to denote by ω. This defines an orientation, and in particular, by integration, we get a positive scalar

$$c = \int_G \omega.$$

Set $dV_G = \frac{1}{c}\omega$ so that

$$\int_G dV_G = 1. \tag{3.4.5}$$

The differential form dV_G is the unique left-invariant n-form $(n = \dim G)$ on G satisfying (3.4.5) (assuming a fixed orientation on G). We claim dV_G is also right invariant.

To prove this, consider the *modular function* $G \ni h \mapsto \Delta(h) \in \mathbb{R}$ defined by

$$R_h^*(dV_G) = \Delta(h)dV_G.$$

The quantity $\Delta(h)$ is independent of h because $R_h^* dV_G$ is a left invariant form, so it has to be a scalar multiple of dV_G. Since $(R_{h_1 h_2})^* = (R_{h_2} R_{h_1})^* = R_{h_1}^* R_{h_2}^*$ we deduce

$$\Delta(h_1 h_2) = \Delta(h_1)\Delta(h_2) \quad \forall h_1,\, h_2 \in G.$$

Hence $h \mapsto \Delta h$ is a *smooth* morphism

$$G \to (\mathbb{R} \setminus \{0\}, \cdot).$$

Since G is connected $\Delta(G) \subset \mathbb{R}_+$, and since G is compact, the set $\Delta(G)$ is bounded. If there exists $x \in G$ such that $\Delta(x) \neq 1$, then either $\Delta(x) > 1$, or $\Delta(x^{-1}) > 1$, and in particular, we would deduce the set $(\Delta(x^n))_{n \in \mathbf{Z}}$ is unbounded. Thus $\Delta \equiv 1$ which establishes the right invariance of dV_G.

The invariant measure dV_G provides a very simple way of producing invariant objects on G. More precisely, if T is tensor field on G, then for each $x \in G$ define

$$\overline{T}^{\ell}{}_x = \int_G ((L_g)_* T)_x dV_G(g).$$

Then $x \mapsto \overline{T}_x$ defines a smooth tensor field on G. We claim that $\overline{T}$ is left invariant. Indeed, for any $h \in G$ we have

$$(L_h)_* \overline{T}^{\ell} = \int_G (L_h)_* ((L_g)_* T) dV_G(g) = \int_G ((L_{hg})_* T) dV_G(g)$$

$$\stackrel{u=hg}{=} \int_G (L_u)_* T L_{h^{-1}}^* dV_G(u) = \overline{T}^{\ell} \quad (L_{h^{-1}}^* dV_G = dV_G).$$

If we average once more on the right we get a tensor

$$G \ni x \mapsto \int_G ((R_g)_* \overline{T}^{\ell})_x dV_G,$$

which is both left and right invariant. □

Exercise 3.4.16. Let G be a Lie group. For any $X \in \mathcal{L}_G$ denote by $\mathrm{ad}(X)$ the linear map $\mathcal{L}_G \to \mathcal{L}_G$ defined by

$$\mathcal{L}_G \ni Y \mapsto [X, Y] \in \mathcal{L}_G.$$

(a) If ω denotes a left invariant volume form prove that $\forall X \in \mathcal{L}_G$

$$L_X \omega = \mathrm{tr}\, \mathrm{ad}(X)\omega.$$

(b) Prove that if G is a compact Lie group, then $\mathrm{tr}\, \mathrm{ad}(X) = 0$, for any $X \in \mathcal{L}_G$. □

3.4.3 *Stokes' formula*

The Stokes' formula is the higher dimensional version of the fundamental theorem of calculus (Leibniz-Newton formula)

$$\int_a^b df = f(b) - f(a),$$

where $f : [a, b] \to \mathbb{R}$ is a smooth function and $df = f'(t)dt$. In fact, the higher dimensional formula will follow from the simplest 1-dimensional situation.

We will spend most of the time finding the correct formulation of the general version, and this requires the concept of *manifold with boundary*. The standard example is the lower half-space

$$\mathbf{H}_-^n = \{ (x^1, ..., x^n) \in \mathbb{R}^n ; x^1 \leq 0 \}.$$

Definition 3.4.17. A smooth *manifold with boundary* of dimension n is a topological space with the following properties.

(a) There exists a smooth n-dimensional manifold $\widetilde{M}$ that contains M as a closed subset.
(b) The interior of M, denoted by M^0, is non empty.
(c) For each point $p \in \partial M := M \setminus M^0$, there exist smooth local coordinates $(x^1, ..., x^n)$ defined on an open neighborhood $\mathcal{N}$ of p in $\widetilde{M}$ such that

(c_1) $M^0 \cap \mathcal{N} = \{ q \in \mathcal{N}; x^1(q) < 0 \}$.
(c_2) $\partial M \cap \mathcal{N} = \{ x^1 = 0 \}$.

The set ∂M is called the *boundary* of M. A manifold with boundary M is called orientable if its interior M^0 is orientable. $\square$

Example 3.4.18. (a) A closed interval $I = [a, b]$ is a smooth 1-dimensional manifold with boundary $\partial I = \{a, b\}$. We can take $\widetilde{M} = \mathbb{R}$.

(b) The closed unit ball $B^3 \subset \mathbb{R}^3$ is an orientable manifold with boundary $\partial B^3 = S^2$. We can take $\widetilde{M} = \mathbb{R}^3$.

(c) Suppose X is a smooth manifold, and $f : X \to \mathbb{R}$ is a smooth function such that $0 \in \mathbb{R}$ is a regular value of f, i.e.,

$$f(x) = 0 \Longrightarrow df(x) \neq 0.$$

Define $M := \{ x \in X; f(x) \leq 0 \}$. A simple application of the implicit function theorem shows that the pair (X, M) defines a manifold with boundary. Note that examples (a) and (b) are special cases of this construction. In the case (a) we take $X = \mathbb{R}$, and $f(x) = (x - a)(x - b)$, while in the case (b) we take $X = \mathbb{R}^3$ and $f(x, y, z) = (x^2 + y^2 + z^2) - 1$. $\square$

Definition 3.4.19. Two manifolds with boundary $M_1 \subset \widetilde{M}_1$, and $M_2 \subset \widetilde{M}_2$ are said to be *diffeomorphic* if, for every $i = 1, 2$ there exists an open neighborhood U_i of M_i in $\widetilde{M}_i$, and a diffeomorphism $F : U_1 \to U_2$ such that $F(M_1) = M_2$. $\square$

Exercise 3.4.20. Prove that any manifold with boundary is diffeomorphic to a manifold with boundary constructed via the process described in Example 3.4.18(c). $\square$

Proposition 3.4.21. *Let M be a smooth manifold with boundary. Then its boundary ∂M is also a smooth manifold of dimension $\dim \partial M = \dim M - 1$. Moreover, if M is orientable, then so is its boundary.* $\square$

The proof is left to the reader as an exercise.

☞**Important convention.** Let M be an orientable manifold with boundary. There is a (*non-canonical*) way to associate to an orientation on M^0 an orientation on the boundary ∂M. This will be the only way in which we will orient boundaries throughout this book. If we do not pay attention to this convention then our results may be off by a sign.

We now proceed to described this *induced orientation* on ∂M. For any $p \in \partial M$ choose local coordinates $(x^1, ..., x^n)$ as in Definition 3.4.17. Then the induced orientation of $T_p \partial M$ is defined by

$$\epsilon dx^2 \wedge \cdots \wedge dx^n \in \det T_p \partial M, \quad \epsilon = \pm 1,$$

where ϵ is chosen so that for $x^1 < 0$, i.e. inside M, the form

$$\epsilon dx^1 \wedge dx^2 \wedge \cdots \wedge dx^n$$

is positively oriented. The differential dx^1 is usually called an *outer conormal* since x^1 increases as we go towards the exterior of M. $-dx^1$ is then the *inner conormal* for analogous reasons. The rule by which we get the induced orientation on the boundary can be rephrased as

{outer conormal}$\wedge${induced orientation on boundary} = {orientation in the interior}.

We may call this rule *"outer (co)normal first"* for obvious reasons. $\square$

Example 3.4.22. The canonical orientation on $S^n \subset \mathbb{R}^{n+1}$ coincides with the induced orientation of S^{n+1} as the boundary of the unit ball B^{n+1}. $\square$

Exercise 3.4.23. Consider the hyperplane $H_i \subset \mathbb{R}^n$ defined by the equation $\{x^i = 0\}$. Prove that the induced orientation of H_i as the boundary of the half-space $\mathbf{H}^{n,i}_+ = \{x^i \geq 0\}$ is given by the $(n-1)$-form $(-1)^i dx^1 \wedge \cdots \wedge \widehat{dx^i} \wedge \cdots dx^n$ where, as usual, the hat indicates a missing term. $\square$

Theorem 3.4.24 (Stokes formula). *Let M be an oriented n-dimensional manifold with boundary ∂M and $\omega \in \Omega^{n-1}(M)$ a compactly supported form. Then*

$$\int_{M^0} d\omega = \int_{\partial M} \omega.$$

In the above formula d denotes the exterior derivative, and ∂M has the induced orientation.

Proof. Via partitions of unity the verification is reduced to the following two situations.

Case 1. *The $(n-1)$-form ω is compactly supported in $\mathbb{R}^n$. We have to show*

$$\int_{\mathbb{R}^n} d\omega = 0.$$

It suffices to consider only the special case

$$\omega = f(x)dx^2 \wedge \cdots \wedge dx^n,$$

where $f(x)$ is a compactly supported smooth function. The general case is a linear combination of these special situations. We compute

$$\int_{\mathbb{R}^n} d\omega = \int_{\mathbb{R}^n} \frac{\partial f}{\partial x^1} dx^1 \wedge \cdots \wedge dx^n = \int_{\mathbb{R}^{n-1}} \left(\int_{\mathbb{R}} \frac{\partial f}{\partial x^1} dx^1 \right) dx^2 \wedge \cdots \wedge dx^n = 0,$$

since

$$\int_{\mathbb{R}} \frac{\partial f}{\partial x^1} dx^1 = f(\infty, x^2, ..., x^n) - f(-\infty, x^2, ..., x^n),$$

and f has compact support.

Case 2. *The $(n-1)$-form ω is compactly supported in $\mathbf{H}^n_-$. Let*

$$\omega = \sum_i f_i(x)dx^1 \wedge \cdots \wedge \widehat{dx^i} \wedge \cdots \wedge dx^n.$$

Then

$$d\omega = \left(\sum_i (-1)^{i+1} \frac{\partial f}{\partial x^i} \right) dx^1 \wedge \cdots \wedge dx^n.$$

One verifies as in **Case 1** that

$$\int_{\mathbf{H}^n_-} \frac{\partial f}{\partial x^i} dx^1 \wedge \cdots \wedge dx^n = 0 \quad \text{for } i \neq 1.$$

For $i = 1$ we have

$$\int_{\mathbf{H}^n_+} \frac{\partial f}{\partial x^1} dx^1 \wedge \cdots \wedge dx^n = \int_{\mathbf{R}^{n-1}} \left(\int_{-\infty}^0 \frac{\partial f}{\partial x^1} dx^1 \right) dx^2 \wedge \cdots \wedge dx^n$$

$$= \int_{\mathbb{R}^{n-1}} \left(f(0, x^2, ..., x^n) - f(-\infty, x^2, ..., x^n) \right) dx^2 \wedge \cdots \wedge dx^n$$

$$= \int_{\mathbb{R}^{n-1}} f(0, x^2, ..., x^n) dx^2 \wedge \cdots \wedge dx^n = \int_{\partial \mathbf{H}^n_-} \omega.$$

The last equality follows from the fact that the induced orientation on $\partial \mathbf{H}^n_-$ is given by $dx^2 \wedge \cdots \wedge dx^n$. This concludes the proof of the Stokes formula. $\quad\square$

Remark 3.4.25. Stokes formula illustrates an interesting global phenomenon. It shows that the integral $\int_M d\omega$ is independent of the behavior of ω inside M. It only depends on the behavior of ω on the boundary. $\quad\square$

Example 3.4.26.

$$\int_{S^2} x dy \wedge dz = \int_{B^3} dx \wedge dy \wedge dz = \mathrm{vol}\,(B^3) = \frac{4\pi}{3}. \qquad\square$$

Remark 3.4.27. The above considerations extend easily to more singular situations. For example, when M is the cube $[0,1]^n$ its topological boundary is no longer a smooth manifold. However, its singularities are inessential as far as integration is concerned. The Stokes formula continues to hold

$$\int_{I^n} d\omega = \int_{\partial I} \omega \quad \forall \omega \in \Omega^{n-1}(I^n).$$

The boundary is smooth outside a set of measure zero and is given the induced orientation: " outer (co)normal first". The above equality can be used to give an explanation for the terminology "exterior derivative" we use to call d. Indeed if $\omega \in \Omega^{n-1}(\mathbb{R}^n)$ and $I_h = [0,h]$ then we deduce

$$d\omega \mid_{x=0} = \lim_{h \to 0} h^{-n} \int_{\partial I^n_h} \omega. \tag{3.4.6}$$

When $n = 1$ this is the usual definition of the derivative. $\quad\square$

Example 3.4.28. We now have sufficient technical background to describe an example of vector bundle which admits no flat connections, thus answering the question raised at the end of Section 3.3.3.

Consider the complex line bundle $L_n \to S^2$ constructed in Example 2.1.36. Recall that L_n is described by the open cover

$$S^2 = U_0 \cup U_1, \quad U_0 = S^2 \setminus \{\text{South pole}\}, \quad U_1 = S^2 \setminus \{\text{North pole}\},$$

and gluing cocycle

$$g_{10} : U_0 \cap U_1 \to \mathbb{C}^*, \quad g_{10}(z) = z^{-n} = g_{01}(z)^{-1},$$

where we identified the overlap $U_0 \cap U_1$ with the punctured complex line $\mathbb{C}^*$.

A connection on L_n is a collection of two complex valued forms $\omega_0 \in \Omega^1(U_1) \otimes \mathbb{C}$, $\omega_1 \in \Omega^1(U_1) \otimes \mathbb{C}$, satisfying a gluing relation on the overlap (see Example 3.3.7)

$$\omega_1 = -\frac{dg_{10}}{g_{10}} + \omega_0 = n\frac{dz}{z} + \omega_0.$$

If the connection is flat, then

$$d\omega_0 = 0 \text{ on } U_0 \text{ and } d\omega_1 = 0 \text{ on } U_1.$$

Let E^+ be the Equator equipped with the induced orientation as the boundary of the northern hemisphere, and E^- the equator with the opposite orientation, as the boundary of the southern hemisphere. The orientation of E^+ coincides with the orientation given by the form $d\theta$, where $z = \exp(i\theta)$.

We deduce from the Stokes formula (which works for complex valued forms as well) that

$$\int_{E^+} \omega_0 = 0 \quad \int_{E^+} \omega_1 = -\int_{E^-} \omega_1 = 0.$$

On the other hand, over the Equator, we have

$$\omega_1 - \omega_0 = n\frac{dz}{z} = nid\theta,$$

from which we deduce

$$0 = \int_{E^+} \omega_0 - \omega_1 = ni \int_{E^+} d\theta = 2n\pi i \text{ !!!}$$

Thus there exist no flat connections on the line bundle L_n, $n \neq 0$, and at fault is the gluing cocycle defining L. In a future chapter we will quantify the measure in which the gluing data obstruct the existence of flat connections. $\quad\square$

3.4.4   *Representations and characters of compact Lie groups*

The invariant integration on compact Lie groups is a very powerful tool with many uses. Undoubtedly, one of the most spectacular application is Hermann Weyl's computation of the characters of representations of compact semi-simple Lie groups. The invariant integration occupies a central place in his solution to this problem.

We devote this subsection to the description of the most elementary aspects of the representation theory of compact Lie groups.

Let G be a Lie group. Recall that a *(linear) representation* of G is a left action on a (finite dimensional) vector space V

$$G \times V \to V \quad (g, v) \mapsto T(g)v \in V,$$

such that the map $T(g)$ is linear for any g. One also says that V has a *structure of G-module*. If V is a real (respectively complex) vector space, then it is said to be a real (respectively complex) *G-module*.

Example 3.4.29. Let $V = \mathbb{C}^n$. Then $G = \mathrm{GL}(n, \mathbb{C})$ acts linearly on V in the tautological manner. Moreover V^*, $V^{\otimes k}$, $\Lambda^m V$ and $S^\ell V$ are complex G-modules. $\quad\square$

Example 3.4.30. Suppose G is a Lie group with Lie algebra $\mathcal{L}_G - T_1 G$. For every $g \in G$, the conjugation

$$C_g : G \to G, \quad h \mapsto C_g(h) = ghg^{-1},$$

sends the identity $1 \in G$ to itself. We denote by $\mathrm{Ad}_g : \mathcal{L}_G \to \mathcal{L}$ the differential of $h \mapsto C_g(h)$ at $h = 1$. The operator Ad_g is a linear isomorphism of $\mathcal{L}_G$, and the resulting map

$$G \mapsto \mathrm{Aut}(\mathcal{L}_G), \quad g \mapsto \mathrm{Ad}_g,$$

is called *adjoint representation* of G.

For example, if $G = SO(n)$, then $\mathcal{L}_G = \underline{so}(n)$, and the adjoint representation $\mathrm{Ad} : SO(n) \to \mathrm{Aut}(\underline{so}(n))$, can be given the more explicit description

$$\underline{so}(n) \ni X \xmapsto{\mathrm{Ad}(g)} gXg^{-1} \in \underline{so}(n), \quad \forall g \in SO(n). \qquad \square$$

Exercise 3.4.31. Let G be a Lie group, with Lie algebra $\mathcal{L}_G$. Prove that for any $X, Y \in \mathcal{L}_G$ we have

$$\frac{d}{dt}\Big|_{t=0} \mathrm{Ad}_{\exp(tX)}(Y) = \mathrm{ad}(X)Y = [X, Y]. \qquad \square$$

Definition 3.4.32. A *morphism of G-modules* V_1 and V_2 is a linear map $L : V_1 \to V_2$ such that, for any $g \in G$, the diagram below is commutative, i.e., $T_2(g)L = LT_1(g)$.

$$
\begin{array}{ccc}
V_1 & \xrightarrow{\ L\ } & V_2 \\
{\scriptstyle T_1(g)}\big\downarrow & & \big\downarrow{\scriptstyle T_2(g)} \\
V_1 & \xrightarrow{\ L\ } & V_2
\end{array}
$$

The space of morphisms of G-modules is denoted by $\mathrm{Hom}_G(V_1, V_2)$. The collection of isomorphisms classes of complex G-modules is denoted by $G - \boldsymbol{Mod}$. $\qquad \square$

If V is a G-module, then an *invariant subspace* (or submodule) is a subspace $U \subset V$ such that $T(g)(U) \subset U$, $\forall g \in G$. A G-module is said to be *irreducible* if it has no invariant subspaces other than $\{0\}$ and V itself.

Proposition 3.4.33. *The direct sum "$\oplus$", and the tensor product "$\otimes$" define a structure of semi-ring with 1 on $G - \boldsymbol{Mod}$. 0 is represented by the null representation $\{0\}$, while 1 is represented by the trivial module $G \to \mathrm{Aut}(\mathbb{C})$, $g \mapsto 1$.* $\qquad \square$

The proof of this proposition is left to the reader.

Example 3.4.34. Let $T_i : G \to \mathrm{Aut}(U_i)$ $(i = 1, 2)$ be two complex G-modules. Then U_1^* is a G-module given by $(g, u^*) \mapsto T_1(g^{-1})^\dagger u^*$. Hence $\mathrm{Hom}(U_1, U_2)$ is also a G-module. Explicitly, the action of $g \in G$ is given by

$$(g, L) \longmapsto T_2(g)LT_1(g^{-1}), \quad \forall L \in \mathrm{Hom}(U_1, U_2).$$

We see that $\mathrm{Hom}_G(U_1, U_2)$ can be identified with the linear subspace in $\mathrm{Hom}\,(U_1, U_2)$ consisting of the linear maps $U_1 \to U_2$ unchanged by the above action of G. $\qquad\square$

Proposition 3.4.35 (Weyl's unitary trick). *Let G be a compact Lie group, and V a complex G-module. Then there exists a Hermitian metric h on V which is G-invariant, i.e., $h(gv_1, gv_2) = h(v_1, v_2)$, $\forall v_1, v_2 \in V$.*

Proof. Let h be an arbitrary Hermitian metric on V. Define its G-average by

$$\overline{h}(u, v) := \int_G h(gu, gv) dV_G(g),$$

where $dV_G(g)$ denotes the normalized bi-invariant measure on G. One can now check easily that $\overline{h}$ is G-invariant. $\qquad\square$

In the sequel, G will always denote a compact Lie group.

Proposition 3.4.36. *Let V be a complex G-module and h a G-invariant Hermitian metric. If U is an invariant subspace of V then so is $U^\perp$, where "$\perp$" denotes the orthogonal complement with respect to h.*

Proof. Since h is G-invariant it follows that, $\forall g \in G$, the operator $T(g)$ is unitary, $T(g)T^*(g) = \mathbb{1}_V$. Hence, $T^*(g) = T^{-1}(g) = T(g^{-1})$, $\forall g \in G$.

If $x \in U^\perp$, then for all $u \in U$, and $\forall g \in G$

$$h(T(g)x, u) = h(x, T^*(g)u) = h(x, T(g^{-1})u) = 0.$$

Thus $T(g)x \in U^\perp$, so that $U^\perp$ is G-invariant. $\qquad\square$

Corollary 3.4.37. *Every G-module V can be decomposed as a direct sum of irreducible ones.* $\qquad\square$

If we denoted by $\boldsymbol{Irr}(G)$ the collection of isomorphism classes of irreducible G-modules, then we can rephrase the above corollary by saying that $\boldsymbol{Irr}(G)$ generates the semigroup $(G - \boldsymbol{Mod}, \oplus)$.

To gain a little more insight we need to use the following remarkable trick due to Isaac Schur.

Lemma 3.4.38 (Schur lemma). *Let V_1, V_2 be two irreducible complex G-modules. Then*

$$\mathrm{dim}_{\mathbb{C}} \, \mathrm{Hom}_G(V_1, V_2) = \begin{cases} 1 \text{ if } V_1 \cong V_2 \\ 0 \text{ if } V_1 \not\cong V_2 \end{cases}.$$

Proof. Let $L \in \mathrm{Hom}_G(V_1, V_2)$. Then $\ker L \subset V_1$ is an invariant subspace of V_1. Similarly, $\mathrm{Range}\,(L) \subset V_2$ is an invariant subspace of V_2. Thus, either $\ker L = 0$ or $\ker L = V_1$.

The first situation forces $\operatorname{Range} L \neq 0$ and, since V_2 is irreducible, we conclude $\operatorname{Range} L = V_2$. Hence, L has to be in isomorphism of G-modules. We deduce that, if V_1 and V_2 are not isomorphic as G-modules, then $\operatorname{Hom}_G(V_1, V_2) = \{0\}$.

Assume now that $V_1 \cong V_2$ and $S : V_1 \to V_2$ is an isomorphism of G-modules. According to the previous discussion, any other nontrivial G-morphism $L : V_1 \to V_2$ has to be an isomorphism. Consider the automorphism $T = S^{-1}L : V_1 \to V_1$. Since V_1 is a *complex* vector space T admits at least one (non-zero) eigenvalue λ.

The map $\lambda \mathbb{1}_{V_1} - T$ is an endomorphism of G-modules, and $\ker(\lambda \mathbb{1}_{V_1} - T) \neq 0$. Invoking again the above discussion we deduce $T \equiv \lambda \mathbb{1}_{V_1}$, i.e. $L \equiv \lambda S$. This shows $\dim \operatorname{Hom}_G(V_1, V_2) = 1$. $\qquad\square$

Schur's lemma is powerful enough to completely characterize $S^1 - \boldsymbol{Mod}$, the representations of S^1.

Example 3.4.39. (The irreducible (complex) representations of S^1). Let V be a complex irreducible S^1-module
$$S^1 \times V \ni (e^{i\theta}, v) \longmapsto T_\theta v \in V,$$
where $T_{\theta_1} \cdot T_{\theta_2} = T_{\theta_1 + \theta_2 \bmod 2\pi}$. In particular, this implies that each T_θ is an S^1-automorphism since it obviously commutes with the action of this group. Hence $T_\theta = \lambda(\theta)\mathbb{1}_V$ which shows that $\dim V = 1$ since any 1-dimensional subspace of V is S^1-invariant. We have thus obtained a smooth map
$$\lambda : S^1 \to \mathbb{C}^*,$$
such that
$$\lambda(e^{i\theta} \cdot e^{i\tau}) = \lambda(e^{i\theta})\lambda(e^{i\theta}).$$
Hence $\lambda : S^1 \to \mathbb{C}^*$ is a group morphism. As in the discussion of the modular function we deduce that $|\lambda| \equiv 1$. Thus, λ looks like an exponential, i.e., there exists $\alpha \in \mathbb{R}$ such that (*verify!*)
$$\lambda(e^{i\theta}) = \exp(i\alpha\theta), \quad \forall \theta \in \mathbb{R}.$$
Moreover, $\exp(2\pi i\alpha) = 1$, so that $\alpha \in \mathbb{Z}$.

Conversely, for any integer $n \in \mathbb{Z}$ we have a representation
$$S^1 \xrightarrow{\rho_n} \operatorname{Aut}(\mathbb{C}) \quad (e^{i\theta}, z) \mapsto e^{in\theta}z.$$
The exponentials $\exp(in\theta)$ are called the *characters* of the representations ρ_n. $\quad\square$

Exercise 3.4.40. Describe the irreducible representations of T^n-the n-dimensional torus. $\qquad\square$

Definition 3.4.41. (a) Let V be a complex G-module, $g \mapsto T(g) \in \operatorname{Aut}(V)$. The *character* of V is the smooth function
$$\chi_V : G \to \mathbb{C}, \quad \chi_V(g) := \operatorname{tr} T(g).$$
(b) A *class function* is a continuous function $f : G \to \mathbb{C}$ such that
$$f(hgh^{-1}) = f(g) \quad \forall g, h \in G.$$
(The character of a representation is an example of class function.) $\qquad\square$

Theorem 3.4.42. *Let G be a compact Lie group, U_1, U_2 complex G-modules and χ_{U_i} their characters. Then the following hold.*

(a)$\chi_{U_1 \oplus U_2} = \chi_{U_1} + \chi_{U_2}$, $\chi_{U_1 \otimes U_2} = \chi_{U_1} \cdot \chi_{U_1}$.

(b) $\chi_{U_i}(1) = \dim U_i$.

(c) $\chi_{U_i^*} = \overline{\chi}_{U_i}$*-the complex conjugate of* χ_{U_i}.

(d)

$$\int_G \chi_{U_i}(g) dV_G(g) = \dim U_i^G,$$

where U_i^G denotes the space of G-invariant elements of U_i,

$$U_i^G = \{x \in U_i \ ; \ x = T_i(g)x \ \ \forall g \in G\}.$$

(e)

$$\int_G \chi_{U_1}(g) \cdot \overline{\chi}_{U_2}(g) dV_G(g) = \dim \operatorname{Hom}_G(U_2, U_1).$$

Proof. The parts (a) and (b) are left to the reader. To prove (c), fix an invariant Hermitian metric on $U = U_i$. Thus, each $T(g)$ is a unitary operator on U. The action of G on U^* is given by $T(g^{-1})^\dagger$. Since $T(g)$ is unitary, we have $T(g^{-1})^\dagger = \overline{T(g)}$. This proves (c).

Proof of (d). Consider

$$P : U \to U, \quad Pu = \int_G T(g)u \, dV_G(g).$$

Note that $PT(h) = T(h)P$, $\forall h \in G$, i.e., $P \in \operatorname{Hom}_G(U, U)$. We now compute

$$T(h)Pu = \int_G T(hg)u \, dV_G(g) = \int_G T(\gamma)u \, R_{h^{-1}}^* dV_G(\gamma),$$

$$\int_G T(\gamma)u \, dV_G(\gamma) = Pu.$$

Thus, each Pu is G-invariant. Conversely, if $x \in U$ is G-invariant, then

$$Px = \int_G T(g)x dg = \int_G x \, dV_G(g) = x,$$

i.e., $U^G = \operatorname{Range} P$. Note also that P is a projector, i.e., $P^2 = P$. Indeed,

$$P^2 u = \int_G T(g)Pu \, dV_G(g) = \int_G Pu \, dV_G(g) = Pu.$$

Hence P is a projection onto U^G, and in particular

$$\dim_{\mathbb{C}} U^G = \operatorname{tr} P = \int_G \operatorname{tr} T(g) \, dV_G(g) = \int_G \chi_U(g) \, dV_G(g).$$

Proof of (e).

$$\int_G \chi_{U_1} \cdot \overline{\chi}_{U_2} dV_G(g) = \int_G \chi_{U_1} \cdot \chi_{U_2^*} dV_G(g) = \int_G \chi_{U_1 \otimes U_2^*} dV_G(g) = \int_G \chi_{\mathrm{Hom}(U_2, U_1)}$$
$$= \dim_{\mathbb{C}} \left(\mathrm{Hom}\,(U_2, U_1) \right)^G = \dim_{\mathbb{C}} \mathrm{Hom}_G(U_2, U_1),$$

since Hom_G coincides with the space of G-invariant morphisms. $\square$

Corollary 3.4.43. *Let U, V be irreducible G-modules. Then*

$$(\chi_U, \chi_V) = \int_G \chi_U \cdot \overline{\chi}_V dg = \delta_{UV} = \begin{cases} 1 \,, U \cong V \\ 0 \,, U \not\cong V \end{cases}.$$

Proof. Follows from Theorem 3.4.42 using Schur's lemma. $\square$

Corollary 3.4.44. *Let U, V be two G-modules. Then $U \cong V$ if and only if $\chi_U = \chi_V$.*

Proof. Decompose U and V as direct sums of irreducible G-modules

$$U = \oplus_1^m (m_i U_i) \quad V = \oplus_1^\ell (n_j V_j).$$

Hence $\chi_U = \sum m_i \chi_{V_i}$ and $\chi_V = \sum n_j \chi_{V_j}$. The equivalence "representation" $\Longleftrightarrow$ "characters" stated by this corollary now follows immediately from Schur's lemma and the previous corollary. $\square$

Thus, the problem of describing the representations of a compact Lie group boils down to describing the characters of its irreducible representations. This problem was completely solved by Hermann Weyl, but its solution requires a lot more work that goes beyond the scope of this book. We will spend the remaining part of this subsection analyzing the equality (d) in Theorem 3.4.42.

Describing the invariants of a group action was a very fashionable problem in the second half of the nineteenth century. Formula (d) mentioned above is a truly remarkable result. It allows (in principle) to compute the maximum number of linearly independent invariant elements.

Let V be a complex G-module and denote by χ_V its character. The complex exterior algebra $\Lambda_c^\bullet V^*$ is a complex G-module, as the space of complex multi-linear skew-symmetric maps

$$V \times \cdots \times V \to \mathbb{C}.$$

Denote by $b_k^c(V)$ the complex dimension of the space of G-invariant elements in $\Lambda_c^k V^*$. One has the equality

$$b_k^c(V) = \int_G \chi_{\Lambda_c^k V^*} dV_G(g).$$

These facts can be presented coherently by considering the $\mathbb{Z}$-graded vector space

$$\mathfrak{I}_c^\bullet(V) := \bigoplus_k \Lambda_{inv}^k V^*.$$

Its Poincaré polynomial is

$$P_{\mathfrak{I}_c^\bullet(V)}(t) = \sum t^k b_k^c(V) = \int_G t^k \chi_{\Lambda_c^k V^*} dV_G(g).$$

To obtain a more concentrated formulation of the above equality we need to recall some elementary facts of linear algebra.

For each endomorphism A of V denote by $\sigma_k(A)$ the trace of the endomorphism

$$\Lambda^k A : \Lambda^k V \to \Lambda^k V.$$

Equivalently, (see Exercise 2.2.25) the number $\sigma_k(A)$ is the coefficient of t^k in the characteristic polynomial

$$\sigma_t(A) = \det(\mathbb{1}_V + tA).$$

Explicitly, $\sigma_k(A)$ is given by the sum

$$\sigma_k(A) = \sum_{1 \leq i_1 \cdots i_k \leq n} \det\left(a_{i_\alpha i_\beta}\right) \quad (n = \dim V).$$

If $g \in G$ acts on V by $T(g)$, then g acts on $\Lambda^k V^*$ by $\Lambda^k T(g^{-1})^\dagger = \Lambda^k \overline{T(g)}$. (We implicitly assumed that each $T(g)$ is unitary with respect to some G-invariant metric on V.) Hence

$$\chi_{\Lambda_c^k V^*} = \sigma_k\left(\overline{T(g)}\right). \tag{3.4.7}$$

We conclude that

$$P_{\mathfrak{I}_c^\bullet(V)}(t) = \int_G \sum t^k \sigma_k\left(\overline{T(g)}\right) dV_G(g) = \int_G \det\left(\mathbb{1}_V + t\overline{T(g)}\right) dV_G(g). \tag{3.4.8}$$

Consider now the following situation. Let V be a *complex* G-module. Denote by $\Lambda_r^\bullet V$ the space of $\mathbb{R}$-multi-linear, skew-symmetric maps

$$V \times \cdots \times V \to \mathbb{R}.$$

The vector space $\Lambda_r^\bullet V^*$ is a *real* G-module. We complexify it, so that $\Lambda_r^\bullet V \otimes \mathbb{C}$ is the space of $\mathbb{R}$-multi-linear, skew-symmetric maps

$$V \times \cdots \times V \to \mathbb{C},$$

and as such, it is a *complex* G-module. The *real* dimension of the subspace $\mathfrak{I}_r^k(V)$ of G-invariant elements in $\Lambda_r^k V^*$ will be denoted by $b_k^r(V)$, so that the Poincaré polynomial of $\mathfrak{I}_r^\bullet(V) = \oplus_k \mathfrak{I}_r^k$ is

$$P_{\mathfrak{I}_r^\bullet(V)}(t) = \sum_k t^k b_k^r(V).$$

On the other hand, $b_k^r(V)$ is equal to the *complex* dimension of $\Lambda_r^k V^* \otimes \mathbb{C}$. Using the results of Subsection 2.2.5 we deduce

$$\Lambda_r^\bullet V \otimes \mathbb{C} \cong \Lambda_c^\bullet V^* \otimes_\mathbb{C} \Lambda_c^\bullet \overline{V}^* = \bigoplus_k \left(\oplus_{i+j=k} \Lambda_c^i V^* \otimes \Lambda_c^j \overline{V}^* \right). \tag{3.4.9}$$

Each of the above summands is a G-invariant subspace. Using (3.4.7) and (3.4.9) we deduce

$$P_{\mathfrak{I}_{\bullet}^{\bullet}(V)}(t) = \sum_k \int_G \sum_{i+j=k} \sigma_i\big(T(g)\big)\sigma_j\big(\overline{T(g)}\big)t^{i+j}\, dV_G(g)$$

$$= \int_G \det\big(\mathbb{1}_V + tT(g)\big)\det\big(\mathbb{1}_{\overline{V}} + t\overline{T}(g)\big)\, dV_G(g) \qquad (3.4.10)$$

$$= \int_G \big|\det\big(\mathbb{1}_V + tT(g)\big)\big|^2\, dV_G(g).$$

We will have the chance to use this result in computing topological invariants of manifolds with a "high degree of symmetry" like, e.g., the complex Grassmannians.

3.4.5 *Fibered calculus*

In the previous section we have described the calculus associated to objects defined on a *single manifold*. The aim of this subsection is to discuss what happens when we deal with an entire family of objects parameterized by some smooth manifold. We will discuss only the fibered version of integration. The exterior derivative also has a fibered version but its true meaning can only be grasped by referring to Leray's spectral sequence of a fibration and so we will not deal with it. The interested reader can learn more about this operation from [40], Chapter 3, Sec.5.

Assume now that, instead of a single manifold F, we have an entire (smooth) family of them $(F_b)_{b \in B}$. In more rigorous terms this means that we are given a smooth fiber bundle $p : E \to B$ with standard fiber F.

On the total space E we will always work with *split coordinates* $(x^i; y^j)$, where (x^i) are local coordinates on the standard fiber F, and (y^j) are local coordinates on the base B (the parameter space).

The model situation is the bundle

$$E = \mathbb{R}^k \times \mathbb{R}^m \xrightarrow{p} \mathbb{R}^m = B, \quad (x, y) \xmapsto{p} y.$$

We will first define a fibered version of integration. This requires a fibered version of orientability.

Definition 3.4.45. Let $p : E \to B$ be a smooth bundle with standard fiber F. The bundle is said to be *orientable* if the following hold.

(a) The manifold F is *orientable*;

(b) There exists an open cover (U_α), and trivializations $p^{-1}(U_\alpha) \xrightarrow{\Psi_\alpha} F \times U_\alpha$, such that the gluing maps

$$\Psi_\beta \circ \Psi_\alpha^{-1} : F \times U_{\alpha\beta} \to F \times U_{\alpha\beta} \quad (U_{\alpha\beta} = U_\alpha \cap U_\beta)$$

are fiberwise orientation preserving, i.e., for each $y \in U_{\alpha\beta}$, the diffeomorphism

$$F \ni f \mapsto \Psi_{\alpha\beta}(f, y) \in F$$

preserves any orientation on F. $\qquad\qquad \square$

Exercise 3.4.46. If the base B of an orientable bundle $p : E \to B$ is orientable, then so is the total space E (as an abstract smooth manifold). $\qquad\square$

☞**Important convention.**　Let $p : E \to B$ be an orientable bundle with oriented basis B. The *natural orientation* of the total space E is defined as follows.

If $E = F \times B$ then the orientation of the tangent space $T_{(f,b)}E$ is given by $\Omega_F \times \omega_B$, where $\omega_F \in \det T_f F$ (respectively $\omega_B \in \det T_b B$) defines the orientation of $T_f F$ (respectively $T_b B$).

The general case reduces to this one since any bundle is locally a product, and the gluing maps are fiberwise orientation preserving. This convention can be briefly described as

orientation total space = orientation fiber $\wedge$ orientation base.

The natural orientation can thus be called the *fiber-first* orientation. In the sequel all orientable bundles will be given the fiber-first orientation. $\qquad\square$

Let $p : E \to B$ be an orientable fiber bundle with standard fiber F.

Proposition 3.4.47. *There exists a linear operator*
$$p_* = \int_{E/B} : \Omega^\bullet_{cpt}(E) \to \Omega^{\bullet-r}_{cpt}(B), \quad r = \dim F,$$
uniquely defined by its action on forms supported on domains D of split coordinates
$$D \cong \mathbb{R}^r \times \mathbb{R}^m \xrightarrow{p} \mathbb{R}^m, \quad (x; y) \mapsto y.$$
If $\omega = f dx^I \wedge dy^J$, $f \in C^\infty_{cpt}(\mathbb{R}^{r+m})$, then
$$\int_{E/B} = \begin{cases} 0, & |I| \neq r \\ \left(\int_{\mathbb{R}^r} f dx^I \right) dy^J, & |I| = r \end{cases}.$$
The operator $\int_{E/B}$ is called the integration-along-fibers *operator.* $\qquad\square$

The proof goes *exactly* as in the non-parametric case, i.e., when B is a point. One shows using partitions of unity that these local definitions can be patched together to produce a well defined map
$$\int_{E/B} : \Omega^\bullet_{cpt}(E) \to \Omega^{\bullet-r}_{cpt}(B).$$
The details are left to the reader.

Proposition 3.4.48. *Let $p : E \to B$ be an orientable bundle with an r-dimensional standard fiber F. Then for any $\omega \in \Omega^*_{cpt}(E)$ and $\eta \in \omega^*_{cpt}(B)$ such that $\deg \omega + \deg \eta = \dim E$ we have*
$$\int_{E/B} d_E \omega = (-1)^r d_B \int_{E/B} \omega.$$
If B is oriented and ω, η are as above then
$$\int_E \omega \wedge p^*(\eta) = \int_B \left(\int_{E/B} \omega \right) \wedge \eta. \tag{Fubini}$$

The last equality implies immediately the *projection formula*

$$p_*(\omega \wedge p^*\eta) = p_*\omega \wedge \eta. \tag{3.4.11}$$

Proof. It suffices to consider only the model case

$$p : E = \mathbb{R}^r \times \mathbb{R}^m \to \mathbb{R}^m = B, \quad (x;y) \xrightarrow{p} y,$$

and $\omega = f dx^I \wedge dy^J$. Then

$$d_E\omega = \sum_i \frac{\partial f}{\partial x^i} dx^i \wedge dx^I \wedge dy^J + (-1)^{|I|} \sum_j \frac{\partial f}{\partial y^j} dx^I \wedge dy^j \wedge dy^J.$$

$$\int_{E/B} d_E\omega = \left(\int_{\mathbb{R}^r} \sum_i \frac{\partial f}{\partial x^i} dx^i \wedge dx^I \right) + (-1)^{|I|} \left(\int_{\mathbb{R}^r} \sum_j \frac{\partial f}{\partial y^j} dx^I \right) \wedge dy^j \wedge dy^J.$$

The above integrals are defined to be zero if the corresponding forms do not have degree r. Stokes' formula shows that the first integral is always zero. Hence

$$\int_{E/B} d_E\omega = (-1)^{|I|} \frac{\partial}{\partial y^j} \left(\int_{\mathbb{R}^r} \sum_j dx^I \right) \wedge dy^j \wedge dy^J = (-1)^r d_B \int_{E/B} \omega.$$

The second equality is left to the reader as an exercise. $\qquad\square$

Exercise 3.4.49. (Gelfand-Leray). Suppose that $p : E \to B$ is an oriented fibration, ω_E is a volume form on E, and ω_B is a volume form on B.

(a) Prove that, for every $b \in B$, there exists a unique volume form $\omega_{E/B}$ on $E_b = p^{-1}(b)$ with the property that, for every $x \in E_b$, we have

$$\omega_E(x) = \omega_{E/B}(x) \wedge (p^*\omega_B)(x) \in \Lambda^{\dim E} T_x^* E.$$

This form is called the *Gelfand-Leray residue* of ω_E rel p.

(b) Prove that for every compactly supported smooth function $f : E \to \mathbb{R}$ we have

$$(p_*(f\omega_E))_b = \left(\int_{E_b} f\omega_{E/B} \right) \omega_B(b), \forall b \in B, \quad \int_E f\omega_E = \int_B \left(\int_{E_b} f\omega_{E/B} \right) \omega_B.$$

(c) Consider the fibration $\mathbb{R}^2 \to \mathbb{R}$, $(x,y) \xrightarrow{p} t = ax + by$, $a^2 + b^2 \neq 0$. Compute the Gelfand-Leray residue $\frac{dx \wedge dy}{dt}$ along the fiber $p(x,y) = 0$. $\qquad\square$

Definition 3.4.50. A ∂-bundle is a collection $(E, \partial E, p, B)$ consisting of the following.

(a) A smooth manifold E with boundary ∂E.

(ii) A smooth map $p : E \to B$ such that the restrictions $p : \text{Int } E \to B$ and $p : \partial E \to B$ are smooth bundles.

The standard fiber of $p : \text{Int } E \to B$ is called the *interior fiber*. $\qquad\square$

One can think of a ∂-bundle as a smooth family of manifolds with boundary.

Example 3.4.51. The projection

$$p : [0,1] \times M \to M \quad (t; m) \mapsto m,$$

defines a ∂-bundle. The interior fiber is the open interval $(0,1)$. The fiber of $p : \partial(I \times M) \to M$ is the disjoint union of two points. $\square$

Standard Models   A ∂-bundle is obtained by gluing two types of local models.

- *Interior models* $\mathbb{R}^r \times \mathbb{R}^m \to \mathbb{R}^m$
- *Boundary models* $\mathbf{H}^r_+ \times \mathbb{R}^m \to \mathbb{R}^m$, where

$$\mathbf{H}^r_+ := \left\{ (x^1, \cdots, x^r) \in \mathbb{R}^r; \ x^1 \geq 0 \right\}.$$

Remark 3.4.52. Let $p : (E, \partial E) \to B$ be a ∂-bundle. If $p : \text{Int } E \to B$ is orientable and the basis B is oriented as well, then on ∂E one can define two orientations.

(i) The fiber-first orientation as the total space of an oriented bundle $\partial E \to B$.
(ii) The induced orientation as the boundary of E.

These two orientations coincide. $\square$

Exercise 3.4.53. Prove that the above orientations on ∂E coincide. $\square$

Theorem 3.4.54.   *Let $p : (E, \partial E) \to B$ be an orientable ∂-bundle with an r-dimensional interior fiber. Then for any $\omega \in \Omega^\bullet_{cpt}(E)$ we have*

$$\int_{\partial E/B} \omega = \int_{E/B} d_E \omega - (-1)^r d_B \int_{E/B} \omega \quad \text{(Homotopy formula)}. \qquad \square$$

The last equality can be formulated as

$$\int_{\partial E/B} = \int_{E/B} d_E - (-1)^r d_B \int_{E/B}.$$

This is "the mother of all homotopy formulæ". It will play a crucial part in Chapter 7 when we embark on the study of DeRham cohomology.

Exercise 3.4.55. Prove the above theorem. $\square$

Chapter 4

Riemannian Geometry

Now we can finally put to work the abstract notions discussed in the previous chapters. Loosely speaking, the Riemannian geometry studies the properties of surfaces (manifolds) "made of canvas". These are manifolds with an extra structure arising naturally in many instances. On such manifolds one can speak of the length of a curve, and the angle between two smooth curves. In particular, we will study the problem formulated in Chapter 1: why a plane (flat) canvas disk cannot be wrapped in an one-to-one fashinon around the unit sphere in $\mathbb{R}^3$. Answering this requires the notion of Riemann curvature which will be the central theme of this chapter.

4.1 Metric properties

4.1.1 *Definitions and examples*

To motivate our definition we will first try to formulate rigorously what do we mean by a "canvas surface".

A "canvas surface" can be deformed in many ways but with some limitations: it cannot be stretched as a rubber surface because the fibers of the canvas are flexible but not elastic. Alternatively, this means that the only operations we can perform are those which do not change the lengths of curves on the surface. Thus, one can think of "canvas surfaces" as those surfaces on which any "reasonable" curve has a well defined length.

Adapting a more constructive point of view, one can say that such surfaces are endowed with a clear procedure of measuring lengths of piecewise smooth curves.

Classical vector analysis describes one method of measuring lengths of smooth paths in $\mathbb{R}^3$. If $\gamma : [0,1] \to \mathbb{R}^3$ is such a path, then its length is given by

$$\text{length}\,(\gamma) = \int_0^1 |\dot{\gamma}(t)| dt,$$

where $|\dot{\gamma}(t)|$ is the Euclidean length of the tangent vector $\dot{\gamma}(t)$.

We want to do the same thing on an abstract manifold, and we are clearly faced with one problem: how do we make sense of the length $|\dot{\gamma}(t)|$? Obviously, this problem can be solved if we assume that there is a procedure of measuring lengths of tangent vectors at any point on our manifold. The simplest way to do achieve this is to assume that each tangent space is endowed with an inner product (which can vary from point to point in a smooth way).

Definition 4.1.1. (a) A *Riemann manifold* is a pair (M, g) consisting of a smooth manifold M and a metric g on the tangent bundle, i.e., a smooth, symmetric positive definite $(0, 2)$–tensor field on M. The tensor g is called a *Riemann metric* on M.
(b) Two Riemann manifolds (M_i, g_i) $(i = 1, 2)$ are said to be *isometric* if there exists a diffeomorphism $\phi : M_1 \to M_2$ such that $\phi^* g_2 = g_1$. $\square$

If (M, g) is a Riemann manifold then, for any $x \in M$, the restriction

$$g_x : T_x M \times T_x M \to \mathbb{R}$$

is an inner product on the tangent space $T_x M$. We will frequently use the alternative notation $(\bullet, \bullet)_x = g_x(\bullet, \bullet)$. The length of a tangent vector $v \in T_x M$ is defined as usual,

$$|v|_x := g_x(v, v)^{1/2}.$$

If $\gamma : [a, b] \to M$ is a piecewise smooth path, then we define its length by

$$l(\gamma) = \int_a^b |\dot{\gamma}(t)|_{\gamma(t)} dt.$$

If we choose local coordinates $(x^1, \ldots, x^n)$ on M, then we get a local description of g as

$$g = g_{ij} dx^i dx^j, \quad g_{ij} = g\left(\frac{\partial}{\partial x_i}, \frac{\partial}{\partial x_j}\right).$$

Proposition 4.1.2. *Let M be a smooth manifold, and denote by $\mathcal{R}_M$ the set of Riemann metrics on M. Then $\mathcal{R}_M$ is a non-empty convex cone in the linear space of symmetric $(0, 2)$–tensors.*

Proof. The only thing that is not obvious is that $\mathcal{R}_M$ is non-empty. We will use again partitions of unity. Cover M by coordinate neighborhoods $(U_\alpha)_{\alpha \in \mathcal{A}}$. Let (x^i_α) be a collection of local coordinates on U_α. Using these local coordinates we can construct by hand the metric g_α on U_α by

$$g_\alpha = (dx^1_\alpha)^2 + \cdots + (dx^n_\alpha)^2.$$

Now, pick a partition of unity $\mathcal{B} \subset C_0^\infty(M)$ subordinated to the cover $(U_\alpha)_{\alpha \in \mathcal{A}}$, i.e., there exits a map $\phi : \mathcal{B} \to \mathcal{A}$ such that $\forall \beta \in \mathcal{B}$ supp $\beta \subset U_{\phi(\beta)}$. Then define

$$g = \sum_{\beta \in \mathcal{B}} \beta g_{\phi(\beta)}.$$

The reader can check easily that g is well defined, and it is indeed a Riemann metric on M. $\square$

Example 4.1.3. (The Euclidean space). The space $\mathbb{R}^n$ has a natural Riemann metric

$$g_0 = (dx^1)^2 + \cdots + (dx^n)^2.$$

The geometry of $(\mathbb{R}^n, g_0)$ is the classical Euclidean geometry. $\square$

Example 4.1.4. (Induced metrics on submanifolds). Let (M, g) be a Riemann manifold and $S \subset M$ a submanifold. If $\imath : S \to M$ denotes the natural inclusion then we obtain by pull back a metric on S

$$g_S = \imath^* g = g|_S .$$

For example, any invertible symmetric $n \times n$ matrix defines a quadratic hypersurface in $\mathbb{R}^n$ by

$$\mathcal{H}_A = \big\{\, x \in \mathbb{R}^n \; ; \; (Ax, x) = 1 \big\},$$

where $(\bullet, \bullet)$ denotes the Euclidean inner product on $\mathbb{R}^n$. $\mathcal{H}_A$ has a natural metric induced by the Euclidean metric on $\mathbb{R}^n$. For example, when $A = I_n$, then $\mathcal{H}_{I_n}$ is the unit sphere in $\mathbb{R}^n$, and the induced metric is called the *round metric* of S^{n-1}. $\square$

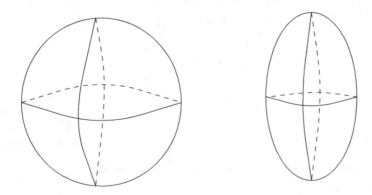

Fig. 4.1   *The unit sphere and an ellipsoid look "different".*

Remark 4.1.5. On any manifold there exist many Riemann metrics, and there is no natural way of selecting one of them. One can visualize a Riemann structure as defining a "shape" of the manifold. For example, the unit sphere $x^2 + y^2 + z^2 = 1$ is diffeomorphic to the ellipsoid $\frac{x^2}{1^2} + \frac{y^2}{2^2} + \frac{z^2}{3^2} = 1$, but they look "different " (see Figure 4.1). However, appearances may be deceiving. In Figure 4.2 it is illustrated the deformation of a sheet of paper to a half cylinder. They look different, but the

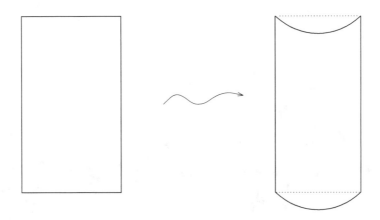

Fig. 4.2   *A plane sheet and a half cylinder are "not so different".*

metric structures are the same since we have not changed the lengths of curves on our sheet. The conclusion to be drawn from these two examples is that we have to be very careful when we use the attribute "different".   □

Example 4.1.6. (The hyperbolic plane).   The Poincaré model of the hyperbolic plane is the Riemann manifold $(\mathbf{D}, g)$ where $\mathbf{D}$ is the unit open disk in the plane $\mathbb{R}^2$ and the metric g is given by

$$g = \frac{1}{1 - x^2 - y^2}(dx^2 + dy^2).$$ □

Exercise 4.1.7.   Let $\mathcal{H}$ denote the upper half-plane
$$\mathcal{H} = \{(u, v) \in \mathbb{R}^2 \; ; \; v > 0\},$$
endowed with the metric
$$h = \frac{1}{4v^2}(du^2 + dv^2).$$
Show that the Cayley transform
$$z = x + iy \mapsto w = -i\frac{z + i}{z - i} = u + iv$$
establishes an isometry $(\mathbf{D}, g) \cong (\mathcal{H}, h)$.   □

Example 4.1.8. (Left invariant metrics on Lie groups).   Consider a Lie group G, and denote by $\mathcal{L}_G$ its Lie algebra. Then any inner product $\langle \cdot, \cdot \rangle$ on $\mathcal{L}_G$ induces a Riemann metric $h = \langle \cdot, \cdot \rangle_g$ on G defined by
$$h_g(X, Y) = \langle X, Y \rangle_g = \langle (L_{g^{-1}})_* X, (L_{g^{-1}})_* Y \rangle, \;\; \forall g \in G, \; X, Y \in T_gG,$$
where $(L_{g^{-1}})_* : T_gG \to T_1G$ is the differential at $g \in G$ of the left translation map $L_{g^{-1}}$. One checks easily that the correspondence
$$G \ni g \mapsto \langle \cdot, \cdot \rangle_g$$
is a smooth tensor field, and it is left invariant, i.e.,
$$L_g^* h = h \;\; \forall g \in G.$$
If G is also compact, we can use the averaging technique of Subsection 3.4.2 to produce metrics which are both left and right invariant.   □

4.1.2 The Levi-Civita connection

To continue our study of Riemann manifolds we will try to follow a close parallel with classical Euclidean geometry. The first question one may ask is whether there is a notion of "straight line" on a Riemann manifold.

In the Euclidean space $\mathbb{R}^3$ there are at least two ways to define a line segment.

(i) A line segment is the shortest path connecting two given points.
(ii) A line segment is a smooth path $\gamma : [0,1] \to \mathbb{R}^3$ satisfying

$$\ddot{\gamma}(t) = 0. \tag{4.1.1}$$

Since we have not said anything about calculus of variations which deals precisely with problems of type (i), we will use the second interpretation as our starting point. We will soon see however that both points of view yield the same conclusion.

Let us first reformulate (4.1.1). As we know, the tangent bundle of $\mathbb{R}^3$ is equipped with a natural trivialization, and as such, it has a natural trivial connection ∇^0 defined by

$$\nabla_i^0 \partial_j = 0 \quad \forall i,j, \quad \text{where } \partial_i := \frac{\partial}{\partial x_i}, \quad \nabla_i := \nabla_{\partial_i},$$

i.e., all the Christoffel symbols vanish. Moreover, if g_0 denotes the Euclidean metric, then

$$\left(\nabla_i^0 g_0\right)(\partial_j, \partial_k) = \nabla_i^0 \delta_{jk} - g_0(\nabla_i^0 \partial_j, \partial_k) - g_0(\partial_j, \nabla_i^0 \partial_k) = 0,$$

i.e., the connection is compatible with the metric. Condition (4.1.1) can be rephrased as

$$\nabla_{\dot{\gamma}(t)}^0 \dot{\gamma}(t) = 0, \tag{4.1.2}$$

so that the problem of defining "lines" in a Riemann manifold reduces to choosing a "natural" connection on the tangent bundle.

Of course, we would like this connection to be compatible with the metric, but even so, there are infinitely many connections to choose from. The following fundamental result will solve this dilemma.

Proposition 4.1.9. *Consider a Riemann manifold (M, g). Then there exists a unique symmetric connection ∇ on TM compatible with the metric g i.e.*

$$T(\nabla) = 0, \quad \nabla g = 0.$$

The connection ∇ is usually called the Levi-Civita connection *associated to the metric g.*

Proof. *Uniqueness.* We will achieve this by producing an *explicit* description of a connection with the above two mproperties.

Let ∇ be such a connection, i.e.,

$$\nabla g = 0 \quad \text{and} \quad \nabla_X Y - \nabla_Y X = [X, Y], \quad \forall X, Y \in \text{Vect}\,(M).$$

For any $X, Y, Z \in \mathrm{Vect}\,(M)$ we have

$$Zg(X,Y) = g(\nabla_Z X, Y) + g(X, \nabla_Z Y)$$

since $\nabla g = 0$. Using the symmetry of the connection we compute

$$Zg(X,Y) - Yg(Z,X) + Xg(Y,X) = g(\nabla_Z X, Y) - g(\nabla_Y Z, X) + g(\nabla_X Y, Z)$$

$$+ g(X, \nabla_Z Y) - g(Z, \nabla_Y X) + g(Y, \nabla_X Z)$$

$$= g([Z,Y], X) + g([X,Y], Z) + g([Z,X], Y) + 2g(\nabla_X Z, Y).$$

We conclude that

$$g(\nabla_X Z, Y) = \frac{1}{2}\{ Xg(Y,Z) - Yg(Z,X) + Zg(X,Y)$$

$$- g([X,Y], Z) + g([Y,Z], X) - g([Z,X], Y) \}. \tag{4.1.3}$$

The above equality establishes the uniqueness of ∇.

Using local coordinates $(x^1, \ldots, x^n)$ on M we deduce from (4.1.3), with $X = \partial_i = \frac{\partial}{\partial x_i}$, $Y = \partial_k = \frac{\partial}{\partial x_k}$, $Z = \partial_j = \frac{\partial}{\partial x_j}$), that

$$g(\nabla_i \partial_j, \partial_k) = g_{k\ell} \Gamma^\ell_{ij} = \frac{1}{2}\left(\partial_i g_{jk} - \partial_k g_{ij} + \partial_j g_{ik} \right).$$

Above, the scalars Γ^ℓ_{ij} denote the *Christoffel symbols* of ∇ in these coordinates, i.e.,

$$\nabla_{\partial_i} \partial_j = \Gamma^\ell_{ij} \partial_\ell.$$

If $(g^{i\ell})$ denotes the inverse of $(g_{i\ell})$ we deduce

$$\Gamma^\ell_{ij} = \frac{1}{2} g^{k\ell} \left(\partial_i g_{jk} - \partial_k g_{ij} + \partial_j g_{ik} \right). \tag{4.1.4}$$

Existence. It boils down to showing that (4.1.3) indeed defines a connection with the required properties. The routine details are left to the reader. $\qquad\square$

We can now define the notion of "straight line" on a Riemann manifold.

Definition 4.1.10. A *geodesic* on a Riemann manifold (M, g) is a smooth path

$$\gamma : (a, b) \to M,$$

satisfying

$$\nabla_{\dot\gamma(t)} \dot\gamma(t) = 0, \tag{4.1.5}$$

where ∇ is the Levi-Civita connection. $\qquad\square$

Using local coordinates $(x^1, ..., x^n)$ with respect to which the Christoffel symbols are (Γ_{ij}^k), and the path γ is described by $\gamma(t) = (x^1(t), ..., x^n(t))$, we can rewrite the geodesic equation as a second order, nonlinear system of ordinary differential equations. Set

$$\frac{d}{dt} = \dot\gamma(t) = \dot{x}^i \partial_i.$$

Then,

$$\nabla_{\frac{d}{dt}} \dot\gamma(t) = \ddot{x}^i \partial_i + \dot{x}^i \nabla_{\frac{d}{dt}} \partial_i = \ddot{x}^i \partial_i + \dot{x}^i \dot{x}^j \nabla_j \partial_i$$

$$= \ddot{x}^k \partial_k + \Gamma_{ji}^k \dot{x}^i \dot{x}^j \partial_k \quad (\Gamma_{ij}^k = \Gamma_{ji}^k),$$

so that the geodesic equation is equivalent to

$$\ddot{x}^k + \Gamma_{ij}^k \dot{x}^i \dot{x}^j = 0 \quad \forall k = 1, ..., n. \tag{4.1.6}$$

Since the coefficients $\Gamma_{ij}^k = \Gamma_{ij}^k(x)$ depend smoothly upon x, we can use the classical Banach-Picard theorem on existence in initial value problems (see e.g. [4]). We deduce the following local existence result.

Proposition 4.1.11. *Let (M, g) be a Riemann manifold. For any compact subset $K \subset TM$ there exists $\varepsilon > 0$ such that for any $(x, X) \in K$ there exists a unique geodesic $\gamma = \gamma_{x,X} : (-\varepsilon, \varepsilon) \to M$ such that $\gamma(0) = x$, $\dot\gamma(0) = X$.* □

One can think of a geodesic as defining a path in the tangent bundle $t \mapsto (\gamma(t), \dot\gamma(t))$. The above proposition shows that the geodesics define a local flow Φ on TM by

$$\Phi^t(x, X) = (\gamma(t), \dot\gamma(t)) \quad \gamma = \gamma_{x,X}.$$

Definition 4.1.12. The local flow defined above is called the *geodesic flow* of the Riemann manifold (M, g). When the geodesic glow is a global flow, i.e., any $\gamma_{x,X}$ is defined at each moment of time t for any $(x, X) \in TM$, then the Riemann manifold is called *geodesically complete*. □

The geodesic flow has some remarkable properties.

Proposition 4.1.13 (Conservation of energy). *If the path $\gamma(t)$ is a geodesic, then the length of $\dot\gamma(t)$ is independent of time.* □

Proof. We have

$$\frac{d}{dt} |\dot\gamma(t)|^2 = \frac{d}{dt} g(\dot\gamma(t), \dot\gamma(t)) = 2g(\nabla_{\dot\gamma(t)}, \dot\gamma(t)) = 0.$$ □

Thus, if we consider the sphere bundles

$$S_r(M) = \{X \in TM ; |X| = r\},$$

we deduce that $S_r(M)$ are invariant subsets of the geodesic flow.

Exercise 4.1.14. Describe the infinitesimal generator of the geodesic flow. □

Example 4.1.15. Let G be a connected Lie group, and let $\mathcal{L}_G$ be its Lie algebra. Any $X \in \mathcal{L}_G$ defines an endomorphism $\operatorname{ad}(X)$ of $\mathcal{L}_G$ by

$$\operatorname{ad}(X)Y := [X, Y].$$

The Jacobi identity implies that

$$\operatorname{ad}([X, Y]) = [\operatorname{ad}(X), \operatorname{ad}(Y)],$$

where the bracket in the right hand side is the usual commutator of two endomorphisms.

Assume that there exists an inner product $\langle \cdot, \cdot \rangle$ on $\mathcal{L}_G$ such that, for any $X \in \mathcal{L}_G$, the operator $\operatorname{ad}(X)$ is skew-adjoint, i.e.,

$$\langle [X, Y], Z \rangle = -\langle Y, [X, Z] \rangle. \tag{4.1.7}$$

We can now extend this inner product to a left invariant metric h on G. We want to describe its geodesics.

First, we have to determine the associated Levi-Civita connection. Using (4.1.3) we get

$$h(\nabla_X Z, Y) = \frac{1}{2}\{Xh(Y, Z) - Y(Z, X) + Zh(X, Y)$$
$$- h([X, Y], Z) + h([Y, Z], X) - h([Z, X], Y)\}.$$

If we take $X, Y, Z \in \mathcal{L}_G$, i.e., these vector fields are left invariant, then $h(Y, Z) = const.$, $h(Z, X) = const.$, $h(X, Y) = const.$ so that the first three terms in the above formula vanish. We obtain the following equality (at $1 \in G$)

$$\langle \nabla_X Z, Y \rangle = \frac{1}{2}\{-\langle [X, Y], Z \rangle + \langle [Y, Z], X \rangle - \langle [Z, X], Y \rangle\}.$$

Using the skew-symmetry of $\operatorname{ad}(X)$ and $\operatorname{ad}(Z)$ we deduce

$$\langle \nabla_X Z, Y \rangle = \frac{1}{2}\langle [X, Z], Y \rangle,$$

so that, at $1 \in G$, we have

$$\nabla_X Z = \frac{1}{2}[X, Z] \quad \forall X, Z \in \mathcal{L}_G. \tag{4.1.8}$$

This formula correctly defines a connection since any $X \in \operatorname{Vect}(G)$ can be written as a linear combination

$$X = \sum \alpha_i X_i \quad \alpha_i \in C^\infty(G) \quad X_i \in \mathcal{L}_G.$$

If $\gamma(t)$ is a geodesic, we can write $\dot{\gamma}(t) = \sum \gamma_i X_i$, so that

$$0 = \nabla_{\dot{\gamma}(t)}\dot{\gamma}(t) = \sum_i \dot{\gamma}_i X_i + \frac{1}{2}\sum_{i,j} \gamma_i \gamma_j [X_i, X_j].$$

Since $[X_i, X_j] = -[X_j, X_i]$, we deduce $\dot{\gamma}_i = 0$, i.e.,

$$\dot{\gamma}(t) = \sum \gamma_i(0) X_i = X.$$

This means that γ is an integral curve of the left invariant vector field X so that the geodesics through the origin with initial direction $X \in T_1 G$ are

$$\gamma_X(t) = \exp(tX). \qquad \square$$

Exercise 4.1.16. Let G be a Lie group and h a bi-invariant metric on G. Prove that its restriction to $\mathcal{L}_G$ satisfies (4.1.7). In particular, on any compact Lie groups there exist metrics satisfying (4.1.7). □

Definition 4.1.17. Let $\mathcal{L}$ be a finite dimensional real Lie algebra. The *Killing pairing* or *form* is the bilinear map

$$\kappa : \mathcal{L} \times \mathcal{L} \to \mathbb{R}, \ \kappa(x,y) := -\mathrm{tr}\left(\mathrm{ad}(x)\,\mathrm{ad}(y)\right) \ x, y \in \mathcal{L}.$$

The Lie algebra $\mathcal{L}$ is said to be *semisimple* if the Killing pairing is a duality. A Lie group G is called *semisimple* if its Lie algebra is semisimple. □

Exercise 4.1.18. Prove that $SO(n)$, $SU(n)$ and $SL(n,\mathbb{R})$ are semisimple Lie groups, but $U(n)$ is not. □

Exercise 4.1.19. Let G be a *compact* Lie group. Prove that the Killing form is positive semi-definite[1] and satisfies (4.1.7).
Hint: Use Exercise 4.1.16. □

Exercise 4.1.20. Show that the parallel transport of X along $\exp(tY)$ is

$$(L_{\exp(\frac{t}{2}Y)})_*(R_{\exp(\frac{t}{2}Y)})_*X. \quad \square$$

Example 4.1.21. (Geodesics on flat tori, and on $SU(2)$). The n-dimensional torus $T^n \cong S^1 \times \cdots \times S^1$ is an Abelian, compact Lie group. If $(\theta^1, ..., \theta^n)$ are the natural angular coordinates on T^n, then the flat metric is defined by

$$g = (d\theta^1)^2 + \cdots + (d\theta^n)^2.$$

The metric g on T^n is bi-invariant, and obviously, its restriction to the origin satisfies the skew-symmetry condition (4.1.7) since the bracket is 0. The geodesics through 1 will be the exponentials

$$\gamma_{\alpha_1,...,\alpha_n}(t) \ t \mapsto (e^{i\alpha_1 t}, ..., e^{i\alpha_n t}) \ \alpha_k \in \mathbb{R}.$$

If the numbers α_k are linearly dependent over $\mathbb{Q}$, then obviously $\gamma_{\alpha_1,...,\alpha_n}(t)$ is a closed curve. On the contrary, when the α's are linearly independent over $\mathbb{Q}$ then a classical result of Kronecker (see e.g. [42]) states that the image of $\gamma_{\alpha_1,...,\alpha_n}$ is dense in T^n!!! (see also Section 7.4 to come)

The special unitary group $SU(2)$ can also be identified with the group of unit quaternions

$$\left\{ a + b\boldsymbol{i} + c\boldsymbol{j} + d\boldsymbol{k} \ ; \ a^2 + b^2 + c^2 + d^2 = 1 \right\},$$

---

[1]The converse of the above exercise is also true, i.e., any semisimple Lie group with positive definite Killing form is compact. This is known as Weyl's theorem. Its proof, which will be given later in the book, requires substantially more work.

so that $SU(2)$ is diffeomorphic with the unit sphere $S^3 \subset \mathbb{R}^4$. The round metric on S^3 is bi-invariant with respect to left and right (unit) quaternionic multiplication (verify this), and its restriction to $(1, 0, 0, 0)$ satisfies (4.1.7). The geodesics of this metric are the 1-parameter subgroups of S^3, and we let the reader verify that these are in fact the great circles of S^3, i.e., the circles of maximal diameter on S^3. Thus, all the geodesics on S^3 are closed. $\square$

4.1.3 *The exponential map and normal coordinates*

We have already seen that there are many differences between the classical Euclidean geometry and the general Riemannian geometry in the large. In particular, we have seen examples in which one of the basic axioms of Euclidean geometry no longer holds: two distinct geodesic (read lines) may intersect in more than one point. The global topology of the manifold is responsible for this "failure".

Locally however, things are not "as bad". Local Riemannian geometry is similar in many respects with the Euclidean geometry. For example, locally, all of the classical incidence axioms hold.

In this section we will define using the metric some special collections of local coordinates in which things are very close to being Euclidean.

Let (M, g) be a Riemann manifold and U an open coordinate neighborhood with coordinates $(x^1, ..., x^n)$. We will try to find a local change in coordinates $(x^i) \mapsto (y^j)$ in which the expression of the metric is as close as possible to the Euclidean metric $g_0 = \delta_{ij} dy^i dy^j$.

Let $q \in U$ be the point with coordinates $(0, ..., 0)$. Via a linear change in coordinates we may as well assume that

$$g_{ij}(q) = \delta_{ij}.$$

We can formulate this by saying that (g_{ij}) is Euclidean up to order zero.

We would like to "spread" the above equality to an entire neighborhood of q. To achieve this we try to find local coordinates (y^j) near q such that in these new coordinates the metric is Euclidean up to order one, i.e.,

$$g_{ij}(q) = \delta_{ij} \quad \frac{\partial g_{ij}}{\partial y^k}(q) = \frac{\partial \delta_{ij}}{\partial y^k}(q) = 0, \quad \forall i, j, k.$$

We now describe a geometric way of producing such coordinates using the geodesic flow.

Denote as usual the geodesic from q with initial direction $X \in T_q M$ by $\gamma_{q,X}(t)$. Note the following simple fact.

$$\forall s > 0 \quad \gamma_{q,sX}(t) = \gamma_{q,X}(st).$$

Hence, there exists a small neighborhood V of $0 \in T_q M$ such that, for any $X \in V$, the geodesic $\gamma_{q,X}(t)$ is defined for all $|t| \leq 1$. We define the *exponential map* at q by

$$\exp_q : V \subset T_q M \to M, \quad X \mapsto \gamma_{q,X}(1).$$

The tangent space T_qM is a Euclidean space, and we can define $\mathbf{D}_q(r) \subset T_qM$, the open "disk" of radius r centered at the origin. We have the following result.

Proposition 4.1.22. *Let (M, g) and $q \in M$ as above. Then there exists $r > 0$ such that the exponential map*

$$\exp_q : \mathbf{D}_q(r) \to M$$

is a diffeomorphism onto. The supremum of all radii r with this property is denoted by $\rho_M(q)$. □

Definition 4.1.23. The positive real number $\rho_M(q)$ is called the *injectivity radius* of M at q. The infimum

$$\rho_M = \inf_q \rho_M(q)$$

is called the *injectivity radius* of M. □

The proof of Proposition 4.1.22 relies on the following key fact.

Lemma 4.1.24. *The Fréchet differential at $0 \in T_qM$ of the exponential map*

$$D_0 \exp_q : T_qM \to T_{\exp_q(0)}M = T_qM$$

is the identity $T_qM \to T_qM$.

Proof. Consider $X \in T_qM$. It defines a line $t \mapsto tX$ in T_qM which is mapped via the exponential map to the geodesic $\gamma_{q,X}(t)$. By definition

$$(D_0 \exp_q)X = \dot{\gamma}_{q,X}(0) = X.$$ □

Proposition 4.1.22 follows immediately from the above lemma using the inverse function theorem. □

Now choose an orthonormal frame $(\boldsymbol{e}_1, ..., \boldsymbol{e}_n)$ of T_qM, and denote by $(\boldsymbol{x}^1, ..., \boldsymbol{x}^n)$ the resulting cartesian coordinates in T_qM. For $0 < r < \rho_M(q)$, any point $p \in \exp_q(\mathbf{D}_q(r))$ can be uniquely written as

$$p = \exp_q(\boldsymbol{x}^i \boldsymbol{e}_i),$$

so that the collection $(\boldsymbol{x}^1, ..., \boldsymbol{x}^n)$ provides a coordinatization of the open set $\exp_q(\mathbf{D}_q(r)) \subset M$. The coordinates thus obtained are called *normal coordinates* at q, the open set $\exp_q(\mathbf{D}_q(r))$ is called a *normal neighborhood*, and will be denoted by $\mathbf{B}_r(q)$ for reasons that will become apparent a little later.

Proposition 4.1.25. *Let $(\boldsymbol{x}^i)$ be normal coordinates at $q \in M$, and denote by $\boldsymbol{g}_{ij}$ the expression of the metric tensor in these coordinates. Then we have*

$$\boldsymbol{g}_{ij}(q) = \delta_{ij} \text{ and } \frac{\partial \boldsymbol{g}_{ij}}{\partial \boldsymbol{x}^k}(q) = 0 \quad \forall i, j, k.$$

Thus, the normal coordinates provide a first order contact between g and the Euclidean metric.

Proof.   By construction, the vectors $e_i = \frac{\partial}{\partial x^i}$ form an orthonormal basis of $T_q M$ and this proves the first equality. To prove the second equality we need the following auxiliary result.

Lemma 4.1.26. *In normal coordinates (x_i) (at q) the Christoffel symbols Γ^i_{jk} vanish at q.*

Proof.   For any $(m^1, ..., m^n) \in \mathbb{R}^n$ the curve $x^i = m^i t$ is the geodesic $t \mapsto \exp_q \left(\sum m^i \frac{\partial}{\partial x^i} \right)$ so that

$$\Gamma^i_{jk}(x(t))m^j m^k = 0.$$

In particular,

$$\Gamma^i_{jk}(0)m^j m^k = 0 \quad \forall m^j \in \mathbb{R}^n$$

from which we deduce the lemma.   □

The result in the above lemma can be formulated as

$$g \left(\nabla_{\frac{\partial}{\partial x^j}} \frac{\partial}{\partial x^i}, \frac{\partial}{\partial x^k} \right) = 0, \quad \forall i, j, k$$

so that,

$$\nabla_{\frac{\partial}{\partial x^j}} \frac{\partial}{\partial x^i} = 0 \quad \text{at } q, \quad \forall i, j. \tag{4.1.9}$$

Using $\nabla g = 0$ we deduce $\frac{\partial g_{ij}}{\partial x^k}(q) = (\frac{\partial}{\partial x^k} g_{ij}) |_q = 0.$ □

The reader may ask whether we can go one step further, and find local coordinates which produce a second order contact with the Euclidean metric. At this step we are in for a big surprise. This thing is in general not possible and, in fact, there is a geometric way of measuring the "second order distance" between an arbitrary metric and the Euclidean metric. This is where the curvature of the Levi-Civita connection comes in, and we will devote an entire section to this subject.

4.1.4   *The length minimizing property of geodesics*

We defined geodesics via a second order equation imitating the second order equation defining lines in a Euclidean space. As we have already mentioned, this is not the unique way of extending the notion of Euclidean straight line to arbitrary Riemann manifolds.

One may try to look for curves of minimal length joining two given points. We will prove that the geodesics defined as in the previous subsection do just that, at least locally. We begin with a technical result which is interesting in its own. Let (M, g) be a Riemann manifold.

Lemma 4.1.27.   *For each $q \in M$ there exists $0 < r < \rho_M(q)$, and $\varepsilon > 0$ such that, $\forall m \in \mathbf{B}_r(q)$, we have $\varepsilon < \rho_M(m)$ and $\mathbf{B}_\varepsilon(m) \supset \mathbf{B}_r(q)$. In particular, any two points of $\mathbf{B}_r(q)$ can be joined by a* unique *geodesic of length $< \varepsilon$.*

We must warn the reader that the above result does not guarantee that the postulated connecting geodesic lies entirely in $\mathbf{B}_r(q)$. This is a different ball game.

Proof. Using the smooth dependence upon initial data in ordinary differential equations we deduce that there exists an open neighborhood V of $(q,0) \in TM$ such that $\exp_m X$ is well defined for all $(m,X) \in V$. We get a smooth map

$$F : V \to M \times M \quad (m,X) \mapsto (m, \exp_m X).$$

We compute the differential of F at $(q,0)$. First, using normal coordinates $(\boldsymbol{x}^i)$ near q we get coordinates $(\boldsymbol{x}^i; \boldsymbol{X}^j)$ near $(q,0) \in TM$. The partial derivatives of F at $(q,0)$ are

$$D_{(q,0)}F(\frac{\partial}{\partial \boldsymbol{x}^i}) = \frac{\partial}{\partial \boldsymbol{x}^i} + \frac{\partial}{\partial \boldsymbol{X}^i}, \quad D_{(q,0)}F(\frac{\partial}{\partial \boldsymbol{X}^i}) = \frac{\partial}{\partial \boldsymbol{X}^i}.$$

Thus, the matrix defining $D_{(q,0)}F$ has the form

$$\begin{bmatrix} \mathbb{1} & 0 \\ * & \mathbb{1} \end{bmatrix},$$

and in particular, it is nonsingular.

It follows from the implicit function theorem that F maps some neighborhood V of $(q,0) \in TM$ diffeomorphically onto some neighborhood U of $(q,q) \in M \times M$. We can choose V to have the form $\{(m,X) \; ; \; |X|_m < \varepsilon, \quad m \in \mathbf{B}_\delta(q)\}$ for some sufficiently small ε and δ. Choose $0 < r < \min(\varepsilon, \rho_M(q))$ such that

$$m_1, m_2 \in \mathbf{B}_r(q) \Longrightarrow (m_1, m_2) \in U.$$

In particular, we deduce that, for any $m \in \mathbf{B}_r(q)$, the map $\exp_m : \mathbf{D}_\varepsilon(m) \subset T_mM \to M$ is a diffeomorphism onto its image, and

$$\mathbf{B}_\varepsilon(m) = \exp_m(\mathbf{D}_\varepsilon(m)) \supset \mathbf{B}_r(q).$$

Clearly, for any $m \in M$, the curve $t \mapsto \exp_m(tX)$ is a geodesic of length $< \varepsilon$ joining m to $\exp_m(X)$. It is the *unique geodesic* with this property since $F : V \to U$ is injective. $\square$

We can now formulate the main result of this subsection.

Theorem 4.1.28. *Let q, r and ε as in the previous lemma, and consider the unique geodesic $\gamma : [0,1] \to M$ of length $< \varepsilon$ joining two points of $\mathbf{B}_r(q)$. If $\omega : [0,1] \to M$ is a piecewise smooth path with the same endpoints as γ then*

$$\int_0^1 |\dot{\gamma}(t)|dt \le \int_0^1 |\dot{\omega}(t)|dt.$$

with equality if and only if $\omega([0,1]) = \gamma([0,1])$. Thus, γ is the shortest path joining its endpoints.

The proof relies on two lemmata. Let $m \in M$ be an arbitrary point, and assume $0 < R < \rho_M(m)$.

Lemma 4.1.29 (Gauss). *In* $\mathbf{B}_R(m) \subset M$, *the geodesics through* m *are orthogonal to the hypersurfaces*

$$\Sigma_\delta = \exp_q(S_\delta(0)) = \left\{ \exp_m(X); \ |X| = \delta \right\}, \ 0 < \delta < R.$$

Proof. Let $t \mapsto X(t)$, $0 \le t \le 1$ denote a smooth curve in $T_m M$ such that $|X(t)|_m = 1$, i.e., $X(t)$ is a curve on the unit sphere $S_1(0) \subset T_m M$. We assume that the map $t \mapsto X(t) \in S_1(0)$ is an embedding, i.e., it is injective, and its differential is nowhere zero. We have to prove that the curve $t \mapsto \exp_m(\delta X(t))$ are orthogonal to the radial geodesics

$$s \mapsto \exp_m(sX(t)), \ 0 \le s \le R.$$

Consider the smooth map

$$f : [0, R] \times [0, 1] \to M, \ f(s, t) = \exp_m(sX(t)) \ (s, t) \in (0, R) \times (0, 1).$$

If we use normal coordinates on $B_R(m)$ we can express f as the embedding

$$(0, R) \times (0, 1) \to T_m M, \ (s, t) \mapsto sX(t).$$

Set

$$\partial_s := f_* \left(\frac{\partial}{\partial s} \right) \in T_{f(s,t)} M.$$

Define ∂_t similarly. The objects ∂_s, and ∂_t are sections of the restriction of TM to the image of f. Using the normal coordinates on $\mathbf{B}_R(m)$ we can think of ∂_s as a vector field on a region in $T_m M$, and as such we have the equality $\partial_s = X(t)$. We have to show

$$\langle \partial_s, \partial_t \rangle = 0 \ \ \forall (s, t),$$

where $\langle \cdot, \cdot \rangle$ denotes the inner product defined by g.

Using the metric compatibility of the Levi-Civita connection along the image of f we compute

$$\partial_s \langle \partial_s, \partial_t \rangle = \langle \nabla_{\partial_s} \partial_s, \partial_t \rangle + \langle \partial_s, \nabla_{\partial_s} \partial_t \rangle.$$

Since the curves $s \mapsto f(s, t = const.)$ are geodesics, we deduce

$$\nabla_{\partial_s} \partial_s = 0.$$

On the other hand, since $[\partial_s, \partial_t] = 0$, we deduce (using the symmetry of the Levi-Civita connection)

$$\langle \partial_s, \nabla_{\partial_s} \partial_t \rangle = \langle \partial_s, \nabla_{\partial_t} \partial_s \rangle = \frac{1}{2} \partial_t |\partial_s|^2 = 0,$$

since $|\partial_s| = |X(t)| = 1$. We conclude that the quantity $\langle \partial_s, \partial_t \rangle$ is independent of s. For $s = 0$ we have $f(0, t) = \exp_m(0)$ so that $\partial_t|_{s=0} = 0$, and therefore

$$\langle \partial_s, \partial_t \rangle = 0 \ \ \forall (s, t),$$

as needed. □

Now consider any continuous, piecewise smooth curve

$$\omega : [a, b] \to \mathbf{B}_R(m) \setminus \{m\}.$$

Each $\omega(t)$ can be uniquely expressed in the form

$$\omega(t) = \exp_m(\rho(t)X(t)) \quad |X(t)| = 1 \quad 0 < |\rho(t)| < R.$$

Lemma 4.1.30. *The length of the curve $\omega(t)$ is $\geq |\rho(b) - \rho(a)|$. The equality holds if and only if $X(t) = const$ and $\dot{\rho}(t) \geq 0$. In other words, the shortest path joining two concentrical shells Σ_δ is a radial geodesic.*

Proof. Let $f(\rho, t) := \exp_m(\rho X(t))$, so that $\omega(t) = f(\rho(t), t)$. Then

$$\dot{\omega} = \frac{\partial f}{\partial \rho}\dot{\rho} + \frac{\partial f}{\partial t}.$$

Since the vectors $\frac{\partial f}{\partial \rho}$ and $\frac{\partial f}{\partial t}$ are orthogonal, and since

$$\left|\frac{\partial f}{\partial \rho}\right| = |X(t)| = 1$$

we get

$$|\dot{\omega}|^2 = |\dot{\rho}|^2 + \left|\frac{\partial f}{\partial t}\right|^2 \geq |\dot{\rho}|^2.$$

The equality holds if and only if $\frac{\partial f}{\partial t} = 0$, i.e. $\dot{X} = 0$. Thus

$$\int_a^b |\dot{\omega}|dt \geq \int_a^b |\dot{\rho}|dt \geq |\rho(b) - \rho(a)|.$$

Equality holds if and only if $\rho(t)$ is monotone, and $X(t)$ is constant. This completes the proof of the lemma. □

The proof of Theorem 4.1.28 is now immediate. Let $m_0, m_1 \in \mathbf{B}_r(q)$, and consider a geodesic $\gamma : [0, 1] \to M$ of length $< \varepsilon$ such that $\gamma(i) = m_i$, $i = 0, 1$. We can write

$$\gamma(t) = \exp_{m_0}(tX) \quad X \in \mathbf{D}_\varepsilon(m_0).$$

Set $R = |X|$. Consider any other piecewise smooth path $\omega : [a, b] \to M$ joining m_0 to m_1. For any $\delta > 0$ this path must contain a portion joining the shell $\Sigma_\delta(m_0)$ to the shell $\Sigma_R(m_0)$ and lying between them. By the previous lemma the length of this segment will be $\geq R - \delta$. Letting $\delta \to 0$ we deduce

$$l(\omega) \geq R = l(\gamma).$$

If $\omega([a, b])$ does not coincide with $\gamma([0, 1])$ then we obtain a strict inequality. □

Any Riemann manifold has a natural structure of metric space. More precisely, we set

$$d(p,q) = \inf\Big\{ l(\omega); \quad \omega : [0,1] \to M \text{ piecewise smooth path joining } p \text{ to } q \Big\}.$$

A piecewise smooth path ω connecting two points p, q such that $l(\omega) = d(p,q)$ is said to be *minimal*. From Theorem 4.1.28 we deduce immediately the following consequence.

Corollary 4.1.31. *The image of any minimal path coincides with the image of a geodesic. In other words, any minimal path can be reparametrized such that it satisfies the geodesic equation.*

Exercise 4.1.32. Prove the above corollary.   □

Theorem 4.1.28 also implies that any two nearby points can be joined by a unique minimal geodesic. In particular we have the following consequence.

Corollary 4.1.33. *Let $q \in M$. Then for all $r > 0$ sufficiently small*

$$\exp_q(\mathbf{D}_r(0))(= \mathbf{B}_r(q)) = \{ p \in M \quad d(p,q) < r \}. \tag{4.1.10}$$

Corollary 4.1.34. *For any $q \in M$ we have the equality*

$$\rho_M(q) = \sup\{ r; \quad r \text{ satisfies } (4.1.10) \}.$$

Proof. The same argument used in the proof of Theorem 4.1.28 shows that for any $0 < r < \rho_M(q)$ the radial geodesics $\exp_q(tX)$ are minimal.   □

Definition 4.1.35. A subset $U \subset M$ is said to be *convex* if any two points in U can be joined by a *unique minimal* geodesic which lies entirely *inside* U.   □

Proposition 4.1.36. *For any $q \in M$ there exists $0 < R < \iota_M(q)$ such that for any $r < R$ the ball $\mathbf{B}_r(q)$ is convex.*

Proof. Choose ax $0 < \varepsilon < \frac{1}{2}\rho_M(q)$, and $0 < R < \varepsilon$ such that any two points m_0, m_1 in $B_R(q)$ can be joined by a unique minimal geodesic $[0,1] \ni t \mapsto \gamma_{m_0,m_1}(t)$ of length $< \varepsilon$, not necessarily contained in $\mathbf{B}_R(q)$. We will prove that $\forall m_0, m_1 \in \mathbf{B}_R(q)$ the map $t \mapsto d(q, \gamma_{m_0,m_1}(t))$ is convex and thus it achieves its maxima at the endpoints $t = 0, 1$. Note that

$$d(q, \gamma(t)) < R + \varepsilon < \rho_M(q).$$

The geodesic $\gamma_{m_0,m_1}(t)$ can be uniquely expressed as

$$\gamma_{m_0,m_1}(t) = \exp_q(r(t)X(t)) \quad X(t) \in T_q M \quad \text{with } r(t) = d(q, \gamma_{m_0,m_1}(t)).$$

It suffices to show $\frac{d^2}{dt^2}(r^2) \geq 0$ for $t \in [0,1]$ if $d(q, m_0)$ and $d(q, m_1)$ are sufficiently small.

At this moment it is convenient to use normal coordinates $(\boldsymbol{x}^i)$ near q. The geodesic γ_{m_0,m_1} takes the form $(\boldsymbol{x}^i(t))$, and we have

$$r^2 = (\boldsymbol{x}^1)^2 + \cdots + (\boldsymbol{x}^n)^2.$$

We compute easily

$$\frac{d^2}{dt^2}(r^2) = 2r^2(\ddot{\boldsymbol{x}}^1 + \cdots + \ddot{\boldsymbol{x}}^n) + |\dot{\boldsymbol{x}}|^2 \tag{4.1.11}$$

where $\dot{\boldsymbol{x}}(t) = \sum \dot{\boldsymbol{x}}^i \mathbf{e}_i \in T_q M$. The path γ satisfies the equation

$$\ddot{\boldsymbol{x}}^i + \Gamma^i_{jk}(\boldsymbol{x})\dot{\boldsymbol{x}}^j \dot{\boldsymbol{x}}^k = 0.$$

Since $\Gamma^i_{jk}(0) = 0$ (normal coordinates), we deduce that there exists a constant $C > 0$ (depending only on the magnitude of the second derivatives of the metric at q) such that

$$|\Gamma^i_{jk}(\boldsymbol{x})| \le C|x|.$$

Using the geodesic equation we obtain

$$\ddot{\boldsymbol{x}}^i \ge -C|\boldsymbol{x}||\dot{\boldsymbol{x}}|^2.$$

We substitute the above inequality in (4.1.11) to get

$$\frac{d^2}{dt^2}(r^2) \ge 2|\dot{\boldsymbol{x}}|^2 \left(1 - nC|\boldsymbol{x}|^3\right). \tag{4.1.12}$$

If we choose from the very beginning

$$R + \varepsilon \le (nC)^{-1/3},$$

then, because along the geodesic we have $|x| \le R+\varepsilon$, the right-hand side of (4.1.12) is nonnegative. This establishes the convexity of $t \mapsto r^2(t)$ and concludes the proof of the proposition. $\qquad \square$

In the last result of this subsection we return to the concept of geodesic completeness. We will see that this can be described in terms of the metric space structure alone.

Theorem 4.1.37 (Hopf-Rinow). *Let M be a Riemann manifold and $q \in M$. The following assertions are equivalent:*
(a) $\exp_q$ is defined on all of $T_q M$.
(b) The closed and bounded (with respect to the metric structure) sets of M are compact.
(c) M is complete as a metric space.
(d) M is geodesically complete.
(e) There exists a sequence of compact sets $K_n \subset M$, $K_n \subset K_{n+1}$ and $\bigcup_n K_n = M$ such that if $p_n \notin K_n$ then $d(q, p_n) \to \infty$.
Moreover, on a (geodesically) complete manifold any two points can be joined by a minimal geodesic. $\qquad \square$

Remark 4.1.38. On a complete manifold there could exist points (sufficiently far apart) which can be joined by more than one minimal geodesic. Think for example of a manifold where there exist closed geodesic, e.g., the tori T^n. □

Exercise 4.1.39. Prove the Hopf-Rinow theorem. □

Exercise 4.1.40. Let (M, g) be a Riemann manifold and let (U_α) be an open cover consisting of *bounded geodesically convex open sets*. Set $d_\alpha = (\text{diameter}\,(U_\alpha))^2$. Denote by g_α the metric on U_α defined by $g_\alpha = d_\alpha^{-1} g$ so that the diameter of U_α in the new metric is 1. Using a partition of unity (φ_i) subordinated to this cover we can form a new metric

$$\tilde{g} = \sum_i \varphi_i g_{\alpha(i)} \quad (\text{supp}\,\varphi_i \subset U_{\alpha(i)}).$$

Prove that $\tilde{g}$ is a complete Riemann metric. Hence, on any manifold there exist complete Riemann metrics. □

4.1.5 *Calculus on Riemann manifolds*

The classical vector analysis extends nicely to Riemann manifolds. We devote this subsection to describing this more general "vector analysis".

Let (M, g) be an *oriented* Riemann manifold. We now have two structures at our disposal: a Riemann metric, and an orientation, and we will use both of them to construct a plethora of operations on tensors.

First, using the *metric* we can construct by duality the *lowering-the-indices* isomorphism $\mathcal{L} : \text{Vect}\,(M) \to \Omega^1(M)$.

Example 4.1.41. Let $M = \mathbb{R}^3$ with the Euclidean metric. A vector field V on M has the form

$$V = P\frac{\partial}{\partial x} + Q\frac{\partial}{\partial y} + R\frac{\partial}{\partial z}.$$

Then

$$W = \mathcal{L}V = P\,dx + Q\,dy + R\,dz.$$

If we think of V as a field of forces in the space, then W is the infinitesimal work of V. □

On a Riemann manifold there is an equivalent way of describing the exterior derivative.

Proposition 4.1.42. *Let*

$$\varepsilon : C^\infty(T^*M \otimes \Lambda^k T^*M) \to C^\infty(\Lambda^{k+1} T^*M)$$

denote the exterior multiplication operator

$$\varepsilon(\alpha \otimes \beta) = \alpha \wedge \beta, \ \ \forall \alpha \in \Omega^1(M), \ \ \beta \in \Omega^k(M).$$

Then the exterior derivative d is related to the Levi-Civita on $\Lambda^k T^* M$ connection via the equality $d = \varepsilon \circ \nabla$.

Proof.　We will use a strategy useful in many other situations. Our discussion about normal coordinates will payoff. Denote temporarily by D the operator $\varepsilon \circ \nabla$.

The equality $d = D$ is a local statement, and it suffices to prove it in any coordinate neighborhood. Choose $(\boldsymbol{x}^i)$ normal coordinates at an arbitrary point $p \in M$, and set $\partial_i := \frac{\partial}{\partial \boldsymbol{x}^i}$. Note that

$$D = \sum_i d\boldsymbol{x}^i \wedge \nabla_i, \quad \nabla_i = \nabla_{\partial_i}.$$

Let $\omega \in \Omega^k(M)$. Near p it can be written as

$$\omega = \sum_I \omega_I d\boldsymbol{x}^I,$$

where as usual, for any ordered multi-index I: $(1 \le i_1 < \cdots < i_k \le n)$, we set

$$d\boldsymbol{x}^I := d\boldsymbol{x}^{i_1} \wedge \cdots \wedge d\boldsymbol{x}^{i_k}.$$

In normal coordinates *at p* we have $(\nabla_i \partial_i)|_p = 0$ from which we get the equalities

$$(\nabla_i d\boldsymbol{x}^j)|_p (\partial_k) = -(d\boldsymbol{x}^j (\nabla_i \partial_k))|_p = 0.$$

Thus, *at p*,

$$D\omega = \sum_I d\boldsymbol{x}^i \wedge \nabla_i (\omega_I d\boldsymbol{x}^I)$$

$$= \sum_I d\boldsymbol{x}^i \wedge (\partial_j \omega_I d\boldsymbol{x}^I + \omega_I \nabla_i(d\boldsymbol{x}^I)) = \sum_I d\boldsymbol{x}^i \wedge \partial_j \omega_I = d\omega.$$

Since the point p was chosen arbitrarily this completes the proof of Proposition 4.1.42.　□

Exercise 4.1.43.　Show that for any k-form ω on the Riemann manifold (M, g) the exterior derivative $d\omega$ can be expressed by

$$d\omega(X_0, X_1, \ldots, X_k) = \sum_{i=0}^k (-1)^i (\nabla_{X_i} \omega)(X_0, \ldots, \hat{X}_i, \ldots, X_k),$$

for all $X_0, \ldots, X_k \in \text{Vect}(M)$. ($\nabla$ denotes the Levi-Civita connection.)　□

The Riemann metric defines a metric in any tensor bundle $\mathcal{T}_s^r(M)$ which we continue to denote by g. Thus, given two tensor fields T_1, T_2 of the same type (r, s) we can form their pointwise scalar product

$$M \ni p \mapsto g(T, S)_p = g_p(T_1(p), T_2(p)).$$

In particular, any such tensor has a pointwise norm

$$M \ni p \mapsto |T|_{g, p} = (T, T)_p^{1/2}.$$

Using the *orientation* we can construct (using the results in subsection 2.2.4) a natural *volume form* on M which we denote by dV_g, and we call it the *metric volume*. This is the positively oriented volume form of pointwise norm $\equiv 1$.

If $(x^1, ..., x^n)$ are local coordinates such that $dx^1 \wedge \cdots \wedge dx^n$ is positively oriented, then

$$dV_g = \sqrt{|g|} dx^1 \wedge \cdots \wedge dx^n,$$

where $|g| := \det(g_{ij})$. In particular, we can integrate (compactly supported) functions on M by

$$\int_{(M,g)} f \overset{def}{=} \int_M f \, dv_g \quad \forall f \in C_0^\infty(M).$$

We have the following not so surprising result.

Proposition 4.1.44. $\nabla_X dV_g = 0$, $\forall X \in \text{Vect}(M)$.

Proof. We have to show that for any $p \in M$

$$(\nabla_X dV_g)(e_1, ..., e_n) = 0, \tag{4.1.13}$$

where $e_1, ..., e_p$ is a basis of $T_p M$. Choose normal coordinates (x^i) near p. Set $\partial_i = \frac{\partial}{\partial x^i}$, $g_{ij} = g(\partial_i, \partial_k)$, and $e_i = \partial_i|_p$. Since the expression in (4.1.13) is linear in X, we may as well assume $X = \partial_k$, for some $k = 1, ..., n$. We compute

$$(\nabla_X dV_g)(e_1, ..., e_n) = X(dV_g(\partial_1, ..., \partial_n))|_p$$
$$- \sum_i dv_g(e_1, ..., (\nabla_X \partial_i)|_p, ..., \partial_n). \tag{4.1.14}$$

We consider each term separately. Note first that $dV_g(\partial_1, ..., \partial_n) = (\det(g_{ij}))^{1/2}$, so that $X(\det(g_{ij}))^{1/2}|_p = \partial_k(\det(g_{ij}))^{1/2}|_p$ is a linear combination of products in which each product has a factor of the form $\partial_k g_{ij}|_p$. Such a factor is zero since we are working in normal coordinates. Thus, the first term in (4.1.14) is zero. The other terms are zero as well since in normal coordinates at p we have the equality

$$\nabla_X \partial_i = \nabla_{\partial_k} \partial_i = 0.$$

Proposition 4.1.44 is proved. □

Once we have an orientation, we also have the *Hodge ∗-operator*

$$* : \Omega^k(M) \to \Omega^{n-k}(M),$$

uniquely determined by

$$\alpha \wedge *\beta = (\alpha, \beta) dV_g, \quad \forall \alpha \, \beta \in \Omega^k(M). \tag{4.1.15}$$

In particular, $*1 = dV_g$.

Example 4.1.45. To any vector field $F = P\partial_x + Q\partial_y + R\partial_z$ on $\mathbb{R}^3$ we associated its *infinitesimal work*

$$W_F = \mathcal{L}(F) = P\,dx + Q\,dy + R\,dz.$$

The *infinitesimal energy flux* of F is the 2-form

$$\Phi_F = *W_F = Pdy \wedge dz + Qdz \wedge dx + Rdx \wedge dy.$$

The exterior derivative of W_F is the infinitesimal flux of the vector field **curl** F

$$dW_F = (\partial_y R - \partial_z Q)dy \wedge dz + (\partial_z P - \partial_x R)dz \wedge dx + (\partial_x Q - \partial_y P)dx \wedge dy$$

$$= \Phi_{\mathbf{curl}\,F} = *W_{\mathbf{curl}\,F}.$$

The divergence of F is the scalar defined as

$$\mathbf{div}\,F = *d * W_F = *d\Phi_F$$

$$= *\{(\partial_x P + \partial_y Q + \partial_z R)dx \wedge dy \wedge dz\} = \partial_x P + \partial_y Q + \partial_z R.$$

If f is a function on $\mathbb{R}^3$, then we compute easily

$$*d * df = \partial_x^2 f + \partial_y^2 f + \partial_x^2 f = \Delta f. \qquad \square$$

Definition 4.1.46. (a) For any smooth function f on the Riemann manifold (M, g) we denote by **grad** f, or **grad**$_g f$, the vector field g-dual to the 1-form df. In other words

$$g(\mathbf{grad}\,f, X) = df(X) = X \cdot f \quad \forall X \in \text{Vect}\,(M).$$

(b) If (M, g) is an *oriented* Riemann manifold, and $X \in \text{Vect}\,(M)$, we denote by **div** X, or **div**$_g X$, the smooth function defined by the equality

$$L_X dV_g = (\mathbf{div}\,X)dV_g. \qquad \square$$

Exercise 4.1.47. Consider the unit sphere

$$S^2 = \{\,(x, y, z) \in \mathbb{R}^3;\ x^2 + y^2 + z^2 = 1\,\},$$

and denote by g the Riemann metric on S^2 induced by the Euclidean metric $g_0 = dx^2 + dy^2 + dz^2$ on $\mathbb{R}^3$.

(a) Express g in the spherical coordinates (r, θ, φ) defined as in Example 3.4.14.

(b) Denote by h the restriction to S^2 of the function $\hat{h}(x, y, z) = z$. Express **grad**$^g f$ in spherical coordinates. $\qquad \square$

Proposition 4.1.48. *Let X be a vector field on the oriented Riemann manifold (M, g), and denote by α the 1-form dual to X. Then*

(a) $\mathbf{div}\,X = \text{tr}\,(\nabla X)$, where we view ∇X as an element of $C^\infty(\text{End}\,(TM))$ via the identifications

$$\nabla X \in \Omega^1(TM) \cong C^\infty(T^*M \otimes TM) \cong C^\infty(\text{End}\,(TM)).$$

(b) $\mathbf{div}\,X = *d * \alpha$.

(c) If $(x^1, ..., x^n)$ are local coordinates such that $dx^1 \wedge \cdots \wedge dx^n$ is positively oriented, then

$$\mathbf{div}\,X = \frac{1}{\sqrt{|g|}}\partial_i(\sqrt{|g|}X^i),$$

where $X = X^i \partial_i$.

The proof will rely on the following technical result which is interesting in its own. For simplicity, we will denote the inner products by $(\bullet, \bullet)$, instead of the more precise $g(\bullet, \bullet)$.

Lemma 4.1.49. *Denote by δ the operator*

$$\delta = *d* : \Omega^k(M) \to \Omega^{k-1}(M).$$

Let α be a $(k-1)$-form and β a k-form such that at least one of them is compactly supported. Then

$$\int_M (d\alpha, \beta) dV_g = (-1)^{\nu_{n,k}} \int_M (\alpha, \delta\beta) dV_g,$$

where $\nu_{n,k} = nk + n + 1$.

Proof. We have

$$\int_M (d\alpha, \beta) dV_g = \int_M d\alpha \wedge *\beta = \int_M d(\alpha \wedge *\beta) + (-1)^k \int_M \alpha \wedge d*\beta.$$

The first integral in the right-hand side vanishes by the Stokes formula since $\alpha \wedge *\beta$ has compact support. Since

$$d*\beta \in \Omega^{n-k+1}(M) \text{ and } *^2 = (-1)^{(n-k+1)(k-1)} \text{ on } \Omega^{n-k+1}(M)$$

we deduce

$$\int_M (d\alpha, \beta) dV_g = (-1)^{k+(n-k+1)(n-k)} \int_M \alpha \wedge *\delta\beta.$$

This establishes the assertion in the lemma since

$$(n-k+1)(k-1) + k \equiv \nu_{n,k} \,(\text{mod } 2). \qquad \square$$

Definition 4.1.50. Define $d^* : \Omega^k(M) \to \Omega^{k-1}(M)$ by

$$d^* := (-1)^{\nu_{n,k}} \delta = (-1)^{\nu_{n,k}} *d* . \qquad \square$$

Proof of the proposition. Set $\Omega := dV_g$, and let $(X_1, ..., X_n)$ be a local moving frame of TM in a neighborhood of some point. Then

$$(L_X\Omega)(X_1, ..., X_n) = X(\Omega(X_1, ..., X_n)) - \sum_i \Omega(X_1, ..., [X, X_i], ..., X_n). \quad (4.1.16)$$

Since $\nabla\Omega = 0$ we get

$$X \cdot (\Omega(X_1, ..., X_n)) = \sum_i \Omega(X_1, ..., \nabla_X X_i, ..., X_n).$$

Using the above equality in (4.1.16) we deduce from $\nabla_X Y - [X, Y] = \nabla_Y X$ that

$$(L_X\Omega)(X_1, ..., X_n) = \sum_i \Omega(X_1, ..., \nabla_{X_i} X, ..., X_n). \quad (4.1.17)$$

Over the neighborhood where the local moving frame is defined, we can find smooth functions f_i^j, such that

$$\nabla_{X_i} X = f_i^j X_j \Rightarrow \text{tr}\,(\nabla X) = f_i^i.$$

Part (a) of the proposition follows after we substitute the above equality in (4.1.17).

Proof of (b) For any $f \in C_0^\infty(M)$ we have

$$L_X(f\omega) = (Xf)\Omega + f(\boldsymbol{div}\, X)\Omega.$$

On the other hand,

$$L_X(f\Omega) = (i_X d + d i_X)(f\Omega) = d i_X(f\Omega).$$

Hence

$$\{(Xf) + f(\boldsymbol{div}\, X)\}\, dV_g = d(i_X f\Omega).$$

Since the form $f\Omega$ is *compactly supported* we deduce from Stokes formula

$$\int_M d(i_X f\Omega) = 0.$$

We have thus proved that for any compactly supported function f we have the equality

$$-\int_M f(\boldsymbol{div}\, X)dV_g = \int_M (Xf)dV_g = \int_M df(X)dV_g$$

$$= \int_M (\boldsymbol{grad}\, f, X)dV_g = \int_M (df, \alpha)dV_g.$$

Using Lemma 4.1.49 we deduce

$$-\int_M f(\boldsymbol{div}\, X)dV_g = -\int_M f\delta\alpha dV_g \quad \forall f \in C_0^\infty(M).$$

This completes the proof of (b).

Proof of (c) We use the equality

$$L_X(\sqrt{|g|}dx^1 \wedge \cdots \wedge dx^n) = \boldsymbol{div}(X)(\sqrt{|g|}dx^1 \wedge \cdots \wedge dx^n).$$

The desired formula follows derivating in the left-hand side. One uses the fact that L_X is an even s-derivation and the equalities

$$L_X dx^i = \partial_i X^i dx^i \quad \text{(no summation)},$$

proved in Subsection 3.1.3. $\qquad\square$

Exercise 4.1.51. Let (M, g) be a Riemann manifold and $X \in \text{Vect}(M)$. Show that the following conditions on X are equivalent.
(a) $L_X g = 0$.
(b) $g(\nabla_Y X, Z) + g(Y, \nabla_Z X) = 0$ for all $Y, Z \in \text{Vect}(M)$.
(A vector field X satisfying the above equivalent conditions is called a *Killing* vector field.) $\qquad\square$

Exercise 4.1.52. Consider a Killing vector field X on the oriented Riemann manifold (M, g), and denote by η the 1-form dual to X. Show that $\delta\eta = 0$, i.e., $\boldsymbol{div}(X) = 0$. $\qquad\square$

Definition 4.1.53. Let (M, g) be an oriented Riemann manifold (possibly with boundary). For any k-forms α, β define

$$\langle \alpha, \beta \rangle = \langle \alpha, \beta \rangle_M = \int_M (\alpha, \beta) dV_g = \int_M \alpha \wedge *\beta,$$

whenever the integrals in the right-hand side are finite.   □

Let (M, g) be an oriented Riemann manifold with boundary ∂M. By definition, M is a closed subset of a boundary-less manifold $\tilde{M}$ of the same dimension. Along ∂M we have a vector bundle decomposition

$$(T\tilde{M}) \mid_{\partial M} = T(\partial M) \oplus \boldsymbol{n}$$

where $\boldsymbol{n} = (T\partial M)^\perp$ is the orthogonal complement of $T\partial M$ in $(TM) \mid_{\partial M}$. Since both M and ∂M are oriented manifolds it follows that $\boldsymbol{\nu}$ is a trivial line bundle. Indeed, over the boundary

$$\det TM = \det(T\partial M) \otimes \boldsymbol{n}$$

so that

$$\boldsymbol{n} \cong \det TM \otimes \det(T\partial M)^*.$$

In particular, $\boldsymbol{n}$ admits nowhere vanishing sections, and each such section defines an orientation in the fibers of $\boldsymbol{n}$.

An *outer normal* is a nowhere vanishing section σ of $\boldsymbol{n}$ such that, for each $x \in \partial M$, and any positively oriented $\omega_x \in \det T_x\partial M$, the product $\sigma_x \wedge \omega_x$ is a positively oriented element of $\det T_x M$. Since $\boldsymbol{n}$ carries a fiber metric, we can select a unique outer normal of pointwise length $\equiv 1$. This will be called the *unit outer normal*, and will be denoted by $\vec{\nu}$. Using partitions of unity we can extend $\vec{\nu}$ to a vector field defined on M.

Proposition 4.1.54 (Integration by parts). *Let (M, g) be a compact, oriented Riemann manifold with boundary, $\alpha \in \Omega^{k-1}(M)$ and $\beta \in \Omega^k(M)$. Then*

$$\int_M (d\alpha, \beta) dV_g = \int_{\partial M} (\alpha \wedge *\beta) \mid_{\partial M} + \int_M (\alpha, d^*\beta) dV_g$$

$$= \int_{\partial M} \alpha\mid_{\partial M} \wedge \hat{*}(i_{\vec{\nu}}\beta)\mid_{\partial M} + \int_M (\alpha, d^*\beta) dV_g$$

where $\hat{*}$ denotes the Hodge $*$-operator on ∂M with the induced metric $\hat{g}$ and orientation.

Using the $\langle \cdot, \cdot \rangle$ notation of Definition 4.1.53 we can rephrase the above equality as

$$\langle d\alpha, \beta \rangle_M = \langle \alpha, i_{\vec{\nu}}\beta \rangle_{\partial M} + \langle \alpha, d^*\beta \rangle_M.$$

Proof. As in the proof of Lemma 4.1.49 we have

$$(d\alpha, \beta)dV_g = d\alpha \wedge *\beta = d(\alpha \wedge *\beta) + (-1)^k \alpha \wedge d * \beta.$$

The first part of the proposition follows from Stokes formula arguing precisely as in Lemma 4.1.49. To prove the second part we have to check that

$$(\alpha \wedge *\beta)|_{\partial M} = \alpha|_{\partial M} \wedge \hat{*}(i_{\vec{\nu}}\beta)|_{\partial M}.$$

This is a local (even a pointwise) assertion so we may as well assume

$$M = \mathbf{H}^n_+ = \{(x^1, ..., x^n) \in \mathbb{R}^n \; ; \; x^1 \geq 0\},$$

and that the metric is the Euclidean metric. Note that $\vec{\nu} = -\partial_1$. Let I be an ordered $(k-1)$-index, and J be an ordered k-index. Denote by J^c the ordered $(n-k)$-index complementary to J so that (as sets) $J \cup J^c = \{1, ..., n\}$. By linearity, it suffices to consider only the cases $\alpha = dx^I$, $\beta = dx^J$. We have

$$*dx^J = \epsilon_J dx^{J^c} \quad (\epsilon_J = \pm 1) \tag{4.1.18}$$

and

$$i_{\vec{\nu}}dx^J = \begin{cases} 0 \, , 1 \notin J \\ -dx^{J'} \, , 1 \in J \end{cases},$$

where $J' = J \setminus \{1\}$. Note that, if $1 \notin J$, then $1 \in J^c$ so that

$$(\alpha \wedge *\beta)|_{\partial M} = 0 = \alpha|_{\partial M} \wedge \hat{*}(i_{\vec{\nu}}\beta)|_{\partial M},$$

and therefore, the only nontrivial situation left to be discussed is $1 \in J$. On the boundary we have the equality

$$\hat{*}(i_{\vec{\nu}}dx^J) = -\hat{*}(dx^{J'}) = -\epsilon'_J dx^{J^c} \quad (\epsilon'_J = \pm 1). \tag{4.1.19}$$

We have to compare the two signs ϵ_J and ϵ'_J. in (4.1.18) and (4.1.19). The sign ϵ_J is the signature of the permutation $J \vec{\cup} J^c$ of $\{1, ..., n\}$ obtained by writing the two increasing multi-indices one after the other, first J and then J^c. Similarly, since the boundary ∂M has the orientation $-dx^2 \wedge \cdots \wedge dx^n$, we deduce that the sign ϵ'_J is $(-1) \times$(the signature of the permutation $J' \vec{\cup} J^c$ of $\{2, ..., n\}$). Obviously

$$\text{sign}\,(J \vec{\cup} J^c) = \text{sign}\,(J' \vec{\cup} J^c),$$

so that $\epsilon_J = -\epsilon'_J$. The proposition now follows from (4.1.18) and (4.1.19). $\square$

Corollary 4.1.55 (Gauss). *Let (M, g) be a compact, oriented Riemann manifold with boundary, and X a vector field on M. Then*

$$\int_M \boldsymbol{div}(X)dV_g = \int_{\partial M} (X, \vec{\nu})dv_{g_\partial},$$

where $g_\partial = g|_{\partial M}$.

Proof. Denote by α the 1-form dual to X. We have

$$\int_M \boldsymbol{div}(X)dV_g = \int_M 1 \wedge *d*\alpha dV_g = \int_M (1, *d*\alpha)dV_g = -\int_M (1, d^*\alpha)dV_g$$

$$= \int_{\partial M} \alpha(\vec{\nu})dv_{g_\partial} = \int_{\partial M} (X, \vec{\nu})dv_{g_\partial}. \qquad \square$$

Remark 4.1.56. The compactness assumption on M can be replaced with an integrability condition on the forms α, β so that the previous results hold for non-compact manifolds as well provided all the integrals are finite. $\qquad \square$

Definition 4.1.57. Let (M, g) be an oriented Riemann manifold. The *geometric Laplacian* is the linear operator $\Delta_M : C^\infty(M) \to C^\infty(M)$ defined by

$$\Delta_M = d^*df = -*d*df = -\boldsymbol{div}(\boldsymbol{grad}\, f).$$

A smooth function f on M satisfying the equation $\Delta_M f = 0$ is called *harmonic*. $\square$

Using Proposition 4.1.48 we deduce that in local coordinates $(x^1, ..., x^n)$, the geometer's Laplacian takes the form

$$\Delta_M = -\frac{1}{\sqrt{|g|}}\partial_i\left(\sqrt{|g|}g^{ij}\partial_j\right),$$

where (g^{ij}) denotes as usual the matrix inverse to (g_{ij}). Note that when g is the Euclidean metric, then the geometers' Laplacian is

$$\Delta_0 = -(\partial_i^2 + \cdots + \partial_n^2),$$

which differs from the physicists' Laplacian by a sign.

Corollary 4.1.58 (Green). *Let (M, g) as in Proposition 4.1.54, and $f, g \in C^\infty(M)$. Then*

$$\langle f, \Delta_M g\rangle_M = \langle df, dg\rangle_M - \langle f, \frac{\partial g}{\partial \vec{\nu}}\rangle_{\partial M},$$

and

$$\langle f, \Delta_M g\rangle_M - \langle \Delta_M f, g\rangle_M = \langle \frac{\partial f}{\partial \vec{\nu}}, g\rangle_{\partial M} - \langle f, \frac{\partial g}{\partial \vec{\nu}}\rangle_{\partial M}.$$

Proof. The first equality follows immediately from the integration by parts formula (Proposition 4.1.54), with $\alpha = f$, and $\beta = dg$. The second identity is now obvious. $\qquad \square$

Exercise 4.1.59. (a) Prove that the only harmonic functions on a compact oriented Riemann manifold M are the constant ones.
(b) If $u, f \in C^\infty(M)$ are such that $\Delta_M u = f$ show that $\int_M f = 0$. $\qquad \square$

Exercise 4.1.60. Denote by $(u^1, \ldots, u^n)$ the coordinates on the round sphere $S^n \hookrightarrow \mathbb{R}^{n+1}$ obtained via the stereographic projection from the south pole.
(a) Show that the round metric g_0 on S^n is given in these coordinates by

$$g_0 = \frac{4}{1+r^2}\{\,(du^1)^2 + \cdots + (du^n)^2\,\},$$

where $r^2 = (u^1)^2 + \cdots + (u^n)^2$.
(b) Show that the n-dimensional "area" of S^n is

$$\sigma_n = \int_{S^n} dv_{g_0} = \frac{2\pi^{(n+1)/2}}{\Gamma(\frac{n+1}{2})},$$

where Γ is Euler's Gamma function

$$\Gamma(s) = \int_0^\infty t^{s-1}e^{-t}dt. \qquad \square$$

Hint: Use the "doubling formula"

$$\pi^{1/2}\Gamma(2s) = 2^{2s-1}\Gamma(s)\Gamma(s+1/2),$$

and the classical Beta integrals (see [39], or [101], Chapter XII)

$$\int_0^\infty \frac{r^{n-1}}{(1+r^2)^n}dr = \frac{(\Gamma(n/2))^2}{2\Gamma(n)}. \qquad \square$$

Exercise 4.1.61. Consider the Killing form on $\underline{su}(2)$ (the Lie algebra of $SU(2)$) defined by

$$\langle X, Y \rangle = -\operatorname{tr} X \cdot Y.$$

(a) Show that the Killing form defines a bi-invariant metric on $SU(2)$, and then compute the volume of the group with respect to this metric. The group $SU(2)$ is given the orientation defined by $e_1 \wedge e_2 \wedge e_3 \in \Lambda^3 \underline{su}(2)$, where $e_i \in \underline{su}(2)$ are the *Pauli matrices*

$$e_1 = \begin{bmatrix} i & 0 \\ 0 & -i \end{bmatrix} \quad e_2 = \begin{bmatrix} 0 & 1 \\ -1 & 0 \end{bmatrix} \quad e_3 = \begin{bmatrix} 0 & i \\ i & 0 \end{bmatrix}$$

(b) Show that the trilinear form on $\underline{su}(2)$ defined by

$$B(X, Y, Z) = \langle [X, Y], Z \rangle,$$

is skew-symmetric. In particular, $B \in \Lambda^3 \underline{su}(2)^*$.
(c) B has a unique extension as a left-invariant 3-form on $SU(2)$ known as the *Cartan form* on $SU(2)$ which we continue to denote by B. Compute $\int_{SU(2)} B$.
Hint: Use the natural diffeomorphism $SU(2) \cong S^3$, and the computations in the previous exercise. $\qquad \square$

4.2 The Riemann curvature

Roughly speaking, a Riemann metric on a manifold has the effect of "giving a shape" to the manifold. Thus, a very short (in diameter) manifold is different from a very long one, and a large (in volume) manifold is different from a small one. However, there is a lot more information encoded in the Riemann manifold than just its size. To recover it, we need to look deeper in the structure, and go beyond the first order approximations we have used so far.

The Riemann curvature tensor achieves just that. It is an object which is very rich in information about the "shape" of a manifold, and loosely speaking, provides a second order approximation to the geometry of the manifold. As Riemann himself observed, we do not need to go beyond this order of approximation to recover all the information.

In this section we introduce the reader to the Riemann curvature tensor and its associates. We will describe some special examples, and we will conclude with the Gauss-Bonnet theorem which shows that the local object which is the Riemann curvature has global effects.

☞ *Unless otherwise indicated, we will use Einstein's summation convention.*

4.2.1 *Definitions and properties*

Let (M, g) be a Riemann manifold, and denote by ∇ the Levi-Civita connection.

Definition 4.2.1. The *Riemann curvature* is the tensor $R = R(g)$, defined as

$$R(g) = F(\nabla),$$

where $F(\nabla)$ is the curvature of the Levi-Civita connection.   □

The Riemann curvature is a tensor $R \in \Omega^2(\text{End}\,(TM))$ explicitly defined by

$$R(X, Y)Z = [\nabla_X, \nabla_Y]Z - \nabla_{[X,Y]}Z.$$

In local coordinates $(x^1, \ldots, x^n)$ we have the description

$$R^\ell_{ijk}\partial_\ell = R(\partial_j, \partial_k)\partial_i.$$

In terms of the Christoffel symbols we have

$$R^\ell_{ijk} = \partial_j\Gamma^\ell_{ik} - \partial_k\Gamma^\ell_{ij} + \Gamma^\ell_{mj}\Gamma^m_{ik} - \Gamma^\ell_{mk}\Gamma^m_{ij}.$$

Lowering the indices we get a new tensor

$$R_{ijk\ell} := g_{im}R^m_{jk\ell} = \big(R(\partial_k, \partial_\ell)\partial_j, \partial_i\big) = \big(\partial_i, R(\partial_k, \partial_\ell)\partial_j\big).$$

Theorem 4.2.2 (The symmetries of the curvature tensor). *The   Riemann curvature tensor R satisfies the following identities $(X, Y, Z, U, V \in \text{Vect}\,(M))$.*

(a) $g(R(X, Y)U, V) = -g(R((Y, X), U, V)$.

(b) $g(R(X,Y)U,V) = -g(R(X,Y)V,U)$.

(c) *(The 1st Bianchi identity)*

$$R(X,Y)Z + R(Z,X)Y + R(Y,Z)X = 0.$$

(d) $g(R(X,Y)U,V) = g(R(U,V)X,Y)$.

(e) *(The 2nd Bianchi identity)*

$$(\nabla_X R)(Y,Z) + (\nabla_Y R)(Z,X) + (\nabla_Z R)(X,Y) = 0.$$

In local coordinates the above identities have the form

$$R_{ijk\ell} = -R_{jik\ell} = -R_{ij\ell k},$$

$$R_{ijk\ell} = R_{k\ell ij},$$

$$R^i_{jk\ell} + R^i_{\ell jk} + R^i_{k\ell j} = 0,$$

$$(\nabla_i R)^j_{mk\ell} + (\nabla_\ell R)^j_{mik} + (\nabla_k R)^j_{m\ell i} = 0.$$

Proof. (a) It follows immediately from the definition of R as an $\mathrm{End}(TM)$-valued skew-symmetric bilinear map $(X,Y) \mapsto R(X,Y)$.

(b) We have to show that the symmetric bilinear form

$$Q(U,V) = g(R(X,Y)U,V) + g(R(X,Y)V,U)$$

is trivial. Thus, it suffices to check $Q(U,U) = 0$. We may as well assume that $[X,Y] = 0$, since (locally) X, Y can be written as linear combinations (over $C^\infty(M)$) of commuting vector fields. (E.g. $X = X^i \partial_i$). Then

$$Q(U,U) = g((\nabla_X \nabla_Y - \nabla_Y \nabla_X)U,U).$$

We compute

$$Y(Xg(U,U)) = 2Yg(\nabla_X U,U) = 2g(\nabla_Y \nabla_X U,U) + 2g(\nabla_X U, \nabla_Y U),$$

and similarly,

$$X(Yg(U,U)) = 2g(\nabla_X \nabla_Y U,U) + 2g(\nabla_X U, \nabla_Y U).$$

Subtracting the two equalities we deduce (b).

(c) As before, we can assume the vector fields X, Y, Z pairwise commute. The 1st Bianchi identity is then equivalent to

$$\nabla_X \nabla_Y Z - \nabla_Y \nabla_X Z + \nabla_Z \nabla_X Y - \nabla_X \nabla_Z Y + \nabla_Y \nabla_Z X - \nabla_Z \nabla_Y X = 0.$$

The identity now follows from the symmetry of the connection: $\nabla_X Y = \nabla_Y X$ etc.

(d) We will use the following algebraic lemma ([60], Chapter 5).

Lemma 4.2.3. *Let* $R : E \times E \times E \times E \to \mathbb{R}$ *be a quadrilinear map on a real vector space* E. *Define*

$$S(X_1, X_2, X_3, X_4) = R(X_1, X_2, X_3, X_4) + R(X_2, X_3, X_1, X_4) + R(X_3, X_1, X_2, X_4).$$

If R satisfies the symmetry conditions

$$R(X_1, X_2, X_3, X_4) = -R(X_2, X_1, X_3, X_4)$$

$$R(X_1, X_2, X_3, X_4) = -R(X_1, X_2, X_4, X_3),$$

then

$$R(X_1, X_2, X_3, X_4) - R(X_3, X_4, X_1, X_2)$$
$$= \frac{1}{2} \Big\{ S(X_1, X_2, X_3, X_4) - S(X_2, X_3, X_4, X_1)$$
$$- S(X_3, X_4, X_1, X_2) + S(X_4, X_3, X_1, X_2) \Big\}.$$

The proof of the lemma is a straightforward (but tedious) computation which is left to the reader. The Riemann curvature $R = g(R(X_1, X_2)X_3, X_4)$ satisfies the symmetries required in the lemma and moreover, the 1st Bianchi identity shows that the associated form S is identically zero. This concludes the proof of (d).

(e) This is the Bianchi identity we established for any linear connection (see Exercise 3.3.23). □

Exercise 4.2.4. Denote by $\mathcal{C}_n$ of n-dimensional curvature tensors, i.e., tensors $(R_{ijkl}) \in (\mathbb{R}^n)^{\otimes 4}$ satisfying the conditions,

$$R_{ijk\ell} = R_{k\ell ij} = -R_{jik\ell}, \quad R_{ijk\ell} + R_{i\ell jk} + R_{ik\ell j} = 0, \quad \forall i, j, k, \ell.$$

Prove that

$$\dim \mathcal{C}_n = \binom{\binom{n}{2} + 1}{2} - \binom{n}{4} = \frac{1}{2} \binom{n}{2} \left(\binom{n}{2} + 1 \right) - \binom{n}{4}.$$

(**Hint:** Consult [13], Chapter 1, Section G.) □

The Riemann curvature tensor is the source of many important invariants associated to a Riemann manifold. We begin by presenting the simplest ones.

Definition 4.2.5. Let (M, g) be a Riemann manifold with curvature tensor R. Any two vector fields X, Y on M define an endomorphism of TM by

$$U \mapsto R(U, X)Y.$$

The *Ricci curvature* is the trace of this endomorphism, i.e.,

$$\text{Ric}(X, Y) = \text{tr}(U \mapsto R(U, X)Y).$$

We view it as a (0,2)-tensor $(X, Y) \mapsto \text{Ric}(X, Y) \in C^\infty(M)$. □

If $(x^1, \ldots, x^n)$ are local coordinates on M and the curvature R has the local expression $R = (R^\ell_{kij})$ then the Ricci curvature has the local description

$$\text{Ric} = (\text{Ric}_{ij}) = \sum_\ell R^\ell_{j\ell i}.$$

The symmetries of the Riemann curvature imply that Ric is a *symmetric* (0,2)-tensor (as the metric).

Definition 4.2.6. The *scalar curvature s* of a Riemann manifold is the trace of the Ricci tensor. In local coordinates, the scalar curvature is described by

$$s = g^{ij} \operatorname{Ric}_{ij} = g^{ij} R^{\ell}_{i\ell j}, \tag{4.2.1}$$

where (g^{ij}) is the inverse matrix of (g_{ij}). □

Let (M, g) be a Riemann manifold and $p \in M$. For any linearly independent $X, Y \in T_p M$ set

$$K_p(X, Y) = \frac{(R(X, Y)Y, X)}{|X \wedge Y|},$$

where $|X \wedge Y|$ denotes the Gramm determinant

$$|X \wedge Y| = \begin{vmatrix} (X, X) & (X, Y) \\ (Y, X) & (Y, Y) \end{vmatrix},$$

which is non-zero since X and Y are linearly independent. ($|X \wedge Y|^{1/2}$ measures the area of the parallelogram in $T_p M$ spanned by X and Y.)

Remark 4.2.7. Given a metric g on a smooth manifold M, and a constant $\lambda > 0$, we obtained a new, rescaled metric $g_\lambda = \lambda^2 g$. A simple computation shows that the Christoffel symbols and Riemann tensor of g_λ are equal with the Christoffel symbols and the Riemann tensor of the metric g. In particular, this implies

$$\operatorname{Ric}_{g_\lambda} = \operatorname{Ric}_g.$$

However, the sectional curvatures are sensitive to metric rescaling.

For example, if g is the canonical metric on the 2-sphere of radius 1 in $\mathbb{R}^3$, then g_λ is the induced metric on the 2-sphere of radius λ in $\mathbb{R}^3$. Intuitively, the larger the constant λ, the less curved is the corresponding sphere.

In general, for any two linearly independent vectors $X, Y \in \operatorname{Vect}(M)$ we have

$$K_{g_\lambda}(X, Y) = \lambda^{-2} K_g(X, Y).$$

In particular, the scalar curvature changes by the same factor upon rescaling the metric.

If we think of the metric as a quantity measured in meter^2, then the sectional curvatures are measured in meter^{-2}. □

Exercise 4.2.8. Let $X, Y, Z, W \in T_p M$ such that $\operatorname{span}(X, Y) = \operatorname{span}(Z, W)$ is a 2-dimensional subspace of $T_p M$ prove that $K_p(X, Y) = K_p(Z, W)$. □

According to the above exercise the quantity $K_p(X, Y)$ depends only upon the 2-plane in T_pM generated by X and Y. Thus K_p is in fact a function on $\mathbf{Gr}_2(p)$ the Grassmannian of 2-dimensional subspaces of T_pM.

Definition 4.2.9. The function $K_p : \mathbf{Gr}_2(p) \to \mathbb{R}$ defined above is called the *sectional curvature* of M at p.   □

Exercise 4.2.10. Prove that

$$\mathbf{Gr}_2(M) = \text{disjoint union of } \mathbf{Gr}_2(p) \quad p \in M$$

can be organized as a smooth fiber bundle over M with standard fiber $\mathbf{Gr}_2(\mathbb{R}^n)$, $n = \dim M$, such that, if M is a Riemann manifold, $\mathbf{Gr}_2(M) \ni (p; \pi) \mapsto K_p(\pi)$ is a smooth map.   □

4.2.2   *Examples*

Example 4.2.11. Consider again the situation discussed in Example 4.1.15. Thus, G is a Lie group, and $\langle \bullet, \bullet \rangle$ is a metric on the Lie algebra $\mathcal{L}_G$ satisfying

$$\langle \text{ad}(X)Y, Z \rangle = -\langle Y, \text{ad}(X)Z \rangle.$$

In other words, $\langle \bullet, \bullet \rangle$ is the restriction of a bi-invariant metric $\mathfrak{m}$ on G. We have shown that the Levi-Civita connection of this metric is

$$\nabla_X Y = \frac{1}{2}[X, Y], \quad \forall X, Y \in \mathcal{L}_G.$$

We can now easily compute the curvature

$$R(X, Y)Z = \frac{1}{4}\left\{[X, [Y, Z]] - [Y, [X, Z]]\right\} - \frac{1}{2}[[X, Y], Z]$$

(Jacobi identity) $= \frac{1}{4}[[X,Y], Z] + \frac{1}{4}[Y, [X, Z]] - \frac{1}{4}[Y, [X, Z]] - \frac{1}{2}[[X, Y], Z] = -\frac{1}{4}[[X, Y], Z].$

We deduce

$$\langle R(X, Y)Z, W \rangle = -\frac{1}{4}\langle [[X, Y], Z], W \rangle = \frac{1}{4}\langle \text{ad}(Z)[X, Y], W \rangle$$

$$= -\frac{1}{4}\langle [X, Y], ad(Z)W \rangle = -\frac{1}{4}\langle [X, Y], [Z, W] \rangle.$$

Now let $\pi \in \mathbf{Gr}_2(T_gG)$ be a 2-dimensional subspace of T_gG, for some $g \in G$. If (X, Y) is an orthonormal basis of π, viewed as left invariant vector fields on G, then the sectional curvature along π is

$$K_g(\pi) = \frac{1}{4}\langle\, [X, Y], [X, Y]\,\rangle \geq 0.$$

Denote the Killing form by $\kappa(X, Y) = -\text{tr}\left(\text{ad}(X)\,\text{ad}(Y)\right)$. To compute the Ricci curvature we pick an orthonormal basis $E_1, \ldots, E_n$ of $\mathcal{L}_G$. For any $X = X^iE_i, Y = Y^jE_j \in \mathcal{L}_G$ we have

$$\text{Ric}\,(X, Y) = \frac{1}{4}\text{tr}\,(Z \mapsto [[X, Z], Y])$$

$$= \frac{1}{4} \sum_i \langle [[X, E_i], Y], E_i) \rangle = -\frac{1}{4} \sum_i \langle \mathrm{ad}(Y)[X, E_i], E_i \rangle$$

$$= \frac{1}{4} \sum_i \langle [X, E_i], [Y, E_i] \rangle = \frac{1}{4} \sum_i \langle \mathrm{ad}(X)E_i, \mathrm{ad}(Y)E_i \rangle$$

$$= -\frac{1}{4} \sum_i \langle \mathrm{ad}(Y)\,\mathrm{ad}(X)E_i, E_i \rangle = -\frac{1}{4}\mathrm{tr}\left(\mathrm{ad}(Y)\,\mathrm{ad}(X)\right) = \frac{1}{4}\kappa(X, Y).$$

In particular, on a compact semisimple Lie group the Ricci curvature is a symmetric positive definite $(0, 2)$-tensor, and more precisely, it is a scalar multiple of the Killing metric.

We can now easily compute the scalar curvature. Using the same notations as above we get

$$s = \frac{1}{4} \sum_i \mathrm{Ric}\,(E_i, E_i) = \frac{1}{4} \sum_i \kappa(E_i, E_i).$$

In particular, if G is a compact semisimple group and the metric is given by the Killing form then the scalar curvature is

$$s(\kappa) = \frac{1}{4} \dim G. \qquad \qquad \square$$

Remark 4.2.12. Many problems in topology lead to a slightly more general situation than the one discussed in the above example namely to metrics on Lie groups which are only left invariant. Although the results are not as "crisp" as in the bi-invariant case many nice things do happen. For details we refer to [73]. $\square$

Example 4.2.13. Let M be a 2-dimensional Riemann manifold (surface), and consider local coordinates on M, (x^1, x^2). Due to the symmetries of R,

$$R_{ijkl} = -R_{ijlk} = R_{klij},$$

we deduce that the only nontrivial component of the Riemann tensor is $R = R_{1212}$. The sectional curvature is simply a function on M

$$K = \frac{1}{|g|} R_{1212} = \frac{1}{2}s(g), \quad \text{where } |g| = \det(g_{ij}).$$

In this case, the scalar K is known as the *total curvature* or the *Gauss curvature* of the surface.

In particular, if M is oriented, and the form $dx^1 \wedge dx^2$ defines the orientation, we can construct a 2-form

$$\varepsilon(g) = \frac{1}{2\pi}K dv_g = \frac{1}{4\pi}s(g)dV_g = \frac{1}{2\pi\sqrt{|g|}}R_{1212}dx^1 \wedge dx^2.$$

The 2-form $\varepsilon(g)$ is called the *Euler form associated to the metric g*. We want to emphasize that this form is defined *only* when M is *oriented*.

We can rewrite this using the pfaffian construction of Subsection 2.2.4. The curvature R is a 2-form with coefficients in the bundle of skew-symmetric endomorphisms of TM so we can write

$$R = A \otimes dV_g, \quad A = \frac{1}{\sqrt{|g|}} \begin{bmatrix} 0 & R_{1212} \\ R_{2112} & 0 \end{bmatrix}$$

Assume for simplicity that (x^1, x^2) are normal coordinates at a point $q \in M$. Thus at q, $|g| = 1$ since ∂_1, ∂_2 is an orthonormal basis of $T_q M$. Hence, at q, $dV_g = dx^1 \wedge dx^2$, and

$$\varepsilon(g) = \frac{1}{2\pi} g\big(R(\partial_1, \partial_2) \partial_2, \partial_1 \big) dx^1 \wedge dx^2 = \frac{1}{2\pi} R_{1212} dx^1 \wedge dx^2.$$

Hence we can write

$$\varepsilon(g) = \frac{1}{2\pi} \boldsymbol{Pf}_g(-A) dv_g =: \frac{1}{2\pi} \boldsymbol{Pf}_g(-R).$$

The Euler form has a very nice interpretation in terms of holonomy. Assume as before that (x^1, x^2) are normal coordinates at q, and consider the square $S_t = [0, \sqrt{t}] \times [0, \sqrt{t}]$ in the (x^1, x^2) plane. Denote the (counterclockwise) holonomy along ∂S_t by $\mathcal{T}_t$. This is an orthogonal transformation of $T_q M$, and with respect to the orthogonal basis (∂_1, ∂_2) of $T_q M$, it has a matrix description as

$$\mathcal{T}_t = \begin{bmatrix} \cos\theta(t) & -\sin\theta(t) \\ \sin\theta(t) & \cos\theta(t) \end{bmatrix}.$$

The result in Subsection 3.3.4 can be rephrased as

$$\sin\theta(t) = -tg(R(\partial_1, \partial_2)\partial_2, \partial_1) + O(t^2),$$

so that

$$R_{1212} = \dot{\theta}(0).$$

Hence R_{1212} is simply the infinitesimal angle measuring the infinitesimal rotation suffered by ∂_1 along S_t. We can think of the Euler form as a "density" of holonomy since it measures the holonomy per elementary parallelogram.   □

4.2.3   *Cartan's moving frame method*

This method was introduced by Élie Cartan at the beginning of the 20th century. Cartan's insight was that the local properties of a manifold equipped with a geometric structure can be very well understood if one knows how the frames of the tangent bundle (compatible with the geometric structure) vary from one point of the manifold to another. We will begin our discussion with the model case of $\mathbb{R}^n$. Throughout this subsection we will use Einstein's convention.

Example 4.2.14. Consider an *orthonormal moving frame* on $\mathbb{R}^n$, $X_\alpha = X_\alpha^i \partial_i$, $\alpha = 1, ..., n$, where $(x^1, ..., x^n)$ are the usual Cartesian coordinates, and $\partial_i := \frac{\partial}{\partial x^i}$. Denote by (θ^α) the *dual coframe*, i.e., the moving frame of $T^*\mathbb{R}^n$ defined by

$$\theta^\alpha(X_\beta) = \delta_\beta^\alpha.$$

The 1-forms θ^α measure the infinitesimal displacement of a point P with respect to the frame (X_α). Note that the TM-valued 1-form $\theta = \theta^\alpha X_\alpha$ is the differential of the identity map $\mathbb{1} : \mathbb{R}^n \to \mathbb{R}^n$ expressed using the given moving frame.

Introduce the 1-forms ω_β^α defined by

$$dX_\beta = \omega_\beta^\alpha X_\alpha, \tag{4.2.2}$$

where we set

$$dX_\alpha := \left(\frac{\partial X_\alpha^i}{\partial x^j}\right) dx^j \otimes \partial_i.$$

We can form the matrix valued 1-form $\omega = (\omega_\beta^\alpha)$ which measures the infinitesimal rotation suffered by the moving frame (X_α) following the infinitesimal displacement $x \mapsto x + dx$. In particular, $\omega = (\omega_\beta^\alpha)$ is a skew-symmetric matrix since

$$0 = d\langle X_\alpha, X_\beta \rangle = \langle \omega \cdot X_\alpha, X_\beta \rangle + \langle X_\alpha, \omega \cdot X_\beta \rangle.$$

Since $\theta = d\mathbb{1}$ we deduce

$$0 = d^2\mathbb{1} = d\theta = d\theta^\alpha \otimes X_\alpha - \theta^\beta \otimes dX_\beta = (d\theta^\alpha - \theta^\beta \wedge \omega_\beta^\alpha) \otimes X_\alpha,$$

and we can rewrite this as

$$d\theta^\alpha = \theta^\beta \wedge \omega_\beta^\alpha, \text{ or } d\theta = -\omega \wedge \theta. \tag{4.2.3}$$

Above, in the last equality we interpret ω as an $n \times n$ matrix whose entries are 1-forms, and θ as a column matrix, or an $n \times 1$ matrix whose entries are 1-forms.

Using the equality $d^2 X_\beta = 0$ in (4.2.2) we deduce

$$d\omega_\beta^\alpha = -\omega_\gamma^\alpha \wedge \omega_\beta^\gamma, \text{ or equivalently, } d\omega = -\omega \wedge \omega. \tag{4.2.4}$$

The equations (4.2.3)–(4.2.4) are called *the structural equations* of the Euclidean space. The significance of these structural equations will become evident in a little while. $\square$

We now try to perform the same computations on an arbitrary Riemann manifold (M, g), $\dim M = n$. We choose a local orthonormal moving frame $(X_\alpha)_{1 \leq \alpha \leq n}$, and we construct similarly its dual coframe $(\theta^\alpha)_{1 \leq \alpha \leq n}$. Unfortunately, there is no natural way to define dX_α to produce the forms ω_β^α entering the structural equations. We will find them using a different (dual) search strategy.

Proposition 4.2.15 (E. Cartan). *There exists a collection of 1-forms* $(\omega_\beta^\alpha)_{1 \leq \alpha, \beta \leq n}$ *uniquely defined by the requirements*

(a) $\omega_\beta^\alpha = -\omega_\alpha^\beta, \forall \alpha, \beta.$

(b) $d\theta^\alpha = \theta^\beta \wedge \omega_\beta^\alpha$, $\forall\alpha$.

Proof.   Since the collection of two forms $(\theta^\alpha \wedge \theta^\beta)_{1\le\alpha<\beta\le n}$ defines a local frame of $\Lambda^2 T^*\mathbb{R}^n$, there exist functions $g_{\beta\gamma}^\alpha$, uniquely determined by the conditions

$$d\theta^\alpha = \frac{1}{2}g_{\beta\gamma}^\alpha\theta^\beta \wedge \theta^\gamma, \quad g_{\beta\gamma}^\alpha = -g_{\gamma\beta}^\alpha.$$

Uniqueness. Suppose that there exist forms ω_β^α satisfy the conditions (a) and (b) above. Then there exist functions $f_{\beta\gamma}^\alpha$ such that

$$\omega_\beta^\alpha = f_{\beta\gamma}^\alpha\theta^\gamma.$$

Then the condition (a) is equivalent to

(a1) $f_{\beta\gamma}^\alpha = -f_{\alpha\gamma}^\beta$,

while (b) gives

(b1) $f_{\beta\gamma}^\alpha - f_{\gamma\beta}^\alpha = g_{\beta\gamma}^\alpha$.

The above two relations uniquely determine the f's in terms of the g's via a cyclic permutation of the indices α, β, γ

$$f_{\beta\gamma}^\alpha = \frac{1}{2}(g_{\beta\gamma}^\alpha + g_{\gamma\alpha}^\beta - g_{\alpha\beta}^\gamma). \tag{4.2.5}$$

Existence. Define $\omega_\beta^\alpha = f_{\beta\gamma}^\alpha\theta^\gamma$, where the f's are given by (4.2.5). We let the reader check that the forms ω_β^α satisfy both (a) and (b).   □

The reader may now ask why go through all this trouble. What have we gained by constructing the forms ω, and after all, what is their significance?

To answer these questions, consider the Levi-Civita connection ∇. Define $\hat{\omega}_\beta^\alpha$ by

$$\nabla X_\beta = \hat{\omega}_\beta^\alpha X_\alpha.$$

Hence

$$\nabla_{X_\gamma} X_\beta = \hat{\omega}_\beta^\alpha(X_\gamma)X_\alpha.$$

Since ∇ is compatible with the Riemann metric, we deduce in standard manner that $\hat{\omega}_\beta^\alpha = -\hat{\omega}_\alpha^\beta$.

The differential of θ^α can be computed in terms of the Levi-Civita connection (see Subsection 4.1.5), and we have

$$d\theta^\alpha(X_\beta, X_\gamma) = X_\beta\theta^\alpha(X_\gamma) - X_\gamma\theta^\alpha(X_\beta) - \theta^\alpha(\nabla_{X_\beta}X_\gamma) + \theta^\alpha(\nabla_{X_\gamma}X_\beta)$$

$$(\text{use } \theta^\alpha(X_\beta) = \delta_\beta^\alpha = \text{const}) = -\theta^\alpha(\hat{\omega}_\gamma^\delta(X_\beta)X_\delta) + \theta^\alpha(\hat{\omega}_\beta^\delta(X_\gamma)X_\delta)$$

$$= \hat{\omega}_\beta^\alpha(X_\gamma) - \hat{\omega}_\gamma^\alpha(X_\beta) = (\theta^\beta \wedge \hat{\omega}_\beta^\alpha)(X_\beta, X_\gamma).$$

Thus the $\hat{\omega}$'s satisfy both conditions (a) and (b) of Proposition 4.2.15 so that we must have

$$\hat{\omega}_\beta^\alpha = \omega_\beta^\alpha.$$

In other words, the matrix valued 1-form (ω_β^α) is the 1-form associated to the Levi-Civita connection in this local moving frame. In particular, using the computation in Example 3.3.12 we deduce that the 2-form

$$\Omega = (d\omega + \omega \wedge \omega)$$

is the Riemannian curvature of g. The *Cartan structural equations* of a Riemann manifold take the form

$$d\theta = -\omega \wedge \theta, \quad d\omega + \omega \wedge \omega = \Omega. \tag{4.2.6}$$

Comparing these with the Euclidean structural equations we deduce another interpretation of the Riemann curvature: It measures "the distance" between the given Riemann metric and the Euclidean one". We refer to [92] for more details on this aspect of the Riemann tensor.

The technique of orthonormal frames is extremely versatile in concrete computations.

Example 4.2.16. We will use the moving frame method to compute the curvature of the *hyperbolic plane*, i.e., the upper half space

$$\mathbf{H}_+ = \{(x,y)\,;\, y > 0\}$$

endowed with the metric $g = y^{-2}(dx^2 + dy^2)$.

The pair $(y\partial_x, y\partial_y)$ is an orthonormal moving frame, and $(\theta^x = \frac{1}{y}dx, \theta^y = \frac{1}{y}dy)$ is its dual coframe. We compute easily

$$d\theta^x = d(\frac{1}{y}dx) = \frac{1}{y^2}dx \wedge dy = (\frac{1}{y}dx) \wedge \theta^y,$$

$$d\theta^y = d(\frac{1}{y}dy) = 0 = (-\frac{1}{y}dx) \wedge \theta^x.$$

Thus the connection 1-form in this local moving frame is

$$\omega = \begin{bmatrix} 0 & -\frac{1}{y} \\ \frac{1}{y} & 0 \end{bmatrix} dx.$$

Note that $\omega \wedge \omega = 0$. Using the structural equations (4.2.6) we deduce that the Riemann curvature is

$$\Omega = d\omega = \begin{bmatrix} 0 & \frac{1}{y^2} \\ -\frac{1}{y^2} & 0 \end{bmatrix} dy \wedge dx = \begin{bmatrix} 0 & -1 \\ 1 & 0 \end{bmatrix} \theta^x \wedge \theta^y.$$

The Gauss curvature is

$$K = \frac{1}{|g|}g(\Omega(\partial_x, \partial_y)\partial_y, \partial_x) = y^4(-\frac{1}{y^4}) = -1. \qquad \square$$

Exercise 4.2.17. Suppose (M, g) is a Riemann manifold, and $u \in C^\infty(M)$. Define a new metric $g_u := e^{2f}g$. Using the moving frames method, describe the scalar curvature of g_u in terms of u and the scalar curvature of g. $\qquad \square$

4.2.4   *The geometry of submanifolds*

We now want to apply Cartan's method of moving frames to discuss the local geometry of submanifolds of a Riemann manifold.

Let (M, g) be a Riemann manifold of dimension m, and S a k-dimensional submanifold in M. The restriction of g to S induces a Riemann metric g_S on S. We want to analyze the relationship between the Riemann geometry of M (assumed to be known) and the geometry of S with the induced metric.

Denote by ∇^M (respectively ∇^S) the Levi-Civita connection of (M, g) (respectively of (S, g_S)). The metric g produces an orthogonal splitting of vector bundles

$$(TM)|_S \cong TS \oplus N_S S.$$

The N_S is called the normal bundle of $S \hookrightarrow M$, and it is the orthogonal complement of TS in $(TM)|_S$. Thus, a section of $(TM)|_S$, that is a vector field X of M along S, splits into two components: a tangential component X^τ, and a normal component, X^ν.

Now choose a local orthonormal moving frame $(X_1, ..., X_k; X_{k+1}, ..., X_m)$ such that the first k vectors $(X_1, ..., X_k)$ are tangent to S. Denote the dual coframe by $(\theta^\alpha)_{1 \leq \alpha \leq m}$. Note that

$$\theta^\alpha|_S = 0 \quad \text{for } \alpha > k.$$

Denote by (μ^α_β), $(1 \leq \alpha, \beta \leq m)$ the connection 1-forms associated to ∇^M by this frame, and let σ^α_β, $(1 \leq \alpha, \beta \leq k)$ be the connection 1-forms of ∇^S corresponding to the frame $(X_1, \ldots, X_k)$. We will analyze the structural equations of M restricted to $S \hookrightarrow M$.

$$d\theta^\alpha = \theta^\beta \wedge \mu^\alpha_\beta \quad 1 \leq \alpha, \beta \leq m. \tag{4.2.7}$$

We distinguish two situations.

A. $1 \leq \alpha \leq k$. Since $\theta^\beta|_S = 0$ for $\beta > k$ the equality (4.2.7) yields

$$d\theta^\alpha = \sum_{\beta=1}^{k} \theta^\beta \wedge \mu^\alpha_\beta, \quad \mu^\alpha_\beta = -\mu^\beta_\alpha \quad 1 \leq \alpha, \beta \leq k.$$

The uniqueness part of Proposition 4.2.15 implies that along S

$$\sigma^\alpha_\beta = \mu^\alpha_\beta \quad 1 \leq \alpha, \beta \leq k.$$

This can be equivalently rephrased as

$$\nabla^S_X Y = (\nabla^M_X Y)^\tau \quad \forall X, Y \in \text{Vect}(S). \tag{4.2.8}$$

B. $k < \alpha \leq m$. We deduce

$$0 = \sum_{\beta=1}^{k} \theta^\beta \wedge \mu^\alpha_\beta.$$

At this point we want to use the following elementary result.

Exercise 4.2.18 (Cartan Lemma). *Let V be a d-dimensional real vector space and consider p linearly independent elements $\omega_1, \ldots, \omega_p \in \Lambda^1 V$, $p \leq d$. If $\theta_1, \ldots, \theta_p \in \Lambda^1 V$ are such that*

$$\sum_{i=1}^{p} \theta_i \wedge \omega_i = 0,$$

then there exist scalars A_{ij}, $1 \leq i, j \leq p$ such that $A_{ij} = A_{ji}$ and

$$\theta_i = \sum_{j=1}^{p} A_{ij} \omega_j. \qquad \square$$

Using Cartan lemma we can find smooth functions $f_{\beta\gamma}^{\lambda}$, $\lambda > k$, $1 \leq \beta, \gamma \leq k$ satisfying

$$f_{\beta\gamma}^{\lambda} = f_{\gamma\beta}^{\lambda}, \text{ and } \mu_{\beta}^{\lambda} = f_{\beta\gamma}^{\lambda} \theta^{\gamma}.$$

Now form

$$\mathcal{N} = f_{\beta\gamma}^{\lambda} \theta^{\beta} \otimes \theta^{\gamma} \otimes X_{\lambda}.$$

We can view $\mathcal{N}$ as a symmetric bilinear map

$$\text{Vect}\,(S) \times \text{Vect}\,(S) \to C^{\infty}(N_S).$$

If $U, V \in \text{Vect}\,(S)$

$$U = U^{\beta} X_{\beta} = \theta^{\beta}(U) X_{\beta} \quad 1 \leq \beta \leq k,$$

and

$$V = V^{\gamma} X_{\gamma} = \theta^{\gamma}(V) X_{\gamma} \quad 1 \leq \gamma \leq k,$$

then

$$\mathcal{N}(U, V) = \sum_{\lambda > k} \left\{ \sum_{\beta} \left(\sum_{\gamma} f_{\beta\gamma}^{\lambda} \theta^{\gamma}(V) \right) \theta^{\beta}(U) \right\} X^{\lambda}$$

$$= \sum_{\lambda > k} \left(\sum_{\beta} \mu_{\beta}^{\lambda}(V) U^{\beta} \right) X_{\lambda}.$$

The last term is precisely the normal component of $\nabla_V^M U$. We have thus proved the following equality, so that we have established

$$\left(\nabla_V^M U \right)^{\nu} = \mathcal{N}(U, V) = \mathcal{N}(V, U) = \left(\nabla_U^M V \right)^{\nu}. \qquad (4.2.9)$$

The map $\mathcal{N}$ is called the *2nd fundamental form*[2] of $S \hookrightarrow M$.

There is an alternative way of looking at $\mathcal{N}$. Choose

$$U, V \in \text{Vect}\,(S), \quad N \in C^{\infty}(N_S).$$

_____
[2]The *first* fundamental form is the induced metric.

If we write $g(\bullet, \bullet) = \langle \bullet, \bullet \rangle$, then

$$\langle \mathcal{N}(U,V), N \rangle = \langle (\nabla_U^M V)^\nu, N \rangle = \langle \nabla_U^M V, N \rangle$$

$$= \nabla_U^M \langle V, N \rangle - \langle V, \nabla_U^M N \rangle = -\langle V, (\nabla_U^M N)^\tau \rangle.$$

We have thus established

$$-\langle V, (\nabla_U^M N)^\tau \rangle = \langle \mathcal{N}(U,V), N \rangle = \langle \mathcal{N}(V,U), N \rangle = -\langle U, (\nabla_V^M N)^\tau \rangle. \quad (4.2.10)$$

The 2nd fundamental form can be used to determine a relationship between the curvature of M and that of S. More precisely we have the following celebrated result.

***Theorema Egregium* (Gauss).**   Let R^M (resp. R^S) denote the Riemann curvature of (M, g) (resp. $(S, g|_S)$). Then for any $X, Y, Z, T \in \text{Vect}\,(S)$ we have

$$\langle R^M(X,Y)Z, T \rangle = \langle R^S(X,Y)Z, T \rangle$$
$$+ \langle \mathcal{N}(X,Z), \mathcal{N}(Y,T) \rangle - \langle \mathcal{N}(X,T), \mathcal{N}(Y,Z) \rangle. \quad (4.2.11)$$

Proof.   Note that

$$\nabla_X^M Y = \nabla_X^S Y + \mathcal{N}(X,Y).$$

We have

$$R^M(X,Y)X = [\nabla_X^M, \nabla_Y^M]Z - \nabla_{[X,Y]}^M Z$$

$$= \nabla_X^M \big(\nabla_Y^S Z + \mathcal{N}(Y,Z) \big) - \nabla_Y^M \big(\nabla_X^S Z + \mathcal{N}(X,Z) \big) - \nabla_{[X,Y]}^S Z - \mathcal{N}([X,Y], Z).$$

Take the inner product with T of both sides above. Since $\mathcal{N}(\bullet, \bullet)$ is N_S-valued, we deduce using (4.2.8)-(4.2.10)

$$\langle R^M(X,Y)Z, T \rangle = \langle \nabla_X^M \nabla_Y^S Z, T \rangle + \langle \nabla_X^M \mathcal{N}(Y,Z), T \rangle$$

$$-\langle \nabla_Y^M \nabla_X^S Z, T \rangle - \langle \nabla_Y^M \mathcal{N}(X,Z), T \rangle - \langle \nabla_{[X,Y]}^S Z, T \rangle$$

$$= \langle [\nabla_X^S, \nabla_Y^S]Z, T \rangle - \langle \mathcal{N}(Y,Z), \mathcal{N}(X,T) \rangle + \langle \mathcal{N}(X,Z), \mathcal{N}(Y,T) \rangle - \langle \nabla_{[X,Y]}^S Z, T \rangle.$$

This is precisely the equality (4.2.11). $\square$

The above result is especially interesting when S is a transversally oriented hypersurface, i.e., S is a codimension 1 submanifold such that the normal bundle N_S is trivial[3]. Pick an orthonormal frame $\boldsymbol{n}$ of N_S, i.e., a length 1 section of N_S, and choose an orthonormal moving frame $(X_1, ..., X_{m-1})$ of TS.

Then $(X_1, ..., X_{m-1}, \boldsymbol{n})$ is an orthonormal moving frame of $(TM)|_S$, and the second fundamental form is completely described by

$$\mathcal{N}_{\boldsymbol{n}}(X,Y) := \langle \mathcal{N}(X,Y), \boldsymbol{n} \rangle.$$

---

[3]Locally, all hypersurfaces are transversally oriented since N_S is locally trivial by definition.

$\mathcal{N}_n$ is a bona-fide symmetric bilinear form, and moreover, according to (4.2.10) we have

$$\mathcal{N}_n(X, Y) = -\langle \nabla_X^M \boldsymbol{n}, Y \rangle = -\langle \nabla_Y^M \boldsymbol{n}, X \rangle.$$

In this case, Gauss formula becomes

$$\langle R^S(X,Y)Z,T \rangle = \langle R^M(X,Y)Z,T \rangle - \begin{vmatrix} \mathcal{N}_n(X,Z) & \mathcal{N}_n(X,T) \\ \mathcal{N}_n(Y,Z) & \mathcal{N}_n(Y,T) \end{vmatrix}.$$

Let us further specialize, and assume $M = \mathbb{R}^m$. Then

$$\langle R^S(X,Y)Z,T \rangle = \begin{vmatrix} \mathcal{N}_n(X,T) & \mathcal{N}_n(X,Z) \\ \mathcal{N}_n(Y,T) & \mathcal{N}_n(Y,Z) \end{vmatrix}. \tag{4.2.12}$$

In particular, the sectional curvature along the plane spanned by X, Y is

$$\langle R^S(X,Y)Y,X \rangle = \mathcal{N}_n(X,X) \cdot \mathcal{N}_n(Y,Y) - |\mathcal{N}_n(X,Y)|^2.$$

This is a truly remarkable result. On the right-hand side we have an extrinsic term (it depends on the "space surrounding S"), while in the left-hand side we have a purely intrinsic term (which is defined entirely in terms of the internal geometry of S). Historically, the extrinsic term was discovered first (by Gauss), and very much to Gauss surprise (?!?) one does not need to look outside S to compute it. This marked the beginning of a new era in geometry. It changed dramatically the way people looked at manifolds and thus it fully deserves the name of The Golden (egregium) Theorem of Geometry.

We can now explain rigorously why we cannot wrap a plane canvas around the sphere. Notice that, when we deform a plane canvas, the only thing that changes is the *extrinsic geometry*, while the *intrinsic geometry* is not changed since the lengths of the "fibers" stays the same. Thus, any intrinsic quantity is invariant under "bending". In particular, no matter how we deform the plane canvas we will always get a surface with Gauss curvature 0 which cannot be wrapped on a surface of constant *positive* curvature! Gauss himself called the total curvature a "bending invariant".

Example 4.2.19. (Quadrics in $\mathbb{R}^3$). Let $A : \mathbb{R}^3 \to \mathbb{R}^3$ be a selfadjoint, invertible linear operator with at least one positive eigenvalue. This implies the quadric

$$Q_A = \{u \in \mathbb{R}^3 \; ; \; \langle Au, u \rangle = 1\},$$

is nonempty and smooth (use implicit function theorem to check this). Let $u_0 \in Q_A$. Then

$$T_{u_0} Q_A = \{x \in \mathbb{R}^3 \; ; \; \langle Au_0, x \rangle = 0\} = (Au_0)^\perp.$$

Q_A is a transversally oriented hypersurface in $\mathbb{R}^3$ since the map $Q_A \ni u \mapsto Au$ defines a nowhere vanishing section of the normal bundle. Set $\boldsymbol{n} = \frac{1}{|Au|} Au$.

Consider an orthonormal frame (e_0, e_1, e_2) of $\mathbb{R}^3$ such that $e_0 = \boldsymbol{n}(u_0)$. Denote the Cartesian coordinates in $\mathbb{R}^3$ with respect to this frame by (x^0, x^1, x^2), and set $\partial_i := \frac{\partial}{\partial x_i}$. Extend (e_1, e_2) to a local moving frame of TQ_A near u_0.

The second fundamental form of Q_A at u_0 is

$$\mathcal{N}_{\boldsymbol{n}}(\partial_i, \partial_j) = \langle\, \partial_i \boldsymbol{n}, \partial_j \,\rangle|_{u_0}.$$

We compute

$$\partial_i \boldsymbol{n} = \partial_i \left(\frac{Au}{|Au|} \right) = \partial_i(\langle\, Au, Au \,\rangle^{-1/2}) Au + \frac{1}{|Au|} A\partial_i u$$

$$= -\frac{\langle\, \partial_i Au, Au \,\rangle}{|Au|^{3/2}} Au + \frac{1}{|Au|} \partial_i Au.$$

Hence

$$\mathcal{N}_{\boldsymbol{n}}(\partial_i, \partial_j)|_{u_0} = \frac{1}{|Au_0|} \langle\, A\partial_i u, e_j \,\rangle|_{u_0}$$

$$= \frac{1}{|Au_0|} \langle\, \partial_i u, Ae_j \,\rangle|_{u_0} = \frac{1}{|Au_0|} \langle\, e_i, Ae_j \,\rangle. \qquad (4.2.13)$$

We can now compute the Gaussian curvature at u_0.

$$K_{u_0} = \frac{1}{|Au_0|^2} \begin{vmatrix} \langle\, Ae_1, e_1 \,\rangle & \langle\, Ae_1, e_2 \,\rangle \\ \langle\, Ae_2, e_1 \,\rangle & \langle\, Ae_2, e_2 \,\rangle \end{vmatrix}.$$

In particular, when $A = r^{-2}I$ so that Q_A is the round sphere of radius r we deduce

$$K_u = \frac{1}{r^2} \quad \forall |u| = r.$$

Thus, the round sphere has constant positive curvature. $\qquad\qquad\square$

Example 4.2.20. (Gauss). Let Σ be a transversally oriented, compact surface in $\mathbb{R}^3$, e.g., a connected sum of a finite number of tori. Note that the Whitney sum $N_\Sigma \oplus T\Sigma$ is the trivial bundle $\mathbb{R}^3_\Sigma$. We orient N_Σ such that

$$\text{orientation } N_\Sigma \wedge \text{orientation } T\Sigma = \text{orientation } \mathbb{R}^3.$$

Let $\boldsymbol{n}$ be the unit section of N_Σ defining the above orientation. We obtain in this way a map

$$\mathcal{G} : \Sigma \to S^2 = \{u \in \mathbb{R}^3 \,;\, |u| = 1\}, \quad \Sigma \ni x \mapsto \boldsymbol{n}(x) \in S^2.$$

The map $\mathcal{G}$ is called the *Gauss map* of $\Sigma \hookrightarrow S^2$. It really depends on how Σ is embedded in $\mathbb{R}^3$ so it is an *extrinsic object*. Denote by $\mathcal{N}_{\boldsymbol{n}}$ the second fundamental form of $\Sigma \hookrightarrow \mathbb{R}^3$, and let (x^1, x^2) be normal coordinates at $q \in \Sigma$ such that

$$\text{orientation } T_q\Sigma = \partial_1 \wedge \partial_2.$$

Consider the Euler form ε_Σ on Σ with the metric induced by the Euclidean metric in $\mathbb{R}^3$. Then, taking into account our orientation conventions, we have

$$2\pi\varepsilon_\Sigma(\partial_1, \partial_2) = R_{1212} = \begin{vmatrix} \mathcal{N}_{\boldsymbol{n}}(\partial_1, \partial_1) & \mathcal{N}_{\boldsymbol{n}}(\partial_1, \partial_2) \\ \mathcal{N}_{\boldsymbol{n}}(\partial_2, \partial_1) & \mathcal{N}_{\boldsymbol{n}}(\partial_2, \partial_2) \end{vmatrix}. \qquad (4.2.14)$$

Now notice that

$$\partial_i \boldsymbol{n} = -\mathcal{N}_{\boldsymbol{n}}(\partial_i, \partial_1)\partial_1 - \mathcal{N}_{\boldsymbol{n}}(\partial_i, \partial_2)\partial_2.$$

We can think of $\boldsymbol{n}, \partial_1 \mid_q$ and $\partial_2 \mid_q$ as defining a (positively oriented) frame of $\mathbb{R}^3$. The last equality can be rephrased by saying that the derivative of the Gauss map

$$\mathcal{G}_* : T_q\Sigma \to T_{\boldsymbol{n}(q)}S^2$$

acts according to

$$\partial_i \mapsto -\mathcal{N}_{\boldsymbol{n}}(\partial_i, \partial_1)\partial_1 - \mathcal{N}_{\boldsymbol{n}}(\partial_i, \partial_2)\partial_2.$$

In particular, we deduce

$$\mathcal{G}_* \text{ preserves (reverses) orientations} \Longleftrightarrow R_{1212} > 0 \ (< 0), \qquad (4.2.15)$$

because the orientability issue is decided by the sign of the determinant

$$\begin{vmatrix} 1 & 0 & 0 \\ 0 & -\mathcal{N}_{\boldsymbol{n}}(\partial_1, \partial_1) & -\mathcal{N}_{\boldsymbol{n}}(\partial_1, \partial_2) \\ 0 & -\mathcal{N}_{\boldsymbol{n}}(\partial_2, \partial_1) & -\mathcal{N}_{\boldsymbol{n}}(\partial_2, \partial_2) \end{vmatrix}.$$

At q, $\partial_1 \perp \partial_2$ so that,

$$\langle \partial_i \boldsymbol{n}, \partial_j \boldsymbol{n} \rangle = \mathcal{N}_{\boldsymbol{n}}(\partial_i, \partial_1)\mathcal{N}_{\boldsymbol{n}}(\partial_j, \partial_1) + \mathcal{N}_{\boldsymbol{n}}(\partial_i, \partial_2)\mathcal{N}_{\boldsymbol{n}}(\partial_j, \partial_2).$$

We can rephrase this coherently as an equality of matrices

$$\begin{bmatrix} \langle \partial_1 \boldsymbol{n}, \partial_1 \boldsymbol{n} \rangle & \langle \partial_1 \boldsymbol{n}, \partial_2 \boldsymbol{n} \rangle \\ \langle \partial_2 \boldsymbol{n}, \partial_1 \boldsymbol{n} \rangle & \langle \partial_2 \boldsymbol{n}, \partial_2 \boldsymbol{n} \rangle \end{bmatrix}$$

$$= \begin{bmatrix} \mathcal{N}_{\boldsymbol{n}}(\partial_1, \partial_1) & \mathcal{N}_{\boldsymbol{n}}(\partial_1, \partial_2) \\ \mathcal{N}_{\boldsymbol{n}}(\partial_2, \partial_1) & \mathcal{N}_{\boldsymbol{n}}(\partial_2, \partial_2) \end{bmatrix} \times \begin{bmatrix} \mathcal{N}_{\boldsymbol{n}}(\partial_1, \partial_1) & \mathcal{N}_{\boldsymbol{n}}(\partial_1, \partial_2) \\ \mathcal{N}_{\boldsymbol{n}}(\partial_2, \partial_1) & \mathcal{N}_{\boldsymbol{n}}(\partial_2, \partial_2) \end{bmatrix}^t.$$

Hence

$$\begin{vmatrix} \mathcal{N}_{\boldsymbol{n}}(\partial_1, \partial_1) & \mathcal{N}_{\boldsymbol{n}}(\partial_1, \partial_2) \\ \mathcal{N}_{\boldsymbol{n}}(\partial_1, \partial_2) & \mathcal{N}_{\boldsymbol{n}}(\partial_2, \partial_2) \end{vmatrix}^2 = \begin{vmatrix} \langle \partial_1 \boldsymbol{n}, \partial_1 \boldsymbol{n} \rangle & \langle \partial_1 \boldsymbol{n}, \partial_2 \boldsymbol{n} \rangle \\ \langle \partial_1 \boldsymbol{n}, \partial_2 \boldsymbol{n} \rangle & \langle \partial_2 \boldsymbol{n}, \partial_2 \boldsymbol{n} \rangle \end{vmatrix}. \qquad (4.2.16)$$

If we denote by dv_0 the metric volume form on S^2 induced by the restriction of the Euclidean metric on $\mathbb{R}^3$, we see that (4.2.14) and (4.2.16) put together yield

$$2\pi|\varepsilon_\Sigma(\partial_1, \partial_2)| = |dv_0(\partial_1 \boldsymbol{n}, \partial_2 \boldsymbol{n})| = |dv_0(\mathcal{G}_*(\partial_1), \mathcal{G}_*(\partial_2))|.$$

Using (4.2.15) we get

$$\varepsilon_\Sigma = \frac{1}{2\pi}\mathcal{G}_\Sigma^* dv_0 = \frac{1}{2\pi}\mathcal{G}_\Sigma^* \varepsilon_{S^2}. \qquad (4.2.17)$$

This is one form of the celebrated Gauss-Bonnet theorem. We will have more to say about it in the next subsection.

Note that the last equality offers yet another interpretation of the Gauss curvature. From this point of view the curvature is a *"distortion factor"*. The Gauss map "stretches" an infinitesimal parallelogram to some infinitesimal region on the unit sphere. The Gauss curvature describes by what factor the area of this parallelogram was changed. In Chapter 9 we will investigate in greater detail the Gauss map of *arbitrary* submanifolds of a Euclidean space. $\qquad \square$

4.2.5   *The Gauss-Bonnet theorem for oriented surfaces*

We conclude this chapter with one of the most beautiful results in geometry. Its meaning reaches deep inside the structure of a manifold and can be viewed as the origin of many fertile ideas.

Recall one of the questions we formulated at the beginning of our study: explain unambiguously why a sphere is "different" from a torus. This may sound like forcing our way in through an open door since everybody can "see" they are different. Unfortunately this is not a conclusive explanation since we can see only 3-dimensional things and possibly there are many ways to deform a surface outside our tight 3D Universe.

The elements of Riemann geometry we discussed so far will allow us to produce an invariant powerful enough to distinguish a sphere from a torus. But it will do more than that.

Theorem 4.2.21. (Gauss-Bonnet Theorem. Preliminary version.)   *Let S be a compact oriented surface without boundary. If g_0 and g_1 are two Riemann metrics on S and $\varepsilon_{g_i}(S)$ ($i = 0, 1$) are the corresponding Euler forms then*

$$\int_S \varepsilon_{g_0}(S) = \int_S \varepsilon_{g_1}(S).$$

Hence the quantity $\int_S \varepsilon_g(S)$ is independent of the Riemann metric g so that it really depends only on the topology of S!!!

The idea behind the proof is very natural. Denote by g_t the metric $g_t = g_0 + t(g_1 - g_0)$. We will show

$$\frac{d}{dt} \int_S \varepsilon_{g_t} = 0 \quad \forall t \in [0, 1].$$

It is convenient to consider a more general problem.

Definition 4.2.22.   Let M be a compact oriented manifold. For any Riemann metric g on E define

$$\mathcal{E}_M(M, g) = \int_M s(g) dV_g,$$

where $s(g)$ denotes the scalar curvature of (M, g). The functional $g \mapsto \mathcal{E}(g)$ is called the *Hilbert-Einstein functional.*   □

We have the following remarkable result.

Lemma 4.2.23.   *Let M be a compact oriented manifold without boundary and $g^t = (g_{ij}^t)$ be a 1-parameter family of Riemann metrics on M depending smoothly upon $t \in \mathbb{R}$. Then*

$$\frac{d}{dt} \mathcal{E}(g^t) = -\int_M \langle \mathrm{Ric}_{g^t} - \frac{1}{2} s(g^t) g^t \, , \, \dot{g}^t \rangle_t dV_{g^t}, \quad \forall t.$$

In the above formula $\langle \cdot, \cdot \rangle_t$ denotes the inner product induced by g^t on the space of symmetric (0,2)-tensors while the dot denotes the t-derivative.

Definition 4.2.24. A Riemann manifold (M, g) of dimension n is said to be *Einstein* if the metric g satisfies *Einstein's equation*

$$\mathrm{Ric}_g = \frac{s(x)}{n} g,$$

where $s(x)$ denotes the scalar curvature.

Example 4.2.25. Observe that if the Riemann metric g satisfies the condition

$$\mathrm{Ric}_g(x) = \lambda(x) g(x) \tag{4.2.18}$$

for some smooth function $\lambda \in C^\infty(M)$, then by taking the traces on both sides of the above equality we deduce

$$\lambda(x) = \frac{s(g)(x)}{n}, \quad n := \dim M.$$

Thus, the Riemann manifold is Einstein if and only if it satisfies (4.2.18).

Using the computations in Example 4.2.11 we deduce that a certain constant multiple of the Killing metric on a compact semisimple Lie group is an Einstein metric.

We refer to [13] for an in-depth study of the Einstein manifolds. $\square$

The $(0, 2)$-tensor

$$\mathcal{E}_{ij} := R_{ij}(x) - \frac{1}{2} s(x) g_{ij}(x)$$

is called the *Einstein tensor* of (M, g).

Exercise 4.2.26. Consider a 3-dimensional Riemann manifold (M, g). Show that

$$R_{ijk\ell} = \mathcal{E}_{ik} g_{j\ell} - \mathcal{E}_{i\ell} g_{jk} + \mathcal{E}_{j\ell} g_{ik} - \mathcal{E}_{jk} g_{i\ell} + \frac{s}{2}(g_{i\ell} g_{jk} - g_{ik} g_{j\ell}).$$

In particular, this shows that on a Riemann 3-manifold the full Riemann tensor is completely determined by the Einstein tensor. $\square$

Exercise 4.2.27. (Schouten-Struik, [87]). Prove that the scalar curvature of an Einstein manifold of dimension ≥ 3 is constant.
Hint: Use the 2nd Bianchi identity. $\square$

Notice that when (S, g) is a compact oriented Riemann surface two very nice things happen.
(i) (S, g) is Einstein (recall that only R_{1212} is nontrivial).
(ii) $\mathcal{E}(g) = 2 \int_S \varepsilon_g$.

Theorem 4.2.21 is thus an immediate consequence of Lemma 4.2.23.

Proof of the lemma We will produce a very explicit description of the integrand

$$\frac{d}{dt}\left(s(g^t) dV_{g^t} \right) = \left(\frac{d}{dt} s(g^t) \right) dV_{g^t} + s(g^t) \left(\frac{d}{dt} dV_{g^t} \right) \tag{4.2.19}$$

of $\frac{d}{dt}\mathcal{E}(g^t)$. We will adopt a "roll up your sleeves, and just do it" strategy reminiscent to the good old days of the tensor calculus frenzy. By this we mean that we will work in a nicely chosen collection of local coordinates, and we will keep track of the zillion indices we will encounter. As we will see, the computations are not as hopeless as they may seem to be.

We will study the integrand (4.2.19) at $t = 0$. The general case is entirely analogous. For typographical reasons we will be forced to introduce new notations. Thus, $\hat{g}$ will denote (g^t) for $t = 0$, while g^t will be denoted simply by g. A hat over a quantity means we think of that quantity at $t = 0$, while a dot means differentiation with respect to t at $t = 0$.

Let q be an arbitrary point on S, and denote by $(x^1, \ldots, x^n)$ a collection of $\hat{g}$-normal coordinates at q. Denote by ∇ the Levi-Civita connection of g and let Γ^i_{jk} denote its Christoffel symbols in the coordinates (x^i).

Many nice things happen   *at* q, and we list a few of them which will be used later.

$$\hat{g}_{ij} = \hat{g}^{ij} = \delta_{ij}, \;\; \partial_k \hat{g}_{ij} = 0. \tag{4.2.20}$$

$$\hat{\nabla}_i \partial_j = 0, \;\; \hat{\Gamma}^i_{jk} = 0. \tag{4.2.21}$$

If $\alpha = \alpha_i dx^i$ is a 1-form then, *at* q,

$$\delta_{\hat{g}}\alpha = \sum_i \partial_i \alpha_i. \text{ where } \delta = *d*. \tag{4.2.22}$$

In particular, for any smooth function u we have

$$(\Delta_{M,\hat{g}} u)(q) = -\sum_i \partial_i^2 u. \tag{4.2.23}$$

Set

$$h = (h_{ij}) := (\dot{g}) = (\dot{g}_{ij}).$$

The tensor h is a symmetric $(0, 2)$-tensor. Its $\hat{g}$-trace is the scalar

$$\text{tr}_{\hat{g}} h = \hat{g}^{ij} h_{ij} = \text{tr}\,\mathcal{L}^{-1}(h),$$

where $\mathcal{L}$ is the lowering the indices operator defined by $\hat{g}$. In particular, *at* q

$$\text{tr}_{\hat{g}} h = \sum_i h_{ii}. \tag{4.2.24}$$

The curvature of g is given by

$$R^\ell_{ikj} = -R^\ell_{ijk} = \partial_k \Gamma^\ell_{ij} - \partial_j \Gamma^\ell_{ik} + \Gamma^\ell_{mk}\Gamma^m_{ij} - \Gamma^\ell_{mj}\Gamma^m_{ik}.$$

The Ricci tensor is

$$R_{ij} = R^k_{ikj} = \partial_k \Gamma^k_{ij} - \partial_j \Gamma^k_{ik} + \Gamma^k_{mk}\Gamma^m_{ij} - \Gamma^k_{mj}\Gamma^m_{ik}.$$

Finally, the scalar curvature is

$$s = \text{tr}_g R_{ij} = g^{ij} R_{ij} = g^{ij}\left(\partial_k \Gamma^k_{ij} - \partial_j \Gamma^k_{ik} + \Gamma^k_{mk}\Gamma^m_{ij} - \Gamma^k_{mj}\Gamma^m_{ik}\right).$$

Differentiating s at $t = 0$, and then evaluating *at* q we obtain

$$\dot{s} = \dot{g}^{ij} \left(\partial_k \hat{\Gamma}^k_{ij} - \partial_j \hat{\Gamma}^k_{ik} \right) + \delta^{ij} \left(\partial_k \dot{\Gamma}^k_{ij} - \partial_j \dot{\Gamma}^k_{ik} \right)$$
$$= \dot{g}^{ij} \hat{R}_{ij} + \sum_i \left(\partial_k \dot{\Gamma}^k_{ii} - \partial_i \dot{\Gamma}^k_{ik} \right). \tag{4.2.25}$$

The term $\dot{g}^{ij}$ can be computed by derivating the equality $g^{ik}g_{jk} = \delta^i_k$ at $t = 0$. We get

$$\dot{g}^{ik}\hat{g}_{jk} + \hat{g}^{ik}h_{jk} = 0,$$

so that

$$\dot{g}^{ij} = -h_{ij}. \tag{4.2.26}$$

To evaluate the derivatives $\dot{\Gamma}$'s we use the known formulæ

$$\Gamma^m_{ij} = \frac{1}{2}g^{km} \left(\partial_i g_{jk} - \partial_k g_{ij} + \partial_j g_{ik} \right),$$

which, upon differentiation at $t = 0$, yield

$$\dot{\Gamma}^m_{ij} = \frac{1}{2} \left(\partial_i \hat{g}_{jk} - \partial_k \hat{g}_{ij} + \partial_j \hat{g}_{ik} \right) + \frac{1}{2}\hat{g}^{km} \left(\partial_i h_{jk} - \partial_k h_{ij} + \partial_j h_{ik} \right)$$
$$= \frac{1}{2} \left(\partial_i h_{jm} - \partial_m h_{ij} + \partial_j h_{im} \right). \tag{4.2.27}$$

We substitute (4.2.26)-(4.2.27) in (4.2.25), and we get, *at* q

$$\dot{s} = -\sum_{i,j} h_{ij}\hat{R}_{ij} + \frac{1}{2}\sum_{i,k}(\partial_k \partial_i h_{ik} - \partial_k^2 h_{ii} + \partial_k \partial_i h_{ik}) - \frac{1}{2}\sum_{i,k}(\partial_i^2 h_{kk} - \partial_i \partial_k h_{ik})$$
$$= -\sum_{i,j} h_{ij}\hat{R}_{ij} - \sum_{i,k}\partial_i^2 h_{kk} + \sum_{i,k}\partial_i \partial_k h_{ik}$$
$$= -\langle \widehat{\mathrm{Ric}}, \dot{g} \rangle_{\hat{g}} + \Delta_{M,\hat{g}}\mathrm{tr}_{\hat{g}}\,\dot{g} + \sum_{i,k}\partial_i \partial_k h_{ik}. \tag{4.2.28}$$

To get a coordinate free description of the last term note that, *at q*,

$$(\hat{\nabla}_k h)(\partial_i, \partial_m) = \partial_k h_{im}.$$

The total covariant derivative $\hat{\nabla}h$ is a $(0,3)$-tensor. Using the $\hat{g}$-trace we can construct a $(0,1)$-tensor, i.e., a 1-form

$$\mathrm{tr}_{\hat{g}}(\hat{\nabla}h) = \mathrm{tr}(\mathcal{L}_{\hat{g}}^{-1}\hat{\nabla}h),$$

where $\mathcal{L}_{\hat{g}}^{-1}$ is the raising the indices operator defined by $\hat{g}$. In the local coordinates (x^i) we have

$$\mathrm{tr}_{\hat{g}}(\hat{\nabla}h) = \hat{g}^{ij}(\hat{\nabla}_i h)_{jk}dx^k.$$

Using (4.2.20), and (4.2.22) we deduce that the last term in (4.2.28) can be rewritten (at q) as

$$\delta\mathrm{tr}_{\hat{g}}\left(\hat{\nabla}h\right) = \delta\mathrm{tr}_{\hat{g}}(\hat{\nabla}\dot{g}).$$

We have thus established that

$$\dot{s} = -\langle \widehat{\mathrm{Ric}}, \dot{g} \rangle_{\hat{g}} + \Delta_{M,\hat{g}} \mathrm{tr}_{\hat{g}} \dot{g} + \delta \mathrm{tr}_{\hat{g}}(\hat{\nabla}\dot{g}). \qquad (4.2.29)$$

The second term of the integrand (4.2.19) is a lot easier to compute.

$$dV_g = \pm\sqrt{|g|}dx^1 \wedge \cdots \wedge dx^n,$$

so that

$$d\dot{V}_g = \pm\frac{1}{2}|\hat{g}|^{-1/2}\frac{d}{dt}|g|dx^1 \wedge \cdots \wedge dx^n.$$

At q the metric is Euclidean, $\hat{g}_{ij} = \delta_{ij}$, and

$$\frac{d}{dt}|g| = \sum_i \dot{g}_{ii} = |\hat{g}| \cdot \mathrm{tr}_{\hat{g}}(\dot{g}) = \langle \hat{g}, \dot{g} \rangle_{\hat{g}}|\hat{g}|.$$

Hence

$$\dot{\mathcal{E}}(g) = \int_M \langle \left(\frac{1}{2}s(\hat{g})\hat{g} - \mathrm{Ric}_{\hat{g}}\right), \dot{g} \rangle_{\hat{g}} dV_{\hat{g}} + \int_M \left(\Delta_{M,\hat{g}}\, \mathrm{tr}_{\hat{g}}\, \dot{g} + \delta \mathrm{tr}_{\hat{g}}(\hat{\nabla}\dot{g})\right) dV_{\hat{g}}.$$

Green's formula shows the last two terms vanish and the proof of the Lemma is concluded.   □

Definition 4.2.28. Let S be a compact, oriented surface without boundary. We define its *Euler characteristic* as the number

$$\chi(S) = \frac{1}{2\pi} \int_S \varepsilon(g),$$

where g is an arbitrary Riemann metric on S. The number

$$g(S) = \frac{1}{2}(2 - \chi(S))$$

is called the *genus* of the surface.   □

Remark 4.2.29. According to the theorem we have just proved, the Euler characteristic is independent of the metric used to define it. Hence, the Euler characteristic is a *topological invariant* of the surface. The reason for this terminology will become apparent when we discuss DeRham cohomology, a $\mathbb{Z}$-graded vector space naturally associated to a surface whose Euler characteristic coincides with the number defined above. So far we have no idea whether $\chi(S)$ is even an integer.   □

Proposition 4.2.30.

$$\chi(S^2) = 2 \text{ and } \chi(T^2) = 0.$$

Proof. To compute $\chi(S^2)$ we use the round metric g_0 for which $K = 1$ so that

$$\chi(S^2) = \frac{1}{2\pi} \int_{S^2} dv_{g_0} = \frac{1}{2\pi} \operatorname{area}_{g_0}(S^2) = 2.$$

To compute the Euler characteristic of the torus we think of it as an Abelian Lie group with a bi-invariant metric. Since the Lie bracket is trivial we deduce from the computations in Subsection 4.2.2 that its curvature is zero. This concludes the proof of the proposition. □

Proposition 4.2.31. *If S_i (i=1,2) are two compact oriented surfaces without boundary then*

$$\chi(S_1 \# S_2) = \chi(S_1) + \chi(S_2) - 2.$$

Thus upon iteration we get

$$\chi(S_1 \# \cdots \# S_k) = \sum_{i=1}^{k} \chi(S_i) - 2(k-1),$$

for any compact oriented surfaces $S_1, \ldots, S_k$. In terms of genera, the last equality can be rephrased as

$$g(S_1 \# \cdots \# S_k) = \sum_{i=1}^{k} g(S_i).$$

In the proof of this proposition we will use special metrics on connected sums of surfaces which require a preliminary analytical discussion.

Consider $f : (-4, 4) \to (0, \infty)$ a smooth, even function such that
(i) $f(x) = 1$ for $|x| \le 2$.
(ii) $f(x) = \sqrt{1 - (x+3)^2}$ for $x \in [-4, -3.5]$.
(iii) $f(x) = \sqrt{1 - (x-3)^2}$ for $x \in [3.5, 4]$.
(iv) f is non-decreasing on $[-4, 0]$.
One such function is shown in Figure 4.3.

Denote by S_f the surface inside $\mathbb{R}^3$ obtained by rotating the graph of f about the x-axis. Because of properties (i)-(iv), S_f is a smooth surface diffeomorphic[4] to S^2. We denote by g the metric on S_f induced by the Euclidean metric in $\mathbb{R}^3$. Since S_f is diffeomorphic to a sphere

$$\int_{S_f} K_g dV_g = 2\pi \chi(S^2) = 4\pi.$$

Set

$$S_f^{\pm} := S_f \cap \{\pm x > 0\}, \quad S_f^{\pm 1} := S_f \cap \{\pm x > 1\}$$

---

[4]One such diffeomorphism can be explicitly constructed projecting along radii starting at the origin.

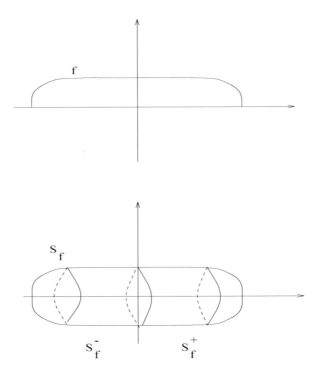

Fig. 4.3   *Generating a hot-dog-shaped surface*

Since f is even we deduce

$$\int_{S_f^\pm} K_g dv_g = \frac{1}{2} \int_{S_f} K_g dv_g = 2\pi. \tag{4.2.30}$$

On the other hand, on the neck $C = \{|x| \leq 2\}$ the metric g is locally Euclidean $g = dx^2 + d\theta^2$, so that over this region $K_g = 0$. Hence

$$\int_C K_g dv_g = 0. \tag{4.2.31}$$

Proof of Proposition 4.2.31   Let $D_i \subset S_i$ $(i = 1, 2)$ be a local coordinate neighborhood diffeomorphic with a disk in the plane. Pick a metric g_i on S_i such that (D_1, g_1) is *isometric* with S_f^+ and (D_2, g_2) is isometric to S_f^-. The connected sum $S_1 \# S_2$ is obtained by chopping off the regions S_f^1 from D_1 and S_f^{-1} from D_2 and (isometrically) identifying the remaining cylinders $S_f^\pm \cap \{|x| \leq 1\} = C$ and call O the overlap created by gluing (see Figure 4.4). Denote the metric thus obtained

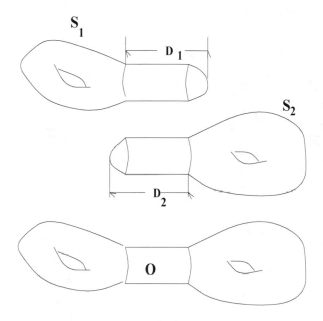

Fig. 4.4 *Special metric on a connected sum*

on $S_1 \# S_2$ by $\hat{g}$. We can now compute

$$\chi(S_1 \# S_2) = \frac{1}{2\pi} \int_{S_1 \# S_2} K_{\hat{g}} dv_{\hat{g}}$$

$$= \frac{1}{2\pi} \int_{S_1 \backslash D_1} K_{g_1} dV_{g_1} + \frac{1}{2\pi} \int_{S_2 \backslash D_2} K_{g_2} dV_{g_2} + \frac{1}{2\pi} \int_O K_g \, dV_g$$

$$\overset{(4.2.31)}{=} \frac{1}{2\pi} \int_{S_1} K_{g_1} dV_{g_1} - \frac{1}{2\pi} \int_{D_1} K_g dV_g$$

$$+ \frac{1}{2\pi} \int_{S_2} K_{g_2} dV_{g_2} - \frac{1}{2\pi} \int_{D_2} K_g dV_g \overset{(4.2.30)}{=} \chi(S_1) + \chi(S_2) - 2.$$

This completes the proof of the proposition. □

Corollary 4.2.32 (Gauss-Bonnet). *Let Σ_g denote the connected sum of g-tori. (By definition $\Sigma_0 = S^2$.) Then*

$$\chi(\Sigma_g) = 2 - 2g \ \text{ and } \ g(\Sigma_g) = g.$$

In particular, a sphere is not diffeomorphic to a torus.

Remark 4.2.33. It is a classical result that the only compact oriented surfaces are the connected sums of g-tori (see [68]), so that the genus of a compact oriented surface is a complete topological invariant. □

Chapter 5

Elements of the Calculus of Variations

This is a very exciting subject lying at the frontier between mathematics and physics. The limited space we will devote to this subject will hardly do it justice, and we will barely touch its physical significance. We recommend to anyone looking for an intellectual feast the Chapter 16 in vol.2 of "The Feynmann Lectures on Physics" [35], which in our opinion is the most eloquent argument for the raison d'être of the calculus of variations.

5.1 The least action principle

5.1.1 *The 1-dimensional Euler-Lagrange equations*

From a very "dry" point of view, the fundamental problem of the calculus of variations can be easily formulated as follows.

Consider a smooth manifold M, and let $L : \mathbb{R} \times TM \to \mathbb{R}$ by a smooth function called the *lagrangian*. Fix two points p_0, $p_1 \in M$. The *action* of a piecewise smooth path $\gamma : [0, 1] \to M$ connecting these points is the real number $S(\gamma) = S_L(\gamma)$ defined by

$$S(\gamma) = S_L(\gamma) := \int_0^1 L(t, \dot{\gamma}(t), \gamma(t)) dt.$$

In the calculus of variations one is interested in those paths as above with minimal action.

Example 5.1.1. Given a smooth function $U : \mathbb{R}^3 \to \mathbb{R}$ called the *potential*, we can form the lagrangian

$$L(\dot{q}, q) : \mathbb{R}^3 \times \mathbb{R}^3 \cong T\mathbb{R}^3 \to \mathbb{R},$$

given by

$$L = Q - U = \text{kinetic energy} - \text{potential energy} = \frac{1}{2}m|\dot{q}|^2 - U(q).$$

The scalar m is called the *mass*. The action of a path (trajectory) $\gamma : [0, 1] \to \mathbb{R}^3$ is a quantity called the *Newtonian action*. Note that, as a physical quantity, the Newtonian action is measured in the same units as the energy. $\square$

Example 5.1.2. To any Riemann manifold (M, g) one can naturally associate two lagrangians L_1, $L_2 : TM \to \mathbb{R}$ defined by

$$L_1(v, q) = g_q(v, v)^{1/2} \quad (v \in T_q M),$$

and

$$L_2(v, q) = \frac{1}{2} g_q(v, v).$$

We see that the action defined by L_1 coincides with the length of a path. The action defined by L_2 is called the *energy* of a path.   □

Before we present the main result of this subsection we need to introduce a bit of notation.

Tangent bundles are very peculiar manifolds. Any collection $(q^1, \dots, q^n)$ of local coordinates on a smooth manifold M automatically induces local coordinates on TM. Any point in TM can be described by a pair (v, q), where $q \in M$, $v \in T_q M$. Furthermore, v has a decomposition

$$v = v^i \partial_i, \quad \text{where} \quad \partial_i := \frac{\partial}{\partial q^i}.$$

We set $\dot{q}^i := v^i$ so that

$$v = \dot{q}^i \partial_i.$$

The collection $(\dot{q}^1, \dots, \dot{q}^n; q^1, \dots, q^n)$ defines local coordinates on TM. These are said to be *holonomic* local coordinates on TM. This will be the only type of local coordinates we will ever use.

Theorem 5.1.3 (The least action principle). *Let* $L : \mathbb{R} \times TM \to \mathbb{R}$ *be a lagrangian, and* p_0, $p_1 \in M$ *two fixed points. Suppose* $\gamma : [0, 1] \to M$ *is a smooth path such that the following hold.*

(i) $\gamma(i) = p_i$, $i = 0, 1$.
(ii) $S_L(\gamma) \leq S_L(\tilde{\gamma})$, *for any smooth path* $\tilde{\gamma} : [0, 1] \to M$ *joining* p_0 *to* p_1.

Then the path γ *satisfies the* Euler-Lagrange equations

$$\frac{d}{dt} \frac{\partial}{\partial \dot{\gamma}} L(t, \dot{\gamma}, \gamma) = \frac{\partial}{\partial \gamma} L(t, \dot{\gamma}, \gamma).$$

More precisely, if $(\dot{q}^j, q^i)$ *are holonomic local coordinates on* TM *such that* $\gamma(t) = (q^i(t))$, *and* $\dot{\gamma} = (\dot{q}^j(t))$, *then* γ *is a solution of the system of nonlinear ordinary differential equations*

$$\frac{d}{dt} \frac{\partial L}{\partial \dot{q}^k}(t, \dot{q}^j, q^i) = \frac{\partial L}{\partial q^k}(t, \dot{q}^j, q^i), \quad k = 1, \dots, n = \dim M.$$

Definition 5.1.4. A path $\gamma : [0,1] \to M$ satisfying the Euler-Lagrange equations with respect to some lagrangian L is said to be an *extremal* of L. $\qquad\square$

To get a better feeling of these equations consider the special case discussed in Example 5.1.1

$$L = \frac{1}{2} m|\dot{q}|^2 - U(q).$$

Then

$$\frac{\partial}{\partial \dot{q}} L = m\dot{q}, \quad \frac{\partial}{\partial q} L = -\nabla U(q),$$

and the Euler-Lagrange equations become

$$m\ddot{q} = -\nabla U(q). \qquad (5.1.1)$$

These are precisely Newton's equation of the motion of a particle of mass m in the force field $-\nabla U$ generated by the potential U.

In the proof of the least action principle we will use the notion of *variation of a path*.

Definition 5.1.5. Let $\gamma : [0,1] \to M$ be a smooth path. A *variation* of γ is a smooth map

$$\alpha = \alpha_s(t) : (-\varepsilon, \varepsilon) \times [0,1] \to M,$$

such that $\alpha_0(t) = \gamma(t)$. If moreover, $\alpha_s(i) = p_i \; \forall s, \; i = 0, 1$, then we say that α is a *variation rel endpoints*. $\qquad\square$

Proof of Theorem 5.1.3. Let α_s be a variation of γ rel endpoints. Then

$$S_L(\alpha_0) \le S_L(\alpha_s) \quad \forall s,$$

so that

$$\frac{d}{ds}\Big|_{s=0} S_L(\alpha_s) = 0.$$

Assume for simplicity that the image of γ is entirely contained in some open coordinate neighborhood U with coordinates $(q^1, \ldots, q^n)$. Then, for very small $|s|$, we can write

$$\alpha_s(t) = (q^i(s,t)) \quad \text{and} \quad \frac{d\alpha_s}{dt} = (\dot{q}^i(s,t)).$$

Following the tradition, we set

$$\delta\alpha := \frac{\partial \alpha}{\partial s}\Big|_{s=0} = \frac{\partial q^i}{\partial s} \partial_i \quad \delta\dot{\alpha} := \frac{\partial}{\partial s}\Big|_{s=0} \frac{d\alpha_s}{dt} = \frac{\partial \dot{q}^j}{\partial s} \frac{\partial}{\partial \dot{q}^j}.$$

The quantity $\delta\alpha$ is a vector field along γ called *infinitesimal variation* (see Figure 5.1). In fact, the pair $(\delta\alpha; \delta\dot{\alpha}) \in T(TM)$ is a vector field along $t \mapsto (\gamma(t), \dot{\gamma}(t)) \in TM$. Note that $\delta\dot{\alpha} = \frac{d}{dt}\delta\alpha$, and at endpoints $\delta\alpha = 0$.

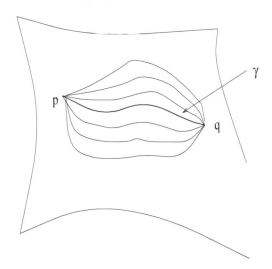

Fig. 5.1   *Deforming a path rel endpoints*

Exercise 5.1.6.   Prove that if $t \mapsto X(t) \in T_{\gamma(t)}M$ is a smooth vector field along γ, such that $X(t) = 0$ for $t = 0, 1$ then there exists at least one variation rel endpoints α such that $\delta\alpha = X$.

Hint: Use the exponential map of some Riemann metric on M.   □

We compute (at $s = 0$)

$$0 = \frac{d}{ds}S_L(\alpha_s) = \frac{d}{ds}\int_0^1 L(t, \dot{\alpha}_s, \alpha_s) = \int_0^1 \frac{\partial L}{\partial q^i}\delta\alpha^i dt + \int_0^1 \frac{\partial L}{\partial \dot{q}^j}\delta\dot{\alpha}_s^j dt.$$

Integrating by parts in the second term in the right-hand side we deduce

$$\int_0^1 \left\{ \frac{\partial L}{\partial q^i} - \frac{d}{dt}\left(\frac{\partial L}{\partial \dot{q}^i} \right) \right\} \delta\alpha^i dt. \tag{5.1.2}$$

The last equality holds *for any variation* α. From Exercise 5.1.6 deduce that it holds for any vector field $\delta\alpha^i\partial_i$ along γ. At this point we use the following classical result of analysis.

If $f(t)$ is a continuous function on [0,1] such that

$$\int_0^1 f(t)g(t)dt = 0 \quad \forall g \in C_0^\infty(0, 1),$$

then f is identically zero.

Using this result in (5.1.2) we deduce the desired conclusion.   □

Remark 5.1.7.   (a) In the proof of the least action principle we used a simplifying assumption, namely that the image of γ lies in a coordinate neighborhood. This is true locally, and for the above arguments to work it suffices to choose only a special type of variations, localized on small intervals of $[0, 1]$. In terms of infinitesimal variations this means we need to look only at vector fields along γ supported in local coordinate neighborhoods. We leave the reader fill in the details.

(b) The Euler-Lagrange equations were described using holonomic local coordinates. The minimizers of the action, if any, are objects independent of any choice of local coordinates, so that the Euler-Lagrange equations have to be independent of such choices. We check this directly.

If (x^i) is another collection of local coordinates on M and $(\dot{x}^j, x^i)$ are the coordinates induced on TM, then we have the transition rules

$$x^i = x^i(q^1, \ldots, q^n), \quad \dot{x}^j = \frac{\partial x^j}{\partial q^k} \dot{q}^k,$$

so that

$$\frac{\partial}{\partial q^i} = \frac{\partial x^j}{\partial q^i} \frac{\partial}{\partial x^j} + \frac{\partial^2 x^j}{\partial q^k \partial q^i} \dot{q}^k \frac{\partial}{\partial \dot{x}^j}$$

$$\frac{\partial}{\partial \dot{q}^j} = \frac{\partial \dot{x}^i}{\partial \dot{q}^j} \frac{\partial}{\partial \dot{x}^i} = \frac{\partial x^j}{\partial q^i} \frac{\partial}{\partial \dot{x}^j}$$

Then

$$\frac{\partial L}{\partial q^i} = \frac{\partial x^j}{\partial q^i} \frac{\partial L}{\partial x^j} + \frac{\partial^2 x^j}{\partial q^k \partial q^i} \dot{q}^k \frac{\partial L}{\partial \dot{x}^j}.$$

$$\frac{d}{dt} \left(\frac{\partial L}{\partial \dot{q}^i} \right) = \frac{d}{dt} \left(\frac{\partial x^j}{\partial q^i} \frac{\partial L}{\partial \dot{x}^j} \right) = \frac{\partial^2 x^j}{\partial q^k \partial q^i} \dot{q}^k \frac{\partial L}{\partial \dot{x}^j} + \frac{\partial x^j}{\partial q^i} \frac{d}{dt} \left(\frac{\partial L}{\partial \dot{x}^j} \right).$$

We now see that the Euler-Lagrange equations in the q-variables imply the Euler-Lagrange in the x-variable, i.e., these equations are *independent of coordinates*.

The æsthetically conscious reader may object to the way we chose to present the Euler-Lagrange equations. These are intrinsic equations we formulated in a coordinate dependent fashion. Is there any way of writing these equation so that the intrinsic nature is visible "on the nose"?

If the lagrangian L satisfies certain nondegeneracy conditions there are two ways of achieving this goal. One method is to consider a natural nonlinear connection ∇^L on TM as in [77]. The Euler-Lagrange equations for an extremal $\gamma(t)$ can then be rewritten as a "geodesics equation"

$$\nabla^L_{\dot{\gamma}} \dot{\gamma}.$$

The example below will illustrate this approach on a very special case when L is the lagrangian L_2 defined in Example 5.1.2, in which the extremals will turn out to be precisely the geodesics on a Riemann manifold.

Another far reaching method of globalizing the formulation of the Euler-Lagrange equation is through the Legendre transform, which again requires a nondegeneracy condition on the lagrangian. Via the Legendre transform the Euler-Lagrange equations become a system of *first order equations* on the cotangent bundle T^*M known as *Hamilton equations*.

These equations have the advantage that can be formulated on manifolds more general than the cotangent bundles, namely on *symplectic manifolds*. These are manifolds carrying a closed 2-form whose restriction to each tangent space defines a symplectic duality (see Subsection 2.2.4).

Much like the geodesics equations on a Riemann manifold, the Hamilton equations carry a lot of information about the structure of symplectic manifolds, and they are currently the focus of very intense research. For more details and examples we refer to the monographs [5, 25]. $\qquad\square$

Example 5.1.8. Let (M, g) be a Riemann manifold. We will compute the Euler-Lagrange equations for the lagrangians L_1, L_2 in Example 5.1.2.

$$L_2(\dot{q}, q) = \frac{1}{2} g_{ij}(q) \dot{q}^i \dot{q}^j,$$

so that

$$\frac{\partial L_2}{\partial \dot{q}^k} = g_{jk} \dot{q}^j \qquad \frac{\partial L_2}{\partial q^k} = \frac{1}{2} \frac{\partial g_{ij}}{\partial q^k} \dot{q}^i \dot{q}^j.$$

The Euler-Lagrange equations are

$$\ddot{q}^j g_{jk} + \frac{\partial g_{jk}}{\partial q^i} \dot{q}^i \dot{q}^j = \frac{1}{2} \frac{\partial g_{ij}}{\partial q^k} \dot{q}^i \dot{q}^j. \qquad (5.1.3)$$

Since $g^{km} g_{jm} = \delta_j^m$, we get

$$\ddot{q}^m + g^{km} \left(\frac{\partial g_{jk}}{\partial q^i} - \frac{1}{2} \frac{\partial g_{ij}}{\partial q^k} \right) \dot{q}^i \dot{q}^j = 0. \qquad (5.1.4)$$

When we derivate with respect to t the equality $g_{ik} \dot{q}^i = g_{jk} \dot{q}^j$ we deduce

$$g^{km} \frac{\partial g_{jk}}{\partial q^i} \dot{q}^i \dot{q}^j = \frac{1}{2} g^{km} \left(\frac{\partial g_{ik}}{\partial q^j} + \frac{\partial g_{jk}}{\partial q^i} \right) \dot{q}^i \dot{q}^j.$$

We substitute this equality in (5.1.4), and we get

$$\ddot{q}^m + \frac{1}{2} g^{km} \left(\frac{\partial g_{ik}}{\partial q^j} + \frac{\partial g_{jk}}{\partial q^i} - \frac{\partial g_{ij}}{\partial q^k} \right) \dot{q}^i \dot{q}^j = 0. \qquad (5.1.5)$$

The coefficient of $\dot{q}^i \dot{q}^j$ in (5.1.5) is none other than the Christoffel symbol Γ_{ij}^m so this equation is precisely the geodesic equation. $\qquad\square$

Example 5.1.9. Consider now the lagrangian $L_1(\dot{q}, q) = (g_{ij} \dot{q}^i \dot{q}^j)^{1/2}$. Note that the action

$$S_L(q(t)) = \int_{p_0}^{p_1} L(\dot{q}, q) dt,$$

is independent of the parametrization $t \mapsto q(t)$ since it computes the length of the path $t \mapsto q(t)$. Thus, when we express the Euler-Lagrange equations for a minimizer γ_0 of this action, we may as well assume it is parametrized by arclength, i.e., $|\dot{\gamma}_0| = 1$. The Euler-Lagrange equations for L_1 are

$$\frac{d}{dt} \frac{g_{kj}\dot{q}^j}{\sqrt{g_{ij}\dot{q}^i\dot{q}^j}} = \frac{\frac{\partial g_{ij}}{\partial q^k}\dot{q}^i\dot{q}^j}{2\sqrt{g_{ij}\dot{q}^i\dot{q}^j}}.$$

Along the extremal we have $g_{ij}\dot{q}^i\dot{q}^j = 1$ (arclength parametrization) so that the previous equations can be rewritten as

$$\frac{d}{dt}\left(g_{kj}\dot{q}^j\right) = \frac{1}{2}\frac{\partial g_{ij}}{\partial q^k}\dot{q}^i\dot{q}^j.$$

We recognize here the equation (5.1.3) which, as we have seen, is the geodesic equation in disguise. This fact almost explains why the geodesics are the shortest paths between two nearby points. $\qquad\square$

5.1.2 *Noether's conservation principle*

This subsection is intended to offer the reader a glimpse at a fascinating subject touching both physics and geometry. We first need to introduce a bit of traditional terminology commonly used by physicists.

Consider a smooth manifold M. The tangent bundle TM is usually referred to as the *space of states* or the *lagrangian phase space*. A point in TM is said to be a *state*. A lagrangian $L : \mathbb{R} \times TM \to \mathbb{R}$ associates to each state several meaningful quantities.

- The *generalized momenta*: $p_i = \frac{\partial L}{\partial \dot{q}^i}$.
- The *energy*: $H = p_i\dot{q}^i - L$.
- The *generalized force*: $F = \frac{\partial L}{\partial q^i}$.

This terminology can be justified by looking at the lagrangian of a classical particle in a potential force field, $F = -\nabla U$,

$$L = \frac{1}{2}m|\dot{q}|^2 - U(q).$$

The momenta associated to this lagrangian are the usual kinetic momenta of the Newtonian mechanics

$$p_i = m\dot{q}^i,$$

while H is simply the total energy

$$H = \frac{1}{2}m|\dot{q}|^2 + U(q).$$

It will be convenient to think of an extremal for an arbitrary lagrangian $L(t, \dot{q}, q)$ as describing the motion of a particle under the influence of the generalized force.

Proposition 5.1.10 (Conservation of energy). *Let $\gamma(t)$ be an extremal of a time independent lagrangian $L = L(\dot{q}, q)$. Then the energy is conserved along γ, i.e.,*

$$\frac{d}{dt} H(\gamma, \dot{\gamma}) = 0.$$

Proof. By direct computation we get

$$\frac{d}{dt} H(\gamma, \dot{\gamma}) = \frac{d}{dt}(p_i \dot{q}^i - L) = \frac{d}{dt}\left(\frac{\partial L}{\partial \dot{q}^i}\right) \dot{q}^i + \frac{\partial L}{\partial \dot{q}^i} \ddot{q}^i - \frac{\partial L}{\partial q^i} \dot{q}^i - \frac{\partial L}{\partial \dot{q}^i} \ddot{q}^i$$

$$= \left\{ \frac{d}{dt}\left(\frac{\partial L}{\partial \dot{q}^i}\right) - \frac{\partial L}{\partial q^i} \right\} \dot{q}^i = 0 \quad (\text{by Euler} - \text{Lagrange}). \qquad \square$$

At the beginning of the 20th century (1918), Emmy Noether discovered that many of the conservation laws of the classical mechanics had a geometric origin: they were, most of them, reflecting a symmetry of the lagrangian!!!

This became a driving principle in the search for conservation laws, and in fact, conservation became almost synonymous with symmetry. It eased the leap from classical to quantum mechanics, and one can say it is a very important building block of quantum physics in general. In the few instances of conservation laws where the symmetry was not apparent the conservation was always "blamed" on a "hidden symmetry". What is then this Noether principle?

To answer this question we need to make some simple observations.

Let X be a vector field on a smooth manifold M defining a global flow Φ^s. This flow induces a flow Ψ^s on the tangent bundle TM defined by

$$\Psi^s(v, x) = \left(\Phi_*^s(v), \Phi^s(x) \right).$$

One can think of Ψ^s as defining an action of the additive group $\mathbb{R}$ on TM. Alternatively, physicists say that X is an *infinitesimal symmetry* of the given mechanical system described by the lagrangian L.

Example 5.1.11. Let M be the unit round sphere $S^2 \subset \mathbb{R}^3$. The rotations about the z-axis define a 1-parameter group of isometries of S^2 generated by $\frac{\partial}{\partial \theta}$, where θ is the longitude on S^2. $\qquad \square$

Definition 5.1.12. Let L be a lagrangian on TM, and X a vector field on M. The lagrangian L is said to be *X-invariant* if

$$L \circ \Psi^s = L, \quad \forall s. \qquad \square$$

Denote by $\mathfrak{X} \in \text{Vect}\,(TM)$ the infinitesimal generator of Ψ^s, and by $\mathcal{L}_\mathfrak{X}$ the Lie derivative on TM along $\mathfrak{X}$. We see that L is X-invariant if and only if

$$\mathcal{L}_\mathfrak{X} L = 0.$$

We describe this derivative using the local coordinates $(\dot{q}^j, q^i)$. Set

$$\left(\dot{q}^j(s), q^i(s) \right) := \Psi^s(\dot{q}^j, q^i).$$

Then

$$\frac{d}{ds}\big|_{s=0}\, q^i(s) = X^k \delta^i_k.$$

To compute $\frac{d}{ds}\big|_{s=0}\, \dot{q}^j(s)\frac{\partial}{\partial q^j}$ we use the definition of the Lie derivative on M

$$-\frac{d}{ds}\dot{q}^j\frac{\partial}{\partial \dot{q}^j} = L_X(\dot{q}^i\frac{\partial}{\partial q^i}) = \left(X^k\frac{\partial \dot{q}^j}{\partial q^k} - \dot{q}^k\frac{\partial X^j}{\partial q^k} \right)\frac{\partial}{\partial q^j} = -\dot{q}^i\frac{\partial X^j}{\partial q^i}\frac{\partial}{\partial q^j},$$

since $\partial \dot{q}^j/\partial q^i = 0$ on TM. Hence

$$\mathfrak{X} = X^i\frac{\partial}{\partial q^i} + \dot{q}^k\frac{\partial X^j}{\partial q^k}\frac{\partial}{\partial \dot{q}^j}.$$

Corollary 5.1.13. *The lagrangian L is X-invariant if and only if*

$$X^i\frac{\partial L}{\partial q^i} + \dot{q}^k\frac{\partial X^j}{\partial q^k}\frac{\partial L}{\partial \dot{q}^j} = 0. \qquad (5.1.6)$$

$\square$

Theorem 5.1.14 (E. Noether). *If the lagrangian L is X-invariant, then the quantity*

$$P_X = X^i\frac{\partial L}{\partial \dot{q}^i} = X^i p_i$$

is conserved along the extremals of L.

Proof. Consider an extremal $\gamma = \gamma(q^i(t))$ of L. We compute

$$\frac{d}{dt}P_X(\gamma,\dot{\gamma}) = \frac{d}{dt}\left\{ X^i(\gamma(t))\frac{\partial L}{\partial \dot{q}^i} \right\} = \frac{\partial X^i}{\partial q^k}\dot{q}^k\frac{\partial L}{\partial \dot{q}^i} + X^i\frac{d}{dt}\left(\frac{\partial L}{\partial \dot{q}^i} \right)$$

$$\overset{\text{Euler-Lagrange}}{=} \frac{\partial X^i}{\partial q^k}\dot{q}^k\frac{\partial L}{\partial \dot{q}^i} + X^i\frac{\partial L}{\partial q^i} \overset{(5.1.6)}{=} 0. \qquad \square$$

The classical conservation-of-momentum law is a special consequence of Noether's theorem.

Corollary 5.1.15. *Consider a lagrangian $L = L(t,\dot{q},q)$ on $\mathbb{R}^n$. If $\frac{\partial L}{\partial q^i} = 0$, i.e., the i-th component of the force is zero, then $\frac{dp_i}{dt} = 0$ along any extremal, i.e., the i-th component of the momentum is conserved.*

Proof. Take $X = \frac{\partial}{\partial q^i}$ in Noether's conservation law. $\square$

The conservation of momentum has an interesting application in the study of geodesics.

Example 5.1.16. (Geodesics on surfaces of revolution). Consider a surface of revolution S in $\mathbb{R}^3$ obtained by rotating about the z-axis the curve $y = f(z)$

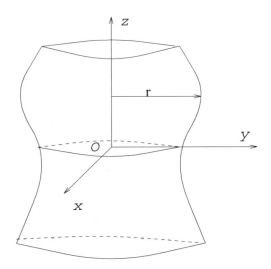

Fig. 5.2 *A surface of revolution*

situated in the yz plane. If we use cylindrical coordinates (r, θ, z) we can describe S as $r = f(z)$.

In these coordinates, the Euclidean metric in $\mathbb{R}^3$ has the form

$$ds^2 = dr^2 + dz^2 + r^2 d\theta^2.$$

We can choose (z, θ) as local coordinates on S, then the induced metric has the form

$$g_S = \{1 + (f'(z))^2\}dz^2 + f^2(z)d\theta^2 = A(z)dz^2 + r^2 d\theta^2, \quad r = f(z).$$

The lagrangian defining the geodesics on S is

$$L = \frac{1}{2}\left(A\dot{z}^2 + r^2\dot{\theta}^2\right).$$

We see that L is independent of θ: $\frac{\partial L}{\partial \theta} = 0$, so that the generalized momentum

$$\frac{\partial L}{\partial \dot{\theta}} = r^2\dot{\theta}$$

is conserved along the geodesics. $\square$

This fact can be given a nice geometric interpretation. Consider a geodesic

$$\gamma(t) = (z(t), \theta(t)),$$

and compute the angle ϕ between $\dot{\gamma}$ and $\frac{\partial}{\partial \theta}$. We get

$$\cos\phi = \frac{\langle \dot{\gamma}, \partial/\partial\theta \rangle}{|\dot{\gamma}| \cdot |\partial/\partial\theta|} = \frac{r^2\dot{\theta}}{r|\dot{\gamma}|},$$

i.e., $r\cos\phi = r^2\dot{\theta}|\dot{\gamma}|^{-1}$. The conservation of energy implies that $|\dot{\gamma}|^2 = 2L = H$ is constant along the geodesics. We deduce the following classical result.

Theorem 5.1.17 (Clairaut). *On a surface of revolution* $r = f(z)$ *the quantity* $r \cos \phi$ *is constant along any geodesic, where* $\phi \in (-\pi, \pi)$ *is the oriented angle the geodesic makes with the parallels* $z = const.$ □

Exercise 5.1.18. Describe the geodesics on the round sphere S^2, and on the cylinder $\{x^2 + y^2 = 1\} \subset \mathbb{R}^3$. □

5.2 The variational theory of geodesics

We have seen that the paths of minimal length between two points on a Riemann manifold are necessarily geodesics.

However, given a geodesic joining two points q_0, q_1 it may happen that it is not a minimal path. This should be compared with the situation in calculus, when a critical point of a function f may not be a minimum or a maximum.

To decide this issue one has to look at the second derivative. This is precisely what we intend to do in the case of geodesics. This situation is a bit more complicated since the action functional

$$S = \frac{1}{2} \int |\dot{\gamma}|^2 dt$$

is not defined on a finite dimensional manifold. It is a function defined on the "space of all paths" joining the two given points. With some extra effort this space can be organized as an infinite dimensional manifold. We will not attempt to formalize these prescriptions, but rather follow the ad-hoc, intuitive approach of [71].

5.2.1 *Variational formulae*

Let M be a connected Riemann manifold, and consider $p, q \in M$. Denote by $\Omega_{p,q} = \Omega_{p,q}(M)$ the space of all *continuous, piecewise smooth* paths $\gamma : [0,1] \to M$ connecting p to q.

An *infinitesimal variation* of a path $\gamma \in \Omega_{p,q}$ is a *continuous, piecewise smooth* vector field V along γ such that $V(0) = 0$ and $V(1) = 0$ and

$$\lim_{h \searrow 0} V(t \pm h)$$

exists for every $t \in [0,1]$, they are vectors $V(t)^{\pm} \in T_{\gamma}(t)M$, and $V(t)^+ = V(t)^-$ for all but finitely many t-s. The space of infinitesimal variations of γ is an infinite dimensional linear space denoted by $T_{\gamma} = T_{\gamma}\Omega_{p,q}$.

Definition 5.2.1. Let $\gamma \in \Omega_{p,q}$. A *variation* of γ is a continuous map

$$\alpha = \alpha_s(t) : (-\varepsilon, \varepsilon) \times [0,1] \to M$$

such that

(i) $\forall s \in (-\varepsilon, \varepsilon)$, $\alpha_s \in \Omega_{p,q}$.

(ii) There exists a partition $0 = t_0 < t_1 \cdots < t_{k-1} < t_k = 1$ of $[0,1]$ such that the restriction of α to each $(-\varepsilon, \varepsilon) \times (t_{i-1}, t_i)$ is a smooth map. □

Every variation α of γ defines an *infinitesimal variation*

$$\delta\alpha := \frac{\partial \alpha_s}{\partial s}\big|_{s=0} .$$

Exercise 5.2.2. Given $V \in T_\gamma$ construct a variation α such that $\delta\alpha = V$. □

Consider now the *energy functional*

$$E : \Omega_{p,q} \to \mathbb{R}, \quad E(\gamma) = \frac{1}{2}\int_0^1 |\dot\gamma(t)|^2 dt.$$

Fix $\gamma \in \Omega_{p,q}$, and let α be a variation of γ. The velocity $\dot\gamma(t)$ has a finite number of discontinuities, so that the quantity

$$\Delta_t \dot\gamma = \lim_{h \to 0^+} (\dot\gamma(t+h) - \dot\gamma(t-h))$$

is nonzero only for finitely many t's.

Theorem 5.2.3 (The first variation formula).

$$E_*(\delta\alpha) := \frac{d}{ds}\big|_{s=0} E(\alpha_s) = -\sum_t \langle (\delta\alpha)(t), \Delta_t \dot\gamma \rangle - \int_0^1 \langle \delta\alpha, \nabla_{\frac{d}{dt}} \dot\gamma \rangle dt, \qquad (5.2.1)$$

where ∇ denotes the Levi-Civita connection. (Note that the right-hand side depends on α only through $\delta\alpha$ so it is really a linear function on T_γ.)

Proof. Set $\dot\alpha_s = \frac{\partial \alpha_s}{\partial t}$. We differentiate under the integral sign using the equality

$$\frac{\partial}{\partial s}|\dot\alpha_s|^2 = 2\langle \nabla_{\frac{\partial}{\partial s}}\dot\alpha_s, \dot\alpha_s \rangle,$$

and we get

$$\frac{d}{ds}\big|_{s=0} E(\alpha_s) = \int_0^1 \langle \nabla_{\frac{\partial}{\partial s}}\dot\alpha_s, \dot\alpha_s \rangle\big|_{s=0} dt.$$

Since the vector fields $\frac{\partial}{\partial s}$ and $\frac{\partial}{\partial t}$ commute we have $\nabla_{\frac{\partial}{\partial s}}\frac{\partial \alpha}{\partial t} = \nabla_{\frac{\partial}{\partial t}}\frac{\partial \alpha}{\partial s}$.

Let $0 = t_0 < t_2 < \cdots < t_k = 1$ be a partition of $[0,1]$ as in Definition 5.2.1. Since $\alpha_s = \gamma$ for $s = 0$ we conclude

$$E_*(\delta\alpha) = \sum_{i=1}^k \int_{t_{i-1}}^{t_i} \langle \nabla_{\frac{\partial}{\partial t}}\delta\alpha, \dot\gamma \rangle.$$

We use the equality

$$\frac{\partial}{\partial t}\langle \delta\alpha, \dot\gamma \rangle = \langle \nabla_{\frac{\partial}{\partial t}}\delta\alpha, \dot\gamma \rangle + \langle \delta\alpha, \nabla_{\frac{\partial}{\partial t}}\dot\gamma \rangle$$

to integrate by parts, and we obtain

$$E_*(\delta\alpha) = \sum_{i=1}^{k} \langle \delta\alpha, \dot\gamma \rangle \Big|_{t_{i-1}}^{t_i} - \sum_{i=1}^{k} \int_{t_{i-1}}^{t_i} \langle \delta\alpha, \nabla_{\frac{\partial}{\partial t}} \dot\gamma \rangle dt.$$

This is precisely equality (5.2.1). □

Definition 5.2.4. A path $\gamma \in \Omega_{p,q}$ is called *critical* if
$$E_*(V) = 0, \quad \forall V \in T_\gamma.$$ □

Corollary 5.2.5. *A path $\gamma \in \Omega_{p,q}$ is critical if and only if it is a geodesic.* □

Exercise 5.2.6. Prove the above corollary. □

Remark 5.2.7. Note that, a priori, a critical path may have a discontinuous first derivative. The above corollary shows that this is not the case: the criticality also implies smoothness. This is a manifestation of a more general analytical phenomenon called *elliptic regularity*. We will have more to say about it in Chapter 11. □

The map $E_* : T_\gamma \to \mathbb{R}$, $\delta\alpha \mapsto E_*(\delta\alpha)$ is called the first derivative of E at $\gamma \in \Omega_{p,q}$. We want to define a second derivative of E in order to address the issue raised at the beginning of this section. We will imitate the finite dimensional case which we now briefly analyze.

Let $f : X \to \mathbb{R}$ be a smooth function on the finite dimensional smooth manifold X. If x_0 is a critical point of f, i.e., $df(x_0) = 0$, then we can define the *Hessian* at x_0

$$f_{**} : T_{x_0}X \times T_{x_0}X \to \mathbb{R}$$

as follows. Given $V_1, V_2 \in T_{x_0}X$, consider a smooth map $(s_1, s_2) \mapsto \alpha(s_1, s_2) \in X$ such that

$$\alpha(0,0) = x_0 \quad \text{and} \quad \frac{\partial\alpha}{\partial s_i}(0,0) = V_i, \quad i = 1,2. \tag{5.2.2}$$

Now set

$$f_{**}(V_1, V_2) = \frac{\partial^2 f(\alpha(s_1, s_2))}{\partial s_1 \partial s_2}\Big|_{(0,0)}.$$

Note that since x_0 is a critical point of f, the Hessian $f_{**}(V_1, V_2)$ is independent of the function α satisfying (5.2.2).

We now return to our energy functional $E : \Omega_{p,q} \to \mathbb{R}$. Let $\gamma \in \Omega_{p,q}$ be a critical path. Consider a 2-parameter variation of γ

$$\alpha := \alpha_{s_1,s_2} : (-\varepsilon, \varepsilon) \times (-\varepsilon, \varepsilon) \times [0,1] \to M, \quad (s_1, s_2, t) \mapsto \alpha_{s_1,s_2}(t).$$

Set $U := (-\varepsilon, \varepsilon) \times (-\varepsilon, \varepsilon) \subset \mathbb{R}^2$ and $\gamma := \alpha_{0,0}$. The map α is continuous, and has second order derivatives everywhere except maybe on finitely many "coordinate" planes $s_i = $ const, or $t = $ const. Set $\delta_i\alpha := \frac{\partial\alpha}{\partial s_i}\big|_{(0,0)}$, $i = 1,2$. Note that $\delta_i\alpha \in T_\gamma$.

Exercise 5.2.8. Given $V_1, V_2 \in T_\gamma$ construct a 2-parameter variation α such that $V_i = \delta_i\alpha$. □

We can now define the *Hessian* of E at γ by

$$E_{**}(\delta_1\alpha, \delta_2\alpha) := \frac{\partial^2 E(\alpha_{s_1,s_2})}{\partial s_1 \partial s_2}\Big|_{(0,0)}.$$

Theorem 5.2.9 (The second variation formula).

$$E_{**}(\delta_1\alpha, \delta_2\alpha) = -\sum_t \langle \delta_2\alpha, \Delta_t\delta_1\alpha \rangle - \int_0^1 \langle \delta_2\alpha, \nabla^2 \frac{\partial}{\partial t}\delta_1\alpha - R(\dot\gamma, \delta_1\alpha)\dot\gamma \rangle dt, \quad (5.2.3)$$

where R denotes the Riemann curvature. In particular, E_{**} is a bilinear functional on T_γ.

Proof.   According to the first variation formula we have

$$\frac{\partial E}{\partial s_2} = -\sum_t \langle \delta_2\alpha, \Delta_t \frac{\partial\alpha}{\partial t}\rangle - \int_0^1 \langle \delta_2\alpha, \nabla_{\frac{\partial}{\partial t}}\frac{\partial\alpha}{\partial t}\rangle dt.$$

Hence

$$\frac{\partial^2 E}{\partial s_1 \partial s_2} = -\sum_t \langle \nabla_{\frac{\partial}{\partial s_1}}\delta_2\alpha, \Delta_1\dot\gamma\rangle - \sum_t \langle \delta_2\alpha, \nabla_{\frac{\partial}{\partial s_1}}\left(\Delta_t\frac{\partial\alpha}{\partial t}\right)\rangle$$

$$-\int_0^1 \langle \nabla_{\frac{\partial}{\partial s_1}}\delta_2\alpha, \nabla_{\frac{\partial}{\partial t}}\dot\gamma\rangle dt - \int_0^1 \langle \delta_2\alpha, \nabla_{\frac{\partial}{\partial s_1}}\nabla_{\frac{\partial}{\partial t}}\frac{\partial\alpha}{\partial t}\rangle dt. \quad (5.2.4)$$

Since γ is a geodesic, we have

$$\Delta_t\dot\gamma = 0 \text{ and } \nabla_{\frac{\partial}{\partial t}}\dot\gamma = 0.$$

Using the commutativity of $\frac{\partial}{\partial t}$ with $\frac{\partial}{\partial s_1}$ we deduce

$$\nabla_{\frac{\partial}{\partial s_1}}\left(\Delta_t\frac{\partial\alpha}{\partial t}\right) = \Delta_t\left(\nabla_{\frac{\partial}{\partial s_1}}\frac{\partial\alpha}{\partial t}\right) = \Delta_t\left(\nabla_{\frac{\partial}{\partial t}}\delta_1\alpha\right).$$

Finally, the definition of the curvature implies

$$\nabla_{\frac{\partial}{\partial s_1}}\nabla_{\frac{\partial}{\partial t}} = \nabla_{\frac{\partial}{\partial t}}\nabla_{\frac{\partial}{\partial s_1}} + R(\delta_1\alpha, \dot\gamma).$$

Putting all the above together we deduce immediately the equality (5.2.3). $\square$

Corollary 5.2.10.

$$E_{**}(V_1, V_2) = E_{**}(V_2, V_1), \ \ \forall V_1, V_2 \in T_\gamma. \qquad\qquad \square$$

5.2.2 *Jacobi fields*

In this subsection we will put to work the elements of calculus of variations presented so far. Let (M, g) be a Riemann manifold and $p, q \in M$.

Definition 5.2.11. Let $\gamma \in \Omega_{p,q}$ be a geodesic. A *geodesic variation* of γ is a smooth map $\alpha_s(t) : (-\varepsilon, \varepsilon) \times [0, 1] \to M$ such that, $\alpha_0 = \gamma$, and $t \mapsto \alpha_s(t)$ is a geodesic for all s. We set as usual $\delta\alpha = \frac{\partial\alpha}{\partial s}|_{s=0}$. □

Proposition 5.2.12. Let $\gamma \in \Omega_{p,q}$ be a geodesic and (α_s) a geodesic variation of γ. Then the infinitesimal variation $\delta\alpha$ satisfies the Jacobi equation

$$\nabla_t^2 \delta\alpha = R(\dot\gamma, \delta\alpha)\dot\gamma \quad (\nabla_t = \nabla_{\frac{\partial}{\partial t}}).$$

Proof.

$$\nabla_t^2 \delta\alpha = \nabla_t\left(\nabla_t \frac{\partial\alpha}{\partial s}\right) = \nabla_t\left(\nabla_s \frac{\partial\alpha}{\partial t}\right)$$

$$= \nabla_s\left(\nabla_t \frac{\partial\alpha}{\partial t}\right) + R(\dot\gamma, \delta\alpha)\frac{\partial\alpha}{\partial t} = R(\dot\gamma, \delta\alpha)\frac{\partial\alpha}{\partial t}.$$ □

Definition 5.2.13. A smooth vector field J along a geodesic γ is called a *Jacobi field* if it satisfies the *Jacobi equation*

$$\nabla_t^2 J = R(\dot\gamma, J)\dot\gamma.$$ □

Exercise 5.2.14. Show that if J is a Jacobi field along a geodesic γ, then there exists a geodesic variation α_s of γ such that $J = \delta\alpha$. □

Exercise 5.2.15. Let $\gamma \in \Omega_{p,q}$, and J a vector field along γ.
(a) Prove that J is a Jacobi field if and only if

$$E_{**}(J, V) = 0, \quad \forall V \in T_\gamma.$$

(b) Show that a vector field J along γ which *vanishes at endpoints*, is a Jacobi field if any only if $E_{**}(J, W) = 0$, for *all* vector fields W along γ. □

Exercise 5.2.16. Let $\gamma \in \Omega_{p,q}$ be a geodesic. Define $\mathfrak{J}_p$ to be the space of Jacobi fields V along γ such that $V(p) = 0$. Show that $\dim \mathfrak{J}_p = \dim M$, and moreover, the evaluation map

$$\mathbf{ev}_q : \mathfrak{J}_p \to T_p M \quad V \mapsto \nabla_t V(p)$$

is a linear isomorphism. □

Definition 5.2.17. Let $\gamma(t)$ be a geodesic. Two points $\gamma(t_1)$ and $\gamma(t_2)$ on γ are said to be *conjugate* along γ if there exists a nontrivial Jacobi field J along γ such that $J(t_i) = 0$, $i = 1, 2$. □

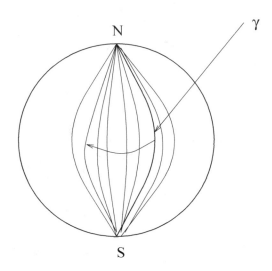

Fig. 5.3   *The poles are conjugate along meridians.*

Example 5.2.18.   Consider $\gamma : [0, 2\pi] \to S^2$ a meridian on the round sphere connection the poles. One can verify easily (using Clairaut's theorem) that γ is a geodesic. The counterclockwise rotation by an angle θ about the z-axis will produce a new meridian, hence a new geodesic γ_θ. Thus (γ_θ) is a geodesic variation of γ with fixed endpoints. $\delta\gamma$ is a Jacobi field vanishing at the poles. We conclude that the poles are conjugate along any meridian (see Figure 5.3). $\square$

Definition 5.2.19.   A geodesic $\gamma \in \Omega_{p,q}$is said to be *nondegenerate* if q is not conjugated to p along γ. $\square$

The following result (partially) explains the geometric significance of conjugate points.

Theorem 5.2.20.   *Let $\gamma \in \Omega_{p,q}$ be a nondegenerate, minimal geodesic. Then p is conjugate with no point on γ other than itself. In particular, a geodesic segment containing conjugate points cannot be minimal !*

Proof.   We argue by contradiction. Let $p_1 = \gamma(t_1)$ be a point on γ conjugate with p. Denote by J_t a Jacobi field along $\gamma|_{[0,t_1]}$ such that $J_0 = 0$ and $J_{t_1} = 0$. Define $V \in T_\gamma$ by

$$V_t = \begin{cases} J_t, & t \in [0, t_1] \\ 0, & t \geq t_1. \end{cases}$$

We will prove that V_t is a Jacobi field along γ which contradicts the nondegeneracy of γ.

Step 1.

$$E_{**}(U, U) \geq 0   \forall U \in T_\gamma. \tag{5.2.5}$$

Indeed, let α_s denote a variation of γ such that $\delta\alpha = U$. One computes easily that

$$\frac{d^2}{ds^2}E(\alpha_{s^2}) = 2E_{**}(U,U).$$

Since γ is *minimal* for any small s we have length $(\alpha_{s^2}) \geq$ length (α_0) so that

$$E(\alpha_{s^2}) \geq \frac{1}{2}\left(\int_0^1 |\dot\alpha_{s^2}|dt\right)^2 = \frac{1}{2}\text{length}\,(\alpha_{s^2})^2 \geq \frac{1}{2}\text{length}\,(\alpha_0)^2$$

$$= \frac{1}{2}\text{length}\,(\gamma)^2 = E(\alpha_0).$$

Hence

$$\frac{d^2}{ds^2}|_{s=0}\,E(\alpha_{s^2}) \geq 0.$$

This proves (5.2.5).

Step 2. $E_{**}(V,V) = 0$. This follows immediately from the second variation formula and the fact that the nontrivial portion of V is a Jacobi field.

Step 3.

$$E_{**}(U,V) = 0, \;\; \forall U \in T_\gamma.$$

From (5.2.5) and Step 2 we deduce

$$0 = E_{**}(V,V) \leq E_{**}(V + \tau U, V + \tau U) = f_U(\tau)\;\;\forall\tau.$$

Thus, $\tau = 0$ is a global minimum of $f_U(\tau)$ so that

$$f_U'(0) = 0.$$

Step 3 follows from the above equality using the bilinearity and the symmetry of E_{**}. The final conclusion (that V is a Jacobi field) follows from Exercise 5.2.15. $\square$

Exercise 5.2.21. Let $\gamma : \mathbb{R} \to M$ be a geodesic. Prove that the set

$$\{\,t \in \mathbb{R}\,;\,\gamma(t)\text{ is conjugate to }\gamma(0)\,\}$$

is discrete. $\square$

Definition 5.2.22. Let $\gamma \in \Omega_{p,q}$ be a geodesic. We define its *index*, denoted by ind (γ), as the cardinality of the set

$$C_\gamma = \{\,t \in (0,1)\,;\,\text{is conjugate to }\gamma(0)\,\}$$

which by Exercise 5.2.21 is finite. $\square$

Theorem 5.2.20 can be reformulated as follows: the index of a nondegenerate minimal geodesic is zero.

The index of a geodesic obviously depends on the curvature of the manifold. Often, this dependence is very powerful.

Theorem 5.2.23. *Let M be a Riemann manifold with non-positive sectional curvature, i.e.,*

$$\langle R(X,Y)Y, X\rangle \leq 0 \;\;\forall X, Y \in T_x M \;\;\forall x \in M. \tag{5.2.6}$$

Then for any $p, q \in M$ and any geodesic $\gamma \in \Omega_{p,q}$, ind $(\gamma) = 0$.

Proof.   It suffices to show that for any geodesic $\gamma : [0,1] \to M$ the point $\gamma(1)$ is not conjugated to $\gamma(0)$.

Let J_t be a Jacobi field along γ vanishing at the endpoints. Thus

$$\nabla_t^2 J = R(\dot{\gamma}, J)\dot{\gamma},$$

so that

$$\int_0^1 \langle \nabla_t^2 J, J \rangle dt = \int_0^1 \langle R(\dot{\gamma}, J)\dot{\gamma}, J \rangle dt = -\int_0^1 \langle R(J, \dot{\gamma})\dot{\gamma}, J \rangle dt.$$

We integrate by parts the left-hand side of the above equality, and we deduce

$$\langle \nabla_t J, J \rangle|_0^1 - \int_0^1 |\nabla_t J|^2 dt = -\int_0^1 \langle R(J, \dot{\gamma})\dot{\gamma}, J \rangle dt.$$

Since $J(\tau) = 0$ for $\tau = 0, 1$, we deduce using (5.2.6)

$$\int_0^1 |\nabla_t J|^2 dt \leq 0.$$

This implies $\nabla_t J = 0$ which coupled with the condition $J(0) = 0$ implies $J \equiv 0$. The proof is complete.   □

The notion of conjugacy is intimately related to the behavior of the exponential map.

Theorem 5.2.24.   *Let (M, g) be a connected, complete, Riemann manifold and $q_0 \in M$. A point $q \in M$ is conjugated to q_0 along some geodesic if and only if it is a critical value of the exponential map*

$$\exp_{q_0} : T_{q_0} M \to M.$$

Proof.   Let $q = \exp_{q_0} v$, $v \in T_{q_0} M$. Assume first that q is a critical value for $\exp_{q_0}$, and v is a critical point. Then $D_v \exp_{q_0}(X) = 0$, for some $X \in T_v(T_{q_0} M)$. Let $v(s)$ be a path in $T_{q_0} M$ such that $v(0) = v$ and $\dot{v}(0) = X$. The map $(s, t) \mapsto \exp_{q_0}(tv(s))$ is a geodesic variation of the radial geodesic $\gamma_v : t \mapsto \exp_{q_0}(tv)$. Hence, the vector field

$$W = \frac{\partial}{\partial s}|_{s=0} \exp_{q_0}(tv(s))$$

is a Jacobi field along γ_v. Obviously $W(0) = 0$, and moreover

$$W(1) = \frac{\partial}{\partial s}|_{s=0} \exp_{q_0}(v(s)) = D_v \exp_{q_0}(X) = 0.$$

On the other hand this is a nontrivial field since

$$\nabla_t W = \nabla_s|_{s=0} \frac{\partial}{\partial t} \exp_{q_0}(tv(s)) = \nabla_s v(s)|_{s=0} \neq 0.$$

This proves q_0 and q are conjugated along γ_v.

Conversely, assume v is not a critical point for $\exp_{q_0}$. For any $X \in T_v(T_{q_0}M)$ denote by J_X the Jacobi field along γ_v such that

$$J_X(q_0) = 0. \tag{5.2.7}$$

The existence of such a Jacobi field follows from Exercise 5.2.16. As in that exercise, denote by $\mathfrak{J}_{q_0}$ the space of Jacobi fields J along γ_v such that $J(q_0) = 0$. The map

$$T_v(T_{q_0}M) \to \mathfrak{J}_{q_0} \quad X \mapsto J_X$$

is a linear isomorphism. Thus, a Jacobi field along γ_v vanishing at both q_0 and q must have the form J_X, where $X \in T_v(T_{q_0}M)$ satisfies $D_v \exp_{q_0}(X) = 0$. Since v is not a critical point, this means $X = 0$ so that $J_X \equiv 0$. □

Corollary 5.2.25. *On a complete Riemann manifold M with non-positive sectional curvature the exponential map $\exp_q$ has no critical values for any $q \in M$.* □

We will see in the next chapter that this corollary has a lot to say about the topology of M.

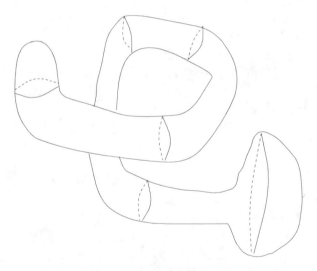

Fig. 5.4 *Lengthening a sphere.*

Consider now the following experiment. Stretch the round two-dimensional sphere of radius 1 until it becomes "very long". A possible shape one can obtain may look like in Figure 5.4. The long tube is very similar to a piece of cylinder so that the total (= scalar) curvature is very close to zero, in other words is very small. The lesson to learn from this intuitive experiment is that the price we have to pay for lengthening the sphere is decreasing the curvature. Equivalently, a highly curved surface cannot have a large diameter. Our next result offers a more quantitative description of this phenomenon.

Theorem 5.2.26 (Myers). *Let M be an n-dimensional complete Riemann manifold. If for all $X \in \mathrm{Vect}\,(M)$*

$$\mathrm{Ric}\,(X, X) \geq \frac{(n-1)}{r^2} |X|^2,$$

then every geodesic of length $\geq \pi r$ has conjugate points and thus is not minimal. Hence

$$\mathrm{diam}\,(M) = \sup\{\mathrm{dist}\,(p, q) \ ; \ p, q \in M\} \leq \pi r,$$

and in particular, Hopf-Rinow theorem implies that M must be compact. □

Proof. Fix a minimal geodesic $\gamma : [0, \ell] \to M$ of length ℓ, and let $e_i(t)$ be an orthonormal basis of vector fields along γ such that $e_n(t) = \dot{\gamma}(t)$. Set $q_0 = \gamma(0)$, and $q_1 = \gamma(\ell)$. Since γ is minimal we deduce

$$E_{**}(V, V) \geq 0 \quad \forall V \in T_\gamma.$$

Set $W_i = \sin(\pi t/\ell)e_i$. Then

$$E_{**}(W_i, W_i) = -\int_0^\ell \langle W_i, \nabla_t W_i + R(W_i, \dot{\gamma})\dot{\gamma}\rangle dt$$

$$= \int_0^\ell \sin^2(\pi t/\ell)\left(\pi^2/\ell^2 - \langle R(e_i, \dot{\gamma})\dot{\gamma}, e_i\rangle\right) dt.$$

We sum over $i = 1, \ldots, n-1$, and we obtain

$$\sum_{i=1}^{n-1} E_{**}(W_i, W_i) = \int_0^\ell \sin^2 \pi t/\ell \left((n-1)\pi^2/\ell^2 - \mathrm{Ric}\,(\dot{\gamma}, \dot{\gamma})\right) dt \geq 0.$$

If $\ell > \pi r$, then

$$(n-1)\pi^2/\ell^2 - \mathrm{Ric}\,(\dot{\gamma}, \dot{\gamma}) < 0,$$

so that,

$$\sum_{i=1}^{n-1} E_{**}(W_i, W_i) < 0.$$

Hence, at least for some W_i, we have $E_{**}(W_i, W_i) < 0$. This contradicts the minimality of γ. The proof is complete. □

We already know that the Killing form of a *compact, semisimple* Lie group is positive definite; see Exercise 4.1.19. The next result shows that the converse is also true.

Corollary 5.2.27. *A semisimple Lie group G with positive definite Killing pairing is compact.*

Proof. The Killing form defines in this case a bi-invariant Riemann metric on G. Its geodesics through the origin $1 \in G$ are the 1-parameter subgroups $\exp(tX)$ which are defined for all $t \in \mathbb{R}$. Hence, by Hopf-Rinow theorem G has to be complete.

On the other hand, we have computed the Ricci curvature of the Killing metric, and we found

$$\text{Ric}\,(X,Y) = \frac{1}{4}\kappa(X,Y) \quad \forall X,Y \in \mathcal{L}_G.$$

The corollary now follows from Myers' theorem. $\qquad\square$

Exercise 5.2.28. Let M be a Riemann manifold and $q \in M$. For the unitary vectors $X,Y \in T_qM$ consider the family of geodesics

$$\gamma_s(t) = \exp_q t(X + sY).$$

Denote by $W_t = \delta\gamma_s$ the associated Jacobi field along $\gamma_0(t)$. Form $f(t) = |W_t|^2$. Prove the following.
(a) $W_t = D_{tX}\exp_q(Y) =$ Frechet derivative of $v \mapsto \exp_q(v)$.
(b) $f(t) = t^2 - \frac{1}{3}\langle R(Y,X)X,Y\rangle_q t^4 + O(t^5)$.
(c) Denote by $\boldsymbol{x}^i$ a collection of normal coordinates at q. Prove that

$$g_{k\ell}(\boldsymbol{x}) = \delta_{k\ell} - \frac{1}{3}R_{kij\ell}\boldsymbol{x}^i\boldsymbol{x}^j + O(3).$$

$$\det g_{ij}(\boldsymbol{x}) = 1 - \frac{1}{3}R_{ij}\boldsymbol{x}^i\boldsymbol{x}^j + O(3).$$

(d) Let

$$\mathbf{D}_r(q) = \{x \in T_qM \ ; \ |x| \le r\}.$$

Prove that if the Ricci curvature is negative definite at q then

$$\text{vol}_0\,(\mathbf{D}_r(q)) \le \text{vol}_g\,(\exp_q(\mathbf{D}_r(q)))$$

for all r sufficiently small. vol_0 denotes the Euclidean volume in T_qM while vol_g denotes the volume on the Riemann manifold M. $\qquad\square$

Remark 5.2.29. The interdependence "curvature-topology" on a Riemann manifold has deep reaching ramifications which stimulate the curiosity of many researchers. We refer to [27] or [71] and the extensive references therein for a presentation of some of the most attractive results in this direction. $\qquad\square$

Chapter 6

The Fundamental Group and Covering Spaces

In the previous chapters we almost exclusively studied local properties of manifolds. This study is interesting only if some additional structure is present since otherwise all manifolds are locally alike.

We noticed an interesting phenomenon: the global "shape" (topology) of a manifold restricts the types of structures that can exist on a manifold. For example, the Gauss-Bonnet theorem implies that on a connected sum of two tori there cannot exist metrics with curvature everywhere positive because the integral of the curvature is a negative universal constant.

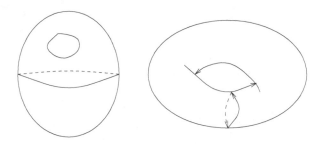

Fig. 6.1   *Looking for unshrinkable loops.*

We used the Gauss-Bonnet theorem in the opposite direction, and we deduced the intuitively obvious fact that a sphere is not diffeomorphic to a torus because they have distinct genera. The Gauss-Bonnet theorem involves a heavy analytical machinery which may obscure the intuition. Notice that S^2 has a remarkable property which distinguishes it from T^2: on the sphere any closed curve can be shrunk to a point while on the torus there exist at least two "independent" unshrinkable curves (see Figure 6.1). In particular, this means the sphere is not diffeomorphic to a torus.

This chapter will set the above observations on a rigorous foundation.

6.1 The fundamental group

6.1.1 *Basic notions*

In the sequel all topological spaces will be locally path connected spaces.

Definition 6.1.1. (a) Let X and Y be two topological spaces. Two continuous maps $f_0, f_1 : X \to Y$ are said to be *homotopic* if there exists a continuous map

$$F : [0, 1] \times X \to Y \quad (t, x) \mapsto F_t(x),$$

such that, $F_i \equiv f_i$ for $i = 0, 1$. We write this as $f_0 \simeq f_1$.
(b) Two topological spaces X, Y are said to be *homotopy equivalent* if there exist maps $f : X \to Y$ and $g : Y \to X$ such that $f \circ g \simeq \mathbb{1}_Y$ and $g \circ f \simeq \mathbb{1}_X$. We write this $X \simeq Y$.
(c) A topological space is said to be *contractible* if it is homotopy equivalent to a point. □

Example 6.1.2. The unit disk in the plane is contractible. The annulus $\{1 \le |z| \le 2\}$ is homotopy equivalent to the unit circle. □

Definition 6.1.3. (a) Let X be a topological space and $x_0 \in X$. A *loop based at* x_0 is a continuous map

$$\gamma : [0, 1] \to X, \quad \text{such that } \gamma(0) = \gamma(1) = x_0.$$

The space of loops in X based at x_0 is denoted by $\Omega(X, x_0)$.
(b) Two loops $\gamma_0, \gamma_1 : I \to X$ based at x_0 are said to be *homotopic rel* x_0 if there exists a continuous map

$$\Gamma : [0, 1] \times I \to X, \quad (t, s) \mapsto \Gamma_t(s),$$

such that

$$\Gamma_i(s) = \gamma_i(s) \quad i = 0, 1,$$

and

$$(s \mapsto \Gamma_t(s)) \in \Omega(X, x_0) \quad \forall t \in [0, 1].$$

We write this as $\gamma_0 \simeq_{x_0} \gamma_1$. □

Note that a loop is more than a closed curve; it is a closed curve + a description of a motion of a point around the closed curve.

Example 6.1.4. The two loops $\gamma_1, \gamma_2 : I \to \mathbb{C}$, $\gamma_k(t) = \exp(2k\pi t)$, $k = 1, 2$ are different though they have the same image. □

Definition 6.1.5. (a) Let γ_1, γ_2 be two loops based at $x_0 \in X$. The *product* of γ_1 and γ_2 is the loop

$$\gamma_1 * \gamma_2(s) = \begin{cases} \gamma_1(2s) \, , \, 0 \le s 1/2 \\ \gamma_2(2s-1) \, , \, 1/2 \le s \le 1 \end{cases} .$$

The *inverse* of a based loop γ is the based loop γ^- defined by

$$\gamma^-(s) = \gamma(1-s).$$

(c) The *identity* loop is the constant loop $\mathfrak{e}_{x_0}(s) \equiv x_0$. $\square$

Intuitively, the product of two loops γ_1 and γ_2 is the loop obtained by first following γ_1 (twice faster), and then γ_2 (twice faster).

The following result is left to the reader as an exercise.

Lemma 6.1.6. *Let* $\alpha_0 \simeq_{x_0} \alpha_1$, $\beta_0 \simeq_{x_0} \beta_1$ *and* $\gamma_0 \simeq_{x_0} \gamma_1$ *be three pairs of homotopic based loops.. Then*
(a) $\alpha_0 * \beta_0 \simeq_{x_0} \alpha_1 * \beta_1$.
(b) $\alpha_0 * \alpha_0^- \simeq_{x_0} \mathfrak{e}_{x_0}$.
(c) $\alpha_0 * \mathfrak{e}_{x_0} \simeq_{x_0} \alpha_0$.
(d) $(\alpha_0 * \beta_0) * \gamma_0 \simeq_{x_0} \alpha_0 * (\beta_0 * \gamma_0)$. $\square$

Hence, the product operation descends to an operation "$\cdot$" on $\Omega(X, x_0)/\simeq_{x_0}$, the set of homotopy classes of based loops. Moreover the induced operation is associative, it has a unit and each element has an inverse. Hence $(\Omega(X, x_0)/\simeq_{x_0}, \cdot)$ is a group.

Definition 6.1.7. The group $(\Omega(X, x_0)/\simeq_{x_0}, \cdot)$ is called the *fundamental group* (or the *Poincaré group*) of the topological space X, and it is denoted by $\pi_1(X, x_0)$. The image of a based loop γ in $\pi_1(X, x_0)$ is denoted by $[\gamma]$. $\square$

The elements of $\pi_1(X, x_0)$ are the "unshrinkable loops" discussed at the beginning of this chapter.

The fundamental group $\pi_1(X, x_0)$ "sees" only the connected component of X which contains x_0. To get more information about X one should study all the groups $\{\pi_1(X, x)\}_{x \in X}$.

Proposition 6.1.8. *Let X and Y be two topological spaces. Fix two points, $x_0 \in X$ and $y_0 \in Y$. Then any continuous map $f : X \to Y$ such that $f(x_0) = y_0$ induces a morphism of groups*

$$f_* : \pi_1(X, x_0) \to \pi_1(Y, y_0),$$

satisfying the following functoriality properties.
(a) $(\mathbb{1}_X)_* = \mathbb{1}_{\pi_1(X, x_0)}$.
(b) If

$$(X, x_0) \xrightarrow{f} (Y, y_0) \xrightarrow{g} (Z, z_0)$$

are continuous maps, such that $f(x_0) = y_0$ *and* $g(y_0) = z_0$, *then* $(g \circ f)_* = g_* \circ f_*$.
(c) Let $f_0, f_1 : (X, x_0) \to (Y, y_0)$ *be two base-point-preserving continuous maps. Assume* f_0 *is homotopic to* f_1 *rel* x_0, *i.e., there exists a continuous map* $F : I \times X \to Y$, $(t, x) \mapsto F_t(x)$ *such that,* $F_i(x) \equiv f_i(x)$. *for* $i = 0, 1$ *and* $F_t(x_0) \equiv y_0$. *Then* $(f_0)_* = (f_1)_*$.

Proof. Let $\gamma \in \Omega(X, x_0)$, Then $f(\gamma) \in \Omega(Y, y_0)$, and one can check immediately that

$$\gamma \simeq_{x_0} \gamma' \Rightarrow f(\gamma) \simeq_{y_0} f(\gamma').$$

Hence the correspondence

$$\Omega(X, x_0) \ni \gamma \mapsto f(\gamma) \in \Omega(Y, y_0)$$

descends to a map $f : \pi_1(X, x_0) \to \pi_1(Y, y_0)$. This is clearly a group morphism. The statements (a) and (b) are now obvious. We prove (c).

Let $f_0, f_1 : (X, x_0) \to (Y, y_0)$ be two continuous maps, and F_t a homotopy rel x_0 connecting them. For any $\gamma \in \Omega(X, x_0)$ we have

$$\beta_0 = f_0(\gamma) \simeq_{y_0} f_1(\gamma) = \beta_1.$$

The above homotopy is realized by $\mathcal{B}_t = F_t(\gamma)$. □

A priori, the fundamental group of a topological space X may change as the base point varies and it almost certainly does if X has several connected components. However, if X is connected, and thus path connected since it is locally so, all the fundamental groups $\pi_1(X, x)$, $x \in X$ are isomorphic.

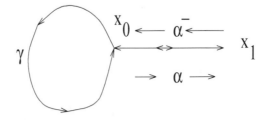

Fig. 6.2   *Connecting base points.*

Proposition 6.1.9. *Let* X *be a connected topological space. Any continuous path* $\alpha : [0, 1] \to X$ *joining* x_0 *to* x_1 *induces an isomorphism*

$$\alpha_* : \pi_1(X, x_0) \to \pi_1(X, x_1),$$

defined by, $\alpha_*([\gamma]) := [\alpha^- * \gamma * \alpha]$; *see Figure 6.2.* □

Exercise 6.1.10. Prove Proposition 6.1.9. □

Thus, the fundamental group of a connected space X is independent of the base point modulo some isomorphism. We will write $\pi_1(X, pt)$ to underscore this weak dependence on the base point.

Corollary 6.1.11. *Two homotopically equivalent connected spaces have isomorphic fundamental groups.*

Example 6.1.12. (a) $\pi_1(\mathbb{R}^n, pt) \sim \pi_1(pt, pt) = \{1\}$.
(b) $\pi_1(\text{annulus}) \sim \pi_1(S^1)$. □

Definition 6.1.13. A connected space X such that $\pi_1(X, pt) = \{1\}$ is said to be *simply connected*. □

Exercise 6.1.14. Prove that the spheres of dimension ≥ 2 are simply connected.□

Exercise 6.1.15. Let G be a connected Lie group. Define a new operation "$\star$ " on $\Omega(G, 1)$ by

$$(\alpha \star \beta)(s) = \alpha(s) \cdot \beta(s),$$

where $\cdot$ denotes the group multiplication.
(a) Prove that $\alpha \star \beta \simeq_1 \alpha \ast \beta$.
(b) Prove that $\pi_1(G, 1)$ is Abelian. □

Exercise 6.1.16. Let $E \to X$ be a rank r complex vector bundle over the smooth manifold X, and let ∇ be a flat connection on E, i.e., $F(\nabla) = 0$. Pick $x_0 \in X$, and identify the fiber E_{x_0} with $\mathbb{C}^r$. For any continuous, piecewise smooth $\gamma \in \Omega(X, x_0)$ denote by $T_\gamma = T_\gamma(\nabla)$ the parallel transport along γ, so that $T_\gamma \in \mathrm{GL}(\mathbb{C}^r)$.
(a) Prove that $\alpha \simeq_{x_0} \beta \Rightarrow T_\alpha = T_\beta$.
(b) $T_{\beta \ast \alpha} = T_\alpha \circ T_\beta$.
  Thus, any flat connection induces a group morphism

$$T : \pi_1(X, x_0) \to \mathrm{GL}(\mathbb{C}^r) \quad \gamma \mapsto T_\gamma^{-1}.$$

This morphism (representation) is called the *monodromy* of the connection. □

Example 6.1.17. We want to compute the fundamental group of the complex projective space $\mathbb{CP}^n$. More precisely, we want to show it is simply connected. We will establish this by induction.
  For $n = 1$, $\mathbb{CP}^1 \cong S^2$, and by Exercise 6.1.14, the sphere S^2 is simply connected. We next assume $\mathbb{CP}^k$ is simply connected for $k < n$ and prove the same is true for n.
  Notice first that the natural embedding $\mathbb{C}^{k+1} \hookrightarrow \mathbb{C}^{n+1}$ induces an embedding $\mathbb{CP}^k \hookrightarrow \mathbb{CP}^n$. More precisely, in terms of homogeneous coordinates this embedding is given by

$$[z_0, \ldots : z_k] \mapsto [z_0, \ldots, z_k, 0, \ldots, 0] \in \mathbb{CP}^n.$$

Choose as base point $\boldsymbol{pt} = [1, 0, \ldots, 0] \in \mathbb{CP}^n$, and let $\gamma \in \Omega(\mathbb{CP}^n, \boldsymbol{pt})$. We may assume γ avoids the point $P = [0, \ldots, 0, 1]$ since we can deform it out of any neighborhood of P.

We now use a classical construction of projective geometry. We project γ from P to the hyperplane $\mathcal{H} = \mathbb{CP}^{n-1} \hookrightarrow \mathbb{CP}^n$. More precisely, if $\zeta = [z_0, \ldots, z_n] \in \mathbb{CP}^n$, we denote by $\pi(\zeta)$ the intersection of the line $P\zeta$ with the hyperplane $\mathcal{H}$. In homogeneous coordinates

$$\pi(\zeta) = [z_0(1 - z_n), \ldots, z_{n-1}(1 - z_n), 0] \ (= [z_0, \ldots, z_{n-1}, 0] \text{ when } z_n \neq 1).$$

Clearly π is continuous. For $t \in [0, 1]$ define

$$\pi_t(\zeta) = [z_0(1 - tz_n), \ldots, z_{n-1}(1 - tz_n), (1 - t)z_n].$$

Geometrically, π_t flows the point ζ along the line $P\zeta$ until it reaches the hyperplane $\mathcal{H}$. Note that $\pi_t(\zeta) = \zeta$, $\forall t$, and $\forall \zeta \in \mathcal{H}$. Clearly, π_t is a homotopy rel $\boldsymbol{pt}$ connecting $\gamma = \pi_0(\gamma)$ to a loop γ_1 in $\mathcal{H} \cong \mathbb{CP}^{n-1}$ based at $\boldsymbol{pt}$. Our induction hypothesis shows that γ_1 can be shrunk to $\boldsymbol{pt}$ inside $\mathcal{H}$. This proves that $\mathbb{CP}^n$ is simply connected.

$\square$

6.1.2 *Of categories and functors*

The considerations in the previous subsection can be very elegantly presented using the language of categories and functors. This brief subsection is a minimal introduction to this language. The interested reader can learn more about it from the monograph [54, 67].

A *category* is a triplet $\mathcal{C} = (\boldsymbol{Ob}(\mathcal{C}), \boldsymbol{Hom}(\mathcal{C}), \circ)$ satisfying the following conditions.

(i) $\boldsymbol{Ob}(\mathcal{C})$ is a set whose elements are called the *objects* of the category.
(ii) $\boldsymbol{Hom}(\mathcal{C})$ is a family of sets $\mathrm{Hom}(X, Y)$, one for each pair of objects X and Y. The elements of $\mathrm{Hom}(X, Y)$ are called the *morphisms* (or arrows) from X to Y.
(iii) $\circ$ is a collection of maps

$$\circ : \mathrm{Hom}(X, Y) \times \mathrm{Hom}(Y, Z) \rightarrow \mathrm{Hom}(X, Z), \quad (f, g) \mapsto g \circ f,$$

which satisfies the following conditions.

(C1) For any object X, there exists a unique element $\mathbb{1}_X \in \mathrm{Hom}(X, X)$ such that,

$$f \circ \mathbb{1}_X = f \quad g \circ \mathbb{1}_X = g \ \ \forall f \in \mathrm{Hom}(X, Y), \ \ \forall g \in \mathrm{Hom}(Z, X).$$

(C2) $\forall f \in \mathrm{Hom}(X, Y), \ \ g \in \mathrm{Hom}(Y, Z), \ h \in \mathrm{Hom}(Z, W)$

$$h \circ (g \circ f) = (h \circ g) \circ f.$$

Example 6.1.18. • **Top** is the category of topological spaces. The objects are topological spaces and the morphisms are the continuous maps. Here we have to be careful about one foundational issue. Namely, the collection of all topological spaces is not a set. To avoid this problem we need to restrict to topological spaces whose subjacent sets belong to a certain Universe. For more about this foundational issue we refer to [54].

• (**Top**, $*$) is the category of marked topological spaces. The objects are pairs $(X, *)$, where X is a topological space, and $*$ is a distinguished point of X. The morphisms

$$(X, *) \xrightarrow{f} (Y, \diamond)$$

are the continuous maps $f : X \to Y$ such that $f(*) = \diamond$.

• $_{\mathbb{F}}$**Vect** is the category of vector spaces over the field $\mathbb{F}$. The morphisms are the $\mathbb{F}$-linear maps.

• **Gr** is the category of groups, while **Ab** denotes the category of Abelian groups. The morphisms are the obvious ones.

• $_R$**Mod** denotes the category of left R-modules, where R is some ring. □

Definition 6.1.19. Let $\mathcal{C}_1$ and $\mathcal{C}_2$ be two categories. A *covariant* (respectively *contravariant*) functor is a map

$$\mathcal{F} : Ob(\mathcal{C}_1) \times Hom(\mathcal{C}_1) \to Ob(\mathcal{C}_2) \times Hom(\mathcal{C}_2),$$

$$(X, f) \mapsto (\mathcal{F}(X), \mathcal{F}(f)),$$

such that, if $X \xrightarrow{f} Y$, then $\mathcal{F}(X) \xrightarrow{\mathcal{F}(f)} \mathcal{F}(Y)$ (respectively $\mathcal{F}(X) \xleftarrow{\mathcal{F}(f)} \mathcal{F}(Y)$), and
(i) $\mathcal{F}(\mathbb{1}_X) = \mathbb{1}_{\mathcal{F}(X)}$,
(ii) $\mathcal{F}(g) \circ \mathcal{F}(f) = \mathcal{F}(g \circ f)$ (respectively $\mathcal{F}(f) \circ \mathcal{F}(g) = \mathcal{F}(g \circ f)$). □

Example 6.1.20. Let V be a real vector space. Then the operation of right tensoring with V is a covariant functor

$$\otimes V : {}_{\mathbb{R}}\mathbf{Vect} \to {}_{\mathbb{R}}\mathbf{Vect}, \quad U \rightsquigarrow U \otimes V, \quad (U_1 \xrightarrow{L} U_2) \rightsquigarrow (U_1 \otimes V \xrightarrow{L \otimes \mathbb{1}_V} U_2 \otimes V).$$

On the other hand, the operation of taking the dual defines a contravariant functor,

$$* : {}_{\mathbb{R}}\mathbf{Vect} \to {}_{\mathbb{R}}\mathbf{Vect}, \quad V \rightsquigarrow V^*, \quad (U \xrightarrow{L} V) \rightsquigarrow (V^* \xrightarrow{L^t} U^*).$$

The fundamental group construction of the previous is a covariant functor

$$\pi_1 : (\mathbf{Top}, *) \to \mathbf{Gr}. \qquad \square$$

In Chapter 7 we will introduce other functors very important in geometry. For more information about categories and functors we refer to [54, 67].

6.2 Covering spaces

6.2.1 *Definitions and examples*

As in the previous section we will assume that all topological spaces are locally path connected.

Definition 6.2.1. (a) A continuous map $\pi : X \to Y$ is said to be a *covering map* if, for any $y \in Y$, there exists an open neighborhood U of y in Y, such that $\pi^{-1}(U)$ is a disjoint union of open sets $V_i \subset X$ each of which is mapped homeomorphically onto U by π. Such a neighborhood U is said to be an *evenly covered* neighborhood. The sets V_i are called the *sheets* of π over U.
(b) Let Y be a topological space. A *covering space* of Y is a topological space X, together with a covering map $\pi : X \to Y$.
(c) If $\pi : X \to Y$ is a covering map, then for any $y \in Y$ the set $\pi^{-1}(y)$ is called the *fiber* over y. □

Example 6.2.2. Let D be a discrete set. Then, for any topological space X, the product $X \times D$ is a covering of X with covering projection $\pi(x, d) = x$. This type of covering space is said to be *trivial*. □

Exercise 6.2.3. Show that a fibration with standard fiber of a discrete space is a covering. □

Example 6.2.4. The exponential map $\exp : \mathbb{R} \to S^1$, $t \mapsto \exp(2\pi i t)$ is a covering map. However, its restriction to $(0, \infty)$ is no longer a a covering map. (Prove this!) □

Exercise 6.2.5. Let (M, g) and $(\tilde{M}, \tilde{g})$ be two Riemann manifolds of the same dimension such that $(\tilde{M}, \tilde{g})$ is complete. Let $\phi : \tilde{M} \to M$ be a surjective local isometry i.e. Φ is smooth and

$$|v|_g = |D\phi(v)|_{\tilde{g}} \quad \forall v \in T\tilde{M}.$$

Prove that ϕ is a covering map. □

The above exercise has a particularly nice consequence.

Theorem 6.2.6 (Cartan-Hadamard). *Let (M, g) be a complete Riemann manifold with non-positive sectional curvature. Then for every point $q \in M$, the exponential map*

$$\exp_q : T_q M \to M$$

is a covering map.

Proof. The pull-back $h = \exp_q^*(g)$ is a symmetric non-negative definite $(0,2)$-tensor field on T_qM. It is in fact *positive definite* since the map $\exp_q$ has no critical points due to the non-positivity of the sectional curvature.

The lines $t \mapsto tv$ through the origin of T_qM are geodesics of h and they are defined for all $t \in \mathbb{R}$. By Hopf-Rinow theorem we deduce that (T_qM, h) is complete. The theorem now follows from Exercise 6.2.5. $\qquad\square$

Exercise 6.2.7. Let $\tilde{G}$ and G be two Lie groups of the same dimension and $\phi : \tilde{G} \to G$ a smooth, surjective group morphism. Prove that ϕ is a covering map. In particular, this explains why $\exp : \mathbb{R} \to S^1$ is a covering map. $\qquad\square$

Exercise 6.2.8. Identify $S^3 \subset \mathbb{R}^4$ with the group of unit quaternions

$$S^3 = \{q \in \mathbb{H} \; ; \; |q| = 1\}.$$

The linear space $\mathbb{R}^3$ can be identified with the space of purely imaginary quaternions

$$\mathbb{R}^3 = \boldsymbol{Im}\,\mathbb{H} = \{x\boldsymbol{i} + y\boldsymbol{j} + z\boldsymbol{k}\}.$$

(a) Prove that $qxq^{-1} \in \boldsymbol{Im}\,\mathbb{H}$, $\forall q \in S^3$.
(b) Prove that for any $q \in S^3$ the linear map

$$T_q : \boldsymbol{Im}\,\mathbb{H} \to \boldsymbol{Im}\mathbb{H} \qquad x \mapsto qxq^{-1}$$

is an isometry so that $T_q \in SO(3)$. Moreover the map

$$S^3 \ni q \mapsto T_q \in SO(3)$$

is a group morphism.
(c) Prove the above group morphism is a covering map. $\qquad\square$

Example 6.2.9. Let M be a smooth manifold. A Riemann metric on M induces a metric on the determinant line bundle $\det TM$. The sphere bundle of $\det TM$ (with respect to this metric) is a $2 : 1$ covering space of M called the *orientation cover* of M. $\qquad\square$

Definition 6.2.10. Let $X_1 \stackrel{\pi_1}{\to} Y$ and $X_2 \stackrel{\pi_2}{\to} Y$ be two covering spaces of Y. A *morphism of covering spaces* is a continuous map $F : X_1 \to X_2$ such that $\pi_2 \circ F = \pi_1$, i.e., the diagram below is commutative.

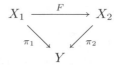

If F is also a homeomorphism we say that F is an isomorphism of covering spaces.

Finally, if $X \stackrel{\pi}{\to} Y$ is a covering space then its automorphisms are called *deck transformations*. The deck transformations form a group denoted by $\mathrm{Deck}\,(X, \pi)$.$\square$

Exercise 6.2.11. Show that Deck $(\mathbb{R} \overset{\exp}{\to} S^1) \cong \mathbb{Z}$.   □

Exercise 6.2.12. (a) Prove that the natural projection $S^n \to \mathbb{RP}^n$ is a covering map.
(b) Denote by $\mathcal{U}_1^{\mathbb{R}}$ the tautological (real) line bundle over $\mathbb{RP}^n$. Using a metric on this line bundle form the associated sphere bundle $S(\mathcal{U}_1^{\mathbb{R}}) \to \mathbb{RP}^n$. Prove that this fiber bundle defines a covering space isomorphic with the one described in part (a).
□

6.2.2   Unique lifting property

Definition 6.2.13. Let $X \overset{\pi}{\to} Y$ be a covering space and $F : Z \to Y$ a continuous map. A *lift* of f is a continuous map $F : Z \to X$ such that $\pi \circ F = f$, i.e. the diagram below is commutative.

$$
\begin{array}{ccc}
 & & X \\
 & \overset{F}{\nearrow} & \downarrow \pi \\
Z & \overset{f}{\longrightarrow} & Y
\end{array}
$$
□

Proposition 6.2.14 (Unique Path Lifting). *Let $X \overset{\pi}{\to} Y$ be a covering map, $\gamma : [0,1] \to Y$ a path in Y and x_0 a point in the fiber over $y_0 = \gamma(0)$, $x_0 \in \pi^{-1}(y_0)$. Then there exists at most one lift of γ, $\Gamma : [0,1] \to Y$ such that $\Gamma(0) = x_0$.*

Proof.   We argue by contradiction. Assume there exist two such lifts, $\Gamma_1, \Gamma_2 : [0,1] \to X$. Set

$$ S := \{ t \in [0,1]; \ \Gamma_1(t) = \Gamma_2(t) \}. $$

The set S is nonempty since $0 \in S$. Obviously S is closed so it suffices to prove that it is also open. We will prove that for every $s \in S$, there exists $\varepsilon > 0$ such that $[s - \varepsilon, s + \varepsilon] \cap [0,1] \subset S$. For simplicity, we consider the case $s = 0$. The general situation is entirely similar.

We will prove that there exists $r_0 > 0$ such that $[0, r_0] \subset S$. Pick a small open neighborhood U of x_0 such that π restricts to a homeomorphism onto $\pi(U)$. There exists $r_0 > 0$ such that $\gamma_i([0, r_0]) \subset U$, $i = 1, 2$. Since $\pi \circ \Gamma_1 = \pi \circ \Gamma_2$, we deduce $\Gamma_1|_{[0,r_0]} = \Gamma_2|_{[0,r_0]}$. The proposition is proved.   □

Theorem 6.2.15.   *Let $X \overset{\pi}{\to} Y$ be a covering space, and $f : Z \to Y$ be a continuous map, where Z is a connected space. Fix $z_0 \in Z$, and $x_0 \in \pi^{-1}(y_0)$, where $y_0 = f(z_0)$. Then there exists at most one lift $F : Z \to X$ of f such that $F(z_0) = x_0$.*

Proof.   For each $z \in Z$ let α_z be a continuous path connecting z_0 to z. If F_1, F_2 are two lifts of f such that $F_1(z_0) = F_2(z_0) = x_0$ then, for any $z \in Z$, the paths

$\Gamma_1 = F_i(\alpha_z)$, and $\Gamma_2 = F_2(\alpha_z)$ are two lifts of $\gamma = f(\alpha_z)$ starting at the same point. From Proposition 6.2.14 we deduce that $\Gamma_1 \equiv \Gamma_2$, i.e., $F_1(z) = F_2(z)$, for any $z \in Z$.

□

6.2.3 Homotopy lifting property

Theorem 6.2.16 (Homotopy lifting property). *Let* $X \xrightarrow{\pi} Y$ *be a covering space,* $f : Z \to Y$ *be a continuous map, and* $F : Z \to X$ *be a lift of* f. *If*

$$h : [0,1] \times Z \to Y \quad (t,z) \mapsto h_t(z)$$

is a homotopy of f *(*$h_0(z) \equiv f(z)$*), then there exists a unique lift of* h

$$H : [0,1] \times Z \to X \quad (t,z) \mapsto H_t(z),$$

such that $H_0(z) \equiv F(z)$.

Proof. For each $z \in Z$ we can find an open neighborhood U_z of $z \in Z$, and a partition $0 = t_0 < t_1 < \ldots < t_n = 1$, depending on z, such that h maps $[t_{i-1}, t_i] \times U_z$ into an evenly covered neighborhood of $h_{t_{i-1}}(z)$. Following this partition, we can now successively lift $h|_{I \times U_z}$ to a continuous map $H = H^z : I \times U_z \to X$ such that $H_0(\zeta) = F(\zeta)$, $\forall \zeta \in U_z$. By unique lifting property, the liftings on $I \times U_{z_1}$ and $I \times U_{z_2}$ agree on $I \times (U_{z_1} \cap U_{z_2})$, for any $z_1, z_2 \in Z$, and hence we can glue all these local lifts together to obtain the desired lift H on $I \times Z$.

□

Corollary 6.2.17 (Path lifting property). *Suppose that* $X \xrightarrow{\pi} Y$ *is a covering map,* $y_0 \in Y$, *and* $\gamma : [0,1] \to Y$ *is a continuous path starting at* y_0. *Then, for every* $x_0 \in \pi^{-1}(y_0)$ *there exists a unique lift* $\Gamma : [0,1] \to X$ *of* γ *starting at* x_0.

Proof. Use the previous theorem with $f : \{pt\} \to Y$, $f(pt) = \gamma(0)$, and $h_t(pt) = \gamma(t)$.

□

Corollary 6.2.18. *Let* $X \xrightarrow{\pi} Y$ *be a covering space, and* $y_0 \in Y$. *If* $\gamma_0, \gamma_1 \in \Omega(Y, y_0)$ *are homotopic rel* y_0, *then any lifts* Γ_0, Γ_1 *which start at the same point also end at the same point, i.e.* $\Gamma_0(1) = \Gamma_1(1)$.

Proof. Lift the homotopy connecting γ_0 to γ_1 to a homotopy in X. By unique lifting property this lift connects Γ_0 to Γ_1. We thus get a continuous path $\Gamma_t(1)$ inside the fiber $\pi^{-1}(y_0)$ which connects $\Gamma_0(1)$ to $\Gamma_1(1)$. Since the fibers are discrete, this path must be constant.

□

Let $X \xrightarrow{\pi} Y$ be a covering space, and $y_0 \in Y$. Then, for every $x \in \pi^{-1}(y_0)$, and any $\gamma \in \Omega(Y, y_0)$, denote by Γ_x the unique lift of γ starting at x. Set

$$x \cdot \gamma := \Gamma_x(1).$$

By Corollary 6.2.18, if $\omega \in \Omega(Y, y_0)$ is homotopic to γ rel y_0, then

$$x \cdot \gamma = x \cdot \omega.$$

Hence, $x \cdot \gamma$ depends only upon the equivalence class $[\gamma] \in \pi_1(Y, y_0)$. Clearly

$$x \cdot ([\gamma] \cdot [\omega]) = (x \cdot [\gamma]) \cdot [\omega],$$

and

$$x \cdot \mathfrak{e}_{y_0} = x,$$

so that the correspondence

$$\pi^{-1}(y_0) \times \pi_1(Y, y_0) \ni (x, \gamma) \mapsto x \cdot \gamma \in x \cdot \gamma \in \pi^{-1}(y_0)$$

defines a right action of $\pi_1(Y, y_0)$ on the fiber $\pi^{-1}(y_0)$. This action is called the *monodromy of the covering*. The map $x \mapsto x \cdot \gamma$ is called the monodromy along γ. Note that when Y is simply connected, the monodromy is trivial. The map π induces a group morphism

$$\pi_* : \pi_1(X, x_0) \to \pi_1(Y, y_0) \quad x_0 \in \pi^{-1}(y_0).$$

Proposition 6.2.19. π_* *is injective.*

Proof. Indeed, let $\gamma \in \Omega(X, x_0)$ such that $\pi(\gamma)$ is trivial in $\pi_1(Y, y_0)$. The homotopy connecting $\pi(\gamma)$ to $\mathfrak{e}_{y_0}$ lifts to a homotopy connecting γ to the unique lift of $\mathfrak{e}_{y_0}$ at x_0, which is $\mathfrak{e}_{x_0}$. $\qquad\qquad\square$

6.2.4 *On the existence of lifts*

Theorem 6.2.20. *Let $X \xrightarrow{\pi} Y$ be a covering space, $x_0 \in X$, $y_0 = \pi(x_0) \in Y$, $f : Z \to Y$ a continuous map and $z_0 \in Z$ such that $f(z_0) = y_0$. Assume the spaces Y and Z are connected (and thus path connected). f admits a lift $F : Z \to X$ such that $F(z_0) = x_0$ if and only if*

$$f_* (\pi_1(Z, z_0)) \subset \pi_* (\pi_1(X, x_0)). \tag{6.2.1}$$

Proof. **Necessity.** If F is such a lift then, using the functoriality of the fundamental group construction, we deduce $f_* = \pi_* \circ F_*$. This implies the inclusion (6.2.1).

Sufficiency. For any $z \in Z$, choose a path γ_z from z_0 to z. Then $\alpha_z = f(\gamma_z)$ is a path from y_0 to $y = f(z)$. Denote by A_z the unique lift of α_z starting at x_0, and set $F(z) = A_z(1)$. We claim that F is a well defined map.

Indeed, let ω_z be another path in Z connecting z_0 to z. Set $\lambda_z := f(\omega_z)$, and denote by Λ_z its unique lift in X starting at x_0. We have to show that $\Lambda_z(1) = A_z(1)$. Construct the loop based at z_0

$$\beta_z = \omega_z * \gamma_z^-.$$

Then $f(\beta_z)$ is a loop in Y based at y_0. From (6.2.1) we deduce that the lift B_z of $f(\beta_z)$ at $x_0 \in X$ is a closed path, i.e., the monodromy along $f(\beta_z)$ is trivial. We now have

$$\Lambda_z(1) = B_z(1/2) = A_z(0) = A_z(1).$$

This proves that F is a well defined map. We still have to show that this map is also continuous.

Pick $z \in Z$. Since f is continuous, for every arbitrarily small, evenly covered neighborhood U of $f(z) \in Y$ there exists a path connected neighborhood V of $z \in Z$ such that $f(V) \subset U$. For any $\zeta \in V$ pick a path $\sigma = \sigma_\zeta$ in V connecting z to ζ. Let ω denote the path $\omega = \gamma_z * \sigma_\zeta$ (go from z_0 to z along γ_z, and then from z to ζ along σ_ζ). Then $F(\zeta) = \Omega(1)$, where Ω is the unique lift of $f(\omega)$ starting at x_0. Since $(f(\zeta) \in U$, we deduce that $\Omega(1)$ belongs to the local sheet Σ, containing $F(z)$, which homeomorphically covers U. We have thus proved $z \in V \subset F^{-1}(\Sigma)$. The proof is complete since the local sheets Σ form a basis of neighborhoods of $F(z)$. $\square$

Definition 6.2.21. Let Y be a connected space. A covering space $X \overset{\pi}{\to} Y$ is said to be *universal* if X is simply connected. $\square$

Corollary 6.2.22. *Let $X_1 \overset{p_i}{\to} Y$ (i=0,1) be two covering spaces of Y. Fix $x_i \in X_i$ such that $p_0(x_0) = p(x_1) = y_0 \in Y$. If X_0 is universal, then there exists a unique covering morphism $F : X_0 \to X_1$ such that $F(x_0) = x_1$.*

Proof. A bundle morphism $F : X_0 \to X_1$ can be viewed as a lift of the map $p_0 : X_0 \to Y$ to the total space of the covering space defined by p_1. The corollary follows immediately from Theorem 6.2.20 and the unique lifting property. $\square$

Corollary 6.2.23. *Every space admits at most one universal covering space (up to isomorphism).*

Theorem 6.2.24. *Let Y be a connected, locally path connected space such that each of its points admits a simply connected neighborhood. Then Y admits an (essentially unique) universal covering space.*

Sketch of proof. Assume for simplicity that Y is a metric space. Fix $y_0 \in Y$. Let $\mathcal{P}_{y_0}$ denote the collection of continuous paths in Y starting at y_0. It can be topologized using the metric of uniform convergence in Y. Two paths in $\mathcal{P}_{y_0}$ are said to be homotopic rel endpoints if they we can deform one to the other while keeping the endpoints fixed. This defines an equivalence relation on $\mathcal{P}_{y_0}$. We denote the space of equivalence classes by $\tilde{Y}$, and we endow it with the quotient topology. Define $p : \tilde{Y} \to Y$ by

$$p([\gamma]) = \gamma(1) \quad \forall \gamma \in \mathcal{P}_{y_0}.$$

Then $(\tilde{Y}, p)$ is a universal covering space of Y. $\square$

Exercise 6.2.25. Finish the proof of the above theorem. $\square$

Example 6.2.26. The map $\mathbb{R} \xrightarrow{\exp} S^1$ is the universal cover of S^1. More generally,
$\exp : \mathbb{R}^n \to T^n$

$$(t_1, \ldots, t_n) \mapsto (\exp(2\pi i t_1), \ldots, \exp(2\pi i t_n)), \quad i = \sqrt{-1},$$

is the universal cover of T^n. The natural projection $p : S^n \to \mathbb{RP}^n$ is the universal cover of $\mathbb{RP}^n$. □

Example 6.2.27. Let (M, g) be a complete Riemann manifold with non-positive sectional curvature. By Cartan-Hadamard theorem, the exponential map $\exp_q : T_q M \to M$ is a covering map. Thus, the universal cover of such a manifold is a linear space of the same dimension. In particular, the universal covering space is *contractible*!!!

We now have another explanation why $\exp : \mathbb{R}^n \to T^n$ is a universal covering space of the torus: the sectional curvature of the (flat) torus is zero. □

Exercise 6.2.28. Let (M, g) be a complete Riemann manifold and $p : \tilde{M} \to M$ its universal covering space.
(a) Prove that $\tilde{M}$ has a natural structure of smooth manifold such that p is a local diffeomorphism.
(b) Prove that the pullback $p^* g$ defines a complete Riemann metric on $\tilde{M}$ locally isometric with g. □

Example 6.2.29. Let (M, g) be a complete Riemann manifold such that

$$\operatorname{Ric}(X, X) \geq const. |X|_g^2, \tag{6.2.2}$$

where *const* denotes a strictly positive constant. By Myers theorem M is compact. Using the previous exercise we deduce that the universal cover $\tilde{M}$ is a complete Riemann manifold locally isometric with (M, g). Hence the inequality (6.2.2) continues to hold on the covering $\tilde{M}$. Myers theorem implies again that the universal cover $\tilde{M}$ is *compact*!! In particular, the universal cover of a semisimple, compact Lie group is compact!!! □

6.2.5 *The universal cover and the fundamental group*

Theorem 6.2.30. *Let $\tilde{X} \xrightarrow{p} X$ be the universal cover of a space X. Then*

$$\pi_1(X, pt) \cong \operatorname{Deck}(\tilde{X} \to X).$$

Proof.   Fix $\xi_0 \in \tilde{X}$ and set $x_0 = p(\xi_0)$. There exists a bijection

$$\mathbf{Ev} : \operatorname{Deck}(\tilde{X}) \to p^{-1}(x_0),$$

given by the **ev**aluation

$$\mathbf{Ev}(F) := F(\xi_0).$$

For any $\xi \in \pi^{-1}(x_0)$, let γ_ξ be a path connecting ξ_0 to ξ. Any two such paths are homotopic rel endpoints since $\tilde{X}$ is simply connected (check this). Their projections on the base X determine identical elements in $\pi_1(X, x_0)$. We thus have a natural map

$$\Psi : \mathrm{Deck}\,(\tilde{X}) \to \pi_1(X, x_0) \quad F \mapsto p(\gamma_{F(\xi_0)}).$$

The map Ψ is clearly a group morphism. (Think monodromy!) The injectivity and the surjectivity of Ψ are consequences of the lifting properties of the universal cover. □

Corollary 6.2.31. *If the space X has a compact universal cover, then $\pi_1(X, pt)$ is finite.*

Proof. Indeed, the fibers of the universal cover have to be both discrete and compact. Hence they must be finite. The map **Ev** in the above proof is a bijection onto Deck $(\tilde{X})$. □

Corollary 6.2.32 (H. Weyl). *The fundamental group of a compact semisimple group is finite.*

Proof. Indeed, we deduce from Example 6.2.29 that the universal cover of such a group is compact. □

Example 6.2.33. From Example 6.2.11 we deduce that $\pi_1(S^1) \cong (\mathbb{Z}, +)$. □

Exercise 6.2.34. (a) Prove that $\pi_1(\mathbb{RP}^n, pt) \cong \mathbb{Z}_2$, $\forall n \geq 2$.
(b) Prove that $\pi_1(T^n) \cong \mathbb{Z}^n$. □

Exercise 6.2.35. Show that the natural inclusion $U(n-1) \hookrightarrow U(n)$ induces an isomorphism between the fundamental groups. Conclude that the map

$$\det : U(n) \to S^1$$

induces an isomorphism

$$\pi_1(U(n), 1) \cong \pi_1(S^1, 1) \cong \mathbb{Z}.$$ □

Chapter 7

Cohomology

7.1 DeRham cohomology

7.1.1 *Speculations around the Poincaré lemma*

To start off, consider the following partial differential equation in the plane. Given two smooth functions P and Q, find a smooth function u such that

$$\frac{\partial u}{\partial x} = P, \quad \frac{\partial u}{\partial y} = Q. \qquad (7.1.1)$$

As is, the formulation is still ambiguous since we have not specified the domains of the functions u, P and Q. As it will turn out, this aspect has an incredible relevance in geometry.

Equation (7.1.1) has another interesting feature: it is overdetermined, i.e., it imposes too many conditions on too few unknowns. It is therefore quite natural to impose some additional restrictions on the data P, Q just like the zero determinant condition when solving overdetermined linear systems.

To see what restrictions one should add it is convenient to introduce the 1-form $\alpha = Pdx + Qdy$. The equality (7.1.1) can be rewritten as

$$du = \alpha. \qquad (7.1.2)$$

If (7.1.2) has at least one solution u, then $0 = d^2u = d\alpha$, so that a necessary condition for existence is

$$d\alpha = 0, \qquad (7.1.3)$$

i.e.,

$$\frac{\partial P}{\partial y} = \frac{\partial Q}{\partial x}.$$

A form satisfying (7.1.3) is said to be *closed*. Thus, if the equation $du = \alpha$ has a solution then α is necessarily closed. Is the converse also true?

Let us introduce a bit more terminology. A form α such that the equation (7.1.2) has a solution is said to be *exact*. The motivation for this terminology comes from

231

the fact that sometimes the differential form du is called the exact differential of u. We thus have an inclusion of vector spaces

$$\{\text{exact forms}\} \subset \{\text{closed forms}\}.$$

Is it true that the opposite inclusion also holds?

Amazingly, the answer to this question *depends on the domain* on which we study (7.1.2). The Poincaré lemma comes to raise our hopes. It says that this is always true, at least *locally*.

Lemma 7.1.1 (Poincaré lemma). *Let C be an open convex set in $\mathbb{R}^n$ and $\alpha \in \Omega^k(C)$. Then the equation*

$$du = \alpha \qquad (7.1.4)$$

has a solution $u \in \Omega^{k-1}(C)$ if and only if α is closed, $d\alpha = 0$.

Proof. The necessity is clear. We prove the sufficiency. We may as well assume that $0 \in C$. Consider the radial vector field on C

$$\vec{r} = x^i \frac{\partial}{\partial x_i},$$

and denote by Φ^t the flow it generates. More explicitly, Φ_t is the linear flow

$$\Phi^t(x) = e^t x, \quad x \in \mathbb{R}^n.$$

The flow lines of Φ^t are half-lines, and since C is *convex*, for every $x \in C$, and any $t \leq 0$ we have $\Phi^t(x) \in C$.

We begin with an a priori study of (7.1.4). Let u satisfy $du = \alpha$. Using the homotopy formula,

$$L_{\vec{r}} = di_{\vec{r}} + i_{\vec{r}}d,$$

we get

$$dL_{\vec{r}}u = d(di_{\vec{r}} + i_{\vec{r}}d)u = di_{\vec{r}}\alpha \Rightarrow d(L_{\vec{r}}u - i_{\vec{r}}\alpha) = 0.$$

This suggests looking for solutions φ of the equation

$$L_{\vec{r}}\varphi = i_{\vec{r}}\alpha, \quad \varphi \in \Omega^{k-1}(C). \qquad (7.1.5)$$

If φ is a solution of this equation, then

$$L_{\vec{r}}d\varphi = dL_{\vec{r}}\varphi = di_{\vec{r}}\alpha = L_{\vec{r}}\alpha - i_{\vec{R}}d\alpha = L_{\vec{r}}\alpha.$$

Hence the form φ also satisfies

$$L_{\vec{r}}(d\varphi - \alpha) = 0.$$

Set $\omega := d\varphi - \alpha = \sum_I \omega_I dx^I$. Using the computations in Subsection 3.1.3 we deduce

$$L_{\vec{r}}dx^i = dx^i,$$

so that

$$L_{\vec{r}}\omega = \sum_I (L_{\vec{r}}\omega_I)dx^I = 0.$$

We deduce that $L_{\vec{r}}\omega_I = 0$, and consequently, the coefficients ω_I are constants along the flow lines of Φ_t which all converge at 0. Thus,

$$\omega_I = c_I = \text{const.}$$

Each monomial $c_I dx^I$ is exact, i.e., there exist $\eta_I \in \Omega^{k-1}(C)$ such that $d\eta_I = c_I dx^I$. For example, when $I = 1 < 2 < \cdots < k$

$$dx^1 \wedge dx^2 \wedge \cdots \wedge dx^k = d(x^1 dx^2 \wedge \cdots \wedge dx^k).$$

Thus, the equality $L_{\vec{r}}\omega = 0$ implies ω is exact. Hence there exists $\eta \in \Omega^{k-1}(C)$ such that

$$d(\varphi - \eta) = \alpha,$$

i.e., the differential form $u := \varphi - \eta$ solves (7.1.4). Conclusion: any solution of (7.1.5) produces a solution of (7.1.4).

We now proceed to solve (7.1.5), and to this aim, we use the flow Φ_t. Define

$$u = \int_{-\infty}^{0} (\Phi^t)^*(i_{\vec{r}}\alpha)dt. \tag{7.1.6}$$

Here the convexity assumption on C enters essentially since it implies that

$$\Phi^t(C) \subset C \quad \forall t \leq 0,$$

so that if the above integral is convergent, then u is a form on C. If we write

$$(\Phi^t)^*(i_{\vec{r}}\alpha) = \sum_{|I|=k-1} \eta_I^t(x)dx^I,$$

then

$$u(x) = \sum_I \left(\int_{-\infty}^{0} \eta_I^t(x)dt \right) dx^I. \tag{7.1.7}$$

We have to check two things.

A. The integral in (7.1.7) is well defined. To see this, we first write

$$\alpha = \sum_{|J|=k} \alpha_J dx^J,$$

and then we set

$$A(x) := \max_{J; 0 \leq \tau \leq 1} |\alpha_J(\tau x)|.$$

Then

$$(\Phi^t)^*(i_{\vec{r}}\alpha) = e^t i_{\vec{r}} \left(\sum_J \alpha_J(e^t x)dx^J \right),$$

so that
$$|\eta_I^t(x)| \le Ce^t |x| A(x) \quad \forall t \le 0.$$
This proves the integral in (7.1.7) converges.

B. The differential form u defined by (7.1.6) is a solution of (7.1.5). Indeed,

$$L_{\vec{r}} u = \lim_{s \to 0} \frac{1}{s} \left((\Phi^s)^* u - u \right)$$

$$= \lim_{s \to 0} \frac{1}{s} \left((\Phi^s)^* \int_{-\infty}^0 (\Phi^t)^* (i_{\vec{r}}\alpha) dt - \int_{-\infty}^0 (\Phi^t)^* (i_{\vec{r}}\alpha) dt \right)$$

$$= \lim_{s \to 0} \left(\int_{-\infty}^s (\Phi^t)^* (i_{\vec{r}}\alpha) dt - \int_{-\infty}^0 (\Phi^t)^* (i_{\vec{r}}\alpha) dt \right)$$

$$= \lim_{s \to 0} \int_0^s (\Phi^t)^* (i_{\vec{r}}\alpha) dt = (\Phi^0)^* (i_{\vec{r}}\alpha) = i_{\vec{r}}\alpha.$$

The Poincaré lemma is proved. $\square$

The local solvability does not in any way implies global solvability. Something happens when one tries to go from local to global.

Example 7.1.2. Consider the form $d\theta$ on $\mathbb{R}^2 \setminus \{0\}$ where (r, θ) denote the polar coordinates in the punctured plane. To write it in cartesian coordinates (x, y) we use the equality

$$\tan \theta = \frac{y}{x}$$

so that

$$(1 + \tan^2 \theta) d\theta = -\frac{y}{x^2} dx + \frac{dy}{x} \quad \text{and} \quad (1 + \frac{y^2}{x^2}) d\theta = \frac{-y dx + x dy}{x^2},$$

i.e.,

$$d\theta = \frac{-y dx + x dy}{x^2 + y^2} = \alpha.$$

Obviously, $d\alpha = d^2\theta = 0$ on $\mathbb{R}^2 \setminus \{0\}$ so that α is closed on the punctured plane. Can we find a smooth function u on $\mathbb{R}^2 \setminus \{0\}$ such that $du = \alpha$?

We know that we can always do this locally. However, we cannot achieve this globally. Indeed, if this was possible, then

$$\int_{S^1} du = \int_{S^1} \alpha = \int_{S^1} d\theta = 2\pi.$$

On the other hand, using polar coordinates $u = u(r, \theta)$ we get

$$\int_{S^1} du = \int_{S^1} \frac{\partial u}{\partial \theta} d\theta = \int_0^{2\pi} \frac{\partial u}{\partial \theta} d\theta = u(1, 2\pi) - u(1, 0) = 0.$$

Hence on $\mathbb{R}^2 \setminus \{0\}$

$$\{\text{exact forms}\} \ne \{\text{closed forms}\}.$$

We see what a dramatic difference a point can make: $\mathbb{R}^2 \setminus \{\text{point}\}$ is structurally very different from $\mathbb{R}^2$. $\square$

The artifice in the previous example simply increases the mystery. It is still not clear what makes it impossible to patch-up local solutions. The next subsection describes two ways to deal with this issue.

7.1.2 Čech vs. DeRham

Let us try to analyze what prevents the "spreading" of local solvability of (7.1.4) to global solvability. We will stay in the low degree range.

The Čech approach. Consider a closed 1-form ω on a smooth manifold. To solve the equation $du = \omega$, $u \in C^\infty(M)$ we first cover M by open sets (U_α) which are geodesically convex with respect to some fixed Riemann metric.

Poincaré lemma shows that we can solve $du = \omega$ on each open set U_α so that we can find a smooth function $f_\alpha \in C^\infty(U_\alpha)$ such that $df_\alpha = \omega$. We get a global solution if and only if

$$f_{\alpha\beta} = f_\alpha - f_\beta = 0 \quad \text{on each } U_{\alpha\beta} = U_\alpha \cap U_\beta \neq \emptyset.$$

For fixed α, the solutions of the equation $du = \omega$ on U_α differ only by additive constants, i.e., closed 0-forms.

The quantities $f_{\alpha\beta}$ satisfy $df_{\alpha\beta} = 0$ on the (*connected*) overlaps $U_{\alpha\beta}$ so they are constants. Clearly they satisfy the conditions

$$f_{\alpha\beta} + f_{\beta\gamma} + f_{\gamma\alpha} = 0 \quad \text{on every } U_{\alpha\beta\gamma} := U_\alpha \cap U_\beta \cap U_\gamma \neq \emptyset. \tag{7.1.8}$$

On each U_α we have, as we have seen, several choices of solutions. Altering a choice is tantamount to adding a constant $f_\alpha \to f_\alpha + c_\alpha$. The quantities $f_{\alpha\beta}$ change according to

$$f_{\alpha\beta} \to f_{\alpha\beta} + c_\alpha - c_\beta.$$

Thus, the global solvability issue leads to the following situation.

Pick a collection of local solutions f_α. The equation $du = \omega$ is globally solvable if we can alter each f_α by a constant c_α such that

$$f_{\alpha\beta} = (c_\beta - c_\alpha) \quad \forall \alpha, \beta \text{ such that } U_{\alpha\beta} \neq \emptyset. \tag{7.1.9}$$

We can start the alteration at some open set U_α, and work our way up from one such open set to its neighbors, always trying to implement (7.1.9). It may happen that in the process we might have to return to an open set whose solution was already altered. Now we are in trouble. (Try this on S^1, and $\omega = d\theta$.) After several attempts one can point the finger to the culprit: the global topology of the manifold may force us to always return to some already altered local solution.

Notice that we replaced the partial differential equation $du = \omega$ with a system of linear equations (7.1.9), where the constants $f_{\alpha\beta}$ are subject to the constraints (7.1.8). This is no computational progress since the complexity of the combinatorics of this system makes it impossible solve in most cases.

The above considerations extend to higher degree, and one can imagine that the complexity increases considerably. This is however the approach Čech adopted in order to study the topology of manifolds, and although it may seem computationally hopeless, its theoretical insights are invaluable.

The DeRham approach. This time we postpone asking why the global solvability is not always possible. Instead, for each smooth manifold M one considers the $\mathbb{Z}$-graded vector spaces

$$B^\bullet(M) = \bigoplus_{k \geq 0} B^k(M), \quad (B^0(M) := \{0\}) \quad Z^\bullet(M) := \bigoplus_{k \geq 0} Z^k(M),$$

where

$$B^k(M) = \{\, d\omega \in \Omega^k(M); \;\; \omega \in \Omega^{k-1}(M) \,\} = \text{exact } k\text{-forms},$$

and

$$Z^k(M = \{\, \eta \in \Omega^k(M); \;\; d\eta = 0 \,\} = \text{closed } k\text{-forms}.$$

Clearly $B^k \subset Z^k$. We form the quotients,

$$H^k(M) := Z^k(M)/B^k(M).$$

Intuitively, this space consists of those closed k-forms ω for which the equation $du = \omega$ has no global solution $u \in \Omega^{k-1}(M)$. Thus, if we can somehow describe these spaces, we may get an idea "who" is responsible for the global nonsolvability.

Definition 7.1.3. For any smooth manifold M the vector space $H^k(M)$ is called the k-th *DeRham cohomology group of M*. $\square$

Clearly $H^k(M) = 0$ for $k > \dim M$.

Example 7.1.4. The Poincaré lemma shows that $H^k(\mathbb{R}^n) = 0$ for $k > 0$. The discussion in Example 7.1.2 shows that $H^1(\mathbb{R}^2 \setminus \{0\}) \neq 0$. $\square$

Proposition 7.1.5. *For any smooth manifold M*

$$\dim H^0(M) = \text{number of connected components of } M.$$

Proof. Indeed

$$H^0(M) = Z^0(M) = \{\, f \in C^\infty(M) \,;\; df = 0 \,\}.$$

Thus, $H^0(M)$ coincides with the linear space of locally constant functions. These are constants on the connected components of M. $\square$

We see that $H^0(M)$, the simplest of the DeRham groups, already contains an important topological information. Obviously the groups H^k are diffeomorphism invariants, and its is reasonable to suspect that the higher cohomology groups may contain more topological information.

Thus, to any smooth manifold M we can now associate the graded vector space

$$H^\bullet(M) := \bigoplus_{k \geq 0} H^k(M).$$

A priori, the spaces $H^k(M)$ may be infinite dimensional. The *Poincaré polynomial of M*, denoted by $P_M(t)$, is defined by

$$P_M(t) = \sum_{k \geq 0} t^k \dim H^k(M),$$

every time the right-hand-side expression makes sense. The number $\dim H^k(M)$ is usually denoted by $b_k(M)$, and it is called the *k-th Betti number* of M. Hence

$$P_M(t) = \sum_k b_k(M) t^k.$$

The alternating sum

$$\chi(M) := \sum_k (-1)^k b_k(M),$$

is called the *Euler characteristic* of M.

Exercise 7.1.6. Show that $P_{S^1}(t) = 1 + t$. □

We will spend the remaining of this chapter trying to understand what is that these groups do and which, if any, is the connection between the two approaches outlined above.

7.1.3 *Very little homological algebra*

At this point it is important to isolate the common algebraic skeleton on which both the DeRham and the Čech approaches are built. This requires a little terminology from homological algebra.

☞ *In the sequel all rings will be assumed commutative with 1.*

Definition 7.1.7. (a) Let R be a ring, and let

$$C^\bullet = \bigoplus_{n \in \mathbb{Z}} C^n, \quad D^\bullet = \bigoplus_{n \in \mathbb{Z}} D^n$$

be two $\mathbb{Z}$-graded left R-modules. A *degree k-morphism* $\phi : C^\bullet \to D^\bullet$ is an R-module morphism such that

$$\phi(C^n) \subset D^{n+k} \quad \forall n \in \mathbb{Z}.$$

(b) Let

$$C^\bullet = \bigoplus_{n \in \mathbb{Z}} C^n$$

be a $\mathbb{Z}$-graded R-module. A *boundary* (respectively *coboundary*) operator is a degree -1 (respectively a degree 1) endomorphism $d : C^\bullet \to C^\bullet$ such that $d^2 = 0$.

A *chain* (respectively *cochain*) complex over R is a pair $(C^\bullet, d)$, where $C^\bullet$ is a $\mathbb{Z}$-graded R-module, and d is a boundary (respectively a coboundary) operator. □

In this book we will be interested mainly in cochain complexes so in the remaining part of this subsection we will stick to this situation. In this case cochain complexes are usually described as $(C^\bullet = \oplus_{n \in \mathbb{Z}} C^n, d)$. Moreover, we will consider only the case $C^n = 0$ for $n < 0$.

Traditionally, a cochain complex is represented as a long sequence of R-modules and morphisms of R-modules

$$(C^\bullet, d): \quad \cdots \to C^{n-1} \xrightarrow{d_{n-1}} C^n \xrightarrow{d_n} C^{n+1} \to \cdots,$$

such that $\mathrm{range}\,(d_{n-1}) \subset \ker\,(d_n)$, i.e., $d_n d_{n-1} = 0$.

Definition 7.1.8. Let

$$\cdots \to C^{n-1} \xrightarrow{d_{n-1}} C^n \xrightarrow{d_n} C^{n+1} \to \cdots$$

be a cochain complex of R-modules. Set

$$Z^n(C) := \ker d_n \quad B^n(C) := \mathrm{range}\,(d_{n-1}).$$

The elements of $Z^n(C)$ are called *cocycles*, and the elements of $B^n(C)$ are called *coboundaries*.

Two cocycles $c, c' \in Z^n(C)$ are said to be *cohomologous* if $c - c' \in B^n(C)$. The quotient module

$$H^n(C) := Z^n(C)/B^n(C)$$

is called the n-th *cohomology group* (module) of C. It can be identified with the set of equivalence classes of cohomologous cocycles. A cochain complex complex C is said to be *acyclic* if $H^n(C) = 0$ for all $n > 0$. □

For a cochain complex $(C^\bullet, d)$ one usually writes

$$H^\bullet(C) = H^\bullet(C, d) = \bigoplus_{n \geq 0} H^n(C).$$

Example 7.1.9. (The DeRham complex). Let M be an m-dimensional smooth manifold. Then the sequence

$$0 \to \Omega^0(M) \xrightarrow{d} \Omega^1 \xrightarrow{d} \cdots \xrightarrow{d} \Omega^m(M) \to 0$$

(where d is the exterior derivative) is a cochain complex of real vector spaces called the *DeRham complex*. Its cohomology groups are the *DeRham cohomology groups* of the manifold. □

Example 7.1.10. Let $(\mathfrak{g}, [\cdot, \cdot])$ be a real Lie algebra. Define

$$d: \Lambda^k \mathfrak{g}^* \to \Lambda^{k+1} \mathfrak{g}^*,$$

by

$$(d\omega)(X_0, X_1, \ldots, X_k) := \sum_{0 \leq i < j \leq k} (-1)^{i+j} \omega([X_i, X_j], X_0, \ldots, \hat{X}_i, \ldots, \hat{X}_j, \ldots, X_k),$$

where, as usual, the hat $\hat{\ }$ indicates a missing argument.

According to the computations in Example 3.2.9 the operator d is a coboundary operator, so that $(\Lambda^\bullet \mathfrak{g}^*, d)$ is a cochain complex. Its cohomology is called the *Lie algebra cohomology*, and it is denoted by $H^\bullet(\mathfrak{g})$. □

Exercise 7.1.11. (a) Let $\mathfrak{g}$ be a real Lie algebra. Show that

$$H^1(\mathfrak{g}) \cong (\mathfrak{g}/[\mathfrak{g},\mathfrak{g}])^*,$$

where $[\mathfrak{g},\mathfrak{g}] = \text{span}\left\{\, [X,Y] \,;\, X,Y \in \mathfrak{g} \,\right\}$.
(b) Compute $H^1(\underline{\mathrm{gl}}(n,\mathbb{R}))$, where $\underline{\mathrm{gl}}(n,\mathbb{R})$ denotes the Lie algebra of $n \times n$ real matrices with the bracket given by the commutator.
(c) (Whitehead) Let $\mathfrak{g}$ be a semisimple Lie algebra, i.e., its Killing pairing is nondegenerate. Prove that $H^1(\mathfrak{g}) = \{0\}$. (**Hint:** Prove that $[\mathfrak{g},\mathfrak{g}]^\perp = 0$, where $\perp$ denotes the orthogonal complement with respect to the Killing pairing.) $\square$

Proposition 7.1.12. *Let*

$$(C^\bullet, d) : \cdots \to C^{n-1} \xrightarrow{d_{n-1}} C^n \xrightarrow{d_n} C^{n+1} \to \cdots,$$

be a cochain complex of R-modules. Assume moreover that C is also a $\mathbb{Z}$-graded R-algebra, i.e., there exists an associative multiplication such that

$$C^n \cdot C^m \subset C^{m+n} \quad \forall m, n.$$

If d is a quasi-derivation, i.e.,

$$d(x \cdot y) = \pm(dx) \cdot y \pm x \cdot (dy) \quad \forall x, y \in C,$$

then $H^\bullet(C)$ inherits a structure of $\mathbb{Z}$-graded R-algebra.

A cochain complex as in the above proposition is called a *differential graded algebra* or *DGA*.

Proof. It suffices to show $Z^\bullet(C) \cdot Z^\bullet(C) \subset Z^\bullet(C)$, and $B^\bullet(C) \cdot B^\bullet(C) \subset B^\bullet(C)$.
If $dx = dy = 0$, then $d(xy) = \pm(dx)y \pm x(dy) = 0$. If $x = dx'$ and $y = dy'$ then, since $d^2 = 0$, we deduce $xy = \pm(dx'dy')$. $\square$

Corollary 7.1.13. *The DeRham cohomology of a smooth manifold has an $\mathbb{R}$-algebra structure induced by the exterior multiplication of differential forms.* $\square$

Definition 7.1.14. Let $(A^\bullet, d)$ and $(B^\bullet, \delta)$ be two cochain complexes of R-modules.
(a) A *cochain morphism*, or *morphism of cochain complexes* is a degree 0 morphism $\phi : A^\bullet \to B^\bullet$ such that $\phi \circ d = \delta \circ \phi$, i.e., the diagram below is commutative for any n.

$$
\begin{CD}
A^n @>{d_n}>> A^{n+1} \\
@V{\phi_n}VV @VV{\phi_{n+1}}V \\
B^n @>{\delta_n}>> B^{n+1}
\end{CD}
$$

(b) Two cochain morphisms $\phi, \psi : A^\bullet \to B^\bullet$ are said to be *cochain homotopic*, and we write this $\phi \simeq \psi$, if there exists a degree -1 morphism $\chi : \{A^n \to B^{n-1}\}$ such that

$$\phi(a) - \psi(a) = \pm\delta \circ \chi(a) \pm \chi \circ d(a).$$

(c) Two cochain complexes $(A^\bullet, d)$, and $(B^\bullet, \delta)$ are said to be *homotopic*, if there exist cochain morphism

$$\phi : A \to B \text{ and } \psi : B \to A,$$

such that $\psi \circ \phi \simeq \mathbb{1}_A$, and $\phi \circ \psi \simeq \mathbb{1}_B$. $\qquad\square$

Example 7.1.15. The commutation rules in Subsection 3.2.1, namely $[L_X, d] = 0$, and $[i_X, d]_s = L_X$, show that for each vector field X on a smooth manifold M, the Lie derivative along X, $L_X : \Omega^\bullet(M) \to \Omega^\bullet(M)$ is a cochain morphism homotopic with the trivial map ($\equiv 0$). The interior derivative i_X is the cochain homotopy achieving this. $\qquad\square$

Proposition 7.1.16. *(a) Any cochain morphism $\phi : (A^\bullet, d) \to (B^\bullet, \delta)$ induces a degree zero morphism in cohomology*

$$\phi_* : H^\bullet(A) \to H^\bullet(B).$$

(b) If the cochain maps $\phi, \psi : A \to B$ are cochain homotopic, then they induce identical morphisms in cohomology, $\phi_* = \psi_*$.

(c) $(\mathbb{1}_A)_* = \mathbb{1}_{H^\bullet(A)}$, and if $(A_0^\bullet, d^0) \xrightarrow{\phi} (A_1^\bullet, d^1) \xrightarrow{\psi} (A_2^\bullet, d^2)$ are cochain morphisms, then $(\psi \circ \phi)_* = \psi_* \circ \phi_*$.

Proof. (a) It boils down to checking the inclusions

$$\phi(Z^n(A)) \subset Z^n(B) \text{ and } \phi(B^n(A)) \subset B^n(B).$$

These follow immediately from the definition of a cochain map.

(b) We have to show that $\phi(\text{cocycle}) - \psi(\text{cocycle}) = \text{coboundary}$. Let $da = 0$. Then

$$\phi(a) - \psi(a) = \pm\delta(\chi(a) \pm \chi(da) = \delta(\pm\chi(a)) = \text{coboundary in } B.$$

(c) Obvious. $\qquad\square$

Corollary 7.1.17. *If two cochain complexes $(A^\bullet, d)$ and $(B^\bullet, \delta)$ are cochain homotopic, then their cohomology modules are isomorphic.* $\qquad\square$

Proposition 7.1.18. *Let*

$$0 \to (A^\bullet, d^A) \xrightarrow{\phi} (B^\bullet, d^B) \xrightarrow{\psi} (C^\bullet, d^C) \to 0$$

be a short exact sequence of cochain complexes of R-modules. This means that we have a commutative diagram

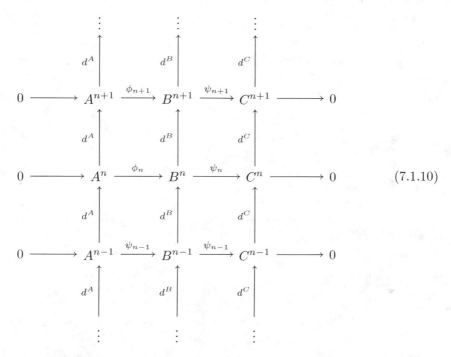

$$(7.1.10)$$

in which the rows are exact. Then there exists a long exact sequence

$$\cdots \to H^{n-1}(C) \overset{\partial_{n-1}}{\to} H^n(A) \overset{\phi_*}{\to} H^n(B) \overset{\psi_*}{\to} H^n(C) \overset{\partial_n}{\to} H^{n+1}(A) \to \cdots. \quad (7.1.11)$$

We will not include a proof of this proposition. We believe this is one proof in homological algebra that the reader should try to produce on his/her own. We will just indicate the construction of the connecting maps ∂_n. This construction, and in fact the entire proof, relies on a simple technique called *diagram chasing*.

Start with $x \in H^n(C)$. The cohomology class x can be represented by some cocycle $c \in Z^n(C)$. Since ψ_n is surjective there exists $b \in B^n$ such that $c = \psi_n(b)$. From the commutativity of the diagram (7.1.10) we deduce $0 = d^C \psi_n(b) = \psi_{n+1} d^B(b)$, i.e., $d^B(b) \in \ker \psi_{n+1} = \operatorname{range} \phi_{n+1}$. In other words, there exists $a \in A^{n+1}$ such that $\phi_{n+1}(a) = d_n^B b$. We claim that a is a cocycle. Indeed,

$$\phi_{n+2} d_{n+1}^A a = d_{n+1}^B \phi_{n+1} a = d_{n+1}^B d_n^B b = 0.$$

Since ϕ_{n+2} is injective we deduce $d_{n+1}^A a = 0$, i.e., a is a cocycle.

If we trace back the path which lead us from $c \in Z^n(C)$ to $a \in Z^{n+1}(A)$, we can write

$$a = \phi_{n+1}^{-1} \circ d^B \circ \psi_n^{-1}(c) = \phi_{n+1}^{-1} \circ d^B b.$$

This description is not entirely precise since a depends on various choices. We let the reader check that the correspondence $Z^n(C) \ni c \mapsto a \in Z^{n+1}(A)$ above induces a well defined map in cohomology, $\partial_n : H^n(C) \to H^{n+1}(A)$ and moreover, the sequence (7.1.11) is exact.

Exercise 7.1.19. [1] Suppose R is a commutative ring with 1. For any cochain complex $(K^\bullet, d_K)$ of R-modules, and any integer n we denote by $K[n]^\bullet$ the complex defined by $K[n]^m = K^{n+m}$, $d_{K[n]} = (-1)^n d_K$. We associate to any cochain map $f : K^\bullet \to L^\bullet$ two new cochain complexes:

(a) *The cone* $\left(C(f)^\bullet, d_{C(f)} \right)$ where

$$C(f)^\bullet = K[1]^\bullet \oplus L^\bullet, \quad d_{C(f)} \begin{bmatrix} k^{i+1} \\ \ell^i \end{bmatrix} = \begin{bmatrix} -d_K & 0 \\ f & d_L \end{bmatrix} \cdot \begin{bmatrix} k^{i+1} \\ \ell^i \end{bmatrix}.$$

(b) *The cylinder* $\left(Cyl(f), d_{Cyl(f)} \right)$

$$Cyl(f)^\bullet \cong K^\bullet \oplus C(f)^\bullet, \quad d_{Cyl(f)} \begin{bmatrix} k^i \\ k^{i+1} \\ \ell^i \end{bmatrix} = \begin{bmatrix} d_K & -\mathbb{1}_{K^{i+1}} & 0 \\ 0 & -d_K & 0 \\ 0 & f & d_L \end{bmatrix} \cdot \begin{bmatrix} k^i \\ k^{i+1} \\ \ell^i \end{bmatrix}.$$

We have canonical inclusions $\alpha : L^\bullet \to Cyl(f)$, $\bar{f} : K^\bullet \to Cyl(f)^\bullet$, a canonical projections $\beta : Cyl(f)^\bullet \to L^\bullet$, $\delta = \delta(f) : C(f)^\bullet \to K[1]^\bullet$, and $\pi : Cyl(f) \to C(f)$.

(i) Prove that $\alpha, \beta\bar{f}, \delta(f)$ are cochain maps, $\beta \circ \alpha = \mathbb{1}_L$ and $\alpha \circ \beta$ is cochain homotopic to $\mathbb{1}_{Cyl(f)}$.

(ii) Show that we have the following commutative diagram of cochain complexes, where the rows are exact.

$$
\begin{array}{ccccccccc}
0 & \longrightarrow & L^\bullet & \xrightarrow{\bar{\pi}} & C(f)^\bullet & \xrightarrow{\delta(f)} & K[1]^\bullet & \longrightarrow & 0 \\
& & \downarrow{\alpha} & & \downarrow{\mathbb{1}_{C(f)}} & & & & \\
0 & \longrightarrow & K^\bullet & \xrightarrow{\bar{f}} & Cyl(f) & \xrightarrow{\pi} & C(f) & \longrightarrow & 0 \\
& & \downarrow{\mathbb{1}_K} & & \downarrow{\beta} & & & & \\
& & K^\bullet & \xrightarrow{f} & L^\bullet & & & &
\end{array}
$$

(iii) Show that the connecting morphism in the long exact sequence corresponding to the short exact sequence

$$0 \to K^\bullet \xrightarrow{\bar{f}} Cyl(f) \xrightarrow{\pi} C(f) \to 0$$

coincides with the morphism induced in cohomology by $\delta(f) : C(f) \to K[1]^\bullet$.

(iv) Prove that f induces an isomorphism in cohomology if and only if the cone of f is acyclic. $\square$

---

[1] This exercise describes additional features of the long exact sequence in cohomology. They are particularly useful in the study of derived categories.

Exercise 7.1.20. (Abstract Morse inequalities). Let $C^\bullet = \bigoplus_{n\geq 0} C^n$ be a cochain complex of vector spaces over the field $\mathbb{F}$. Assume each of the vector spaces C^n is finite dimensional. Form the Poincaré series

$$P_C(t) = \sum_{n\geq 0} t^n \dim_\mathbb{F} C^n,$$

and

$$P_{H^\bullet(C)}(t) = \sum_{n\geq 0} t^n \dim_\mathbb{F} H^n(C).$$

Prove that there exists a formal series $R(t) \in \mathbb{Z}[[t]]$ with *non-negative* coefficients such that

$$P_{C^\bullet}(t) = P_{H^\bullet(C)}(t) + (1+t)R(t).$$

In particular, whenever it makes sense, the graded spaces C^* and H^* have identical Euler characteristics

$$\chi(C^\bullet) = P_C(-1) = P_{H^\bullet(C)}(-1) = \chi(H^\bullet(C)). \qquad \square$$

Exercise 7.1.21. (Additivity of Euler characteristic). Let

$$0 \to A^\bullet \to B^\bullet \to C^\bullet \to 0$$

be a short exact sequence of cochain complexes of vector spaces over the field $\mathbb{F}$. Prove that if at least two of the cohomology modules $H^\bullet(A)$, $H^\bullet(B)$ and $H^\bullet(C)$ have finite dimension over $\mathbb{F}$, then the same is true about the third one, and moreover

$$\chi(H^\bullet(B)) = \chi(H^\bullet(A)) + \chi(H^\bullet(C)). \qquad \square$$

Exercise 7.1.22. (Finite dimensional Hodge theory). Let

$$(V^\bullet, d) := \bigoplus_{n\geq 0} (V^n, d_n),$$

be a cochain complex of real vector spaces such that $\dim V^n < \infty$, for all n. Assume that each V^n is equipped with a Euclidean metric, and denote by $d_n^* : V^{n+1} \to V^n$ the adjoint of d_n. We can now form the *Laplacians*

$$\Delta_n : V^n \to V^n, \quad \Delta_n := d_n^* d_n + d_{n-1} d_{n-1}^*.$$

(a) Prove that $\bigoplus_{n\geq 0} \Delta_n = (d + d^*)^2$.

(b) Prove that $\Delta_n c = 0$ if and only if $d_n c = 0$ and $d_{n-1}^* c = 0$. In particular, $\ker \Delta_n \subset Z^n(V^\bullet)$.

(c) Let $c \in Z^n(C)$. Prove that there exists a unique $\bar{c} \in Z^n(C)$ cohomologous to c such that

$$|\bar{c}| = \min\{\, |c'|; \;\; c - c' \in B^n(C) \,\},$$

where $|\cdot|$ denotes the Euclidean norm in V^n.

(d) Prove that $\bar{c}$ determined in part (b) satisfies $\Delta_n \bar{c} = 0$. Deduce from all the above that the natural map

$$\ker \Delta_n \to H^n(V^\bullet)$$

is a linear isomorphism. $\qquad \square$

Exercise 7.1.23. Let V be a finite dimensional real vector space, and $v_0 \in V$. Define

$$d_k = d_k(v_0) : \Lambda^k V \to \Lambda^{k+1}V, \quad \omega \longmapsto v_0 \wedge \omega.$$

(a) Prove that

$$\cdots \xrightarrow{d_{k-1}} \Lambda^k V \xrightarrow{d_k} \Lambda^{k+1}V \xrightarrow{d_{k+1}} \cdots$$

is a cochain complex. (It is known as the *Koszul complex.*)
(b) Use the finite dimensional Hodge theory described in previous exercise to prove that the Koszul complex is acyclic if $v_0 \neq 0$, i.e.,

$$H^k\big(\Lambda^\bullet V, d(v_0)\big) = 0, \quad \forall k \geq 0. \qquad \square$$

7.1.4 *Functorial properties of the DeRham cohomology*

Let M and N be two smooth manifolds. For any smooth map $\phi : M \to N$ the pullback

$$\phi^* : \Omega^\bullet(N) \to \Omega^\bullet(M)$$

is a cochain morphism, i.e., $\phi^* d_N = d_M \phi^*$, where d_M and respectively d_N denote the exterior derivative on M, and respectively N. Thus, ϕ^* induces a morphism in cohomology which we continue to denote by ϕ^*;

$$\phi^* : H^\bullet(N) \to H^\bullet(M).$$

In fact, we have a more precise statement.

Proposition 7.1.24. *The DeRham cohomology construction is a contravariant functor from the category of smooth manifolds and smooth maps to the category of $\mathbb{Z}$-graded vector spaces with degree zero morphisms.* $\qquad \square$

Note that the pullback is an algebra morphism $\phi^* : \Omega^\bullet(N) \to \Omega^\bullet(M)$,

$$\phi^*(\alpha \wedge \beta) = (\phi^*\alpha) \wedge (\phi^*\beta), \quad \forall \alpha, \beta \in \Omega^\bullet(N),$$

and the exterior differentiation is a quasi-derivation, so that the map it induces in cohomology will also be a ring morphism.

Definition 7.1.25. (a) Two smooth maps $\phi_0, \phi_1 : M \to N$ are said to be (smoothly) homotopic, and we write this $\phi_0 \simeq_{sh} \phi_1$, if there exists a smooth map

$$\Phi : I \times M \to N \quad (t, m) \mapsto \Phi_t(m),$$

such that $\Phi_i = \phi_i$, for $i = 0, 1$.
(b) A smooth map $\phi : M \to N$ is said to be a (smooth) homotopy equivalence if there exists a smooth map $\psi : N \to M$ such that $\phi \circ \psi \simeq_{sh} \mathbb{1}_N$, and $\psi \circ \phi \simeq_{sh} \mathbb{1}_M$.
(c) Two smooth manifolds M and N are said to be homotopy equivalent if there exists a homotopy equivalence $\phi : M \to N$. $\qquad \square$

Proposition 7.1.26. *Let $\phi_0, \phi_1 : M \to N$ be two homotopic smooth maps. Then they induce* identical *maps in cohomology*

$$\phi_0^* = \phi_1^* : H^\bullet(N) \to H^\bullet(M).$$

Proof. According to the general results in homological algebra, it suffices to show that the pullbacks

$$\phi_0^*, \phi_1^* : \Omega^\bullet(N) \to \Omega^\bullet(M)$$

are cochain homotopic. Thus, we have to produce a map

$$\chi : \Omega^\bullet(N) \to \Omega^{*\bullet-1}(M),$$

such that

$$\phi_1^*(\omega) - \phi_0^*(\omega) = \pm\chi(d\omega) \pm d\chi\omega, \ \ \forall \omega \in \Omega^\bullet(M).$$

At this point, our discussion on the fibered calculus of Subsection 3.4.5 will pay off.

The projection $\Phi : I \times M \to M$ defines an oriented ∂-bundle with standard fiber I. For any $\omega \in \Omega^\bullet(N)$ we have the equality

$$\phi_1^*(\omega) - \phi_0^*(\omega) = \Phi^*(\omega)|_{1 \times M} - \Phi^*(\omega)|_{0 \times M} = \int_{(\partial I \times M)/M} \Phi^*(\omega).$$

We now use the homotopy formula in Theorem 3.4.54 of Subsection 3.4.5, and we deduce

$$\int_{(\partial I \times M)/M} \Phi^*(\omega) = \int_{(I \times M)/M} d_{I \times M} \Phi^*(\omega) - d_M \int_{(I \times M)/M} \Phi^*(\omega)$$

$$= \int_{(I \times M)/M} \Phi^*(d_N\omega) - d_M \int_{(I \times M)/M} \Phi^*(\omega).$$

Thus

$$\chi(\omega) := \int_{(I \times M)/M} \Phi^*(\omega)$$

is the sought for cochain homotopy. $\qquad\square$

Corollary 7.1.27. *Two homotopy equivalent spaces have isomorphic cohomology rings.* $\qquad\square$

Consider a smooth manifold M, and U, V two open subsets of M such that $M = U \cup V$. Denote by $\imath_U$ (respectively $\imath_V$) the inclusions $U \hookrightarrow M$ (resp. $V \hookrightarrow M$). These induce the restriction maps

$$\imath_U^* : \Omega^\bullet(M) \to \Omega^\bullet(U), \ \ \omega \mapsto \omega|_U,$$

and

$$\imath_V^* : \Omega^\bullet(M) \to \Omega^\bullet(V), \ \ \omega \mapsto \omega|_V .$$

We get a cochain morphism

$$r : \Omega^\bullet(M) \to \Omega^\bullet(U) \oplus \Omega^\bullet(V), \quad \omega \mapsto (i_U^* \omega, i_V^* \omega).$$

There exists another cochain morphism

$$\delta : \Omega^\bullet(U) \oplus \Omega^\bullet(V) \to \Omega^\bullet(U \cap V), \quad (\omega, \eta) \mapsto -\omega|_{U \cap V} + \eta|_{U \cap V} .$$

Lemma 7.1.28.   *The* short Mayer-Vietoris sequence

$$0 \to \Omega^\bullet(M) \xrightarrow{r} \Omega^\bullet(U) \oplus \Omega^\bullet(V) \xrightarrow{\delta} \Omega^*(U \cap V) \to 0$$

is exact.

Proof.   Obviously r is injective. The proof of the equality $\mathrm{Range}\, r = \ker \delta$ can be safely left to the reader. The surjectivity of δ requires a little more effort.

The collection $\{U, V\}$ is an open cover of M, so we can find a partition of unity $\{\varphi_U, \varphi_V\} \subset C^\infty(M)$ subordinated to this cover, i.e.,

$$\mathrm{supp}\, \varphi_U \subset U, \ \ \mathrm{supp}\, \varphi_V \subset V, \ \ 0 \le \varphi_U, \ \varphi_V \le 1, \ \ \varphi_U + \varphi_V = 1.$$

Note that for any $\omega \in \Omega^*(U \cap V)$ we have

$$\mathrm{supp}\, \varphi_V \omega \subset \mathrm{supp}\, \varphi_V \subset V,$$

and thus, upon extending $\varphi_V \omega$ by 0 outside V, we can view it as a form on U. Similarly, $\varphi_U \omega \in \Omega^*(V)$. Note that

$$\delta(-\varphi_V \omega, \varphi_U \omega) = (\varphi_V + \varphi_U)\omega = \omega.$$

This establishes the surjectivity of δ. $\square$

Using the abstract results in homological algebra we deduce from the above lemma the following fundamental result.

Theorem 7.1.29 (Mayer-Vietoris).   *Let* $M = U \cup V$ *be an open cover of the smooth manifold M. Then there exists a long exact sequence*

$$\cdots \to H^k(M) \xrightarrow{r} H^k(U) \oplus H^k(V) \xrightarrow{\delta} H^k(U \cap V) \xrightarrow{\partial} H^{k+1}(M) \to \cdots,$$

called the long Mayer-Vietoris sequence. $\square$

The *connecting morphisms* ∂ can be explicitly described using the prescriptions following Proposition 7.1.18 in the previous subsection. Let us recall that construction.

Start with $\omega \in \Omega^k(U \cap V)$ such that $d\omega = 0$. Writing as before

$$\omega = \varphi_V \omega + \varphi_U \omega,$$

we deduce

$$d(\varphi_V \omega) = d(-\varphi_U \omega) \quad \text{on } U \cap V.$$

Thus, we can find $\eta \in \Omega^{k+1}(M)$ such that

$$\eta|_U = d(\varphi_V \omega) \quad \eta|_V = d(-\varphi_U \omega).$$

Then

$$\partial \omega := \eta.$$

The reader can prove directly that the above definition is independent of the various choices.

The Mayer-Vietoris sequence has the following functorial property.

Proposition 7.1.30. *Let $\phi : M \to N$ be a smooth map and $\{U, V\}$ an open cover of N. Then $U' = \phi^{-1}(U), V' = \phi^{-1}(V)$ form an open cover of M and moreover, the diagram below is commutative.*

$$
\begin{array}{ccccccc}
H^k(N) & \xrightarrow{\ r\ } & H^k(U) \oplus H^k(V) & \xrightarrow{\ \delta\ } & H^k(U \cap V) & \xrightarrow{\ \partial\ } & H^{k+1}(N) \\
\downarrow{\scriptstyle \phi^*} & & \downarrow{\scriptstyle \phi^*} & & \downarrow{\scriptstyle \phi^*} & & \downarrow{\scriptstyle \phi^*} \\
H^k(M) & \xrightarrow{\ r'\ } & H^k(U') \oplus H^k(V') & \xrightarrow{\ \delta'\ } & H^k(U' \cap V') & \xrightarrow{\ \partial'\ } & H^{k+1}(M)
\end{array}
$$

$\square$

Exercise 7.1.31. Prove the above proposition. $\square$

7.1.5 *Some simple examples*

The Mayer-Vietoris theorem established in the previous subsection is a very powerful tool for computing the cohomology of manifolds. In principle, it allows one to recover the cohomology of a manifold decomposed into simpler parts, knowing the cohomologies of its constituents. In this subsection we will illustrate this principle on some simple examples.

Example 7.1.32. (The cohomology of spheres). The cohomology of S^1 can be easily computed using the definition of DeRham cohomology. We have $H^0(S^1) = \mathbb{R}$ since S^1 is connected. A 1-form $\eta \in \Omega^1(S^1)$ is automatically closed, and it is exact if and only if

$$\int_{S^1} \eta = 0.$$

Indeed, if $\eta = dF$, where $F : \mathbb{R} \to \mathbb{R}$ is a smooth 2π-periodic function, then

$$\int_{S^1} \eta = F(2\pi) - F(0) = 0.$$

Conversely if $\eta = f(\theta)d\theta$, where $f : \mathbb{R} \to \mathbb{R}$ is a smooth 2π-periodic function and

$$\int_0^{2\pi} f(\theta)d\theta = 0,$$

then the function

$$F(t) = \int_0^t f(s)ds$$

is smooth, 2π-periodic, and $dF = \eta$.

Thus, the map

$$\int_{S^1} : \Omega^1(S^1) \to \mathbb{R}, \quad \eta \mapsto \int_{S^1} \eta,$$

induces an isomorphism $H^1(S^1) \to \mathbb{R}$, and we deduce

$$P_{S^1}(t) = 1 + t.$$

To compute the cohomology of higher dimensional spheres we use the Mayer-Vietoris theorem.

The $(n+1)$-dimensional sphere S^{n+1} can be covered by two open sets

$$U_{south} = S^{n+1} \setminus \{\text{North pole}\} \quad \text{and} \quad U_{north} = S^{n+1} \setminus \{\text{South pole}\}.$$

Each is diffeomorphic to $\mathbb{R}^{n+1}$. Note that the overlap $U_{north} \cap U_{south}$ is homotopically equivalent with the equator S^n. The Poincaré lemma implies that

$$H^{k+1}(U_{north}) \oplus H^{k+1}(U_{south}) \cong 0$$

for $k \geq 0$. The Mayer-Vietoris sequence gives

$$H^k(U_{north}) \oplus H^k(U_{south}) \to H^k(U_{north} \cap U_{south}) \to H^{k+1}(S^{n+1}) \to 0.$$

For $k > 0$ the group on the left is also trivial, so that we have the isomorphisms

$$H^k(S^n) \cong H^k(U_{north} \cap U_{south}) \cong H^{k+1}(S^{n+1}) \quad k > 0.$$

Denote by $P_n(t)$ the Poincaré polynomial of S^n and set $Q_n(t) = P_n(t) - P_n(0) = P_n(t) - 1$. We can rewrite the above equality as

$$Q_{n+1}(t) = tQ_n(t) \quad n > 0.$$

Since $Q_1(t) = t$ we deduce $Q_n(t) = t^n$, i.e.,

$$P_{S^n}(t) = 1 + t^n. \qquad \qquad \square$$

Example 7.1.33. Let $\{U, V\}$ be an open cover of the smooth manifold M. We assume that all the Betti numbers of U, V and $U \cap V$ are finite. Using the Mayer-Vietoris short exact sequence, and the Exercise 7.1.21 in Subsection 7.1.3, we deduce that all the Betti numbers of M are finite, and moreover

$$\chi(M) = \chi(U) + \chi(V) - \chi(U \cap V). \tag{7.1.12}$$

This resembles very much the classical inclusion-exclusion principle in combinatorics. We will use this simple observation to prove that the Betti numbers of a connected sum of g tori is finite, and then compute its Euler characteristic.

Let Σ be a surface with finite Betti numbers. From the decomposition

$$\Sigma = (\Sigma \setminus \text{disk}) \cup \text{disk},$$

we deduce (using again Exercise 7.1.21)

$$\chi(\Sigma) = \chi(\Sigma \setminus \text{disk}) + \chi(\text{disk}) - \chi\left((\Sigma \setminus \text{disk}) \cap (\text{disk})\right).$$

Since $(\Sigma \setminus \text{disk}) \cap \text{disk}$ is homotopic to a circle, and $\chi(\text{disk}) = 1$, we deduce

$$\chi(\Sigma) = \chi(\Sigma \setminus \text{disk}) + 1 - \chi(S^1) = \chi(\Sigma \setminus \text{disk}) + 1.$$

If now Σ_1 and Σ_2 are two surfaces with finite Betti numbers, then

$$\Sigma_1 \# \Sigma_2 = (\Sigma_1 \setminus \text{disk}) \cup (\Sigma_2 \setminus \text{disk}),$$

where the two holed surfaces intersect over an entire annulus, which is homotopically a circle. Thus

$$\chi(\Sigma_1 \# \Sigma_2) = \chi(\Sigma_1 \setminus \text{disk}) + \chi(\Sigma_2 \setminus \text{disk}) - \chi(S^1)$$
$$= \chi(\Sigma_1) + \chi(\Sigma_2) - 2.$$

This equality is identical with the one proved in Proposition 4.2.31 of Subsection 4.2.5.

We can decompose a torus as a union of two cylinders. The intersection of these cylinders is the disjoint union of two annuli so homotopically, this overlap is a disjoint union of two circles. In particular, the Euler characteristic of the intersection is zero. Hence

$$\chi(\text{torus}) = 2\chi(\text{cylinder}) = 2\chi(\text{circle}) = 0.$$

We conclude as in Proposition 4.2.31 that

$$\chi(\text{connected sum of } g \text{ tori}) = 2 - 2g.$$

This is a pleasant, surprising connection with the Gauss-Bonnet theorem. And the story is not over. $\qquad\square$

7.1.6 *The Mayer-Vietoris principle*

We describe in this subsection a "patching" technique which is extremely versatile in establishing general homological results about arbitrary manifolds building up from elementary ones.

Definition 7.1.34. A smooth manifold M is said to be of *finite type* if it can be covered by finitely many open sets $U_1, \dots, U_m$ such that any nonempty intersection $U_{i_1} \cap \dots \cap U_{i_k}$ ($k \geq 1$) is diffeomorphic to $\mathbb{R}^{\dim M}$. Such a cover is said to be a *good cover*. $\qquad\square$

Example 7.1.35. (a) All compact manifolds are of finite type. To see this, it suffices to cover such a manifold by finitely many open sets which are geodesically convex with respect to some Riemann metric.

(b) If M is a finite type manifold, and $U \subset M$ is a closed subset homeomorphic with the closed unit ball in $\mathbb{R}^{\dim M}$, then $M \setminus U$ is a finite type non-compact manifold. (It suffices to see that $\mathbb{R}^n \setminus$ closed ball is of finite type).

(c) The connected sums, and the direct products of finite type manifolds are finite type manifolds. $\qquad\square$

Proposition 7.1.36. *Let $p : E \to B$ be a smooth vector bundle. If the base B is of finite type, then so is the total space E.*

In the proof of this proposition we will use the following fundamental result.

Lemma 7.1.37. *Let $p : E \to B$ be a smooth vector bundle such that B is diffeomorphic to $\mathbb{R}^n$. Then $p : E \to B$ is a* trivializable *bundle.* $\square$

Proof of Proposition 7.1.36.   Denote by F the standard fiber of E. The fiber F is a vector space. Let $(U_i)_{1 \leq i \leq \nu}$ be a good cover of B. For each ordered multi-index $I := \{i_1 < \cdots < i_k\}$ denote by U_I the multiple overlap $U_{i_1} \cap \cdots \cap U_{i_k}$. Using the previous lemma we deduce that each $E_I = E|_{U_I}$ is a product $F \times U_i$, and thus it is diffeomorphic with some vector space. Hence (E_i) is a good cover. $\square$

Exercise 7.1.38.   Prove Lemma 7.1.37.
Hint: Assume that E is a vector bundle over the unit open ball $B \subset \mathbb{R}^n$. Fix a connection ∇ on E, and then use the ∇-parallel transport along the half-lines L_x, $[0, \infty) \ni t \mapsto tx \in \mathbb{R}^n$, $x \in \mathbb{R}^n \setminus \{0\}$. $\square$

$\clubsuit$ *We denote by $\mathfrak{M}_n$ the category finite type smooth manifolds of dimension n. The morphisms of this category are the smooth embeddings, i.e., the one-to-one immersions $M_1 \hookrightarrow M_2$, $M_i \in \mathfrak{M}_n$.*

Definition 7.1.39. Let R be a commutative ring with 1. A *contravariant Mayer-Vietoris functor* (or *MV-functor* for brevity) is a contravariant functor from the category $\mathfrak{M}_n$, to the category of $\mathbb{Z}$-graded R-modules

$$\mathcal{F} = \oplus_{n \in \mathbb{Z}} \mathcal{F}^n \to \mathbf{Grad}_R\mathbf{Mod}, \quad M \mapsto \bigoplus_n \mathcal{F}^n(M),$$

with the following property. If $\{U, V\}$ is a *MV-cover* of $M \in \mathfrak{M}_n$, i.e., U, V, $U \cap V \in \mathfrak{M}_n$, then there exist morphisms of R-modules

$$\partial_n : \mathcal{F}^n(U \cap V) \to \mathcal{F}^{n+1}(M),$$

such that the sequence below is exact

$$\cdots \to \mathcal{F}^n(M) \overset{r^*}{\to} \mathcal{F}^n(U) \oplus \mathcal{F}^n(V) \overset{\delta}{\to} \mathcal{F}^n(U \cap V) \overset{\partial_n}{\to} \mathcal{F}^{n+1}(M) \to \cdots,$$

where r^* is defined by

$$r^* = \mathcal{F}(\imath_U) \oplus \mathcal{F}(\imath_V),$$

and δ is defined by

$$\delta(x \oplus y) = \mathcal{F}(\imath_{U \cap V})(y) - \mathcal{F}(\imath_{U \cap V})(x).$$

(The maps $\imath_\bullet$ denote natural embeddings.) Moreover, if $N \in \mathcal{M}_n$ is an open submanifold of N, and $\{U, V\}$ is an MV-cover of M such that $\{U \cap N, V \cap N\}$ is an MV-cover of N, then the diagram below is commutative.

$$
\begin{array}{ccc}
\mathcal{F}^n(U \cap V) & \xrightarrow{\ \partial_n\ } & \mathcal{F}^{n+1}(M) \\
\downarrow & & \downarrow \\
\mathcal{F}^n(U \cap V \cap N) & \xrightarrow{\ \partial_n\ } & \mathcal{F}^{n+1}(N)
\end{array}
$$

The vertical arrows are the morphisms $\mathcal{F}(\imath_\bullet)$ induced by inclusions. $\qquad\square$

The covariant MV-functors are defined in the dual way, by reversing the orientation of all the arrows in the above definition.

Definition 7.1.40. Let $\mathcal{F}$, $\mathcal{G}$ be two contravariant MV-functors,

$$
\mathcal{F},\ \mathcal{G} : \mathcal{M}_n \to \mathbf{Grad}_R\mathbf{Mod}.
$$

A *correspondence*[2] between these functors is a collection of R-module morphisms

$$
\phi_M = \bigoplus_{n \in \mathbb{Z}} \phi_M^n : \bigoplus \mathcal{F}^n(M) \longrightarrow \bigoplus \mathcal{G}^n(M),
$$

one morphism for each $M \in \mathcal{M}_n$, such that, for any embedding $M_1 \xhookrightarrow{\varphi} M_2$, the diagram below is commutative

$$
\begin{array}{ccc}
\mathcal{F}^n(M_2) & \xrightarrow{\ \mathcal{F}(\varphi)\ } & \mathcal{F}^n(M_1) \\
\phi_{M_2} \downarrow & & \downarrow \phi_{M_1} \\
\mathcal{G}^n(M_2) & \xrightarrow{\ \mathcal{G}(\varphi)\ } & \mathcal{G}^n(M_1)
\end{array} \ ,
$$

and, for any $M \in \mathcal{M}_n$ and any MV-cover $\{U, V\}$ of M, the diagram below is commutative.

$$
\begin{array}{ccc}
\mathcal{F}^n(U \cap V) & \xrightarrow{\ \partial_n\ } & \mathcal{F}^{n+1}(M) \\
\phi_{U \cap V} \downarrow & & \downarrow \phi_M \\
\mathcal{G}^n(U \cap V) & \xrightarrow{\ \partial_n\ } & \mathcal{G}^{n+1}(M)
\end{array}
$$

The correspondence is said to be a *natural equivalence* if all the morphisms ϕ_M are isomorphisms. $\qquad\square$

---

[2] Our notion of correspondence corresponds to the categorical notion of morphism of functors.

Theorem 7.1.41 (Mayer-Vietoris principle). *Let $\mathcal{F}$, $\mathcal{G}$ be two (contravariant) Mayer-Vietoris functors on $\mathcal{M}_n$ and $\phi : \mathcal{F} \to \mathcal{G}$ a correspondence. If*

$$\phi_{\mathbb{R}^n}^k : \mathcal{F}^k(\mathbb{R}^n) \to \mathcal{G}^k(\mathbb{R}^n)$$

is an isomorphism for any $k \in \mathbb{Z}$, then ϕ is a natural equivalence. $\square$

Proof.   The family of finite type manifolds $\mathcal{M}_n$ has a natural filtration

$$\mathcal{M}_n^1 \subset \mathcal{M}_n^2 \subset \cdots \subset \mathcal{M}_n^r \subset \cdots,$$

where $\mathcal{M}_n^r$ is the collection of all smooth manifolds which admit a good cover consisting of at most r open sets. We will prove the theorem using an induction over r.

The theorem is clearly true for $r = 1$ by hypothesis. Assume ϕ_M^k is an isomorphism for all $M \in \mathcal{M}_n^{r-1}$. Let $M \in \mathcal{M}_n^r$ and consider a good cover $\{U_1, \ldots, U_r\}$ of M. Then

$$\{U = U_1 \cup \cdots \cup U_{r-1}, U_r\}$$

is an MV-cover of M. We thus get a commutative diagram

$$
\begin{array}{ccccccc}
\mathcal{F}^n(U) \oplus \mathcal{F}^n(U_r) & \longrightarrow & \mathcal{F}^n(U \cap U_r) & \overset{\partial}{\longrightarrow} & \mathcal{F}^{n+1}(M) & \longrightarrow & \mathcal{F}^{n+1}(U) \oplus \mathcal{G}^{n+1}(U_r) \\
\downarrow & & \downarrow & & \downarrow & & \downarrow \\
\mathcal{G}^n(U) \oplus \mathcal{G}^n(U_r) & \longrightarrow & \mathcal{G}^n(U \cap U_r) & \overset{\partial}{\longrightarrow} & \mathcal{G}^{n+1}(M) & \longrightarrow & \mathcal{F}^{n+1}(U) \oplus \mathcal{G}^{n+1}(U_r)
\end{array}
$$

The vertical arrows are defined by the correspondence ϕ. Note the inductive assumption implies that in the above infinite sequence only the morphisms ϕ_M may not be isomorphisms. At this point we invoke the following technical result.

Lemma 7.1.42. (The five lemma). *Consider the following commutative diagram of R-modules.*

$$
\begin{array}{ccccccccc}
A_{-2} & \longrightarrow & A_{-1} & \longrightarrow & A_0 & \longrightarrow & A_1 & \longrightarrow & A_2 \\
f_{-2}\downarrow & & f_{-1}\downarrow & & f_0\downarrow & & f_1\downarrow & & f_2\downarrow \\
B_{-2} & \longrightarrow & B_{-1} & \longrightarrow & B_0 & \longrightarrow & B_1 & \longrightarrow & B_2
\end{array}
$$

If f_i is an isomorphism for any $i \neq 0$, then so is f_0. $\square$

Exercise 7.1.43. Prove the five lemma. $\square$

The five lemma applied to our situation shows that the morphisms ϕ_M must be isomorphisms. $\square$

Remark 7.1.44. (a) The Mayer-Vietoris principle is true for covariant MV-functors as well. The proof is obtained by reversing the orientation of the horizontal arrows in the above proof.

(b) The Mayer-Vietoris principle can be refined a little bit. Assume that $\mathcal{F}$ and $\mathcal{G}$ are functors from $\mathcal{M}_n$ to the category of $\mathbb{Z}$-graded R-algebras, and $\phi : \mathcal{F} \to \mathcal{G}$ is a correspondence compatible with the multiplicative structures, i.e., each of the R-module morphisms ϕ_M are in fact morphisms of R-algebras. Then, if $\phi_{\mathbb{R}^n}$ are isomorphisms of $\mathbb{Z}$-graded R-algebras, then so are the ϕ_M's, for any $M \in \mathcal{M}_n$.

(c) Assume R is a field. The proof of the Mayer-Vietoris principle shows that if $\mathcal{F}$ is a MV-functor and $\dim_R \mathcal{F}^*(\mathbb{R}^n) < \infty$ then $\dim \mathcal{F}^*(M) < \infty$ for all $M \in \mathcal{M}_n$.

(d) The Mayer-Vietoris principle is a baby case of the very general technique in algebraic topology called the *acyclic models principle*. $\square$

Corollary 7.1.45. *Any finite type manifold has finite Betti numbers.* $\square$

7.1.7 The Künneth formula

We learned in principle how to compute the cohomology of a "union of manifolds". We will now use the Mayer-Vietoris principle to compute the cohomology of products of manifolds.

Theorem 7.1.46 (Künneth formula). *Let $M \in \mathcal{M}_m$ and $N \in \mathcal{M}_n$. Then there exists a natural isomorphism of graded $\mathbb{R}$-algebras*

$$H^\bullet(M \times N) \cong H^\bullet(M) \otimes H^*(N) = \bigoplus_{n \geq 0} \left(\bigoplus_{p+q=n} H^p(M) \otimes H^q(N) \right).$$

In particular, we deduce

$$P_{M \times N}(t) = P_M(t) \cdot P_N(t).$$

Proof. We construct two functors

$$\mathcal{F}, \mathcal{G} : \mathcal{M}_m \to \mathbf{Grad}_{\mathbb{R}}\mathbf{Alg},$$

$$\mathcal{F} : M \mapsto \bigoplus_{r \geq 0} \mathcal{F}^r(M) = \bigoplus_{r \geq 0} \left\{ \bigoplus_{p+q=r} H^p(M) \otimes H^q(N) \right\},$$

and

$$\mathcal{G} : M \mapsto \bigoplus_{r \geq 0} \mathcal{G}^r(M) = \bigoplus_{r \geq 0} H^r(M \times N),$$

where

$$\mathcal{F}(f) = \bigoplus_{r \geq 0} \left(\bigoplus_{p+q=r} f^*|_{H^p(M_2)} \otimes \mathbb{1}_{H^q(N)} \right), \quad \forall f : M_1 \hookrightarrow M_2,$$

and

$$\mathcal{G}(f) = \bigoplus_{r \geq 0} (f \times \mathbb{1}_N)^*|_{H^r(M_2 \times N)}, \quad \forall f : M_1 \hookrightarrow M_2.$$

We let the reader check the following elementary fact.

Exercise 7.1.47. $\mathcal{F}$ and $\mathcal{G}$ are contravariant MV-functors. $\square$

For $M \in \mathcal{M}_M$, define $\phi_M : \mathcal{F}(M) \to \mathcal{G}(N)$ by

$$\phi_M(\omega \otimes \eta) = \omega \times \eta := \pi_M^* \omega \wedge \pi_N^* \eta \quad (\omega \in H^\bullet(M), \ \eta \in H^\bullet(M)),$$

where π_M (respectively π_N) are the canonical projections $M \times N \to M$ (respectively $M \times N \to N$). The operation

$$\times : H^\bullet(M) \otimes H^\bullet(N) \to H^\bullet(M \times N) \quad (\omega \otimes \eta) \mapsto \omega \times \eta,$$

is called the *cross product*. The Künneth formula is a consequence of the following lemma.

Lemma 7.1.48. *(a) ϕ is a correspondence of MV-functors.*
(b) $\phi_{\mathbb{R}^m}$ is an isomorphism.

Proof.   The only nontrivial thing to prove is that for any MV-cover $\{U, V\}$ of $M \in \mathcal{M}_m$ the diagram below is commutative.

$$\begin{array}{ccc}
\oplus_{p+q=r} H^p(U \cap V) \otimes H^q(N) & \xrightarrow{\ \partial\ } & \oplus_{p+q=r} H^{p+1}(M) \otimes H^q(N) \\
\Big\downarrow {\scriptstyle \phi_{U \cap V}} & & \Big\downarrow {\scriptstyle \phi_M} \\
H^r((U \times N) \cap (V \times N)) & \xrightarrow{\ \partial'\ } & H^{r+1}(M \times N)
\end{array}$$

We briefly recall the construction of the connecting morphisms ∂ and ∂'.

One considers a partition of unity $\{\varphi_U, \varphi_V\}$ subordinated to the cover $\{U, V\}$. Then, the functions $\psi_U = \pi_M^* \varphi_U$ and $\psi_V = \pi_M^* \varphi_V$ form a partition of unity subordinated to the cover

$$\{\, U \times N, V \times N \,\}$$

of $M \times N$. If $\omega \otimes \eta \in H^\bullet(U \cap V) \otimes H^\bullet(N)$,

$$\partial(\omega \otimes \eta) = \hat{\omega} \otimes \eta,$$

where

$$\hat{\omega}\,|_U = -d(\varphi_V \omega) \quad \hat{\omega} = d(\varphi_U \omega).$$

On the other hand, $\phi_{U \cap V}(\omega \otimes \eta) = \omega \times \eta$, and $\partial'(\omega \times \eta) = \hat{\omega} \times \eta$. This proves (a).

To establish (b), note that the inclusion

$$\jmath : N \hookrightarrow \mathbb{R}^m \times N, \quad x \mapsto (0, x)$$

is a homotopy equivalence, with π_N a homotopy inverse. Hence, by the homotopy invariance of the DeRham cohomology we deduce

$$\mathcal{G}(\mathbb{R}^m) \cong H^\bullet(N).$$

Using the Poincaré lemma and the above isomorphism we can identify the morphism $\phi_{\mathbb{R}^m}$ with $\mathbb{1}_{\mathbb{R}^m}$. $\qquad\qquad\qquad\qquad\qquad\qquad\qquad\qquad\qquad\qquad\qquad\qquad\Box$

Example 7.1.49. Consider the n-dimensional torus, T^n. By writing it as a direct product of n circles we deduce from Künneth formula that

$$P_{T^n}(t) = \{P_{S^1}(t)\}^n = (1+t)^n.$$

Thus

$$b_k(T^n) = \binom{n}{k}, \quad \dim H^*(T^n) = 2^n,$$

and $\chi(T^n) = 0$.

One can easily describe a basis of $H^*(T^n)$. Choose angular coordinates $(\theta^1, \ldots, \theta^n)$ on T^n. For each ordered multi-index $I = (1 \leq i_1 < \cdots < i_k \leq n)$ we have a closed, non-exact form $d\theta^I$. These monomials are linearly independent (over $\mathbb{R}$) and there are 2^n of them. Thus, they form a basis of $H^*(T^n)$. In fact, one can read the multiplicative structure using this basis. We have an isomorphism of $\mathbb{R}$-algebras

$$H^\bullet(T^n) \cong \Lambda^\bullet \mathbb{R}^n. \qquad \square$$

Exercise 7.1.50. Let $M \in \mathcal{M}_m$ and $N \in \mathcal{M}_n$. Show that for any $\omega_i \in H^*(M)$, $\eta_j \in H^*(N)$ $(i, j = 0, 1)$ the following equality holds.

$$(\omega_0 \times \eta_0) \wedge (\omega_1 \times \eta_1) = (-1)^{\deg \eta_0 \deg \omega_1} (\omega_0 \wedge \omega_1) \times (\eta_0 \wedge \eta_1). \qquad \square$$

Exercise 7.1.51. (Leray-Hirsch). Let $p : E \to M$ be smooth bundle with standard fiber F. We assume the following:

(a) Both M and F are of finite type.
(b) There exist cohomology classes $e_1, \ldots, e_r \in H^\bullet(E)$ such that their restrictions to any fiber generate the cohomology algebra of that fiber.

The projection p induces a $H^\bullet(M)$-module structure on $H^\bullet(E)$ by

$$\omega \cdot \eta = p^* \omega \wedge \eta \quad \forall \omega \in H^\bullet(M), \ \eta \in H^\bullet(E).$$

Show that $H^\bullet(E)$ is a free $H^\bullet(M)$-module with generators $e_1, \ldots, e_r$. $\qquad \square$

7.2 The Poincaré duality

7.2.1 *Cohomology with compact supports*

Let M be a smooth n-dimensional manifold. Denote by $\Omega^k_{cpt}(M)$ the space of smooth compactly supported k-forms. Then

$$0 \to \Omega^0_{cpt}(M) \xrightarrow{d} \cdots \xrightarrow{d} \Omega^n_{cpt}(M) \to 0$$

is a cochain complex. Its cohomology is denoted by $H^\bullet_{cpt}(M)$, and it is called the *DeRham cohomology with compact supports*. Note that when M is compact this cohomology coincides with the usual DeRham cohomology.

Although it looks very similar to the usual DeRham cohomology, there are many important differences. The most visible one is that if $\phi : M \to N$ is a smooth map, and $\omega \in \Omega^\bullet_{cpt}(N)$, then the pull-back $\phi^*\omega$ may not have compact support, so this new construction is no longer a contravariant functor from the category of smooth manifolds and smooth maps, to the category of graded vector spaces.

On the other hand, if $\dim M = \dim N$, and ϕ is an embedding, we can identify M with an open subset of N, and then any $\eta \in \Omega^\bullet_{cpt}(M)$ can be extended by 0 outside $M \subset N$. This extension by zero defines a *push-forward* map

$$\phi_* : \Omega^\bullet_{cpt}(M) \to \Omega^\bullet_{cpt}(N).$$

One can verify easily that ϕ_* is a cochain map so that it induces a morphism

$$\phi_* : H^\bullet_{cpt}(M) \to H^\bullet_{cpt}(N).$$

In terms of our category $\mathcal{M}_n$ we see that $H^\bullet_{cpt}$ is a *covariant functor* from the category $\mathcal{M}_n$, to the category of graded real vector spaces. As we will see, it is a rather nice functor.

Theorem 7.2.1. *$H^\bullet_{cpt}$ is a covariant MV-functor, and moreover*

$$H^k_{cpt}(\mathbb{R}^n) = \begin{cases} 0 \,, k < n \\ \mathbb{R} \,, k = n \end{cases}.$$

The last assertion of this theorem is usually called the Poincaré lemma for compact supports.

We first prove the Poincaré lemma for compact supports. The crucial step is the following technical result that we borrowed from [11].

Lemma 7.2.2. *Let $E \xrightarrow{p} B$ be a rank r real vector bundle which is orientable in the sense described in Subsection 3.4.5. Denote by p_* the integration-along-fibers map*

$$p_* : \Omega^\bullet_{cpt}(E) \to \Omega^{\bullet - r}_{cpt}(B).$$

Then there exists a smooth bilinear map

$$\mathfrak{m} : \Omega^i_{cpt}(E) \times \Omega^j_{cpt}(E) \to \Omega^{i+j-r-1}_{cpt}(E)$$

such that,

$$p^* p_* \alpha \wedge \beta - \alpha \wedge p^* p_* \beta = (-1)^r d(\mathfrak{m}(\alpha, \beta)) - \mathfrak{m}(d\alpha, \beta) + (-1)^{\deg \alpha} \mathfrak{m}(\alpha, d\beta).$$

Proof. Consider the ∂-bundle

$$\pi : \mathcal{E} = I \times (E \oplus E) \to E, \quad \pi : (t; v_0, v_1) \mapsto (t; v_0 + t(v_1 - v_0)).$$

Note that

$$\partial \mathcal{E} = (\{0\} \times (E \oplus E)) \sqcup (\{1\} \times (E \oplus E)).$$

Define $\pi^t : E \oplus E \to E$ as the composition

$$E \oplus E \cong \{t\} \times E \oplus E \hookrightarrow I \times (E \oplus E) \xrightarrow{\pi} E.$$

Observe that

$$\partial\pi = \pi|_{\partial\mathcal{E}} = (-\pi^0) \sqcup \pi^1.$$

For $(\alpha, \beta) \in \Omega^\bullet_{cpt}(E) \times \Omega^\bullet_{cpt}(E)$ define $\alpha \odot \beta \in \Omega^\bullet_{cpt}(E \oplus E)$ by

$$\alpha \odot \beta := (\pi^0)^* \alpha \wedge (\pi^1)^* \beta.$$

(Verify that the support of $\alpha \odot \beta$ is indeed compact.) For $\alpha \in \Omega^i_{cpt}(E)$, and $\beta \in \Omega^j_{cpt}(E)$ we have the equalities

$$p^* p_* \alpha \wedge \beta = \pi^1_*(\alpha \odot \beta) \in \Omega^{i+j-r}_{cpt}(E), \quad \alpha \wedge p^* p_* \beta = \pi^0_*(\alpha \odot \beta) \in \Omega^{i+j-r}_{cpt}(E).$$

Hence

$$D(\alpha, \beta) = p^* p_* \alpha \wedge \beta - \alpha \wedge p^* p_* \beta = \pi^1_*(\alpha \odot \beta) - \pi^0_*(\alpha \odot \beta) = \int_{\partial\mathcal{E}/E} \alpha \odot \beta.$$

We now use the fibered Stokes formula to get

$$D(\alpha, \beta) = \int_{\mathcal{E}/E} d_\mathcal{E} \mathfrak{T}^*(\alpha \odot \beta) + (-1)^r d_E \int_{\mathcal{E}/E} \mathfrak{T}^*(\alpha \odot \beta),$$

where $\mathfrak{T}$ is the natural projection $\mathcal{E} = I \times (E \oplus E) \to E \oplus E$. The lemma holds with

$$\mathfrak{m}(\alpha, \beta) = \int_{\mathcal{E}/E} \mathfrak{T}^*(\alpha \odot \beta). \qquad \square$$

Proof of the Poincaré lemma for compact supports. Consider $\delta \in C_0^\infty(\mathbb{R}^n)$ such that

$$0 \le \delta \le 1, \quad \int_{\mathbb{R}^n} \delta(x) dx = 1.$$

Define the, closed, compactly supported n-form

$$\tau := \delta(x) dx^1 \wedge \cdots \wedge dx^n.$$

We want to use Lemma 7.2.2 in which E is the rank n bundle over a point, i.e., $E = \{pt\} \times \mathbb{R}^n \xrightarrow{p} \{pt\}$. The integration along fibers is simply the integration map.

$$p_* : \Omega^\bullet_{cpt}(\mathbb{R}^n) \to \mathbb{R}, \quad \omega \mapsto p_* \omega = \begin{cases} 0, & \deg \omega < n \\ \int_{\mathbb{R}^n} \omega, & \deg \omega = n. \end{cases}$$

If now ω is a closed, compactly supported form on $\mathbb{R}^n$, we have

$$\omega = (p^* p_* \tau) \wedge \omega.$$

Using Lemma 7.2.2 we deduce

$$\omega - \tau \wedge p^* p_* \omega = (-1)^n d\mathfrak{m}(\tau, \omega).$$

Thus any closed, compactly supported form ω on $\mathbb{R}^n$ is cohomologous to $\tau \wedge p^* p_* \omega$. The latter is always zero if $\deg \omega < n$. When $\deg \omega = n$ we deduce that ω is cohomologous to $(\int \omega)\tau$. This completes the proof of the Poincaré lemma. $\quad \square$

To finish the proof of Theorem 7.2.1 we must construct a Mayer-Vietoris sequence. Let M be a smooth manifold decomposed as a union of two open sets $M = U \cup V$, the sequence of inclusions

$$U \cap V \hookrightarrow U, V \hookrightarrow M,$$

induces a short sequence

$$0 \to \Omega^\bullet_{cpt}(U \cap V) \xrightarrow{i} \Omega^\bullet_{cpt}(U) \oplus \Omega^\bullet_{cpt}(V) \xrightarrow{j} \Omega^\bullet_{cpt}(M) \to 0,$$

where

$$i(\omega) = (\hat{\omega}, \hat{\omega}), \quad j(\omega, \eta) = \hat{\eta} - \hat{\omega}.$$

The hat $\hat{\ }$ denotes the extension by zero outside the support. This sequence is called the Mayer-Vietoris short sequence for compact supports.

Lemma 7.2.3. *The above Mayer-Vietoris sequence is exact.*

Proof. The morphism i is obviously injective. Clearly, $\text{Range}\,(i) = \ker(j)$. We have to prove that j is surjective.

Let (φ_U, φ_V) a partition of unity subordinated to the cover $\{U, V\}$. Then, for any $\eta \in \Omega^*_{cpt}(M)$, we have

$$\varphi_U \eta \in \Omega^\bullet_{cpt}(U) \ \text{and} \ \varphi_V \eta \in \Omega^\bullet_{cpt}(V).$$

In particular, $\eta = j(-\varphi_U \eta, \varphi_V \eta)$, which shows that j is surjective. $\qquad\square$

We get a long exact sequence called the long Mayer-Vietoris sequence for compact supports.

$$\cdots \to H^k_{cpt}(U \cap V) \to H^k_{cpt}(U) \oplus H^k_{cpt}(V) \to H^k_{cpt}(M) \xrightarrow{\delta} H^{k+1}_{cpt} \to \cdots$$

The connecting homomorphism can be explicitly described as follows. If $\omega \in \Omega^k_{cpt}(M)$ is a closed form then

$$d(\varphi_U \omega) = d(-\varphi_V \omega) \quad \text{on } U \cap V.$$

We set $\delta\omega := d(\varphi_U \omega)$. The reader can check immediately that the cohomology class of $\delta\omega$ is independent of all the choices made.

If $\phi : N \hookrightarrow M$ is a morphism of $\mathcal{M}_n$ then for any MV-cover of $\{U, V\}$ of M $\{\phi^{-1}(U), \phi^{-1}(V)\}$ is an MV-cover of N. Moreover, we almost tautologically get a commutative diagram

$$
\begin{array}{ccc}
H^k_{cpt}(N) & \xrightarrow{\delta} & H^{k+1}_{cpt}(\phi^{-1}(U \cap V)) \\
\phi_* \downarrow & & \phi_* \downarrow \\
H^k_{cpt}(M) & \xrightarrow{\delta} & H^{k+1}_{cpt}(U \cap V)
\end{array}
$$

This proves $H^\bullet_{cpt}$ is a covariant Mayer-Vietoris sequence. $\qquad\square$

Remark 7.2.4. To be perfectly honest (from a categorical point of view) we should have considered the *chain complex*

$$\bigoplus_{k \leq 0} \tilde{\Omega}^k = \bigoplus_{k \leq 0} \Omega_{cpt}^{-k},$$

and correspondingly the associated *homology*

$$\tilde{H}_\bullet := H_{cpt}^{-\bullet}.$$

This makes the sequence

$$0 \leftarrow \tilde{\Omega}^{-n} \leftarrow \cdots \leftarrow \tilde{\Omega}^{-1} \leftarrow \tilde{\Omega}^0 \leftarrow 0$$

a *chain* complex, and its *homology* $\tilde{H}_\bullet$ is a bona-fide *covariant* Mayer-Vietoris functor since the connecting morphism δ goes in the right direction $\tilde{H}_\bullet \to \tilde{H}_{\bullet-1}$. However, the simplicity of the original notation is worth the small formal ambiguity, so we stick to our upper indices. □

From the proof of the Mayer-Vietoris principle we deduce the following.

Corollary 7.2.5. *For any $M \in \mathcal{M}_n$, and any $k \leq n$ we have* $\dim H_{cpt}^k(M) < \infty$.□

7.2.2 *The Poincaré duality*

Definition 7.2.6. Denote by $\mathcal{M}_n^+$ the category of n-dimensional, finite type, *oriented* manifolds. The morphisms are the embeddings of such manifold. The MV functors on $\mathcal{M}_n^+$ are defined exactly as for $\mathcal{M}_n$. □

Given $M \in \mathcal{M}_n^+$, there is a natural pairing

$$\langle \bullet, \bullet \rangle_\kappa : \Omega^k(M) \times \Omega_{cpt}^{n-k}(M) \to \mathbb{R},$$

defined by

$$\langle \omega, \eta \rangle_\kappa := \int_M \omega \wedge \eta.$$

This pairing is called the *Kronecker pairing*. We can extend this pairing to any $(\omega, \eta) \in \Omega^\bullet \times \Omega_{cpt}^\bullet$ as

$$\langle \omega, \eta \rangle_\kappa = \begin{cases} 0 & , \deg \omega + \deg \eta \neq n \\ \int_M \omega \wedge \eta & , \deg \omega + \deg \eta = n \end{cases}.$$

The Kronecker pairing induces maps

$$\mathcal{D} = \mathcal{D}^k : \Omega^k(M) \to (\Omega_{cpt}^{n-k}(M))^*, \quad \langle \mathcal{D}(\omega), \eta \rangle = \langle \omega, \eta \rangle_\kappa.$$

Above, $\langle \bullet, \bullet \rangle$ denotes the natural pairing between a vector space V and its dual V^*,

$$\langle \bullet, \bullet \rangle : V^* \times V \to \mathbb{R}.$$

If ω is closed, the restriction of $\mathcal{D}(\omega)$ to the space Z_{cpt}^{n-k} of closed, compactly supported $(n-k)$-forms vanishes on the subspace $B_{cpt}^{n-k} = d\Omega_{cpt}^{n-k-1}(M)$. Indeed, if $\eta = d\eta'$, $\eta' \in \Omega_{cpt}^{n-k-1}(M)$ then

$$\langle \mathcal{D}(\omega), \eta \rangle = \int_M \omega \wedge d\eta' \overset{Stokes}{=} \pm \int_M d\omega \wedge \eta' = 0.$$

Thus, if ω is closed, the linear functional $\mathcal{D}(\omega)$ defines an element of $(H_{cpt}^{n-k}(M))^*$. If moreover ω is exact, a computation as above shows that $\mathcal{D}(\omega) = 0 \in (H_{cpt}^{n-k}(M))^*$. Hence $\mathcal{D}$ descends to a map in cohomology

$$\mathcal{D} : H^k(M) \to (H_{cpt}^{n-k}(M))^*.$$

Equivalently, this means that the Kronecker pairing descends to a pairing in cohomology,

$$\langle \bullet, \bullet \rangle_\kappa : H^k(M) \times H_{cpt}^{n-k}(M) \to \mathbb{R}.$$

Theorem 7.2.7 (Poincaré duality). *The Kronecker pairing in cohomology is a duality for all $M \in \mathcal{M}_n^+$.*

Proof.   The functor $\mathcal{M}_n^+ \to$ **Graded Vector Spaces** defined by

$$M \to \bigoplus_k \tilde{H}^k(M) = \bigoplus_k \left(H_{cpt}^{n-k}(M) \right)^*$$

is a contravariant MV-functor. (The exactness of the Mayer-Vietoris sequence is preserved by transposition. This is where the fact that all the cohomology groups are finite dimensional vector spaces plays a very important role.)

For purely formal reasons which will become apparent in a little while, we re-define the connecting morphism

$$\tilde{H}^k(U \cap V) \overset{\tilde{d}}{\to} \tilde{H}^{k+1}(U \cup V),$$

to be $(-1)^k \delta^\dagger$, where $H_{cpt}^{n-k-1}(U \cap V) \overset{\delta}{\to} H_{cpt}^{n-k}(U \cup V)$ denotes the connecting morphism in the DeRham cohomology with compact supports, and $\delta^\dagger$ denotes its transpose.

The Poincaré lemma for compact supports can be rephrased

$$\tilde{H}^k(\mathbb{R}^n) = \begin{cases} \mathbb{R} \ , \ k = 0 \\ 0 \ , \ k > 0 \end{cases}.$$

The Kronecker pairing induces linear maps

$$\mathcal{D}_M : H^k(M) \to \tilde{H}^k(M).$$

Lemma 7.2.8. $\bigoplus_k \mathcal{D}^k$ *is a correspondence of MV functors.*

Proof. We have to check two facts.

Fact A. Let $M \overset{\phi}{\hookrightarrow} N$ be a morphism in $\mathcal{M}_n^+$. Then the diagram below is commutative.

$$
\begin{array}{ccc}
H^k(N) & \overset{\phi^*}{\longrightarrow} & H^k(M) \\
\mathcal{D}_N \downarrow & & \mathcal{D}_M \downarrow \\
\tilde{H}^k(N) & \overset{\tilde{\phi}^*}{\longrightarrow} & \tilde{H}^k(M)
\end{array} \;\;.
$$

Fact B. If $\{U, V\}$ is an MV-cover of $M \in \mathcal{M}_n^+$, then the diagram bellow is commutative

$$
\begin{array}{ccc}
H^k(U \cap V) & \overset{\partial}{\longrightarrow} & H^{k+1}(M) \\
\mathcal{D}_{U \cap V} \downarrow & & \mathcal{D}_M \downarrow \\
\tilde{H}^k(U \cap V) & \overset{(-1)^k \delta^\dagger}{\longrightarrow} & \tilde{H}^{k+1}(M)
\end{array} \;\;.
$$

Proof of Fact A. Let $\omega \in H^k(N)$. Denoting by $\langle \bullet, \bullet \rangle$ the natural duality between a vector space and its dual we deduce that for any $\eta \in H_{cpt}^{n-k}(M)$ we have

$$
\langle \tilde{\phi}^* \circ \mathcal{D}_N(\omega), \eta \rangle = \langle (\phi_*)^\dagger \mathcal{D}_N(\omega), \eta \rangle = \langle \mathcal{D}_N(\omega), \phi_* \eta \rangle
$$

$$
= \int_N \omega \wedge \phi_* \eta = \int_{M \hookrightarrow N} \omega |_M \wedge \eta = \int_M \phi^* \omega \wedge \eta = \langle \mathcal{D}_M(\phi^* \omega), \eta \rangle.
$$

Hence $\tilde{\phi}^* \circ \mathcal{D}_N = \mathcal{D}_M \circ \phi^*$.

Proof of Fact B. Let φ_U, φ_V be a partition of unity subordinated to the MV-cover $\{U, V\}$ of $M \in \mathcal{M}_n^+$. Consider a closed k-form $\omega \in \Omega^k(U \cap V)$. Then the connecting morphism in usual DeRham cohomology acts as

$$
\partial \omega = \begin{cases} d(-\varphi_V \omega) \text{ on } U \\ d(\varphi_U \omega) \text{ on } V \end{cases} .
$$

Choose $\eta \in \Omega_{cpt}^{n-k-1}(M)$ such that $d\eta = 0$. We have

$$
\langle \mathcal{D}_M \partial \omega, \eta \rangle = \int_M \partial \omega \wedge \eta = \int_U \partial \omega \wedge \eta + \int_V \partial \omega \wedge \eta - \int_{U \cap V} \partial \omega \wedge \eta
$$

$$
= -\int_U d(\varphi_V \omega) \wedge \eta + \int_V d(\varphi_U \omega) \wedge \eta + \int_{U \cap V} d(\varphi_V \omega) \wedge \eta.
$$

Note that the first two integrals vanish. Indeed, over U we have the equality

$$
\big(d(\varphi_V \omega)\big) \wedge \eta = d\big(\varphi_V \omega \wedge \eta\big),
$$

and the vanishing now follows from Stokes formula. The second term is dealt with in a similar fashion. As for the last term, we have

$$
\int_{U \cap V} d(\varphi_V \omega) \wedge \eta = \int_{U \cap V} d\varphi_V \wedge \omega \wedge \eta = (-1)^{\deg \omega} \int_{U \cap V} \omega \wedge (d\varphi_V \wedge \eta)
$$

$$= (-1)^k \int_{U \cap V} \omega \wedge \delta\eta = (-1)^k \langle \mathcal{D}_{U \cap V}\omega, \delta\eta \rangle = \langle (-1)^k \delta^\dagger \mathcal{D}_{U \cap V}\omega, \eta \rangle.$$

This concludes the proof of Fact B. The Poincaré duality now follows from the Mayer-Vietoris principle. $\qquad \square$

Remark 7.2.9. Using the Poincaré duality we can associate to any smooth map $f : M \to N$ between compact oriented manifolds of dimensions m and respectively n a natural *push-forward* or *Gysin map*

$$f_* : H^\bullet(M) \to H^{\bullet+q}(N), \quad q := \dim N - \dim M = n - m$$

defined by the composition

$$H^\bullet(M) \overset{\mathcal{D}_M}{\to} (H^{m-\bullet}(M))^* \overset{(f^*)^\dagger}{\to} (H^{m-\bullet}(N))^* \overset{\mathcal{D}_N^{-1}}{\to} H^{\bullet+n-m}(N),$$

where $(f^*)^\dagger$ denotes the transpose of the pullback morphism. $\qquad \square$

Corollary 7.2.10. *If $M \in \mathcal{M}_n^+$ then*

$$H^\bullet_{cpt}(M) \cong (H^{n-k}(M))^\bullet.$$

Proof. Since $H^k_{cpt}(M)$ is finite dimensional, the transpose

$$\mathcal{D}_M^t : (H^{n-k}_{cpt}(M))^{**} \to (H^k(M))^*$$

is an isomorphism. On the other hand, for any finite dimensional vector space there exists a natural isomorphism

$$V^{**} \cong V. \qquad \square$$

Corollary 7.2.11. *Let M be a compact, connected, oriented, n-dimensional manifold. Then the pairing*

$$H^k(M) \times H^{n-k}(M) \to \mathbb{R} \quad (\omega, \eta) \mapsto \int_M \omega \wedge \eta$$

is a duality. In particular, $b_k(M) = b_{n-k}(M)$, $\forall k$. $\qquad \square$

If M is connected $H^0(M) \cong H^n(M) \cong \mathbb{R}$ so that $H^n(M)$ is generated by any volume form defining the orientation.

The symmetry of Betti numbers can be translated in the language of Poincaré polynomials as

$$t^n P_M(\frac{1}{t}) = P_M(t). \qquad (7.2.1)$$

Example 7.2.12. Let Σ_g denote the connected sum of g tori. We have shown that

$$\chi(\Sigma_g) = b_0 - b_1 + b_2 = 2 - 2g.$$

Since Σ_g is connected, the Poincaré duality implies $b_2 = b_0 = 1$. Hence $b_1 = 2g$ i.e.

$$P_{\Sigma_g}(t) = 1 + 2gt + t^2. \qquad \square$$

Consider now a compact oriented smooth manifold such that $\dim M = 2k$. The Kronecker pairing induces a *non-degenerate* bilinear form

$$\mathfrak{I} : H^k(M) \times H^k(M) \to \mathbb{R} \quad \mathfrak{I}(\omega, \eta) = \int_M \omega \wedge \eta.$$

The bilinear form $\mathfrak{I}$ is called the *cohomological intersection form* of M.

When k is even (so that n is divisible by 4) $\mathfrak{I}$ is a symmetric form. Its signature is called the *signature* of M, and it is denoted by $\sigma(M)$.

When k is odd, $\mathfrak{I}$ is skew-symmetric, i.e., it is a symplectic form. In particular, $H^{2k+1}(M)$ must be an even dimensional space.

Corollary 7.2.13. *For any compact manifold $M \in \mathcal{M}^+_{4k+2}$ the middle Betti number $b_{2k+1}(M)$ is even.* □

Exercise 7.2.14. (a) Let $P \in \mathbb{Z}[t]$ be an odd degree polynomial with non-negative integer coefficients such that $P(0) = 1$. Show that if P satisfies the symmetry condition (7.2.1) there exists a compact, connected, oriented manifold M such that $P_M(t) = P(t)$.

(b) Let $P \in \mathbb{Z}[t]$ be a polynomial of degree $2k$ with non-negative integer coefficients. Assume $P(0) = 1$ and P satisfies (7.2.1). If the coefficient of t^k is even then there exists a compact connected manifold $M \in \mathcal{M}^+_{2k}$ such that $P_M(t) = P(t)$.
Hint: Describe the Poincaré polynomial of a connect sum in terms of the polynomials of its constituents. Combine this fact with the Künneth formula. □

Remark 7.2.15. The result in the above exercise is sharp. Using his intersection theorem F. Hirzebruch showed that there exist no *smooth* manifolds M of dimension 12 or 20 with Poincaré polynomials $1 + t^6 + t^{12}$ and respectively $1 + t^{10} + t^{20}$. Note that in each of these cases both middle Betti numbers are odd. For details we refer to J. P. Serre, *"Travaux de Hirzebruch sur la topologie des variétés"*, Seminaire Bourbaki 1953/54,$n°$ **88**. □

7.3 Intersection theory

7.3.1 *Cycles and their duals*

Suppose M is a smooth manifold.

Definition 7.3.1. A k-dimensional *cycle* in M is a pair (S, ϕ), where S is a compact, oriented k-dimensional manifold without boundary, and $\phi : S \to M$ is a smooth map. We denote by $\mathcal{C}_k(M)$ the set of k-dimensional cycles in M. □

Definition 7.3.2. (a) Two cycles $(S_0, \phi_0), (S_1, \phi_1) \in \mathcal{C}_k(M)$ are said to be *cobordant*, and we write this $(S_0, \phi_0) \sim_c (S_1, \phi_1)$, if there exists a compact, oriented manifold with boundary Σ, and a smooth map $\Phi : \Sigma \to M$ such that the following hold.

(a1) $\partial \Sigma = (-S_0) \sqcup S_1$ where $-S_0$ denotes the oriented manifold S_0 equipped with the opposite orientation, and "$\sqcup$" denotes the disjoint union.

(a2) $\Phi|_{S_i} = \phi_i$, $i = 0, 1$.

(b) A cycle $(S, \phi) \in \mathcal{C}_k(M)$ is called *trivial* if there exists a $(k+1)$-dimensional, oriented manifold Σ with (oriented) boundary S, and a smooth map $\Phi : \Sigma \to M$ such that $\Phi|_{\partial \Sigma} = \phi$. We denote by $\mathcal{T}_k(M)$ the set of trivial cycles.

(c) A cycle $(S, \phi) \in \mathcal{C}_k(M)$ is said to be *degenerate* if it is cobordant to a constant cycle, i.e., a cycle (S', ϕ') such that ϕ' is map constant on the components of S'. We denote by $\mathcal{D}_k(M)$ the set of degenerate cycles. □

Exercise 7.3.3. Let $(S_0, \phi_0) \sim_c (S_1, \phi_1)$. Prove that $(-S_0 \sqcup S_1, \phi_0 \sqcup \phi_1)$ is a trivial cycle. □

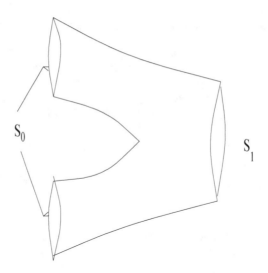

S_0

S_1

Fig. 7.1   *A cobordism in* $\mathbb{R}^3$

The cobordism relation on $\mathcal{C}_k(M)$ is an equivalence relation[3], and we denote by $\mathcal{Z}_k(M)$ the set of equivalence classes. For any cycle $(S, \phi) \in \mathcal{C}_k(M)$ we denote by $[S, \phi]$ its image in $\mathcal{Z}_k(M)$. Since M is connected, all the trivial cycles are cobordant, and they define an element in $\mathcal{Z}_k(M)$ which we denote by $[0]$.

Given $[S_0, \phi_0], [S_1, \phi_1] \in \mathcal{Z}_k(M)$ we define

$$[S_0, \phi_0] + [S_1, \phi_1] := [S_0 \sqcup S_1, \phi_0 \sqcup \phi_1],$$

where "$\sqcup$" denotes the disjoint union.

Proposition 7.3.4. *Suppose M is a smooth manifold.*

---
[3]We urge the reader to supply a proof of this fact.

(a) Let $[S_i, \phi_i], [S_i', \phi_i'] \in \mathcal{C}_k(M)$, $i = 0, 1$. If $[S_i, \phi_i] \sim_c [S_i', \phi_i']$, for $i = 0, 1$, then

$$[S_0 \sqcup S_1, \phi_0 \sqcup \phi_1] \sim_c [S_0' \sqcup S_1', \phi_0' \sqcup \phi_1']$$

so that the above map $+ : \mathcal{Z}_k(M) \times \mathcal{Z}_k(M) \to \mathcal{Z}_k(M)$ is well defined.

(b) The binary operation $+$ induces a structure of Abelian group on $\mathcal{Z}_k(M)$. The trivial element is represented by the trivial cycles. Moreover,

$$-[S, \phi] = [-S, \phi] \in \mathcal{H}_k(M). \qquad \square$$

Exercise 7.3.5. Prove the above proposition. $\qquad \square$

We denote by $\mathcal{H}_k(M)$ the quotient of $\mathcal{Z}_k(M)$ modulo the subgroup generated by the degenerate cycles. Let us point out that any trivial cycle is degenerate, but the converse is not necessarily true.

Suppose $M \in \mathcal{M}_n^+$. Any k-cycle (S, ϕ) defines a linear map $H^k(M) \to \mathbb{R}$ given by

$$H^k(M) \ni \omega \mapsto \int_S \phi^* \omega.$$

Stokes formula shows that this map is well defined, i.e., it is independent of the closed form representing a cohomology class.

Indeed, if ω is exact, i.e., $\omega = d\omega'$, then

$$\int_S \phi^* d\omega' = \int_S d\phi^* \omega' = 0.$$

In other words, each cycle defines an element in $(H^k(M))^*$. Via the Poincaré duality we identify the vector space $(H^k(M))^*$ with $H_{cpt}^{n-k}(M)$. Thus, there exists $\delta_S \in H_{cpt}^{n-k}(M)$ such that

$$\int_M \omega \wedge \delta_S = \int_S \phi^* \omega \quad \forall \omega \in H^k(M).$$

The compactly supported cohomology class δ_S is called the *Poincaré dual* of (S, ϕ).

There exist many closed forms $\eta \in \Omega_{cpt}^{n-k}(M)$ representing δ_S. When there is no risk of confusion, we continue denote any such representative by δ_S.

Example 7.3.6. Let $M = \mathbb{R}^n$, and S is a point, $S = \{pt\} \subset \mathbb{R}^n$. pt is canonically a 0-cycle. Its Poincaré dual is a compactly supported n-form ω such that for any constant λ (i.e. closed 0-form)

$$\int_{\mathbb{R}^n} \lambda \omega = \int_{pt} \lambda = \lambda,$$

i.e.,

$$\int_{\mathbb{R}^n} \omega = 1.$$

Thus δ_{pt} can be represented by any compactly supported n-form with integral 1. In particular, we can choose representatives with arbitrarily small supports. Their "profiles" look like in Figure 7.2. "At limit" they approach Dirac's delta distribution. $\qquad \square$

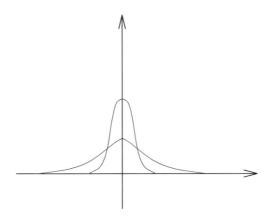

Fig. 7.2   *The dual of a point is Dirac's distribution*

Example 7.3.7. Consider an n-dimensional, compact, connected, oriented manifold M. We denote by $[M]$ the cycle $(M, \mathbb{1}_M)$. Then $\delta_{[M]} = 1 \in H^0(M)$.   □

☞ *For any differential form ω, we set (for typographical reasons), $|\omega| := \deg \omega$.*

Example 7.3.8. Consider the manifolds $M \in \mathcal{M}_m^+$ and $N \in \mathcal{M}_n^+$. (The manifolds M and N need not be compact.) To any pair of cycles $(S, \phi) \in \mathcal{C}_p(M)$, and $(T, \psi) \in \mathcal{C}_q(N)$ we can associate the cycle $(S \times T, \phi \times \psi) \in \mathcal{C}_{p+q}(M \times N)$. We denote by π_M (respectively π_N) the natural projection $M \times N \to M$ (respectively $M \times N \to N$). We want to prove the equality

$$\delta_{S \times T} = (-1)^{(m-p)q} \delta_S \times \delta_T, \tag{7.3.1}$$

where

$$\omega \times \eta := \pi_M^* \omega \wedge \pi_N^* \eta, \ \ \forall (\omega, \eta) \in \Omega^\bullet(M) \times \Omega^\bullet(N).$$

Pick $(\omega, \eta) \in \Omega^\bullet(M) \times \Omega^\bullet(N)$ such that, $\deg \omega + \deg \eta = (m + n) - (p + q)$. Then, using Exercise 7.1.50

$$\int_{M \times N} (\omega \times \eta) \wedge (\delta_S \times \delta_T) = (-1)^{(m-p)|\eta|} \int_{M \times N} (\omega \wedge \delta_S) \times (\eta \wedge \delta_T).$$

The above integral should be understood in the generalized sense of Kronecker pairing. The only time when this pairing does not vanish is when $|\omega| = p$ and $|\eta| = q$. In this case the last term equals

$$(-1)^{q(m-p)} \left(\int_S \omega \wedge \delta_S \right) \left(\int_T \eta \wedge \delta_T \right) = (-1)^{q(m-p)} \left(\int_S \phi^* \omega \right) \left(\int_T \psi^* \eta \right)$$

$$= (-1)^{q(m-p)} \int_{S \times T} (\phi \times \psi)^* (\omega \wedge \eta).$$

This establishes the equality (7.3.1).   □

Example 7.3.9. Consider a *compact* manifold $M \in \mathcal{M}_n^+$. Fix a basis (ω_i) of $H^\bullet(M)$ such that each ω_i is homogeneous of degree $|\omega_i| = d_i$, and denote by ω^i the basis of $H^\bullet(M)$ dual to (ω_i) with respect to the Kronecker pairing, i.e.,

$$\langle \omega^i, \omega_j \rangle_\kappa = (-1)^{|\omega^i| \cdot |\omega_j|} \langle \omega_j, \omega^i \rangle_\kappa = \delta_j^i.$$

In $M \times M$ there exists a remarkable cycle, the *diagonal*

$$\Delta = \Delta_M : M \to M \times M, \quad x \mapsto (x, x).$$

We claim that the Poincaré dual of this cycle is

$$\delta_\Delta = \boldsymbol{\delta}_M = \sum_i (-1)^{|\omega^i|} \omega^i \times \omega_i. \tag{7.3.2}$$

Indeed, for any homogeneous forms $\alpha, \beta \in \Omega^\bullet(M)$ such that $|\alpha| + |\beta| = n$, we have

$$\int_{M \times M} (\alpha \times \beta) \wedge \boldsymbol{\delta}_M = \sum_i (-1)^{|\omega^i|} \int_{M \times M} (\alpha \times \beta) \wedge (\omega^i \times \omega_i)$$

$$= \sum_i (-1)^{|\omega^i|} (-1)^{|\beta| \cdot |\omega^i|} \int_{M \times M} (\alpha \wedge \omega^i) \times (\beta \wedge \omega_i)$$

$$= \sum_i (-1)^{|\omega^i|} (-1)^{|\beta| \cdot |\omega^i|} \left(\int_M \alpha \wedge \omega^i \right) \left(\int_M \beta \wedge \omega_i \right).$$

The i-th summand is nontrivial only when $|\beta| = |\omega^i|$, and $|\alpha| = |\omega_i|$. Using the equality $|\omega^i| + |\omega^i|^2 \equiv 0 \pmod 2$ we deduce

$$\int_{M \times M} (\alpha \times \beta) \wedge \boldsymbol{\delta}_M = \sum_i \left(\int_M \alpha \wedge \omega^i \right) \left(\int_M \beta \wedge \omega_i \right) = \sum_i \langle \alpha, \omega^i \rangle_\kappa \langle \beta, \omega_i \rangle_\kappa.$$

From the equalities

$$\alpha = \sum_i \omega_i \langle \omega^i, \alpha \rangle_\kappa \quad \text{and} \quad \beta = \sum_j \langle \beta, \omega_j, \rangle_\kappa \omega^j$$

we conclude

$$\int_M \Delta^*(\alpha \times \beta) = \int_M \alpha \wedge \beta = \int_M \left(\sum_i \omega_i \langle \omega^i, \alpha \rangle_\kappa \right) \wedge \left(\sum_j \langle \beta, \omega_j \rangle_\kappa \omega^j \right)$$

$$= \int_M \sum_{i,j} \langle \omega^i, \alpha \rangle_\kappa \langle \beta, \omega_j \rangle_\kappa \omega_i \wedge \omega^j = \sum_{i,j} \langle \omega^i, \alpha \rangle_\kappa \langle \beta, \omega_j \rangle_\kappa \langle \omega_i, \omega^j \rangle_\kappa$$

$$= \sum_i (-1)^{|\omega_i|(n-|\omega_i|)} \langle \alpha, \omega^i \rangle_\kappa (-1)^{|\omega_i|(n-|\omega_i|)} \langle \beta, \omega_i \rangle_\kappa = \sum_i \langle \alpha, \omega^i \rangle_\kappa \langle \beta, \omega_i \rangle_\kappa.$$

Equality (7.3.2) is proved. $\qquad\qquad\qquad\qquad\qquad\qquad\qquad\qquad\qquad\qquad\qquad\square$

Proposition 7.3.10. *Let $M \in \mathcal{M}_n^+$ be a manifold, and suppose that $(S_i, \phi_i) \in \mathcal{C}_k(M)$ ($i = 0, 1$) are two k-cycles in M.*

(a) *If $(S_0, \phi_0) \sim_c (S_1, \phi_1)$, then $\delta_{S_0} = \delta_{S_1}$ in $H_{cpt}^{n-k}(M)$.*
(b) *If (S_0, ϕ_0) is trivial, then $\delta_{S_0} = 0$ in $H_{cpt}^{n-k}(M)$.*
(c) *$\delta_{S_0 \sqcup S_1} = \delta_{S_0} + \delta_{S_1}$ in $H_{cpt}^{n-k}(M)$.*
(d) *$\delta_{-S_0} = -\delta_{S_0}$ in $H_{cpt}^{n-k}(M)$.*

Proof. (a) Consider a compact manifold Σ with boundary $\partial\Sigma = -S_0 \sqcup S_1$ and a smooth map $\Phi : \Sigma \to M$ such that $\Phi \mid_{\partial\Sigma} = \phi_0 \sqcup \phi_1$. For any closed k-form $\omega \in \Omega^k(M)$ we have

$$0 = \int_\Sigma \Phi^*(d\omega) = \int_\Sigma d\Phi^*\omega \overset{Stokes}{=} \int_{\partial\Sigma} \omega = \int_{S_1} \phi_1^*\omega - \int_{S_0} \phi_0^*\omega.$$

Part (b) is left to the reader. Part (c) is obvious. To prove (d) consider $\Sigma = [0, 1] \times S_0$ and

$$\Phi : [0, 1] \times S_0 \to M, \quad \Phi(t, x) = \phi_0(x), \quad \forall(t, x) \in \Sigma.$$

Note that $\partial\Sigma = (-S_0) \sqcup S_0$ so that

$$\delta_{-S_0} + \delta_{S_0} = \delta_{-S_0 \sqcup S_0} = \delta_{\partial\Sigma} = 0. \qquad \square$$

The above proposition shows that the correspondence

$$\mathcal{C}_k(M) \ni (S, \phi) \mapsto \delta_S \in H_{cpt}^{n-k}(M)$$

descends to a map

$$\delta : \mathcal{H}_k(M) \to H_{cpt}^{n-k}(M).$$

This is usually called the *homological Poincaré duality*. We are not claiming that δ is an isomorphism.

7.3.2 *Intersection theory*

Consider $M \in \mathcal{M}_n^+$, and a k-dimensional compact oriented submanifold S of M. We denote by $i : S \hookrightarrow M$ inclusion map so that (S, i) is a k-cycle.

Definition 7.3.11. A smooth map $\phi : T \to M$ from an $(n-k)$-dimensional, oriented manifold T is said to be *transversal* to S, and we write this $S \pitchfork \phi$, if the following hold.

(a) $\phi^{-1}(S)$ is a finite subset of T;
(b) for every $x \in \phi^{-1}(S)$ we have

$$\phi_*(T_x T) + T_{\phi(x)}S = T_{\phi(x)}M \quad \text{(direct sum)}.$$

If $S \pitchfork \Phi$, then for each $x \in \phi^{-1}(S)$ we define the *local intersection number* at x to be ($\boldsymbol{or}$ = orientation)

$$i_x(S, T) = \begin{cases} 1 \ , & \boldsymbol{or}\,(T_{\phi(x)}S) \wedge \boldsymbol{or}(\phi_* T_x T) = \boldsymbol{or}(T_{\phi(x)}M) \\ -1 \ , & \boldsymbol{or}(T_{\phi(x)}S) \wedge \boldsymbol{or}(\phi_* T_x T) = -\boldsymbol{or}(T_{\phi(x)}M) \end{cases} .$$

Finally, we define the *intersection number* of S with T to be

$$S \cdot T := \sum_{x \in \phi^{-1}(S)} i_x(S, T). \qquad \qquad \square$$

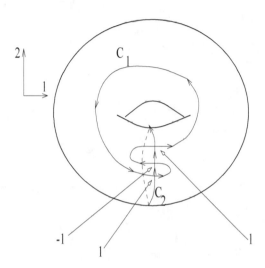

Fig. 7.3   *The intersection number of the two cycles on T^2 is 1*

Our next result offers a different description of the intersection number indicating how one can drop the transversality assumption from the original definition.

Proposition 7.3.12.   *Let $M \in \mathcal{M}_n^+$. Consider a compact, oriented, k-dimensional submanifold $S \hookrightarrow M$, and $(T, \phi) \in \mathcal{C}_{n-k}(M)$ a $(n-k)$-dimensional cycle intersecting S transversally, i.e., $S \pitchfork \phi$. Then*

$$S \cdot T = \int_M \delta_S \wedge \delta_T, \qquad \qquad (7.3.3)$$

where $\delta_\bullet$ denotes the Poincaré dual of $\bullet$.

The proof of the proposition relies on a couple of technical lemmata of independent interest.

Lemma 7.3.13 (Localization lemma).   *Let $M \in \mathcal{M}_n^+$ and $(S, \phi) \in \mathcal{C}_k(M)$. Then, for any open neighborhood $\mathcal{N}$ of $\phi(S)$ in M there exists $\delta_S^{\mathcal{N}} \in \Omega_{cpt}^{n-k}(M)$ such that*
(a) $\delta_S^{\mathcal{N}}$ represents the Poincaré dual $\delta_S \in H_{cpt}^{n-k}(M)$;
(b) $\operatorname{supp} \delta_S^{\mathcal{N}} \subset \mathcal{N}$.

Proof.   Fix a Riemann metric on M. Each point $p \in \phi(S)$ has a geodesically convex open neighborhood entirely contained in $\mathcal{N}$. Cover $\phi(S)$ by finitely many such neighborhoods, and denote by N their union. Then $N \in \mathcal{M}_n^+$, and $(S, \phi) \in \mathcal{C}_k^+(N)$.

Denote by δ_S^N the Poincaré dual of S in N. It can be represented by a closed form in $\Omega_{cpt}^{n-k}(N)$ which we continue to denote by δ_S^N. If we pick a closed form $\omega \in \Omega^k(M)$, then $\omega|_N$ is also closed, and

$$\int_M \delta_S^N \wedge \omega = \int_N \delta_S^N \wedge \omega = \int_S \phi^* \omega.$$

Hence, δ_S^N represents the Poincaré dual of S in $H_{cpt}^{n-k}(M)$, and moreover, $\operatorname{supp} \delta_S^N \subset \mathcal{N}$. □

Definition 7.3.14.   Let $M \in \mathcal{M}_n^+$, and $S \hookrightarrow M$ a compact, k-dimensional, oriented submanifold of M. A *local transversal* at $p \in S$ is an embedding

$$\phi : B \subset \mathbb{R}^{n-k} \to M, \quad B = \text{open ball centered at } 0 \in \mathbb{R}^{m-k},$$

such that $S \pitchfork \phi$ and $\phi^{-1}(S) = \{0\}$. □

Lemma 7.3.15.   *Let $M \in \mathcal{M}_n^+$, $S \hookrightarrow M$ a compact, k-dimensional, oriented submanifold of M and (B, ϕ) a local transversal at $p \in S$. Then for any sufficiently "thin" closed neighborhood $\mathcal{N}$ of $S \subset M$ we have*

$$S \cdot (B, \phi) = \int_B \phi^* \delta_S^N.$$

Proof.   Using the transversality $S \pitchfork \phi$, the implicit function theorem, and eventually restricting ϕ to a smaller ball, we deduce that, for some sufficiently "thin" neighborhood $\mathcal{N}$ of S, there exist local coordinates $(x^1, \ldots, x^n)$ defined on some neighborhood U of $p \in M$ diffeomorphic with the cube

$$\{|x^i| < 1, \ \forall i\}$$

such that the following hold.
(i) $S \cap U = \{x^{k+1} = \cdots = x^n = 0\}$, $p = (0, \ldots, 0)$.
(ii) The orientation of $S \cap U$ is defined by $dx^1 \wedge \cdots \wedge dx^k$.
(iii) The map $\phi : B \subset \mathbb{R}^{n-k}_{(y^1, \ldots, y^{n-k})} \to M$ is expressed in these coordinates as

$$x^1 = 0, \ldots, x^k = 0, x^{k+1} = y^1, \ldots, x^n = y^{n-k}.$$

(iv) $\mathcal{N} \cap U = \{|x^j| \le 1/2 \ ; \ j = 1, \ldots, n\}$.

Let $\epsilon = \pm 1$ such that $\epsilon dx^1 \wedge \ldots \wedge dx^n$ defines the orientation of TM. In other words,

$$\epsilon = S \cdot (B, \phi).$$

For each $\xi = (x^1, \ldots, x^k) \in S \cap U$ denote by P_ξ the $(n-k)$-"plane"

$$P_\xi = \{(\xi; x^{k+1}, \ldots, x^n) \ ; \ |x^j| < 1 \ j > k\}.$$

We orient each P_ξ using the $(n-k)$-form $dx^{k+1} \wedge \cdots \wedge dx^n$, and set

$$v(\xi) := \int_{P_\xi} \delta_S^N.$$

Equivalently,

$$v(\xi) = \int_B \phi_\xi^* \delta_S^N,$$

where $\phi_\xi : B \to M$ is defined by

$$\phi_\xi(y^1, \ldots, y^{n-k}) = (\xi; y^1, \ldots, y^{n-k}).$$

To any function $\varphi = \varphi(\xi) \in C^\infty(S \cap U)$ such that

$$\operatorname{supp}\varphi \subset \{|x^i| \leq 1/2 \; ; \; i \leq k\},$$

we associate the k-form

$$\omega_\varphi := \varphi dx^1 \wedge \cdots \wedge dx^k = \varphi d\xi \in \Omega_{cpt}^k(S \cap U).$$

Extend the functions $x^1, \ldots, x^k \in C^\infty(U \cap N)$ to smooth compactly supported functions

$$\tilde{x}^i \in C_0^\infty(M) \to [0,1].$$

The form ω_φ is then the restriction to $U \cap S$ of the closed compactly supported form

$$\tilde{\omega}_\varphi = \varphi(\tilde{x}^1, \ldots, \tilde{x}^k) d\tilde{x}^1 \wedge \cdots \wedge d\tilde{x}^k.$$

We have

$$\int_M \tilde{\omega}_\varphi \wedge \delta_S^N = \int_U \omega_\varphi \wedge \delta_S^N = \int_S \omega_\varphi = \int_{\mathbb{R}^k} \varphi(\xi) d\xi. \qquad (7.3.4)$$

The integral over U can be evaluated using the Fubini theorem. Write

$$\delta_S^N = f dx^{k+1} \wedge \cdots \wedge dx^n + \varrho,$$

where ϱ is an $(n-k)$-form not containing the monomial $dx^{k+1} \wedge \cdots \wedge dx^n$. Then

$$\int_U \omega_\varphi \wedge \delta_S^N = \int_U f\varphi dx^1 \wedge \cdots \wedge dx^n$$

$$= \epsilon \int_U f\varphi |dx^1 \wedge \cdots \wedge dx^n| \quad (|dx^1 \wedge \cdots \wedge dx^n| = \text{Lebesgue density})$$

$$\overset{Fubini}{=} \epsilon \int_{S \cap U} \varphi(\xi) \left(\int_{P_\xi} f|dx^{k+1} \wedge \cdots \wedge dx^n| \right) |d\xi|$$

$$= \epsilon \int_S \varphi(\xi) \left(\int_{P_\xi} \delta_S^N \right) |d\xi| = \epsilon \int_S \varphi(\xi) v(\xi) |d\xi|.$$

Comparing with (7.3.4), and taking into account that φ was chosen arbitrarily, we deduce that

$$\int_B \phi^* \delta_S^{\mathcal{N}} = v(0) = \epsilon.$$

The local transversal lemma is proved. □

Proof of Proposition 7.3.12   Let

$$\phi^{-1}(S) = \{p_1, \ldots, p_m\}.$$

The transversality assumption implies that each p_i has an open neighborhood B_i diffeomorphic to an open ball such that $\phi_i = \phi|_{B_i}$ is a local transversal at $y_i = \phi(p_i)$. Moreover, we can choose the neighborhoods B_i to be mutually disjoint. Then

$$S \cdot T = \sum_i S \cdot (B_i, \phi_i).$$

The compact set $K := \phi(T \setminus \cup B_i)$ does not intersect S so that we can find a "thin", closed neighborhood" $\mathcal{N}$ of $S \hookrightarrow M$ such that $K \cap \mathcal{N} = \emptyset$. Then, $\phi^* \delta_S^{\mathcal{N}}$ is compactly supported in the union of the B_i's and

$$\int_M \delta_S^{\mathcal{N}} \wedge \delta_T = \int_T \phi^* \delta_S^{\mathcal{N}} = \sum_i \int_{B_i} \phi_i^* \delta_S^{\mathcal{N}}.$$

From the local transversal lemma we get

$$\int_{B_i} \phi^* \delta_S^{\mathcal{N}} = S \cdot (B_i, \phi) = i_{p_i}(S, T). \qquad \square$$

Equality (7.3.3) has a remarkable feature. Its right-hand side is an integer which is defined only for cycles S, T such that S is embedded and $S \pitchfork T$.

The left-hand side makes sense for any cycles of complementary dimensions, but a priori it may not be an integer. In any event, we have a remarkable consequence.

Corollary 7.3.16. *Let $(S_i, \phi_i) \in \mathcal{C}_k(M)$ and $((T_i, \psi_i) \in \mathcal{C}_{n-k}(M)$ where $M \in \mathcal{M}_n^+$, $i = 0, 1$. If*

(a) $S_0 \sim_c S_1$, $T_0 \sim_c T_1$,
(b) the cycles S_i are embedded, and
(c) $S_i \pitchfork T_i$,

then

$$S_0 \cdot T_0 = S_1 \cdot T_1. \qquad \square$$

Definition 7.3.17. The *homological intersection pairing* is the $\mathbb{Z}$-bilinear map

$$\mathcal{I} = \mathcal{I}_M : \mathcal{H}_k(M) \times \mathcal{H}_{n-k}(M) \to \mathbb{R},$$

$(M \in \mathcal{M}_n^+)$ defined by

$$\mathcal{I}(S, T) = \int_M \delta_S \wedge \delta_T. \qquad \square$$

We have proved that in some special instances $\mathfrak{I}(S,T) \in \mathbb{Z}$. We want to prove that, when M is *compact*, this is *always* the case.

Theorem 7.3.18. *Let $M \in \mathcal{M}_n^+$ be compact manifold. Then, for any $(S,T) \in \mathcal{H}_k(M) \times \mathcal{H}_{n-k}(M)$, the intersection number $\mathfrak{I}(S,T)$ is an integer.*

The theorem will follow from two lemmata. The first one will show that it suffices to consider only the situation when one of the two cycles is embedded. The second one will show that, if one of the cycles is embedded, then the second cycle can be deformed so that it intersects the former transversally. (This is called a *general position result.*)

Lemma 7.3.19 (Reduction-to-diagonal trick). *Let M, S and T as in Theorem 7.3.18. Then*

$$\mathfrak{I}(S,T) = (-1)^{n-k} \int_{M \times M} \delta_{S \times T} \wedge \delta_{\Delta},$$

where Δ is the diagonal cycle $\Delta : M \to M \times M$, $x \mapsto (x,x)$. (It is here where the compactness of M is essential, since otherwise Δ would not be a cycle.)

Proof. We will use the equality (7.3.1)

$$\delta_{S \times T} = (-1)^{n-k} \delta_S \times \delta_T.$$

Then

$$(-1)^{n-k} = \int_{M \times M} \delta_{S \times T} \wedge \delta_{\Delta} = \int_{M \times M} (\delta_S \times \delta_T) \wedge \delta_{\Delta}$$

$$= \int_M \Delta^*(\delta_S \times \delta_T) = \int_M \delta_S \wedge \delta_T. \qquad \square$$

Lemma 7.3.20 (Moving lemma). *Let $S \in \mathcal{C}_k(M)$, and $T \in \mathcal{C}_{n-k}(M)$ be two cycles in $M \in \mathcal{M}_n^+$. If S is embedded, then T is cobordant to a cycle $\tilde{T}$ such that $S \pitchfork \tilde{T}$.*

The proof of this result relies on Sard's theorem. For details we refer to [45], Chapter 3.

Proof of Theorem 7.3.18 Let $(S,T) \in \mathcal{C}_k(M) \times \mathcal{C}_{n-k}(M)$. Then,

$$\mathfrak{I}(S,T) = (-1)^{n-k}\mathfrak{I}(S \times T, \Delta).$$

Since Δ is embedded we may assume by the moving lemma that $(S \times T) \pitchfork \Delta$, so that $\mathfrak{I}(S \times T, \Delta) \in \mathbb{Z}$. $\qquad \square$

7.3.3   *The topological degree*

Consider two compact, connected, oriented smooth manifolds M, N having the same dimension n. Any smooth map $F : M \to N$ canonically defines an n-dimensional cycle in $M \times N$

$$\Gamma_F : M \to M \times N \quad x \mapsto (x, F(x)).$$

Γ_F is called the *graph of F*.

Any point $y \in N$ defines an n-dimensional cycle $M \times \{y\}$. Since N is connected all these cycles are cobordant so that the integer $\Gamma_F \cdot (M \times \{y\})$ is independent of y.

Definition 7.3.21. The *topological degree* of the map F is defined by

$$\deg F := (M \times \{y\}) \cdot \Gamma_F. \qquad \square$$

Note that the intersections of Γ_F with $M \times \{y\}$ correspond to the solutions of the equation $F(x) = y$. Thus the topological degree counts these solutions (with sign).

Proposition 7.3.22. *Let $F : M \to N$ as above. Then for any n-form $\omega \in \Omega^n(N)$*

$$\int_M F^*\omega = \deg F \int_N \omega.$$

Remark 7.3.23. The map F induces a morphism

$$\mathbb{R} \cong H^n(N) \xrightarrow{F^*} H^n(M) \cong \mathbb{R}$$

which can be identified with a real number. The above proposition guarantees that this number is an *integer*. $\qquad \square$

Proof of the proposition   Note that if $\omega \in \Omega^n(N)$ is exact, then

$$\int_N \omega = \int_M F^*\omega = 0.$$

Thus, to prove the proposition it suffices to check it for any particular form which generates $H^n(N)$. Our candidate will be the Poincaré dual δ_y of a point $y \in N$. We have

$$\int_N \delta_y = 1,$$

while equality (7.3.1) gives

$$\delta_{M \times \{y\}} = \delta_M \times \delta_y = 1 \times \delta_y.$$

We can then compute the degree of F using Theorem 7.3.18

$$\deg F = \deg F \int_N \delta_y = \int_{M \times N} (1 \times \delta_y) \wedge \delta_{\Gamma_F} = \int_M \Gamma_F^*(1 \times \delta_y) = \int_M F^*\delta_y. \quad \square$$

Corollary 7.3.24 (Gauss-Bonnet). *Consider a connected sum of g-tori $\Sigma = \Sigma_g$ embedded in $\mathbb{R}^3$ and let $\mathcal{G}_\Sigma : \Sigma \to S^2$ be its Gauss map. Then*

$$\deg \mathcal{G}_\Sigma = \chi(\Sigma) = 2 - 2g. \qquad \square$$

This corollary follows immediately from the considerations at the end of Subsection 4.2.4.

Exercise 7.3.25. Consider the compact, connected manifolds $M_0, M_1, N \in \mathcal{M}_n^+$ and the smooth maps $F_i : M_i \to N$, $i = 0, 1$. Show that if F_0 is cobordant to F_1 then $\deg F_0 = \deg F_1$. In particular, homotopic maps have the same degree. $\qquad \square$

Exercise 7.3.26. Let A be a nonsingular $n \times n$ real matrix. It defines a smooth map

$$F_A : S^{n-1} \to S^{n-1}, \quad x \mapsto \frac{Ax}{|Ax|}.$$

Prove that $\deg F_A = \operatorname{sign} \det A$.
Hint: Use the polar decomposition $A = P \cdot O$ (where P is a positive symmetric matrix and O is an orthogonal one) to deform A inside $GL(n, \mathbb{R})$ to a diagonal matrix. $\qquad \square$

Exercise 7.3.27. Let $M \xrightarrow{F} N$ be a smooth map (M, N are smooth, compact oriented of dimension n). Assume $y \in N$ is a regular value of F, i.e., for all $x \in F^{-1}(y)$ the derivative

$$D_x F : T_x M \to T_y N$$

is invertible. For $x \in F^{-1}(y)$ define

$$\deg(F, x) = \begin{cases} 1, & D_x F \text{ preserves orientations} \\ -1, & \text{otherwise} \end{cases}$$

Prove that

$$\deg F = \sum_{F(x)=y} \deg(F, x). \qquad \square$$

Exercise 7.3.28. Let M denote a compact oriented manifold, and consider a smooth map $F : M \to M$ Regard $H^\bullet(M)$ as a superspace with the obvious $\mathbb{Z}_2$-grading

$$H^\bullet(M) = H^{even}(M) \oplus H^{odd}(M),$$

and define the *Lefschetz number* $\lambda(F)$ of F as the supertrace of the pull back

$$F^* : H^\bullet(M) \to H^\bullet(M).$$

Prove that $\lambda(F) = \Delta \cdot \Gamma_F$, and deduce from this the Lefschetz fixed point theorem: $\lambda(F) \neq 0 \Rightarrow F$ has a fixed point. $\qquad \square$

Remark 7.3.29. For an elementary approach to degree theory, based only on Sard's theorem, we refer to the beautiful book of Milnor, [74]. $\qquad \square$

7.3.4 *Thom isomorphism theorem*

Let $p : E \to B$ be an orientable fiber bundle (in the sense of Definition 3.4.45) with standard fiber F and compact, oriented basis B. Let $\dim B = m$ and $\dim F = r$. The total space E is a compact orientable manifold which we equip with the *fiber-first orientation*.

In this subsection we will extensively use the techniques of fibered calculus described in Subsection 3.4.5. The integration along fibers

$$\int_{E/B} = p_* : \Omega^\bullet_{cpt}(E) \to \Omega^{\bullet - r}(B)$$

satisfies

$$p_* d_E = (-1)^r d_B p_*,$$

so that it induces a map in cohomology

$$p_* : H^\bullet_{cpt}(E) \to H^{\bullet - r}(B).$$

This induced map in cohomology is sometimes called the *Gysin map*.

Remark 7.3.30. If the standard fiber F is compact, then the total space E is also compact. Using Proposition 3.4.48 we deduce that for any $\omega \in \Omega^\bullet(E)$, and any $\eta \in \Omega^\bullet(B)$ such that

$$\deg \omega + \deg \eta = \dim E,$$

we have

$$\langle p_*(\omega), \eta \rangle_\kappa = \langle \omega, p^*(\eta) \rangle_k$$

If we denote by $\mathcal{D}_M : \Omega^\bullet(M) \to \Omega^{\dim M - \bullet}(M)^*$ the Poincaré duality isomorphism on a compact oriented smooth manifold M, then we can rewrite the above equality as

$$\langle \mathcal{D}_B p_*(\omega), \eta \rangle = \langle \mathcal{D}_E \omega, p^* \eta \rangle = \langle (p^*)^\dagger \mathcal{D}_E \omega, \eta \rangle \implies \mathcal{D}_B p_*(\omega) = (p^*)^\dagger \mathcal{D}_E \omega.$$

Hence

$$p_* = \mathcal{D}_B^{-1} (p^*)^\dagger \mathcal{D}_E,$$

so that, in this case, the Gysin map coincides with the pushforward (Gysin) map defined in Remark 7.2.9. $\qquad\qquad\square$

Exercise 7.3.31. Consider a smooth map $f : M \to N$ between compact, oriented manifolds M, N of dimensions m and respectively n. Denote by i_f the embedding of M in $M \times N$ as the graph of f

$$M \ni x \mapsto (x, f(x)) \in M \times N.$$

The natural projection $M \times N \to N$ allows us to regard $M \times N$ as a trivial fiber bundle over N.

Show that the push-forward map $f_* : H^\bullet(M) \to H^{\bullet+n-m}(N)$ defined in Remark 7.2.9 can be equivalently defined by

$$f_* = \pi_* \circ (i_f)_*,$$

where $\pi_* : H^\bullet(M \times N) \to H^{\bullet-m}(N)$ denotes the integration along fibers while

$$(i_f)_* : H^\bullet(M) \to H^{\bullet+n}(M \times N),$$

is the pushforward morphism defined by the *embedding* i_f. □

Let us return to the fiber bundle $p : E \to B$. Any smooth section $\sigma : B \to E$ defines an embedded cycle in E of dimension $m = \dim B$. Denote by δ_σ its Poincaré dual in $H^r_{cpt}(E)$.

Using the properties of the integration along fibers we deduce that, for any $\omega \in \Omega^m(B)$, we have

$$\int_E \delta_\sigma \wedge p^*\omega = \int_B \left(\int_{E/B} \delta_\sigma \right) \omega.$$

On the other hand, by Poincaré duality we get

$$\int_E \delta_\sigma \wedge p^*\omega = (-1)^{rm} \int_E p^*\omega \wedge \delta_\sigma$$

$$= (-1)^{rm} \int_B \sigma^* p^*\omega = (-1)^{rm} \int_B (p\sigma)^*\omega = (-1)^{rm} \int_B \omega.$$

Hence

$$p_*\delta_\sigma = \int_{E/B} \delta_\sigma = (-1)^{rm} \in \Omega^0(B).$$

Proposition 7.3.32. *Let $p : E \to B$ a bundle as above. If it admits at least one section, then the Gysin map*

$$p_* : H^\bullet_{cpt}(E) \to H^{\bullet-r}(B)$$

is surjective.

Proof. Denote by τ_σ the map

$$\tau_\sigma : H^\bullet(B) \to H^{\bullet+r}_{cpt}(E) \quad \omega \mapsto (-1)^{rm} \delta_\sigma \wedge p^*\omega = p^*\omega \wedge \delta_\sigma.$$

Then τ_σ is a right inverse for p_*. Indeed

$$\omega = (-1)^{rm} p_*\delta_\sigma \wedge \omega = (-1)^{rm} p_*(\delta_\sigma \wedge p^*\omega) = p_*(\tau_\sigma\omega). \qquad \square$$

The map p_* is not injective in general. For example, if (S, ϕ) is a k-cycle in F, then it defines a cycle in any fiber $\pi^{-1}(b)$, and consequently in E. Denote by δ_S its Poincaré dual in $H^{m+r-k}_{cpt}(E)$. Then for any $\omega \in \Omega^{m-k}(B)$ we have

$$\int_B (p_*\delta_S) \wedge \omega = \int_E \delta_S \wedge p^*\omega = \pm \int_D \phi^* p^*\omega = \int_S (p \circ \phi)^*\omega = 0,$$

since $p \circ \phi$ is constant. Hence $p_* \delta_S = 0$, and we conclude that if F carries nontrivial cycles $\ker p_*$ may not be trivial.

The simplest example of standard fiber with only trivial cycles is a vector space.

Definition 7.3.33. Let $p : E \to B$ be an orientable vector bundle over the compact oriented manifold B ($\dim B = m$, $\mathrm{rank}(E) = r$). The *Thom class* of E, denoted by τ_E is the Poincaré dual of the cycle defined by the zero section $\zeta_0 : B \to E$, $b \mapsto 0 \in E_b$. Note that $\tau_E \in H^r_{cpt}(E)$. $\square$

Theorem 7.3.34 (Thom isomorphism). *Let $p : E \to B$ as in the above definition. Then the map*

$$\tau : H^\bullet(B) \to H^{\bullet+r}_{cpt}(E) \quad \omega \mapsto \tau_E \wedge p^* \omega$$

is an isomorphism called the Thom isomorphism. *Its inverse is the Gysin map*

$$(-1)^{rm} p_* : H^\bullet_{cpt}(E) \to H^{\bullet-r}(B).$$

Proof. We have already established that $p_* \tau = (-1)^{rm}$. To prove the reverse equality, $\tau p_* = (-1)^{rm}$, we will use Lemma 7.2.2 of Subsection 7.2.1. For $\beta \in \Omega^\bullet_{cpt}(E)$ we have

$$(p^* p_* \tau_E) \wedge \beta - \tau_E \wedge (p^* p_* \beta) = (-1)^r d(\mathfrak{m}(\tau_E, \beta)),$$

where $\mathfrak{m}(\tau_E, \beta) \in \Omega^\bullet_{cpt}(E)$. Since $p^* p_* \tau_E = (-1)^{rm}$, we deduce

$$(-1)^{rm} \beta = (\tau_E \wedge p^*(p_* \beta)) + \text{exact form} \Rightarrow (-1)^{rm} \beta = \tau_E \circ p_*(\beta) \text{ in } H^\bullet_{cpt}(E). \quad \square$$

Exercise 7.3.35. Show that $\tau_E = \zeta_* 1$, where $\zeta_\bullet : H^\bullet(M) \to H^{\bullet+\dim M}_{cpt}(E)$ is the push-forward map defined by a section $\zeta : M \to E$. $\square$

Remark 7.3.36. Suppose $M \in \mathcal{M}^+_n$, and $S \hookrightarrow M$ is a compact, oriented, k-dimensional manifold. Fix a Riemann metric g on M, and denote by N_S the normal bundle of the embedding $S \hookrightarrow M$, i.e., the orthogonal complement of TN in $(TM)|_S$.

The exponential map defined by the metric g defines a smooth map

$$\exp : N_S \to M,$$

which induces a diffeomorphism from an open neighborhood $\mathcal{O}$ of S in N_S, to an open (tubular) neighborhood $\mathcal{N}$ of S in M. Fix a closed form $\delta^N_S \in \Omega^{n-k}(M)$ with compact support contained in $\mathcal{N}$, and representing the Poincaré dual of (S, i), where $i : S \hookrightarrow M$ denotes the canonical inclusion.

Then $\exp^* \delta^N_S$ is the Poincaré dual of the cycle (S, ζ) in N_S, where $\zeta : S \to N_S$ denotes the zero section. This shows that $\exp^* \delta^N_S$ is a compactly supported form representing the Thom class of the normal bundle N_S.

Using the identification between $\mathcal{O}$ and $\mathcal{N}$ we obtain a natural submersive projection $\pi : \mathcal{N} \to S$ corresponding to the natural projection $N_S \to S$. In more intuitive terms, π associates to each $x \in \mathcal{N}$ the unique point in S which is closest to x. One can prove that the Gysin map

$$i_* : H^\bullet(S) \to H_{cpt}^{\bullet+(n-k)}(M),$$

is given by,

$$H^\bullet(S) \ni \omega \longmapsto \exp^* \delta_S^{\mathcal{N}} \wedge \pi^*\omega \in H_{cpt}^{\bullet+(n-k)}(M). \qquad (7.3.5)$$

$\square$

Exercise 7.3.37. Prove the equality (7.3.5). $\square$

7.3.5 *Gauss-Bonnet revisited*

We now examine a very special type of vector bundle: the tangent bundle of a compact, oriented, smooth manifold M. Note first the following fact.

Exercise 7.3.38. Prove that M is orientable if and only if TM is orientable as a bundle. $\square$

Definition 7.3.39. Let $E \to M$ be a real orientable vector bundle over the compact, oriented, n-dimensional, smooth manifold M. Denote by $\tau_E \in H_{cpt}^n(E)$ the Thom class of E. The *Euler class* of E is defined by

$$e(E) := \zeta_0^* \tau_E \in H^n(M),$$

where $\zeta_0 : M \to E$ denotes the zero section. $e(TM)$ is called the *Euler class* of M, and it is denoted by $e(M)$. $\square$

Note that the sections of TM are precisely the vector fields on M. Moreover, any such section $\sigma : M \to TM$ tautologically defines an n-dimensional cycle in TM, and in fact, any two such cycles are homotopic: try a homotopy, affine along the fibers of TM.

Any two sections $\sigma_0, \sigma_1 : M \to TM$ determine cycles of complementary dimension, and thus the intersection number $\sigma_0 \cdot \sigma_1$ is a well defined integer, independent of the two sections. It is a number reflecting the topological structure of the manifold.

Proposition 7.3.40. *Let $\sigma_0, \sigma_1 : M \to TM$ be two sections of TM. Then*

$$\int_M e(M) = \sigma_0 \cdot \sigma_1.$$

In particular, if $\dim M$ *is odd then*

$$\int_M e(M) = 0.$$

Proof.   The sections σ_0, σ_1 are cobordant, and their Poincaré dual in $H^n_{cpt}(TM)$ is the Thom class τ_M. Hence

$$\sigma_0 \cdot \sigma_1 = \int_{TM} \delta_{\sigma_0} \wedge \delta_{\sigma_1} = \int_{TM} \tau_M \wedge \tau_M$$

$$= \int_{TM} \tau_M \wedge \delta_{\zeta_0} = \int_M \zeta_0^* \tau_M = \int_M e(M).$$

If $\dim M$ is odd then

$$\int_M e(M) = \sigma_0 \cdot \sigma_1 = -\sigma_1 \cdot \sigma_0 = -\int_M e(M). \qquad \square$$

Theorem 7.3.41. *Let M be a compact oriented n-dimensional manifold, and denote by $e(M) \in H^n(M)$ its Euler class. Then the integral of $e(M)$ over M is equal to the Euler characteristic of M,*

$$\int_M e(M) = \chi(M) = \sum_{k=0}^{n} (-1)^k b_k(M).$$

In the proof we will use an equivalent description of $\chi(M)$.

Lemma 7.3.42. *Denote by Δ the diagonal cycle $\Delta \in \mathfrak{H}_n(M \times M)$. Then $\chi(M) = \Delta \cdot \Delta$.*

Proof.   Denote by $\delta_M \in H^n_{cpt}(M \times M)$ the Poincaré dual of Δ. Consider a basis (ω_j) of $H^\bullet(M)$ consisting of homogeneous elements. We denote by (ω^i) the dual basis, i.e.,

$$\langle \omega^i, \omega_j \rangle_\kappa = \delta^i_j, \quad \forall i, j.$$

According to (7.3.2), we have

$$\delta_M = \sum_i (-1)^{|\omega^i|} \omega^i \times \omega_i.$$

Similarly, if we start first with the basis (ω^i), then its dual basis is

$$(-1)^{|\omega^i| \cdot |\omega_i|} \omega_i,$$

so, taking into account that $|\omega_j| + |\omega^j| \cdot |\omega_j| \equiv n|\omega_j| \pmod 2$, we also have

$$\delta_M = \sum_i (-1)^{n \cdot |\omega_j|} \omega_j \times \omega^j.$$

Using Exercise 7.1.50 we deduce

$$\Delta \cdot \Delta = \int_{M \times M} \delta_M \wedge \delta_M = \int_{M \times M} \left(\sum_i (-1)^{|\omega^i|} \omega^i \times \omega_i \right) \wedge \left(\sum_j (-1)^{n|\omega_j|} \omega_j \times \omega^j \right)$$

$$= \int_{M \times M} \left(\sum_{i,j} (-1)^{|\omega^i|+n \cdot |\omega_j|} (-1)^{|\omega_i| \cdot |\omega_j|} \omega^i \wedge \omega_j \right) \times \left(\omega_i \times \omega^j \right)$$

$$= \sum_{i,j} (-1)^{|\omega^i|+n \cdot |\omega_j|} (-1)^{|\omega_i| \cdot |\omega_j|} \langle \omega^i, \omega_j \rangle_\kappa \langle \omega_i, \omega^j \rangle_\kappa.$$

In the last expression we now use the duality equations

$$\langle \omega_i, \omega^j \rangle_\kappa = (-1)^{|\omega_i| \cdot |\omega^j|} \delta_i^j,$$

and the congruences

$$n \equiv 0 \mod 2, \ |\omega^i| + n|\omega_i| + |\omega_i|^2 + |\omega_i| \cdot |\omega^i| \equiv |\omega^i| \mod 2,$$

to conclude that

$$\Delta \cdot \Delta = \sum_{\omega^i} (-1)^{|\omega^i|} = \chi(M). \qquad \square$$

Proof of theorem 7.3.41   The tangent bundle of $M \times M$ restricts to the diagonal Δ as a rank $2n$ vector bundle. If we choose a Riemann metric on $M \times M$ then we get an orthogonal splitting

$$T(M \times M)|_\Delta = N_\Delta \oplus T\Delta.$$

The diagonal map $M \to M \times M$ identifies M with Δ so that $T\Delta \cong TM$. We now have the following remarkable result.

Lemma 7.3.43. $N_\Delta \cong TM$.

Proof   Use the isomorphisms

$$T(M \times M)|_\Delta \cong T\Delta \oplus N_\Delta \cong TM \oplus N_\Delta$$

and

$$T(M \times M)|_\Delta = TM \oplus TM. \qquad \square$$

From this lemma we immediately deduce the equality of Thom classes

$$\tau_{N_\Delta} = \tau_M. \tag{7.3.6}$$

At this point we want to invoke a technical result whose proof is left to the reader as an exercise in Riemann geometry.

Lemma 7.3.44.   *Denote by* $\exp$ *the exponential map of a Riemann metric* g *on* $M \times M$. *Regard* Δ *as a submanifold in* N_Δ *via the embedding given by the zero section. Then there exists an open neighborhood* $\mathcal{U}$ *of* $\Delta \subset N_\Delta \subset T(M \times M)$ *such that*

$$\exp|_\mathcal{U} \colon \mathcal{U} \to M \times M$$

is an embedding. $\qquad \square$

Let $\mathcal{U}$ be a neighborhood of $\Delta \subset N_\Delta$ as in the above lemma, and set $\mathcal{N} := \exp(\mathcal{U})$. Denote by $\delta_\Delta^{\mathcal{U}}$ the Poincaré dual of Δ in $\mathcal{U}$, $\delta_\Delta^{\mathcal{U}} \in H_{cpt}^n(\mathcal{U})$, and by $\delta_\Delta^{\mathcal{N}}$ the Poincaré dual of Δ in $\mathcal{N}$, $\delta_\Delta^{\mathcal{N}} \in H_{cpt}^n(\mathcal{N})$. Then

$$\int_{\mathcal{U}} \delta_\Delta^{\mathcal{U}} \wedge \delta_\Delta^{\mathcal{U}} = \int_{\mathcal{N}} \delta_\Delta^{\mathcal{N}} \wedge \delta_\Delta^{\mathcal{N}} = \int_{M \times M} \boldsymbol{\delta}_M \wedge \boldsymbol{\delta}_M = \chi(M).$$

The cohomology class $\delta_\Delta^{\mathcal{U}}$ is the Thom class of the bundle $N_\Delta \to \Delta$, which in view of (7.3.6) means that $\delta_\Delta^{\mathcal{U}} = \tau_{N_\Delta} = \tau_M$. We get

$$\int_{\mathcal{U}} \delta_\Delta^{\mathcal{U}} \wedge \delta_\Delta^{\mathcal{U}} = \int_\Delta \delta_\Delta^{\mathcal{U}} = \int_\Delta \zeta_0^* \tau_{N_\Delta} = \int_M \zeta_0^* \tau_M = \int_M e(M).$$

Hence

$$\int_M \boldsymbol{e}(M) = \chi(M). \qquad \square$$

If M is a connected sum of g tori then we can rephrase the Gauss-Bonnet theorem as follows.

Corollary 7.3.45. *For any Riemann metric h on a connected sum of g-tori Σ_g we have*

$$\frac{1}{2\pi} \varepsilon(h) = \frac{1}{4\pi} s_h dv_h = \boldsymbol{e}(\Sigma_g) \text{ in } H^*(\Sigma_g). \qquad \square$$

The remarkable feature of the Gauss-Bonnet theorem is that, once we choose a metric, we can *explicitly* describe a representative of the Euler class in terms of the Riemann curvature.

The same is true for any compact, oriented, even-dimensional Riemann manifold. In this generality, the result is known as Gauss-Bonnet-Chern, and we will have more to say about it in the next two chapters.

We now have a new interpretation of the Euler characteristic of a compact oriented manifold M.

Given a smooth vector field X on M, its "graph" in TM,

$$\Gamma_X = \{(x, X(x)) \in T_x M \; ; \; x \in M\},$$

is an n-dimensional submanifold of TM. The Euler characteristic is then the intersection number

$$\chi(M) = \Gamma_X \cdot M,$$

where we regard M as a submanifold in TM via the embedding given by the zero section. In other words, the Euler characteristic counts (with sign) the zeroes of the vector fields on M. For example if $\chi(M) \neq 0$ this means that any vector field on M *must have a zero !* We have thus proved the following result.

Corollary 7.3.46. *If $\chi(M) \neq 0$ then the tangent bundle TM is nontrivial.* $\qquad \square$

The equality $\chi(S^{2n}) = 2$ is particularly relevant in the vector field problem discussed in Subsection 2.1.4. Using the notations of that subsection we can write

$$v(S^{2n}) = 0.$$

We have thus solved "half" the vector field problem.

Exercise 7.3.47. Let X be a vector field over the compact oriented manifold M. A point $x \in M$ is said to be a *non-degenerate* zero of X if $X(0) = 0$ and

$$\det \left(\frac{\partial X_i}{\partial x^j} \right) |_{x=x_0} \neq 0$$

for some local coordinates (x^i) near x_0 such that the orientation of $T^*_{x_0} M$ is given by $dx^1 \wedge \cdots \wedge dx^n$. Prove that the local intersection number of Γ_X with M at x_0 is given by

$$i_{x_0}(\Gamma_X, M) = \text{sign } \det \left(\frac{\partial X_i}{\partial x^j} \right) |_{x=x_0} .$$

This is sometimes called the *local index* of X at x_0, and it is denoted by $i(X, x_0)$. □

From the above exercise we deduce the following celebrated result.

Corollary 7.3.48 (Poincaré-Hopf). *If X is a vector field along a compact, oriented manifold M, with only non-degenerated zeros $x_1, \ldots, x_k$, then*

$$\chi(M) = \sum_j i(X, x_j).$$ □

Exercise 7.3.49. Let X be a vector field on $\mathbb{R}^n$ and having a non-degenerate zero at the origin.
(a) prove that for all $r > 0$ sufficiently small X has no zeros on $S_r = \{|x| = r\}$.
(b) Consider $F_r : S_r \to S^{n-1}$ defined by

$$F_r(x) = \frac{1}{|X(x)|} X(x).$$

Prove that $i(X, 0) = \deg F_r$ for all $r > 0$ sufficiently small.
Hint: Deform X to a linear vector field. □

7.4 Symmetry and topology

The symmetry properties of a manifold have a great impact on its global (topological) structure. We devote this section to a more in depth investigation of the correlation symmetry-topology.

7.4.1 *Symmetric spaces*

Definition 7.4.1. A *homogeneous space* is a smooth manifold M acted transitively by a Lie group G called the *symmetry group*.   □

Recall that a smooth left action of a Lie group G, on a smooth manifold M

$$G \times M \to M \quad (g, m) \mapsto g \cdot m,$$

is called transitive if, for any $m \in M$, the map

$$\Psi_m : G \ni g \mapsto g \cdot m \in M$$

is surjective. For any point x of a homogeneous space M we define the *isotropy group* at x by

$$\mathcal{I}_x = \{g \in G \; ; \; g \cdot x = x\}.$$

Lemma 7.4.2. *Let M be a homogeneous space with symmetry group G and $x, y \in M$. Then*
(a) $\mathcal{I}_x$ is a closed subgroup of G;
(b) $\mathcal{I}_x \cong \mathcal{I}_y$.

Proof. (a) is immediate. To prove (b), choose $g \in G$ such that $y = g \cdot x$. Then note that $\mathcal{I}_y = g\mathcal{I}_x g^{-1}$.   □

Remark 7.4.3. It is worth mentioning some fundamental results in the theory of Lie groups which will shed a new light on the considerations of this section. For proofs we refer to [44, 95].

Fact 1. Any closed subgroup of a Lie group is also a Lie group (see Remark 1.2.29). In particular, the isotropy groups $\mathcal{I}_x$ of a homogeneous space are all Lie groups. They are smooth submanifolds of the symmetry group.

Fact 2. Let G be a Lie group, and H a closed subgroup. Then the space of left cosets,

$$G/H := \{g \cdot H; \; g \in G\},$$

can be given a smooth structure such that the map

$$G \times (G/H) \to G/H \quad (g_1, g_2 H) \mapsto (g_1 g_2) \cdot H$$

is smooth. The manifold G/H becomes a homogeneous space with symmetry group G. All the isotropy groups are subgroups of G conjugate to H.

Fact 3. If M is a homogeneous space with symmetry group G, and $x \in M$, then M is equivariantly diffeomorphic to $G/\mathcal{I}_x$, i.e., there exists a diffeomorphism

$$\phi : M \to G/\mathcal{I}_x,$$

such that $\phi(g \cdot y) = g \cdot \phi(y)$.   □

We will be mainly interested in a very special class of homogeneous spaces.

Definition 7.4.4. A *symmetric space* is a collection of data $(M, h, G, \sigma, \mathfrak{i})$ satisfying the following conditions.
(a) (M, h) is a Riemann manifold.
(b) G is a connected Lie group acting isometrically and transitively on M

$$G \times M \ni (g, m) \mapsto g \cdot m \in M.$$

(c) $\sigma : M \times M \to M$ is a smooth map $(m_1, m_2) \mapsto \sigma_{m_1}(m_2)$ such that the following hold.

(c1) $\forall m \in M$ $\sigma_m : M \to M$ is an isometry, and $\sigma_m(m) = m$.
(c2) $\sigma_m \circ \sigma_m = \mathbb{1}_M$.
(c3) $D\sigma_m|_{T_m M} = -\mathbb{1}_{T_m M}$.
(c4) $\sigma_{gm} = g\sigma_m g^{-1}$.

(d) $\mathfrak{i} : M \times G \to G$, $(m, g) \mapsto \mathfrak{i}_m g$ is a smooth map such that the following hold.

(d1) $\forall m \in M$, $\mathfrak{i}_m : G \to G$ is a homomorphism of G.
(d2) $\mathfrak{i}_m \circ \mathfrak{i}_m = \mathbb{1}_G$.
(d3) $\mathfrak{i}_{gm} = g\mathfrak{i}_m g^{-1}$, $\forall (m, g) \in M \times G$.

(e) $\sigma_m g \sigma_m^{-1}(x) = \sigma_m g \sigma_m(x) = \mathfrak{i}_m(g) \cdot x$, $\forall m, x \in M$, $g \in G$. $\qquad\square$

Remark 7.4.5. This may not be the most elegant definition of a symmetric space, and certainly it is not the minimal one. As a matter of fact, a Riemann manifold (M, g) is a symmetric space if and only if there exists a smooth map $\sigma : M \times M \to M$ satisfying the conditions (c1), (c2) and (c3). We refer to [44, 50] for an extensive presentation of this subject, including a proof of the equivalence of the two descriptions. Our definition has one academic advantage: it lists all the properties we need to establish the topological results of this section. $\qquad\square$

The next exercise offers the reader a feeling of what symmetric spaces are all about. In particular, it describes the geometric significance of the family of involutions σ_m.

Exercise 7.4.6. Let (M, h) be a symmetric space. Denote by ∇ the Levi-Civita connection, and by R the Riemann curvature tensor.
(a) Prove that $\nabla R = 0$.
(b) Fix $m \in M$, and let $\gamma(t)$ be a geodesic of M such that $\gamma(0) = m$. Show that

$$\sigma_m \gamma(t) = \gamma(-t).$$
$\qquad\square$

Example 7.4.7. Perhaps the most popular example of symmetric space is the round sphere $S^n \subset \mathbb{R}^{n+1}$. The symmetry group is $SO(n+1)$, the group of orientation preserving "rotations" of $\mathbb{R}^{n+1}$. For each $m \in S^{n+1}$ we denote by σ_m

the orthogonal reflection through the 1-dimensional space determined by the radius Om. We then set $i_m(T) = \sigma_m T \sigma_m^{-1}$, $\forall T \in SO(n+1)$. We let the reader check that σ and i satisfy all the required axioms.   □

Example 7.4.8. Let G be a connected Lie group and $\mathfrak{m}$ a bi-invariant Riemann metric on G. The direct product $G \times G$ acts on G by

$$(g_1, g_2) \cdot h = g_1 h g_2^{-1}.$$

This action is clearly transitive and since $\mathfrak{m}$ is bi-invariant it is also isometric. Define

$$\sigma : G \times G \to G \quad \sigma_g h = g h^{-1} g^{-1}$$

and

$$i : G \times (G \times G) \to G \times G \quad i_g(g_1, g_2) = (g g_1 g^{-1}, g g_2 g^{-1}).$$

We leave the reader to check that these data do indeed define a symmetric space structure on $(G, \mathfrak{m})$. The symmetry group is $G \times G$.   □

Example 7.4.9. Consider the complex Grassmannian $M = \mathbf{Gr}_k(\mathbb{C}^n)$. Recall that in Example 1.2.20 we described $\mathbf{Gr}_k(\mathbb{C}^n)$ as a submanifold of $\mathrm{End}^+(V)$ – the linear space of selfadjoint $n \times n$ complex matrices, via the map which associates to each complex subspace S, the orthogononal projection $P_S : \mathbb{C}^n \to \mathbb{C}^n$ onto S.

The linear space $\mathrm{End}^+(\mathbb{C}^n)$ has a natural metric $g_0(A, B) = \frac{1}{2}\mathbf{Re}\,\mathrm{tr}\,(AB^*)$ that restricts to a Riemann metric g on M. The unitary group $U(n)$ acts on $\mathrm{End}^+(\mathbb{C}^n)$ by conjugation,

$$U(n) \times \mathrm{End}^+(\mathbb{C}^n) \ni (T, A) \longmapsto T \star A := TAT^*.$$

Note that $U(n) \star M = M$, and g_0 is $U(n)$-invariant. Thus $U(n)$ acts transitively, and isometrically on M.

For each subspace $S \in M$ define

$$R_S := P_S - P_{S^\perp} = 2P_S - 1.$$

The operator R_S is the orthogonal reflection through $S^\perp$. Note that $R_S \in U(n)$, and $R_S^2 = 2$. The map $A \mapsto R_S \star A$ is an involution of $\mathrm{End}^+(\mathbb{C}^n)$. It descends to an involution of M. We thus get an entire family of involutions

$$\sigma : M \times M \to M, \quad (P_{S_1}, P_{S_2}) \mapsto R_{S_1} \star P_{S_2}.$$

Define

$$i : M \times U(n) \to U(n), \quad i_S T = R_S T R_S.$$

We leave the reader to check that the above collection of data defines a symmetric space structure on $\mathbf{Gr}_k(\mathbb{C}^n)$.   □

Exercise 7.4.10. Fill in the details left out in the above example.   □

7.4.2 *Symmetry and cohomology*

Definition 7.4.11. Let M be a homogeneous space with symmetry group G. A differential form $\omega \in \Omega^*(M)$ is said to be (left) *invariant* if $\ell_g^* \omega = \omega \ \forall g \in G$, where we denoted by

$$\ell_g^* : \Omega^\bullet(M) \to \Omega^\bullet(M)$$

the pullback defined by the left action by g: $m \mapsto g \cdot m$. □

Proposition 7.4.12. *Let M be a compact homogeneous space with compact, connected symmetry group G. Then any cohomology class of M can be represented by a (not necessarily unique) invariant form.*

Proof. Denote by $dV_G(g)$ the normalized bi-invariant volume form on G. For any form $\omega \in \Omega^*(M)$ we define its G-average by

$$\overline{\omega} = \int_G \ell_g^* \omega \, dV_G(g).$$

The form $\overline{\omega}$ is an invariant form on M. The proposition is a consequence of the following result.

Lemma 7.4.13. *If ω is a closed form on M then $\overline{\omega}$ is closed and cohomologous to ω.*

Proof. The form $\overline{\omega}$ is obviously closed so we only need to prove it is cohomologous to ω. Consider a bi-invariant Riemann metric $\mathfrak{m}$ on G. Since G is connected, the exponential map

$$\exp : \mathcal{L}_G \to G \quad X \mapsto \exp(tX)$$

is surjective. Choose $r > 0$ sufficiently small so that the map

$$\exp : D_r = \big\{ |X|_{\mathfrak{m}} = r \; ; \; X \in \mathcal{L}_G \big\} \to G$$

is an embedding. Set $B_r = \exp D_r$. We can select finitely many $g_1, \ldots, g_m \in G$ such that

$$G = \bigcup_{j=1}^m B_j, \quad B_j := g_j B_r.$$

Now pick a partition of unity $(\alpha_j) \subset C^\infty(G)$ subordinated to the cover (B_j), i.e.,

$$0 \le \alpha_j \le 1, \quad \operatorname{supp} \alpha_j \subset B_j, \quad \sum_j \alpha_j = 1.$$

Set

$$a_j := \int_G \alpha_j \, dV_G(g).$$

Since the volume of G is normalized to 1, and $\sum_j \alpha_j = 1$, we deduce that $\sum_j a_j = 1$. For any $j = 1, \ldots, m$ define $T_j : \Omega^*(M) \to \Omega^*(M)$ by

$$T_j\omega := \int_G \alpha_j(g)\ell_g^*\omega\, dV_G(g).$$

Note that

$$\overline{\omega} = \sum_j T_j\omega \quad \text{and} \quad dT_j\omega = T_j d\omega.$$

Each T_j is thus a cochain morphism. It induces a morphism in cohomology which we continue to denote by T_j. The proof of the lemma will be completed in several steps.

Step 1. $\ell_g^* = \mathbb{1}$ on $H^*(M)$, for all $g \in G$. Let $X \in \mathcal{L}_G$ such that $g = \exp X$. Define

$$f : I \times M \to M \quad f_t(m) := \exp(tX) \cdot m = \ell_{\exp(tX)}m.$$

The map f is a homotopy connecting $\mathbb{1}_M$ with ℓ_g^*. This concludes Step 1.

Step 2.

$$T_j = a_j \mathbb{1}_{H^*(M)}.$$

For $t \in [0, 1]$, consider $\phi_{j,t} : B_j \to G$ defined as the composition

$$B_j \xrightarrow{g_j^{-1}} B_r \xrightarrow{\exp^{-1}} D_r \xrightarrow{t\cdot-} D_{tr} \xrightarrow{\exp} B_{tr} \xhookrightarrow{g_j} G.$$

Define $T_{j,t} : \Omega^*(M) \to \Omega^*(M)$ by

$$T_{j,t}\omega = \int_G \alpha_j(g)\ell_{\phi_{j,t}(t)}^*\omega\, dV_G(g) = T_j\phi_{j,t}^*\omega. \tag{7.4.1}$$

We claim that $T_{j,0}$ is cochain homotopic to $T_{j,1}$.

To verify this claim set $t := e^s$, $-\infty < s \le 0$ and

$$g_s = \exp(e^s \exp^{-1}(g)) \quad \forall g \in B_r.$$

Then

$$U_s\omega \overset{def}{=} T_{j,e^s}\omega = \int_{B_r} \alpha_j(g_j g)\ell_{g_j g_s}^*\omega = \int_{B_r} \alpha(g_j g)\ell_{g_s}^*\ell_{g_j}^*\omega\, dV_G(g).$$

For each $g \in B_r$ the map $(s, m) \mapsto \Psi_s(m) = g_s(m)$ defines a local flow on M. We denote by X_g its infinitesimal generator. Then

$$\frac{d}{ds}(U_s\omega) = \int_{B_r} \alpha_j(g_j g)L_{X_g}\ell_{g_s}^*\ell_{g_j}^*\omega\, dV_G(g)$$

$$= \int_{B_r} \alpha_j(g_j g)(d_M i_{X_g} + i_{X_g} d_M)(\ell_{g_s}^*\ell_{g_j}^*\omega)\, dV_G(g).$$

Consequently,

$$T_{j,0}\omega - T_{j,1}\omega = U_{-\infty}\omega - U_0\omega$$

$$= -\int_{-\infty}^{0} \left(\int_{B_r} \alpha_j(g_j g)(di_{X_g} + i_{X_g}d)(\ell_{g_s}^* \ell_{g_j}^* \omega) \, dV_G(g) \right) ds.$$

(An argument entirely similar to the one we used in the proof of the Poincaré lemma shows that the above improper integral is pointwise convergent.) From the above formula we immediately read a cochain homotopy $\chi : \Omega^\bullet(M) \to \Omega^{\bullet-1}(M)$ connecting $U_{-\infty}$ to U_0. More precisely

$$\{\chi(\omega)\}|_{x \in M} := -\int_{-\infty}^{0} \left(\int_{B_r} \alpha_j(g_j g) \left\{ i_{X_g} \ell_{g_s}^* \ell_{g_j}^* \omega \right\}|_x \, dV_G(g) \right) ds.$$

Now notice that

$$T_{j,0}\omega = \left(\int_{B_j} \alpha_j(g) dV_G(g) \right) = a_j \ell_{g_j}^* \omega,$$

while $T_{j,1}\omega = T_j \omega$. Taking into account Step 1, we deduce $T_{j,0} = a_j \cdot \mathbb{1}$. Step 2 is completed.

The lemma and hence the proposition follow from the equality

$$\mathbb{1}_{H^*(M)} = \sum_j a_j \mathbb{1}_{H^*(M)} = \sum_j T_j = G - \text{average.} \qquad \square$$

The proposition we have just proved has a greater impact when M is a symmetric space.

Proposition 7.4.14. *Let (M, h) be an, oriented symmetric space with symmetry group G. Then the following are true.*
(a) Every invariant form on M is closed.
(b) If moreover M is compact, then the only invariant form cohomologous to zero is the trivial one.

Proof. (a) Consider an invariant k-form ω. Fix $m_0 \in M$ and set $\hat\omega = \sigma_{m_0}^* \omega$. We claim $\hat\omega$ is invariant. Indeed, $\forall g \in G$

$$\ell_g^* \hat\omega = \ell_g^* \sigma_{m_0}^* \omega = (\sigma_{m_0} g)^* \omega = (g i_g \sigma_{m_0} g)^* \omega = \sigma_{m_0}^* \ell_{i(g)}^* \omega = \sigma_{m_0}^* \omega = \hat\omega.$$

Since $D\sigma_{m_0}|_{T_{m_0}M} = -\mathbb{1}_{T_{m_0}M}$, we deduce that, at $m_0 \in M$, we have $\hat\omega = (-1)^k \omega$.

Both ω and $\hat\omega$ are G-invariant, and we deduce that the above equality holds at any point $m = g \cdot m_0$. Invoking the transitivity of the G-action we conclude that

$$\hat\omega = (-1)^k \omega \quad \text{on } M.$$

In particular, $d\hat\omega = (-1)^k d\omega$ on M.

The $(k+1)$-forms $d\hat\omega = \sigma_{m_0}^* d\omega = \widehat{d\omega}$, and $d\omega$ are both invariant and, arguing as above, we deduce

$$d\hat\omega = \widehat{d\omega} = (-1)^{k+1} d\omega.$$

The last two inequalities imply $d\omega = 0$.

(b) Let ω be an invariant form cohomologous to zero, i.e. $\omega = d\alpha$. Denote by $*$ the Hodge $*$-operator corresponding to the invariant metric h. Since G acts by isometries, the form $\eta = *\omega$ is also invariant, so that $d\eta = 0$. We can now integrate (M is compact), and use Stokes theorem to get

$$\int_M \omega \wedge *\omega = \int_M d\alpha \wedge = \pm \int_M \alpha \wedge d\eta = 0.$$

This forces $\omega \equiv 0$. $\square$

From Proposition 7.4.12 and the above theorem we deduce the following celebrated result of Élie Cartan ([20]).

Corollary 7.4.15 (É. Cartan). *Let (M, h) be a compact, oriented symmetric space with compact, connected symmetry group G. Then the cohomology algebra $H^*(M)$ of M is isomorphic with the graded algebra $\Omega^*_{inv}(M)$ of invariant forms on M.* $\square$

In the coming subsections we will apply this result to the symmetric spaces discussed in the previous subsection: the Lie groups and the complex Grassmannians.

7.4.3   *The cohomology of compact Lie groups*

Consider a compact, connected Lie group G, and denote by $\mathcal{L}_G$ its Lie algebra. According to Proposition 7.4.12, in computing its cohomology, it suffices to restrict our considerations to the subcomplex consisting of left invariant forms. This can be identified with the exterior algebra $\Lambda^\bullet \mathcal{L}_G^*$. We deduce the following result.

Corollary 7.4.16. $H^\bullet(G) \cong H^\bullet(\mathcal{L}_G) \cong \Lambda^\bullet_{inv} \mathcal{L}_G$, *where $\Lambda^\bullet_{inv} \mathcal{L}_G$ denotes the algebra of bi-invariant forms on G, while $H^\bullet(\mathcal{L}_G)$ denotes the Lie algebra cohomology introduced in Example 7.1.10.* $\square$

Using Exercise 7.1.11 we deduce the following consequence.

Corollary 7.4.17. *If G is a compact semisimple Lie group then $H^1(G) = 0$.* $\square$

Proposition 7.4.18. *Let G be a compact semisimple Lie group. Then $H^2(G) = 0$.*

Proof.   A closed bi-invariant 2-form ω on G is uniquely defined by its restriction to $\mathcal{L}_G$, and satisfies the following conditions.

$$d\omega = 0 \iff \omega([X_0, X_1], X_2) - \omega([X_0, X_2], X_1) + \omega([X_1, X_2], X_0) = 0,$$

and (right-invariance)

$$(L_{X_0}\omega)(X_1, X_2) = 0 \ \forall X_0 \in \mathcal{L}_G \iff \omega([X_0, X_1], X_2) - \omega([X_0, X_2], X_1) = 0.$$

Thus

$$\omega([X_0, X_1], X_2) = 0 \ \ \forall X_0, X_1, X_2 \in \mathcal{L}_G.$$

On the other hand, since $H^1(\mathcal{L}_G) = 0$ we deduce (see Exercise 7.1.11) $\mathcal{L}_G = [\mathcal{L}_G, \mathcal{L}_G]$, so that the last equality can be rephrased as

$$\omega(X, Y) = 0 \quad \forall X, Y \in \mathcal{L}_G. \qquad \square$$

Definition 7.4.19. A Lie algebra is called *simple* if it has no nontrivial ideal. A Lie group is called *simple* if its Lie algebra is simple. $\qquad \square$

Exercise 7.4.20. Prove that $SU(n)$ and $SO(m)$ are simple. $\qquad \square$

Proposition 7.4.21. *Let G be a compact, simple Lie group. Then $H^3(G) \cong \mathbb{R}$. Moreover, $H^3(G)$ is generated by the Cartan form*

$$\alpha(X, Y, Z) = \kappa([X, Y], Z),$$

where κ denotes the Killing pairing. $\qquad \square$

The proof of the proposition is contained in the following sequence of exercises.

Exercise 7.4.22. Prove that a simple Lie algebra is necessarily semi-simple. $\qquad \square$

Exercise 7.4.23. Let ω be a closed, bi-invariant 3-form on a Lie group G. Then

$$\omega(X, Y, [Z, T]) = \omega([X, Y], Z, T) \quad \forall X, Y, Z, T \in \mathcal{L}_G. \qquad \square$$

Exercise 7.4.24. Let ω be a closed, bi-invariant 3-form on a compact, semisimple Lie group.
(a) Prove that for any $X \in \mathcal{L}_G$ there exists a unique left-invariant form $\eta_X \in \Omega^1(G)$ such that

$$(i_X \omega)(Y, Z) = \eta_X([Y, Z]).$$

Moreover, the correspondence $X \mapsto \eta_X$ is linear. **Hint:** Use $H^1(G) = H^2(G) = 0$.
(b) Denote by A the linear operator $\mathcal{L}_G \to \mathcal{L}_G$ defined by

$$\kappa(AX, Y) = \eta_X(Y).$$

Prove that A is selfadjoint with respect to the Killing metric.
(c) Prove that the eigenspaces of A are ideals of $\mathcal{L}_G$. Use this to prove Proposition 7.4.21. $\qquad \square$

Exercise 7.4.25. Compute

$$\int_{SU(2)} \alpha \quad \text{and} \quad \int_{SO(3)} \alpha,$$

where α denotes the Cartan form. (These groups are oriented by their Cartan forms.)
Hint: Use the computation in Exercise 4.1.61 and the double cover $SU(2) \to SO(3)$ described in Subsection 6.2.1. Pay very much attention to the various constants. $\qquad \square$

7.4.4 *Invariant forms on Grassmannians and Weyl's integral formula*

We will use the results of Subsection 7.4.2 to compute the Poincaré polynomial of the complex Grassmannian $\mathbf{Gr}_k(\mathbb{C}^n)$. Set $\ell = n - k$.

As we have seen in the previous subsection, the Grassmannian $\mathbf{Gr}_k(\mathbb{C}^n)$ is a symmetric space with symmetry group $U(n)$. It is a complex manifold so that it is orientable (cf. Exercise 3.4.13). Alternatively, the orientability of $\mathbf{Gr}_k(\mathbb{C}^n)$ is a consequence of the following fact.

Exercise 7.4.26. If M is a homogeneous space with connected isotropy groups, then M is orientable. $\qquad\square$

We have to describe the $U(n)$-invariant forms on $\mathbf{Gr}_k(\mathbb{C}^n)$. These forms are completely determined by their values at a particular point in the Grassmannian. We choose this point to correspond to the subspace S_0 determined by the canonical inclusion $\mathbb{C}^k \hookrightarrow \mathbb{C}^n$.

The isotropy of S_0 is the group $H = U(k) \times U(\ell)$. The group H acts linearly on the tangent space $V_0 = T_{S_0} \mathbf{Gr}_k(\mathbb{C}^n)$. If ω is an $U(n)$-invariant form, then its restriction to V_0 is an H-invariant skew-symmetric, multilinear map

$$V_0 \times \cdots \times V_0 \to \mathbb{R}.$$

Conversely, any H-invariant element of $\Lambda^\bullet V_0^*$ extends via the transitive action of $U(n)$ to an invariant form on $\mathbf{Gr}_k(\mathbb{C}^n)$. Denote by $\Lambda^\bullet_{inv}$ the space of H-invariant elements of $\Lambda^\bullet V_0^*$. We have thus established the following result.

Proposition 7.4.27. *There exists an isomorphism of graded $\mathbb{R}$-algebras:*

$$H^\bullet(\mathbf{Gr}_k(\mathbb{C}^n)) \cong \Lambda^\bullet_{inv}. \qquad\square$$

We want to determine the Poincaré polynomial of the complexified $\mathbb{Z}$-graded space, $\Lambda^\bullet_{inv} \otimes \mathbb{C}$

$$P_{k,\ell}(t) = \sum_j t^j \dim_{\mathbb{C}} \Lambda^j_{inv} \otimes \mathbb{C} = P_{\mathbf{Gr}_k(\mathbb{C}^n)}(t).$$

Denote the action of H on V_0 by

$$H \ni h \mapsto T_h \in \mathrm{Aut}(V_0).$$

Using the equality (3.4.10) of Subsection 3.4.4 we deduce

$$P_{k,\ell}(t) = \int_H |\det(\mathbb{1}_{V_0} + tT_h)|^2 dV_H(h), \qquad (7.4.2)$$

where dV_H denotes the normalized bi-invariant volume form on H.

At this point, the above formula may look hopelessly complicated. Fortunately, it can be dramatically simplified using a truly remarkable idea of H. Weyl.

Note first that the function
$$H \ni h \mapsto \varphi(h) = |\det(\mathbb{1}_{V_0} + tT_h)|^2$$
is a class function, i.e., $\varphi(ghg^{-1}) = \varphi(h)$, $\forall g, h \in H$.

Inside H sits the *maximal torus* $\mathbb{T} = \mathbb{T}^k \times \mathbb{T}^\ell$, where
$$\mathbb{T}^k = \left\{ \operatorname{diag}(e^{i\theta_1}, \ldots, e^{i\theta_k}) \in U(k) \right\}, \quad \text{and} \quad \mathbb{T}^\ell = \{\operatorname{diag}(e^{i\phi_1}, \ldots, e^{i\phi_\ell}) \in U(\ell)\}.$$
Each $h \in U(k) \times U(\ell)$ is conjugate to diagonal unitary matrix, i.e., there exists $g \in H$ such that $ghg^{-1} \in \mathbb{T}$.

We can rephrase this fact in terms of the conjugation action of H on itself
$$\boldsymbol{C} : H \times H \to H \quad (g, h) \mapsto \boldsymbol{C}_g(h) = ghg^{-1}.$$
The class functions are constant along the orbits of this conjugation action, and each such orbit intersects the maximal torus $\mathbb{T}$. In other words, a class function is completely determined by its restriction to the maximal torus. Hence, it is reasonable to expect that we ought to be able to describe the integral in (7.4.2) as an integral over $\mathbb{T}$. This is achieved in a very explicit manner by the next result.

Define $\Delta_k : \mathbb{T}^k \to \mathbb{C}$ by
$$\Delta_k(\theta^1, \ldots, \theta^k) = \prod_{1 \le i < j \le n} (e^{i\theta^i} - e^{i\theta^j}), \quad \boldsymbol{i} = \sqrt{-1}.$$
On the unitary group $U(k)$ we fix the bi-invariant metric $\mathfrak{m} = \mathfrak{m}_k$ such that at $T_1 U(k)$ we have
$$\mathfrak{m}(X, Y) := \boldsymbol{Re} \operatorname{tr}(XY^*).$$
If we think of X, Y as $k \times k$ matrices, then
$$\mathfrak{m}(X, Y) = \boldsymbol{Re} \sum_{i,j} x_{ij} \bar{y}_{ij}.$$
We denote by dv_k the volume form induced by this metric, and by V_k the total volume
$$V_k := \int_{U(k)} dv_k.$$

Proposition 7.4.28 (Weyl's integration formula). *Consider a class function φ on the group $G = U(k_1) \times \cdots \times U(k_s)$, denote by dg the volume form*
$$dg = dv_{k_1} \wedge \cdots dv_{k_s},$$
and by V the volume of G, $V = V_{k_1} \cdots V_{k_s}$. Then
$$\frac{1}{V} \int_G \varphi(g) dg = \frac{1}{\prod_{j=1}^s k_j!} \frac{1}{\operatorname{vol}(\mathbb{T})} \int_{\mathbb{T}} \varphi(\boldsymbol{\theta}_1, \ldots, \boldsymbol{\theta}_s) \prod_{j=1}^s |\Delta_{k_j}(\boldsymbol{\theta}_j)|^2 d\boldsymbol{\theta}_1 \wedge \cdots \wedge d\boldsymbol{\theta}_s,$$
$$= \frac{1}{\prod_{j=1}^s (2\pi)^{k_j} k_j!} \int_{\mathbb{T}} \varphi(\boldsymbol{\theta}_1, \ldots, \boldsymbol{\theta}_s) \prod_{j=1}^s |\Delta_{k_j}(\boldsymbol{\theta}_j)|^2 d\boldsymbol{\theta}_1 \wedge \cdots \wedge d\boldsymbol{\theta}_s$$
Above, for every $j = 1, \ldots, s$, we denoted by $\boldsymbol{\theta}_j$ the angular coordinates on $\mathbb{T}^{k_j}$,
$$\boldsymbol{\theta}_j := (\theta_j^1, \ldots, \theta_j^{k_j}),$$
while $d\boldsymbol{\theta}_j$ denotes the bi-invariant volume on $\mathbb{T}^{k_j}$,
$$d\boldsymbol{\theta}_j := d\theta_j^1 \wedge \cdots \wedge d\theta_j^{k_j}.$$

The remainder of this subsection is devoted to the proof of this proposition. The reader may skip this part at the first lecture, and go directly to Subsection 7.4.5 where this formula is used to produce an explicit description of the Poincaré polynomial of a complex Grassmannian.

Proof of Proposition 7.4.28   To keep the main ideas as transparent as possible, we will consider only the case $s = 1$. The general situation is entirely similar. Thus, $G = U(k)$ and $\mathbb{T} = \mathbb{T}^k$. In this case, the volume form dg is the volume form associated to metric $\mathfrak{m}_k$.

We denote by $\mathrm{Ad} : G \to \mathrm{Aut}(\mathcal{L}_G)$ the adjoint representation of G; see Example 3.4.30. In this concrete case, Ad can be given the explicit description

$$\mathrm{Ad}_g(X) = gXg^{-1}, \quad \forall g \in U(k), \quad X \in \underline{u}(k).$$

Denote the angular coordinates $\boldsymbol{\theta}$ on $\mathbb{T}$ by $\boldsymbol{\theta} = (\theta^1, \dots, \theta^k)$. The restriction of the metric $\mathfrak{m}$ to $\mathbb{T}$ is described in these coordinates by

$$\mathfrak{m}|_{\mathbb{T}} = (d\theta^1)^2 + \cdots + (d\theta^k)^2.$$

To any class function $\varphi : G \to \mathbb{C}$ we associate the complex valued differential form

$$\omega_\varphi := \varphi(g)dg.$$

Consider the homogeneous space $G/\mathbb{T}$, and the smooth map

$$q : \mathbb{T} \times G/\mathbb{T} \to G \quad (t, gT) = gtg^{-1}.$$

Note that if $g_1\mathbb{T} = g_2\mathbb{T}$, then $g_1tg_1^{-1} = g_1tg_2^{-1}$, so the map q is well defined.

We have a $\mathfrak{m}$-orthogonal splitting of the Lie algebra $\mathcal{L}_G$

$$\mathcal{L}_G = \mathcal{L}_{\mathbb{T}} \oplus \mathcal{L}_G^\perp.$$

The tangent space to $G/\mathbb{T}$ at $1 \cdot \mathbb{T} \in G/\mathbb{T}$ can be identified with $\mathcal{L}_{\mathbb{T}}^\perp$. For this reason we will write $\mathcal{L}_{G/\mathbb{T}}$ instead of $\mathcal{L}_{\mathbb{T}}^\perp$.

Fix $x \in G/\mathbb{T}$. Any $g \in G$ defines a linear map $L_g : T_xG/\mathbb{T} \to T_{gx}G/\mathbb{T}$. Moreover, if $gx = hx = y$, then L_g and L_h differ by an element in the stabilizer of $x \in G/\mathbb{T}$. This stabilizer is isomorphic to $\mathbb{T}$, and in particular it is connected.

Hence, if $\omega \in \det T_xG/\mathbb{T}$, then $L_g\omega \in \det T_yG/\mathbb{T}$ and $L_h\omega \in \det T_yG/\mathbb{T}$ define the same orientation of $T_yG/\mathbb{T}$. In other words, an orientation in one of the tangent spaces of $G/\mathbb{T}$ "spreads" via the action of G to an orientation of the entire manifold. Thus, we can orient $G/\mathbb{T}$ by fixing an orientation on $\mathcal{L}_{G/\mathbb{T}}$.

We fix an orientation on $\mathcal{L}_G$, and we orient $\mathcal{L}_{\mathbb{T}}$ using the form $d\boldsymbol{\theta} = d\theta^1 \wedge \cdots \wedge d\theta^k$. The orientation on $\mathcal{L}_{G/\mathbb{T}}$ will be determined by the condition

$$\boldsymbol{or}(\mathcal{L}_G) = \boldsymbol{or}(\mathcal{L}_{\mathbb{T}}) \wedge \boldsymbol{or}(\mathcal{L}_{G/\mathbb{T}}).$$

The proof of Weyl's integration formula will be carried out in three steps.

Step 1.

$$\int_G \omega_\varphi = \frac{1}{k!} \int_{\mathbb{T} \times G/\mathbb{T}} q^* \omega_\varphi \quad \forall \omega.$$

Step 2. We prove that there exists a positive constant C_k such that for any class function φ on G we have

$$k! \int_G \omega_\varphi = \int_{\mathbb{T} \times G/\mathbb{T}} q^* \omega_\varphi = C_k \int_\mathbb{T} \varphi(\boldsymbol{\theta}) |\Delta_k(\boldsymbol{\theta})|^2 d\boldsymbol{\theta}.$$

Step 3. We prove that

$$C_k = \frac{V_k}{\mathrm{vol}\,(\mathbb{T})} = \frac{V_k}{(2\pi)^k}.$$

Step 1. We use the equality

$$\int_{\mathbb{T} \times G/\mathbb{T}} q^* \omega_\varphi = \deg q \int_G \omega_\varphi,$$

so it suffices to compute the degree of q.

Denote by $N(\mathbb{T})$ the normalizer of $\mathbb{T}$ in G, i.e.,

$$N(\mathbb{T}) = \{\, g \in G \,;\, g\mathbb{T}g^{-1} \subset \mathbb{T} \,\},$$

and then form the *Weyl group*

$$\mathcal{W} := N(\mathbb{T})/\mathbb{T}.$$

Lemma 7.4.29. *The Weyl group $\mathcal{W}$ is isomorphic to the group $\mathcal{S}_k$ of permutations of k symbols.*

Proof. This is a pompous rephrasing of the classical statement in linear algebra that two unitary matrices are similar if and only if they have the same spectrum, multiplicities included. The adjoint action of $N(\mathbb{T})$ on $\mathbb{T} =$ diagonal unitary matrices simply permutes the entries of a diagonal unitary matrix. This action descends to an action on the quotient $\mathcal{W}$ so that $\mathcal{W} \subset \mathcal{S}_k$.

Conversely, any permutation of the entries of a diagonal matrix can be achieved by a conjugation. Geometrically, this corresponds to a reordering of an orthonormal basis. $\qquad\square$

Lemma 7.4.30. *Let $(\alpha^1, \ldots, \alpha^k) \in \mathbb{R}^k$ such that $1, \frac{\alpha^1}{2\pi}, \ldots, \frac{\alpha^k}{2\pi}$ are linearly independent over $\mathbb{Q}$. Set $\tau := (\exp(i\alpha^1), \ldots, \exp(i\alpha^k)) \in \mathbb{T}^k$. Then the sequence $(\tau^n)_{n \in \mathbb{Z}}$ is dense in $\mathbb{T}^k$. (The element τ is said to be a generator of $\mathbb{T}^k$.)*

For the sake of clarity, we defer the proof of this lemma to the end of this subsection.

Lemma 7.4.31. *Let $\tau \in \mathbb{T}^k \subset G$ be a generator of $\mathbb{T}^k$. Then $q^{-1}(\tau) \subset \mathbb{T} \times G/\mathbb{T}$ consists of $|\mathcal{W}| = k!$ points.*

Proof.

$$q(s, gT) = \tau \iff gsg^{-1} = \tau \iff g\tau g^{-1} = s \in \mathbb{T}.$$

In particular, $g\tau^n g^{-1} = s^n \in \mathbb{T}$, $\forall n \in \mathbb{Z}$. Since (τ^n) is dense in $\mathbb{T}$, we deduce

$$gTg^{-1} \subset \mathbb{T} \Rightarrow g \in N(\mathbb{T}).$$

Hence

$$q^{-1}(\tau) = \big\{ (g^{-1}\tau g, g\mathbb{T}) \in \mathbb{T} \times G/\mathbb{T} \ ; \ g \in N(\mathbb{T}) \big\},$$

and thus $q^{-1}(\tau)$ has the same cardinality as the Weyl group $\mathcal{W}$. $\qquad\square$

The metric $\mathfrak{m}$ on $\mathcal{L}_{G/\mathbb{T}}$ extends to a G-invariant metric on $G/\mathbb{T}$. It defines a left-invariant volume form $d\mu$ on $G/\mathbb{T}$. Observe that the volume form on $\mathbb{T}$ induced by the metric $\mathfrak{m}$ is precisely

$$d\boldsymbol{\theta} = \theta^1 \wedge \cdots \wedge d\theta^k.$$

Lemma 7.4.32. $q^* dg = |\Delta_k(\theta)|^2 d\theta \wedge d\mu$. *In particular, any generator τ of $\mathbb{T}^k \subset G$ is a regular value of q since $\Delta_k(\sigma) \neq 0$, $\forall \sigma \in q^{-1}(\tau)$.*

Proof. Fix a point $x_0 = (\boldsymbol{t}_0, g_0 \mathbb{T}) \in \mathbb{T} \times G/\mathbb{T}$. We can identify $T_{x_0}(\mathbb{T} \times G/\mathbb{T})$ with $\mathcal{L}_{\mathbb{T}} \oplus \mathcal{L}_{G/\mathbb{T}}$ using the isometric action of $\mathbb{T} \times G$ on $\mathbb{T} \times G/\mathbb{T}$.

Fix $X \in \mathcal{L}_{\mathbb{T}}$, $Y \in \mathcal{L}_{G/\mathbb{T}}$. For every real number s consider

$$h_s = q\big(\boldsymbol{t}_0 \exp(sX) \,, \, g_0 \exp(sY)\mathbb{T} \big) = g_0 \exp(sY)\boldsymbol{t}_0 \exp(sX) \exp(-sY)g_0^{-1} \in G.$$

We want to describe

$$\frac{d}{ds}\big|_{s=0} h_0^{-1} h_s \in T_1 G = \mathcal{L}_G.$$

Using the Taylor expansions

$$\exp(sX) = 1 + sX + O(s^2), \ \text{ and } \ \exp(sY) = 1 + sY + O(s^2),$$

we deduce

$$h_0^{-1} h_s = g_0 \boldsymbol{t}_0^{-1}(1 + sY)\boldsymbol{t}_0(1 + sX)(1 - sY)g_0^{-1} + O(s^2)$$

$$= 1 + s\big(g_0 \boldsymbol{t}_0^{-1} Y \boldsymbol{t}_0 g_0^{-1} + g_0 X g_0^{-1} - g_0 Y g_0^{-1}\big) + O(s^2).$$

Hence, the differential

$$D_{x_0} : T_{x_0}(\mathbb{T} \times G/\mathbb{T}) \cong \mathcal{L}_{\mathbb{T}} \oplus \mathcal{L}_{G/\mathbb{T}} \to \mathcal{L}_{\mathbb{T}} \oplus \mathcal{L}_{G/\mathbb{T}} \cong \mathcal{L}_G,$$

can be rewritten as

$$D_{x_0}q(X \oplus Y) = \mathrm{Ad}_{g_0}(\mathrm{Ad}_{\boldsymbol{t}_0^{-1}} - \mathbb{1})Y + \mathrm{Ad}_{g_0}X,$$

or, in block form,

$$D_{x_0}q = \mathrm{Ad}_{g_0} \begin{bmatrix} \mathbb{1}_{\mathcal{L}_{\mathbb{T}}} & 0 \\ 0 & \mathrm{Ad}_{\boldsymbol{t}_0^{-1}} - \mathbb{1}_{\mathcal{L}_{G/\mathbb{T}}} \end{bmatrix}.$$

The linear operator Ad_g is an $\mathfrak{m}$-orthogonal endomorphism of $\mathcal{L}_G$, so that $\det \mathrm{Ad}_g = \pm 1$. On the other hand, since $G = U(k)$ is connected, $\det \mathrm{Ad}_g = \det \mathrm{Ad}_1 = 1$. Hence,

$$\det D_{x_0} q = \det(\mathrm{Ad}_{t^{-1}} - \mathbb{1}_{\mathcal{L}_{G/\mathbb{T}}}).$$

Now observe that $\mathcal{L}_{G/\mathbb{T}} = \mathcal{L}_{\mathbb{T}}^{\perp}$ is equal to

$$\big\{ X \in \underline{u}(k) \; ; \; X_{jj} = 0 \; \forall j = 1, \ldots, k \big\} \subset \underline{u}(k) = \mathcal{L}_G.$$

Given $t = \mathrm{diag}\,(\exp(i\theta^1), \ldots, \exp(i\theta^k)) \in \mathbb{T}^k \subset U(k)$, we can explicitly compute the eigenvalues of $\mathrm{Ad}_{t^{-1}}$ acting on $\mathcal{L}_{G/\mathbb{T}}$. More precisely, they are

$$\big\{ \exp(-i(\theta_i - \theta_j)) \; ; \; 1 \leq i \neq j \leq k \big\}.$$

Consequently

$$\det D_{x_0} q = \det(\mathrm{Ad}_{t^{-1}} - 1) = |\Delta_k(t)|^2. \qquad \square$$

Lemma 7.4.32 shows that q is an orientation preserving map. Using Lemma 7.4.31 and Exercise 7.3.27 we deduce $\deg q = |\mathcal{W}| = k!$. Step 1 is completed.

Step 2 follows immediately from Lemma 7.4.32. More precisely, we deduce

$$\int_G \omega_\varphi = \frac{1}{k!} \underbrace{\left(\int_{G/\mathbb{T}} d\mu \right)}_{=:C_k} \int_{\mathbb{T}^k} \varphi(\theta) \Delta_k(\theta) d\theta.$$

To complete **Step 3**, that is, to find the constant C_k, we apply the above equality in the special case $\varphi = 1$, so that $\omega_\varphi = dv_k$. We deduce

$$V_k = \frac{C_k}{k!} \int_{\mathbb{T}^k} \Delta_k(\theta) d\theta,$$

so that

$$C_k = \frac{V_k k!}{\int_{\mathbb{T}^k} \Delta_k(\theta) d\theta}.$$

Thus, we have to show that

$$\int_{\mathbb{T}^k} \Delta_k(\theta) d\theta = k! \mathrm{vol}\,(\mathbb{T}) = (2\pi)^k k!.$$

To compute the last integral we observe that $\Delta_k(\theta)$ can be expressed as a Vandermonde determinant

$$\Delta_k(\theta) = \begin{vmatrix} 1 & 1 & \cdots & 1 \\ e^{i\theta^1} & e^{i\theta^2} & \cdots & e^{i\theta^k} \\ e^{2i\theta^1} & e^{2i\theta^2} & \cdots & e^{2i\theta^k} \\ \vdots & \vdots & \vdots & \vdots \\ e^{i(k-1)\theta^1} & e^{i(k-1)\theta^2} & \cdots & e^{i(k-1)\theta^k} \end{vmatrix}.$$

This shows that we can write Δ_k as a trigonometric polynomial

$$\Delta_k(\boldsymbol{\theta}) = \sum_{\sigma \in \mathcal{S}_k} \epsilon(\sigma) e_\sigma(\boldsymbol{\theta}),$$

where, for any permutation σ of $\{1, 2, \ldots, k\}$ we denoted by $\epsilon(\sigma)$ its signature, and by $e_\sigma(\boldsymbol{\theta})$ the trigonometric monomial

$$e\sigma(\boldsymbol{\theta}) = e_\sigma(\theta^1, \ldots, \theta^k) = \frac{\prod_{j=1}^k e^{i\sigma(j)\theta^j}}{\prod_{j=1}^k e^{i\theta^j}}.$$

The monomials e_σ are orthogonal with respect to the L^2-metric on $\mathbb{T}_k$, and we deduce

$$\int_{\mathbb{T}_k} |\Delta_k(\boldsymbol{\theta})|^2 d\boldsymbol{\theta} = \sum_{\sigma \in \mathcal{S}_k} \int_{\mathbb{T}}^k |e_\sigma(\boldsymbol{\theta})|^2 d\boldsymbol{\theta} = (2\pi)^k k!.$$

This completes the proof of Weyl's integration formula. $\qquad\qquad\square$

Proof of Lemma 7.4.30 We follow Weyl's original approach ([96, 97]) in a modern presentation.

Let $X = C(\mathbb{T}, \mathbb{C})$ denote the Banach space of continuous complex valued functions on $\mathbb{T}$. We will prove that

$$\lim_{n \to \infty} \frac{1}{n+1} \sum_{j=0}^n f(\tau^j) = \int_T f dt, \quad \forall f \in X. \tag{7.4.3}$$

If $U \subset \mathbb{T}$ is an open subset and f is a continuous, non-negative function supported in U ($f \not\equiv 0$) then, for very large n,

$$\frac{1}{n+1} \sum_{j=0}^n f(\tau^j) \approx \int_T f dt \neq 0.$$

This means that $f(\tau^j) \neq 0$, i.e., $\tau^j \in U$ for some j.

To prove the equality (7.4.3) consider the continuous linear functionals $L_n, L : X \to \mathbb{C}$

$$L_n(f) = \frac{1}{n+1} \sum_{j=0}^n f(\tau^j), \quad \text{and} \quad L(f) = \int_T f dt.$$

We have to prove that

$$\lim_{n \to \infty} L_n(f) = L(f) \quad f \in X. \tag{7.4.4}$$

It suffices to establish (7.4.4) for any $f \in \mathcal{S}$, where $\mathcal{S}$ is a subset of X spanning a dense subspace. We let $\mathcal{S}$ be the subset consisting of the trigonometric monomials

$$e_\zeta(\theta^1, \ldots, \theta^k) = \exp(i\zeta_1\theta^1) \cdots \exp(i\zeta_k\theta^k), \quad \zeta = (\zeta_1, \ldots, \zeta_k) \in \mathbb{Z}^k.$$

The Weierstrass approximation theorem guarantees that this S spans a dense subspace. We compute easily

$$L_n(e_\zeta) = \frac{1}{n+1} \sum_{j=0}^{n} e_{j\zeta}(\alpha) = \frac{1}{n+1} \frac{e_\zeta(\alpha)^{n+1} - 1}{e_\zeta(\alpha) - 1}.$$

Since $1, \frac{1}{2\pi}\alpha_1, \ldots, \frac{1}{2\pi}\alpha_k$ are linearly independent over $\mathbb{Q}$ we deduce that $e_\zeta(\alpha) \neq 1$ for all $\zeta \in \mathbb{Z}^k$. Hence

$$\lim_{n \to \infty} L_n(e_\zeta) = 0 = \int_T e_\zeta dt = L(e_\zeta).$$

Lemma 7.4.30 is proved. $\square$

7.4.5 *The Poincaré polynomial of a complex Grassmannian*

After this rather long detour we can continue our search for the Poincaré polynomial of $\mathbf{Gr}_k(\mathbb{C}^n)$.

Let S_0 denote the canonical subspace $\mathbb{C}^k \hookrightarrow \mathbb{C}^n$. The tangent space of $\mathbf{Gr}_k(\mathbb{C}^n)$ at S_0 can be identified with the linear space $\mathcal{E}$ of complex linear maps $\mathbb{C}^k \to \mathbb{C}^\ell$, $\ell = n - k$. The isotropy group at S_0 is $H = U(k) \times U(\ell)$.

Exercise 7.4.33. Prove that the isotropy group H acts on $\mathcal{E} = \{L : \mathbb{C}^k \to \mathbb{C}^\ell\}$ by

$$(T, S) \cdot L = SLT^* \quad \forall L \in \mathcal{E}, \; T \in U(k), \; S \in U(\ell). \qquad \square$$

Consider the maximal torus $T^k \times T^\ell \subset H$ formed by the diagonal unitary matrices. We will denote the elements of T^k by

$$\underline{\varepsilon} := (\varepsilon_1, \ldots, \varepsilon_k), \quad \varepsilon_\alpha = e^{2\pi i \tau^\alpha},$$

and the elements of T^ℓ by

$$\underline{e} := (e_1, \ldots, e_\ell), \quad e_j = \exp(2\pi i \theta^j).$$

The *normalized* measure on T^k is then

$$d\boldsymbol{\tau} = d\tau^1 \wedge \cdots \wedge d\tau^k,$$

and the *normalized* measure on T^ℓ is

$$d\boldsymbol{\theta} = d\theta^1 \wedge \cdots \wedge d\theta^\ell.$$

The element $(\underline{\varepsilon}, \underline{e}) \in T^k \times T^\ell$ viewed as a linear operator on $\mathcal{E}$ has eigenvalues

$$\{\bar{\varepsilon}_\alpha e_j \; ; \; 1 \leq \alpha \leq k \; 1 \leq j \leq \ell\}.$$

Using the Weyl integration formula we deduce that the Poincaré polynomial of $\mathbf{Gr}_k(\mathbb{C}^n)$ is

$$P_{k,\ell}(t) = \frac{1}{k!\ell!} \int_{T^k \times T^\ell} \prod_{\alpha,j} |1 + t\bar{\varepsilon}_\alpha e_j|^2 |\Delta_k(\underline{\varepsilon}(\boldsymbol{\tau}))|^2 |\Delta_\ell(\underline{e}(\boldsymbol{\theta}))|^2 d\boldsymbol{\tau} \wedge d\boldsymbol{\theta}$$

$$= \frac{1}{k!\ell!} \int_{T^k \times T^\ell} \prod_{\alpha,j} |\varepsilon_\alpha + te_j|^2 |\Delta_k(\underline{\varepsilon}(\boldsymbol{\tau}))|^2 |\Delta_\ell(\underline{e}(\boldsymbol{\theta}))|^2 d\boldsymbol{\tau} \wedge d\boldsymbol{\theta}.$$

We definitely need to analyze the integrand in the above formula. Set

$$I_{k,\ell}(t) := \prod_{\alpha,j} (\varepsilon_\alpha + te_j)\Delta_k(\underline{\varepsilon})\Delta_\ell(\underline{e}),$$

so that

$$P_{k,\ell}(t) = \frac{1}{k!\ell!} \int_{T^k \times T^\ell} I_{k,\ell}(t)\overline{I_{k,\ell}(t)} d\boldsymbol{\tau} \wedge d\boldsymbol{\theta}. \tag{7.4.5}$$

We will study in great detail the formal expression

$$J_{k,\ell}(t; x; y) = \prod_{\alpha,j} (x_\alpha + ty_j).$$

The Weyl group $\mathcal{W} = \mathcal{S}_k \times \mathcal{S}_\ell$ acts on the variables $(x; y)$ by separately permuting the x-components and the y-components. If $(\sigma, \varphi) \in \mathcal{W}$ then

$$J_{k,\ell}(t; \sigma(x); \varphi(y)) = J_{k,\ell}(t; x; y).$$

Thus we can write $J_{k,\ell}$ as a sum

$$J_{k,\ell}(t) = \sum_{d \geq 0} t^d Q_d(x) R_d(y)$$

where $Q_d(x)$ and $R_d(y)$ are symmetric polynomials in x and respectively y.

To understand the nature of these polynomials we need to introduce a very useful class of symmetric polynomials, namely the *Schur polynomials*. This will require a short trip in the beautiful subject of symmetric polynomials. An extensive presentation of this topic is contained in the monograph [66].

A *partition* is a compactly supported, decreasing function

$$\lambda : \{1, 2, \ldots\} \to \{0, 1, 2, \ldots\}.$$

We will describe a partition by an ordered finite collection $(\lambda_1, \lambda_2, \ldots, \lambda_n)$, where $\lambda_1 \geq \cdots \geq \lambda_n \geq \lambda_{n+1} = 0$. The *length* of a partition λ is the number

$$L(\lambda) := \max\{n \; ; \; \lambda_n \neq 0\}.$$

The *weight* of a partition λ is the number

$$|\lambda| := \sum_{n \geq 1} \lambda_n.$$

Traditionally, one can visualize a partition using *Young diagrams*. A Young diagram is an array of boxes arranged in left justified rows (see Figure 7.4). Given a partition $(\lambda_1 \geq \cdots \geq \lambda_n)$, its Young diagram will have λ_1 boxes on the first row, λ_2 boxes on the second row etc.

Any partition λ has a *conjugate* $\hat{\lambda}$ defined by

$$\hat{\lambda}_n := \#\{j \geq 0 \; ; \; \lambda_j \geq n\}.$$

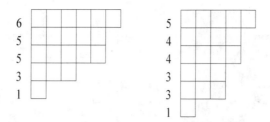

Fig. 7.4   *The conjugate of (6,5,5,3,1) is (5,4,4,3,3,1)*

The Young diagram of $\hat{\lambda}$ is the transpose of the Young diagram of λ (see Figure 7.4).

A *strict partition* is a partition which is *strictly decreasing* on its support. Denote by $\mathcal{P}_n$ the set of partitions of length $\leq n$, and by $\mathcal{P}_n^*$ the set of *strict* partitions λ of length $n - 1 \leq L(\lambda) \leq n$. Clearly $\mathcal{P}_n^* \subset \mathcal{P}_n$. Denote by $\delta = \delta_n \in \mathcal{P}_n^*$ the partition $(n-1, n-2, \ldots, 1, 0, \ldots)$.

Remark 7.4.34. The correspondence $\mathcal{P}_n \ni \lambda \mapsto \lambda + \delta_n \in \mathcal{P}_n^*$, is a bijection. $\qquad\square$

To any $\lambda \in \mathcal{P}_n^*$ we can associate a skew-symmetric polynomial

$$a_\lambda(x_1, \ldots, x_n) := \det(x_j^{\lambda_i}) = \sum_{\sigma \in \mathcal{S}_n} \epsilon(\sigma) x_{\sigma(i)}^{\lambda_i}.$$

Note that a_{δ_n} is the Vandermonde determinant

$$a_\delta(x_1, \ldots, x_n) = \det(x_i^{n-1-i}) = \prod_{i<j}(x_i - x_j) = \Delta_n(x).$$

For each $\lambda \in \mathcal{P}_n$ we have $\lambda + \delta \in \mathcal{P}_n^*$, so that $a_{\lambda+\delta}$ is well defined and nontrivial.

Note that $a_{\lambda+\delta}$ vanishes when $x_i = x_j$, so that the polynomial $a_{\lambda+\delta}(x)$ is divisible by each of the differences $(x_i - x_j)$ and consequently, it is divisible by a_δ. Hence

$$S_\lambda(x) := \frac{a_{\lambda+\delta}(x)}{a_\delta(x)}$$

is a well defined polynomial. It is a symmetric polynomial since each of the quantities $a_{\lambda+\delta}$ and a_δ is skew-symmetric in its arguments. The polynomial $S_\lambda(x)$ is called the *Schur polynomial* corresponding to the partition λ. Note that each Schur polynomial S_λ is homogeneous of degree $|\lambda|$. We have the following remarkable result.

Lemma 7.4.35.

$$J_{k,\ell}(t) = \sum_{\lambda \in \mathcal{P}_{k,\ell}} t^{|\bar{\lambda}|} S_{\hat{\lambda}}(x) S_{\bar{\lambda}}(y),$$

where

$$\mathcal{P}_{k,\ell} := \{\, \lambda \,;\, \lambda_1 \leq k \ \ L(\lambda) \leq \ell \,\}.$$

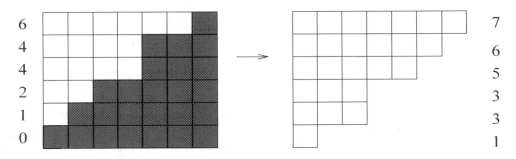

Fig. 7.5   *The complementary of* $(6, 4, 4, 2, 1, 0) \in \mathcal{P}_{7,6}$ *is* $(7, 6, 5, 3, 3, 1) \in \mathcal{P}_{6,7}$.

For each $\lambda \in \mathcal{P}_{k,\ell}$ *we denoted by* $\bar{\lambda}$ *the complementary partition*

$$\bar{\lambda} = (k - \lambda_\ell, k - \lambda_{\ell-1}, \ldots, k - \lambda_1).$$

Geometrically, the partitions in $\mathcal{P}_{k,\ell}$ are precisely those partitions whose Young diagrams fit inside a $\ell \times k$ rectangle. If λ is such a partition then the Young diagram of the complementary of $\bar{\lambda}$ is (up to a $180°$ rotation) the complementary of the diagram of λ in the $\ell \times k$ rectangle (see Figure 7.5).

For a proof of Lemma 7.4.35 we refer to [66], Section I.4, Example 5. The true essence of the Schur polynomials is however representation theoretic, and a reader with a little more representation theoretic background may want to consult the classical reference [65], Chapter VI, Section 6.4, Theorem V, for a very exciting presentation of the Schur polynomials, and the various identities they satisfy, including the one in Lemma 7.4.35.

Using (7.4.5), Lemma 7.4.35, and the definition of the Schur polynomials, we can describe the Poincaré polynomial of $\mathbf{Gr}_k(\mathbb{C}^n)$ as

$$P_{k,\ell}(t) = \frac{1}{k!\ell!} \int_{T^k \times T^\ell} \left| \sum_{\lambda \in \mathcal{P}_{k,\ell}} t^{|\bar{\lambda}|} a_{\hat{\lambda} + \delta_k}(\underline{\varepsilon}) a_{\overline{\lambda} + \delta_\ell}(\underline{e}) \right|^2 d\boldsymbol{\tau} \wedge d\boldsymbol{\theta}. \qquad (7.4.6)$$

The integrand in (7.4.6) is a linear combination of trigonometric monomials $\varepsilon_1^{r_1} \cdots \varepsilon_k^{r_k} \cdot e_1^{s_1} \cdots e_\ell^{s_\ell}$, where the r-s and s-s are nonnegative integers.

Note that if $\lambda, \mu \in \mathcal{P}_{k,\ell}$ are distinct partitions, then the terms $a_{\hat{\lambda}+\delta}(\underline{\varepsilon})$ and $a_{\hat{\mu}+\delta}(\underline{\varepsilon})$ have no monomials in common. Hence

$$\int_{T^k} a_{\hat{\lambda}+\delta}(\underline{\varepsilon}) \overline{a_{\hat{\mu}+\delta}(\underline{\varepsilon})} \, d\boldsymbol{\tau} = 0.$$

Similarly,

$$\int_{T^\ell} a_{\overline{\lambda}+\delta}(\underline{e}) \overline{a_{\overline{\mu}+\delta}(\underline{e})} d\boldsymbol{\theta} = 0, \ \text{ if } \lambda \neq \mu.$$

On the other hand, a simple computation shows that

$$\int_{T^k} |a_{\hat{\lambda}+\delta}(\underline{\varepsilon})|^2 d\boldsymbol{\tau} = k!,$$

and

$$\int_{T^\ell} |a_{\overline{\lambda}+\delta}(\underline{e})|^2 d\boldsymbol{\theta} = \ell!.$$

In other words, the terms

$$\left(\frac{1}{k!\ell!}\right)^{1/2} a_{\hat{\lambda}+\delta}(\underline{\varepsilon}) a_{\overline{\lambda}+\delta}(\underline{e})$$

form an *orthonormal* system in the space of trigonometric (Fourier) polynomials endowed with the L^2 inner product. We deduce immediately from (7.4.6) that

$$P_{k,\ell}(t) = \sum_{\lambda \in \mathcal{P}_{k,\ell}} t^{2|\overline{\lambda}|}. \tag{7.4.7}$$

The map

$$\mathcal{P}_{k,\ell} \ni \lambda \mapsto \overline{\lambda} \in \mathcal{P}_{\ell,k},$$

is a bijection so that

$$P_{k,\ell}(t) = \sum_{\lambda \in \mathcal{P}_{\ell,k}} t^{2|\lambda|} = P_{\ell,k}(t). \tag{7.4.8}$$

Computing the Betti numbers, i.e., the number of partitions in $\mathcal{P}_{k,\ell}$ with a given weight is a very complicated combinatorial problem, and currently there are no exact general formulæ. We will achieve the next best thing, and rewrite the Poincaré polynomial as a "fake" rational function.

Denote by $b_{k,\ell}(w)$ the number of partitions $\lambda \in \mathcal{P}_{\ell,k}$ with weight $|\lambda| = w$. Hence

$$P_{k,\ell}(t) = \sum_{w=1}^{k\ell} b_{k,\ell}(w) t^{2w}.$$

Alternatively, $b_{k,\ell}(w)$ is the number of Young diagrams of weight w which fit inside a $k \times \ell$ rectangle.

Lemma 7.4.36.

$$b_{k+1,\ell+1}(w) = b_{k,\ell+1}(w) + b_{k+1,\ell}(w - \ell - 1).$$

Proof. Look at the $(k+1) \times \ell$ rectangle R^{k+1} inside the $(k+1) \times (l+1)$-rectangle $R^{k+1}_{\ell+1}$ (see Figure 7.6). Then

$$b_{k+1,\ell+1}(w) = \# \left\{ \text{diagrams of weight } w \text{ which fit inside } R^{k+1}_{\ell+1} \right\}$$

$$= \#\left\{ \text{diagrams which fit inside } R^{k+1} \right\}$$

$$+\#\left\{ \text{diagrams which do not fit inside } R^{k+1} \right\}.$$

On the other hand,

$$b_{k+1,\ell} = \#\left\{ \text{diagrams which fit inside } R^{k+1} \right\}.$$

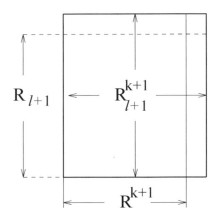

Fig. 7.6   *The rectangles $R_{\ell+1}$ and R^{k+1} are "framed" inside $R_{\ell+1}^{k+1}$*

If a diagram does not fit inside R^{k+1} this means that its first line consists of $\ell + 1$ boxes. When we drop this line, we get a diagram of weight $w - \ell - 1$ which fits inside the $k \times (\ell + 1)$ rectangle $R_{\ell+1}$ of Figure 7.6. Thus, the second contribution to $b_{k+1,\ell+1}(w)$ is $b_{k,\ell+1}(w - \ell - 1)$. $\qquad\square$

The result in the above lemma can be reformulated as

$$P_{k+1,\ell+1}(t) = P_{k+1,\ell}(t) + t^{2(\ell+1)} P_{k,\ell+1}(t).$$

Because the roles of k and ℓ are symmetric (cf. (7.4.8)) we also have

$$P_{k+1,\ell+1}(t) = P_{k,\ell+1}(t) + t^{2(k+1)} P_{k+1,\ell}(t).$$

These two equalities together yield

$$P_{k,\ell+1}(1 - t^{2(\ell+1)}) = P_{k+1,\ell}(1 - t^{2(k+1)}).$$

Let $m = k + \ell + 1$ and set $Q_{d,m}(t) = P_{d,m-d}(t) = P_{G_d(m,\mathbb{C})}(t)$. The last equality can be rephrased as

$$Q_{k+1,m}(t) = Q_{k,m} \cdot \frac{1 - t^{m-k}}{1 - t^{2(k+1)}},$$

so that

$$Q_{k+1,m}(t) = \frac{1 - t^{2(m-k)}}{1 - t^{2(k+1)}} \cdot \frac{1 - t^{2(m-k+1)}}{1 - t^{2k}} \cdots \frac{1 - t^{2(n-1)}}{1 - t^4}.$$

Now we can check easily that $b_{1,m-1}(w) = 1$, i.e.,

$$Q_{1,m}(t) = 1 + t^2 + t^4 + \cdots + t^{2(m-1)} = \frac{1 - t^{2m}}{1 - t^2}.$$

Hence

$$P_{G_k(m,\mathbb{C})}(t) = Q_{k,m}(t) = \frac{(1 - t^{2(m-k+1)}) \cdots (1 - t^{2m})}{(1 - t^2) \cdots (1 - t^{2k})}$$

$$= \frac{(1-t^2)\cdots(1-t^{2m})}{(1-t^2)\cdots(1-t^{2k})(1-t^2)\cdots(1-t^{2(m-k)})}.$$

Remark 7.4.37. (a) The invariant theoretic approach in computing the cohomology of $\mathbf{Gr}_k(\mathbb{C}^n)$ was used successfully for the first time by C. Ehresmann [31]. His method was then extended to arbitrary compact, oriented symmetric spaces by H. Iwamoto [49]. However, we followed a different avenue which did not require Cartan's maximal weight theory.

(b) We borrowed the idea of using the Weyl's integration formula from Weyl's classical monograph [100]. In turn, Weyl attributes this line of attack to R. Brauer. Our strategy is however quite different from Weyl's. Weyl uses an equality similar to (7.4.5) to produce an upper estimate for the Betti numbers (of $U(n)$ in his case) and then produces by hand sufficiently many invariant forms. The upper estimate is then used to establish that these are the only ones. We refer also to [92] for an explicit description of the invariant forms on Grassmannians. $\qquad\square$

Exercise 7.4.38. Show that the cohomology *algebra* of $\mathbb{CP}^n$ is isomorphic to the truncated ring of polynomials

$$\mathbb{R}[x]/(x^{n+1}),$$

where x is a formal variable of degree 2 while (x^{n+1}) denotes the ideal generated by x^{n+1}.

Hint: Describe $\Lambda^\bullet_{inv}\mathbb{CP}^n$ explicitly. $\qquad\square$

7.5 Čech cohomology

In this last section we return to the problem formulated in the beginning of this chapter: what is the relationship between the Čech and the DeRham approach. We will see that these are essentially two equivalent facets of the same phenomenon. Understanding this equivalence requires the introduction of a new and very versatile concept namely, the concept of sheaf. This is done in the first part of the section. The second part is a fast paced introduction to Čech cohomology. A concise yet very clear presentation of these topics can be found in [46]. For a very detailed presentation of the classical aspects of subject we refer to [37]. For a modern presentation, from the point of view of derived categories we refer to the monographs [48, 53].

7.5.1 *Sheaves and presheaves*

Consider a topological space X. The topology $\mathfrak{T}_X$ on X, i.e., the collection of open subsets, can be organized as a category. The morphisms are the inclusions $U \hookrightarrow V$. A *presheaf* of Abelian groups on X is a *contravariant* functor $\mathcal{S} : \mathfrak{T}_X \to \mathbf{Ab}$.

In other words, $\mathcal{S}$ associates to each open set an Abelian group $\mathcal{S}(U)$, and to each inclusion $U \hookrightarrow V$, a group morphism $r_V^U : \mathcal{S}(V) \to \mathcal{S}(U)$ such that, if $U \hookrightarrow V \hookrightarrow W$, then $r_W^U = r_V^U \circ r_W^V$. If $s \in \mathcal{S}(V)$ then, for any $U \hookrightarrow V$, we set

$$s|_U := r_V^U(s) \in \mathcal{S}(U).$$

If $f \in \mathcal{S}(U)$ then we define dom $f := U$.

The presheaves of rings, modules, vector spaces are defined in an obvious fashion.

Example 7.5.1. Let X be a topological space. For each open set $U \subset X$ we denote by $C(U)$ the space of continuous functions $U \to \mathbb{R}$. The assignment $U \mapsto C(U)$ defines a presheaf of $\mathbb{R}$-algebras on X. The maps r_V^U are determined by the restrictions $|_U : C(V) \to C(U)$.

If X is a smooth manifold we get another presheaf $U \mapsto C^\infty(U)$. More generally, the differential forms of degree k can be organized in a presheaf $\Omega^k(\bullet)$. If E is a smooth vector bundle, then the E-valued differential forms of degree k can be organized as a presheaf of vector spaces

$$U \mapsto \Omega_E^k(U) = \Omega^k(E|_U).$$

If G is an Abelian group equipped with the discrete topology, then the G-valued continuous functions $C(U, G)$ determine a presheaf called the *constant G-presheaf* which is denoted by $\underline{G}_X$.   □

Definition 7.5.2. A presheaf $\mathcal{S}$ on a topological space X is said to be a *sheaf* if the following hold.

(a) If (U_α) is an open cover of the open set U, and $f, g \in \mathcal{S}(U)$ satisfy $f|_{U_\alpha} = g|_{U_\alpha}$, $\forall \alpha$, then $f = g$.

(b) If (U_α) is an open cover of the open set U, and $f_\alpha \in \mathcal{S}(U_\alpha)$ satisfy

$$f_\alpha|_{U_\alpha \cap U_\beta} = f_\beta|_{U_\alpha \cap U_\beta} \quad \forall U_\alpha \cap U_\beta \neq \emptyset$$

then there exists $f \in \mathcal{S}(U)$ such that, $f|_{U_\alpha} = f_\alpha$, $\forall \alpha$.   □

Example 7.5.3. All the presheaves discussed in Example 7.5.1 are sheaves.   □

Example 7.5.4. Consider the presheaf $\mathcal{S}$ over $\mathbb{R}$ defined by

$$\mathcal{S}(U) := \text{continuous, bounded functions } f : U \to \mathbb{R}.$$

We let the reader verify this is not a sheaf since the condition (b) is violated. The reason behind this violation is that in the definition of this presheaf we included a *global* condition namely the boundedness assumption.   □

Definition 7.5.5. Let X be a topological space, and R a commutative ring with 1. We equip R with the discrete topology

(a) A *space of germs* over X (*"espace étalé"* in the French literature) is a topological space $\mathcal{E}$, together with a continuous map $\pi : \mathcal{E} \to X$ satisfying the following conditions.

(a1) The map π is a local homeomorphism that is, each point $e \in \mathcal{E}$ has a neighborhood U such that $\pi \mid_U$ is a homeomorphism onto the open subset $\pi(U) \subset X$. For every $x \in X$, the set $\mathcal{E}_x := \pi^{-1}(x)$, is called the *stalk* at x.

(a2) There exist continuous maps

$$\cdot : R \times \mathcal{E} \to \mathcal{E},$$

and

$$+ : \big\{(u,v) \in \mathcal{E} \times \mathcal{E}; \ \pi(u) = \pi(v)\big\} \to \mathcal{E},$$

such that

$$\forall r \in R, \ \ \forall x \in X, \ \ \forall u, v \in \mathcal{E}_x \ \text{ we have } \ r \cdot u \in \mathcal{E}_x, \ \ u + v \in \mathcal{E}_x,$$

and with respect to the operations $+$, $\cdot$, the stalk $\mathcal{E}_x$ is an R-module, $\forall x \in X$.

(b) A *section* of a space of germs $\pi : \mathcal{E} \to X$ over a subset $Y \subset X$ is a continuous function $s : Y \to \mathcal{E}$ such that $s(y) \in \mathcal{E}_y$, $\forall y \in Y$. The spaces of sections defined over Y will be denoted by $\mathcal{E}(Y)$. $\qquad \square$

Example 7.5.6. (The space of germs associated to a presheaf). Let $\mathcal{S}$ be a presheaf of Abelian groups over a topological space X. For each $x \in X$ define an equivalence relation $\sim_x$ on

$$\bigsqcup_{U \ni x} \mathcal{S}(U), \ \ U \text{ runs through the open neighborhoods of } X,$$

by

$$f \sim_x g \iff \exists \text{ open } U \ni x \text{ such that } f\mid_U = g\mid_U .$$

The equivalence class of $f \in \bigsqcup_{U \ni x} \mathcal{S}(U)$ is denoted by $[f]_x$, and it is called the *germ* of f at x. Set

$$\mathcal{S}_x := \{[f]_x \ ; \ \operatorname{dom} f \ni x\}, \ \text{ and } \ \widehat{\mathcal{S}} := \bigsqcup_{x \in X} \mathcal{S}_x.$$

There exists a natural projection $\pi : \widehat{\mathcal{S}} \to X$ which maps $[f]_x$ to x. The "fibers" of this map are $\pi^{-1}(x) = \mathcal{S}_x$ – the germs at $x \in X$. Any $f \in \mathcal{S}(U)$ defines a subset

$$U^f = \big\{ [f]_u \ ; \ u \in U \big\} \subset \widehat{\mathcal{S}}.$$

We can define a topology in $\widehat{\mathcal{S}}$ by indicating a basis of neighborhoods. A basis of *open* neighborhoods of $[f]_x \in \widehat{\mathcal{S}}$ is given by the collection

$$\big\{ U^g \ ; \ U \ni x, \ g \in \mathcal{S}(U) \ [g]_x = [f]_x \big\}.$$

We let the reader check that this collection of sets satisfies the axioms of a basis of neighborhoods as discussed e.g. in [57].

With this topology, each $f \in \mathcal{S}(U)$ defines a *continuous* section of π over U

$$[f] : \ U \ni u \mapsto [f]_u \in \mathcal{S}_u.$$

Note that each fiber S_x has a well defined structure of Abelian group

$$[f]_x + [g]_x = [(f + g)|_U]_x \quad U \ni x \text{ is open and } U \subset \text{dom} f \cap \text{dom} g.$$

(Check that this addition is independent of the various choices.) Since $\pi : f(U) \to U$ is a homeomorphism, it follows that $\pi : \widehat{S} \to X$ is a space of germs. It is called the *space of germs associated to the presheaf* S.   □

If the space of germs associated to a sheaf S is a covering space, we say that S is a sheaf of locally constant functions (valued in some discrete Abelian group). When the covering is trivial, i.e., it is isomorphic to a product $X \times \{ \text{ discrete set } \}$, then the sheaf is really the constant sheaf associated to a discrete Abelian group.

Example 7.5.7. (The sheaf associated to a space of germs).   Consider a space of germs $\mathcal{E} \xrightarrow{\pi} X$ over the topological space X. For each open subset $U \subset X$, we denote by $\overline{\mathcal{E}}(U)$ the space of continuous sections $U \to \mathcal{E}$. The correspondence $U \mapsto \overline{\mathcal{E}}(U)$ clearly a sheaf. $\overline{\mathcal{E}}$ is called the *sheaf associated to the space of germs.* □

Proposition 7.5.8.   *(a) Let* $\mathcal{E} \xrightarrow{\pi} X$ *be a space of germs. Then* $\widehat{\overline{\mathcal{E}}} = \mathcal{E}$.

(b) A presheaf S *over the topological space* X *is a sheaf if and only if* $\overline{\widehat{S}} = S$.   □

Exercise 7.5.9.   Prove the above proposition.   □

Definition 7.5.10.   If S is a presheaf over the topological space X, then the sheaf $\overline{\widehat{S}}$ is called the *sheaf associated to* S, or the *sheafification of* S.   □

Definition 7.5.11.   (a) Let A *morphism* between the (pre)sheaves of Abelian groups (modules etc.) S and $\tilde{S}$ over the topological space X is a collection of morphisms of Abelian groups (modules etc.) $h_U : S(U) \to \tilde{S}(U)$, one for each open set $U \subset X$, such that, when $V \subset U$, we have $h_V \circ r_U^V = \tilde{r}_U^V \circ h_U$. Above, r_U^V denotes the restriction morphisms of S, while the $\tilde{r}_U^V$ denotes the restriction morphisms of $\tilde{S}$. A morphism h is said to be *injective* if each h_U is injective.

(b) Let S be a presheaf over the topological space X. A *sub-presheaf* of S is a pair $(\mathcal{T}, \imath)$, where $\mathcal{T}$ is a presheaf over X, and $\imath : \mathcal{T} \to S$ is an injective morphism. The morphism $\imath$ is called the *canonical inclusion* of the sub-presheaf.   □

Let $h : S \to \mathcal{T}$ be a morphism of presheaves. The correspondence

$$U \mapsto \ker h_U \subset S(U),$$

defines a presheaf called the *kernel* of the morphism h. It is a sub-presheaf of S.

Proposition 7.5.12.   *Let* $h : S \to \mathcal{T}$ *be a morphism of presheaves. If both* S *and* $\mathcal{T}$ *are sheaves, then so is the kernel of* h.   □

The proof of this proposition is left to the reader as an exercise.

Definition 7.5.13. (a) Let $\mathcal{E}_i \overset{\pi_i}{\to} X$ $(i = 0, 1)$ be two spaces of germs over the same topological space X. A morphism of spaces of germs is a continuous map $h : \mathcal{E}_0 \to \mathcal{E}_1$ satisfying the following conditions.

(a1) $\pi_1 \circ h = \pi_0$, i.e., $h(\pi_0^{-1}(x)) \subset \pi_1^{-1}(x)$, $\forall x \in X$.
(a2) For any $x \in X$, the induced map $h_x : \pi_0^{-1}(x) \to \pi_1^{-1}(x)$ is a morphism of Abelian groups (modules etc.).

The morphism h is called injective if each h_x is injective.

(b) Let $\mathcal{E} \overset{\pi}{\to} X$ be a space of germs. A *subspace of germs* is a pair $(\mathcal{F}, \jmath)$, where $\mathcal{F}$ is a space of germs over X and $\jmath : \mathcal{F} \to \mathcal{E}$ is an injective morphism. $\square$

Proposition 7.5.14. *Let $h : \mathcal{E}_0 \to \mathcal{E}_1$ be a morphism between two spaces of germs over X. Then $h(\mathcal{E}_0) \overset{\pi_1}{\to} X$ is a space of germs over X called the image of h. It is denoted by $\operatorname{Im} h$, and it is a subspace of $\mathcal{E}_1$.* $\square$

Exercise 7.5.15. Prove the above proposition. $\square$

Lemma 7.5.16. *Consider two sheaves $\mathcal{S}$ and $\mathcal{T}$, and let $h : \mathcal{S} \to \mathcal{T}$ be a morphism. Then h induces a morphism between the associated spaces of germs $\hat{h} : \hat{\mathcal{S}} \to \hat{\mathcal{T}}$.* $\square$

The definition of $\hat{h}$ should be obvious. If $f \in \mathcal{S}_U$, and $x \in U$, then $\hat{h}([f]_x) = [h(f)]_x$, where $h(f)$ is now an element of $\mathcal{T}(U)$. We let the reader check that $\hat{h}$ is independent of the various choices, and that it is a *continuous* map $\hat{\mathcal{S}} \to \hat{\mathcal{T}}$ with respect to the topologies described in Example 7.5.6.

The sheaf associated to the space of germs $\operatorname{Im} \hat{h}$ is a subsheaf of $\mathcal{T}$ called the *image* of h, and denoted by $\operatorname{Im} h$.

Exercise 7.5.17. Consider a morphism of sheaves over X, $h : \mathcal{S} \to \mathcal{T}$. Let $U \subset X$ be an open set. Show that a section $g \in \mathcal{T}(U)$ belongs to $(\operatorname{Im} h)(U)$ if and only if, for every $x \in X$, there exists an open neighborhood $V_x \subset U$, such that $g|_{V_x} = h(f)$, for some $f \in \mathcal{S}(V_x)$. $\square$

☞ *Due to Proposition 7.5.8(a), in the sequel we will make no distinction between sheaves and spaces of germs.*

Definition 7.5.18. (a) A sequence of sheaves and morphisms of sheaves,

$$\cdots \to \mathcal{S}_n \overset{h_n}{\to} \mathcal{S}_{n+1} \overset{h_{n+1}}{\to} \mathcal{S}_{n+2} \to \cdots,$$

is said to be exact if $\operatorname{Im} h_n = \ker h_{n+1}$, $\forall n$.

(b) Consider a sheaf $\mathcal{S}$ over the space X. A *resolution* of $\mathcal{S}$ is a long exact sequence

$$0 \hookrightarrow \mathcal{S} \overset{\imath}{\hookrightarrow} \mathcal{S}^0 \overset{d_0}{\to} \mathcal{S}^1 \overset{d_1}{\to} \cdots \to \mathcal{S}^n \overset{d_n}{\to} \mathcal{S}^{n+1} \to \cdots.$$ $\square$

Exercise 7.5.19. Consider a short exact sequence of sheaves

$$0 \to \mathcal{S}_{-1} \to \mathcal{S}_0 \to \mathcal{S}_1 \to 0.$$

For each open set U define $\mathcal{S}(U) = \mathcal{S}_0(U)/\mathcal{S}_{-1}(U)$.
(a) Prove that $U \mapsto \mathcal{S}(U)$ is a presheaf.
(b) Prove that $\mathcal{S}_1 \cong \hat{\mathcal{S}}$ = the sheaf associated to the presheaf $\mathcal{S}$.   □

Example 7.5.20. (The DeRham resolution). Let M be a smooth n-dimensional manifold. Using the Poincaré lemma and Exercise 7.5.17 we deduce immediately that the sequence

$$0 \hookrightarrow \underline{\mathbb{R}}_M \hookrightarrow \Omega_M^0 \xrightarrow{d} \Omega_M^1 \xrightarrow{d} \cdots \xrightarrow{d} \Omega_M^n \to 0$$

is a resolution of the constant sheaf $\underline{\mathbb{R}}_M$. Ω_M^k denotes the sheaf of k-forms on M while d denotes the exterior differentiation.   □

7.5.2   Čech cohomology

Let $U \mapsto \mathcal{S}(U)$ be a presheaf of Abelian groups over a topological space X. Consider an open cover $\mathcal{U} = (U_\alpha)_{\alpha \in \mathcal{A}}$ of X. A *simplex* associated to $\mathcal{U}$ is a nonempty subset $A = \{\alpha_0, \ldots, \alpha_q\} \subset \mathcal{A}$, such that

$$U_A := \bigcap_0^q U_{\alpha_i} \neq \emptyset.$$

We define $\dim A$ to be one less the cardinality of A, $\dim A = |A| - 1$. The set of all q-dimensional simplices is denoted by $\mathcal{N}_q(\mathcal{U})$. Their union,

$$\bigcup_q \mathcal{N}_q(\mathcal{U})$$

is denoted by $\mathcal{N}(\mathcal{U})$, and it is called the *nerve* of the cover.

For every nonnegative integer q we set $\Delta_q := \{0, \ldots, q\}$, and we define an *ordered q-simplex* to be a map $\sigma : \Delta_q \to \mathcal{A}$ with the property that $\sigma(\Delta_q)$ is a simplex of $\mathcal{N}(\mathcal{U})$, possibly of dimension $< q$. We will denote by $\vec{\mathcal{N}}_q(\mathcal{U})$ the set of ordered q-simplices, and we will use the symbol $(\alpha_0, \ldots, \alpha_q)$ to denote an ordered q-simplex $\sigma : \Delta_q \to \mathcal{A}$ such that $\sigma(k) = \alpha_k$.

Define

$$C^q(\mathcal{S}, \mathcal{U}) := \prod_{\sigma \in \vec{\mathcal{N}}_q(\mathcal{U})} \mathcal{S}_\sigma, \quad \mathcal{S}_\sigma := \mathcal{S}(U_{\sigma(\Delta_q)}).$$

The elements of $C^q(\mathcal{S}, \mathcal{U})$ are called Čech q-cochains subordinated to the cover $\mathcal{U}$. In other words, a q-cochain c associates to each ordered q-simplex σ an element $\langle c, \sigma \rangle \in \mathcal{S}_\sigma$.

We have *face* operators

$$\partial^j = \partial^j : \vec{\mathcal{N}}_q(\mathcal{U}) \to \vec{\mathcal{N}}_{q-1}(\mathcal{U}), \quad j = 0, \ldots, q,$$

where for an ordered q-simplex $\sigma = (\alpha_0, \ldots, \alpha_q)$, we set

$$\partial^j \sigma = \partial^j_q \sigma = (\alpha_0, \ldots, \hat{\alpha}_j, \ldots, \alpha_q) \in \mathcal{U}_{(q-1)},$$

and where a hat "ˆ" indicates a missing entry.

Exercise 7.5.21. Prove that $\partial^i_{q-1} \partial^j_q = \partial^{j-1}_{q-1} \partial^i_q$, for all $j > i$. □

We can now define an operator

$$\delta : C^{q-1}(\mathcal{S}, \mathcal{U}) \to C^q(\mathcal{S}, \mathcal{U}),$$

which assigns to each $(q-1)$-cochain c, a q-cochain δc whose value on an ordered q-simplex σ is given by

$$\langle \delta c, \sigma \rangle := \sum_{j=0}^{q} (-1)^j \langle c, \partial^j \sigma \rangle |_{U_\sigma} .$$

Using Exercise 7.5.21 above we deduce immediately the following result.

Lemma 7.5.22. $\delta^2 = 0$ *so that*

$$0 \hookrightarrow C^0(\mathcal{S}, \mathcal{U}) \xrightarrow{\delta} C^1(\mathcal{S}, \mathcal{U}) \xrightarrow{\delta} \cdots \xrightarrow{\delta} C^q(\mathcal{S}, \mathcal{U}) \xrightarrow{\delta} \cdots$$

is a cochain complex.

The cohomology of this cochain complex is called the *Čech cohomology* of the cover $\mathcal{U}$ with coefficients in the pre-sheaf $\mathcal{S}$.

Example 7.5.23. Let $\mathcal{U}$ and $\mathcal{S}$ as above. A 0-cochain is a correspondence which associates to each open set $U_\alpha \in \mathcal{U}$ an element $c_\alpha \in \mathcal{S}(U_\alpha)$. It is a cocycle if, for any pair (α, β) such that $U_{\alpha \cap \beta} \neq \emptyset$, we have

$$c_\beta - c_\alpha = 0.$$

Observe that if $\mathcal{S}$ is a *sheaf*, then the collection (c_α) determines a unique section of $\mathcal{S}$.

A 1-cochain associates to each pair (α, β) such that $U_{\alpha\beta} \neq \emptyset$ an element

$$c_{\alpha\beta} \in \mathcal{S}(U_{\alpha\beta}).$$

This correspondence is a cocycle if, for any ordered 2-simplex (α, β, γ) we have

$$c_{\beta\gamma} - c_{\alpha\gamma} + c_{\alpha\beta} = 0.$$

For example, if X is a smooth manifold, and $\mathcal{U}$ is a good cover, then we can associate to each closed 1-form $\omega \in \Omega^1(M)$ a Čech 1-cocycle valued in $\mathbb{R}_X$ as follows.

First, select for each U_α a solution $f_\alpha \in C^\infty(U_\alpha)$ of

$$df_\alpha = \omega.$$

Since $d(f_\alpha - f_\beta) \equiv 0$ on $U_{\alpha\beta}$, we deduce there exist constants $c_{\alpha\beta}$ such that $f_\alpha - f_\beta = c_{\alpha\beta}$. Obviously this is a cocycle, and it is easy to see that its cohomology class is independent of the initial selection of local solutions f_α. Moreover, if ω is exact, this cocycle is a coboundary. In other words we have a natural map

$$H^1(X) \to H^1\big(\mathcal{N}(\mathcal{U}), \mathbb{R}_X\big).$$

We will see later that this is an isomorphism. □

Definition 7.5.24. Consider two open covers $\mathcal{U} = (U_\alpha)_{\alpha \in \mathcal{A}}$, and $\mathcal{V} = (V_\beta)_{\beta \in \mathcal{B}}$ of the same topological space X. We say $\mathcal{V}$ is finer than $\mathcal{U}$, and we write this $\mathcal{U} \prec \mathcal{V}$, if there exists a map $\varrho : \mathcal{B} \to \mathcal{A}$, such that

$$V_\beta \subset U_{\varrho(\beta)} \quad \forall \beta \in \mathcal{B}.$$

The map ϱ is said to be a *refinement map*. $\square$

Proposition 7.5.25. *Consider two open covers $\mathcal{U} = (U_\alpha)_{\alpha \in \mathcal{A}}$ and $\mathcal{V} = (V_\beta)_{\beta \in \mathcal{B}}$ of the same topological space X such that $\mathcal{U} \prec \mathcal{V}$. Fix a sheaf of Abelian groups $\mathcal{S}$. Then the following are true.*

(a) Any refinement map ϱ induces a cochain morphism

$$\varrho_* : \bigoplus_q C^q(\mathcal{U}, \mathcal{S}) \to \bigoplus_q C^q(\mathcal{V}, \mathcal{S}).$$

(b) If $r : \mathcal{A} \to \mathcal{B}$ is another refinement map, then ϱ_* is cochain homotopic to r_*. In particular, any relation $\mathcal{U} \prec \mathcal{V}$ defines a unique morphism in cohomology

$$\imath_{\mathcal{U}}^{\mathcal{V}} : H^\bullet(\mathcal{U}, \mathcal{S}) \to H^\bullet(\mathcal{V}, \mathcal{S}).$$

(c) If $\mathcal{U} \prec \mathcal{V} \prec \mathcal{W}$, then

$$\imath_{\mathcal{U}}^{\mathcal{W}} = \imath_{\mathcal{V}}^{\mathcal{W}} \circ \imath_{\mathcal{U}}^{\mathcal{V}}.$$

Proof. (a) We define $\varrho_* : C^q(\mathcal{U}) \to C^q(\mathcal{V})$ by

$$\mathcal{S}(V_\sigma) \ni \langle \varrho_*(c), \sigma \rangle := \langle c, \varrho(\sigma) \rangle |_{V_\sigma} \quad \forall c \in C^q(\mathcal{U})\ \sigma \in \mathcal{V}_{(q)},$$

where, by definition, $\varrho(\sigma) \in \mathcal{A}^q$ is the ordered q-simplex $(\varrho(\beta_0), \dots, \varrho(\beta_q))$. The fact that ϱ_* is a cochain morphism follows immediately from the obvious equality

$$\varrho \circ \partial_q^j = \partial_q^j \circ \varrho.$$

(b) We define $h_j : \vec{N}_{q-1}(\mathcal{V}) \to \vec{N}_q(\mathcal{U})$ by

$$h_j(\beta_0, \dots, \beta_{q-1}) = (\varrho(\beta_0), \dots, \varrho(\beta_j), r(\beta_j), \cdots, r(\beta_{q-1})).$$

The reader should check that $h_j(\sigma)$ is indeed an ordered simplex of $\mathcal{U}$ for any ordered simplex σ of $\mathcal{V}$. Note that $V_\sigma \subset U_{h_j(\sigma)} \ \forall j$. Now define

$$\chi := \chi_q : C^q(\mathcal{U}) \to C^{q-1}(\mathcal{V})$$

by

$$\langle \chi_q(c), \sigma \rangle := \sum_{j=0}^{q-1} (-1)^j \langle c, h_j(\sigma) \rangle |_{V_\sigma} \quad \forall c \in C^q(\mathcal{U})\ \forall \sigma \in \mathcal{V}_{(q-1)}.$$

We will show that

$$\delta \circ \chi_q(c) + \chi_{q+1} \circ \delta(\sigma) = \varrho_*(c) - r_*(c) \quad \forall c \in C^q(\mathcal{U}).$$

Let $\sigma = (\beta_0, \dots, \beta_q) \in \vec{N}_q(\mathcal{V})$, and set

$$\varrho(\sigma) := (\lambda_0, \dots, \lambda_q), \ r(\sigma) = (\mu_0, \dots, \mu_q) \in \mathcal{U}_{(q)},$$

so that,

$$h_j(\sigma) = (\lambda_0, \ldots, \lambda_j, \mu_j, \ldots, \mu_q).$$

Then

$$\langle \chi \circ \delta(c), \sigma \rangle = \sum (-1)^j \langle \delta c, h_j(\sigma) \rangle |_{V_\sigma}$$

$$= \sum_{j=0}^{q} (-1)^j \left(\sum_{k=0}^{q+1} (-1)^k \langle c, \partial_{q+1}^k h_j(\sigma) \rangle |_{V_\sigma} \right)$$

$$= \sum_{j=0}^{q} (-1)^j \left(\sum_{k=0}^{j} (-1)^k \langle c, (\lambda_0, \ldots, \hat{\lambda}_j, \ldots, \lambda_j, \mu_j, \ldots, \mu_q) \rangle |_{V_\sigma} \right.$$

$$\left. + \sum_{\ell=j}^{q} (-1)^{\ell+1} \langle c, (\lambda_0, \ldots, \lambda_j, \mu_j, \ldots, \hat{\mu}_k, \ldots, \mu_q) \rangle |_{V_\sigma} \right)$$

$$= \sum_{j=0}^{q} (-1)^j \left(\sum_{k=0}^{j-1} (-1)^k \langle c, (\lambda_0, \ldots, \hat{\lambda}_j, \ldots, \lambda_j, \mu_j, \ldots, \mu_q) \rangle |_{V_\sigma} \right.$$

$$\left. + \sum_{\ell=j+1}^{q} (-1)^{\ell+1} \langle c, (\lambda_0, \ldots, \lambda_j, \mu_j, \ldots, \hat{\mu}_k, \ldots, \mu_q) \rangle |_{V_\sigma} \right)$$

$$+ \sum_{j=0}^{q} (-1)^j \left\{ \langle c, (\lambda_0, \ldots, \lambda_{j-1}, \mu_j, \ldots, \mu_q) \rangle |_{V_\sigma} + \langle c, (\lambda_0, \ldots, \lambda_j, \mu_{j+1}, \ldots, \mu_q) \rangle |_{V_\sigma} \right\}.$$

The last term is a telescopic sum which is equal to

$$\langle c, (\lambda_0, \ldots, \lambda_q) \rangle |_{V_\sigma} - \langle c, (\mu_0, \ldots, \mu_q) \rangle |_{V_\sigma} = \langle \varrho_* c, \sigma \rangle - \langle r_* c, \sigma \rangle.$$

If we change the order of summation in the first two terms we recover the term $\langle -\delta \chi c, \sigma \rangle$.

Part (c) is left to the reader as an exercise. $\qquad \square$

We now have a collection of graded groups

$$\{ H^\bullet(\mathcal{U}, \mathcal{S}) \; ; \; \mathcal{U} - \text{open cover of } X \},$$

and morphisms

$$\left\{ \imath_{\mathcal{U}}^{\mathcal{V}} : H^\bullet(\mathcal{U}, \mathcal{S}) \to H^\bullet(\mathcal{V}, \mathcal{S}); \; \mathcal{U} \prec \mathcal{V} \right\},$$

such that,

$$\imath_{\mathcal{U}}^{\mathcal{U}} = \mathbb{1} \text{ and } \imath_{\mathcal{U}}^{\mathcal{W}} = \imath_{\mathcal{V}}^{\mathcal{W}} \circ \imath_{\mathcal{U}}^{\mathcal{V}},$$

whenever $\mathcal{U} \prec \mathcal{V} \prec \mathcal{W}$. We can thus define the inductive (direct) limit

$$H^\bullet(X, \mathcal{S}) := \varinjlim_{\mathcal{U}} H^\bullet(\mathcal{U}, \mathcal{S}).$$

The group $H^\bullet(X, \mathcal{S})$ is called the *Čech cohomology* of the space X with coefficients in the pre-sheaf $\mathcal{S}$.

Let us briefly recall the definition of the direct limit. One defines an equivalence relation on the disjoint union

$$\coprod_{\mathcal{U}} H^\bullet(\mathcal{U}, \mathcal{S}),$$

by

$$H^\bullet(\mathcal{U}) \ni f \sim g \in H^\bullet(\mathcal{V}) \Longleftrightarrow \exists \mathcal{W} \succ \mathcal{U}, \mathcal{V} : \imath_{\mathcal{U}}^{\mathcal{W}} f = \imath_{\mathcal{V}}^{\mathcal{W}} g.$$

We denote the equivalence class of f by $\overline{f}$. Then

$$\varinjlim_{\mathcal{U}} H^\bullet(\mathcal{U}) = \left(\coprod_{\mathcal{U}} H^\bullet(\mathcal{U}, \mathcal{S}) \right) / \sim .$$

Note that we have canonical morphisms,

$$\imath_{\mathcal{U}} : H^\bullet(\mathcal{U}, \mathcal{S}) \to H^\bullet(X, \mathcal{S}).$$

Example 7.5.26. Let $\mathcal{S}$ be a *sheaf* over the space X. For any open cover $\mathcal{U} = (U_\alpha)$, a 0-cycle subordinated to $\mathcal{U}$ is a collection of sections $f_\alpha \in \mathcal{S}(U_\alpha)$ such that, every time $U_\alpha \cap U_\beta \neq \emptyset$, we have

$$f_\alpha |_{U_{\alpha\beta}} = f_\beta |_{U_{\alpha\beta}} .$$

According to the properties of a sheaf, such a collection defines a unique *global* section $f \in \mathcal{S}(X)$. Hence, $H^0(X, \mathcal{S}) \cong \mathcal{S}(X)$. $\qquad\square$

Proposition 7.5.27. *Any morphism of pre-sheaves $h : \mathcal{S}_0 \to \mathcal{S}_1$ over X induces a morphism in cohomology*

$$h_* : H^\bullet(X, \mathcal{S}_0) \to H^\bullet(X, \mathcal{S}_1).$$

Sketch of proof. Let $\mathcal{U}$ be an open cover of X. Define

$$h_* : C^q(\mathcal{U}, \mathcal{S}_0) \to C^q(\mathcal{U}, \mathcal{S}_1),$$

by

$$\langle h_* c, \sigma \rangle = h_U(\langle c, \sigma \rangle) \quad \forall c \in C^q(\mathcal{U}, \mathcal{S}_0) \ \sigma \in \mathcal{U}_{(q)}.$$

The reader can check easily that h_* is a cochain map so it induces a map in cohomology

$$h_*^{\mathcal{U}} : H^\bullet(\mathcal{U}, \mathcal{S}_0) \to H^\bullet(\mathcal{U}, \mathcal{S}_1)$$

which commutes with the refinements $i_{\mathcal{U}}^{\mathcal{V}}$. The proposition follows by passing to direct limits. $\square$

Theorem 7.5.28. *Let*

$$0 \to \mathcal{S}_{-1} \xrightarrow{\jmath} \mathcal{S}_0 \xrightarrow{p} \mathcal{S}_1 \to 0,$$

be an exact sequence of sheaves over a paracompact *space X. Then there exists a natural long exact sequence*

$$\cdots \to H^q(X, \mathcal{S}_{-1}) \xrightarrow{\jmath_*} H^q(X, \mathcal{S}_0) \xrightarrow{p_*} H^q(X, \mathcal{S}_1) \xrightarrow{\delta_*} H^{q+1}(X, \mathcal{S}_{-1}) \to \cdots .$$

Sketch of proof. For each open set $U \subset X$ define $\mathcal{S}(U) := \mathcal{S}_0(U)/\mathcal{S}_{-1}(U)$. Then the correspondence $U \mapsto \mathcal{S}(U)$ defines a pre-sheaf on X. Its associated sheaf is isomorphic with $\mathcal{S}_1$ (see Exercise 7.5.19). Thus, for each open cover $\mathcal{U}$ we have a short exact sequence

$$0 \to C^q(\mathcal{U}, \mathcal{S}_{-1}) \xrightarrow{\jmath} C^q(\mathcal{U}, \mathcal{S}_0) \xrightarrow{\pi} C^q(\mathcal{U}, \mathcal{S}) \to 0.$$

We obtain a long exact sequence in cohomology

$$\cdots \to H^q(\mathcal{U}, \mathcal{S}_{-1}) \to H^q(\mathcal{U}, \mathcal{S}_0) \to H^q(\mathcal{U}, \mathcal{S}) \to H^{q+1}(\mathcal{U}, X) \to \cdots .$$

Passing to direct limits we get a long exact sequence

$$\cdots \to H^q(X, \mathcal{S}_{-1}) \to H^q(X, \mathcal{S}_0) \to H^q(X, \mathcal{S}) \to H^{q+1}(X, \mathcal{S}_{-1}) \to \cdots .$$

To conclude the proof of the proposition we invoke the following technical result. Its proof can be found in [88].

Lemma 7.5.29. *If two pre-sheaves $\mathcal{S}$, $\mathcal{S}'$ over a* paracompact *topological space X have isomorphic associated sheaves then*

$$H^\bullet(X, \mathcal{S}) \cong H^\bullet(X, \mathcal{S}'). \qquad \square$$

Definition 7.5.30. A sheaf $\mathcal{S}$ is said to be *fine* if, for any locally finite open cover $\mathcal{U} = (U_\alpha)_{\alpha \in \mathcal{A}}$ there exist morphisms $h_\alpha : \mathcal{S} \to \mathcal{S}$ with the following properties.

(a) For any $\alpha \in \mathcal{A}$ there exists a closed set $C_\alpha \subset U_\alpha$, and $h_\alpha(\mathcal{S}_x) = 0$ for $x \notin C_\alpha$, where $\mathcal{S}_x$ denotes the stalk of $\mathcal{S}$ at $x \in X$.
(b) $\sum_\alpha h_\alpha = \mathbb{1}_\mathcal{S}$. This sum is well defined since the cover $\mathcal{U}$ is locally finite. $\square$

Example 7.5.31. Let X be a smooth manifold. Using partitions of unity we deduce that the sheaf Ω_X^k of smooth k-forms is fine. More generally, if E is a smooth vector bundle over X then the space Ω_E^k of E-valued k-forms is fine. $\square$

Proposition 7.5.32. *Let $\mathcal{S}$ be a fine sheaf over a paracompact space X. Then $H^q(X, \mathcal{S}) \cong 0$ for $q \geq 1$.*

Proof.   Because X is paracompact, any open cover admits a locally finite refinement. Thus, it suffices to show that, for each locally finite open cover $\mathcal{U} = (U_\alpha)_{\alpha \in \mathcal{A}}$, the cohomology groups $H^q(\mathcal{U}, \mathcal{S})$ are trivial for $q \geq 1$. We will achieve this by showing that the identity map $C^q(\mathcal{U}, \mathcal{S}) \to C^q(\mathcal{U}, \mathcal{S})$ is cochain homotopic with the trivial map. We thus need to produce a map

$$\chi^q : C^q(\mathcal{U}, \mathcal{S}) \to C^{q-1}(\mathcal{U}, \mathcal{S}),$$

such that

$$\chi^{q+1}\delta^q + \delta^{q-1}\chi^q = \mathbb{1}. \tag{7.5.1}$$

Consider the morphisms $h_\alpha : \mathcal{S} \to \mathcal{S}$ associated to the cover $\mathcal{U}$ postulated by the definition of a fine sheaf. For every $\alpha \in \mathcal{A}$, $\sigma \in \vec{\mathcal{N}}_q(\mathcal{U})$, and every $f \in C^q(\mathcal{U}, \mathcal{S})$, we construct $\langle t_\alpha(f), \sigma \rangle \in \mathcal{S}(U_\sigma)$ as follows. Consider the open cover of U_σ

$$\{\, V = U_\alpha \cap U_\sigma, \ \ W := U_\sigma \setminus C_\alpha \,\}, \quad (\operatorname{supp} h_\alpha \subset C_\alpha).$$

Note that $h_\alpha f|_{V \cap W} = 0$ and, according to the axioms of a sheaf, the section $h_\alpha(f|_V)$ can be extended by zero to a section $\langle t_\alpha(f), \sigma \rangle \in \mathcal{S}(U_\sigma)$. Now, for every $f \in C^q(\mathcal{U}, \mathcal{S})$, define $\chi^q f \in C^{q-1}(\mathcal{U}, \mathcal{S})$ by

$$\langle \chi^q(f), \sigma \rangle := \sum_\alpha \langle t_\alpha(f), \sigma \rangle.$$

The above sum is well defined since the cover $\mathcal{U}$ is locally finite. We let the reader check that χ^q satisfies (7.5.1).   ☐

Definition 7.5.33.   Let $\mathcal{S}$ be a sheaf over a space X. A *fine resolution* is a resolution

$$0 \to \mathcal{S} \hookrightarrow \mathcal{S}^0 \xrightarrow{d} \mathcal{S}^1 \xrightarrow{d} \cdots,$$

such that each of the sheaves $\mathcal{S}_j$ is fine.   ☐

Theorem 7.5.34 (Abstract DeRham theorem).   *Let*

$$0 \hookrightarrow \mathcal{S} \to \mathcal{S}^0 \xrightarrow{d^0} \mathcal{S}^1 \xrightarrow{d^1} \cdots$$

be a fine resolution of the sheaf $\mathcal{S}$ over the paracompact space X. Then

$$0 \to \mathcal{S}^0(X) \xrightarrow{d_0} \mathcal{S}^1(X) \xrightarrow{d_1} \cdots$$

is a cochain complex, and there exists a natural isomorphism

$$H^q(X, \mathcal{S}) \cong H^q(\mathcal{S}_q(X)).$$

Proof.   The first statement in the theorem can be safely left to the reader. For $q \geq 1$, denote by $\mathcal{Z}^q$ the kernel of the sheaf morphism d_q. We set for uniformity $\mathcal{Z}^0 := \mathcal{S}$. We get a short exact sequence of sheaves

$$0 \to \mathcal{Z}^q \to \mathcal{S}^q \to \mathcal{Z}^{q+1} \to 0 \quad q \geq 0. \tag{7.5.2}$$

We use the associated long exact sequence in which $H^k(X, \mathcal{S}_q) = 0$, for $k \geq 1$, since $\mathcal{S}^q$ is a fine sheaf. This yields the isomorphisms

$$H^{k-1}(X, \mathcal{Z}^{q+1}) \cong H^k(X, \mathcal{Z}^q) \quad k \geq 2.$$

We deduce inductively that

$$H^m(X, \mathcal{Z}^0) \cong H^1(X, \mathcal{Z}^{m-1}) \quad m \geq 1. \tag{7.5.3}$$

Using again the long sequence associated to (7.5.2), we get an exact sequence

$$H^0(X, \mathcal{Z}^{m-1}) \xrightarrow{d_*^{m-1}} H^0(X, \mathcal{Z}^m) \to H^1(X, \mathcal{Z}^{m-1}) \to 0.$$

We apply the computation in Example 7.5.26, and we get

$$H^1(X, \mathcal{Z}^{m-1}) \cong \mathcal{Z}^m(X)/d_*^{m-1}\left(\mathcal{S}^{m-1}(X)\right).$$

This is precisely the content of the theorem. $\qquad\square$

Corollary 7.5.35. *Let M be a smooth manifold. Then*

$$H^\bullet(M, \underline{\mathbb{R}}_M) \cong H^\bullet(M).$$

Proof. The manifold M is paracompact. We conclude using the fine resolution

$$0 \to \underline{\mathbb{R}}_M \hookrightarrow \Omega^0_M \xrightarrow{d} \Omega^1_M \to \cdots. \qquad\square$$

Exercise 7.5.36. Describe explicitly the isomorphisms

$$H^1(M) \cong H^1(M, \underline{\mathbb{R}}_M) \quad \text{and} \quad H^2(M) \cong H^2(M, \underline{\mathbb{R}}_M). \qquad\square$$

Remark 7.5.37. The above corollary has a surprising implication. Since the Čech cohomology is obviously a *topological* invariant, so must be the DeRham cohomology which is defined in terms of a smooth structure. Hence if two *smooth* manifolds are *homeomorphic* they must have isomorphic DeRham groups, even if the manifolds may not be diffeomorphic.

Such exotic situations do exist. In a celebrated paper [70], John Milnor has constructed a family of nondiffeomorphic manifolds all homeomorphic to the sphere S^7. More recently, the work of Simon Donaldson in gauge theory was used by Michael Freedman to construct a smooth manifold homeomorphic to $\mathbb{R}^4$ but not diffeomorphic with $\mathbb{R}^4$ equipped with the natural smooth structure. (This is possible only for 4-dimensional vector spaces!) These three mathematicians, J. Milnor, S. Donaldson and M. Freedman were awarded Fields medals for their contributions. $\square$

Theorem 7.5.38. (Leray) *Let M be a smooth manifold and $\mathcal{U} = (U_\alpha)_{\alpha \in A}$ a good cover of M, i.e.,*

$$U_\sigma \cong \mathbb{R}^{\dim M}.$$

Then

$$H^\bullet(\mathcal{U}, \underline{\mathbb{R}}_M) \cong H^\bullet(M).$$

Proof. Let $\mathcal{Z}^k$ denote the sheaf of closed k-forms on M. Using the Poincaré lemma we deduce

$$\mathcal{Z}^k(U_\sigma) = d\Omega^{k-1}(U_\sigma).$$

We thus have a short exact sequence

$$0 \to C^q(\mathcal{U}, \mathcal{Z}^k) \to C^q(\mathcal{U}, \Omega^{k-1}) \xrightarrow{d_*} C^q(\mathcal{U}, \mathcal{Z}^k) \to 0.$$

Using the associated long exact sequence and the fact that Ω^{k-1} is a fine sheaf, we deduce as in the proof of the abstract DeRham theorem that

$$H^q(\mathcal{U}, \underline{\mathbb{R}}_M) \cong H^{q-1}(\mathcal{U}, \mathcal{Z}^1) \cong \cdots$$

$$\cong H^1(\mathcal{U}, \mathcal{Z}^{q-1}) \cong H^0(\mathcal{U}, \mathcal{Z}^q)/d_* H^0(\mathcal{U}, \Omega^{q-1}) \cong \mathcal{Z}^q(M)/d\Omega^{q-1}(M) \cong H^\bullet(M). \quad \square$$

Remark 7.5.39. The above result is a special case of a theorem of Leray: if $\mathcal{S}$ is a sheaf on a *paracompact* space X, and $\mathcal{U}$ is an open cover such that

$$H^q(U_A, \mathcal{S}) = 0, \quad \forall q \geq 1 \quad A \in \mathcal{N}(\mathcal{U}),$$

then $H^\bullet(\mathcal{U}, \mathcal{S}) = H^\bullet(X, \mathcal{S})$. For a proof we refer to [37].

When $\underline{G}_M$ is a constant sheaf, where G is an arbitrary Abelian group, we have a Poincaré lemma (see [32], Chapter IX, Thm. 5.1),

$$H^q(\mathbb{R}^n, \underline{G}) = 0 \quad q \geq 1.$$

Hence, for any good cover $\mathcal{U}$

$$H^\bullet(M, \underline{G}_M) = H^\bullet(\mathcal{U}, \underline{G}_M). \quad\quad\quad\quad \square$$

Remark 7.5.40. A *combinatorial simplicial complex*, or a *simplicial scheme*, is a collection $\mathcal{K}$ of nonempty, finite subsets of a set V with the property that any nonempty subset B of a subset $A \in \mathcal{K}$ must also belong to the collection $\mathcal{K}$, i.e.,

$$A \in \mathcal{K}, \quad B \subset A, \quad B \neq \emptyset \Longrightarrow B \in \mathcal{K}.$$

The set V is called the vertex set of $\mathcal{K}$, while the subsets in $\mathcal{K}$ are called the ((open) faces) of the simplicial scheme. The nerve of an open cover is an example of simplicial scheme.

To any simplicial scheme $\mathcal{K}$, with vertex set V, we can associate a topological space $|\mathcal{K}|$ in the following way.

Denote by $\mathbb{R}^V$ the vector space of all maps $f : V \to \mathbb{R}$ with the property that $f(v) = 0$, for all but finitely many v's. In other words,

$$\mathbb{R}^V = \bigoplus_{v \in V} \mathbb{R}.$$

Note that $\mathbb{R}^V$ has a canonical basis given by the Dirac functions

$$\delta_v : V \to \mathbb{R}, \quad \delta_v(u) = \begin{cases} 1 & u = v \\ 0 & u \neq v. \end{cases}$$

We topologize $\mathbb{R}^V$ by declaring a subset $C \subset \mathbb{R}^V$ closed if the intersection of C with any finite dimensional subspace of $\mathbb{R}^V$ is a closed subset with respect to the Euclidean topology of that finite dimensional subspace.

For any face $F \subset \mathcal{K}$ we denote by $\Delta_F \subset \mathbb{R}^V$ the convex hull of the set $\{\delta_v; \ v \in F\}$. Note that Δ_F is a simplex of dimension $|F| - 1$. Now set

$$|\mathcal{K}| = \bigcup_{F \in \mathcal{K}} \Delta_F.$$

We will say that $|\mathcal{K}|$ is the geometric realization of the simplicial scheme $\mathcal{K}$.

A result of Borsuk-Weil (see [14]) states that if $\mathcal{U} = (U_\alpha)_{\alpha \in \mathcal{A}}$ is a good cover of a compact manifold M, then the geometric realization of the nerve $\mathcal{N}(\mathcal{U})$ is homotopy equivalent to M. Thus, the nerve of a good cover contains all the homotopy information about the manifold. Since the DeRham cohomology is homotopy invariant, Leray's theorem comes as no suprise. What is remarkable is the explicit way in which one can extract the cohomological information from the combinatorics of the nerve.

While at this point, we should remark that any compact manifold admits *finite* good covers. This shows that the homotopy type is determined by a finite, albeit very large, set of data. □

Example 7.5.41. Let M be a smooth manifold and $\mathcal{U} = (U_\alpha)_{\alpha \in \mathcal{A}}$ a good cover of M. A 1-cocycle of $\underline{\mathbb{R}}_M$ is a collection of real numbers $f_{\alpha\beta}$ - one for each pair $(\alpha, \beta) \in \mathcal{A}^2$ such that $U_{\alpha\beta} \neq \emptyset$ satisfying

$$f_{\alpha\beta} + f_{\beta\gamma} + f_{\gamma\alpha} = 0,$$

whenever $U_{\alpha\beta\gamma} \neq \emptyset$. The collection is a coboundary if there exist the constants f_α such that $f_{\alpha\beta} = f_\beta - f_\alpha$. This is precisely the situation encountered in Subsection 7.1.2. The abstract DeRham theorem explains why the Čech approach is equivalent with the DeRham approach. □

Remark 7.5.42. Often, in concrete applications it is convenient to work with skew-symmetric Čech cochains. A cochain $c \in C^q(\mathcal{S}, \mathcal{U})$ is skew-symmetric if for any ordered q-simplex $\sigma = (\alpha_0, \ldots, \alpha_q)$, and for any permutation φ of Δ_q we have

$$\langle c, (\alpha_0, \ldots, \alpha_q) \rangle = \epsilon(\varphi) \langle c, (\alpha_{\varphi(0)}, \ldots, \alpha_{\varphi(q)}) \rangle.$$

One can then define a "skew-symmetric" Čech cohomology following the same strategy. The resulting cohomology coincides with the cohomology described in this subsection. For a proof of this fact we refer to [88]. □

Exercise 7.5.43. A sheaf of Abelian groups $\mathcal{S}$ on a paracompact space X is called *soft* if for every closed subset $C \subset X$ the restriction map $\mathcal{S}(X) \to \mathcal{S}(C)$ is surjective. The sheaf $\mathcal{S}$ is called *flabby* if, for any open subset $U \subset X$, the restriction map $\mathcal{S}(X) \to \mathcal{S}(U)$ is surjective.

(a) Prove that any flabby sheaf is also soft.

(b) Suppose $0 \to \mathcal{S}_0 \to \mathcal{S}_1 \to \mathcal{S}_2 \to 0$ is a short exact sequence of sheaves on X. Show that if $\mathcal{S}_0$ is a flabby sheaf, then the sequence of Abelian groups

$$0 \to \mathcal{S}_0(X) \to \mathcal{S}_1(X) \to \mathcal{S}_2(X) \to 0$$

is exact. ☐

Exercise 7.5.44. Suppose X is a topological space and $\mathcal{S}$ is a sheaf of Abelian groups on X.

(a) Prove that $\mathcal{S}$ admits a *flabby resolution*, i.e., there exists a sequence of flabby sheaves over X, $\mathcal{S}_k$, $k \geq 0$, and morphisms of sheaves $f_k : \mathcal{S}_{k-1} \to \mathcal{S}_k$, $k = 0, 1, \dots$, $\mathcal{S}_{-1} = \mathcal{S}$, such that the sequence below is exacft

$$0 \to \mathcal{S} \xrightarrow{f_0} \mathcal{S}_0 \xrightarrow{f_1} \mathcal{S}_1 \longrightarrow \cdots .$$

(b) Prove that if

$$0 \to \mathcal{S} \xrightarrow{f_0} \mathcal{S}_0 \xrightarrow{f_1} \mathcal{S}_1 \longrightarrow \cdots$$

is a flabby resolution of $\mathcal{S}$, then the cohomology groups $H^k(X, \mathcal{S})$ are isomorphic to the cohomology groups of the cochain complex

$$\mathcal{S}_0(X) \xrightarrow{f_0} \mathcal{S}_1(X) \longrightarrow \cdots .$$

☐

Chapter 8

Characteristic Classes

We now have sufficient background to approach a problem formulated in Chapter 2: find a way to measure the "extent of nontriviality" of a given vector bundle. This is essentially a topological issue but, as we will see, in the context of smooth manifolds there are powerful differential geometric methods which will solve a large part of this problem. Ultimately, only topological techniques yield the best results.

8.1 Chern-Weil theory

8.1.1 *Connections on principal G-bundles*

In this subsection we will describe how to take into account the possible symmetries of a vector bundle when describing a connection.

All the Lie groups we will consider will be assumed to be *matrix Lie groups*, i.e., Lie subgroups of a general linear group $GL(n, \mathbb{K}) = GL(\mathbb{K}^n)$.

This restriction is neither severe, nor necessary. It is not severe since, according to a nontrivial result (Peter-Weyl theorem), any compact Lie group is isomorphic with a matrix Lie group, and these groups are sufficient for most applications in geometry. It is not necessary since all the results of this subsection are true for any Lie group. We stick with this assumption since most proofs are easier to "swallow" in this context.

The Lie algebra $\mathfrak{g}$ of a matrix Lie group G is a Lie algebra of matrices in which the bracket is the usual commutator.

Let M be a smooth manifold. Recall that a principal G-bundle P over M can be defined by an open cover (U_α) of M and a gluing cocycle

$$g_{\alpha\beta} : U_{\alpha\beta} \to G.$$

The Lie group G operates on its Lie algebra $\mathfrak{g}$ via the *adjoint action*

$$\mathrm{Ad} : G \to GL(\mathfrak{g}), \quad g \mapsto \mathrm{Ad}(g) \in GL(\mathfrak{g}),$$

where

$$\mathrm{Ad}(g)X := gXg^{-1}, \quad \forall X \in \mathfrak{g}, \ g \in G.$$

We denote by $\mathrm{Ad}(P)$ the vector bundle with standard fiber $\mathfrak{g}$ associated to P via the adjoint representation. In other words, $\mathrm{Ad}(P)$ is the vector bundle defined by the open cover (U_α), and gluing cocycle

$$\mathrm{Ad}(g_{\alpha\beta}) : U_{\alpha\beta} \to \mathrm{GL}(\mathfrak{g}).$$

The bracket operation in the fibers of $\mathrm{Ad}(P)$ induces a bilinear map

$$[\cdot,\cdot] : \Omega^k\big(\mathrm{Ad}(P)\big) \times \Omega^\ell\big(\mathrm{Ad}(P)\big) \to \Omega^{k+\ell}\big(\mathrm{Ad}(P)\big),$$

defined by

$$[\omega^k \otimes X, \eta^\ell \otimes Y] := (\omega^k \wedge \eta^\ell) \otimes [X,Y], \tag{8.1.1}$$

for all $\omega^k \in \Omega^k(M)$, $\eta^\ell \in \Omega^\ell(M)$, and $X, Y \in \Omega^0\big(\mathrm{Ad}(P)\big)$.

Exercise 8.1.1. Prove that for any $\omega, \eta, \phi \in \Omega^*(\mathrm{Ad}(P))$ the following hold.

$$[\omega,\eta] = -(-1)^{|\omega|\cdot|\eta|}[\eta,\omega], \tag{8.1.2}$$

$$[[\omega,\eta],\phi] = [[\omega,\phi],\eta] + (-1)^{|\omega|\cdot|\phi|}[\omega,[\eta,\phi]]. \tag{8.1.3}$$

In other words, $\big(\Omega^\bullet\big(\mathrm{Ad}(P)\big), [\,,\,]\big)$ is a super Lie algebra. $\qquad\square$

Using Proposition 3.3.5 as inspiration we introduce the following fundamental concept.

Definition 8.1.2. (a) A *connection* on the principal bundle P defined by the open cover $\mathcal{U} = (U_\alpha)_{\alpha\in A}$, and the gluing cocycle $g_{\beta\alpha} : U_{\alpha\beta} \to G$ is a collection

$$A_\alpha \in \Omega^1(U_\alpha) \otimes \mathfrak{g},$$

satisfying the transition rules

$$
\begin{aligned}
A_\beta(x) &= g_{\alpha\beta}^{-1}(x)dg_{\alpha\beta}(x) + g_{\alpha\beta}^{-1}(x)A_\alpha(x)g_{\alpha\beta}(x) \\
&= -(dg_{\beta\alpha}(x))g_{\beta\alpha}^{-1}(x) + g_{\beta\alpha}(x)A_\alpha(x)g_{\beta\alpha}^{-1}(x), \quad \forall x \in U_{\alpha\beta}.
\end{aligned}
$$

We will denote by $\mathcal{A}(\mathcal{U}, g_{\bullet\bullet})$ the set of connections defined by the open cover $\mathcal{U}$ and the gluing cocycle $g_{\bullet\bullet} : U_{\bullet\bullet} \to G$.

(b) The *curvature* of a connection $A \in \mathcal{A}(\mathcal{U}, g_{\bullet\bullet})$ is defined as the collection $F_\alpha \in \Omega^2(U_\alpha) \otimes \mathfrak{g}$ where

$$F_\alpha = dA_\alpha + \frac{1}{2}[A_\alpha, A_\alpha]. \qquad\square$$

Remark 8.1.3. Given an open cover (U_α) of M, then two gluing cocycles

$$g_{\beta\alpha}, h_{\beta\alpha} : U_{\alpha\beta} \to G,$$

define isomorphic principal bundles if and only if there exist smooth maps

$$T_\alpha : U_\alpha \to G,$$

such that

$$h_{\beta\alpha}(x) = T_\beta g_{\beta\alpha}(x) T_\alpha(x)^{-1}, \quad \forall \alpha, \beta, \quad \forall x \in U_{\alpha\beta}.$$

If this happens, we say that the cocycles $g_{\bullet\bullet}$ and $h_{\bullet\bullet}$ are cohomologous.

Suppose $(g_{\alpha\beta})$, $(h_{\alpha\beta})$ are two such cohomologous cocycles. Then the map T_α induces a well defined correspondence

$$\mathcal{T} : \mathcal{A}(\mathcal{U}, g_{\bullet\bullet}) \to \mathcal{A}(\mathcal{U}, h_{\bullet\bullet})$$

given by

$$\mathcal{A}(\mathcal{U}, g_{\bullet\bullet}) \ni (A_\alpha) \xmapsto{\mathcal{T}} \left(B_\alpha := -(dT_\alpha)T_\alpha^{-1} + T_\alpha A_\alpha T_\alpha^{-1} \right) \in \mathcal{A}(\mathcal{U}, h_{\bullet\bullet}).$$

The correspondence T is a *bijection*.

Suppose $\mathcal{U} = (U_\alpha)_{\alpha \in A}$ and $\mathcal{V} = (V_i)_{i \in I}$ are open covers, and $g_{\beta\alpha} : U_{\alpha\beta} \to G$ is a gluing cocycle. Suppose that the open cover $(V_i)_{i \in I}$ is finer than the cover $(U_\alpha)_{\alpha \in A}$, i.e., there exists a map $\varphi : I \to A$ such that

$$V_i \subset U_{\varphi(i)}, \quad \forall i \in I.$$

We obtain a new gluing cocycle $g_{ij}^\varphi : V_{ij} \to G$ given by

$$g_{ij}^\varphi(x) = g_{\varphi(i)\varphi(j)}(x), \quad \forall i, j \in I, \quad x \in V_{ij}.$$

This new cocycle defines a principal bundle isomorphic to the principal bundle defined by the cocycle $(g_{\beta\alpha})$. If (A_α) is a connection defined by the open cover (U_α) and the cocycle $(g_{\beta\alpha})$, then it induces a connection

$$A_i^\varphi = A_{\varphi(i)}|_{V_i}$$

defined by the open cover (V_i), and the cocycle g_{ij}. We say that the connection A^φ is the $\mathcal{V}$-refinement of the connection A. The correspondence

$$\mathcal{T}_{\mathcal{V}\mathcal{U}} : \mathcal{A}(\mathcal{U}, g_{\bullet\bullet}) \to \mathcal{A}(\mathcal{V}, g_{\bullet\bullet}^\varphi), \quad A \mapsto A^\varphi$$

is also a bijection.

Finally, given two pairs (open cover, gluing cocycle), $(\mathcal{U}, g_{\bullet\bullet})$ and $(\mathcal{V}, h_{\bullet\bullet})$, both describing the principal bundle $P \to M$, there exists an open cover $\mathcal{W}$, finer that both $\mathcal{U}$ and $\mathcal{W}$, and cohomologous gluing cocycles $\bar{g}_{\bullet\bullet}, \bar{h}_{\bullet\bullet} : \mathcal{W}_{\bullet\bullet} \to G$ refining $g_{\bullet\bullet}$ and respectively $h_{\bullet\bullet}$, such that the induced maps

$$\mathcal{A}(\mathcal{U}, g_{\bullet\bullet}) \longrightarrow \mathcal{A}(\mathcal{W}, \bar{g}_{\bullet\bullet}) \longrightarrow \mathcal{A}(\mathcal{W}, \bar{h}_{\bullet\bullet}) \longleftarrow \mathcal{A}(\mathcal{V}, h_{\bullet\bullet})$$

are bijections.

For simplicity, we will denote by $\mathcal{A}(P)$, any of these isomorphic spaces $\mathcal{A}(\mathcal{U}, g_{\bullet\bullet})$.

☞ *All the constructions that we will perform in the remainder of this section are compatible with the above isomorphisms but, in order to keep the presentation as transparent as possible, in our proofs we will not keep track of these isomorphisms. We are convinced that the reader can easily supply the obvious, and repetitious missing details.* □

Proposition 8.1.4. *Suppose $P \to M$ is a principal G-bundle described by an open cover $\mathcal{U} = (U_\alpha)_{\alpha \in A}$, and gluing cocycle $g_{\beta\alpha} : U_{\alpha\beta} \to G$.*
(a) The set $\mathcal{A}(P)$ of connections on P is an affine space modelled by $\Omega^1(\operatorname{Ad}(P))$.
(b) For any connection $A = (A_\alpha) \in \mathcal{A}(\mathcal{U}, g_{\bullet\bullet})$, the collection (F_α) defines a global $\operatorname{Ad}(P)$-valued 2-form. We will denote it by F_A or $F(A)$, and we will refer to it as the curvature of the connection A.
(c) **(The Bianchi identity.)**

$$dF_\alpha + [A_\alpha, F_\alpha] = 0, \quad \forall \alpha. \tag{8.1.4}$$

Proof.   (a) If $(A_\alpha)_{\alpha \in A}$, $(B_\alpha)_{\alpha \in A} \in \mathcal{A}(P)$, then their difference $C_\alpha = A_\alpha - B_\alpha$ satisfies the gluing rules

$$C_\beta = g_{\beta\alpha} C_\alpha g_{\beta\alpha}^{-1},$$

so that it defines an element of $\Omega^1(\operatorname{Ad}(P))$. Conversely, if $(A_\alpha) \in \mathcal{A}(P)$, and $\omega \in \Omega^1(\operatorname{Ad}(P))$, then ω is described by a collection of $\mathfrak{g}$-valued 1-forms $\omega_\alpha \in \Omega^1(U_\alpha) \otimes \mathfrak{g}$, satisfying the gluing rules

$$\omega_\beta|_{U_{\alpha\beta}} = g_{\beta\alpha} \omega_\alpha|_{U_{\alpha\beta}} g_{\beta\alpha}^{-1}.$$

The collection $A'_\alpha := A_\alpha + \omega_\alpha$ is then a connection on P. This proves that if $\mathcal{A}(P)$ is nonempty, then it is an affine space modelled by $\Omega^1(\operatorname{Ad}(P))$.

To prove that $\mathcal{A}(P) \neq \emptyset$ we consider a partition of unity subordinated to the cover U_α. More precisely, we consider a family of nonnegative smooth functions $u_\alpha : M \to \mathbb{R}$, $\alpha \in A$, such that $\operatorname{supp} u_\alpha \subset U_\alpha$, $\forall \alpha$, and

$$\sum_\alpha u_\alpha = 1.$$

For every $\alpha \in A$ we set

$$B_\alpha := \sum_\gamma u_\gamma g_{\gamma\alpha}^{-1} dg_{\gamma\alpha} \in \Omega^1(U_\alpha) \otimes \mathfrak{g}.$$

Since $g_{\gamma\alpha} = g_{\alpha\gamma}^{-1}$ we deduce $g_{\gamma\alpha}^{-1} dg_{\gamma\alpha} = -(dg_{\alpha\gamma}) g_{\alpha\gamma}^{-1}$, so that

$$B_\alpha = -\sum_\gamma u_\gamma (dg_{\alpha\gamma}) g_{\alpha\gamma}^{-1}.$$

Then, on the overlap $U_{\alpha\beta}$, we have the equality

$$B_\beta - g_{\beta\alpha} B_\alpha g_{\beta\alpha}^{-1} = -\sum_\gamma u_\gamma (dg_{\beta\gamma}) g_{\beta\gamma}^{-1} + \sum_\gamma u_\gamma g_{\beta\alpha} (dg_{\alpha\gamma}) g_{\alpha\gamma}^{-1} g_{\beta\alpha}^{-1}.$$

The cocycle condition implies that

$$dg_{\beta\gamma} = (dg_{\beta\alpha}) g_{\alpha\gamma} + g_{\beta\alpha} (dg_{\alpha\gamma}),$$

so that

$$(dg_{\beta\gamma}) g_{\beta\gamma}^{-1} = (dg_{\beta\gamma}) g_{\gamma\beta} = (dg_{\beta\alpha}) g_{\alpha\beta} + g_{\beta\alpha} (dg_{\alpha\gamma}) g_{\gamma\beta}.$$

Hence

$$-\sum_\gamma u_\gamma(dg_{\beta\gamma})g_{\beta\gamma}^{-1} = -(dg_{\beta\alpha})g_{\alpha\beta} - \sum_\gamma u_\gamma g_{\beta\alpha}(dg_{\alpha\gamma})g_{\gamma\beta}.$$

Using the cocycle condition again, we deduce $g_{\alpha\gamma}^{-1}g_{\beta\alpha}^{-1} = g_{\gamma\beta}$, so that

$$\sum_\gamma u_\gamma g_{\beta\alpha}(dg_{\alpha\gamma})g_{\alpha\gamma}^{-1}g_{\beta\alpha}^{-1} = \sum_\gamma u_\gamma g_{\beta\alpha}(dg_{\alpha\gamma})g_{\gamma\beta}.$$

Hence, on the overlap $U_{\alpha\beta}$ we have

$$B_\beta = -(dg_{\beta\alpha})g_{\beta\alpha}^{-1}g_{\beta\alpha}B_\alpha g_{\beta\alpha}^{-1} = g_{\alpha\beta}dg_{\alpha\beta} + g_{\alpha\beta}^{-1}B_\alpha g_{\alpha\beta}.$$

This shows that the collection (B_α) defines a connection on P.

(b) We need to check that the forms F_α satisfy the gluing rules

$$F_\beta = g^{-1}F_\alpha g,$$

where $g = g_{\alpha\beta} = g_{\beta\alpha}^{-1}$. We have

$$F_\beta = dA_\beta + \frac{1}{2}[A_\beta, A_\beta]$$

$$= d(g^{-1}dg + g^{-1}A_\alpha g) + \frac{1}{2}[g^{-1}dg + g^{-1}A_\alpha g, g^{-1}dg + g^{-1}A_\alpha g].$$

Set $\varpi := g^{-1}dg$. Using (8.1.2) we get

$$F_\beta = d\varpi + \frac{1}{2}[\varpi, \varpi]$$

$$+ d(g^{-1}A_\alpha g) + [\varpi, g^{-1}A_\alpha g] + \frac{1}{2}[g^{-1}A_\alpha g, g^{-1}A_\alpha g]. \qquad (8.1.5)$$

We will check two things.

A. The *Maurer-Cartan* structural equations.

$$d\varpi + \frac{1}{2}[\varpi, \varpi] = 0.$$

B.

$$d(g^{-1}A_\alpha g) + [\varpi, g^{-1}A_\alpha g] = g^{-1}(dA_\alpha)g.$$

Proof of A. Let us first introduce a new operation. Let $\underline{gl}(n, \mathbb{K})$ denote the associative algebra of $\mathbb{K}$-valued $n \times n$ matrices. There exists a natural operation

$$\wedge : \Omega^k(U_\alpha) \otimes \underline{gl}(n, \mathbb{K}) \times \Omega^\ell(U_\alpha) \otimes \underline{gl}(n, \mathbb{K}) \to \Omega^{k+\ell}(U_\alpha) \otimes \underline{gl}(n, \mathbb{K}),$$

uniquely defined by

$$(\omega^k \otimes A) \wedge (\eta^\ell \otimes B) = (\omega^k \wedge \eta^\ell) \otimes (A \cdot B), \qquad (8.1.6)$$

where $\omega^k \in \Omega^k(U_\alpha)$, $\eta^\ell \in \Omega^\ell(U_\alpha)$ and $A, B \in \underline{gl}(n, \mathbb{K})$ (see also Example 3.3.12). The space $\underline{gl}(n, \mathbb{K})$ is naturally a Lie algebra with respect to the commutator of two matrices. This structure induces a bracket

$$[\bullet, \bullet] : \Omega^k(U_\alpha) \otimes \underline{gl}(n, \mathbb{K}) \times \Omega^\ell(U_\alpha) \otimes \underline{gl}(n, \mathbb{K}) \to \Omega^{k+\ell}(U_\alpha) \otimes \underline{gl}(n, \mathbb{K})$$

defined as in (8.1.1). A very simple computation yields the following identity.

$$\omega \wedge \eta = \frac{1}{2}[\omega, \eta] \quad \forall \omega, \eta \in \Omega^1(U_\alpha) \otimes \underline{gl}(n, \mathbb{K}). \tag{8.1.7}$$

The Lie group lies inside $GL(n, \mathbb{K})$, so that its Lie algebra $\mathfrak{g}$ lies inside $\underline{gl}(n, \mathbb{K})$. We can think of the map $g_{\alpha\beta}$ as a matrix valued map, so that we have

$$
\begin{aligned}
d\varpi &= d(g^{-1}dg) = (dg^{-1}) \wedge dg \\
&= -(g^{-1} \cdot dg \cdot g^{-1})dg = -(g^{-1}dg) \wedge (g^{-1}dg) \\
&= -\varpi \wedge \varpi \overset{(8.1.7)}{=} -\frac{1}{2}[\varpi, \varpi].
\end{aligned}
$$

Proof of B. We compute

$$
\begin{aligned}
d(g^{-1}A_\alpha g) &= (dg^{-1}A_\alpha) \cdot g + g^{-1}(dA_\alpha)g + g^{-1}A_\alpha dg \\
&= -g^{-1} \cdot dg \cdot g^{-1} \wedge A_\alpha \cdot g + g^{-1}(dA_\alpha)g + (g^{-1}A_\alpha g) \wedge g^{-1}dg \\
&= -\varpi \wedge g^{-1}A_\alpha g + g^{-1}A_\alpha g \wedge \varpi + g^{-1}(dA_\alpha)g \\
&\overset{(8.1.7)}{=} -\frac{1}{2}[\varpi, g^{-1}A_\alpha g] + \frac{1}{2}[\varpi, g^1 A_\alpha g] + g^{-1}(dA_\alpha)g \\
&\overset{(8.1.2)}{=} -[\varpi, g^{-1}A_\alpha g] + g^{-1}(dA_\alpha)g.
\end{aligned}
$$

Part (b) of the proposition now follows from **A**, **B** and (8.1.5).
(c) First, we let the reader check the following identity

$$d[\omega, \eta] = [d\omega, \eta] + (-1)^{|\omega|}[\omega, d\eta], \tag{8.1.8}$$

where $\omega, \eta \in \Omega^*(U_\alpha) \otimes \mathfrak{g}$. Using the above equality we get

$$
\begin{aligned}
d(F_\alpha) &= \frac{1}{2}\{[dA_\alpha, A_\alpha] - [A_\alpha, dA_\alpha]\} \overset{(8.1.2)}{=} [dA_\alpha, A_\alpha] \\
&= [F_\alpha, A_\alpha] - \frac{1}{2}[[A_\alpha, A_\alpha], A_\alpha] \overset{(8.1.3)}{=} [F_\alpha, A_\alpha].
\end{aligned}
$$

The proposition is proved. $\qquad\square$

Exercise 8.1.5. Let $\omega_\alpha \in \Omega^k(U_\alpha) \otimes \mathfrak{g}$ satisfy the gluing rules

$$\omega_\beta = g_{\beta\alpha}\omega_\alpha g_{\beta\alpha}^{-1} \quad \text{on } U_{\alpha\beta}.$$

In other words, the collection ω_α defines a global k-form $\omega \in \Omega^k(\text{Ad}(P))$. Prove that the collection

$$d\omega_\alpha + [A_\alpha, \omega_\alpha],$$

defines a global $\text{Ad}(P)$-valued $(k+1)$-form on M which we denote by $d_A\omega$. Thus, the Bianchi identity can be rewritten as $d_A F(A) = 0$, for any $A \in \mathcal{A}(P)$. $\qquad\square$

Remark 8.1.6. Suppose $P \to M$ is a principal G-bundle given by the open cover $\mathcal{U} = (U_\alpha)$ and gluing cocycle $g_{\beta\alpha} : U_{\alpha\beta} \to G$. A *gauge transformation* of P is by definition, a collection of smooth maps

$$T_\alpha : U_\alpha \to G$$

satisfying the gluing rules

$$T_\beta(x) = g_{\beta\alpha}(x)T_\alpha(x)g_{\beta\alpha}(x)^{-1}, \quad \forall x \in U\alpha\beta.$$

Observe that a gauge transformation is a section of the G-fiber bundle

$$G \hookrightarrow C(P) \twoheadrightarrow M,$$

(see Definition 2.3.15) with standard fiber G, symmetry group G, where the symmetry group G acts on itself by conjugation

$$G \times G \ni (g, h) \overset{C}{\longmapsto} C_g(h) := ghg^1 \in G.$$

The set of gauge transformations forms a group with respect to the operation

$$(S_\alpha) \cdot (T_\alpha) := (S_\alpha T_\alpha).$$

We denote by $\mathcal{G}(\mathcal{U}, g_{\bullet\bullet})$ this group. One can verify that if P is described the (open cover, gluing cocycle)-pair $(\mathcal{V}, h_{\bullet\bullet})$, then the groups $\mathcal{G}(\mathcal{U}, g_{\bullet\bullet})$ and $\mathcal{G}(\mathcal{V}, h_{\bullet\bullet})$ are isomorphic. We denote by $\mathcal{G}(P)$ the isomorphism class of all these groups.

The group $\mathcal{G}(\mathcal{U}, g_{\bullet\bullet})$ acts on $\mathcal{A}(\mathcal{U}, g_{\bullet\bullet})$ according to

$$\mathcal{G}(P) \times \mathcal{A}(P) \ni (T, A) \mapsto TAT^{-1} := \left(-(dT_\alpha)T_\alpha^{-1} + T_\alpha A_\alpha T_\alpha^{-1} \right) \in \mathcal{A}(P).$$

The group $\mathcal{G}(P)$ also acts on the vector spaces $\Omega^\bullet\big(\operatorname{Ad}(P)\big)$ and, for any $A \in \mathcal{A}(P)$, $T \in \mathcal{G}(P)$ we have

$$F_{TAT^{-1}} = TF_A T^{-1}.$$

We say that two connections $A^0, A^1 \in \mathcal{A}(P)$ are gauge equivalent if there exists $T \in \mathcal{G}(P)$ such that $A^1 = TA^0T^{-1}$. $\qquad\square$

8.1.2 *G-vector bundles*

Definition 8.1.7. Let G be a Lie group, and $E \to M$ a vector bundle with standard fiber a vector space V. A *G-structure* on E is defined by the following collection of data.

(a) A representation $\rho : G \to GL(V)$.

(b) A principal G-bundle P over M such that E is associated to P via ρ. In other words, there exists an open cover (U_α) of M, and a gluing cocycle $g_{\alpha\beta} : U_{\alpha\beta} \to G$, such that the vector bundle E can be defined by the cocycle

$$\rho(g_{\alpha\beta}) : U_{\alpha\beta} \to GL(V).$$

We denote a G-structure by the pair (P, ρ).

Two G-structures (P_i, ρ_i) on E, $i = 1, 2$, are said to be isomorphic, if the representations ρ_i are isomorphic, and the principal G-bundles P_i are isomorphic.   □

Example 8.1.8. Let $E \to M$ be a rank r real vector bundle over a smooth manifold M. A metric on E allows us to talk about orthonormal moving frames. They are easily produced from arbitrary ones via the Gram-Schmidt orthonormalization technique. In particular, two different orthonormal local trivializations are related by a transition map valued in the orthogonal group $O(r)$, so that a metric on a bundle allows one to replace an arbitrary collection of gluing data by an equivalent (cohomologous) one with transitions in $O(r)$. In other words, a metric on a bundle induces an $O(r)$ structure. The representation ρ is in this case the natural injection $O(r) \hookrightarrow \mathrm{GL}(r, \mathbb{R})$.

Conversely, an $O(r)$ structure on a rank r real vector bundle is tantamount to choosing a metric on that bundle.

Similarly, a Hermitian metric on a rank k complex vector bundle defines an $U(k)$-structure on that bundle.   □

Let $E = (P, \rho, V)$ be a G-vector bundle. Assume P is defined by an open cover (U_α), and gluing cocycle

$$g_{\alpha\beta} : U_{\alpha\beta} \to G.$$

If the collection $\{A_\alpha \in \Omega^1(U_\alpha) \otimes \mathfrak{g}\}$ defines a connection on the principal bundle P, then the collection $\rho_*(A_\alpha)$ defines a connection on the vector bundle E. Above, $\rho_* : \mathfrak{g} \to \mathrm{End}(V)$ denotes the derivative of ρ at $1 \in G$. A connection of E obtained in this manner is said to be compatible with the G-structure. Note that if $F(A_\alpha)$ is the curvature of the connection on P, then the collection $\rho_*(F(A_\alpha))$ coincides with the curvature $F(\rho_*(A_\alpha))$ of the connection $\rho_*(A_\alpha)$.

For example, a connection compatible with some metric on a vector bundle is compatible with the orthogonal/unitary structure of that bundle. The curvature of such a connection is skew-symmetric which shows the infinitesimal holonomy is an infinitesimal orthogonal/unitary transformation of a given fiber.

8.1.3   *Invariant polynomials*

Let V be a vector space over $\mathbb{K} = \mathbb{R}, \mathbb{C}$. Consider the symmetric power

$$S^k(V^*) \subset (V^*)^{\otimes k},$$

which consists of symmetric, multilinear maps

$$\varphi : V \times \cdots \times V \to \mathbb{K}.$$

Note that any $\varphi \in S^k(V^*)$ is completely determined by

$$P_\varphi(v) = \varphi(v, \ldots, v).$$

This follows immediately from the *polarization formula*

$$\varphi(v_1, \ldots, v_k) = \frac{1}{k!} \frac{\partial^k}{\partial t_1 \cdots \partial t_k} P_\varphi(t_1 v_1 + \cdots + t_k v_k).$$

If $\dim V = n$ then, fixing a basis of V, we can identify $S^k(V^*)$ with the space of degree k homogeneous polynomials in n variables.

Assume now that $\mathcal{A}$ is a $\mathbb{K}$-algebra with 1. Starting with $\varphi \in S^k(V^*)$ we can produce a $\mathbb{K}$-multilinear map

$$\varphi = \varphi_{\mathcal{A}} : (\mathcal{A} \otimes V) \times \cdots \times (\mathcal{A} \otimes V) \to \mathcal{A},$$

uniquely determined by

$$\varphi(a_1 \otimes v_1, \ldots, a_k \otimes v_k) = \varphi(v_1, \ldots, v_k) a_1 a_2 \cdots a_k \in \mathcal{A}.$$

If moreover the algebra $\mathcal{A}$ is *commutative*, then $\varphi_{\mathcal{A}}$ is uniquely determined by the polynomial

$$P_\varphi(x) = \varphi_{\mathcal{A}}(x, \ldots, x) \quad x \in \mathcal{A} \otimes V.$$

Remark 8.1.9. Let us emphasize that when $\mathcal{A}$ *is not commutative*, then the above function *is not* symmetric in its variables. For example, if $a_1 a_2 = -a_2 a_1$, then

$$P(a_1 X_1, a_2 X_2, \cdots) = -P(a_2 X_2, a_1 X_1, \cdots).$$

For applications to geometry, $\mathcal{A}$ will be the algebra $\Omega^\bullet(M)$ of complex valued differential forms on a smooth manifold M. When restricted to the commutative subalgebra

$$\Omega^{even}(M) = \bigoplus_{k \geq 0} \Omega^{2k}(M) \otimes \mathbb{C},$$

we do get a symmetric function. $\quad\square$

Example 8.1.10. Let $V = \underline{gl}(n, \mathbb{C})$. For each matrix $T \in V$ we denote by $c_k(T)$ the coefficient of λ^k in the characteristic polynomial

$$c_\lambda(T) := \det\left(\mathbb{1} - \frac{\lambda}{2\pi i} T\right) = \sum_{k \geq 0} c_k(T) \lambda^k, \quad (i = \sqrt{-1}).$$

Then, $c_k(T)$ is a degree k homogeneous polynomial in the entries of T. For example,

$$c_1(T) = -\frac{1}{2\pi i} \operatorname{tr} T, \quad c_n(T) = \left(-\frac{1}{2\pi i}\right)^n \det T.$$

Via polarization, $c_k(T)$ defines an element of $S^k(\underline{gl}(n, \mathbb{C})^*)$.

If $\mathcal{A}$ is a commutative $\mathbb{C}$-algebra with 1, then $\mathcal{A} \otimes \underline{gl}(n, \mathbb{C})$ can be identified with the space $\underline{gl}(n, \mathcal{A})$ of $n \times n$ matrices with entries in $\mathcal{A}$. For each $T \in \underline{gl}(n, \mathcal{A})$ we have

$$\det\left(\mathbb{1} - \frac{\lambda}{2\pi i} T\right) \in \mathcal{A}[\lambda],$$

and $c_k(T)$ continues to be the coefficient of λ^k in the above polynomial. $\quad\square$

Consider now a matrix Lie group G. The adjoint action of G on its Lie algebra $\mathfrak{g}$ induces an action on $S^k(\mathfrak{g}^*)$ still denoted by Ad. We denote by $I^k(G)$ the Ad-invariant elements of $S^k(\mathfrak{g}^*)$. It consists of those $\varphi \in S^k(\mathfrak{g}^*)$ such that

$$\varphi(gX_1g^{-1}, \ldots, gX_kg^{-1}) = \varphi(X_1, \ldots, X_k),$$

for all $X_1, \ldots, X_k \in \mathfrak{g}$. Set

$$I^\bullet(G) := \bigoplus_{k \geq 0} I^k(G) \text{ and } I^{\bullet\bullet}(G) := \prod_{k \geq 0} I^k(G).$$

The elements of $I^\bullet(G)$ are usually called *invariant polynomials*. The space $I^{\bullet\bullet}(G)$ can be identified (as vector space) with the space of Ad-invariant formal power series with variables from $\mathfrak{g}^*$.

Example 8.1.11. Let $G = \mathrm{GL}(n, \mathbb{C})$, so that $\mathfrak{g} = \underline{\mathrm{gl}}(n, \mathbb{C})$. The map

$$\underline{\mathrm{gl}}(n, \mathbb{C}) \ni X \mapsto \mathrm{tr}\, \exp(X),$$

defines an element of $I^{\bullet\bullet}(GL(n, \mathbb{C}))$. To see this, we use the "Taylor expansion"

$$\exp(X) = \sum_{k \geq 0} \frac{1}{k!} X^k,$$

which yields

$$\mathrm{tr}\, \exp(X) = \sum_{k \geq 0} \frac{1}{k!} \mathrm{tr}\, X^k.$$

For each k, $\mathrm{tr}\, X^k \in I^k(GL(n, \mathbb{C}))$ since

$$\mathrm{tr}\, (gXg^{-1})^k = \mathrm{tr}\, gX^kg^{-1} = \mathrm{tr}\, X^k. \qquad \square$$

Proposition 8.1.12. *Let $\varphi \in I^k(G)$. Then for any $X, X_1, \ldots, X_k \in \mathfrak{g}$ we have*

$$\varphi([X, X_1], X_2, \ldots, X_k) + \cdots + \varphi(X_1, X_2, \ldots, [X, X_k]) = 0. \qquad (8.1.9)$$

Proof. The proposition follows immediately from the equality

$$\frac{d}{dt}\Big|_{t=0} \varphi(e^{tX}X_1e^{-tX}, \ldots, e^{tX}X_ke^{-tX}) = 0. \qquad \square$$

Let us point out a useful identity. If $P \in I^k(\mathfrak{g})$, U is an open subset of $\mathbb{R}^n$, and

$$F_i = \omega_i \otimes X_i \in \Omega^{d_i}(U) \otimes \mathfrak{g}, \quad A = \omega \otimes X \in \Omega^d(U) \otimes \mathfrak{g}$$

then

$$P(F_1, \ldots, F_{i-1}, [A, F_i], F_{i+1} \ldots, F_k)$$
$$= (-1)^{d(d_1 + \cdots d_{i-1})} \omega\omega_1 \cdots \omega_k P(X_1, \cdots, [X, X_i], \cdots X_k).$$

In particular, if $F_1, \ldots, F_{k-1}$ have even degree, we deduce that for every $i = 1, \ldots, k$ we have

$$P(F_1, \ldots, F_{i-1}, [A, F_i], F_{i+1}, \ldots, F_k) = \omega\omega_1 \cdots \omega_k P(X_1, \ldots, [X, X_i], \ldots, X_k).$$

Summing over i, and using the Ad-invariance of the polynomial P, we deduce

$$\sum_{i=1}^k P(F_1, \ldots, F_{i-1}, [A, F_i], F_{i+1}, \ldots, F_k) = 0, \qquad (8.1.10)$$

$\forall F_1, \ldots, F_{k-1} \in \Omega^{even}(U) \otimes \mathfrak{g}, \, F_k, A \in \Omega^\bullet(U) \otimes \mathfrak{g}.$

8.1.4 The Chern-Weil theory

Let G be a matrix Lie group with Lie algebra $\mathfrak{g}$, and $P \to M$ be a principal G-bundle over the smooth manifold M.

Assume P is defined by an open cover (U_α), and a gluing cocycle

$$g_{\alpha\beta} : U_{\alpha\beta} \to G.$$

Pick $A \in \mathcal{A}(P)$ defined by the collection $A_\alpha \in \Omega^1(U_\alpha) \otimes \mathfrak{g}$. Its curvature is then defined by the collection

$$F_\alpha = dA_\alpha + \frac{1}{2}[A_\alpha, A_\alpha].$$

Given $\phi \in I^k(G)$, we can define as in the previous section (with $\mathcal{A} = \Omega^{even}(U_\alpha)$, $V = \mathfrak{g}$)

$$P_\phi(F_\alpha) := \phi(F_\alpha, \dots, F_\alpha) \in \Omega^{2k}(U_\alpha).$$

Because ϕ is Ad-invariant, and $F_\beta = g_{\beta\alpha} F_\alpha g_{\beta\alpha}^{-1}$, we deduce

$$P_\phi(F_\alpha) = P_\phi(F_\beta) \text{ on } U_{\alpha\beta},$$

so that the locally defined forms $P_\phi(F_\alpha)$ patch-up to a global $2k$-form on M which we denote by $\phi(F_A)$.

Theorem 8.1.13 (Chern-Weil). *(a) The form $\phi(F_A)$ is closed, $\forall A \in \mathcal{A}(P)$.*
(b) If A^0, $A^1 \in \mathcal{A}(P)$, then the forms $\phi(F_{A^0})$, and $\phi(F_{A^1})$ are cohomologous. In other words, the closed form $\phi(F_A)$ defines a cohomology class in $H^{2k}(M)$ which is independent of the connection $A \in \mathcal{A}(P)$.

Proof. We use the Bianchi identity $dF_\alpha = -[A_\alpha, F_\alpha]$. The Leibniz' rule yields

$$d\phi(F_\alpha, \dots, F_\alpha) = \phi(dF_\alpha, F_\alpha, \dots, F_\alpha) + \dots + \phi(F_\alpha, \dots, F_\alpha, dF_\alpha)$$

$$= -\phi([A_\alpha, F_\alpha], F_\alpha, \dots, F_\alpha) - \dots - \phi(F_\alpha, \dots, F_\alpha, [A_\alpha, F_\alpha]) \overset{(8.1.9)}{=} 0.$$

(b) Let $A^i \in \mathcal{A}(P)$ $(i = 0, 1)$ be defined by the collections

$$A_\alpha^i \in \Omega^1(U_\alpha) \otimes \mathfrak{g}.$$

Set $C_\alpha := A_\alpha^1 - A_\alpha^0$. For $0 \leq t \leq 1$ we define $A_\alpha^t \in \Omega^1(U_\alpha) \otimes \mathfrak{g}$ by $A_\alpha^t := A_\alpha^0 + tC_\alpha$.

The collection (A_α^t) defines a connection $A^t \in \mathcal{A}(P)$, and $t \mapsto A^t \in \mathcal{A}(P)$ is an (affine) path connecting A^0 to A^1. Note that

$$C = (C_\alpha) = \dot{A}^t.$$

We denote by $F^t = (F_\alpha^t)$ the curvature of A^t. A simple computation yields

$$F_\alpha^t = F_\alpha^0 + t(dC_\alpha + [A_\alpha^0, C_\alpha]) + \frac{t^2}{2}[C_\alpha, C_\alpha]. \tag{8.1.11}$$

Hence,

$$\dot{F}_\alpha^t = dC_\alpha + [A_\alpha^0, C_\alpha] + t[C_\alpha, C_\alpha] = dC_\alpha + [A_\alpha^t, C_\alpha].$$

Consequently,

$$\phi(F_\alpha^1) - \phi(F_\alpha^0) = \int_0^1 \left\{ \phi(\dot{F}_\alpha^t, F_\alpha^t, \dots, F_\alpha^t) + \cdots + \phi(F_\alpha^t, \dots, F_\alpha^t, \dot{F}_\alpha^t) \right\} dt$$

$$= \int_0^1 \left\{ \phi(dC_\alpha, F_\alpha^t, \dots, F_\alpha^t) + \cdots + \phi(F_\alpha^t, \dots, F_\alpha^t, dC_\alpha) \right\} dt$$

$$+ \int_0^1 \left\{ \phi([A_\alpha^t, C_\alpha], F_\alpha^t, \dots, F_\alpha^t) + \cdots + \phi(F_\alpha^t, \dots, F_\alpha^t, [A_\alpha^t, C_\alpha]) \right\} dt.$$

Because the algebra $\Omega^{even}(U_\alpha)$ is commutative, we deduce

$$\phi(\omega_{\sigma(1)}, \dots, \omega_{\sigma(k)}) = \phi(\omega_1, \dots, \omega_k),$$

for all $\sigma \in S_k$ and any $\omega_1, \dots, \omega_k \in \Omega^{even}(U_\alpha) \otimes \mathfrak{g}$. Hence

$$\phi(F_\alpha^1) - \phi(F_\alpha^0) = k \int_0^1 \phi(F_\alpha^t, \dots, F_\alpha^t, dC_\alpha + [A_\alpha^t C_\alpha]) dt.$$

We claim that

$$\phi(F_\alpha^t, \dots, F_\alpha^t, dC_\alpha + [A_\alpha^t, C_\alpha]) = d\phi(F_\alpha^t, \dots, F_\alpha^t, C_\alpha).$$

Using the Bianchi identity we get

$$d\phi(F_\alpha^t, \dots, F_\alpha^t, C_\alpha)$$

$$= \phi(F_\alpha^t, \cdots, F_\alpha^t, dC_\alpha) + \phi(dF_\alpha^t, \cdots, F_\alpha^t, C_\alpha) + \cdots + \phi(F_\alpha^t, \cdots, dF_\alpha^t, C_\alpha)$$

$$= \phi(F_\alpha^t, \dots, F_\alpha^t, dC_\alpha)$$

$$-\phi(C_\alpha, [A_\alpha^t, F_\alpha^t], F_\alpha^t, \dots, F_\alpha^t) - \cdots - \phi(C_\alpha, F_\alpha^t, \dots, F_\alpha^t, [A_\alpha^t, F_\alpha^t])$$

$$= \phi(F_\alpha^t, \dots, F_\alpha^t, dC_\alpha + [A_\alpha^t, C_\alpha]) - \phi(F_\alpha^t, \dots, F_\alpha^t, [A_\alpha^t, C_\alpha])$$

$$-\phi([A_\alpha^t, F_\alpha^t], F_\alpha^t, \dots, F_\alpha^t, C_\alpha) - \cdots - \phi(F_\alpha^t, \dots, F_\alpha^t, [A_\alpha^t, F_\alpha^t], C_\alpha)$$

$$\stackrel{(8.1.10)}{=} \phi(F_\alpha^t, \dots, F_\alpha^t, dC_\alpha + [A_\alpha^t, C_\alpha]) = \phi(dC_\alpha + [A_\alpha^t, C_\alpha], F_\alpha^t, \dots, F_\alpha^t).$$

Hence

$$\phi(F_\alpha^1) - \phi(F_\alpha^0) = d \int_0^1 k\phi(\dot{A}_\alpha^t, F_\alpha^t, \dots, F_\alpha^t) dt. \tag{8.1.12}$$

We set

$$T_\phi(A_\alpha^1, A_\alpha^0) := \int_0^1 k\phi(\dot{A}_\alpha^t, F_\alpha^t, \dots, F_\alpha^t) dt.$$

Since $C_\beta = g_{\beta\alpha} C_\alpha g_{\beta\alpha}^{-1}$, and $F_\beta = g_{\beta\alpha} F_\alpha g_{\beta\alpha}^{-1}$ on $U_{\alpha\beta}$, we conclude from the Ad-invariance of ϕ that the collection $T_\phi(A_\alpha^1, A_\alpha^0)$ defines a global $(2k-1)$-form on M which we denote by $T_\phi(A^1, A^0)$, and we name it the *ϕ-transgression* from A^0 to A^1. We have thus established the *transgression formula*

$$\phi(F_{A^1}) - \phi(F_{A^0}) = dT_\phi(A^1, A^0). \tag{8.1.13}$$

The Chern-Weil theorem is proved. $\square$

Remark 8.1.14. Observe that for every Ad-invariant polynomial $\phi \in I^k(\mathfrak{g})$, any principal G-bundle P, any connection $A \in \mathcal{A}(P)$, and any gauge transformation $T \in \mathcal{G}(P)$ we have

$$\phi(F_A) = \phi(F_{TAT^{-1}}).$$

We say that the Chern-Weil construction is *gauge invariant*. $\square$

Example 8.1.15. Consider a matrix Lie group G with Lie algebra $\mathfrak{g}$, and denote by P_0 the trivial principal G-bundle over G, $P_0 = G \times G$. Denote by ϖ the tautological 1-form $\varpi = g^{-1}dg \in \Omega^1(G) \otimes \mathfrak{g}$. Note that for every left invariant vector field $X \in \mathfrak{g}$ we have

$$\varpi(X) = X.$$

Denote by d the trivial connection on P_0. Clearly d is a flat connection. Moreover, the Maurer-Cartan equation implies that[1] $d + \varpi$ is also a flat connection. Thus, for any $\phi \in I^k(G)$

$$\phi(F_d) = \phi(F_{(d+\varpi)}) = 0.$$

The transgression formula implies that the form

$$\tau_\phi = T_\phi(d + \varpi, d) = k \int_0^1 \phi(\varpi, F_{d+t\varpi}, \ldots, F_{d+t\varpi})dt \in \Omega^{2k-1}(G)$$

is closed.

A simple computation using the Maurer-Cartan equations shows that

$$\tau_\phi = \frac{k}{2^{k-1}} \left(\int_0^1 (t^2 - 1)^{k-1} dt \right) \cdot \phi(\varpi, [\varpi, \varpi], \cdots, [\varpi, \varpi])$$

$$= (-1)^{k-1} \frac{k}{2^{k-1}} \frac{2^{2k-1}k!(k-1)!}{(2k)!} \phi(\varpi, [\varpi, \varpi], \cdots, [\varpi, \varpi])$$

$$= (-1)^{k-1} \frac{2^k}{\binom{2k}{k}} \cdot \phi(\varpi, [\varpi, \varpi], \cdots, [\varpi, \varpi]).$$

We thus have a natural map $\tau : I^\bullet(G) \to H^{odd}(G)$ called *transgression*. The elements in the range of τ are called *transgressive*. When G is compact and connected, then a nontrivial result due to the combined efforts of H. Hopf, C. Chevalley, H. Cartan, A. Weil and L. Koszul states that the cohomology of G is generated as an $\mathbb{R}$-algebra by the transgressive elements. We refer to [21] for a beautiful survey of this subject. $\square$

Exercise 8.1.16. Let $G = SU(2)$. The Killing form κ is a degree 2 Ad-invariant polynomial on $\underline{su}(2)$. Describe $\tau_\kappa \in \Omega^3(G)$ and then compute

$$\int_G \tau_\kappa.$$

Compare this result with the similar computations in Subsection 7.4.3. $\square$

Let us now analyze the essentials of the Chern-Weil construction.
Input: (a) A principal G-bundle P over a smooth manifold M, defined by an open cover (U_α), and gluing cocycle $g_{\alpha\beta} : U_{\alpha\beta} \to G$.
(b) A connection $A \in \mathcal{A}(P)$ defined by the collection

$$A_\alpha \in \Omega^1(U_\alpha) \otimes \mathfrak{g},$$

---

[1]The connections d and $d + \varpi$ are in fact gauge equivalent.

satisfying the transition rules

$$A_\beta = g_{\alpha\beta}^{-1} dg_{\alpha\beta} + g_{\alpha\beta}^{-1} A_\alpha g_{\alpha\beta} \quad \text{on } U_{\alpha\beta}.$$

(c) $\phi \in I^k(G)$.

Output: A closed form $\phi(F(A)) \in \Omega^{2k}(M)$, whose cohomology class is independent of the connection A. We denote this cohomology class by $\phi(P)$.

Thus, the principal bundle P defines a map, called the *Chern-Weil correspondence*

$$\mathfrak{c}w_P : I^\bullet(G) \to H^\bullet(M) \quad \phi \mapsto \phi(P).$$

One can check easily that the map $\mathfrak{c}w_P$ is a morphism of $\mathbb{R}$-algebras.

Definition 8.1.17. Let M and N be two smooth manifolds, and $F : M \to N$ be a smooth map. If P is a principal G-bundle over N defined by an open cover (U_α), and gluing cocycle $g_{\alpha\beta} : U_{\alpha\beta} \to G$, then the *pullback* of P by F is the principal bundle $F^*(P)$ over M defined by the open cover $F^{-1}(U_\alpha)$, and gluing cocycle

$$F^{-1}(U_{\alpha\beta}) \xrightarrow{F} U_{\alpha\beta} \xrightarrow{g_{\alpha\beta}} G. \qquad \square$$

The pullback of a connection on P is defined similarly. The following result should be obvious.

Proposition 8.1.18. *(a) If P is a trivial G-bundle over the smooth manifold M, then $\phi(P) = 0 \in H^\bullet(M)$, for any $\phi \in I^\bullet(G)$.*
(b) Let $M \xrightarrow{F} N$ be a smooth map between the smooth manifolds M and N. Then, for every principal G-bundle over N, and any $\phi \in I^\bullet(G)$ we have

$$\phi(F^*(P)) = F^*(\phi(P)).$$

Equivalently, this means the diagram below is commutative.

$$
\begin{array}{ccc}
I^\bullet(G) & \xrightarrow{\mathfrak{c}w_P} & H^\bullet(N) \\
& \searrow{\scriptstyle \mathfrak{c}w_{F^\bullet(P)}} & \downarrow{\scriptstyle F^\bullet} \\
& & H^\bullet(M)
\end{array}
\qquad \square
$$

Denote by $\mathcal{P}_G$ the collection of smooth principal G-bundles (over smooth manifolds). For each $P \in \mathcal{P}_G$ we denote by $\mathcal{B}_P$ the base of P. Finally, we denote by $\mathcal{F}$ a *contravariant* functor from the category of smooth manifolds (and smooth maps) to the category of Abelian groups.

Definition 8.1.19. An $\mathcal{F}$-valued G-*characteristic class* is a correspondence

$$\mathcal{P}_G \ni P \mapsto c(P) \in \mathcal{F}(\mathcal{B}_P),$$

such that the following hold.

(a) $c(P) = 0$ if P is trivial.

(b) $\mathcal{F}(F)(c(P)) = c(F^*(P))$, for any smooth map $F : M \to N$, and any principal G-bundle $P \to N$. □

Hence, the Chern-Weil construction is just a method of producing G-characteristic classes valued in the DeRham cohomology.

Remark 8.1.20. (a) We see that each characteristic class provides a way of measuring the nontriviality of a principal G-bundle.

(b) A very legitimate question arises. Do there exist characteristic classes (in the DeRham cohomology) not obtainable via the Chern-Weil construction?

The answer is negative, but the proof requires an elaborate topological technology which is beyond the reach of this course. The interested reader can find the details in the monograph [75] which is the ultimate reference on the subject of characteristic classes.

(b) There exist characteristic classes valued in contravariant functors other then the DeRham cohomology. E.g., for each Abelian group A, the Čech cohomology with coefficients in the constant sheaf $\underline{A}$ defines a contravariant functor $H^\bullet(-, A)$, and using topological techniques, one can produce $H^\bullet(-, A)$-valued characteristic classes. For details we refer to [75], or the classical [93]. □

8.2 Important examples

We devote this section to the description of some of the most important examples of characteristic classes. In the process we will describe the invariants of some commonly encountered Lie groups.

8.2.1 *The invariants of the torus T^n*

The n-dimensional torus $T^n = U(1) \times \cdots \times U(1)$ is an Abelian Lie group, so that the adjoint action on its Lie algebra $\mathfrak{t}^n$ is trivial. Hence

$$I^\bullet(T^n) = S^\bullet((\mathfrak{t}^n)^*).$$

In practice one uses a more explicit description obtained as follows. Pick angular coordinates $0 \leq \theta^i \leq 2\pi$, $1 \leq i \leq n$, and set

$$x_j := -\frac{1}{2\pi i} d\theta^j.$$

The x_j's form a basis of $(\mathfrak{t}^n)^*$, and now we can identify

$$I^\bullet(T^n) \cong \mathbb{R}[x_1, \ldots, x_n].$$

8.2.2 *Chern classes*

Let E be a rank r complex vector bundle over the smooth manifold M. We have seen that a Hermitian metric on E induces an $U(r)$-structure (P, ρ), where ρ is the tautological representation

$$\rho : U(r) \hookrightarrow \mathrm{GL}(r, \mathbb{C}).$$

Exercise 8.2.1. Prove that different Hermitian metrics on E define isomorphic $U(r)$-structures. $\square$

Thus, we can identify such a bundle with the tautological principal $U(r)$-bundle of unitary frames. A connection on this $U(r)$-bundle is then equivalent with a linear connection ∇ on E compatible with a Hermitian metric $\langle \bullet, \bullet \rangle$, i.e.,

$$\nabla_X \langle \lambda u, v \rangle = \lambda \{ \langle \nabla_X u, v \rangle + \langle u, \nabla_X v \rangle \},$$

$\forall \lambda \in \mathbb{C}$, $u, v \in C^\infty(E)$, $X \in \mathrm{Vect}\,(M)$.

The characteristic classes of E are by definition the characteristic classes of the tautological principal $U(r)$-bundle. To describe these characteristic classes we need to elucidate the structure of the ring of invariants $I^\bullet(U(r))$.

The ring $I^\bullet(U(r))$ consists of symmetric, r-linear maps

$$\phi : \underline{u}(r) \times \cdots \times \underline{u}(r) \to \mathbb{R},$$

invariant with respect to the adjoint action

$$\underline{u}(r) \ni X \mapsto TXT^{-1} \in \underline{u}(r), \quad T \in U(r).$$

It is convenient to identify such a map with its polynomial form

$$P_\phi(X) = \phi(X, \dots, X).$$

The Lie algebra $\underline{u}(r)$ consists of $r \times r$ complex skew-hermitian matrices. Classical results of linear algebra show that, for any $X \in \underline{u}(r)$, there exists $T \in U(r)$, such that TXT^{-1} is diagonal

$$TXT^{-1} = i \, \mathrm{diag}(\lambda_1, \dots, \lambda_r).$$

The set of diagonal matrices in $\underline{u}(r)$ is called the *Cartan algebra* of $\underline{u}(r)$, and we will denote it by $\mathcal{C}_{\underline{u}(r)}$. It is a (maximal) Abelian Lie subalgebra of $\underline{u}(r)$. Consider the stabilizer

$$\mathcal{S}_{U(r)} := \big\{ T \in U(r) \,;\, TXT^{-1} = X, \ \forall X \in \mathcal{C}_{\underline{u}(r)} \big\},$$

and the normalizer

$$\mathcal{N}_{U(r)} := \big\{ T \in U(r); \ T\mathcal{C}_{\underline{u}(r)}T^{-1} \subset \mathcal{C}_{\underline{u}(r)} \big\}.$$

The stabilizer $\mathcal{S}_{U(r)}$ is a normal subgroup of $\mathcal{N}_{U(r)}$, so we can form the quotient

$$\mathcal{W}_{U(r)} := \mathcal{N}_{U(r)} / \mathcal{S}_{U(r)}$$

called the *Weyl group* of $U(r)$. As in Subsection 7.4.4, we see that the Weyl group is isomorphic with the symmetric group $\mathcal{S}_r$ because two diagonal skew-Hermitian matrices are unitarily equivalent if and only if they have the same eigenvalues, including multiplicities.

We see that P_ϕ is Ad-invariant if and only if its restriction to the Cartan algebra is invariant under the action of the Weyl group.

The Cartan algebra is the Lie algebra of the (maximal) torus T^n consisting of diagonal unitary matrices. As in the previous subsection we introduce the variables

$$x_j := -\frac{1}{2\pi i}d\theta_j.$$

The restriction of P_ϕ to $\mathcal{C}_{\underline{u}(r)}$ is a polynomial in the variables $x_1, \ldots, x_r$. The Weyl group $\mathcal{S}_r$ permutes these variables, so that P_ϕ is Ad-invariant if and only if $P_\phi(x_1, \ldots, x_r)$ is a symmetric polynomial in its variables. According to the fundamental theorem of symmetric polynomials, the ring of these polynomials is generated (as an $\mathbb{R}$-algebra) by the elementary ones

$$
\begin{aligned}
c_1 &= \sum_j x_j \\
c_2 &= \sum_{i<j} x_i x_j \\
&\vdots \quad \vdots \qquad \vdots \\
c_r &= x_1 \cdots x_r
\end{aligned}
$$

Thus

$$I^\bullet(U(r)) = \mathbb{R}[c_1, c_2, \ldots, c_r].$$

In terms of matrices $X \in \underline{u}(r)$ we have

$$\sum_k c_k(X)t^k = \det\left(\mathbb{1} - \frac{t}{2\pi i}X\right) \in I^\bullet(U(r))[t].$$

The above polynomial is known as the *universal rank r Chern polynomial*, and its coefficients are called the *universal, rank r Chern classes*.

Returning to our rank r vector bundle E, we obtain the *Chern classes of E*

$$c_k(E) := c_k(F(\nabla)) \in H^{2k}(M),$$

and the *Chern polynomial of E*

$$c_t(E) := \det\left(\mathbb{1} - \frac{t}{2\pi i}F(\nabla)\right) \in H^\bullet(M)[t].$$

Above, ∇ denotes a connection compatible with a Hermitian metric $\langle \bullet, \bullet \rangle$ on E, while $F(\nabla)$ denotes its curvature.

Remark 8.2.2. The Chern classes produced via the Chern-Weil method capture only a part of what topologists usually refer to characteristic classes of complex bundles. To give the reader a feeling of what the Chern-Weil construction is unable to capture we will sketch a different definition of the first Chern class of a complex line bundle. The following facts are essentially due to Kodaira and Spencer [61]; see also [40] for a nice presentation.

Let $L \to M$ be a smooth complex Hermitian line bundle over the smooth manifold M. Upon choosing a good open cover (U_α) of M we can describe L by a collection of smooth maps $z_{\alpha\beta}: U_{\alpha\beta} \to U(1) \cong S^1$ satisfying the cocycle condition

$$z_{\alpha\beta}z_{\beta\gamma}z_{\gamma\alpha} = \mathbb{1} \quad \forall \alpha, \beta, \gamma. \tag{8.2.1}$$

If we denote by $C^\infty(\cdot, S^1)$ the sheaf of multiplicative groups of smooth S^1-valued functions, we see that the family of complex line bundles on M can be identified with the Čech group $H^1(M, C^\infty(\cdot, S^1))$. This group is called the *smooth Picard group* of M. The group multiplication is precisely the tensor product of two line bundles. We will denote it by $\mathrm{Pic}^\infty(M)$.

If we write $z_{\alpha\beta} = \exp(2\pi i\theta_{\alpha\beta})$ ($\theta_{\beta\alpha} = -\theta_{\alpha\beta} \in C^\infty(U_{\alpha\beta}, \mathbb{R})$) we deduce from (8.2.1) that $\forall U_{\alpha\beta\gamma} \neq \emptyset$

$$\theta_{\alpha\beta} + \theta_{\beta\gamma} + \theta_{\gamma\alpha} = n_{\alpha\beta\gamma} \in \mathbb{Z}.$$

It is not difficult to see that, $\forall U_{\alpha\beta\gamma\delta} \neq \emptyset$, we have

$$n_{\beta\gamma\delta} - n_{\alpha\gamma\delta} + n_{\alpha\beta\delta} - n_{\alpha\beta\gamma} = 0.$$

In other words, the collection $n_{\alpha\beta\gamma}$ defines a Čech 2-cocycle of the constant sheaf $\mathbb{Z}$.

On a more formal level, we can capture the above cocycle starting from the exact sequence of sheaves

$$0 \to \mathbb{Z} \hookrightarrow C^\infty(\cdot, \mathbb{R}) \xrightarrow{\exp(2\pi i\cdot)} C^\infty(\cdot, S^1) \to 0.$$

The middle sheaf is a fine sheaf so its cohomology vanishes in positive dimensions. The long exact sequence in cohomology then gives

$$0 \to \mathrm{Pic}^\infty(M) \xrightarrow{\delta} H^2(M, \mathbb{Z}) \to 0.$$

The cocycle ($n_{\alpha\beta\gamma}$) represents precisely the class $\delta(L)$.

The class $\delta(L)$, $L \in \mathrm{Pic}^\infty(M)$ is called the topological first Chern class and is denoted by $c_1^{top}(L)$. This terminology is motivated by the following result of Kodaira and Spencer, [61]:

The image of $c_1^{top}(L)$ in the DeRham cohomology via the natural morphism

$$H^*(M, \mathbb{Z}) \to H^*(M, \mathbb{R}) \cong H^*_{DR}(M)$$

coincides with the first Chern class obtained via the Chern-Weil procedure.

The Chern-Weil construction misses precisely the torsion elements in $H^2(M, \mathbb{Z})$. For example, if a line bundle admits a flat connection then its first Chern class is trivial. This may not be the case with the topological one, because line bundle may not be topologically trivial. $\qquad\square$

8.2.3 *Pontryagin classes*

Let E be a rank r real vector bundle over the smooth manifold M. An Euclidean metric on E induces an $O(r)$ structure (P, ρ). The representation ρ is the tautological one

$$\rho : O(r) \hookrightarrow \mathrm{GL}(r, \mathbb{R}).$$

Exercise 8.2.3. Prove that two metrics on E induce isomorphic $O(r)$-structures.

$\qquad\square$

Hence, exactly as in the complex case, we can naturally identify the rank r-real vector bundles equipped with metric with principal $O(r)$-bundles. A connection on the principal bundle can be viewed as a metric compatible connection in the associated vector bundle. To describe the various characteristic classes we need to understand the ring of invariants $I^\bullet(O(r))$.

As usual, we will identify the elements of $I^k(O(r))$ with the degree k, Ad-invariant polynomials on the Lie algebra $\underline{o}(r)$ consisting of skew-symmetric $r \times r$ real matrices. Fix $P \in I^k(O(r))$. Set $m = [r/2]$, and denote by J the 2×2 matrix

$$J := \begin{bmatrix} 0 & -1 \\ 1 & 0 \end{bmatrix}.$$

Consider the *Cartan algebra*

$$\mathcal{C}_{\underline{o}(r)} = \begin{cases} \{\lambda_1 J \oplus \cdots \oplus \lambda_m J \in \underline{o}(r) \, ; \, \lambda_j \in \mathbb{R}\}, & r = 2m \\ \{\lambda_1 J \oplus \cdots \oplus \lambda_m J \oplus 0 \in \underline{o}(r) \, ; \, \lambda_j \in \mathbb{R}\}, & r = 2m+1 \end{cases}$$

The Cartan algebra $\mathcal{C}_{\underline{o}(r)}$ is the Lie algebra of the (maximal) torus

$$T^m = \begin{cases} R_{\theta_1} \oplus \cdots \oplus R_{\theta_m} \in O(r), & r = 2m \\ R_{\theta_1} \oplus \cdots \oplus R_{\theta_m} \oplus \mathbb{1}_\mathbb{R} \in O(r), & r = 2m+1 \end{cases},$$

where for each $\theta \in [0, 2\pi]$ we denoted by R_θ the 2×2 rotation

$$R_\theta := \begin{bmatrix} \cos\theta & -\sin\theta \\ \sin\theta & \cos\theta \end{bmatrix}.$$

As in Subsection 8.2.1 we introduce the variables

$$x_j = -\frac{1}{2\pi} d\theta_j.$$

Using standard results concerning the normal Jordan form of a skew-symmetric matrix, we deduce that, for every $X \in \underline{o}(r)$, there exists $T \in O(r)$ such that $TXT^{-1} \in \mathcal{C}_{\underline{o}(r)}$. Consequently, any Ad-invariant polynomial on $\underline{o}(r)$ is uniquely defined by its restriction to the Cartan algebra.

Following the approach in the complex case, we consider

$$\mathcal{S}_{O(r)} = \left\{ T \in O(r); \ TXT^{-1} = X, \ \forall X \in \mathcal{C}_{\underline{o}(r)} \right\},$$

$$\mathcal{N}_{O(r)} = \left\{ T \in O(r); \ T\mathcal{C}_{\underline{o}(r)}T^{-1} \subset \mathcal{C}_{\underline{o}(r)} \right\}.$$

The stabilizer $\mathcal{S}_{O(r)}$ is a normal subgroup in $\mathcal{N}(O(r))$, so we can form the *Weyl group*

$$\mathcal{W}_{O(r)} := \mathcal{N}_{O(r)}/\mathcal{S}_{O(r)}.$$

Exercise 8.2.4. Prove that $\mathcal{W}_{O(r)}$ is the subgroup of $\mathrm{GL}(m, \mathbb{R})$ generated by the involutions

$$\sigma_{ij} : (x_1, \ldots, x_i, \ldots, x_j, \ldots, x_m) \mapsto (x_1, \ldots, x_j, \ldots, x_i, \ldots, x_m)$$

$$\varepsilon_j : (x_1, \ldots, x_j, \ldots, x_m) \mapsto (x_1, \ldots, -x_j, \ldots, x_m). \qquad \square$$

The restriction of $P \in I^k(O(r))$ to $\mathcal{C}_{\underline{o}(r)}$ is a degree k homogeneous polynomial in the variables $x_1, \ldots, x_m$ invariant under the action of the Weyl group. Using the above exercise we deduce that P must be a symmetric polynomial $P = P(x_1, \ldots, x_m)$, separately even in each variable. Invoking once again the fundamental theorem of symmetric polynomials we conclude that P must be a polynomial in the elementary symmetric ones

$$
\begin{aligned}
p_1 &= \sum_j x_j^2 \\
p_2 &= \sum_{i<j} x_i^2 x_j^2 \\
&\vdots \quad \vdots \qquad \vdots \\
p_m &= x_1^2 \cdots x_m^2
\end{aligned}.
$$

Hence

$$I^\bullet(O(r)) = \mathbb{R}[p_1, \ldots, p_{\lfloor r/2 \rfloor}].$$

In terms of $X \in \underline{o}(r)$ we have

$$p_t(X) = \sum_j p_j(X) t^{2j} = \det\left(\mathbb{1} - \frac{t}{2\pi} X\right) \in I^\bullet(O(r))[t].$$

The above polynomial is called the *universal rank r Pontryagin polynomial*, while its coefficients $p_j(X)$ are called the *universal rank r Pontryagin classes*.

The *Pontryagin classes* $p_1(E), \ldots, p_m(E)$ of our real vector bundle E are then defined by the equality

$$p_t(E) = \sum_j p_j(E) t^{2j} = \det\left(\mathbb{1} - \frac{t}{2\pi} F(\nabla)\right) \in H^\bullet(M)[t],$$

where ∇ denotes a connection compatible with some (real) metric on E, while $F(\nabla)$ denotes its curvature. Note that $p_j(E) \in H^{4j}(M)$. The polynomial $p_t(E) \in H^\bullet(M)[t]$ is called the *Pontryagin polynomial of E*.

8.2.4 *The Euler class*

Let E be a rank r, real *oriented* vector bundle. A metric on E induces an $O(r)$-structure, but the existence of an orientation implies the existence of a finer structure, namely an $SO(r)$-symmetry.

The groups $O(r)$ and $SO(r)$ share the same Lie algebra $\underline{so}(r) = \underline{o}(r)$. The inclusion

$$\imath : SO(r) \hookrightarrow O(r),$$

induces a morphism of $\mathbb{R}$-algebras

$$\imath^* : I^\bullet(O(r)) \to I^\bullet(SO(r)).$$

Because $\underline{so}(r) = \underline{o}(r)$ one deduces immediately that $\imath^*$ is injective.

Lemma 8.2.5. *When r is odd then $\imath^* : I^\bullet(O(r)) \to I^\bullet(SO(r))$ is an isomorphism.*

□

Exercise 8.2.6. Prove the above lemma. □

The situation is different when r is even, $r = 2m$. To describe the ring of invariants $I^\bullet(SO(2m))$, we need to study in greater detail the Cartan algebra

$$\mathcal{C}_{\underline{o}(2m)} = \left\{ \lambda_1 J \oplus \cdots \oplus \lambda_m J \in \underline{o}(2m) \right\}$$

and the corresponding Weyl group action. The Weyl group $\mathcal{W}_{SO(2m)}$, defined as usual as the quotient

$$\mathcal{W}_{SO(2m)} = \mathcal{N}_{SO(2m)}/\mathcal{S}_{SO(2m)},$$

is isomorphic to the subgroup of $GL(\mathcal{C}_{\underline{o}(2m)})$ generated by the involutions

$$\sigma_{ij} : (\lambda_1, \ldots, \lambda_i, \ldots, \lambda_j, \ldots, \lambda_m) \mapsto (\lambda_1, \ldots, \lambda_j, \ldots, \lambda_i, \ldots, \lambda_m),$$

and

$$\varepsilon : (\lambda_1, \ldots, \lambda_m) \mapsto (\varepsilon_1 \lambda_1, \ldots, \varepsilon_m \lambda_m),$$

where $\varepsilon_1, \ldots, \varepsilon_m = \pm 1$, and $\varepsilon_1 \cdots \varepsilon_m = 1$. (*Check this!*)

Set as usual $x_i := -\lambda_i/2\pi$. The Pontryagin $O(2m)$-invariants

$$p_j(x_1, \ldots, x_m) = \sum_{1 \leq i_1 < \cdots < i_j \leq m} (x_{i_1} \cdots x_{i_j})^2,$$

continue to be $\mathcal{W}_{SO(2m)}$ invariants. There is however a new invariant,

$$\Delta(x_1, \ldots, x_m) := \prod_j x_j.$$

In terms of

$$X = \lambda_1 J \oplus \cdots \oplus \lambda_m J \in \mathcal{C}_{SO(2m)},$$

we can write

$$\Delta(X) = \left(\frac{-1}{2\pi} \right)^m \boldsymbol{Pf}(X),$$

where $\boldsymbol{Pf}$ denotes the pfaffian of the skewsymmetric matrix X viewed as a linear map $\mathbb{R}^{2m} \to \mathbb{R}^{2m}$, when $\mathbb{R}^{2m}$ is endowed with the canonical orientation. Note that $p_m = \Delta^2$.

Proposition 8.2.7.

$$I^\bullet(SO(2m)) \cong \mathbb{R}[Z_1, Z_2, \ldots, Z_m; Y]/(Y^2 - Z_m) \quad (Z_j = p_j, \ Y = \Delta)$$

where $(Y^2 - Z_m)$ denotes the ideal generated by the polynomial $Y^2 - Z_m$.

Proof. We follow the approach used by H. Weyl in describing the invariants of the alternate group ([100], Sec. II.2). The isomorphism will be established in two steps.

Step 1. The space $I^\bullet(SO(2m))$ is generated as an $\mathbb{R}$-algebra by the polynomials $p_1, \cdots, p_m, \Delta$.

Step 2. The kernel of the morphism

$$\mathbb{R}[Z_1, \ldots, Z_m, ; Y] \overset{\psi}{\to} I^\bullet(SO(2m))$$

defined by $Z_j \mapsto p_j$, $Y \mapsto \Delta$ is the ideal $(Y^2 - Z_m)$.

Proof of Step 1. Note that $\mathcal{W}_{SO(2m)}$ has index 2 as a subgroup in $\mathcal{W}_{O(2m)}$. Thus $\mathcal{W}_{SO(2m)}$ is a normal subgroup, and

$$\mathcal{G} = \mathcal{W}_{O(2m)}/\mathcal{W}SO(2m) \cong \mathbb{Z}_2.$$

The group $\mathcal{G} = \{\mathbb{1}, \mathfrak{e}\}$ acts on $I^\bullet(SO(2m))$ by

$$(\mathfrak{e}F)(x_1, x_2, \ldots, x_m) = F(-x_1, x_2, \ldots, x_m) = \cdots = F(x_1, x_2, \ldots, -x_m),$$

and moreover,

$$I^\bullet(O(2m)) = \ker(\mathbb{1} - \mathfrak{e}).$$

For each $F \in I^\bullet(SO(2m))$ we define $F^+ := (\mathbb{1} + \mathfrak{e})F$, and we observe that $F^+ \in \ker(\mathbb{1} - \mathfrak{e})$. Hence

$$F^+ = P(p_1, \ldots, p_m).$$

On the other hand, the polynomial $F^- := (\mathbb{1} - \mathfrak{e})F$ is separately odd in each of its variables. Indeed,

$$F^-(-x_1, x_2, \ldots, x_m) = \mathfrak{e}F^-(x_1, \ldots, x_m)$$

$$= \mathfrak{e}(\mathbb{1} - \mathfrak{e})F(x_1, \ldots, x_m) = -(\mathbb{1} - \mathfrak{e})F(x_1, \ldots, x_m) = -F^-(x_1, \ldots, x_m).$$

Hence, F^- vanishes when any of its variables vanishes so that F^- is divisible by their product $\Delta = x_1 \cdots x_m$,

$$F^- = \Delta \cdot G.$$

Since $\mathfrak{e}F^- = -F^-$, and $\mathfrak{e}\Delta = -\Delta$ we deduce $\mathfrak{e}G = G$, i.e., $G \in I^\bullet(O(2m))$. Consequently, G can be written as

$$G = Q(p_1, \ldots, p_m),$$

so that

$$F^- = \Delta \cdot Q(p_1, \ldots, p_m).$$

Step 1 follows from

$$F = \frac{1}{2}(F^+ + F^-) = \frac{1}{2}(P(p_1, \ldots, p_m) + \Delta \cdot Q(p_1, \ldots, p_m)).$$

Proof of Step 2. From the equality

$$\det X = \boldsymbol{Pf}(X)^2 \quad \forall X \in \underline{\mathrm{so}}(2m),$$

we deduce

$$(Y^2 - Z_m) \subset \ker \psi,$$

so that we only need to establish the opposite inclusion.

Let $P = P(Z_1, Z_2, \ldots, Z_m; Y) \in \ker \psi$. Consider P as a polynomial in Y with coefficients in $\mathbb{R}[Z_1, \ldots, Z_m]$. Divide P by the quadratic polynomial (in Y) $Y^2 - Z_m$. The remainder is linear

$$R = A(Z_1, \ldots, Z_m)Y + B(Z_1, \ldots, Z_m).$$

Since $Y^2 - Z_m$, $P \in \ker \psi$, we deduce $R \in \ker \psi$. Thus

$$A(p_1, \ldots, p_m)\Delta + B(p_1, \ldots, p_m) = 0.$$

Applying the morphism $\mathfrak{e}$ we get

$$-A(p_1, \ldots, p_m)\Delta + B(p_1, \ldots, p_m) = 0.$$

Hence $A \equiv B \equiv 0$ so that P is divisible by $Y^2 - Z_m$. $\qquad\qquad$ $\square$

Let E be a rank $2m$, real, oriented vector bundle over the smooth manifold M. As in the previous subsection we deduce that we can use a metric to naturally identify E with a principal $SO(2m)$-bundle and in fact, this principal bundle is independent of the metric. Finally, choose a connection ∇ compatible with some metric on E.

Definition 8.2.8. (a) The *universal Euler class* is defined by

$$e = e(X) = \frac{1}{(2\pi)^m} \boldsymbol{Pf}(-X) \in I^m(SO(2m)).$$

(b) The *Euler class* of E, is the cohomology class $e(E) \in H^{2m}(M)$ represented by the *Euler form*

$$e(\nabla) = \frac{1}{(2\pi)^m} \boldsymbol{Pf}(-F(\nabla)) \in \Omega^{2m}(M).$$

(According to the Chern-Weil theorem this cohomology class is independent of the metric and the connection.) $\qquad\qquad$ $\square$

Example 8.2.9. Let (Σ, g) be a compact, oriented, Riemann surface and denote by ∇^g the Levi-Civita connection. Then the Euler form

$$\varepsilon(g) = \frac{1}{4\pi} s(g) dv_g$$

coincides with the Euler form $e(\nabla^g)$ obtained via the Chern-Weil construction. $\quad \square$

Remark 8.2.10.   Let E be a rank $2m$, real, oriented vector bundle over the smooth, compact, oriented manifold M. We now have two apparently conflicting notions of Euler classes.

A *topological Euler class* $e_{top}(E) \in H^{2m}(M)$ defined as the pullback of the Thom class via an arbitrary section of E.

A *geometric Euler class* $e_{geom}(E) \in H^{2m}(M)$ defined via the Chern-Weil construction.

The most general version of the Gauss-Bonnet theorem, which will be established later in this chapter, will show that these two notions coincide! $\square$

8.2.5   *Universal classes*

In each of the situations discussed so far we defined characteristic classes for vector bundles with a given rank. In this subsection we show how one can coherently present these facts all at once, irrespective of rank. The algebraic machinery which will achieve this end is called *inverse limit*. We begin by first describing a special example of inverse limit.

A *projective sequence* of rings is a a sequence of rings $\{R_n\}_{n\geq 0}$, together with a sequence of ring morphisms $R_n \xleftarrow{\phi_n} R_{n+1}$. The inverse limit of a projective system (R_n, ϕ_n) is the subring

$$\varprojlim R_n \subset \prod_{n\geq 0} R_n$$

consisting of the sequences $(x_1, x_2, \ldots)$ such that $\phi_n(x_{n+1}) = x_n$, $\forall n \geq 0$.

Example 8.2.11.   Let $R_n = \mathcal{R}[[X_1, \ldots, X_n]]$ be the ring of formal power series in n variables with coefficients in the commutative ring with unit $\mathcal{R}$. ($R_0 = \mathcal{R}$.) Denote by $\phi_n : R_{n+1} \to R_n$ the natural morphism defined by setting $X_{n+1} = 0$. The inverse limit of this projective system is denoted by $\mathcal{R}[[X_1, X_2, \ldots]]$.

Given a sequence $F_n \in \mathcal{R}[[X]]$ ($n \geq 1$) such that $F_n(0) = 1$, we can form the sequence of products

$$(1, F_1(X_1), F_1(X_1)F_2(X_2), \ldots, F_1(X_1)\cdots F_n(X_n), \ldots),$$

which defines an element in $\mathcal{R}[[X_1, X_2, \ldots]]$ denoted by $F_1(X_1)F_2(X_2)\cdots$. When $F_1 = F_2 = \cdots F_n = \cdots = F$ the corresponding elements is denoted by $(F)^\infty$.

Exercise 8.2.12.   (a) Let $\mathcal{R}[[x]]^\flat$ denote the set of formal power series $F \in \mathcal{R}[[x]]$ such that $F(0) = 1$. Prove that $(\mathcal{R}[[x]]^\flat, \cdot)$ is an abelian group.

(b) Prove that $\forall F, G \in \mathcal{R}[[x]]^\flat$

$$(F \cdot G)^\infty = (F)^\infty \cdot (G)^\infty.$$ $\square$

Similarly, given $G_n \in \mathcal{R}[[x]]$, $(n \geq 1)$ such that $F_n(0) = 0$, we can form the sequence of sums

$$(0, G_1(X_1), G_1(X_1) + G_2(X_2), \ldots, G_1(X_1) + \cdots \oplus G_n(X_n), \ldots),$$

which defines an element in $\mathcal{R}[[X_1, X_2, \ldots]]$ denoted by $G_1(X_1) + G_2(X_2) + \cdots$. When $G_1 = G_2 = \cdots = F_n = \cdots = F$ we denote the corresponding element $(G)_\infty$. $\qquad\square$

In dealing with characteristic classes of vector bundles one naturally encounters the increasing sequences

$$U(1) \hookrightarrow U(2) \hookrightarrow \cdots \qquad (8.2.2)$$

(in the complex case) and (in the real case)

$$O(1) \hookrightarrow O(2) \hookrightarrow \cdots . \qquad (8.2.3)$$

We will discuss these two situations separately.

The complex case. The sequence in (8.2.2) induces a projective sequence of rings

$$\mathbb{R} \leftarrow I^{\bullet\bullet}(U(1)) \leftarrow I^{\bullet\bullet}(U(2)) \leftarrow \cdots . \qquad (8.2.4)$$

We know that $I^{\bullet\bullet}(U(n)) = \mathcal{S}^n[[x_j]] =$ the ring of symmetric formal power series in n variables with coefficients in $\mathbb{R}$. Set

$$\mathcal{S}^\infty[[x_j]] := \varprojlim \mathcal{S}^n.$$

Given a rank n vector bundle E over a smooth manifold M, and $\phi \in I^{\bullet\bullet}(U(n))$, the characteristic class $\phi(E)$ is well defined since $u^{\dim M + 1} = 0$, for any $u \in \Omega^*(M)$. Thus we can work with the ring $I^{\bullet\bullet}(U(n))$ rather than $I^\bullet(U(n))$ as we have done so far. An element

$$\phi = (\phi_1, \phi_2, \ldots) \in I^{\bullet\bullet}(U(\infty)) := \varprojlim I^{\bullet\bullet}(U(n)) = \mathcal{S}^\infty[[x_j]]$$

is called *universal characteristic class*.

If E is a complex vector bundle, we set $\phi(E)" = \phi_r(E)$, where $r = \operatorname{rank} E$. More precisely, to define $\phi_r(E)$ we need to pick a connection ∇ compatible with some Hermitian metric on E, and then set

$$\phi(E) = \phi_r(F(\nabla)).$$

Example 8.2.13. We denote by $c_k^{(n)}$ $(n \geq k)$ the elementary symmetric polynomial in n variables

$$c_k^{(n)} = \sum_{1 \leq i_1 < \cdots < i_k \leq n} x_{i_1} \cdots x_{i_k}.$$

Then the sequence

$$(0, \ldots, 0, c_k^{(k)}, c_k^{(k+1)}, \ldots)$$

defines an element in $\mathcal{S}^\infty[[x_j]]$ denoted by c_k, which we call the *universal k-th Chern class*. Formally, we can write

$$c_k = \sum_{1 \le i_1 < \cdots < i_k < \infty} x_{i_1} x_{i_2} \cdots x_{i_k}.$$

We can present the above arguments in a more concise form as follows. Consider the function $F(x) = (1 + tx) \in \mathcal{R}[[x]]$, where $\mathcal{R} = \mathbb{R}[[t]]$ is the ring of formal power series in the variable t. Then $(F)^\infty$ defines an element in $\mathcal{R}[[X_1, X_2, \ldots]]$. One sees immediately that in fact, $(F)^\infty \in \mathcal{S}^\infty[[x_j]][[t]]$ and moreover

$$(F)^\infty(x_1, x_2, \ldots) = (1 + tx_1)(1 + tx_2) \cdots = 1 + c_1 t + c_2 t^2 + \cdots.$$

We see that

$$c_t := \prod_{j \ge 1} (1 + tx_j)$$

is the *universal Chern polynomial*. $\square$

We can perform a similar operation with any $F \in \mathbb{R}[[x]]$ such that $F(0) = 1$. We get a semigroup morphism

$$(\mathbb{R}[[x]]^\flat, \cdot) \ni F \mapsto (F)^\infty \in (\mathcal{S}^\infty[[x_j]], \cdot)$$

(see Exercise 8.2.12). One very important example is

$$F(x) = \frac{x}{1 - e^{-x}} = 1 + \frac{1}{2}x + \frac{1}{12}x^2 + \cdots = 1 + \frac{1}{2}x + \sum_{k=1}^\infty (-1)^{k-1} \frac{B_k}{(2k)!} x^{2k} \in \mathbb{R}[[x]].$$

The coefficients B_k are known as the *Bernoulli numbers*. The product

$$(F)^\infty = \left(\frac{x_1}{1 - e^{-x_1}} \right) \cdot \left(\frac{x_2}{1 - e^{-x_2}} \right) \cdots,$$

defines an element in $\mathcal{S}^\infty[[x_j]]$ called the *universal Todd class*. We denote it by **Td**. Using the fundamental theorem of symmetric polynomials we can write

$$\mathbf{Td} = 1 + \mathbf{Td}_1 + \mathbf{Td}_2 + \cdots,$$

where $\mathbf{Td}_n \in \mathcal{S}^\infty[[x_j]]$ is a universal symmetric, homogeneous "polynomial" of degree n, hence expressible as a combination of the elementary symmetric "polynomials" $c_1, c_2, \ldots$. By an universal "polynomial" we understand element in the inverse limit

$$\mathbb{R}[x_1, x_2, \ldots] = \varprojlim \mathbb{R}[x_1, x_2, \ldots].$$

A universal "polynomial" P is said to be homogeneous of degree d, if it can be represented as a sequence

$$P = (P_1, P_2, \ldots),$$

where P_m is a homogeneous polynomial of degree d in m variables, and

$$P_{m+1}(x_1, x_2, \ldots, x_m, 0) = P_m(x_1, \ldots, x_m).$$

For example,

$$\mathbf{Td}_1 = \frac{1}{2}c_1, \quad \mathbf{Td}_2 = \frac{1}{12}(c_1^2 + c_2), \quad \mathbf{Td}_3 = \frac{1}{24}c_1 c_2 \text{ etc.}$$

Analogously, any function $G \in \mathbb{R}[[x]]$ such that $G(0) = 0$ defines an element

$$(G)_\infty = G(x_1) + G(x_2) + \cdots \in \mathcal{S}^\infty[[x_j]].$$

We have two examples in mind. First, consider $G(x) = x^k$. We get the symmetric function

$$s_k = (x^k)_\infty = (G)_\infty = x_1^k + x_2^k + \cdots ,$$

called the *universal k-th power sum*. We will denote these by s_k.

Next, consider

$$G(x) = e^x - 1 = \sum_{m \geq 1} \frac{x^m}{m!}.$$

We have

$$(e^x - 1)_\infty = \sum_{m \geq 1} \frac{1}{m!}(x^m)_\infty = \sum_{m=1} \frac{1}{m!} s_m \in \mathcal{S}^\infty[[x_j]].$$

Given a complex vector bundle E we define

$$\mathbf{ch}(E) = \operatorname{rank} E + (e^x - 1)_\infty(E).$$

The cohomology class $\mathbf{ch}(E)$ is called the *Chern character* of the bundle E. If ∇ is a connection on E compatible with some Hermitian metric, then we can express the Chern character of E as

$$\mathbf{ch}(E) = \operatorname{tr}\left(e^{F(\nabla)}\right) = \operatorname{rank}(E) + \sum_{k=1}^{\infty} \frac{1}{k!} \operatorname{tr}\left(F(\nabla)^{\wedge k}\right),$$

where $\wedge$ is the bilinear map

$$\Omega^i(\operatorname{End}(E)) \times \Omega^j(\operatorname{End}(E)) \to \Omega^{i+j}(\operatorname{End}(E))$$

defined in Example 3.3.12.

Proposition 8.2.14. *Consider two complex vector bundles E_1, E_2 over the same manifold M. Then*

$$\mathbf{ch}(E_1 \oplus E_2) = \mathbf{ch}(E_1) + \mathbf{ch}(E_2),$$

and

$$\mathbf{ch}(E_1 \otimes E_2) = \mathbf{ch}(E_1) \wedge \mathbf{ch}(E_2) \in H^\bullet(M).$$

Proof. Consider a connection ∇^i on E_i compatible with some Hermitian metric h_i, $i = 1, 2$. Then $\nabla^1 \oplus \nabla^2$ is a connection on $E_1 \oplus E_2$ compatible with the metric $h_1 \oplus h_2$, and moreover

$$F(\nabla^1 \oplus \nabla^2) = F(\nabla_1) \oplus F(\nabla^2).$$

Hence

$$\exp(F(\nabla^1 \oplus \nabla^2)) = \exp(F(\nabla^1)) \oplus \exp(F(\nabla^2)),$$

from which we deduce the first equality.

As for the second equality, consider the connection ∇ on $E_1 \otimes E_2$ uniquely defined by the product rule

$$\nabla(s_1 \otimes s_2) = (\nabla^1 s_1) \otimes s_2 + s_1 \otimes (\nabla^2 s_2), \quad s_i \in C^\infty(E_i),$$

where the operation

$$\otimes : \Omega^k(E_1) \times \Omega^\ell(E_2) \to \Omega^{k+\ell}(E_1 \otimes E_2)$$

is defined by

$$(\omega^k \otimes s_1) \otimes (\eta^\ell s_2) = (\omega^k \wedge \eta^\ell) \otimes (s_1 \otimes s_2) \tag{8.2.5}$$

$\forall s_i \in C^\infty(E_i)$, $\forall \omega^k \in \Omega^k(M)$, and $\forall \eta^\ell \in \Omega^\ell(M)$.

We compute the curvature of ∇ using the equality $F(\nabla) = (d^\nabla)^2$. If $s_i \in C^\infty(E_i)$, then

$$F(\nabla)(s_1 \otimes s_2) = d^\nabla \{(\nabla^1 s_1) \otimes s_2 + s_1 \otimes (\nabla^2 s_2)\}$$

$$= \{(F(\nabla^1)s_1) \otimes s_2 - (\nabla^1 s_1) \otimes (\nabla^2 s_2) + (\nabla^1 s_1) \otimes (\nabla^2 s_2) + s_1 \otimes (F(\nabla^2))\}$$

$$= F(\nabla^1) \otimes \mathbb{1}_{E_2} + \mathbb{1}_{E_1} \otimes F(\nabla^2).$$

The second equality in the proposition is a consequence of the following technical lemma.

Lemma 8.2.15. *Let A (respectively B) be a skew-adjoint, $n \times n$ (respectively $m \times m$) complex matrix. Then*

$$\mathrm{tr}\,(\exp(A \otimes \mathbb{1}_{\mathbb{C}^m} + \mathbb{1}_{\mathbb{C}^n} \otimes B)) = \mathrm{tr}\,(\exp(A)) \cdot \mathrm{tr}\,(\exp(B)).$$

Proof. Pick an orthonormal basis (e_i) of $\mathbb{C}^n$ and an orthonormal basis (f_j) of $\mathbb{C}^m$ such that, with respect to these bases $A = \mathrm{diag}(\lambda_1, \ldots, \lambda_n)$ and $B = \mathrm{diag}(\mu_1, \ldots, \mu_m)$. Then, with respect to the basis $(e_i \otimes f_j)$ of $\mathbb{C}^n \otimes \mathbb{C}^m$, we have

$$(A \otimes \mathbb{1}_{\mathbb{C}^m} + \mathbb{1}_{\mathbb{C}^n} \otimes B) = \mathrm{diag}(\lambda_i + \mu_j).$$

Hence

$$\exp(A \otimes \mathbb{1}_{\mathbb{C}^m} + \mathbb{1}_{\mathbb{C}^n} \otimes B) = \mathrm{diag}(e^{\lambda_i} e^{\mu_j}),$$

so that,

$$\operatorname{tr}\left(A \otimes 1_{\mathbb{C}^m} + 1_{\mathbb{C}^n} \otimes B\right) = \sum e^{\lambda_i} e^{\mu_j} = \operatorname{tr}\left(e^A\right) \cdot \operatorname{tr}\left(e^B\right). \qquad \square$$

The proposition is proved. $\qquad \square$

Exercise 8.2.16. (Newton's formulæ). Consider the symmetric polynomials

$$c_k = \sum_{1 \le i_1 \le \cdots i_k \le n} x_{i_1} \cdots x_{i_k} \in \mathbb{R}[x_1, \ldots, x_n],$$

and $(r \ge 0)$

$$s_r = \sum_j x_j^r \in \mathbb{R}[x_1, \ldots, x_n].$$

Set

$$f(t) := \prod_{j=1}^{n} (1 - x_j t).$$

(a) Show that

$$\frac{f'(t)}{f(t)} = -\sum_r s_r t^{r-1}.$$

(b) Prove the Newton formulæ

$$\sum_{j=1}^{r} (-1)^j s_{r-j} c_j = 0 \quad \forall 1 \le r \le n.$$

(c) Deduce from the above formulae the following identities between universal symmetric polynomials.

$$s_1 = c_1, \quad s_2 = c_1^2 - 2c_2, \quad s_3 = c_1^3 - 3c_1 c_2 + 3c_3. \qquad \square$$

The real case. The sequence (8.2.3) induces a projective system

$$I^{\bullet\bullet}(O(1)) \leftarrow I^{\bullet\bullet}(O(2)) \leftarrow \cdots .$$

We have proved that $I^{\bullet\bullet}(O(n)) = \mathcal{S}^{\lfloor n/2 \rfloor}[[x_j^2]] =$ the ring of even, symmetric power series in $\lfloor n/2 \rfloor$ variables. The inverse limit of this system is

$$I^{\bullet} \bullet (O(\infty)) = \varprojlim I^{\bullet\bullet}(O(r)) = \mathcal{S}^{\infty}[[x_j^2]] := \varprojlim \mathcal{S}^m[[x_j^2]].$$

As in the complex case, any element of this ring is called a *universal characteristic class*. In fact, for any real vector bundle E and any $\phi \in \mathcal{S}^{\infty}[[x_j^2]]$ there is a well defined characteristic class $\phi(E)$ which can be expressed exactly as in the complex case, using metric compatible connections. If $F \in \mathbb{R}[[x]]^{\flat}$ then $(F(x^2))^{\infty}$ defines an element of $\mathcal{S}^{\infty}[[x_j^2]]$.

In topology, the most commonly encountered situations are the following.

A.

$$F(x) = \frac{\sqrt{x}/2}{\sinh(\sqrt{x}/2)} = 1 + \sum_{k \geq 1}(-1)^k \frac{2^{2k-1}-1}{2^{2k-1}(2k)!}B_k x^k.$$

The universal characteristic class $(F(x))^\infty$ is denoted by $\widehat{A}$, and it is called the $\widehat{A}$-genus. We can write

$$\widehat{A} = 1 + \widehat{A}_1 + \widehat{A}_2 + \cdots,$$

where A_k are universal, symmetric, even, homogeneous "polynomials", and as such they can be described using universal Pontryagin classes

$$p_m = \sum_{1 \leq j_1 < \cdots j_m} x_{j_1}^2 \cdots x_{j_m}^2.$$

The first couple of terms are

$$\widehat{A}_1 = -\frac{p_1}{24}, \quad \widehat{A}_2 = \frac{1}{2^7 \cdot 3^2 \cdot 5}(-4p_2 + 7p_1^2) \text{ etc.}$$

B. Consider

$$F(x) = \frac{\sqrt{x}}{\tanh\sqrt{x}} = 1 + \frac{1}{3}x + \frac{1}{45}x^2 + \cdots = 1 + \sum_{k \geq 1}(-1)^{k-1}\frac{2^{2k}}{(2k)!}B_k x^k.$$

The universal class $(F(x^2))^\infty$ is denoted by L, and it is called the L-genus. As before, we can write

$$L = 1 + L_1 + L_2 + \cdots,$$

where the L_j's are universal, symmetric, even, homogeneous "polynomials". They can be expressed in terms of the universal Pontryagin classes. The first few terms are

$$L_1 = \frac{1}{3}p_1, \quad L_2 = \frac{1}{45}(7p_2 - p_1^2) \text{ etc.}$$

8.3   Computing characteristic classes

The theory of characteristic classes is as useful as one's ability to compute them. In this section we will describe some methods of doing this.

Most concrete applications require the aplication of a combination of techniques from topology, differential and algebraic geometry and Lie group theory that go beyond the scope of this book. We will discuss in some detail a few invariant theoretic methods and we will present one topological result more precisely the Gauss-Bonnet-Chern theorem.

8.3.1 *Reductions*

In applications, the symmetries of a vector bundle are implicitly described through topological or geometric properties.

For example, if a rank r complex vector bundle E splits as a Whitney sum $E = E_1 \oplus E_2$ with rank $E_i = r_i$, then E, which has a natural $U(r)$-symmetry, can be given a finer structure of $U(r_1) \times U(r_2)$ vector bundle.

More generally, assume that a given rank r complex vector bundle E admits a G-structure (P, ρ), where G is a Lie group, P is a principal G bundle, and $\rho : G \to U(r)$ is a representation of G. Then we can perform two types of Chern-Weil constructions: using the $U(r)$ structure, and using the G structure. In particular, we obtain two collections of characteristic classes associated to E. One natural question is whether there is any relationship between them.

In terms of the Whitney splitting $E = E_1 \oplus E_2$ above, the problem takes a more concrete form: compute the Chern classes of E in terms of the Chern classes of E_1 and E_2. Our next definition formalizes the above situations.

Definition 8.3.1. Let $\varphi : H \to G$ be a smooth morphism of (matrix) Lie groups.
(a) If P is a principal H-bundle over the smooth manifold M defined by the open cover (U_α), and gluing cocycle

$$h_{\alpha\beta} : U_{\alpha\beta} \to H,$$

then the principal G-bundle defined by the gluing cocycle

$$g_{\alpha\beta} = \varphi \circ h_{\alpha\beta} : U_{\alpha\beta} \to G$$

is said to be the φ-associate of P, and it is denoted by $\varphi(P)$.
(b) A principal G-bundle Q over M is said to be φ-reducible, if there exists a principal H-bundle $P \to M$ such that $Q = \varphi(P)$. $\square$

The morphism $\varphi : H \to G$ in the above definition induces a morphism of $\mathbb{R}$-algebras

$$\varphi^* : I^\bullet(G) \to I^\bullet(H).$$

The elements of $\ker \varphi^* \subset I^\bullet(G)$ are called *universal identities*.

The following result is immediate.

Proposition 8.3.2. Let P be a principal G-bundle which can be reduced to a principal H-bundle Q. Then for every $\eta \in \ker \varphi^*$ we have

$$\eta(P) = 0 \text{ in } H^*(M).$$

Proof. Denote by $\mathcal{L}_G$ (respectively $\mathcal{L}_H$) the Lie algebra of G (respectively H), and by φ_* the differential of φ at $1 \in H$,

$$\varphi_* : \mathcal{L}_H \to \mathcal{L}_G.$$

Pick a connection (A_α) on Q, and denote by (F_α) its curvature. Then the collection $\varphi_*(A_\alpha)$ defines a connection on P with curvature $\varphi_*(F_\alpha)$. Now

$$\eta(\varphi_*(F_\alpha)) = (\varphi^*\eta)(F_\alpha) = 0. \qquad \square$$

The above result should be seen as a guiding principle in proving identities between characteristic classes, rather than a rigid result. What is important about this result is the simple argument used to prove it.

We conclude this subsection with some simple, but very important applications of the above principle.

Example 8.3.3. Let E and F be two complex vector bundles over the same smooth manifold M of ranks r and respectively s. Then the Chern polynomials of E, F and $E \oplus F$ are related by the identity

$$c_t(E \oplus F) = c_t(E) \cdot c_t(F), \qquad (8.3.1)$$

where the "$\cdot$" denotes the $\wedge$-multiplication in $H^{even}(M)$. Equivalently, this means

$$c_k(E \oplus F) = \sum_{i+j=k} c_i(E) \cdot c_j(F).$$

To check this, pick a Hermitian metric g on E and a Hermitian metric h on F. $g \oplus h$ is a Hermitian metric on $E \oplus F$. Hence, $E \oplus F$ has a $U(r + s)$ structure reducible to a $U(r) \times U(s)$ structure.

The Lie algebra of $U(r) \times U(s)$ is the direct sum $\underline{u}(r) \oplus \underline{u}(s)$. Any element X in this algebra has a block decomposition X_E

$$X = \begin{bmatrix} X_r & 0 \\ 0 & X_s \end{bmatrix} = X_r \oplus X_s,$$

where X_r (respectively X_s) is an $r \times r$ (respectively $s \times s$) complex, skew-hermitian matrix. Let $\imath$ denote the natural inclusion $\underline{u}(r) \oplus \underline{u}(s) \hookrightarrow \underline{u}(r + s)$ and denote by $c_t^{(\nu)} \in I^*(U(\nu))[t]$ the Chern polynomial.

We have

$$\imath^*(c_t^{(r+s)})(X_r \oplus X_s) = \det\left(\mathbb{1}_{r+s} - \frac{t}{2\pi\boldsymbol{i}} X_r \oplus X_s\right)$$

$$= \det\left(\mathbb{1}_r - \frac{t}{2\pi\boldsymbol{i}} X_r\right) \cdot \det\left(\mathbb{1}_s - \frac{t}{2\pi\boldsymbol{i}} X_s\right) = c_t^{(r)} \cdot c_t^{(s)}(s).$$

The equality (8.3.1) now follows using the argument in the proof of Proposition 8.3.2. $\qquad \square$

Remark 8.3.4. Consider the Grassmannian $\mathbf{Gr}_k(\mathbb{C}^n)$ of complex k-dimensional subspaces in $\mathbb{C}^n$. The universal complex vector bundle $\mathcal{U}_{k,n} \to \mathbf{Gr}_k(\mathbb{C}^n)$ is a sub-bundle of the trivial bundle $\underline{\mathbb{C}}^n \to \mathbf{Gr}_k(\mathbb{C}^n)$. The trivial bundle $\underline{\mathbb{C}}^n$ is equipped with a canonical Hermitian metric. We denote by $\mathcal{Q}_{k,n}$ the orthogonal complement of $\mathcal{U}_{k,n}$ in $\underline{\mathbb{C}}^n$ so that we have an isomorphism

$$\underline{\mathbb{C}}^n \cong \mathcal{U}_{k,n} \oplus \mathcal{Q}_{k,n}.$$

From the above example we deduce that

$$c_t(\mathcal{U}_{k,n})c_t(\mathcal{Q}_{k,n}) = c_t(\underline{\mathbb{C}}^n) = 1.$$

Denote by u_j the j-th Chern class of $\mathcal{U}_{k,n}$ and by v_ℓ the ℓ-th Chern class of $\mathcal{Q}_{k,n}$. Then

$$c_t(\mathcal{U}_{k,n}) = \sum_{j=0}^{k} u_j t^j, \quad c_t(\mathcal{Q}_{k,n}) = \sum_{\ell=0}^{n-k} v_\ell t^\ell.$$

One can then prove that the cohomology ring $H^\bullet(\mathbf{Gr}_k(\mathbb{C}^n), \mathbb{R})$ is generated by the classes u_i, v_j, which are subject to the single relation above. More formally

$$H^\bullet(\mathbf{Gr}_k(\mathbb{C}^n), \mathbb{R}) \cong \mathbb{R}[u_i, v_j; \ 0 \le i \le k, \ 0 \le j \le n-k]/\left((\sum_i u_i)(\sum_j v_j) = 1\right),$$

$\deg u_i = 2i$, $\deg v_j = 2j$. The proof requires a more sophisticated topological machinery. For details we refer to [76], Chapter v3, Section 6. $\qquad\square$

Exercise 8.3.5. Let E and F be two complex vector bundles over the same manifold M. Show that

$$\mathbf{Td}(E \oplus F) = \mathbf{Td}(E) \cdot \mathbf{Td}(F). \qquad\square$$

Exercise 8.3.6. Let E and F be two real vector bundles over the same manifold M. Prove that

$$p_t(E \oplus F) = p_t(E) \cdot p_t(F), \tag{8.3.2}$$

where p_t denotes the Pontryagin polynomials. $\qquad\square$

Exercise 8.3.7. Let E and F be two real vector bundles over the same smooth manifold M. Show that

$$\boldsymbol{L}(E \oplus F) = \boldsymbol{L}(E) \cdot \boldsymbol{L}(F)$$

$$\widehat{\boldsymbol{A}}(E \oplus F) = \widehat{\boldsymbol{A}}(E) \cdot \widehat{\boldsymbol{A}}(F). \qquad\square$$

Example 8.3.8. The natural inclusion $\mathbb{R}^n \hookrightarrow \mathbb{C}^n$ induces an embedding $\imath : O(n) \hookrightarrow U(n)$. (An orthogonal map $T : \mathbb{R}^n \to \mathbb{R}^n$ extends by complexification to a unitary map $T_{\mathbb{C}} : \mathbb{C}^n \to \mathbb{C}^n$.) This is mirrored at the Lie algebra level by an inclusion

$$\underline{o}(n) \hookrightarrow \underline{u}(n).$$

We obtain a morphism $\imath^* : I^\bullet(U(n)) \to I^\bullet(O(n))$, and we claim that

$$\imath^*(c_{2k+1}) = 0, \quad \text{and} \quad \imath^*(c_{2k}) = (-1)^k p_k.$$

Indeed, for $X \in \underline{o}(n)$ we have

$$\imath^*(c_{2k+1})(X) = \left(-\frac{1}{2\pi i}\right)^{2k+1} \sum_{1 \le i_1 < \cdots < i_{2k+1} \le n} \lambda_{i_1}(X) \cdots \lambda_{i_{2k+1}}(X),$$

where $\lambda_j(X)$ are the eigenvalues of X over $\mathbb{C}$. Since X is, in effect, a real skew-symmetric matrix we have

$$\lambda_j(\overline{X}) = \overline{\lambda_j(X)} = -\lambda_j(X).$$

Consequently,

$$\imath^*(c_{2k+1})(X) = \imath^*(c_{2k+1}(\overline{X})) = \left(-\frac{1}{2\pi\boldsymbol{i}}\right)^{2k+1} \sum_{1 \leq i_1 < \cdots < i_{2k+1} \leq n} \overline{\lambda_{i_1}(X) \cdots \lambda_{i_{2k+1}}(X)}$$

$$= (-1)^{2k+1} \left(-\frac{1}{2\pi\boldsymbol{i}}\right)^{2k+1} \sum_{1 \leq i_1 < \cdots < i_{2k+1} \leq n} \lambda_{i_1}(X) \cdots \lambda_{i_{2k+1}}(X) = -\imath^*(c_{2k+1})(X).$$

The equality $\imath^*(c_{2k}) = (-1)^k p_k$ is proved similarly. $\qquad\square$

From the above example we deduce immediately the following consequence.

Proposition 8.3.9. If $E \to M$ is a real vector bundle and $E \otimes \mathbb{C}$ is its complexification then

$$p_k(E) = (-1)^k c_{2k}(E \otimes \mathbb{C}), \quad k = 1, 2, \ldots. \tag{8.3.3}$$

In a more concentrated form, this means

$$p_t(E) = p_{-t}(E) = c_{\boldsymbol{i}t}(E \otimes \mathbb{C}). \qquad\square$$

Exercise 8.3.10. Let $E \to M$ be a complex vector bundle of rank r.
(a) Show that $c_k(E^*) = (-1)^k c_k(E)$, i.e.,

$$c_t(E^*) = c_{-t}(E).$$

(b) One can also regard E as a **real, oriented** vector bundle $E_{\mathbb{R}}$. Prove that

$$p_{\boldsymbol{i}t}(E_{\mathbb{R}}) = \sum_k (-1)^k t^{2k} p_k(E_{\mathbb{R}}) = c_t(E) \cdot c_{-t}(E),$$

and

$$c_r(E) = \boldsymbol{e}(E_{\mathbb{R}}). \qquad\square$$

Exercise 8.3.11. (a) The natural morphisms $I^{**}(U(r)) \to I^{**}(O(r))$ described above induce a morphism

$$\Phi_\infty : I^{\bullet\bullet}(U(\infty)) \to I^{\bullet\bullet}(O(\infty)).$$

As we already know, for any $F \in \mathbb{R}[[x]]^\flat$, the product $(F)^\infty$ is an element of $I^{\bullet\bullet}(U(\infty))$, and if moreover F is **even**, then $(F)^\infty$ can be regarded as an element of $I^{**}(O(\infty))$. Show that for every $F \in \mathbb{R}[[x]]^\flat$

$$\Phi_\infty((F)^\infty) = (F \cdot F^-)^\infty \in I^{\bullet\bullet}(O(\infty)),$$

where $F^-(x) = F(-x)$.
(b) Let E be a real vector bundle. Deduce from part (a) that

$$\mathbf{Td}(E \otimes \mathbb{C}) = \widehat{A}(E)^2. \qquad\square$$

Exercise 8.3.12. For every square matrix X we set[2]

$$(\mathbb{1} + X)^{1/2} = \sum_{k=0}^{\infty} (-1)^k \binom{1/2}{k} X^k, \quad \binom{r}{k} := \frac{r(r-1)\cdots(r-k+1)}{k!},$$

and $\det^{1/2}(\mathbb{1} + X) := \det(\mathbb{1} + X)^{1/2}$.

Suppose that $E \to M$ is a rank r real vector bundle, equipped with a metric and a compatible connection ∇. The connection ∇ induces a Hermitian connection ∇^c on the complexification $E_c := E \otimes_{\mathbb{R}} \mathbb{C}$. Prove that

$$\boldsymbol{L}(\nabla) = \det^{1/2}\left(\frac{\frac{i}{2\pi}F(\nabla^c)}{\tanh(\frac{i}{2\pi}F(\nabla^c))} \right), \quad \widehat{\boldsymbol{A}}(\nabla) = \det^{1/2}\left(\frac{\frac{i}{4\pi}F(\nabla^c)}{\sinh(\frac{i}{4\pi}F(\nabla^c))} \right). \qquad \square$$

Example 8.3.13. Consider the inclusion

$$\imath : SO(2k) \times SO(2\ell) \hookrightarrow SO(2k + 2\ell).$$

This induces a ring morphism

$$\imath^* : I^{\bullet}(SO(2k + 2\ell)) \to I^{\bullet}(SO(2k) \times SO(2\ell)).$$

Note that

$$I^{\bullet}(SO(2k) \times SO(2\ell)) \cong I^{\bullet}(SO(2k)) \otimes_{\mathbb{R}} I^{\bullet}(SO(2\ell)).$$

Denote by $\boldsymbol{e}^{(\nu)}$ the Euler class in $I^{\bullet}(SO(2\nu))$. We want to prove that

$$\imath^*(\boldsymbol{e}^{(k+\ell)}) = \boldsymbol{e}^{(k)} \otimes \boldsymbol{e}^{(\ell)}.$$

Let $X = X_k \oplus X_\ell \in \underline{so}(2k) \oplus \underline{so}(2\ell)$. Modulo a conjugation by $(S, T) \in SO(2k) \times SO(2\ell)$ we may assume that

$$X_k = \lambda_1 J \oplus \cdots \oplus \lambda_k J \text{ and } X_\ell = \mu_1 J \oplus \cdots \oplus \mu_\ell J,$$

where J denotes the 2×2 matrix

$$J = \begin{bmatrix} 0 & -1 \\ 1 & 0 \end{bmatrix}.$$

We have

$$\imath^*(\boldsymbol{e}^{(k+\ell)})(X) = \left(-\frac{1}{2\pi} \right)^{k+\ell} \lambda_1 \cdots \lambda_k \cdot \mu_1 \cdots \mu_\ell = \boldsymbol{e}^{(k)}(X_k) \cdot \boldsymbol{e}^{(\ell)}(X_\ell)$$

$$= \boldsymbol{e}^{(k)} \otimes \boldsymbol{e}^{(\ell)}(X_k \oplus X_\ell). \qquad \square$$

The above example has an interesting consequence.

Proposition 8.3.14. Let E and F be two real, oriented vector bundle of even ranks over the same manifold M. Then

$$\boldsymbol{e}(E \oplus F) = \boldsymbol{e}(E) \cdot \boldsymbol{e}(F),$$

where $\cdot$ denotes the $\wedge$-multiplication in $H^{even}(M)$. $\qquad \square$

---

[2]We are not worried about convergence issues because the matrices for which we intend to apply the formula have nilpotent entries.

Example 8.3.15. Let E be a rank $2k$ real, oriented vector bundle over the smooth manifold M. We claim that *if E admits a nowhere vanishing section ξ then $e(E) = 0$.*

To see this, fix a Euclidean metric on E so that E is now endowed with an $SO(2k)$-structure. Denote by L the real line subbundle of E generated by the section ξ. Clearly, L is a trivial line bundle, and E splits as an orthogonal sum

$$E = L \oplus L^{\perp}.$$

The orientation on E, and the orientation on L defined by ξ induce an orientation on $L^{\perp}$, so that $L^{\perp}$ has an $SO(2k-1)$-structure.

In other words, the $SO(2k)$ structure of E can be reduced to an $SO(1) \times SO(2k-1) \cong SO(2k-1)$-structure. Denote by $\imath^*$ the inclusion induced morphism

$$I^*(SO(2k)) \to I^*(SO(2k-1)).$$

Since $\imath^*(e^{(k)}) = 0$, we deduce from the Proposition 8.3.2 that $e(E) = 0$. $\qquad\square$

The result proved in the above example can be reformulated more suggestively as follows.

Corollary 8.3.16. Let E be a real oriented vector bundle of even rank over the smooth manifold M. If $e(E) \neq 0$, then any section of E must vanish somewhere on M! $\qquad\square$

8.3.2 The Gauss-Bonnet-Chern theorem

If $E \to M$ is a real oriented vector bundle over a smooth, compact, oriented manifold M then there are two apparently conflicting notions of Euler class naturally associated to E.

- *The topological Euler class*

$$e_{top}(E) = \zeta_0^* \tau_E,$$

where τ_E is the Thom class of E and $\zeta_0 : M \to E$ is the zero section.

- *The geometric Euler class*

$$e_{geom}(E) = \begin{cases} \frac{1}{(2\pi)^r} \, \boldsymbol{Pf}(-F(\nabla)) & \text{if rank}\,(E) \text{ is even} \\ 0 & \text{if rank}\,(E) \text{ is odd} \end{cases},$$

where ∇ is a connection on E compatible with some metric and $2r = \text{rank}\,(E)$. The next result, which generalizes the Gauss-Bonnet theorem, will show that these two notions of Euler class coincide.

Theorem 8.3.17 (Gauss-Bonnet-Chern). Let $E \xrightarrow{\pi} M$ be a real, oriented vector bundle over the compact oriented manifold M. Then

$$e_{top}(E) = e_{geom}(E).$$

Proof.   We will distinguish two cases.

A. $\operatorname{rank}(E)$ is odd. Consider the automorphism of E

$$\mathbf{i} : E \to E \quad u \mapsto -u \ \ \forall u \in E.$$

Since the fibers of E are odd dimensional, we deduce that $\mathbf{i}$ reverses the orientation in the fibers. In particular, this implies

$$\pi_* \mathbf{i}^* \tau_E = -\pi_* \tau_E = \pi_*(-\tau_E),$$

where π_* denotes the integration along fibers. Since π_* is an isomorphism (Thom isomorphism theorem), we deduce

$$\mathbf{i}^* \tau_E = -\tau_E.$$

Hence

$$e_{top}(E) = -\zeta_0^* \mathbf{i}^* \tau_E. \tag{8.3.4}$$

On the other hand, notice that

$$\zeta_0^* \mathbf{i}^* = \zeta_0.$$

Indeed,

$$\zeta_0^* \mathbf{i}^* = (\mathbf{i}\zeta_0)^* = (-\zeta_0)^* = (\zeta)^* \quad (\zeta_0 = -\zeta_0).$$

The equality $e_{top} = e_{geom}$ now follows from (8.3.4).

B. $\operatorname{rank}(E) = 2k$. We will use a variation of the original argument of Chern, [22]. Let ∇ denote a connection on E compatible with a metric g. The strategy of proof is very simple. We will explicitly construct a closed form $\omega \in \Omega^{2k}_{cpt}(E)$ such that

(i) $\pi_* \omega = 1 \in \Omega^0(M)$.
(ii) $\zeta_0^* \omega = e(\nabla) = (2\pi)^{-k} \, \boldsymbol{Pf}(-F(\nabla))$.

The Thom isomorphism theorem coupled with (i) implies that ω represents the Thom class in $H^{2k}_{cpt}(E)$. The condition (ii) simply states the sought for equality $e_{top} = e_{geom}$.

Denote by $S(E)$ the unit sphere bundle of E,

$$S(E) = \{u \in E \ ; \ |u|_g = 1\}.$$

Then $S(E)$ is a compact manifold, and

$$\dim S(E) = \dim M + 2k - 1.$$

Denote by π_0 the natural projection $S(E) \to M$, and by $\pi_0^*(E) \to S(E)$ the pullback of E to $S(E)$ via the map π_0. The vector bundle $\pi_0^*(E)$ has an $SO(2k)$-structure, and moreover, it admits a tautological, nowhere vanishing section

$$\Upsilon : S(E_x) \ni e \mapsto e \in E_x \equiv (\pi_0^*(E)_x)_e \quad (x \in M).$$

Thus, according to Example 8.3.15 we must have

$$e_{geom}(\pi_0^*(E)) = 0 \in H^{2k}(S(E)),$$

where $e_{geom}(\pi_0^* E)$ denotes the differential form

$$e_{geom}(\pi_0^* \nabla) = \frac{1}{(2\pi)^k} \, \boldsymbol{Pf}(-F(\pi_0^* \nabla)).$$

Hence there must exist $\psi \in \Omega^{2k-1}(S(E))$ such that

$$d\psi = e_{geom}(\pi_0^* E).$$

The decisive step in the proof of Gauss-Bonnet-Chern theorem is contained in the following lemma.

Lemma 8.3.18. There exists $\Psi = \Psi(\nabla) \in \Omega^{2k-1}(S(E))$ such that

$$d\Psi(\nabla) = e_{geom}(\pi_0^*(E)) \qquad (8.3.5)$$

and

$$\int_{S(E)/M} \Psi(\nabla) = -1 \in \Omega^0(M). \qquad (8.3.6)$$

The form $\Psi(\nabla)$ is sometimes referred to as the *global angular form* of the pair (E, ∇). For the clarity of the exposition we will conclude the proof of the Gauss-Bonnet-Chern theorem assuming Lemma 8.3.18 which will be proved later on.

Denote by $r : E \to \mathbb{R}_+$ the norm function

$$E \ni e \mapsto |e|_g.$$

If we set $E^0 = E \setminus \{\text{ zero section }\}$, then we can identify

$$E^0 \cong (0, \infty) \times S(E) \quad e \mapsto (|e|, \frac{1}{|e|} e).$$

Consider the smooth cutoff function

$$\rho = \rho(r) : [0, \infty) \to \mathbb{R},$$

such that $\rho(r) = -1$ for $r \in [0, 1/4]$, and $\rho(r) = 0$ for $r \geq 3/4$. Finally, define

$$\omega = \omega(\nabla) = -\rho'(r) dr \wedge \Psi(\nabla) - \rho(r)\pi^*(e(\nabla)).$$

The differential form ω is well defined since $\rho'(r) \equiv 0$ near the zero section. Obviously ω has compact support on E, and satisfies the condition (ii) since

$$\zeta_0^* \omega = -\rho(0)\zeta_0^* \pi^* e(\nabla) = e(\nabla).$$

From the equality

$$\int_{E/M} \rho(r)\pi^* e(\nabla) = 0,$$

we deduce

$$\int_{E/M} \omega = - \int_{E/M} \rho'(r) dr \wedge \Psi(\nabla) = - \int_0^\infty \rho'(r) dr \cdot \int_{S(E)/M} \Psi(\nabla)$$

$$= -(\rho(1) - \rho(0)) \int_{S(E)/M} \Psi(\nabla) \stackrel{(8.3.6)}{=} 1.$$

To complete the program outlined at the beginning of the proof we need to show that ω is closed.

$$d\omega = \rho'(r)dr \wedge d\Psi(\nabla) - \rho'(r) \wedge \pi^* e(\nabla)$$

$$\stackrel{(8.3.5)}{=} \rho'(r)dr \wedge \{\pi_0^* e(\nabla) - \pi^* e(\nabla)\}.$$

The above form is identically zero since $\pi_0^* e(\nabla) = \pi^* e(\nabla)$ on the support of ρ'. Thus ω is closed and the theorem is proved. □

Proof of Lemma 8.3.18 We denote by $\overline{\nabla}$ the pullback of ∇ to $\pi_0^* E$. The tautological section $\Upsilon : S(E) \to \pi_0^* E$ can be used to produce an orthogonal splitting

$$\pi_0^* E = L \oplus L^\perp,$$

where L is the real line bundle spanned by Υ, while $L^\perp$ is its orthogonal complement in $\pi_0^* E$ with respect to the pullback metric g. Denote by

$$P : \pi_0^* E \to \pi_0^* E$$

the orthogonal projection onto $L^\perp$. Using P, we can produce a new metric compatible connection $\hat{\nabla}$ on $\pi_0^* E$ by

$$\hat{\nabla} = \text{(trivial connection on } L) \oplus P\overline{\nabla}P.$$

We have an equality of differential forms

$$\pi_0^* e(\nabla) = e(\overline{\nabla}) = \frac{1}{(2\pi)^k} \boldsymbol{Pf}(-F(\overline{\nabla})).$$

Since the curvature of $\hat{\nabla}$ splits as a direct sum

$$F(\hat{\nabla}) = 0 \oplus F'(\hat{\nabla}),$$

where $F'(\hat{\nabla})$ denotes the curvature of $\hat{\nabla}|_{L^\perp}$, we deduce

$$\boldsymbol{Pf}(F(\hat{\nabla})) = 0.$$

We denote by ∇^t the connection $\hat{\nabla} + t(\overline{\nabla} - \hat{\nabla})$, so that $\nabla^0 = \hat{\nabla}$, and $\nabla^1 = \overline{\nabla}$. If F^t is the curvature of ∇^t, we deduce from the transgression formula (8.1.12) that

$$\pi_0^* e(\nabla) = e(\overline{\nabla}) - e(\hat{\nabla}) = d\left\{ \left(\frac{-1}{2\pi}\right)^k k \int_0^1 \boldsymbol{Pf}(\overline{\nabla} - \hat{\nabla}, F^t, \ldots, F^t)dt \right\}.$$

We claim that the form

$$\Psi(\nabla) = \left(\frac{-1}{2\pi}\right)^k k \int_0^1 \boldsymbol{Pf}(\overline{\nabla} - \hat{\nabla}, F^t, \ldots, F^t)dt$$

satisfies all the conditions in Lemma 8.3.18.

By construction,

$$d\Psi(\nabla) = \pi_0^* e(\nabla),$$

so all that we need to prove is

$$\int_{S(E)/M} \Psi(\nabla) = -1 \in \Omega^0(M).$$

It suffices to show that for each fiber E_x of E we have

$$\int_{E_x} \Psi(\nabla) = -1.$$

Along this fiber $\pi_0^* E$ is naturally isomorphic with a trivial bundle

$$\pi_0^* E|_{E_x} \cong (E_x \times E_x \to E_x).$$

Moreover, the connection $\overline{\nabla}$ restricts as the trivial connection. By choosing an orthonormal basis of E_x we can identify $\pi_0^* E|_{E_x}$ with the trivial bundle $\mathbb{R}^{2k}$ over $\mathbb{R}^{2k}$. The unit sphere $S(E_x)$ is identified with the unit sphere $S^{2k-1} \subset \mathbb{R}^{2k}$. The splitting $L \oplus L^\perp$ over $S(E)$ restricts over $S(E_x)$ as the splitting

$$\mathbb{R}^{2k} = \nu \oplus TS^{2k-1},$$

where ν denotes the normal bundle of $S^{2k-1} \hookrightarrow \mathbb{R}^{2k}$. The connection $\hat{\nabla}$ is then the direct sum between the trivial connection on ν, and the Levi-Civita connection on TS^{2k-1}.

Fix a point $p \in S^{2k-1}$, and denote by $(x^1, \ldots, x^{2k-1})$ a collection of normal coordinates near p, such that the basis $(\frac{\partial}{\partial x_i}|_p)$ is positively oriented. Set $\partial_i := \frac{\partial}{\partial x_i}$ for $i = 1, \ldots, 2k-1$. Denote the unit outer normal vector field by ∂_0. For $\alpha = 0, 1, \ldots, 2k-1$ set $\boldsymbol{f}_\alpha = \partial_\alpha|_p$. The vectors $\boldsymbol{f}_\alpha$ form a positively oriented orthonormal basis of $\mathbb{R}^{2k}$.

We will use Latin letters to denote indices running from 1 to $2k-1$, and Greek letters indices running from 0 to $2k-1$.

$$\overline{\nabla}_i \partial_\alpha = (\overline{\nabla}_i \partial_\alpha)^\nu + (\overline{\nabla}_i \partial_\alpha)^\tau,$$

where the superscript ν indicates the normal component, while the superscript τ indicates the tangential component. Since at p

$$0 = \hat{\nabla}_i \partial_j = (\overline{\nabla}_i \partial_j)^\tau,$$

we deduce

$$\overline{\nabla}_i \partial_j = (\overline{\nabla}_i \partial_j)^\nu \quad \text{at } p.$$

Hence

$$\overline{\nabla}_i \partial_j = (\overline{\nabla}_i \partial_j, \partial_0) \partial_0 = -(\partial_j, \overline{\nabla}_i \partial_0) \partial_0.$$

Recalling that $\overline{\nabla}$ is the trivial connection in $\mathbb{R}^{2k}$, we deduce

$$\overline{\nabla}_i \partial_0|_p = \left(\frac{\partial}{\partial \boldsymbol{f}_i} \partial_0\right)|_p = \boldsymbol{f}_i = \partial_i|_p.$$

Consequently,

$$\overline{\nabla}_i \partial_j = -\delta_{ji}\partial_0, \quad \text{at} \ \ p.$$

If we denote by θ^i the local frame of $T^* S^{2k-1}$ dual to (∂_i), then we can rephrase the above equality as

$$\overline{\nabla}\partial_j = -(\theta^1 + \cdots + \theta^{2k-1}) \otimes \partial_0.$$

On the other hand, $\overline{\nabla}_i \partial_0 = \partial_i$, i.e.,

$$\overline{\nabla}\partial_0 = \theta^1 \otimes \partial_1 + \cdots + \theta^{2k-1} \otimes \partial_{2k-1}.$$

Since $(x^1, \cdots, x^{2k-1})$ are normal coordinates with respect to the Levi-Civita connection $\hat{\nabla}$, we deduce that $\hat{\nabla}\partial_\alpha = 0$, $\forall \alpha$ so that

$$A = (\overline{\nabla} - \hat{\nabla})|_p = \begin{bmatrix} 0 & -\theta^1 & \cdots & -\theta^{2k-1} \\ \theta^1 & 0 & \cdots & 0 \\ \theta^2 & 0 & \cdots & 0 \\ \vdots & \vdots & \vdots & \vdots \\ \theta^{2k-1} & 0 & \cdots & 0 \end{bmatrix}.$$

Denote by F^0 the curvature of $\nabla^0 = \hat{\nabla}$ at p. Then $F^0 = 0 \oplus R$, where R denotes the Riemann curvature of $\hat{\nabla}$ at p.

The computations in Example 4.2.19 show that the second fundamental form of the embedding

$$S^{2k-1} \hookrightarrow \mathbb{R}^{2k},$$

coincides with the induced Riemann metric (which is the first fundamental form). Using Teorema Egregium we get

$$\langle R(\partial_i, \partial_j)\partial_k, \partial_\ell \rangle = \delta_{i\ell}\delta_{jk} - \delta_{ik}\delta_{j\ell}.$$

In matrix format we have

$$F^0 = 0 \oplus (\Omega_{ij}),$$

where $\Omega_{ij} = \theta^i \wedge \theta^j$. The curvature F^t at p of $\nabla^t = \hat{\nabla} + tA$ can be computed using the equation (8.1.11) of subsection 8.1.4, and we get

$$F^t = F^0 + t^2 A \wedge A = 0 \oplus (1 - t^2)F^0.$$

We can now proceed to evaluate $\Psi(\nabla)$.

$$\Psi(\nabla)|_p = \left(\frac{-1}{2\pi}\right)^k k \int_0^1 \boldsymbol{Pf}(A, (1-t^2)F^0, \cdots, (1-t^2)F^0)dt$$

$$= \left(\frac{-1}{2\pi}\right)^k k \left(\int_0^1 (1-t^2)^{k-1}dt\right) \boldsymbol{Pf}(A, F^0, F^0, \ldots, F^0). \qquad (8.3.7)$$

We need to evaluate the pfaffian in the right-hand-side of the above formula. Set for simplicity $F := F^0$.

Using the polarization formula and Exercise 2.2.65 in Subsection 2.2.4, we get

$$\boldsymbol{Pf}(A, F, F, \cdots, F) = \frac{(-1)^k}{2^k k!} \sum_{\sigma \in \mathcal{S}_{2k}} \epsilon(\sigma) A_{\sigma_0 \sigma_1} F_{\sigma_2 \sigma_3} \cdots F_{\sigma_{2k-2} \sigma_{2k-1}}.$$

For $i = 0, 1$, define

$$\mathcal{S}^i = \{\sigma \in \mathcal{S}_{2k} \; ; \; \sigma_i = 0\}.$$

We deduce

$$2^k k! \, \boldsymbol{Pf}(A, F, \cdots, F) = (-1)^k \sum_{\sigma \in \mathcal{S}^0} \epsilon(\sigma)(-\theta^{\sigma_1}) \wedge \theta^{\sigma_2} \wedge \theta^{\sigma_3} \wedge \cdots \wedge \theta^{\sigma_{2k-2}} \wedge \theta^{\sigma_{2k-1}}$$

$$+(-1)^k \sum_{\sigma \in \mathcal{S}^1} \epsilon(\sigma) \theta^{\sigma_0} \wedge \theta^{\sigma_2} \wedge \theta^{\sigma_3} \wedge \cdots \wedge \theta^{\sigma_{2k-2}} \wedge \theta^{\sigma_{2k-1}}.$$

For each $\sigma \in \mathcal{S}^0$ we get a permutation

$$\phi : (\sigma_1, \sigma_2, \cdots, \sigma_{2k-1}) \in \mathcal{S}_{2k-1},$$

such that $\epsilon(\sigma) = \epsilon(\phi)$. Similarly, for $\sigma \in \mathcal{S}^1$, we get a permutation

$$\phi = (\sigma_0, \sigma_2, \cdots, \sigma_{2k-1}) \in \mathcal{S}_{2k-1},$$

such that $\epsilon(\sigma) = \epsilon(\phi)$. Hence

$$2^k k! \, \boldsymbol{Pf}(A, F, \cdots, F) = 2(-1)^{k+1} \sum_{\phi \in \mathcal{S}_{2k-1}} \epsilon(\phi) \theta^{\phi_1} \wedge \cdots \cdots \theta^{\phi_{2k-1}}$$

$$= 2(-1)^{k+1}(2k-1)! \theta^1 \wedge \cdots \wedge \theta^{2k-1} = 2(-1)^{k+1} dV_{S^{2k-1}},$$

where $dV_{S^{2k-1}}$ denotes the Riemannian volume form on the unit sphere S^{2k-1}. Using the last equality in (8.3.7) we get

$$\Psi(\nabla)|_p = \left(\frac{-1}{2\pi}\right)^k k \left(\int_0^1 (1-t^2)^{k-1} dt\right) \cdot 2(-1)^{k+1}(2k-1)! dV_{S^{2k-1}}$$

$$= -\frac{(2k)!}{(4\pi)^k k!} \left(\int_0^1 (1-t^2)^{k-1} dt\right) dV_{S^{2k-1}}.$$

Using Exercise 4.1.60 we get that

$$\boldsymbol{\omega}_{2k-1} = \int_{S^{2k-1}} dV_{S^{2k-1}} = \frac{2\pi^k}{(k-1)!},$$

and consequently

$$\int_{S^{2k-1}} \Psi(\nabla) = -\boldsymbol{\omega}_{2k-1} \frac{(2k)!}{(4\pi)^k k!} \left(\int_0^1 (1-t^2)^{k-1} dt\right)$$

$$= -\frac{(2k)!}{2^{2k-1} k!(k-1)!} \left(\int_0^1 (1-t^2)^{k-1} dt\right).$$

The above integral can be evaluated inductively using the substitution $t = \cos\varphi$. We have

$$I_k = \left(\int_0^1 (1 - t^2)^{k-1} dt \right) = \int_0^{\pi/2} (\cos\varphi)^{2k-1} d\varphi$$

$$= (\cos\varphi)^{2k-2} \sin\varphi\big|_0^{\pi/2} + (2k - 2) \int_0^{\pi/2} (\cos\varphi)^{2k-3} (\sin\varphi)^2 d\varphi$$

$$= (2k - 1) I_{k-1} - (2k - 2) I_k$$

so that

$$I_k = \frac{2k - 3}{2k - 2} I_{k-1}.$$

One now sees immediately that

$$\int_{S^{2k-1}} \Psi(\nabla) = -1.$$

Lemma 8.3.18 is proved. □

Corollary 8.3.19 (Chern). Let (M, g) be a compact, oriented Riemann Manifold of dimension $2n$. If R denotes the Riemann curvature then

$$\chi(M) = \frac{1}{(2\pi)^n} \int_M \boldsymbol{Pf}(-R). \qquad \square$$

The next exercises provide another description of the Euler class of a real oriented vector bundle $E \to M$ over the compact oriented manifold M in terms of the homological Poincaré duality. Set $r = \operatorname{rank} E$, and let τ_E be a compactly supported form representing the Thom class.

Exercise 8.3.20. Let $\Phi : N \to M$ be a smooth map, where N is compact and oriented. We denote by $\Phi^\#$ the bundle map $\Phi^* E \to E$ induced by the pullback operation. Show that $\Phi^* \tau_E := (\Phi^\#)^* \tau_E \in \Omega^r(\Phi^* E)$ is compactly supported, and represents the Thom class of $\Phi^* E$. □

In the next exercise we will also assume M is endowed with a Riemann structure.

Exercise 8.3.21. Consider a nondegenerate smooth section s of E, i.e.,

(i) the set $\mathcal{Z} = s^{-1}(0)$ is a codimension r smooth submanifold of M, and
(ii) there exists a connection ∇^0 on E such that

$$\nabla^0_X s \neq 0 \quad \text{along } \mathcal{Z}$$

for any vector field X normal to $\mathcal{Z}$ (along $\mathcal{Z}$).

(a) (**Adjunction formula.**) Let ∇ be an **arbitrary** connection on E, and denote by N_Z the normal bundle of the embedding $Z \hookrightarrow M$. Show that the **adjunction map**

$$\mathfrak{a}_\nabla : N_Z \to E|_Z,$$

defined by

$$X \mapsto \nabla_X s \quad X \in C^\infty(N_Z),$$

is a bundle **isomorphism**. Conclude that Z is endowed with a natural orientation.
(b) Show that there exists an open neighborhood N of $Z \hookrightarrow N_Z \hookrightarrow TM$ such that

$$\exp : N \to \exp(N)$$

is a diffeomorphism. (Compare with Lemma 7.3.44.) Deduce that the Poincaré dual of Z in M can be identified (via the above diffeomorphism) with the Thom class of N_Z.
(c) Prove that the Euler class of E coincides with the Poincaré dual of Z .   □

The part (c) of the above exercise generalizes the Poincaré-Hopf theorem (see Corollary 7.3.48). In that case, the section was a nondegenerate vector field, and its zero set was a finite collection of points. The local index of each zero measures the difference between two orientations of the normal bundle of this finite collection of points: one is the orientation obtained if one tautologically identifies this normal bundle with the restriction of TM to this finite collection of points while, and the other one is obtained via the adjunction map.

In many instances one can explicitly describe a nondegenerate section and its zero set, and thus one gets a description of the Euler class which is satisfactory for most topological applications.

Example 8.3.22. Let $\mathcal{U}$ denote the tautological line bundle over the complex projective space $\mathbb{CP}^n$. According to Exercise 8.3.10

$$c_1(\mathcal{U}) = e(\mathcal{U}),$$

when we view $\mathcal{U}$ as a rank 2 oriented, real vector bundle. Denote by $[H]$ the $(2n-2)$-cycle defined by the natural inclusion

$$\imath : \mathbb{CP}^{n-1} \hookrightarrow \mathbb{CP}^n, \quad [z_0, \ldots, z_{n-1}] \mapsto [z_0, \ldots, z_{n-1}, 0] \in \mathbb{CP}^n.$$

We claim that the (homological) Poincaré dual of $c_1(\tau_n)$ is $-[H]$. We will achieve this by showing that $c_1(\mathcal{U}^*) = -c_1(\mathcal{U})$ is the Poincaré dual of $[H]$.

Let P be a degree 1 homogeneous polynomial $P \in \mathbb{C}[z_0, \ldots, z_n]$. For each complex line $L \hookrightarrow \mathbb{C}^{n+1}$, the polynomial P defines a complex linear map $L \to \mathbb{C}$, and hence an element of L^*, which we denote by $P|_L$. We thus have a well defined map

$$\mathbb{CP}^n \ni L \mapsto P|_L \in L^* = \mathcal{U}^*|_L,$$

and the reader can check easily that this is a *smooth* section of $\mathcal{U}^*$, which we denote by $[P]$. Consider the special case $P_0 = z_n$. The zero set of the section $[P_0]$ is precisely the image of $[H]$. We let the reader keep track of all the orientation conventions in Exercise 8.3.21, and conclude that $c_1(\mathcal{U}^*)$ is indeed the Poincaré dual of $[H]$.　□

Exercise 8.3.23. Let $U_S = S^2 \setminus \{\text{north pole}\}$, and $U_N = S^2 \setminus \{\text{south pole}\} \cong \mathbb{C}$. The overlap $U_N \cap U_S$ is diffeomorphic with the punctured plane $\mathbb{C}^* = \mathbb{C} \setminus \{0\}$. For each map $g : \mathbb{C}^* \to U(1) \cong S^1$ denote by L_g the line bundle over S^2 defined by the gluing map

$$g_{NS} : U_N \cap U_S \to U(1), \quad g_{NS}(z) = g(z).$$

Show that

$$\int_{S^2} c_1(L_g) = \deg g,$$

where $\deg g$ denotes the degree of the smooth map $g|_{S^1 \subset \mathbb{C}^*} \to S^1$.　□

Chapter 9

Classical Integral Geometry

Ultimately, mathematics is about solving problems, and we can trace the roots of most remarkable achievements in mathematics to attempts of solving concrete problems which, for various reasons, were deemed very interesting by the mathematical community.

In this chapter, we will present some very beautiful applications of the techniques developed so far. In a sense, we are going against the natural course of things, since the problems we will solve in this chapter were some of the catalysts for the discoveries of many of the geometric results discussed in the previous chapters.

Integral geometry, also known as geometric probability, is a mixed breed subject. The question it addresses have purely geometric formulations, but the solutions borrow ideas from many mathematical areas, such as representation theory and probability. This chapter is intended to wet the reader's appetite for unusual questions, and make him/her appreciate the power of the technology developed in the previous chapters.

9.1 The integral geometry of real Grassmannians

9.1.1 *Co-area formulæ*

As Gelfand and his school pointed out, the main trick of classical integral geometry is a very elementary one, namely the change of order of summation. Let us explain the bare bones version of this trick, unencumbered by various technical assumptions.

Consider a "roof"

$$
\begin{array}{ccc}
 & X & \\
{}^{\alpha}\swarrow & & \searrow{}^{\beta} \\
A & & B
\end{array}
$$

where $X \xrightarrow{\alpha} A$ and $X \xrightarrow{\beta} B$ are maps between finite sets. One should think of α and β as defining two different fibrations with the same total space X.

Suppose we are given a function $f : X \to \mathbb{R}$. We define its "average" over X as the "integral"

$$\langle f \rangle_X := \sum_{x \in X} f(x).$$

We can compute the average $\langle f \rangle_X$ in two different ways, either summing first over the fibers of α, or summing first over the fibers of β, i.e.

$$\sum_{a \in A} \left(\sum_{\alpha(x)=a} f(x) \right) = \langle f \rangle_X = \sum_{b \in B} \left(\sum_{\beta(x)=b} f(x) \right). \tag{9.1.1}$$

We can reformulate the above equality in a more conceptual way by introducing the functions

$$\alpha_*(f) : A \to \mathbb{C}, \quad \alpha_*(f)(a) = \sum_{\alpha(x)=a} f(x),$$

and

$$\beta_*(f) : B \to \mathbb{C}, \quad \beta_*(f)(b) = \sum_{\beta(x)=b} f(x).$$

The equality (9.1.1) implies that

$$\big\langle \alpha_*(f) \big\rangle_A = \big\langle \beta_*(f) \big\rangle_B.$$

All the integral geometric results that we will prove in this chapter will be of this form. The reason such formulæ have captured the imagination of geometers comes from the fact that the two sides $\langle \alpha_*(f) \rangle_A$ and $\langle \beta_*(f) \rangle_B$ have rather unrelated interpretations.

The geometric situation we will investigate is a bit more complicated than the above baby case, because in the geometric case the sets X, A, B are smooth manifolds, the maps α and β define smooth fibrations, and the function f is not really a function, but an object which can be integrated over X, i.e., a density. It turns out that the push forward operations do make sense for densities resulting in formulas of the type

$$\langle f \rangle_X = \big\langle \alpha_*(f) \big\rangle_A.$$

These formulæ generalize the classical Fubini theorem and they are often referred to as *coarea formulæ* for reasons which will become apparent a bit later. To produce such results we need to gain a deeper understanding of the notion of density introduced in Subsection 3.4.1.

Suppose V is a finite dimensional real vector space. We denote by $\det V$ the top exterior power of V. Given a real number s we define an *s-density on V* to be a map $\lambda : \det V \to \mathbb{R}$ such that

$$\lambda(t\Omega) = |t|^s \lambda(\Omega), \quad \forall t \in \mathbb{R}^*, \ \Omega \in \det V.$$

We denote by $|\Lambda|_s(V)$ the one dimensional space of s-densities. Note that we have a canonical identification $|\Lambda|_0(E) = \mathbb{R}$. We will refer to 1-densities simply as *densities*, and we denote the corresponding space by $|\Lambda|(V)$.

We say that an s-density $\lambda : \det V \to \mathbb{R}$ is *positive* if

$$\lambda(\det V \setminus 0) \subset (0, \infty).$$

We denote by $|\Lambda|_s^+(V)$ the cone of positive densities.

Note that any basis $(v_1, \ldots, v_n)$ of V defines linear isomorphisms

$$|\Lambda|_s V \to \mathbb{R}, \quad \lambda \mapsto \lambda(v_1 \wedge \cdots \wedge v_n).$$

In particular, we have a canonical identification

$$|\Lambda|_s(\mathbb{R}^n) \cong \mathbb{R}, \quad \lambda \mapsto \lambda(e_1 \wedge \cdots \wedge e_n),$$

where $(e_1, \ldots, e_n)$ is the canonical basis of $\mathbb{R}^n$.

If V_0 and V_1 are vector spaces of the same dimension n, and $g : V_0 \to V_1$ is a linear isomorphism then we get a linear map

$$g^* : |\Lambda|_s(V_1) \to |\Lambda|_s(V_0), \quad |\Lambda|_s(V_1) \ni \lambda \mapsto g^*\lambda,$$

where

$$(g^*\lambda)\,(\wedge_i v_i) = \lambda\big(\wedge_i(gv_i)\big), \quad \forall v_1, \cdots, v_n \in V_0.$$

If $V_0 = V_1 = V$ so that $g \in \mathrm{Aut}(V)$, then

$$g^*\lambda = |\det g|^s \lambda.$$

For every $g, h \in \mathrm{Aut}(V)$ we gave $(gh)^* = h^*g^*$, and thus we have a *left* action of $\mathrm{Aut}(V)$ on $|\Lambda|_s(V)$

$$\mathrm{Aut}(V) \times |\Lambda|_s(V) \to |\Lambda|_s(V),$$

$$\mathrm{Aut}(V) \times |\Lambda|_s(V) \ni (g, \lambda) \mapsto g_*\lambda = (g^{-1})^*\lambda = |\det g|^{-s}\lambda.$$

We have bilinear maps

$$|\Lambda|_s(V) \otimes |\Lambda|_t(V) \to |\Lambda|_{s+t}(V), \quad (\lambda, \mu) \mapsto \lambda \cdot \mu.$$

To any short exact sequence of vector spaces

$$0 \to U \xrightarrow{\alpha} V \xrightarrow{\beta} W \to 0, \quad \dim U = m, \quad \dim V = m, \quad \dim W = p$$

we can associate maps

$$\backslash : |\Lambda|_s^+(U) \times |\Lambda|_s(V) \to |\Lambda|_s(W),$$

$$/ : |\Lambda|_s(V) \times |\Lambda|_s^+(W) \to |\Lambda|_s(U),$$

and

$$\times : |\Lambda|_s(U) \times |\Lambda|_s(W) \to |\Lambda|_s(V)$$

as follows.

- Let $\mu \in |\Lambda|_s^+(U)$, $\lambda \in |\Lambda|_s(V)$, and suppose $(w_j)_{1 \leq j \leq p}$ is a basis of W. Now choose lifts $v_j \in V$ of w_j, such that $\beta(v_j) = w_j$, and a basis $(u_i)_{1 \leq i \leq m}$ of U such that

$$\left\{ \alpha(u_1), \ldots, \alpha(u_m), v_1, \ldots, v_p \right\}$$

is a basis of V. We set

$$(\mu \backslash \lambda)\left(\wedge_j w_j \right) := \frac{\lambda\left(\left(\wedge_i \alpha(u_i) \right) \wedge \left(\wedge_j v_j \right) \right)}{\mu\left(\wedge_i u_i \right)}.$$

It is easily seen that the above definition is independent of the choices of v's and u's.

- Let $\lambda \in |\Lambda|_s(V)$ and $\nu \in |\Lambda|_s^+(W)$. Given a basis $(u_i)_{1 \leq i \leq m}$ of U, extend the linearly independent set $\left(\alpha(u_i) \right) \subset V$ to a basis

$$\left\{ \alpha(u_1), \ldots, \alpha(u_m), v_1, \cdots, v_p \right\}$$

of V and now define

$$(\lambda / \nu)\left(\wedge_i u_i \right) := \frac{\lambda\left(\left(\wedge_i \alpha(u_i) \right) \wedge \left(\wedge_j v_j \right) \right)}{\nu\left(\wedge_j \beta(v_j) \right)}.$$

Again it is easily verified that the above definition is independent of the various choices.

- Let $\mu \in |\Lambda|_s(U)$ and $\nu \in |\Lambda|_s(W)$. To define $\mu \times \nu : \det V \to \mathbb{R}$ it suffices to indicate its value on a single nonzero vector of the line $\det V$. Fix a basis $(u_i)_{1 \leq i \leq m}$ of U and a basis $(w_j)_{1 \leq j \leq p}$ of W. Choose lifts (v_j) of w_j to V. Then we set

$$(\mu \times \nu)\left(\left(\wedge_i \alpha(u_i) \right) \wedge \left(\wedge_j w_j \right) \right) = \mu(\wedge_i u_i) \nu(\wedge_j v_j).$$

Again one can check that this is independent of the various bases (u_i) and (w_j).

Exercise 9.1.1. Prove that the above constructions are indeed independent of the various choices of bases. $\square$

Remark 9.1.2. The constructions $\backslash$, $/$, and $\times$ associated to a short exact sequence of vector spaces

$$0 \to U \xrightarrow{\alpha} V \xrightarrow{\beta} W \to 0,$$

do depend on the maps α and β! For example if we replace α by $\alpha_t = t\alpha$ and β by $\beta_\tau = \tau\beta$, $t, \tau > 0$, and if

$$\mu \in |\Lambda|_s(U), \quad \lambda \in |\Lambda|_s(V), \quad \nu \in |\Lambda|_s(W),$$

then

$$\mu \backslash_{\alpha_t, \beta_\tau} \lambda = (t/\tau)^s \mu \backslash_{\alpha, \beta} \lambda, \quad \lambda /_{\alpha_t, \beta_\tau} \nu = (t/\tau)^s \lambda /_{\alpha, \beta} \nu,$$

$$\lambda \times_{\alpha_t, \beta_\tau} \nu = (t/\tau)^{-s} \mu \times_{\alpha, \beta} \nu.$$

The next example illustrates this. $\square$

Example 9.1.3. Consider the short exact sequence

$$0 \to U = \mathbb{R} \xrightarrow{\alpha} V = \mathbb{R}^2 \xrightarrow{\beta} W = \mathbb{R} \to 0$$

given by

$$\alpha(s) = (4s, 10s), \quad \beta(x, y) = 5x - 2y.$$

Denote by e the canonical basis of U, by (e_1, e_2) the canonical basis of V and by f the canonical basis of W. We obtain canonical densities λ_U on U, λ_V on V and λ_W on W given by

$$\lambda_U(e) = \lambda_V(e_1 \wedge e_2) = \lambda_W(f) = 1.$$

We would like to describe the density $\lambda_V / \beta^* \lambda_W$ on V. Set

$$f_1 = \alpha(e) = (4, 10).$$

We choose $f_2 \in V$ such that $\beta(f_2) = f$, for example, $f_2 = (1, 2)$. Then

$$\lambda_V / \beta^* \lambda_W(e) = \lambda_V(f_1 \wedge f_2) / \lambda_W(f) = \left| \det \begin{bmatrix} 4 & 1 \\ 10 & 2 \end{bmatrix} \right| = 2.$$

Hence $\lambda_V / \beta^* \lambda_W = 2\lambda_U$. □

Suppose now that $E \to M$ is a real vector bundle of rank n over the smooth manifold M. Assume that it is given by the open cover (U_α) and gluing cocycle

$$g_{\beta\alpha} : U_{\alpha\beta} \to \mathrm{Aut}(V),$$

where V is a *fixed* real vector space of dimension n. Then the bundle of s-densities associated to E is the real line bundle $|\Lambda|_s E$ given by the open cover (U_α) and gluing cocycle

$$|\det g_{\beta\alpha}|^{-s} : U_{\alpha\beta} \to \mathrm{Aut}(|\Lambda|_s(V)) \cong \mathbb{R}^*.$$

We denote by $C^\infty(|\Lambda|_s E)$ the space of smooth sections of $|\Lambda|_s E$. Such a section is given by a collection of smooth functions $\lambda_\alpha : U_\alpha \to |\Lambda|_s(V)$ satisfying the gluing conditions

$$\lambda_\beta(x) = (g_{\beta\alpha}^{-1})^* \lambda_\alpha(x) = |\det g_{\beta\alpha}|^{-s} \lambda_\alpha(x), \quad \forall \alpha, \beta, \ x \in U_{\alpha\beta}.$$

Let us point out that if $V = \mathbb{R}^n$, then we have a canonical identification $|\Lambda|_s(\mathbb{R}^n) \to \mathbb{R}$, and in this case a density can be regarded as a collection of smooth functions $\lambda_\alpha : U_\alpha \to \mathbb{R}$ satisfying the above gluing conditions.

An s-density $\lambda \in C^\infty(|\Lambda|_s E)$ is called *positive* if for every $x \in M$ we have $\lambda(x) \in |\Lambda|_s^+(E_x)$.

If $\phi : N \to M$ is a smooth map, and $E \to M$ is a smooth real vector bundle, then we obtain the pullback bundle $\pi^* E \to N$. We have canonical isomorphisms

$$|\Lambda|_s \pi^* E \cong \pi^* |\Lambda|_s E,$$

and a natural pullback map

$$\phi^* : C^\infty(|\Lambda|_s E) \to C^\infty(\pi^* |\Lambda|_s E) \cong C^\infty(|\Lambda|_s \pi^* E).$$

Given a short exact sequence of vector bundles

$$0 \to E_0 \to E_1 \to E_2 \to 0$$

over M, we obtain maps

$$\backslash : C^\infty(|\Lambda|_s^+ E_0) \times C^\infty(|\Lambda|_s E_1) \to C^\infty(|\Lambda|_s E_2),$$

$$/ : C^\infty(|\Lambda|_s E_1) \times C^\infty(|\Lambda|_s^+ E_2) \to C^\infty(|\Lambda|_s E_0),$$

and

$$\times : C^\infty(|\Lambda|_s E_0) \times C^\infty(|\Lambda|_s E_2) \to C^\infty(|\Lambda|_s E_1).$$

Observe that for every positive smooth function $f : M \to (0, \infty)$ we have

$$(f\mu)\backslash\lambda = (f^{-1})(\mu\backslash\lambda), \quad \lambda/(f\nu) = (f^{-1})(\mu/\lambda).$$

In the sequel we will almost exclusively use a special case of the above construction, when E is the tangent bundle of the smooth manifold M. We will denote by $|\Lambda|_s(M)$ the line bundle $|\Lambda|_s(TM)$, and we will refer to its sections as (smooth) s-densities on M. When $s = 1$, we will use the simpler notation $|\Lambda|_M$ to denote $|\Lambda|_1(M)$.

As explained in Subsection 3.4.1, an s-density on M is described by a coordinate atlas $\left(U_\alpha, \ (x^i_\alpha)\right)$, and smooth functions $\lambda_\alpha : U_\alpha \to \mathbb{R}$ satisfying the conditions

$$\lambda_\beta = |d_{\beta\alpha}|^{-s}\lambda_\alpha, \quad \text{where} \ \ d_{\beta\alpha} = \det\left(\frac{\partial x^j_\beta}{\partial x^i_\alpha}\right)_{1 \le i,j \le n}, \quad n = \dim M. \qquad (9.1.2)$$

We deduce that the smooth 0-densities on M are precisely the smooth functions.

Example 9.1.4. (a) Suppose $\omega \in \Omega^n(M)$ is a top degree differential form on M. Then in a coordinate atlas $(U_\alpha, \ (x^i_\alpha))$ this form is described by a collection of forms

$$\omega_\alpha = \lambda_\alpha dx^1_\alpha \wedge \cdots \wedge dx^n_\alpha.$$

The functions λ_α satisfy the gluing conditions $\lambda_\beta = d_{\beta\alpha}^{-1}\lambda_\alpha$, and we conclude that the collection of functions $|\lambda_\alpha|^s$ defines an s-density on M which we will denote by $|\omega|^s$. Because of this fact, the s-densities are traditionally described as collections

$$\lambda_\alpha |dx_\alpha|^s, \quad dx_\alpha := dx^1_\alpha \wedge \cdots \wedge dx^n_\alpha.$$

(b) Suppose M is an orientable manifold. By fixing an orientation we choose an atlas $(U_\alpha, \ (x^i_\alpha))$ so that all the determinants $d_{\beta\alpha}$ are positive. If ω is a top dimensional form on M described locally by forms $\omega_\alpha = \lambda_\alpha dx_\alpha$, then the collection of functions λ_α defines a density on M. Thus, a choice of orientation produces a linear map

$$\Omega^n(M) \to C^\infty(|\Lambda|(M)).$$

As explained in Subsection 3.4.2, this map is a bijection.

(c) Any Riemann metric g on M defines a canonical density on M denoted by $|dV_g|$ and called the *volume density*. It is locally described by the collection

$$\sqrt{|g_\alpha|}|dx_\alpha|,$$

where $|g_\alpha|$ denotes the determinant of the symmetric matrix representing the metric g in the coordinates (x_α^i). $\qquad\square$

The densities on a manifold serve one major purpose: they can be integrated. We denote by $C_0(|\Lambda|(M))$ the space of continuous densities with compact support. Consider the integration map defined in Subsection 3.4.1

$$\int_M : C_0(|\Lambda|(M)) \to \mathbb{R}, \quad |d\mu| \mapsto \int_M |d\mu|.$$

Note that if f is a continuous, compactly supported function on M, and $|d\mu|$ is a density, then $f|d\mu|$ is a continuous compactly supported density, and thus there is a well defined integral

$$\int_M f|d\mu|.$$

We obtain in this fashion a natural pairing

$$C_0(M) \times C(|\Lambda|_M), \quad (f, |d\mu|) \mapsto \int_M f|d\mu|.$$

Let us observe that if $|d\rho|$ and $|d\tau|$ are two positive densities, then there exists a positive function f such that

$$|d\rho| = f|d\tau|.$$

The existence of this function follows from the Radon-Nicodym theorem. For every $x \in M$ we have

$$f(x) = \lim_{U \to \{x\}} \frac{\int_U |d\rho|}{\int_U |d\tau|},$$

where the above limit is taken over open sets shrinking to x. We will use the notation

$$f := \frac{|d\rho|}{|d\tau|},$$

and we will refer to f as the *jacobian of $|d\rho|$ relative to $|d\tau|$*. If $\phi : M \to N$ is a diffeomorphism and $|d\rho| = (U_\alpha, \rho_\alpha|dy_\alpha|)$ is a density on N, then we define the *pullback* of $|d\rho|$ by ϕ to be the density $\phi^*|d\rho|$ on M defined by

$$\phi^*|d\rho| = (\phi^{-1}(U_\alpha), \rho_\alpha|dy_\alpha|), \quad y_\alpha^i = x_\alpha^i \circ \phi.$$

The classical change in variables formula now takes the form

$$\int_N |d\rho| = \int_M \phi^*|d\rho|.$$

Example 9.1.5. Suppose $\phi : M \to N$ is a diffeomorphism between two smooth m-dimensional manifolds, $\omega \in \Omega^m(M)$, and $|\omega|$ is the associated density. Then

$$\phi^*|\omega| = |\phi^*\omega|.$$

$\qquad\square$

Suppose a Lie group G acts smoothly on M. Then for every $g \in G$, and any density $|d\rho|$, we get a new density $g^*|d\rho|$. The density $|d\rho|$ is called G-invariant if

$$g^*|d\rho| = |d\rho|, \quad \forall g \in G.$$

Note that a density is G-invariant if and only if the associated Borel measure is G-invariant. A positive density is invariant if the jacobian $\frac{g^*|d\rho|}{|d\rho|}$ is identically equal to 1, $\forall g \in G$.

Proposition 9.1.6. *Suppose $|d\rho|$ and $|d\tau|$ are two G-invariant positive densities. Then the jacobian $J = \frac{|d\rho|}{|d\tau|}$ is a G-invariant smooth, positive function on G.*

Proof. Let $x \in M$ and $g \in G$. Then for every open neighborhood U of x we have

$$\int_U |d\rho| = \int_{g(U)} |d\rho|, \quad \int_U |d\tau| = \int_{g(U)} |d\tau| \implies \frac{\int_U |d\tau|}{\int_U |d\rho|} = \frac{\int_{g(U)} |d\tau|}{\int_{g(U)} |d\rho|},$$

and then letting $U \to \{x\}$ we deduce

$$J(x) = J(gx), \quad \forall x \in M, g \in G. \qquad \square$$

Corollary 9.1.7. *If G acts smoothly and transitively on the smooth manifold M then, up to a positive multiplicative constant there exists at most one invariant positive density.* $\qquad \square$

Suppose $\Phi : M \to B$ is a submersion. The kernels of the differentials of Φ form a vector subbundle $T^V M \hookrightarrow TM$ consisting of the planes tangent to the fibers of Φ. We will refer to it as the *vertical tangent bundle*. Since Φ is a submersion, we have a short exact sequence of bundles over M.

$$0 \to T^V M \hookrightarrow TM \xrightarrow{D\Phi} \Phi^* TB \to 0.$$

Observe that any (positive) density $|d\nu|$ on B defines by pullback a (positive) density $\Phi^*|d\nu|$ associated to the bundle $\Phi^* TB \to M$. If λ is a density associated to the vertical tangent bundle $T^V M$, then we obtain a density $\lambda \times \Phi^*|d\nu|$ on M.

Suppose $|d\mu|$ is a density on M such that Φ is proper on the support of $|d\mu|$. Set $k := \dim B$, $r := \dim M - \dim B$. We would like to describe a density $\Phi_*|d\mu|$ on B called the *pushforward* of $|d\mu|$ by Φ. Intuitively, $\Phi_*|d\mu|$ is the unique density on B such that for any open subset $U \subset B$ we have

$$\int_U \Phi_*|d\mu| = \int_{\Phi^{-1}(U)} |d\mu|.$$

Proposition 9.1.8 (The pushforward of a density). *There exists a smooth density $\Phi_*|d\mu|$ on B uniquely characterized by the following condition. For every density $|d\nu|$ on B we have*

$$\Phi_*|d\mu| = V_\nu |d\nu|,$$

where $V_\nu \in C^\infty(B)$ is given by

$$V_\nu(b) := \int_{\Phi^{-1}(b)} |d\mu|/\Phi^*|d\nu|.$$

Proof. Fix a positive density $|d\nu|$ on B. Along every fiber $M_b = \Phi^{-1}(b)$ we have a density $|d\mu|_b/\Phi^*|d\nu| \in C^\infty\big(|\Lambda|(M_b)\big)$ corresponding to the short exact sequence

$$0 \to TM_b \to (TM)|_{M_b} \xrightarrow{D\Phi} (\Phi^* TB)|_{M_b} \to 0.$$

To understand this density fix $x \in M_b$. Then we can find local coordinates $(y_j)_{1 \le j \le k}$ near $b \in B$, and smooth functions $(x^i)_{1 \le i \le r}$ defined in a neighborhood V of x in M such that the collection of functions (x^i, y^j) defines local coordinates near x on M, and in these coordinates the map Φ is given by the projection $(x, y) \mapsto y$.

In the coordinates y on B we can write

$$|d\nu| = \rho_B(y)|dy| \quad \text{and} \quad |d\mu| = \rho_M(x, y)|dx \wedge dy|.$$

Then along the fibers $y = \text{const}$ we have

$$|d\mu|_b/\Phi^*|d\nu| = \frac{\rho_M(x, y)}{\rho_B(y)}|dx|.$$

We set

$$V_\nu(b) := \int_{M_b} |d\mu|_b/\Phi^*|d\nu|.$$

The function $V_\nu : B \to \mathbb{R}$ is smooth function. We can associate to V_ν the density $V_\nu|d\nu|$ which a priori depends on ν.

Observe that if $|d\hat{\nu}|$ is another density, then there exists a positive smooth function $w : B \to \mathbb{R}$ such that

$$|d\hat{\nu}| = w|d\nu|.$$

Then

$$|d\mu|_b/\Phi^*|d\hat{\nu}| = w^{-1}|d\mu|_b/\Phi^*|d\nu|, \quad V_{\hat{\nu}} = w^{-1}V_\nu$$

so that

$$V_{\hat{\nu}}|d\hat{\nu}| = V_\nu|d\nu|.$$

In other words, the density $V_\nu|d\nu|$ on B is independent of ν. It depends only on $|d\mu|$. $\square$

Using partitions of unity and the classical Fubini theorem we obtain the *Fubini formula for densities*

$$\int_{\Phi^{-1}(U)} |d\mu| = \int_U \Phi_*|d\mu|, \quad \text{for any open subset } U \subset B. \qquad (9.1.3)$$

Remark 9.1.9. Very often the submersion $\Phi : M \to B$ satisfies the following condition.

For every point on the base $b \in B$ there exist an open neighborhood U of b in B, a nowhere vanishing form $\omega \in \Omega^k(U)$, a nowhere vanishing form $\Omega \in \Omega^{k+r}(M_U)$, $(M_U := \Phi^{-1}(U))$, and a form $\eta \in \Omega^r(M_U)$ such that

$$\Omega = \eta \wedge \pi^* \omega.$$

Then we can write $|d\mu| = \rho|\Omega|$, for some $\rho \in C^{\infty}(M_U)$. The form ω defines a density $|\omega|$ on U, and we conclude that

$$|d\mu|/f^*|\omega| = \rho|\eta|, \quad \Phi_*|d\mu| = f|\omega|,$$

where for every $u \in U \subset B$ we have

$$f(u) = \int_{M_u} |d\mu|/f^*|\omega| = \int_{M_u} \rho|\Omega|/|f^*\omega| = \int_{M_u} \rho|\eta|.$$

In particular, using (9.1.3) we obtain the *generalized coarea formula*

$$\int_{M_U} |d\mu| = \int_{M_U} \rho|\eta \wedge \pi^*\omega| = \int_U f(u)|\omega| = \int_U \left(\int_{M_u} \rho|\eta|\right)|\omega|. \qquad \square$$

Example 9.1.10. (The classical coarea formula). Suppose (M, g) is a Riemann manifold of dimension $m + 1$, and $f : M \to \mathbb{R}$ is a smooth function without critical points. For simplicity, we also assume that the level sets $M_t = f^{-1}(t)$ are compact.

On M we have a volume density $|dV_g|$. We would like to compute the pushforward density $f_*|dV_g|$ on $\mathbb{R}$. We seek $f_*|dV_g|$ of the form

$$f_*|dV_g| = v(t)|dt|,$$

where $|dt|$ is the Euclidean volume density on $\mathbb{R}$, and v is a smooth function.

For $t \in \mathbb{R}$, the fiber M_t is a compact codimension 1 submanifold of M. We denote by $|dV_t|$ the volume density on M_t defined by the induced metric $g_t := g|_{M_t}$. We denote by ∇f the g-gradient of f, and we set $\boldsymbol{n} := \frac{1}{|\nabla f|}\nabla f$.

Fix $t_0 \in \mathbb{R}$. For every point $p \in M_{t_0}$ we have $df(p) \neq 0$, and from the implicit function theorem we deduce that we can find an open neighborhood U, and smooth function $x^1, \ldots, x^m$ such that $(f, x^1, \ldots, x^m)$ are local coordinates on U. Then along U we can write

$$|dV_g| = \rho|df \wedge dx^1 \wedge \cdots \wedge dx^m|,$$

where ρ is a smooth function. The vector field $\boldsymbol{n}$ is a unit *normal* vector field along $M_{t_0} \cap U$, so that we have the equality

$$|dV_{t_0}|\,|_{U \cap M_{t_0}} = \rho\big|\boldsymbol{n} \lrcorner (df \wedge dx^1 \wedge \cdots \wedge dx^m)\,|_{U \cap M_{t_0}}\big|$$

$$= \rho|\nabla f| \cdot \big|(dx^1 \wedge \cdots \wedge dx^m)\,|_{U \cap M_{t_0}}\big|.$$

Now observe that along U we have

$$\rho(df \wedge dx^1 \wedge \cdots \wedge dx^m) = \rho(f^*dt \wedge dx^1 \wedge \cdots \wedge dx^m).$$

Hence

$$|dV_g|/f^*|dt| = \rho|(dx^1 \wedge \cdots \wedge dx^m)|,$$

so that

$$|dV_{t_0}|\,|_{U \cap M_{t_0}} = |\nabla f| \cdot |dV_g|/f^*|dt| \quad \text{and} \quad |dV_g|/f^*|dt| = \frac{1}{|\nabla f|}|dV_{t_0}|\,|_{U \cap M_{t_0}}.$$

Hence

$$f_*|dV_g| = v(t)|dt|, \quad v(t) = \int_{M_t} \frac{1}{|\nabla f|} |dV_t|, \tag{9.1.4}$$

and we obtain in this fashion the *classical coarea formula*

$$\int_M |dV_g| = \int_{\mathbb{R}} \left(\int_{M_t} \frac{1}{|\nabla f|} |dV_t| \right) |dt|. \tag{9.1.5}$$

To see how this works in practice, consider the unit sphere $S^n \subset \mathbb{R}^{n+1}$. We denote the coordinates in $\mathbb{R}^{n+1}$ by $(t, x^1, \ldots, x^n)$. We let $P_\pm \in S^n$ denote the poles given $t = \pm 1$.

We denote by $|dV_n|$ the volume density on S^n, and by $\pi : S^n \to \mathbb{R}$ the natural projection given by

$$(t, x^1, \ldots, x^n) \mapsto t.$$

The map π is a submersion on the complement of the poles, $M = S^n \setminus \{P_\pm\}$, and $\pi(M) = (-1, 1)$. We want to compute $\pi_*|dV_n|$.

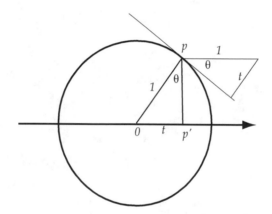

Fig. 9.1   *Slicing a sphere by hyperplanes*

Observe that $\pi^{-1}(t)$ is the $(n-1)$-dimensional sphere of radius $(1 - t^2)^{1/2}$. To find the gradient $\nabla\pi$ observe that for every $p \in S^n$ the tangent vector $\nabla\pi(p)$ is the projection of the vector ∂_t on the tangent space $T_p S^n$, because ∂_t is the gradient with respect to the Euclidean metric on $\mathbb{R}^{n+1}$ of the linear function $\pi : \mathbb{R}^{n+1} \to \mathbb{R}$, $\pi(t, x^i) = t$.

We denote by θ the angle between ∂_t and $T_p S^n$, set $p' = \pi(p)$, and by t the coordinate of p' (see Figure 9.1). Then θ is equal to the angle at p between the radius $[0, p]$ and the segment $[p, p']$. We deduce

$$\cos\theta = \text{length } [p, p'] = (1 - t^2)^{1/2}.$$

Hence

$$|\nabla\pi(p)| = (1 - t^2)^{1/2},$$

and consequently,

$$\int_{\pi^{-1}(t)} \frac{1}{|\nabla \pi|} |dV_t| = (1 - t^2)^{-1/2} \int_{\pi^{-1}(t)} |dV_t| = \sigma_{n-1}(1 - t^2)^{\frac{n-2}{2}},$$

where σ_m denotes the m-dimensional area of the unit m-dimensional sphere S^m. The last formula implies

$$\sigma_n = 2\sigma_n \int_0^1 (1 - t^2)^{\frac{n-2}{2}} |dt| = \sigma_{n-1} \int_0^1 (1 - s)^{\frac{n-2}{2}} s^{-1/2} |ds| = \sigma_{n-1} B\left(\frac{1}{2}, \frac{n}{2}\right),$$

where B denotes the *Beta function*

$$B(p, q) := \int_0^1 s^{p-1}(1 - s)^{q-1} ds, \quad p, q > 0. \tag{9.1.6}$$

It is known (see [101], Section 12.41) that

$$B(p, q) = \int_0^\infty \frac{x^p}{(1 + x)^{p+q}} \cdot \frac{dx}{x} = \frac{\Gamma(p)\Gamma(q)}{\Gamma(p + q)}, \tag{9.1.7}$$

where $\Gamma(x)$ denotes Euler's Gamma function

$$\Gamma(x) = \int_0^\infty e^{-t} t^{x-1} dt.$$

We deduce

$$\sigma_n = \frac{2\Gamma(\frac{1}{2})^{n+1}}{\Gamma(\frac{n+1}{2})}. \tag{9.1.8}$$

$\square$

9.1.2 *Invariant measures on linear Grassmannians*

Suppose that V is a Euclidean real vector space of dimension n. We denote by $\bullet$ the inner product on V. For every subspace $U \subset V$ we denote by P_U the orthogonal projection onto U. We would like to investigate a certain natural density on $\mathbf{Gr}_k(V)$, the Grassmannian of k dimensional subspaces of V. We are forced into considering densities rather than forms because very often the Grassmannians $\mathbf{Gr}_k(V)$ are not orientable. More precisely, $\mathbf{Gr}_k(V)$ is orientable if and only if $\dim V$ is even.

In Example 1.2.20 we have shown that, via the (orthogonal) projection map $U \mapsto P_U$, we can regard $\mathbf{Gr}_k(V)$ as a submanifold of $\mathrm{End}^+(V)$, the vector space of symmetric endomorphisms of V. The Euclidean metric (1.2.2) on $\mathrm{End}(V)$, defined by

$$\langle A, B \rangle = \frac{1}{2} \mathrm{tr}(AB^\dagger), \quad \forall A, B \in \mathrm{End}(V),$$

induces a metric $h = h_{n,k}$ on $\mathbf{Gr}_k(V)$.

Denote by $O(V)$ the group of orthogonal transformations of V. The group $O(V)$ acts smoothly and transitively on $\mathbf{Gr}_k(V)$

$$O(V) \times \mathbf{Gr}_k(V) \ni (g, L) \mapsto g(L) \in \mathbf{Gr}_k(V).$$

Note that

$$P_{gL} = gP_L g^{-1}.$$

The action of $O(V)$ on $\text{End}^+(V)$ by conjugation preserves the inner product on $\text{End}^+(V)$, and we deduce that the action of $O(V)$ on $\mathbf{Gr}_k(V)$ preserves the metric h.

We would like to express this metric in the graph coordinates introduced in Example 1.2.20. Consider $L \in \mathbf{Gr}_k(V)$, and $S \in \text{Hom}(L, L^\perp)$. Then, for every $t \in \mathbb{R}$, we denote by U_t the graph of the map $tS : L \to L^\perp$,

$$U_t := \Gamma_{tS} \in \mathbf{Gr}_k(V, L).$$

If $\dot{U}_0$ denotes the tangent to the path $t \mapsto U_t$ at $t = 0$, then

$$h(\dot{U}_0, \dot{U}_0) = \frac{1}{2}\text{tr}(\dot{P}_0^2), \quad \dot{P}_0 := \frac{d}{dt}|_{t=0}P_{tS},$$

If we write $P_t := P_{\Gamma_{tS}}$ we deduce from (1.2.4) that

$$P_t = \begin{bmatrix} (\mathbb{1}_L + t^2 S^*S)^{-1} & t(\mathbb{1}_L + t^2 S^*S)^{-1}S^* \\ tS(\mathbb{1}_L + t^2 S^*S)^{-1} & t^2 S(\mathbb{1}_L + t^2 S^*S)^{-1}S^* \end{bmatrix}.$$

Hence

$$\dot{P}_{t=0} = \begin{bmatrix} 0 & S^* \\ S & 0 \end{bmatrix} = S^*P_{L^\perp} + SP_L, \tag{9.1.9}$$

so that

$$h(\dot{U}_0, \dot{U}_0) = \frac{1}{2}\big(\text{tr}(SS^*) + \text{tr}(S^*S)\big) = \text{tr}(SS^*).$$

We can be even more concrete.

Let us choose an orthonormal basis $(\vec{e}_i)_{1 \le i \le k}$ of L, and an orthonormal basis $(e_\alpha)_{k < \alpha \le n}$ of $L^\perp$. With respect to these bases, the map $S : L \to L^\perp$ is described by a matrix $(s_{\alpha i})_{1 \le i \le k < \alpha \le n}$, and then

$$\text{tr}(SS^*) = \text{tr}(S^*S) = \sum_{i,\alpha} |s_{\alpha i}|^2.$$

We can think of the collection $(s_{\alpha i})$ as defining local coordinates on the open subset $\mathbf{Gr}_k(V, L) \subset \mathbf{Gr}_k(V)$ consisting of k-planes intersecting $L^\perp$ transversally. Hence

$$h(\dot{U}_0, \dot{U}_0) = \sum_{i,\alpha} |s_{\alpha i}|^2. \tag{9.1.10}$$

In integral geometric computations we will find convenient to relate the above coordinates to the classical language of moving frames. In the sequel we make the following notational conventions.

- We will use lower case Latin letters $i, j, \dots$ to denote indices in the range $\{1, \dots, k\}$.

- We will use lower case Greek letters $\alpha, \beta, \gamma, \ldots$ to denote indices in the range $\{k+1, \ldots, n\}$.
- We will use upper case Latin letters $A, B, C, \ldots$ to denote indices in the range $\{1, \cdots, n\}$.

Suppose we have a smooth 1-parameter family of orthonormal frames

$$(e_A) = (e_A(t)), \quad |t| \ll 1.$$

This defines a smooth path

$$t \mapsto L_t = \text{span}\,(e_i(t)\,) \in \mathbf{Gr}_k(V).$$

We would like to compute $\hat{h}(\dot{L}_0, \dot{L}_0)$.

Observe that we have a smooth path $t \mapsto g_t \in O(V)$, defined by

$$g_t e_A(0) = e_A(t).$$

With respect to the fixed frame $(e_A(0)\,)$, the orthogonal transformation g_t is given by a matrix $(s_{AB}(t)\,)$, where

$$s_{AB} = e_A(0) \bullet (g_t e_B(0)\,).$$

Observe that $g_0 = \mathbb{1}_V$. Let P_t denote the projection onto L_t. Then

$$P_t = g_t P_0 g_t^{-1},$$

so that, if we set $X = \frac{d}{dt}|_{t=0} g_t$, we have

$$\dot{P}_0 = [X, P_0].$$

With respect to the decomposition $V = L_0 + L_0^\perp$ the projector P_0 has the block decomposition

$$P_0 = \begin{bmatrix} \mathbb{1}_{L_0} & 0 \\ 0 & 0 \end{bmatrix}.$$

The operator X is represented by a skew-symmetric matrix with entries

$$x_{AB} = \dot{s}_{AB} = e_A \bullet \dot{e}_B,$$

which has the block form

$$X = \begin{bmatrix} X_{L_0, L_0} & -X^*_{L_0^\perp, L_0} \\ X_{L_0^\perp, L_0} & X_{L_0^\perp, L_0^\perp} \end{bmatrix},$$

where $X_{L_0^\perp, L_0}$ denotes a map $L_0 \to L_0^\perp$ etc. We deduce

$$[X, P_0] = \begin{bmatrix} 0 & X^*_{L_0^\perp, L_0} \\ X_{L_0^\perp, L_0} & 0 \end{bmatrix}.$$

Hence

$$h(\dot{L}_0, \dot{L}_0) = \frac{1}{2}\,\text{tr}(\dot{P}_0, \dot{P}_0) = \text{tr}(X_{L_0^\perp, L_0} X^*_{L_0^\perp, L_0}) = \sum_{\alpha, i} |\dot{s}_{\alpha i}|^2. \tag{9.1.11}$$

We want to interpret this in the language of moving frames.

Suppose M is a smooth m-dimensional manifold and $L : M \to \mathbf{Gr}_k(V)$ is a smooth map. Fix a point $p_0 \in M$ and local coordinates $(u^i)_{1 \leq i \leq m}$ near p_0 such that $u^i(p_0) = 0$.

The map L_t can be described near p_0 via a moving frame, i.e., an orthonormal frame (e_A) of V, varying smoothly with (u^i), such that

$$L(u) = \mathrm{span}\,(\,e_i(u)\,).$$

The above computations show that the differential of L at p_0 is described by the $(n - k) \times k$ matrix of 1-forms on M

$$D_{p_0} L = (\theta_{\alpha i}), \quad \theta_{\alpha i} := e_\alpha \bullet de_i.$$

More precisely, this means that if $X = (\dot{u}^a) \in T_{p_0} M$, and we let " $\cdot$ " denote the differentiation along the flow lines of X, then

$$D_{p_0} L(X) = (x_{\alpha i}) \in T_{L(0)} \mathbf{Gr}_k(V), \quad x_{\alpha i} = e_\alpha \bullet \dot{e}_i = \sum_a e_\alpha \bullet de_i(X). \quad (9.1.12)$$

If M happens to be an open subset of $\mathbf{Gr}_k(V)$, then we can use the forms $\theta_{\alpha i}$ to describe the metric h. More precisely, the equalities (9.1.10) and (9.1.11) show that

$$h = \sum_{\alpha, i} \theta_{\alpha i} \otimes \theta_{\alpha i}. \quad (9.1.13)$$

In other words, the collection $(\theta_{\alpha i})$ is an orthonormal frame of the cotangent bundle $T^* \mathbf{Gr}_k(V)$.

The metric $h = h_{n,k}$ is $O(V)$-invariant so that the associated Riemannian volume defines an invariant density. We will denote this invariant density by $|d\gamma_{n,k}|$, where $n = \dim V$, and we will refer to it as the *kinematic density* on $\mathbf{Gr}_k(V)$. Since the action of the group $O(V)$ is transitive, we deduce that any other invariant density is equivalent to a constant multiple of this metric density. We would like to give a local description of $|d\gamma_{n,k}|$.

Set $n := \dim V$. If $\mathcal{O}$ is a sufficiently small open subset of $\mathbf{Gr}_k(V)$ then we can find smooth maps

$$e_A : \mathcal{O} \to V, \quad A = 1, \cdots, n,$$

with the following properties.

- For every $L \in \mathcal{O}$ the collection $\big(e_A(L)\big)_{1 \leq A \leq n}$ is an orthonormal frame of V.
- For every $L \in \mathcal{O}$ the collection $\big(e_i(L)\big)_{1 \leq i \leq k}$ is an orthonormal frame of L.

For every $1 \leq i \leq k$, and every $k + 1 \leq \alpha \leq 1$, we have a 1-form

$$\theta_{\alpha i} \in \Omega^1(\mathcal{O}), \quad \theta_{\alpha i} = e_\alpha \bullet de_i = -de_\alpha \bullet e_i.$$

As explained above, the metric h is described along $\mathcal{O}$ by the symmetric tensor

$$h = \sum_{\alpha, i} \theta_{\alpha i} \otimes \theta_{\alpha i},$$

and the associated volume density is described by

$$|d\gamma_{n,k}| = \left|\prod_{\alpha,i}\theta_{\alpha i}\right| := \left|\bigwedge_{\alpha,i}\theta_{\alpha i}\right|.$$

Example 9.1.11. To understand the above construction it is helpful to consider the special case of the Grassmannian $\mathbf{Gr}_1(\mathbb{R}^2)$ of lines through the origin in $\mathbb{R}^2$. This space is diffeomorphic to the real projective line $\mathbb{R}\mathbb{P}^1$, which in turn, it is diffeomorphic to a circle.

A line L in $\mathbb{R}^2$ is uniquely determined by the angle $\theta \in [0, \pi]$ that it forms with the x-axis. For such an angle θ, we denote by L_θ the corresponding line. The line L_θ is also represented by the the orthonormal frame

$$\boldsymbol{e}_1(\theta) = (\cos\theta, \sin\theta), \quad \boldsymbol{e}_2(\theta) = (-\sin\theta, \cos\theta), \quad L_\theta = \mathrm{span}\,(\,\boldsymbol{e}_1(\theta)\,),$$

where the first vector $\boldsymbol{e}_1(\theta)$ gives the direction of the line. Then $\theta_{21} = \boldsymbol{e}_2 \bullet d\boldsymbol{e}_1 = d\theta$, and $|d\gamma_{2,1}| = |d\theta|$. $\square$

We would like to compute the volumes of the Grassmannians $\mathbf{Gr}_k(V)$, $\dim V = n$ with respect to the kinematic density $|d\gamma_{n,k}|$, i.e., we would like to compute

$$C_{n,k} := \int_{\mathbf{Gr}_k(V)} |d\gamma_{n,k}|.$$

Denote by $\boldsymbol{\omega}_n$ the volume of the unit ball $\boldsymbol{B}^n \subset \mathbb{R}^n$, and by $\boldsymbol{\sigma}_{n-1}$ the $(n-1)$-dimensional "surface area" of the unit sphere $\boldsymbol{S}^{n-1}$, so that

$$\boldsymbol{\sigma}_{n-1} = n\boldsymbol{\omega}_n, \quad \text{and} \quad \boldsymbol{\omega}_n = \frac{\Gamma(1/2)^n}{\Gamma(1+n/2)} = \begin{cases} \dfrac{\pi^k}{k!} & n = 2k \\[2ex] \dfrac{2^{2k+1}\pi^k k!}{(2k+1)!} & n = 2k+1 \end{cases},$$

where $\Gamma(x)$ is the Gamma function. We list below the values of $\boldsymbol{\omega}_n$ for small n.

n	0	1	2	3	4
$\boldsymbol{\omega}_n$	1	2	π	$\frac{4\pi}{3}$	$\frac{\pi^2}{2}$

To compute the volume of the Grassmannians we need to use yet another description of the Grassmannians, as homogeneous spaces.

Note first that the group $O(V)$ acts transitively on $\mathbf{Gr}_k(V)$. Fix $L_0 \in \mathbf{Gr}_k(V)$. Then the stabilizer of L_0 with respect to the action of $O(V)$ on $\mathbf{Gr}_k(V)$ is the subgroup $O(L_0) \times O(L_0^\perp)$ and thus we can identify $\mathbf{Gr}_k(V)$ with the homogeneous space of left cosets $O(V)/O(L_0) \times O(L_0^\perp)$.

The computation of $C_{n,k}$ is carried out in three steps.

Step 1. We equip the orthogonal groups $O(\mathbb{R}^n)$ with a canonical invariant density $|d\gamma_n|$ called the *kinematic density* on $O(n)$. Set

$$C_n := \int_{O(\mathbb{R}^n)} |d\gamma_n|.$$

Step 2. We show that

$$C_{n,k} = \frac{C_n}{C_k C_{n-k}}.$$

Step 3. We show that

$$C_{n,1} = \frac{1}{2}\sigma_{n-1} = \frac{n\omega_n}{2},$$

and then compute C_n inductively using the recurrence relation from Step 2

$$C_{n+1} = (C_1 C_{n,1}) C_n,$$

and the initial condition

$$C_2 = \mathrm{vol}\,(\,O(2)\,) = 2\sigma_1.$$

Step 1. The group $O(V)$ is a submanifold of $\mathrm{End}(V)$ consisting of endomorphisms S satisfying $SS^* = S^*S = \mathbb{1}_V$. We equip $\mathrm{End}(V)$ with the inner product

$$\langle A, B \rangle = \frac{1}{2}\mathrm{tr}(AB^*).$$

This metric induces an invariant metric h on $O(V)$. We would like to give a more concrete description of this metric.

Denote by $\mathrm{End}^-(V)$ the subspace of $\mathrm{End}(V)$ consisting of skew-symmetric operators. For any $S_0 \in O(V)$ we have a map

$$\exp_{S_0} : \mathrm{End}^-(V) \to O(V), \quad \mathrm{End}^-(V) \longmapsto S_0 \cdot \exp(X).$$

This defines a diffeomorphism from a neighborhood of 0 in $\mathrm{End}^-(V)$ to a neighborhood of S_0 in $O(V)$. Two skew-symmetric endomorphisms $X, Y \in \mathrm{End}^-(V)$ define paths

$$\gamma_X, \gamma_Y : \mathbb{R} \to O(V), \quad \gamma_X(t) = S_0\exp(tX), \quad \gamma_Y(t) = S_0\exp(tY),$$

originating at S_0. We set

$$\dot{X} := \dot{\gamma}_X(0) \in T_{S_0}O(V) \subset \mathrm{End}(V), \quad \dot{Y} := \dot{\gamma}_Y(0) \in T_{S_0}O(V) \subset \mathrm{End}(V).$$

Then $\dot{X} = S_0 X$, $\dot{Y} = S_0 Y$, and

$$h(\dot{X}, \dot{Y}) = \frac{1}{2}\mathrm{tr}\big((S_0 X)(S_0 Y)^*\big) = \frac{1}{2}\mathrm{tr}\big(S_0 XY^* S_0^*\big) = \frac{1}{2}\mathrm{tr}\big(S_0^* S_0 XY^*\big)$$

$$= \frac{1}{2}\mathrm{tr}\big(XY^*\big).$$

If we choose an orthonormal basis (e_A) of V so that X and Y are given by the skew symmetric matrices (x_{AB}), (y_{AB}), then we deduce

$$h(\dot{X}, \dot{Y}) = \sum_{A>B} x_{AB} y_{AB}.$$

If we set $\boldsymbol{f}_A(t) := \exp(tX)e_A$, then we deduce

$$x_{AB} = e_A \bullet \dot{\boldsymbol{f}}_B(0) = \boldsymbol{f}_A(0) \bullet \dot{\boldsymbol{f}}_B(0).$$

More generally, if we define

$$\boldsymbol{f}_A : O(V) \to V, \quad \boldsymbol{f}_A(S) := S\boldsymbol{e}_A,$$

then we obtain the angular forms $\theta_{AB} := \boldsymbol{f}_A \bullet d\boldsymbol{f}_B$. The above metric metric has the description

$$h = \sum_{A>B} \theta_{AB} \otimes \theta_{AB}.$$

The associated volume density is

$$|d\gamma_n| = \left| \bigwedge_{A>B} \theta_{AB} \right|.$$

Step 2. *Fix an orthonormal frame* $(\boldsymbol{e}_A)$ *of* V *such that* $L_0 = \mathrm{span}\,(\boldsymbol{e}_i;\ 1 \leq i \leq k)$. We can identify V with $\mathbb{R}^n$, $O(V)$ with $O(n)$, and L_0 with the subspace

$$\mathbb{R}^k \oplus 0_{n-k} \subset \mathbb{R}^n.$$

An orthogonal $n \times n$ matrix T is uniquely determined by the orthonormal frame $(T\boldsymbol{e}_A)$ via the equalities

$$T_{AB} = \boldsymbol{e}_A \bullet T\boldsymbol{e}_B.$$

Define

$$p : O(n) \to \mathbf{Gr}_k(\mathbb{R}^n), \quad p(T) = T(L_0).$$

More explicitly, we have

$$p(T) = \mathrm{span}\,(T\boldsymbol{e}_i)_{1 \leq i \leq k}.$$

We will prove that we have a principal fibration

$$O(k) \times O(n-k) \lhook\joinrel\longrightarrow O(n)$$
$$\downarrow p \qquad ,$$
$$\mathbf{Gr}_k(\mathbb{R}^n)$$

and that

$$p_* |d\gamma_n| = C_k C_{n-k} |d\gamma_{n,k}|.$$

Once we have this we deduce from the Fubini formula (9.1.3) that

$$C_n = C_k C_{n-k} C_{n,k}.$$

Let us prove the above facts.

For every sufficiently small open subset $U \subset \mathbf{Gr}_k(V)$ we can find a smooth section

$$\phi : U \to O(n)$$

of the map $p : O(n) \to \mathbf{Gr}_k(\mathbb{R}^n)$. The section can be identified with a smooth family of orthonormal frames $(\boldsymbol{\phi}_A(L),\ L \in U)_{1 \leq A \leq n}$ of $\mathbb{R}^n$, such that

$$L = \mathrm{span}\,(\boldsymbol{\phi}_i(L);\ 1 \leq i \leq k).$$

To such a frame we associate the orthogonal matrix $\phi(L) \in O(n)$ which maps the fixed frame (e_A) to the frame (ϕ_B). It is given by a matrix with entries

$$\phi(L)_{AB} = e_A \bullet \phi_B.$$

Then we have a smooth map

$$\Psi : O(k) \times O(n-k) \times U \to p^{-1}(U),$$

defined as follows.

- Given $(s, t, L) \in O(k) \times O(n-k) \times U$, express s as a $k \times k$ matrix $s = (s^i_j)$, and t as an $(n-k) \times (n-k)$ matrix (t^α_β).
- Define the orthonormal frame of $\mathbb{R}^n$

$$(f_A) := (\phi_B) * (s, t),$$

  via the equalities

$$f_i = f_i(s, L) = \sum_j s^j_i \phi_j(L) \in L, \;\; 1 \leq i \leq k, \tag{9.1.14}$$

$$f_\alpha = f_\alpha(t, L) = \sum_\beta t^\beta_\alpha \phi_\beta(L) \in L^\perp, \;\; k+1 \leq \alpha \leq n. \tag{9.1.15}$$

- Now define $\Psi = \Psi(s, t, L)$ to be the orthogonal transformation of $\mathbb{R}^n$ which maps the frame (e_A) to the frame (f_B), i.e.

$$f_A = \Psi e_A, \;\; \forall A.$$

The map Ψ is a homeomorphism with inverse

$$O(n) \ni T \mapsto \Psi^{-1}(T) = (s, t; L) \in O(k) \times O(n-k) \times L$$

defined as follows. We set $f_A = f_A(T) := T e_A, 1 \leq A \leq n$. Then

$$L = L_T = \mathrm{span}\,(f_i)_{1 \leq i \leq k},$$

while the matrices (s^i_j) and (t^α_β) are obtained via (9.1.14) and (9.1.15). More precisely, we have

$$s^i_j = \phi_i(L_T) \bullet f_j, \;\; s^\alpha_\beta = \phi_\alpha(L_T) \bullet f_\beta.$$

Observe that, $\forall s_0, s_1 \in O(k), \;\; t_0, t_1 \in O(n-k)$, we have

$$\big((\phi_B) * (s_0, t_0)\big) * (s_1, t_1) = (\phi_B) * (s_0 s_1, t_0 t_1).$$

This means that Ψ is equivariant with respect to the right actions of $O(k) \times O(n-k)$ on $O(k) \times O(n-k) \times U$ and $O(n)$. We have a commutative diagram

$$
\begin{array}{ccc}
O(k) \times O(n-k) \times U & \xrightarrow{\;\;\;\;\Psi\;\;\;\;} & p^{-1}(U) \\
& \searrow{\scriptstyle \pi} \qquad {\scriptstyle p} \swarrow & \\
& U \subset \mathbf{Gr}_k(\mathbb{R}^n) &
\end{array}
$$

In particular, this shows that p defines a principal $O(k) \times O(n-k)$-bundle.

Observe now that $p_* |d\gamma_n|$ is an invariant density on $\mathbf{Gr}_k(\mathbb{R}^n)$, and thus there exists a constant c such that

$$p_* |d\gamma_n| = c |d\gamma_{n,k}|.$$

This constant is given by the integral of the density $|d\gamma_n|/p^* |d\gamma_{n,k}|$ along the fiber $p^{-1}(L_0)$.

Hence

$$C_n = \int_{O(n)} |d\gamma_n| = \int_{\mathbf{Gr}_k(\mathbb{R}^n)} p_* |d\gamma_n| = c \int_{\mathbf{Gr}_k(\mathbb{R}^n)} |d\gamma_{n,k}| = c C_{n,k}.$$

Recall that if we define $\boldsymbol{f}_A : O(n) \to \mathbb{R}^n$ by $\boldsymbol{f}_A(T) := T e_A$, and $\theta_{AB} := \boldsymbol{f}_A \bullet d\boldsymbol{f}_B$, then

$$|d\gamma_n| = \Big| \bigwedge_{A>B} \theta_{AB} \Big|.$$

We write this as

$$\Big| \Big(\bigwedge_{i>j} \theta_{ij} \Big) \wedge \Big(\bigwedge_{\alpha>\beta} \theta_{\alpha\beta} \Big) \wedge \Big(\bigwedge_{\alpha,i} \theta_{\alpha,i} \Big) \Big|.$$

The form $\Big(\bigwedge_{\alpha,i} \theta_{\alpha,i} \Big)$ is the pullback of a nowhere vanishing form defined in a neighborhood of L_0 in $\mathbf{Gr}_k(\mathbb{R}^n)$, whose associated density is $|d\gamma_{n,k}|$. We now find ourselves in the situation described in Remark 9.1.9. We deduce

$$c = \int_{p^{-1}(L_0)} \Big| \Big(\bigwedge_{i>j} \theta_{ij} \Big) \wedge \Big(\bigwedge_{\alpha>\beta} \theta_{\alpha\beta} \Big) \Big|$$

$$= \Big(\int_{O(k)} |d\gamma_k| \Big) \Big(\int_{O(n-k)} |d\gamma_{n-k}| \Big) = C_k C_{n-k}.$$

Hence

$$C_{n,k} = \frac{C_n}{C_k C_{n-k}}.$$

Step 3. Fix an orthonormal basis $\{e_A\}$ of V, and denote by $\boldsymbol{S}_+^{n-1}$ the open hemisphere

$$\boldsymbol{S}_+^{n-1} = \{ \vec{v} \in V; \ |\vec{v}| = 1, \ \vec{v} \bullet e_1 > 0 \}.$$

Note that $\mathbf{Gr}_1(V) \cong \mathbb{RP}^{n-1}$ is the Grassmannian of lines in V. The set of lines that do not intersect $\boldsymbol{S}_+^{n-1}$ is a smooth hypersurface of $\mathbf{Gr}_1(V)$ diffeomorphic to $\mathbb{RP}^{n-2}$ and thus has kinematic measure zero. We denote by $\mathbf{Gr}_1(V)^*$ the open subset consisting of lines intersecting $\boldsymbol{S}_+^{n-1}$. We thus have a map

$$\psi : \mathbf{Gr}_1^*(V) \to \boldsymbol{S}_+^{n-1}, \quad \ell \mapsto \ell \cap \boldsymbol{S}_+^{n-1}.$$

This map is a diffeomorphism, and we have

$$C_{n,1} = \int_{\mathbf{Gr}_1(V)} |d\gamma_{n,1}| = \int_{\mathbf{Gr}_1^*(V)} |d\gamma_{n,1}| = \int_{\boldsymbol{S}_+^{n-1}} (\psi^{-1})^* |d\gamma_{n,1}|.$$

Now observe that ψ is in fact an *isometry*, and thus we deduce

$$C_{n,1} = \frac{1}{2} \text{area}(\boldsymbol{S}^{n-1}) = \frac{\boldsymbol{\sigma}_{n-1}}{2} = \frac{n\boldsymbol{\omega}_n}{2}.$$

Hence

$$C_{n+1} = C_n C_1 C_{n+1,1} = \boldsymbol{\sigma}_n C_n,$$

which implies inductively that

$$C_n = \boldsymbol{\sigma}_{n-1} \cdots \boldsymbol{\sigma}_2 C_2 = 2 \prod_{k=1}^{n-1} \boldsymbol{\sigma}_k = \prod_{j=0}^{n-1} \boldsymbol{\sigma}_j.$$

In particular, we deduce the following result.

Proposition 9.1.12. *For every* $1 \le k < n$ *we have*

$$C_{n,k} = \int_{\mathbf{Gr}_k(\mathbb{R}^n)} |d\gamma_{n,k}| = \frac{\prod_{j=0}^{n-1} \boldsymbol{\sigma}_j}{\left(\prod_{i=0}^{k-1} \boldsymbol{\sigma}_i\right) \cdot \left(\prod_{j=0}^{n-k-1} \boldsymbol{\sigma}_j\right)}$$

$$= \binom{n}{k} \frac{\prod_{j=1}^{n} \boldsymbol{\omega}_j}{\left(\prod_{i=1}^{k} \boldsymbol{\omega}_i\right) \cdot \left(\prod_{j=1}^{n-k} \boldsymbol{\omega}_j\right)}.$$

$\square$

Following [59], we set

$$[n] := \frac{1}{2} \frac{\boldsymbol{\sigma}_{n-1}}{\boldsymbol{\omega}_{n-1}} = \frac{n\boldsymbol{\omega}_n}{2\boldsymbol{\omega}_{n-1}}, \quad [n]! := \prod_{k=1}^{n} [k] = \frac{\boldsymbol{\omega}_n n!}{2^n},$$

$$\begin{bmatrix} n \\ k \end{bmatrix} := \frac{[n]!}{([k]!)([n-k]!)} = \binom{n}{k} \frac{\boldsymbol{\omega}_n}{\boldsymbol{\omega}_k \boldsymbol{\omega}_{n-k}}. \tag{9.1.16}$$

Denote by $|d\nu_{n,k}|$ the unique invariant density on $\mathbf{Gr}_k(V)$, $\dim V = n$ such that

$$\int_{\mathbf{Gr}_k(V)} |d\nu_{n,k}| = \begin{bmatrix} n \\ k \end{bmatrix}. \tag{9.1.17}$$

We have

$$|d\nu_{n,k}| = \frac{\begin{bmatrix} n \\ k \end{bmatrix}}{C_{n,k}} |d\gamma_{n_k}|.$$

Example 9.1.13. Using the computation in Example 9.1.11 we deduce

$$|d\gamma_{2,1}| = |d\theta|, \quad 0 \le \theta < \pi,$$

which implies that

$$C_{2,1} = \int_0^\pi |d\theta| = 2\frac{\omega_2}{\omega_1^2},$$

as predicted by Proposition 9.1.12. We have

$$\begin{bmatrix} 2 \\ 1 \end{bmatrix} = 2\frac{\omega_2}{\omega_1^2} = C_{2,1},$$

so that $|d\nu_{2,1}| = |d\gamma_{2,1}|$.

$\square$

9.1.3 Affine Grassmannians

We denote by $\mathbf{Graff}_k(V)$ the set of k-dimensional affine subspaces of V. We would like to describe a natural structure of smooth manifold on $\mathbf{Graff}_k(V)$.

We have the tautological vector bundle $\mathcal{U} = \mathcal{U}_{n,k} \to \mathbf{Gr}_k(V)$. It is naturally a subbundle of the trivial vector bundle $\underline{V} = V \times \mathbf{Gr}_k(V) \to \mathbf{Gr}_k(V)$ whose fiber over $L \in \mathbf{Gr}_k(V)$ is the vector subspace L. The trivial vector bundle $\underline{V}$ is equipped with a natural metric, and we denote by $\mathcal{U}^\perp \to \mathbf{Gr}_k(V)$ the orthogonal complement of $\mathcal{U}$ in $\underline{V}$.

The fiber of $\mathcal{U}^\perp$ over $L \in \mathbf{Gr}_k(V)$ is canonically identified with the orthogonal complement $L^\perp$ of L in V. The points of $\mathcal{U}^\perp$ are pairs $(\vec{c}, L)$, where $L \in \mathbf{Gr}_k(V)$, and $\vec{c}$ is a vector in $L^\perp$.

We have a natural map $\mathcal{A} : \mathcal{U}^\perp \to \mathbf{Graff}_k(V)$ given by

$$(\vec{c}, L) \longmapsto \vec{c} + L.$$

This map is a bijection with inverse

$$\mathbf{Graff}_k(V) \ni S \mapsto (S \cap [S]^\perp, [S]),$$

where $[S] \in \mathbf{Gr}_k(V)$ denotes the affine subspace through the origin parallel to S,

$$[S] = S - S = \big\{ s_1 - s_2; \ s_1, s_2 \in S \big\}.$$

We set

$$\vec{c}(S) := S \cap [S]^\perp,$$

and we say that $\vec{c}(S)$ is the *center* of the affine plane S.

We equip $\mathbf{Graff}_k(V)$ with the structure of smooth manifold which makes $\mathcal{A}$ a diffeomorphism. Thus, we identify $\mathbf{Graff}_k(V)$ with a vector subbundle of the *trivial bundle* $V \times \mathbf{Gr}_k(V)$ described by

$$\mathcal{U}^\perp = \big\{ (\vec{c}, L) \in V \times \mathbf{Gr}_k(V); \ P_L \vec{c} = 0 \big\},$$

where P_L denotes the orthogonal projection onto L.

The projection $\pi : \mathcal{U}^\perp \to \mathbf{Gr}_k(V)$ is a submersion. The fiber of this submersion over $L \in \mathbf{Gr}_k(V)$ is canonically identified with the vector subspace $L^\perp \subset V$. As such, it is equipped with a volume density $|dV_{L^\perp}|$. We obtain in this fashion a density $|dV_{L^\perp}|$ associated to the vertical subbundle $\ker D\pi \subset T\mathcal{U}^\perp$.

The base $\mathbf{Gr}_k(V)$ of the submersion $\pi : \mathbf{Graff}_k(V) \to \mathbf{Gr}_k(V)$ is equipped with a density $|d\gamma_{n,k}|$, and thus we obtain a density $\pi^* |d\gamma_{n,k}|$ associated to the bundle

$$\pi^* T \mathbf{Gr}_k(V) \to \mathbf{Graff}_k(V).$$

Using the short exact sequence of vector bundles over $\mathbf{Graff}_k(V)$,

$$0 \to \ker D\pi \longrightarrow T \mathbf{Graff}_k(V) \longrightarrow \pi^* T \mathbf{Gr}_k(V) \to 0,$$

we obtain a density $|\widetilde{\gamma}_{n,k}|$ on $\mathbf{Graff}_k(V)$ given by

$$|d\widetilde{\gamma}_{n,k}| = |dV_{L^\perp}| \times \pi^* |d\gamma_{n,k}|.$$

Let us provide a local description for this density.

Fix a small open subset $\mathcal{O} \subset \mathbf{Gr}_k(V)$, and denote by $\widetilde{\mathcal{O}}$ its preimage in $\mathbf{Graff}_k(V)$ via the projection π. Then we can find smooth maps

$$e_A : \mathcal{O} \to V, \ \vec{r} : \widetilde{\mathcal{O}} \to V$$

with the following properties.

• For every $S \in \widetilde{\mathcal{O}}$, $(e_A(S))$ is an orthonormal frame of V, and

$$[L] = \text{span}\left(e_i([L]) \right).$$

• For every $S \in \widetilde{\mathcal{O}}$ we have

$$S = \vec{r}(S) + [S].$$

We rewrite the last equality as

$$S = S(\vec{r}, e_i).$$

Observe that the center of this affine plane is the projection of $\vec{r}$ onto $[S]^{\perp}$

$$\vec{c}(S) = \sum_{\alpha} (e_{\alpha} \bullet \vec{r}) e_{\alpha}.$$

Following the tradition we introduce the (locally defined) 1-forms

$$\theta_{\alpha} := e_{\alpha} \bullet d\vec{r}, \ \ \theta_{\alpha i} := e_{\alpha} \bullet de_i.$$

For fixed $L \in \mathbf{Gr}_k(V)$ the density on the fiber $\mathcal{U}_L^{\perp} = L^{\perp}$ is given by

$$|dV_{L^{\perp}}| = \left| \bigwedge \theta_{\alpha} \right|.$$

The volume density on $\mathbf{Graff}_k(V)$ is described along $\widetilde{\mathcal{O}}$ by

$$|d\widetilde{\gamma}_{n,k}| = \left| \left(\bigwedge_{\alpha} \theta_{\alpha} \right) \wedge \left(\bigwedge_{\alpha,i} \theta_{\alpha i} \right) \right|.$$

The Fubini formula for densities (9.1.3) implies the following result.

Theorem 9.1.14. *Suppose* $f : \mathbf{Graff}_k(V) \to \mathbb{R}$ *is a compactly supported* $|d\hat{\gamma}_{n,k}|$-*integrable function. Then*

$$\int_{\mathbf{Graff}_k(V)} f(S)|d\widetilde{\gamma}_{n,k}(S)| = \int_{\mathbf{Gr}_k(V)} \left(\int_{L^{\perp}} f(p + L)|dV_{L^{\perp}}(p)| \right) |d\gamma_{n,k}(L)|,$$

where $|dV_{L^{\perp}}|$ *denotes the Euclidean volume density on* $L^{\perp}$. $\qquad\qquad\square$

Denote by $\mathrm{Iso}(V)$ the group of affine isometries of V, i.e., the subgroup of the group of affine transformations generated by translations, and by the rotations about a fixed point. Any affine isometry $T : V \to V$ is described by a unique pair $(t, S) \in V \times O(V)$ so that

$$T(v) = Sv + t, \quad \forall v \in V.$$

The group $\mathrm{Iso}(V)$ acts in an obvious fashion on $\mathbf{Graff}_k(V)$, and a simple computation shows that the associated volume density $|d\tilde{\gamma}_{n,k}|$ is $\mathrm{Iso}(V)$ invariant.

If instead of the density $|d\gamma_{n,k}|$ on $\mathbf{Gr}_k(V)$ we use the density $|d\nu_{n,k}|$, we obtain a density $|d\tilde{\nu}_{n,k}|$ on $\mathbf{Graff}_k(V)$ which is a constant multiple of $|d\tilde{\gamma}_{n,k}|$.

$$|d\tilde{\nu}_{n,k}| = \frac{\left[\begin{smallmatrix} n \\ k \end{smallmatrix}\right]}{C_{n,k}} |d\tilde{\gamma}_{n_k}|. \tag{9.1.18}$$

Example 9.1.15. Let us unravel the above definition in the special case $\mathbf{Graff}_1(\mathbb{R}^2)$, the Grassmannians of affine lines in $\mathbb{R}^2$. Such a line L is determined by two quantities: the angle $\theta \in [0, \pi)$ is makes with the x-axis, and the *signed* distance $\rho \in (-\infty, \infty)$ from the origin. More precisely, for every $\rho \in \mathbb{R}$ and $\theta \in [0, \pi)$ we denote by $L_{\theta,\rho}$ the line is given in Euclidean coordinates by the equation

$$x \sin\theta - y \cos\theta = \rho.$$

As a manifold, the Grassmannian $\mathbf{Graff}_1(\mathbb{R}^2)$ is diffeomorphic to the interior of the Möbius band. The Fubini formula in Theorem 9.1.14 can now be rewritten

$$\int_{\mathbf{Graff}_1(\mathbb{R}^2)} f(L) \, |d\tilde{\gamma}_{2,1}|(L) = \int_{-\infty}^{\infty} \left(\int_0^{\pi} f(L_{\theta,\rho}) |d\theta| \right) |d\rho|,$$

$\forall f \in C^{\infty}_{cpt}(\mathbf{Graff}_1(\mathbb{R}^2))$. $\square$

9.2 Gauss-Bonnet again?!?

The Grassmannians enter into the study of the geometry of a submanifold via a fundamental observation going back to Gauss. To understand the shape of a submanifold $M \hookrightarrow \mathbb{R}^n$, $m = \dim M$, it is very productive to understand how a tangent plane $T_p M$ varies, as p varies along M.

Observe that if M is flat, i.e., it is an open subset of an affine subspace, then the tangent plane $T_p M$ does not vary as p moves around M. However, if M is curved, this tangent plane changes its location in the Grassmannian $\mathbf{Gr}_m(\mathbb{R}^n)$, and in fact, "the more curved is M, the faster will the tangent plane move inside this Grassmannian".

In Example 4.2.20 we have seen this heuristic principle at work in the case of a hypersurface in $\mathbb{R}^3$, where the variation of the unit normal along the surface detects the Riemann curvature of the surface. Note that the tangent plane is uniquely determined by the unit normal.

The goal of this section is to analyze in great detail the map

$$M \ni p \mapsto T_p M \in \mathbf{Gr}_m(\mathbb{R}^n),$$

which, for obvious reasons, is named the *Gauss map* of the submanifold M.

9.2.1 *The shape operator and the second fundamental form of a submanifold in $\mathbb{R}^n$*

Suppose M is a smooth submanifold M of $\mathbb{R}^n$, and set $m := \dim M$. To minimize the notational burden we will use the following conventions.

- We will use lower case Latin letters $i, j, \ldots$ to denote indices in the range

$$1 \leq i, j, \ldots \leq m = \dim M.$$

- We will use lower case Greek letters $\alpha, \beta, \ldots$ to denote indices in the range

$$m < \alpha, \beta, \ldots \leq n.$$

- We will use the capital Latin letters A, B, C to denote indices in the range

$$1 \leq A, B, C \leq n.$$

- We will use Einstein's summation convention.

The Euclidean metric of $\mathbb{R}^n$ induces a metric g on M. In Subsection 4.2.4 we explained how to determine the Levi-Civita connection ∇^M of g using Cartan's moving frame method. Let us recall this construction using notations more appropriate for the applications in this chapter.

Denote by $\boldsymbol{D}$ the Levi-Civita connection of the Euclidean metric on $\mathbb{R}^n$. Every vector field on $\mathbb{R}^n$ can be regarded as an n-uple of functions

$$X = \begin{bmatrix} X^1 \\ \vdots \\ X^n \end{bmatrix}.$$

Then

$$\boldsymbol{D}\, X = \begin{bmatrix} dX^1 \\ \vdots \\ dX^n \end{bmatrix} = dX.$$

The restriction to M of the tangent bundle $T\mathbb{R}^n$ admits an orthogonal decomposition

$$(T\mathbb{R}^n)|_M = TM \oplus (TM)^\perp.$$

Correspondingly, a section X of $(T\mathbb{R}^n)|_M$ decomposes into a tangential part X^τ, and a normal part X^ν. Fix a a point $p_0 \in M$, an open neighborhood U of p_0 in $\mathbb{R}^n$, and a local orthonormal frame $(\boldsymbol{e}_A)$ of $T\mathbb{R}^n$ along U such that the collection

$(e_i|_M)$ is a local orthonormal frame of TM along $U \cap M$. We denote by $(\boldsymbol{\theta}_A)$ the dual coframe of (e_A), i.e.,

$$\boldsymbol{\theta}_A(e_B) = e_A \bullet e_B = \delta_{AB}.$$

If X is a section of $T\mathbb{R}^n|_M$ then

$$X^\tau = \boldsymbol{\theta}^i(X)e_i, \quad X^\nu = \boldsymbol{\theta}^\alpha(X)e_\alpha.$$

We denote by Θ^A_B the 1-forms associated to $\boldsymbol{D}$ by the frame (e_A). They satisfy Cartan's equations

$$d\boldsymbol{\theta}^A = -\Theta^A_B \wedge \boldsymbol{\theta}^B, \quad \boldsymbol{D}e_B = \Theta^A_B e_A, \quad \Theta^A_B = -\Theta^B_A.$$

If we write

$$\phi^A := \boldsymbol{\theta}^A|_M, \quad \Phi^A_B := \Theta^A_B|_M,$$

then, as explained Subsection 4.2.4, we deduce that (Φ^i_j) are the 1-forms associated to the Levi-Civita connection ∇^M by the local orthonormal frame $(e_i|_M)$. This implies that

$$\nabla^M e_j = \Phi^i_j e_i = \text{the tangential component of } \Phi^A_j e_A = \boldsymbol{D}e_j,$$

and thus,

$$\nabla^M_X Y = (\boldsymbol{D}_X Y)^\tau, \quad \forall X, Y \in \text{Vect}(M).$$

Consider the *Gauss map*

$$\mathcal{G} = \mathcal{G}_M : M \to \mathbf{Gr}_m(\mathbb{R}^n), \quad x \mapsto T_x M.$$

The *shape operator* of the submanifold $M \hookrightarrow \mathbb{R}^n$ is, by definition, the differential of the Gauss map. Intuitively, the shape operator measures how fast the tanget plane changes its location as a point in a Grassmannian.

We denote the shape operator by S^M, and we would like to relate it to the structural coefficients Φ^A_B.

As explained in Subsection 9.1.2, in the neighborhood U of p_0, the "moving plane" $x \mapsto T_x M$ can be represented by the orthonormal frame (e_A) which has the property that the first m vectors $e_1(x), \ldots, e_m(x)$ span $T_x M$. The differential of the Gauss map at $x \in U \cap M$ is a linear map

$$D\mathcal{G} : T_x M \to T_{\mathcal{G}(x)} \mathbf{Gr}_m(\mathbb{R}^m) = \text{Hom}(T_x M, (T_x M)^\perp).$$

As explained in (9.1.12), this differential described by the $(n-m) \times m$ matrix of 1-forms

$$\left(e_\alpha \bullet de_j,\right)_{\alpha,i} = \left(e_\alpha \bullet \boldsymbol{D}e_j,\right)_{\alpha,i}.$$

On the other hand, $\boldsymbol{D}e_j = \Phi^A_j e_A$, so that $e_\alpha \bullet \boldsymbol{D}e_j = \Phi^\alpha_j$. Define

$$\Phi^A_{ij} := e_i \,\lrcorner\, \Phi^A_j \in \Omega^0(M \cap U),$$

so that

$$\Phi_j^A = \Phi_{ij}^A \wedge \phi^i.$$

We have thus obtained the following result.

Proposition 9.2.1. *The shape operator of M, that is the differential of the Gauss map, is locally described by the matrix of 1-forms $(\Phi_i^\alpha)_{1 \le i \le m < \alpha < n}$. More precisely the operator*

$$S^M(e_i) \in \mathrm{Hom}(T_x M, (T_x M)^\perp)$$

is given by

$$S^M(e_i)e_j = (D_{e_i} e_j)^\nu = \Phi_{ij}^\alpha e_\alpha, \quad \forall i, j. \qquad \square$$

In Subsection 4.2.4 we have defined the second fundamental form of the submanifold M to be the $C^\infty(M)$-bilinear and symmetric map

$$S_M : \mathrm{Vect}\,(M) \times \mathrm{Vect}\,(M) \to C^\infty\big((TM)^\perp\big), \quad (X, Y) \mapsto (D_X Y)^\nu.$$

Note that

$$S_M(e_i, e_j) = \phi^\alpha(D_{e_i} e_j)e_\alpha = \Phi_{ij}^\alpha e_\alpha = S^M(e_i)e_j. \qquad (9.2.1)$$

The equality (4.2.10) can be rewritten as

$$e_j \bullet (D_{e_i} e_\alpha) = -(D_{e_i} e_j) \bullet e_\alpha = -e_\alpha \bullet S_M(e_i, e_j), \quad \forall i, j, \alpha. \qquad (9.2.2)$$

Theorema Egregium states that the Riemann curvature tensor of M is completely determined by the second fundamental form. We recall this result below.

Theorem 9.2.2. *Suppose M is a submanifold of $\mathbb{R}^n$. We denote by g the induced metric on M and by S_M the second fundamental form of the embedding $M \hookrightarrow \mathbb{R}^n$. Denote by R the Riemann curvature of M with the induced metric. Then for any $X_1, \ldots, X_4 \in \mathrm{Vect}\,(M)$ we have*

$$g(X_1, R(X_3, X_4)X_2) = S_M(X_3, X_1) \bullet S_M(X_4, X_2) - S_M(X_3, X_2) \bullet S_M(X_4, X_1),$$

where $\bullet$ denotes the inner product in $\mathbb{R}^n$. $\qquad \square$

9.2.2 The Gauss-Bonnet theorem for hypersurfaces of a Euclidean space

The results in the previous subsection have very surprising consequences. Suppose M is a compact, *orientable*[1] hypersurface of $\mathbb{R}^{m+1}$, i.e., an orientable submanifold

---

[1] The orientability assumption is superfluous. It follows from the Alexander duality theorem with $\mathbb{Z}/2$ coefficients that a compact hypersurface of a vector space is the boundary of a compact domain, and in particular, it is orientable.

of codimension 1. If we fix an orientation on M, then we obtain a normal vector field

$$\boldsymbol{n} : M \to \mathbb{R}^{m+1}, \ \ \boldsymbol{n}(x) \perp T_x M, \ \ |\boldsymbol{n}(x)| = 1, \ \ \forall x \in M.$$

If we choose an *oriented*, local orthonormal frame $\boldsymbol{e}_1, \ldots, \boldsymbol{e}_m$ of TM, then the ordered set $\{\boldsymbol{n}(x), \boldsymbol{e}_1(x), \ldots, \boldsymbol{e}_m\}$ is an oriented, local orthonormal frame of $\mathbb{R}^{m+1}$. In this case we can identify the second fundamental form with a genuine symmetric bilinear form

$$S_M \in C^\infty\big(T^*M^{\otimes 2}\big), \ \ S_M(X, Y) = \boldsymbol{n} \bullet (\boldsymbol{D}_X Y).$$

The Gauss map $\mathcal{G}_M : M \to \mathbf{Gr}_m(\mathbb{R}^{m+1})$ of the embedding $M \hookrightarrow \mathbb{R}^{m+1}$ can be given the description,

$$M \ni x \mapsto \langle \boldsymbol{n}(x) \rangle^\perp := \text{the vector subspace orthogonal to } \boldsymbol{n}(x).$$

On the other hand, we have an *oriented* Gauss map

$$\vec{\mathcal{G}}_M : M \to \boldsymbol{S}^m, \ \ x \to \boldsymbol{n}(x),$$

and a double cover

$$\pi : \boldsymbol{S}^m \to \mathbf{Gr}_m(\mathbb{R}^{m+1}), \ \ \boldsymbol{S}^m \ni \vec{u} \mapsto \langle \vec{u} \rangle^\perp,$$

so that the diagram below is commutative

$$
\begin{array}{ccc}
M & \xrightarrow{\ \vec{\mathcal{G}}\ } & \boldsymbol{S}^m \\
& \searrow{\scriptstyle \mathcal{G}} & \downarrow{\scriptstyle \pi} \\
& & \mathbf{Gr}_m(\mathbb{R}^{m+1})
\end{array}
$$

We fix an oriented orthonormal frame $(\vec{f}_0, \vec{f}_1, \ldots, \vec{f}_m)$ of $\mathbb{R}^{m+1}$, and we orient the unit sphere $S^m \subset \mathbb{R}^{m+1}$ so that the orientation of $T_{\vec{f}_0} S^m$ is given by the ordered frame

$$(\vec{f}_1, \ldots, \vec{f}_m).$$

In the remainder of this subsection we will assume that m is even, $m = 2h$.

Denote by $dA_m \in \Omega^m(\boldsymbol{S}^m)$ the "area" *form* on the unit m-dimensional sphere $\boldsymbol{S}^m$. Recall that σ_m denotes the "area" of $\boldsymbol{S}^m$,

$$\sigma_{2h} = (2h+1)\omega_{2r+1} \implies \frac{\sigma_{2h}}{2} = \frac{2^{2h}\pi^h h!}{(2h)!}. \tag{9.2.3}$$

Hence

$$\int_{\boldsymbol{S}^m} \frac{1}{\sigma_m} dA_m = 1,$$

so that

$$\frac{1}{\sigma_m} \int_M \vec{\mathcal{G}}_M^* dA_m = \deg \mathcal{G}_M.$$

Denote by g the induced metric on M, and by R the curvature of g. We would like to prove that the integrand $\vec{\mathcal{G}}_M^* dA_m$ has the form

$$\vec{\mathcal{G}}_M^* dA_m = P(R_M) dV_M,$$

where dV_M denotes the metric volume form on M and $P(R_M)$ is a universal polynomial of degree $\frac{m}{2}$ in the curvature R of M.

Fix a positively oriented orthonormal frame $\vec{e} = (e_1, \ldots, e_m)$ of TM defined on some open set $U \subset M$, and denote by $\vec{\theta} = (\theta^1, \ldots, \theta^m)$ the dual coframe. Observe that

$$dV_M = \theta^1 \wedge \cdots \wedge \theta^m.$$

We set

$$S_{ij} := S_M(e_i, e_j) \bullet n, \quad R_{ijk\ell} := g\big(e_i, R(e_k, e_\ell)e_j\big).$$

Theorem 9.2.2 implies that

$$R_{ijk\ell} = S_{ik}S_{j\ell} - S_{i\ell}S_{jk} = \begin{vmatrix} S_{ik} & S_{i\ell} \\ S_{jk} & S_{j\ell} \end{vmatrix}. \qquad (9.2.4)$$

Observe that $R_{ijk\ell} \neq 0 \Longrightarrow i \neq j, \ k \neq \ell$, and in this case the matrix

$$\begin{bmatrix} S_{ik} & S_{i\ell} \\ S_{jk} & S_{j\ell} \end{bmatrix}$$

is the 2×2 submatrix of $S = (S_{ij})_{1 \leq i,j \leq m}$ lying at the intersection of the i, j-rows with the k, ℓ-columns. We can rephrase the equality (9.2.4) in a more convenient form.

First, we regard the curvature $R_{ijk\ell}$ at a point $x \in M$ as a linear map

$$\Lambda^2 T_x M \to \Lambda^2 T_x^* M, \quad R(e_k \wedge e_\ell) = \sum_{i<j} R_{ijk\ell} \theta^i \wedge \theta^j.$$

Next, regard S as a linear map

$$S : T_x M \to T^* M, \quad Se_j = S_{ij}\theta^i.$$

Then S induces linear maps

$$\Lambda^p S : \Lambda^k T_x M \to \Lambda^p T^* M,$$

defined by

$$S(e_{i_1} \wedge \cdots \wedge e_{i_p}) = (Se_{i_1}) \wedge \cdots \wedge (Se_{i_p}), \quad \forall 1 \leq i_1 < \cdots < i_p \leq m.$$

The equality (9.2.4) can now be rephrased as

$$R = \Lambda^2 S. \qquad (9.2.5)$$

Along U we have the equality

$$(\vec{\mathcal{G}}_M^* dA_M)\,|_U = \Lambda^m S(e_1 \wedge \cdots \wedge e_m) = (\det S)\theta^1 \wedge \cdots \wedge \theta^m = (\det S)dV_M.$$

We want to prove that $\det S$ can be described in terms of $\Lambda^2 S$. To see this observe that

$$(\Lambda^{2h} S)(e_1 \wedge e_2 \wedge \cdots e_{2h-1} \wedge e_{2h}) = (\Lambda^2 S)(e_1 \wedge e_2) \wedge \cdots \wedge (\Lambda^2 S)(e_{2h-1} \wedge e_{2h})$$

$$= \bigwedge_{s=1}^{h} \left(\sum_{i<j} R_{ij,2s-1,2s} \boldsymbol{\theta}^i \wedge \boldsymbol{\theta}^j \right) = \sum_{\varphi \in \mathcal{S}'_m} \epsilon(\varphi) \left(\prod_{s=1}^{h} R_{\varphi(2s-1)\varphi(2s),2s-1,2s} \right) dV_M,$$

where $\mathcal{S}'_m$ denotes the set of permutations φ of $\{1, 2, \ldots, m = 2h\}$ such that

$$\varphi(1) < \varphi(2), \ldots, \varphi(2h-1) < \varphi(2h),$$

and $\epsilon(\varphi) = \pm 1$ denotes the signature of a permutation. Observe that

$$\#\mathcal{S}'_m = \binom{2h}{2} \cdot \binom{2h-2}{2} \cdots \binom{2}{2} = \frac{(2h)!}{2^h}. \tag{9.2.6}$$

We would like to give an alternate description of $\det S$ using the concept of *pfaffian*.

The Riemann curvature, like the curvature of any metric connection, can be regarded as a 2-form with coefficients in the bundle of skew-symmetric endomorphisms of TM. It is locally described by an $m \times m$ skew-symmetric matrix

$$\Theta = \Theta_g := \big(\Theta_{ij} \big)_{1 \leq i,j \leq m}$$

whose entries are 2-forms on U,

$$\Theta_{ij} = \sum_{k<\ell} R_{ijk\ell} \boldsymbol{\theta}^k \wedge \boldsymbol{\theta}^\ell = \frac{1}{2} \sum_{k,\ell} R_{ijk\ell} \boldsymbol{\theta}^k \wedge \boldsymbol{\theta}^\ell = \Lambda^2 S(e_i \wedge e_j) \in \Omega^2(U).$$

Recall from Subsection 8.2.4 that the pfaffian of Θ is the differential form

$$\boldsymbol{Pf}(\Theta) := \frac{(-1)^h}{2^h h!} \sum_{\varphi \in \mathcal{S}_m} \epsilon(\varphi) \Theta_{\varphi(1)\varphi(2)} \wedge \cdots \wedge \Theta_{\varphi(2h-1)\varphi(2h)} \in \Omega^{2h}(U),$$

where $\mathcal{S}_m$ denotes the group of permutations of $\{1, 2, \ldots, m\}$. Observe that

$$\boldsymbol{Pf}(\Theta) = \frac{(-1)^h}{h!} \sum_{\varphi \in \mathcal{S}'_m} \epsilon(\varphi) \Theta_{\varphi(1)\varphi(2)} \wedge \cdots \wedge \Theta_{\varphi(2h-1)\varphi(2h)}.$$

We can simplify this some more if we introduce the set $\mathcal{S}''_m$ consisting of permutations $\varphi \in \mathcal{S}'_m$ such that

$$\varphi(1) < \varphi(3) < \cdots < \varphi(2h-1).$$

Observe that

$$\#\mathcal{S}'_m = (\#\mathcal{S}''_m)h! \implies \#\mathcal{S}''_m = \frac{\mathcal{S}'_M}{h!} = \frac{(2h)!}{2^h h!} = 1 \cdot 3 \cdots (2h-1) =: \gamma(2h).$$

Then

$$\boldsymbol{Pf}(\Theta) = (-1)^h \sum_{\varphi \in \mathcal{S}''_m} \epsilon(\varphi) \Theta_{\varphi(1)\varphi(2)} \wedge \cdots \wedge \Theta_{\varphi(2h-1)\varphi(2h)}.$$

We have

$$\boldsymbol{Pf}(-\Theta) = \frac{1}{h!} \sum_{(\sigma,\varphi)\in\mathcal{S}'_m\times\mathcal{S}'_m} \epsilon(\sigma\varphi)\Big(\prod_{j=1}^{h} R_{\varphi(2j-1)\varphi(2j)\sigma(2j-1)\sigma(2j)}\Big)dV_M.$$

On the other hand,

$$(\Lambda^{2h}S)(\boldsymbol{e}_1\wedge\boldsymbol{e}_2\wedge\cdots\wedge\boldsymbol{e}_{2h-1}\wedge\boldsymbol{e}_{2h})$$

$$= \frac{1}{\#\mathcal{S}'_m} \sum_{\varphi\in\mathcal{S}'_m} \epsilon(\varphi)\Lambda^m S(\boldsymbol{e}_{\varphi(1)}\wedge\boldsymbol{e}_{\varphi(2)}\wedge\cdots\wedge\boldsymbol{e}_{\varphi(2h-1)}\wedge\boldsymbol{e}_{\varphi(2h)})$$

$$= \frac{1}{\#\mathcal{S}'_m} \sum_{\varphi\in\mathcal{S}'_m} \epsilon(\varphi)\Theta_{\varphi(1)\varphi(2)}\wedge\cdots\wedge\Theta_{\varphi(2h-1)\varphi(2h)}$$

$$= \frac{1}{(\#\mathcal{S}'_m)} \sum_{(\sigma,\varphi)\in\mathcal{S}'_m\times\mathcal{S}'_m} \epsilon(\sigma\varphi)\Big(\prod_{j=1}^{h} R_{\varphi(2j-1)\varphi(2j)\sigma(2j-1)\sigma(2j)}\Big)dV_M = \frac{h!}{\#\mathcal{S}'_m}\boldsymbol{Pf}(-\Theta).$$

Hence

$$\vec{\mathcal{G}}_M^* dA_M = (\Lambda^{2h}S)(\boldsymbol{e}_1\wedge\boldsymbol{e}_2\wedge\cdots\wedge\boldsymbol{e}_{2h-1}\wedge\boldsymbol{e}_{2h}) = \frac{h!}{\#\mathcal{S}'_m}\boldsymbol{Pf}(-\Theta),$$

so that

$$\vec{\mathcal{G}}_M^*\Big(\frac{2}{\sigma_{2h}}dA_M\Big) = \frac{2}{\sigma_{2h}}\frac{h!}{\#\mathcal{S}'_m}\boldsymbol{Pf}(-\Theta) \overset{(9.2.3)}{=} \frac{h!}{\#\mathcal{S}'_m}\frac{(2h)!}{2^{2h}\pi^h h!}\boldsymbol{Pf}(-\Theta)$$

$$\overset{(9.2.6)}{=} \frac{1}{(2\pi)^h}\boldsymbol{Pf}(-\Theta).$$

The last form is the Euler form $e(\nabla^M)$ associated to the Levi-Civita connection ∇^M. Using the Gauss-Bonnet-Chern theorem we obtain the following result.

Theorem 9.2.3. *If $M^{2h}\subset\mathbb{R}^{2h+1}$ is a compact, oriented hypersurface, and g denotes the induced metric, then*

$$\frac{2}{\sigma_m}\vec{\mathcal{G}}_M^* dA_m = e(\nabla^M) = \frac{1}{(2\pi)^h}\boldsymbol{Pf}(-\Theta_g),$$

and

$$2\deg\vec{\mathcal{G}}_M = \frac{1}{(2\pi)^h}\int_M e(\nabla^M) = \chi(M).$$

Moreover, if $(\boldsymbol{e}_1,\ldots,\boldsymbol{e}_{2h})$ is a local, positively oriented orthonormal frame of TM, then

$$\boldsymbol{Pf}(-\Theta_g) = \frac{1}{h!} \sum_{(\sigma,\varphi)\in\mathcal{S}'_{2h}\times\mathcal{S}'_m} \epsilon(\sigma\varphi)\Big(\prod_{j=1}^{h} R_{\varphi(2j-1)\varphi(2j)\sigma(2j-1)\sigma(2j)}\Big)dV_M \qquad (9.2.7a)$$

$$= \sum_{\varphi \in \mathcal{S}''_{2h}} \epsilon(\varphi) \Theta_{\varphi(1)\varphi(2)} \wedge \cdots \wedge \Theta_{\varphi(2h-1)\varphi(2h)}, \qquad (9.2.7b)$$

where $\mathcal{S}'_{2h}$ denotes the set of permutations φ of $\{1, \ldots, 2h\}$ such that

$$\varphi(2j-1) < \varphi(2j), \quad \forall 1 \le j \le h,$$

$\mathcal{S}''_m$ *denotes the set of permutations $\varphi \in \mathcal{S}'_m$ such that*

$$\varphi(1) < \varphi(3) < \cdots < \varphi(2h-1),$$

and

$$\Theta_{ij} = \sum_{k<\ell} R_{ijk\ell} \boldsymbol{\theta}^k \wedge \boldsymbol{\theta}^\ell. \qquad \square$$

Example 9.2.4. (a) If dim $M = 2$ then

$$\boldsymbol{Pf}(-\Theta_g) = R_{1212} dV_M = (\text{the Gaussian curvature of } M) \times dV_M$$

(b) If dim $M = 4$, then $\mathcal{S}''_4$ consists of 3 permutations

$$1, 2, 3, 4, \ \epsilon = 1 \longrightarrow \Theta_{1212}\Theta_{3434},$$

$$1, 3, 2, 4, \ \epsilon = -1 \longrightarrow -\Theta_{1312}\Theta_{2434}$$

$$1, 4, 2, 3, \ \epsilon = 1 \longrightarrow \Theta_{1412}\Theta_{2334}$$

We deduce

$$\boldsymbol{Pf}(-\Theta_g) = \Theta_{12} \wedge \Theta_{34} - \Theta_{13} \wedge \Theta_{24} + \Theta_{14} \wedge \Theta_{23}$$

$$= \frac{1}{4}(R_{12ij}\boldsymbol{\theta}^i \wedge \boldsymbol{\theta}^j) \wedge (R_{34k\ell}\boldsymbol{\theta}^k \wedge \boldsymbol{\theta}^\ell) - \frac{1}{4}(R_{13ij}\boldsymbol{\theta}^i \wedge \boldsymbol{\theta}^j) \wedge (R_{24k\ell}\boldsymbol{\theta}^k \wedge \boldsymbol{\theta}^\ell)$$

$$+ \frac{1}{4}(R_{14ij}\boldsymbol{\theta}^i \wedge \boldsymbol{\theta}^j) \wedge (R_{23k\ell}\boldsymbol{\theta}^k \wedge \boldsymbol{\theta}^\ell)$$

Each of the three exterior products above in involves only six nontrivial terms, because $\theta^i \wedge \theta^j = 0$, when $i = j$. Thus the total number of terms in the above expression is 36. However, there are many repetitions due to the symmetries of the Riemann tensor

$$R_{ijk\ell} = R_{k\ell ij} = -R_{ij\ell k}.$$

We have

$$(R_{12ij}\boldsymbol{\theta}^i \wedge \boldsymbol{\theta}^j) \wedge (R_{34k\ell}\boldsymbol{\theta}^k \wedge \boldsymbol{\theta}^\ell)$$

$$= \Big(R_{1212}R_{3434} + R_{1213}R_{3442} + R_{1214}R_{3423}$$

$$+ R_{1223}R_{3414} + R_{1242}R_{3413} + R_{1234}R_{3412} \Big) dV_M, \ \text{etc.}$$

(c) We can choose the *positively oriented* local orthonormal frame $(\boldsymbol{e}_1, \ldots, \boldsymbol{e}_m)$ so that it diagonalizes S_M at a given point $x \in M$. Then the eigenvalues of S_M at x are called the *principal curvatures* at x and are denoted by $\kappa_1(x), \ldots, \kappa_m(x)$. Then

$$\boldsymbol{Pf}(-\Theta) = \rho dV_M, \ \rho \in C^\infty(M),$$

where

$$\rho(x) = (2h-1)!! \prod_{k=1}^m \kappa_i(x), \ \forall x \in M. \qquad \square$$

9.2.3 Gauss-Bonnet theorem for domains of a Euclidean space

Suppose D is a relatively compact open subset of an Euclidean space $\mathbb{R}^{m+1}$ with smooth boundary ∂D. We denote by $\boldsymbol{n}$ the outer normal vector field along the boundary. It defines an *oriented Gauss map*

$$\vec{\mathcal{G}}_D : \partial D \to \boldsymbol{S}^m.$$

We denote by dA_m the area form on the unit sphere $\boldsymbol{S}^m$ so that

$$\deg \vec{\mathcal{G}}_D = \frac{1}{\sigma_m} \int_{\partial D} \vec{\mathcal{G}}_D^* dA_m.$$

If m is even then the Gauss-Bonnet theorem for the hypersurface ∂D implies

$$\frac{1}{\sigma_m} \int_{\partial D} \vec{\mathcal{G}}_D^* dA_m = \frac{1}{2} \chi(\partial D).$$

Using the Poincaré duality for the oriented manifold with boundary D we deduce $\chi(\partial D) = 2\chi(D)$, so that

$$\frac{1}{\sigma_m} \int_{\partial D} \vec{\mathcal{G}}_D^* dA_m = \chi(D), \quad m \in 2\mathbb{Z}.$$

We want to prove that the above equality holds even when m is odd. Therefore in the remainder of this section we assume *m is odd*.

We first describe the integrand $\vec{\mathcal{G}}_D^* dA_m$. Let S_D denote the second fundamental form of the hypersurface

$$S_D(X,Y) = \boldsymbol{n} \bullet (\boldsymbol{D}_X Y) = -X \bullet (\boldsymbol{D}_Y \boldsymbol{n}), \quad \forall X, Y \in \mathrm{Vect}\,(\partial D).$$

We deduce

$$\frac{1}{\sigma_m} \vec{\mathcal{G}}_D^* dA_m = \frac{1}{\sigma_m} \det(-S_D) dV_{\partial D},$$

where $dV_{\partial D}$ denotes the volume form on ∂D.

A smooth vector field on $\bar{D}$,

$$X : \bar{D} \to \mathbb{R}^{m+1}$$

is called *admissible* if along the boundary it points towards the exterior of D,

$$X \bullet \boldsymbol{n} > 0, \quad \text{on } \partial D.$$

For an admissible vector field X we define

$$\bar{X} : \partial D \to \boldsymbol{S}^m, \quad \bar{X}(p) := \frac{1}{|X(p)|} X(p), \quad \forall p \in \partial D.$$

Let us observe that the map $\bar{X}$ is homotopic to the map $\vec{\mathcal{G}}_D$. Indeed, for $t \in [0,1]$, define

$$Y_t : \partial D \to \boldsymbol{S}^m, \quad Y_t(p) = \frac{1}{|(1-t)\boldsymbol{n} + t\bar{X}|} (1-t)\boldsymbol{n} + t\bar{X}$$

Observe that this map is well defined since

$$|(1-t)\boldsymbol{n} + t\bar{X}|^2 = t^2 + (1-t)^2 + 2t(1-t)(\boldsymbol{n} \bullet \bar{X}) > 0.$$

Hence

$$\deg \vec{\mathcal{G}}_D = \deg \bar{X},$$

for any admissible vector field X.

Suppose X is a *nondegenerate* admissible vector field. This means that X has a finite number of stationary points,

$$\mathcal{Z}_X = \{p_1, \ldots, p_\nu\}, \quad X(p_i) = 0,$$

and all of them are nondegenerate, i.e., for any $p \in \mathcal{Z}_X$, the linear map

$$A_{X,p} : T_p\mathbb{R}^{m+1} \to T_p\mathbb{R}^{m+1}, \quad T_p\mathbb{R}^{m+1} \ni v \mapsto (\boldsymbol{D}_v X)(p)$$

is invertible. Define

$$\epsilon_X : \mathcal{Z}_X \to \{\pm 1\}, \quad \epsilon(p) = \operatorname{sign} \det A_{X,p}.$$

For any $\varepsilon > 0$ sufficiently small the closed balls of radius ε centered at the points in $\mathcal{Z}_X$ are disjoint. Set

$$D_\varepsilon = D \setminus \bigcup_{p \in \mathcal{Z}_X} B_\varepsilon(p).$$

The vector field X does not vanish on D_ε, and we obtain a map

$$\bar{X} : \bar{D}_\varepsilon \to S^{m-1}, \quad \bar{X} = \frac{1}{|X|}X.$$

Set

$$\Omega := \frac{1}{\sigma_m}\bar{X}^* dA_m.$$

Observe that

$$d\Omega = \frac{1}{\sigma_m}\bar{X}^* d(dA_m) = 0 \text{ on } D_\varepsilon.$$

Stokes' theorem then implies that

$$\int_{\partial D_\varepsilon} \Omega = \int_{D_\varepsilon} d\Omega = 0 \implies \deg \vec{\mathcal{G}}_D = \int_{\partial D} \Omega = \sum_{p \in \mathcal{Z}_X} \int_{\partial B_\varepsilon(p)} \Omega,$$

where the spheres $\partial B_\varepsilon(p)$ are oriented as boundaries of the balls $B_\varepsilon(p)$. If we let $\varepsilon \to 0$ we deduce

$$\deg \vec{\mathcal{G}}_D = \sum_{p \in \mathcal{Z}_X} \epsilon_X(p), \tag{9.2.8}$$

for any nondegenerate admissible vector field X.

To give an interpretation of the right-hand side of the above equality, consider the *double of D*. This is the smooth manifold $\widehat{D}$ obtained by gluing D along ∂D to a copy of itself equipped with the opposite orientation,

$$\widehat{D} = D \cup_{\partial D} (-D).$$

The manifold without boundary $\widehat{D}$ is equipped with an orientation reversing involution $\varphi : \widehat{D} \to \widehat{D}$ whose fixed point set is ∂D. In particular, along $\partial D \subset \widehat{D}$ we have a φ-invariant decomposition

$$T\widehat{D}|_{\partial D} = T\partial D \oplus L,$$

where L is a real line bundle along which the differential of φ acts as $-\mathbb{1}_L$. The normal vector field $\boldsymbol{n}$ defines a basis of L. If X is a vector field on D which is equal to $\boldsymbol{n}$ along ∂D, then we obtain a vector field $\widehat{X}$ on $\widehat{D}$ by setting

$$\widehat{X} := \begin{cases} X & \text{on } D \\ -\varphi_*(X) & \text{on } -D. \end{cases}$$

If X is nondegenerate, then so is $\widehat{X}$, where the nondegeneracy of $\widehat{X}$ is defined in terms of an arbitrary connection on $T\widehat{D}$. More precisely, if ∇ is a connection on $T\widehat{D}$, and $q \in \mathcal{Z}_{\widehat{X}}$, then q is nondegenerate if the map

$$A_{\widehat{X},q} : T_q\widehat{D} \to T_q\widehat{D}, \quad v \mapsto \nabla_v \widehat{X}$$

is an isomorphism. This map is independent of the connection ∇, and we denote by $\epsilon_{\widehat{X}}(q)$ the sign of its determinant. Moreover

$$\mathcal{Z}_{\widehat{X}} = \mathcal{Z}_X \cup \varphi(\mathcal{Z}_X),$$

and, *because m is odd*, the map

$$\epsilon_{\widehat{X}} : \mathcal{Z}_{\widehat{X}} \to \{\pm 1\}$$

satisfies

$$\epsilon_X(p) = \epsilon_{\widehat{X}}(\varphi(p)).$$

Hence

$$\sum_{q\in\mathcal{Z}_{\widehat{X}}} \epsilon_{\widehat{X}}(q) = 2 \sum_{p\in\mathcal{Z}_X} \epsilon_X(q) \stackrel{(9.2.8)}{=} 2\deg \vec{\mathcal{G}}_D.$$

On the other hand, the general Poincaré-Hopf theorem (see Corollary 7.3.48) implies that

$$\sum_{q\in\mathcal{Z}_{\widehat{X}}} \epsilon_{\widehat{X}}(q) = \chi(\widehat{D}).$$

Using the Mayer-Vietoris theorem we deduce

$$\chi(\widehat{D}) = 2\chi(D) - \chi(\partial D).$$

Since ∂D is *odd dimensional and oriented* we deduce that $\chi(\partial D) = 0$, and therefore

$$2\chi(D) = \chi(\widehat{D}) = 2 \sum_{p\in\mathcal{Z}_X} \epsilon_X(q) = 2\deg \vec{\mathcal{G}}_D.$$

We have thus proved the following result.

Theorem 9.2.5 (Gauss-Bonnet for domains). *Suppose D is a relatively compact open subset of $\mathbb{R}^{m+1}$ with smooth boundary ∂D. We denote by $\vec{\mathcal{G}}_D$ the oriented Gauss map*

$$\vec{\mathcal{G}}_D : \partial D \to S^m, \ \ \partial D \ni p \mapsto \boldsymbol{n}(p) = \textit{unit outer normal,}$$

and by S_D the second fundamental form of ∂D,

$$S_D(X,Y) = \boldsymbol{n} \bullet (\boldsymbol{D}_X Y), \ \ \forall X, Y \in \mathrm{Vect}\,(\partial D).$$

Then

$$\frac{1}{\boldsymbol{\sigma}_m} \int_{\partial D} \det(-S_D) dV_{\partial D} = \deg \vec{\mathcal{G}}_D = \chi(D). \qquad \qquad \square$$

9.3   Curvature measures

In this section we introduce the main characters of classical integral geometry, the so called *curvature measures*, and we describe several probabilistic interpretations of them known under the common name of *Crofton formulæ*.

We will introduce these quantities trough the back door, via the beautiful tube formula of Weyl, which describes the volume of a tube of small radius r around a compact submanifold of an Euclidean space as a polynomial in r whose coefficients are these geometric measures. Surprisingly, these coefficients depend only on the Riemann curvature of the induced metric on the submanifold.

The Crofton Formulæ state that these curvature measures can be computed by slicing the submanifold with affine subspaces of various codimensions, and averaging the Euler characteristics of such slices. To keep the geometric ideas as transparent as possible, and the analytical machinery to a minimum, we chose to describe these facts in the slightly more restrictive context of *tame geometry*.

9.3.1   *Tame geometry*

The category of tame spaces and tame maps is sufficiently large to include all the compact triangulable spaces, yet sufficiently restrictive to rule out pathological situations such as Cantor sets, Hawaiian rings, or nasty functions such as $\sin(1/t)$.

We believe that the subject of tame geometry is one mathematical gem which should be familiar to a larger geometric audience, but since this is not yet the case, we devote this section to a brief introduction to this topic. Unavoidably, we will have to omit many interesting details and contributions, but we refer to [23, 28, 29] for more systematic presentations. For every set X we will denote by $\mathcal{P}(X)$ the collection of all subsets of X.

An $\mathbb{R}$-*structure*[2] is a collection $\mathcal{S} = \{\mathcal{S}^n\}_{n\geq 1}$, $\mathcal{S}^n \subset \mathcal{P}(\mathbb{R}^n)$, with the following properties.

$\mathbf{E}_1$. $\mathcal{S}^n$ contains all the real algebraic subsets of $\mathbb{R}^n$, i.e., the subsets described by finitely many polynomial equations.

$\mathbf{E}_2$. $\mathcal{S}^n$ contains all the closed affine half-spaces of $\mathbb{R}^n$.

$\mathbf{P}_1$. $\mathcal{S}^n$ is closed under boolean operations, $\cup$, $\cap$ and complement.

$\mathbf{P}_2$. If $A \in \mathcal{S}^m$, and $B \in \mathcal{S}^n$, then $A \times B \in \mathcal{S}^{m+n}$.

$\mathbf{P}_3$. If $A \in \mathcal{S}^m$, and $T : \mathbb{R}^m \to \mathbb{R}^n$ is an affine map, then $T(A) \in \mathcal{S}^n$.

Example 9.3.1. (Semialgebraic sets). Denote by $\mathcal{S}^n_{alg}$ the collection of semialgebraic subsets of $\mathbb{R}^n$, i.e., the subsets $S \subset \mathbb{R}^n$ which are finite unions

$$S = S_1 \cup \cdots \cup S_\nu,$$

where each S_j is described by finitely many polynomial inequalities. The *Tarski-Seidenberg theorem* (see [15]) states that $\mathcal{S}_{alg}$ is a structure.

To appreciate the strength of this theorem we want to discuss one of its many consequences. Consider the real algebraic set

$$\mathcal{Z}_n := \left\{ (x, a_0, \dots, a_{n-1}) \in \mathbb{R}^{n+1}; \ a_0 + a_1 x + \cdots a_{n-1} x^{n-1} + x^n = 0 \right\}.$$

The Tarski-Seidenberg theorem implies that the projection of $\mathcal{Z}_n$ on the subspace with coordinates (a_i) is a semialgebraic set

$$\mathcal{R}_n := \left\{ \vec{a} = (a_0, \dots, a_{n-1}) \in \mathbb{R}^n; \ \exists x \in \mathbb{R}, \ P_{\vec{a}}(x) = 0 \right\},$$

where

$$P_{\vec{a}}(x) = a_0 + a_1 x + \cdots a_{n-1} x^{n-1} + x^n.$$

In other words, the polynomial $P_{\vec{a}}$ has a real root if and only if the coefficients $\vec{a}$ satisfy at least one system of polynomial inequalities from a *finite, universal, collection of systems of polynomial inequalities*.

For example, when $n = 2$, the resolvent set $\mathcal{R}_2$ is described by the well known inequality $a_1^2 - 4a_0 \geq 0$.

When $n = 3$, we can obtain a similar conclusion using Cardano's formulæ. For $n \geq 5$, we know that there cannot exist any algebraic formulæ describing the roots of a degree n polynomial, yet we can find algebraic inequalities which can decide if such a polynomial has at least one real root. □

Suppose $\mathcal{S}$ is a structure. We say that a set is $\mathcal{S}$-*definable* if it belongs to one of the $\mathcal{S}^n$'s. If A, B are $\mathcal{S}$-definable then a function $f : A \to B$ is called $\mathcal{S}$-*definable* if its graph

$$\Gamma_f := \left\{ (a, b) \in A \times B; \ b = f(a) \right\}$$

---

[2]This is a highly condensed and special version of the traditional definition of structure. The model theoretic definition allows for ordered fields, other than $\mathbb{R}$, such as extensions of $\mathbb{R}$ by "infinitesimals". This can come in handy even if one is interested only in the field $\mathbb{R}$.

is S-definable. The reason these sets are called definable has to do with mathematical logic.

A *formula*[3] is a property defining a certain set. For example, the two different looking formulas

$$\{x \in \mathbb{R};\ x \geq 0\}, \quad \{x \in \mathbb{R};\ \exists y \in \mathbb{R}:\ x = y^2\},$$

describe the same set, $[0, \infty)$.

Given a collection of formulas, we can obtain new formulas, using the logical operations $\wedge, \vee, \neg$, and quantifiers $\exists, \forall$. If we start with a collection of formulas, each describing an S-definable set, then any formula obtained from them by applying the above logical transformations will describe a definable set.

To see this, observe that the operators $\wedge, \vee, \neg$ correspond to the boolean operations, $\cap, \cup$, and taking the complement. The existential quantifier corresponds to taking a projection. For example, suppose we are given a formula $\phi(a, b)$, $(a, b) \in A \times B$, A, B definable, describing a definable set $C \subset A \times B$. Then the formula

$$\{\, a \in A;\ \exists b \in B:\ \phi(a, b) \,\}$$

describes the image of the subset $C \subset A \times B$ via the canonical projection $A \times B \to A$. If $A \subset \mathbb{R}^m$, $B \subset \mathbb{R}^n$, then the projection $A \times B \to A$ is the restriction to $A \times B$ of the linear projection $\mathbb{R}^m \times \mathbb{R}^n \to \mathbb{R}^m$ and $\mathbf{P}_3$ implies that the image of C is also definable. Observe that the universal quantifier can be replaced with the operator $\neg \exists \neg$.

Example 9.3.2. (a) The composition of two definable functions $A \xrightarrow{f} B \xrightarrow{g} C$ is a definable function because

$$\Gamma_{g \circ f} = \{\, (a, c) \in A \times C; \exists b \in B : (a, b) \in \Gamma_f,\ (b, c) \in \Gamma_g \,\}.$$

Note that any polynomial with real coefficients is a definable function.

(b) The image and the preimage of a definable set via a definable function is a definable set.

(c) Observe that S_{alg} is contained in any structure S. In particular, the Euclidean norm

$$|\bullet| : \mathbb{R}^n \to \mathbb{R}, \quad |(x_1, \ldots, x_n)| = \left(\sum_{i=1}^{n} x_i^2\right)^{1/2}$$

is S-definable.

Observe that any Grassmannian $\mathbf{Gr}_k(\mathbb{R}^n)$ is a semialgebraic subset of the Euclidean space $\mathrm{End}^+(\mathbb{R})$ of symmetric operators $\mathbb{R}^n \to \mathbb{R}^n$ because it is defined by the system of algebraic (in)equalities

$$\mathbf{Gr}_k(\mathbb{R}^n) = \{P \in \mathrm{End}^+(\mathbb{R}^n);\ P^2 = P,\ \Lambda^k P \neq 0,\ \Lambda^m P = 0,\ \forall m > k\},$$

---

[3]We are deliberately vague on the meaning of formula.

where $\Lambda^j P$ denotes the endomorphism of $\Lambda^j \mathbb{R}^n$ induced by the linear map $P \in \mathrm{End}(\mathbb{R}^n)$. In the above description, to the orthogonal projection P of rank k, we associate its range, which is a vector subspace of dimension k.

Similarly, the affine Grassmannian $\mathbf{Graff}_k(\mathbb{R}^n)$ is a semialgebraic set because its is described by

$$\mathbf{Graff}_k(\mathbb{R}^n) = \big\{ (\boldsymbol{x}, P) \in \mathbb{R}^n \times \mathbf{Gr}_k(\mathbb{R}^n); \;\; P\boldsymbol{x} = 0 \big\}.$$

In the above description, to the pair $(\boldsymbol{x}, P)$ we associate the affine plane $\boldsymbol{x} + \mathrm{Range}\,(P)$.

(d) Any compact, affine simplicial complex $K \subset \mathbb{R}^n$ is $\mathcal{S}$-definable because it is a finite union of affine images of finite intersections of half-spaces.

(e) Suppose $A \subset \mathbb{R}^n$ is $\mathcal{S}$-definable. Then its closure $\boldsymbol{cl}(A)$ is described by the formula

$$\big\{ x \in \mathbb{R}^n; \;\; \forall \varepsilon > 0, \;\; \exists a \in A : \;\; |x - a| < \varepsilon \big\},$$

and we deduce that $\boldsymbol{cl}(A)$ is also $\mathcal{S}$-definable. Let us examine the correspondence between the operations on formulas and operations on sets on this example.

We rewrite this formula as

$$\forall \varepsilon \Big((\varepsilon > 0) \Rightarrow \exists a (a \in A) \wedge (x \in \mathbb{R}^n) \wedge (|x - a| < \varepsilon) \Big).$$

In the above formula we see one free variable x, and the set described by this formula consists of those x for which that formula is a true statement.

The above formula is made of the "atomic" formulæ,

$$(a \in A), \;\; (x \in \mathbb{R}^n), \;\; (|x - a| < \varepsilon), \;\; (\varepsilon > 0),$$

which all describe definable sets. The logical connector $\Rightarrow$ can be replaced by $\vee\neg$. Finally, we can replace the universal quantifier to rewrite the formula as a transform of atomic formulas via the basic logical operations.

$$\neg \Big\{ \exists \varepsilon \neg \Big((\varepsilon > 0) \Rightarrow \exists a (a \in A) \wedge (x \in \mathbb{R}^n) \wedge (|x - a| < \varepsilon) \Big) \Big\}.$$

Arguing in a similar fashion we deduce that the interior of an $\mathcal{S}$-definable set is an $\mathcal{S}$-definable set. $\qquad \square$

Given an $\mathbb{R}$-structure $\mathcal{S}$, and a collection $\mathcal{A} = (\mathcal{A}_n)_{n \geq 1}$, $\mathcal{A}_n \subset \mathcal{P}(\mathbb{R}^n)$, we can form a new structure $\mathcal{S}(\mathcal{A})$, which is the smallest structure containing $\mathcal{S}$ and the sets in $\mathcal{A}_n$. We say that $\mathcal{S}(\mathcal{A})$ is obtained from $\mathcal{S}$ by *adjoining the collection* $\mathcal{A}$.

Definition 9.3.3. An $\mathbb{R}$-structure $\mathcal{S}$ is called *tame*, or *o-minimal* (order minimal) if it satisfies the property

T. Any set $A \in \mathcal{S}^1$ is a *finite* union of open intervals (a, b), $-\infty \leq a < b \leq \infty$, and singletons $\{r\}$. $\qquad \square$

Example 9.3.4. (a) The collection of real semialgebraic sets $\mathcal{S}_{alg}$ is a tame structure.

(b)(*Gabrielov-Hironaka-Hardt*) A *restricted* real analytic function is a function $f : \mathbb{R}^n \to \mathbb{R}$ with the property that there exists a real analytic function $\tilde{f}$ defined in an open neighborhood U of the cube $C_n := [-1, 1]^n$ such that

$$f(x) = \begin{cases} \tilde{f}(x) & x \in C_n \\ 0 & x \in \mathbb{R}^n \setminus C_n. \end{cases}$$

we denote by $\mathcal{S}_{an}$ the structure obtained from $\mathcal{S}_{alg}$ by adjoining the graphs of all the restricted real analytic functions. For example, all the compact, real analytic submanifolds of $\mathbb{R}^n$ belong to $g\mathcal{S}_{an}$. The structure $\mathcal{S}_{an}$ is tame.

(c)(*Wilkie, van den Dries, Macintyre, Marker*) The structure obtained by adjoining to $\mathcal{S}_{an}$ the graph of the exponential function $\mathbb{R} \to \mathbb{R}$, $t \mapsto e^t$, is a tame structure.

(d)(*Khovanski-Speissegger*) There exists a tame structure $\mathcal{S}'$ with the following properties

(d_1) $\mathcal{S}_{an} \subset \mathcal{S}'$.

(d_2) If $U \subset \mathbb{R}^n$ is open, connected and $\mathcal{S}'$-definable, $F_1, \ldots, F_n : U \times \mathbb{R} \to \mathbb{R}$ are $\mathcal{S}'$-definable and C^1, and $f : U \to \mathbb{R}$ is a C^1 function satisfying

$$\frac{\partial f}{\partial x_i} = F_i(x, f(x)), \quad \forall x \in \mathbb{R}, , \quad i = 1, \ldots, n, \qquad (9.3.1)$$

then f is $\mathcal{S}'$-definable.

The smallest structure satisfying the above two properties, is called the *pfaffian closure*[4] of $\mathcal{S}_{an}$, and we will denote it by $\widehat{\mathcal{S}}_{an}$.

Observe that if $f : (a, b) \to \mathbb{R}$ is C^1, $\widehat{\mathcal{S}}_{an}$-definable, and $x_0 \in (a, b)$ then the antiderivative $F : (a, b) \to \mathbb{R}$

$$F(x) = \int_{x_0}^{x} f(t)dt, \quad x \in (a, b),$$

is also $\widehat{\mathcal{S}}_{an}$-definable.   □

The definable sets and functions of a tame structure have rather remarkable *tame* behavior which prohibits many pathologies. It is perhaps instructive to give an example of function which is not definable in any tame structure. For example, the function $x \mapsto \sin x$ is not definable in a tame structure because the intersection of its graph with the horizontal axis is the countable set $\pi\mathbb{Z}$ which violates the *o*-minimality condition **T**.

We will list below some of the nice properties of the sets and function definable in a tame structure $\mathcal{S}$. Their proofs can be found in [23, 28].

• (*Piecewise smoothness of tame functions and sets.*) Suppose A is an $\mathcal{S}$-definable set, p is a positive integer, and $f : A \to \mathbb{R}$ is a definable function. Then A can

---

[4]Our definition of pfaffian closure is more restrictive than the original one in [58, 91], but it suffices for many geometric applications.

be partitioned into finitely many S definable sets $S_1, \ldots, S_k$, such that each S_i is a C^p-manifold, and each of the restrictions $f|_{S_i}$ is a C^p-function. The dimension of A is then defined as $\max \dim S_i$.

• (*Dimension of the boundary.*) If A is an S-definable set, then $\dim(\boldsymbol{cl}(A) \setminus A) < \dim A$.

• (*Closed graph theorem.*) Suppose X is a tame set and $f : X \to \mathbb{R}^n$ is a tame bounded function. Then f is continuous if and only if its graph is closed in $X \times \mathbb{R}^n$.

• (*Curve selection.*) If A is an S-definable set, $k > 0$ an integer, and $x \in \boldsymbol{cl}(A) \setminus A$, then there exists an S-definable C^k-map $\gamma : (0,1) \to A$ such that $x = \lim_{t \to 0} \gamma(t)$.

• (*Triangulability.*) For every compact definable set A, and any finite collection of definable subsets $\{S_1, \ldots, S_k\}$, there exists a compact simplicial complex K, and a definable homeomorphism

$$\Phi : K \to A$$

such that all the sets $\Phi^{-1}(S_i)$ are unions of relative interiors of faces of K.

• Any definable set has finitely many connected components, and each of them is definable.

• (*Definable selection.*) Suppose A, Λ are S-definable. Then a *definable* family of subsets of A parameterized by Λ is a subset

$$S \subset A \times \Lambda.$$

We set

$$S_\lambda := \big\{ a \in A; \ (a, \lambda) \in S \big\},$$

and we denote by Λ_S the projection of S on Λ. Then there exists a definable function $s : \Lambda_S \to S$ such that

$$s(\lambda) \in S_\lambda, \ \ \forall \lambda \in \Lambda_S.$$

• (*Definability of dimension.*) If $(S_\lambda)_{\lambda \in \Lambda}$ is a definable family of definable sets, then the function

$$\Lambda \ni \lambda \mapsto \dim S_\lambda \in \mathbb{R}$$

is definable. In particular, its range *must be a finite subset of* $\mathbb{Z}$.

• (*Definability of Euler characteristic.*) Suppose $(S_\lambda)_{\lambda \in \Lambda}$ is a definable family of compact tame sets. Then the map

$$\Lambda \ni \lambda \mapsto \chi(S_\lambda) = \text{the Euler characteristic of } S_\lambda \in \mathbb{Z}$$

is definable. In particular, the set $\big\{ \chi(S_\lambda); \ \lambda \in \Lambda, \big\} \subset \mathbb{Z}$ is finite.

• (*Scissor equivalence.*) If A and B are two compact definable sets, then there exists a definable bijection $\varphi : A \to B$ if and only if A and B have the same dimensions and the same Euler characteristics. (The map φ need not be continuous.)

• (*Definable triviality of tame maps.*) We say that a tame map $\Phi : X \to S$ is *definably trivial* if there exists a definable set F, and a definable homeomorphism $\tau : X \to F \times S$ such that the diagram below is commutative

$$
\begin{array}{ccc}
X & \xrightarrow{\ \tau\ } & S \times F \\
& \Phi \searrow \quad \swarrow \pi_S & \\
& S &
\end{array}
$$

If $\Psi : X \to Y$ is a definable map, and p is a positive integer, then there exists a partition of Y into definable C^p-manifolds $Y_1, \ldots, Y_k$ such that each the restrictions

$$
\Psi : \Psi^{-1}(Y_k) \to Y_k
$$

is definably trivial.   □

Definition 9.3.5. A subset A of some Euclidean space $\mathbb{R}^n$ is called *tame* if it is definable within a tame structure $\mathcal{S}$.   □

Exercise 9.3.6. Suppose $\mathcal{S}$ is a tame structure and Suppose $\varphi : [0,1] \to \mathbb{R}^2$ is an $\mathcal{S}$-definable map.
(a) Prove that φ is differentiable on the complement of a finite subset of $[0,1]$.
(b) Prove that the set

$$
\left\{\, (t,m) \in [0,1] \times \mathbb{R}^2; \ \varphi \text{ is differentiable at } t \text{ and } m = \varphi'(t) \,\right\}
$$

is $\mathcal{S}$-definable.
(c) Prove that the curve

$$
C = \left\{ (t, \varphi(t)); \ t \in [0,1] \right\}
$$

has finitely many components, and each of them has finite length.   □

9.3.2   Invariants of the orthogonal group

In the proof of the tube formula we will need to use H.Weyl's characterization of polynomials invariant under the orthogonal group.

Suppose V is a finite dimensional Euclidean space with metric $(-,-)$. We denote by $\langle -, - \rangle$ the canonical pairing

$$
\langle -, - \rangle : V^* \times V \to \mathbb{R}, \ \ \langle \boldsymbol{\lambda}, v \rangle = \boldsymbol{\lambda}(v), \ \ \forall v \in V, \ \ \lambda \in V^* = \mathrm{Hom}(V, \mathbb{R}).
$$

We denote by $O(V)$ the group of orthogonal transformations of the Euclidean space V.

Recall that an $O(V)$-module is a pair (E, ρ), where E is a finite dimensional real vector space, while ρ is a group morphism

$$
\rho : O(V) \to \mathrm{Aut}(E), \ \ g \mapsto \rho(g).
$$

A morphism of $O(V)$-modules (E_i, ρ_i), $i = 0, 1$, is a linear map $A : E_0 \to E_1$ such that for every $g \in O(V)$ the diagram below is commutative

$$
\begin{array}{ccc}
E_0 & \xrightarrow{\ A\ } & E_1 \\
{\scriptstyle \rho_0(g)} \downarrow & & \downarrow {\scriptstyle \rho_1(g)} \\
E_0 & \xrightarrow[\ A\ ]{} & E_1
\end{array}
$$

We will denote by $\mathrm{Hom}_{O(V)}(E_0, E_1)$ the spaces of morphisms of $O(V)$-modules.

The vector space V has a tautological structure of $O(V)$-module given by

$$
\tau : O(V) \to \mathrm{Aut}(V), \quad \tau(g)v = gv, \quad \forall g \in O(V), \quad v \in V.
$$

It induces a structure of $O(V)$-module on $V^* = \mathrm{Hom}(V, \mathbb{R})$ given by

$$
\rho_\dagger : O(V) \to \mathrm{Aut}(V^*), \quad g \mapsto \rho_\dagger(g),
$$

where

$$
\langle \rho_\dagger(g)\boldsymbol{\lambda}, \boldsymbol{v} \rangle = \langle \boldsymbol{\lambda}, g^{-1}\boldsymbol{v} \rangle, \forall \boldsymbol{\lambda} \in V^*, \boldsymbol{v} \in V.
$$

Equivalently, $\rho_\dagger(g) = \rho(g^{-1})^\dagger$, where $\rho(h)^\dagger$ denotes the transpose of $\rho(h)$. We obtain an action on $(V^*)^{\otimes n}$,

$$
(\rho_\dagger)^{\otimes n} : O(V) \to \mathrm{Aut}\big((V^*)^{\otimes n}\big), \quad g \mapsto \rho_\dagger(g)^{\otimes n}.
$$

We denote by $(V^*)^{\otimes n}_{O(V)}$ the subspace consisting of invariant tensors,

$$
\omega \in (V^*)^{\otimes n}_{O(V)} \iff \big(\rho_\dagger(g)\big)^{\otimes n} \omega = \omega, \quad \forall g \in O(V).
$$

Observe that $(V^*)^{\otimes n}$ can be identified with the vector space of multi-linear maps

$$
\omega : V^n = \underbrace{V \times \cdots \times V}_{n} \to \mathbb{R},
$$

so that $(V^*)^{\otimes n}_{O(V)}$ can be identified with the subspace of $O(V)$-invariant multilinear maps $V^n \to \mathbb{R}$.

Hermann Weyl has produced in his classic monograph [100] an explicit description of $(V^*)^{\otimes n}_{O(V)}$. We would like to present here, without proof, this beautiful result of Weyl since it will play an important role in the future. We follow the elegant and more modern presentation in Appendix I of [6] to which we refer for proofs.

Observe first that the metric duality defines a natural isomorphism of vector spaces

$$
D : V \to V^*, \quad \boldsymbol{v} \mapsto \boldsymbol{v}^\dagger,
$$

defined by

$$
\langle \boldsymbol{v}^\dagger, \boldsymbol{u} \rangle = (\boldsymbol{v}, \boldsymbol{u}), \quad \forall \boldsymbol{u}, \boldsymbol{v} \in V.
$$

This isomorphism induces an *isomorphism of $O(V)$-modules*

$$
D : (V, \rho) \to (V^*, \rho_\dagger).
$$

We conclude that for any nonnegative integers r, s we have isomorphisms of G-modules

$$(V^*)^{\otimes(r+s)} \cong (V^* \otimes r) \otimes V^{\otimes s} \cong \operatorname{Hom}(V^{\otimes r}, V^{\otimes s}).$$

In particular,

$$\left((V^*)^{\otimes(r+s)}\right)_{O(V)} \cong \left(\operatorname{Hom}(V^{\otimes r}, V^{\otimes s})\right)_{O(V)} = \operatorname{Hom}_{O(V)}(V^{\otimes r}, V^{\otimes s}).$$

Let us observe that if we denote by $\mathcal{S}_r$ the group of permutations of $\{1, \dots, r\}$, then for every $\varphi \in \mathcal{S}_r$ we obtain a morphism of $O(V)$-modules

$$T_\phi \in \operatorname{Hom}_{O(V)}(V^{\otimes r}, V^{\otimes r}), \quad T_\varphi(v_1 \otimes \cdots \otimes v_r) = v_{\varphi(1)} \otimes \cdots \otimes v_{\varphi(r)}.$$

Weyl's *First Main Theorem of Invariant Theory* states that

$$\operatorname{Hom}_{O(V)}(V^{\otimes r}, V^{\otimes s}) \neq 0 \Longleftrightarrow r = s,$$

and that

$$\operatorname{Hom}_{O(V)}(V^{\otimes r}, V^{\otimes r}) = \mathbb{R}[\mathcal{S}_r] := \left\{ \sum_{\varphi \in \mathcal{S}_r} c_\varphi T_\varphi; \;\; c_\varphi \in \mathbb{R} \right\}.$$

We can translate this result in terms of invariant multi-linear forms. Thus

$$(V^*)_{O(V)}^{\otimes n} \neq 0 \Longleftrightarrow n = 2r, \;\; r \in \mathbb{Z}_{\geq 0},$$

and $(V^*)_{O(V)}^{\otimes 2r}$ is spanned by the multilinear forms

$$P_\varphi : V^{2r} \to \mathbb{R}, \;\; (\varphi \in \mathcal{S}_r),$$

defined by

$$P_\varphi(\boldsymbol{u}_1, \dots, \boldsymbol{u}_r, \boldsymbol{v}_1, \dots, \boldsymbol{v}_r) = \left(\boldsymbol{u}_1, \boldsymbol{v}_{\varphi(1)}\right) \cdots \left(\boldsymbol{u}_r, \boldsymbol{v}_{\varphi(r)}\right).$$

The above result has an immediate consequence. Suppose we have a map

$$f : \underbrace{V \times \cdots \times V}_{n} \to \mathbb{R}, \;\; (\boldsymbol{v}_1, \dots, \boldsymbol{v}_n) \mapsto f(\boldsymbol{v}_1, \dots, \boldsymbol{v}_n),$$

which is a homogeneous polynomial of degree d_i in the variable $\boldsymbol{v}_i$, $\forall i = 1, \dots, n$. This form determines a multilinear form

$$\operatorname{Pol}_f : V^{d_1} \times \cdots \times V^{d_n} \to \mathbb{R}$$

obtained by polarization in each variables separately,

$$\operatorname{Pol}_f(\boldsymbol{u}_1^1, \dots, \boldsymbol{u}_1^{d_1}; \dots; \boldsymbol{v}_n^1, \dots, \boldsymbol{v}_n^{d_n})$$

$=$ the coefficient of the monomial $t_{11}t_{12} \cdots t_{1d_1} \cdots t_{n1} \cdots t_{nd_n}$ in the polynomial

$$P_f(t_{11}, t_{12}, \dots, t_{1d_1}, \dots, t_{n1}, \dots, t_{nd_n}) = f\left(\sum_{j=1}^{d_1} t_{ij} \boldsymbol{u}_1^j, \dots, \sum_{j=1}^{d_n} t_{nj} u_n^j\right).$$

Observe that

$$f(\boldsymbol{v}_1, \ldots, \boldsymbol{v}_n) = \mathrm{Pol}_f(\underbrace{\boldsymbol{v}_1, \ldots, \boldsymbol{v}_1}_{d_1}, \ldots, \underbrace{\boldsymbol{v}_n, \ldots, \boldsymbol{v}_n}_{d_n}),$$

and f is $O(V)$-invariant if and only if Pol_f is $O(V)$ invariant.

Note that every function

$$f : \underbrace{V \times \cdots \times V}_{n} \to \mathbb{R}$$

which is polynomial in each of the variables is a linear combination of functions which is polynomial and homogeneous in each of the variables. For every $1 \le i \le j \le n$ we define

$$q_{ij} : \underbrace{V \times \cdots \times V}_{n} \to \mathbb{R}, \quad q_{ij}(\boldsymbol{v}_1, \ldots, \boldsymbol{v}_n) := (\boldsymbol{v}_i, \boldsymbol{v}_j).$$

Theorem 9.3.7 (Weyl). *If $f : V \times \cdots \times V \to \mathbb{R}$ is a polynomial map then f is $O(V)$-invariant if and only if there exists a polynomial P in the $\binom{n+1}{2}$ variables q_{ij} such that*

$$f(\boldsymbol{v}_1, \ldots, \boldsymbol{v}_n) = P(\,q_{ij}(\boldsymbol{v}_1, \ldots, \boldsymbol{v}_n)_{1 \le i \le j \le n}\,). \qquad \square$$

Example 9.3.8. (a) Consider the space $E = V^{\otimes k}$. Observe that a degree n homogeneous polynomial P on E can by identified with an element in the symmetric tensor product

$$\mathrm{Sym}_d(E^*) \subset (V^*)^{\otimes 2kn}.$$

The polynomial P is called a *degree n orthogonal invariant* of tensors $T \in V^{\otimes k}$ if it is invariant as an element of $(V^*)^{\otimes kn}$. For example, Weyl's theorem implies that the only degree 1 invariant of a tensor

$$T = \sum_{i,j} T_{ij} \boldsymbol{e}_i \otimes \boldsymbol{e}_j \in V^{\otimes 2}$$

is the *trace*

$$\mathrm{tr}(T) = \sum_{i,j} T_{i,j} \boldsymbol{e}_i \otimes \boldsymbol{e}_j = \sum_i T_{ii}.$$

The space of degree 2 invariants is spanned by the polynomials

$$(\mathrm{tr}(T))^2, \quad Q(T) = \sum_{ij} T_{ij}^2, \quad \tilde{Q}(T) = \sum_{i,j} T_{ij} T_{ji}. \qquad \square$$

Here is briefly how to find a basis of the space of degree k invariant polynomials

$$P : V^{\otimes h} \to \mathbb{R}.$$

First of all, such polynomials exist if and only if hk is even, $hk = 2m$. Fix an orthonormal basis $\{e_1, \cdots, e_d\}$ of V. A tensor $T \in V^{\otimes h}$ decomposes as

$$T = \sum_{1 \le i_1, \dots, i_h \le d} T_{i_1 \dots i_h} e_{i_1} \otimes \cdots \otimes e_{i_h}.$$

A degree k polynomial $P : V^{\otimes h} \to \mathbb{R}$ is then a polynomial in the d^h variables $T_{i_1 \dots i_h}$.

We will construct a bijection between a basis of invariant polynomials and certain combinatorial structures called *matchings*.

A matching on the set $I_{hk} := \{1, \dots, hk\}$ is an equivalence relation $\sim$ with the property that each equivalence class consists of two elements. Thus to produce a matching we need to choose a subset $A \subset I_{hk}$ of cardinality $m = \frac{hk}{2}$, and then a bijection $\varphi : A \to I_{hk} \setminus A$. Observing that the matching associated to (A, φ) is the same as the matching associated to $(I_{hk} \setminus A, \varphi^{-1})$, we deduce that the number of matchings is

$$\frac{1}{2}(m!) \cdot \binom{2m}{m} = \frac{(2m)!}{2(m!)}.$$

Denote by $\mathcal{F}_\sim$ the set of functions

$$\mu : I_{hk} \to \{1, 2, \dots, d\}$$

such that $i \sim j \Longrightarrow \mu(i) = \mu(j)$. Equivalently, $\mathcal{F}_\sim$, can be identified with the set of functions $\mu : I_{hk}/\sim \to \{1, \dots, d\}$, so there are d^m such functions.

For any matching $\sim$ define

$$P_\sim(T) = \sum_{\mu \in \mathcal{F}_\sim} T_{\mu(1) \dots \mu(h)} \cdots T_{\mu(hk-h+1) \dots \mu(hk)}.$$

The collection

$$\{P_\sim; \ \sim \ \text{is a matching of } I_{hk}\}$$

is a basis of the space of degree k invariant polynomials $V^{\otimes h} \to \mathbb{R}$.

For example, the space of degree 1 invariant polynomials in tensors of order 4 has dimension $\frac{4!}{2(2!)} = 6$. Each polynomial in a basis constructed as above is a sum of d^2 terms. We see that things are getting "hairy" pretty fast. $\qquad\square$

9.3.3 *The tube formula and curvature measures*

Suppose that M is an m-dimensional submanifold of $\mathbb{R}^n$. We set

$$c := \operatorname{codim} M = n - m.$$

In this section we will assume that M is *compact and without boundary*, but *we will not assume that it is orientable.*

For $r > 0$ we define the tube of radius r around M to be the closed set

$$\mathbb{T}_r(M) := \{x \in M; \ \operatorname{dist}(x, M) \le r\},$$

and we denote by $V(M, r)$ its volume.

Note that a first approximation for $V(M, r)$ is

$$V(M, r) \approx \text{vol}(M) \cdot \boldsymbol{\omega}_c r^c,$$

where $\boldsymbol{\omega}_c r^c$ is the volume of a ball of dimension c and radius r.

We want to prove that there exists a polynomial

$$P_M(r) = p_c r^c + \cdots + p_n r^n,$$

such that, for all sufficiently small r, we have

- $V(M, r) = P_M(r)$,
- $p_c = \boldsymbol{\omega}_c \cdot \text{vol}(M)$,
- all the coefficients of P_M are described, up to some universal multiplicative constants, by certain integral, intrinsic geometric invariants of M.

Let $\mathcal{N}(M)$ denote the orthogonal complement of TM in $(T\mathbb{R}^n)|_M$, and we will call it the *normal bundle* of $M \hookrightarrow \mathbb{R}^n$. We define

$$\mathbb{D}_r(\mathbb{R}^n) := \{(v, p); \; p \in \mathbb{R}^n, \; v \in T_p\mathbb{R}^n, \; |v| \leq r\} \subset T\mathbb{R}^n,$$

and we set

$$\mathcal{N}_r(M) := \mathcal{N}(M) \cap \mathbb{D}_r(\mathbb{R}^n).$$

The manifold with boundary $\mathbb{D}_r(\mathbb{R}^n)$ is a bundle of n-dimensional disks over $\mathbb{R}^n$, while $\mathcal{N}_r(M)$ is a bundle of c-dimensional disks over M.

The exponential map $\mathbb{E} : T\mathbb{R}^n \to \mathbb{R}^n$ restricts to an exponential map

$$\mathbb{E}_M : \mathcal{N}(M) \to \mathbb{R}^n.$$

Observe that because M is compact, there exists $r_0 = r_0(M) > 0$ such that, for every $r \in (0, r_0)$, the exponential map $\mathbb{E}_M$ induces a diffeomorphism

$$\mathbb{E}_M : \mathcal{N}_r(M) \to \mathbb{T}_r(M).$$

If we denote by $|dV_n|$ the Euclidean volume density on $\mathbb{R}^n$ we deduce

$$V(M, r) = \text{vol}(\mathbb{T}_r(M)) = \int_{\mathbb{T}_r(M)} |dV_n| = \int_{\mathcal{N}_r(M)} \mathbb{E}_M^* |dV_n|.$$

If $\pi : \mathcal{N}_r(M) \to M$ denotes the canonical projection, then we deduce from Fubini's theorem that

$$V(M, r) = \int_M \pi_* \mathbb{E}_M^* |dV_n|. \tag{9.3.2}$$

We want to give a more explicit description of the density $\pi_* \mathbb{E}_M^* |dV_n|$. We will continue to use the indexing conventions we have used in Subsection 9.2.1.

Fix a local orthonormal frame $(\boldsymbol{e}_A)$ of $(T\mathbb{R}^n)|_M$ defined in a neighborhood $U \subset M$ of a point $p_0 \in M$ such that for all $1 \leq i \leq m$ vector field $\boldsymbol{e}_i$ is tangent to U. We define

$$D_r^c := \{\vec{t} = (t^\alpha) = (t^{m+1}, \ldots, t^n) \in \mathbb{R}^c; \; \sum_\alpha |t^\alpha|^2 \leq r\}.$$

Note that we have a diffeomorphism

$$D_r^c \times U \longrightarrow \mathcal{N}_r(U) := \mathcal{N}_r(M)|_U, \quad (\vec{t}, x) \mapsto (t^\alpha e_\alpha(x), x) \in \mathcal{N}_r(M),$$

and thus we can identify $D_r^c \times U$ with the open subset $\pi^{-1}(U) \subset \mathcal{N}_r(M)$, and we can use $x \in M$ and $\vec{t} \in D_r^c$ as local coordinates on $\pi^{-1}(U)$. Define

$$\mathbb{T}_r(U) := \mathbb{E}_M(\mathcal{N}_r(U)) \subset \mathbb{R}^n,$$

and

$$\tilde{e}_A : \mathbb{T}_r(U) \to \mathbb{R}^n \text{ by } \tilde{e}_A(x + t^\alpha e_\alpha) = e_A(x).$$

We have thus extended in a special way the local frame (e_A) of $(T\mathbb{R}^n)|_U$ to a local frame of $(T\mathbb{R}^n)|_{\mathbb{T}_r(U)}$ so that

$$\boldsymbol{D}_{\tilde{e}_\alpha} \tilde{e}_A = 0, \quad \forall \alpha, A. \tag{9.3.3}$$

We denote by (θ^A) the coframe of $\mathbb{T}_r(U)$ dual to $\tilde{e}_A$.

Over $D_r^c \times U$ we have a local frame $(\partial_{t^\alpha}, e_i)$ with dual coframe (ϕ^A) defined by

$$\phi^i = \pi^* \boldsymbol{\theta}^i, \quad \phi_\alpha = dt^\alpha.$$

Consider the 1-forms $\Theta_B^A \in \Omega^1(\mathbb{T}_r(U))$ associated to the Levi-Civita connection $\boldsymbol{D}$ by the frame $(\tilde{e}_A)$ on $\mathbb{T}_r(U)$, and set

$$\Theta_{CB}^A = \tilde{e}_C \lrcorner \Theta_B^A, \quad \forall i,$$

so that

$$\boldsymbol{D}_{e_C} \boldsymbol{e}_B = \Theta_{CB}^A \boldsymbol{e}_A.$$

Using (9.3.3) we deduce

$$\Theta_{\alpha B}^A = 0, \quad \forall \alpha \implies \Theta_B^A = \Theta_{iB}^A \boldsymbol{\theta}^i. \tag{9.3.4}$$

Finally set

$$\Phi_B^A = \pi^*(\Theta_B^A|_M) \in \Omega^1(\mathcal{N}_r(U)), \quad \Phi_{iB}^A := \pi^*(\Theta_{iB}^A|_U) \in C^\infty(\mathcal{N}_r(U)).$$

The equalities (9.3.4) imply

$$\Phi_B^A = \Phi_{iB}^A \phi^i.$$

On $\mathcal{N}_r(M)|_U$ we use $(\vec{t}, x)$ as coordinates and we have

$$\mathbb{E}_M(t^\alpha e_\alpha(x), x) = x + t^\alpha e_\alpha.$$

We have

$$\mathbb{E}_M^* \boldsymbol{\theta}^A = \sum_i (e_A \bullet \boldsymbol{D}_{e_i} \mathbb{E}_M) \phi^i + \sum_\alpha (e_A \bullet \partial_{t^\alpha} \mathbb{E}_M) dt^\alpha$$

$$= \sum_i \boldsymbol{e}_A \bullet (\boldsymbol{e}_i + t^\alpha \boldsymbol{D}_{e_i} \boldsymbol{e}_\alpha) \phi^i + \sum_\alpha \delta_{A\alpha} dt^\alpha = \delta_{Ai} \phi^i + t^\alpha \Phi_{i\alpha}^A \phi^i + \delta_{A\alpha} dt^\alpha.$$

Hence

$$\mathbb{E}_M^*\boldsymbol{\theta}^j = \phi^j + t^\alpha\Phi^j_{i\alpha}\phi^i = \phi^j - \sum_\alpha t^\alpha\Phi^\alpha_{ij}\phi^i, \ \ \mathbb{E}_M^*\boldsymbol{\theta}^\beta = dt^\beta.$$

We find it convenient to set

$$\Phi_{ij} = (\Phi_{ij}^{m+1}, \ldots, \Phi_{ij}^n) : U \to \mathbb{R}^c,$$

so that

$$\mathbb{E}_M^*\boldsymbol{\theta}^j = \phi^j - \sum_i (\vec{t} \bullet \Phi_{ij})\phi^i.$$

Define the $m \times m$ symmetric matrix

$$S := S(\vec{t}, x) = (\vec{t} \bullet \Phi_{ij}(x))_{1 \le i, j \le m}.$$

Note that the volume density on $\mathbb{R}^n$ is

$$|dV_n| = |\theta^{m+1} \wedge \cdots \wedge \theta^n \wedge \theta^1 \wedge \cdots \wedge \theta^m|,$$

$$\mathbb{E}_M^* dV_n = |\det(\mathbb{1} - S(\vec{t}, x))| \, |d\vec{t} \wedge d\phi| = \det(\mathbb{1} - S(\vec{t}, x)) \, |d\vec{t} \wedge d\phi|, \qquad (9.3.5)$$

$$d\vec{t} = dt^{m+1} \wedge \cdots \wedge dt^n, \ \ d\phi = d\phi^1 \wedge \cdots \wedge d\phi^m.$$

Recalling that $|\wedge_i d\boldsymbol{\theta}^i|_M$ is the metric volume density $|dV_M|$ on M, we deduce

$$\mathbb{E}_M^*|dV_n| = \det(\mathbb{1} - S(\vec{t}, x)) |d\vec{t}| \times \pi^*|dV_M|,$$

where $|d\vec{t}|$ denotes the volume density on $\mathbb{R}^c$. For simplicity we write $|dV_M|$ instead of $\pi^*|dV_M|$. Now set

$$\rho := |\vec{t}|, \ \ \omega := \frac{1}{\rho}\vec{t},$$

and denote by $|d\omega|$ the area density on the unit sphere in $\mathbb{R}^c$. Then

$$\mathbb{E}_M^*|dV_n| = \det(\mathbb{1} - \rho S(\omega, x)) \rho^{c-1}|d\rho| \times |d\omega| \times |dV_M|. \qquad (9.3.6)$$

Observe that

$$\det(\mathbb{1} - \rho S(\omega, x)) = \sum_{\nu=0}^m (-1)^\nu \rho^\nu P_\nu(\Phi_{ij}(x) \bullet \omega),$$

where P_ν denotes a homogeneous polynomial of degree ν in the m^2 variables

$$\boldsymbol{u}_{ij} \in \mathbb{R}^c, \ \ 1 \le i, j \le m.$$

We set

$$\overline{P}_\nu(\Phi_{ij}(x)) := \int_{S^{c-1}} P_\nu(\Phi_{ij}(x) \bullet \omega)|d\omega|.$$

Above, $\overline{P}_\nu(\boldsymbol{u}_{ij})$ is an $O(c)$-invariant, homogeneous polynomial of degree ν in the variables $\boldsymbol{u}_{ij} \in \mathbb{R}^c$, $1 \le i, j \le m$. We conclude,

$$\pi_*\mathbb{E}_M^*|dV_n| = \sum_{\nu=0}^m \frac{(-1)^\nu}{c+\nu} r^{c+\nu}\overline{P}_\nu(\Phi_{ij}(x))|dV_M(x)|. \qquad (9.3.7)$$

We would like to determine the invariant polynomials $\overline{P}_\nu(\boldsymbol{u}_{ij})$.

Theorem 9.3.7 on invariants of the orthogonal group $O(c)$ implies that $\overline{P}_\nu$ must be a polynomial in the quantities

$$q_{i,j,k,\ell} := \boldsymbol{u}_{ik} \bullet \boldsymbol{u}_{j\ell}.$$

Because these quantities are homogeneous of degree 2 in the variables $\boldsymbol{u}_{ij}$ we deduce $\overline{P}_\nu = 0$ if ν is odd. Assume therefore $\nu = 2h$, $h \in \mathbb{Z}_{\geq 0}$.

For every $\vec{t} \in \mathbb{R}^c = \operatorname{span}(\boldsymbol{e}_{m+1}, \dots, \boldsymbol{e}_{m+c})$ we form the linear operator

$$U(\vec{t}) = U(\boldsymbol{u}_{ij}, \vec{t}) : \mathbb{R}^m \to \mathbb{R}^m$$

given by the $m \times m$ matrix $(\boldsymbol{u}_{ij} \bullet \vec{t})_{1 \leq i,j \leq m}$. We deduce that

$$(-1)^\nu P_\nu(\boldsymbol{u}_{ij} \bullet \vec{t}) = \operatorname{tr} \Lambda^\nu U(\boldsymbol{u}_{ij}, \vec{t})$$

= the sum of all the $\nu \times \nu$ minors of $U(\vec{t})$ symmetric with respect to the diagonal.

These minors are parametrized by the subsets $I \subset \{1, \dots, m\}$ of cardinality $\#I = \nu$. For every $\omega \in \mathbb{R}^c$ we denote by $\mu_I(\boldsymbol{u}_{ij} \bullet \omega)$ the corresponding minor of $U(\omega)$, and by $\overline{\mu}_I$ its average,

$$\overline{\mu}_I(\boldsymbol{u}_{ij}) := \int_{S^{c-1}} \mu_I(\boldsymbol{u}_{ij} \bullet \omega) d\omega.$$

Note that $\overline{\mu}_I$ is an $O(c)$-invariant polynomial in the variables $\{\boldsymbol{u}_{ij}\}_{i,j \in I}$.

Let

$$I = \{1 \leq i_1 < i_2 < \cdots < i_{2h} \leq m\} \subset \{1, \dots, m\},$$

and denote by $\mathcal{S}_I$ the group of permutations of I. For $\varphi \in \mathcal{S}_I$ we set

$$\varphi_j := \varphi(i_j), \quad \forall j = 1, \dots, 2h.$$

For any $\sigma, \varphi \in \mathcal{S}_I$ we denote by $\epsilon(\sigma, \varphi)$ the signature of the permutation $\sigma \circ \varphi^{-1}$, and by $Q_{\sigma,\varphi}$ the invariant polynomial

$$Q_{I,\sigma,\varphi} = \prod_{j=1}^h q_{\varphi_{2j-1}, \varphi_{2j}, \sigma_{2j-1}, \sigma_{2j}} = \prod_{j=1}^h \boldsymbol{u}_{\varphi_{2j-1} \sigma_{2j-1}} \bullet \boldsymbol{u}_{\varphi_{2j} \sigma_{2j}}.$$

Lemma 9.3.9. *There exists a constant* $\xi = \xi_{m,\nu,c}$ *depending only on* m, ν *and* c *such that*

$$\overline{\mu}_I = \xi Q_I, \quad Q_I := \sum_{\varphi, \sigma \in \mathcal{S}_I} \epsilon(\sigma, \varphi) Q_{I,\sigma,\varphi}.$$

Proof. We regard $\overline{\mu}_I$ as a function on the vector space of $(2h) \times (2h)$ matrices U with entries in $\mathbb{R}^c$

$$U = [\boldsymbol{u}_{ij}]_{i,j \in I}.$$

We observe that $\overline{\mu}_I$ satisfies the following determinant like properties.

- $\overline{\mu}_I$ changes sign if we switch two rows (or columns).
- $\overline{\mu}_I$ is separately linear in each of the variables $\boldsymbol{u}_{ij}$.
- $\overline{\mu}_I$ is a homogeneous polynomial of degree h in the variables $q_{i,j,k,\ell}$.

We deduce that $\overline{\mu}_I$ is a linear combination of monomials of the form

$$q_{k_1,k_2,\ell_1,\ell_2} \cdots q_{k_{2h-1}k_{2h},\ell_{2h-1},\ell_{2h}},$$

where

$$\{k_1,\ldots,k_{2h}\} \text{ and } \{\ell_1,\ldots,\ell_{2h}\}$$

are permutations of I. The skew-symmetry of $\overline{\mu}_I$ with respect to the permutations of rows and columns now implies that $\overline{\mu}_I$ must be a multiple of Q_I.

$\square$

The constant ξ satisfies

$$\xi_{m,\nu,c} = \frac{\overline{\mu}_I(\boldsymbol{u}_{ij})}{Q_I(\boldsymbol{u}_{ij})}, \quad \forall \boldsymbol{u}_{ij} \in \mathbb{R}^c$$

so it suffices to compute the numerator and denominator of the above fraction for some special values of $\boldsymbol{u}_{ij}$. We can assume $I = \{1,2,\ldots,2h\}$ and we choose

$$\boldsymbol{u}_{ij} = \begin{bmatrix} 1 \\ 0 \\ \vdots \\ 0 \end{bmatrix} \in \mathbb{R}^c.$$

Then, if we set

$$\vec{t} = \begin{bmatrix} t^1 \\ t^2 \\ \vdots \\ t^c \end{bmatrix} \in \mathbb{R}^c, \quad \omega = \frac{1}{|\vec{t}|}\vec{t}$$

we deduce

$$U(\boldsymbol{u}_{ij},\vec{t}) = \begin{bmatrix} t^1 & 0 & \cdots & 0 \\ 0 & t^1 & \cdots & 0 \\ \vdots & \vdots & \ddots & \vdots \\ 0 & 0 & \cdots & t^1 \end{bmatrix}, \quad \mu_I(\boldsymbol{u}_{ij} \bullet \vec{t}) = |t^1|^\nu.$$

Hence

$$\overline{\mu}_I(\boldsymbol{u}_{ij}) = \int_{S^{c-1}} |\omega^1|^{2h} |d\omega|. \tag{9.3.8}$$

On the other hand, we have

$$Q_{I,\sigma,\varphi} = \prod_{j=1}^{h} \boldsymbol{u}_{\varphi_{2j-1}\sigma_{2j-1}} \bullet \boldsymbol{u}_{\varphi_{2j}\sigma_{2j}},$$

which is nonzero if and only if $\sigma = \varphi$. We conclude that for this particular choice of $\boldsymbol{u}_{ij}$ we have

$$Q_I = (2h)!.$$

Hence

$$\xi_{m,\nu,c} = \frac{1}{(2h)!} \int_{S^{c-1}} |\omega^1|^{2h} |d\omega|.$$

At this point we invoke the following result whose proof is deferred to the end of this subsection.

Lemma 9.3.10. *For any even, nonnegative integers $2h_1, \ldots, 2h_c$ we have*

$$\int_{S^{c-1}} |\omega^1|^{2h_1} \cdots |\omega^c|^{2h_c} |d\omega| = \frac{2\Gamma(\frac{2h_1+1}{2}) \cdots \Gamma(\frac{2h_c+1}{2})}{\Gamma(\frac{c+2h}{2})},$$

where $h = h_1 + \cdots + h_c$.

We deduce

$$\xi_{m,2h,c} = \frac{2\Gamma(\frac{2h+1}{2})\Gamma(1/2)^{c-1}}{(2h)!\Gamma(\frac{c+2h}{2})}, \tag{9.3.9}$$

and

$$\overline{P}_\nu(\boldsymbol{u}_{ij}) = \xi_{m,2h,c} \sum_{\#I=\nu} Q_I(\boldsymbol{u}_{ij}). \tag{9.3.10}$$

We denote by $\mathcal{S}_2$ the group of permutations of a linearly ordered set with two elements. We observe that every element

$$\tau = (\tau_1, \ldots, \tau_h) \in G = \underbrace{\mathcal{S}_2 \times \cdots \times \mathcal{S}_2}_{h}$$

defines a permutation of I by regarding τ_1 as a permutation of $\{i_1, i_2\}$, τ_2 as a permutation of $\{i_3, i_4\}$ etc. Thus G is naturally a subgroup of $\mathcal{S}_I$. The space $\mathcal{S}_k/G$ of left cosets of this group can be identified with the subset $\mathcal{S}'_I \subset \mathcal{S}_I$ consisting of bijections $\varphi : I \to I$ satisfying the conditions

$$\varphi_1 < \varphi_2, \quad \varphi_3 < \varphi_4, \ldots, \quad \varphi_{2h-1} < \varphi_{2h}.$$

We deduce that if $\boldsymbol{u}_{ij} = S_{ij}(x)$, then for every $\sigma \in \mathcal{S}_I$ we have

$$\sum_{\varphi \in \mathcal{S}_I} \epsilon(\sigma, \varphi) Q_{I,\sigma,\varphi} = \sum_{\varphi \in \mathcal{S}'_I, \tau \in G} \epsilon(\sigma, \varphi\tau) Q_{I,\sigma,\varphi\tau}$$

$$= \sum_{\varphi \in \mathcal{S}'_I} \epsilon(\sigma, \varphi) \prod_{j=1}^{h} \left(q_{\varphi_{2j-1}, \varphi_{2j}, \sigma_{2j-1}, \sigma_{2j}} - q_{\varphi_{2j}, \varphi_{2j-1}, \sigma_{2j-1}, \sigma_{2j}} \right)$$

$$= \sum_{\varphi \in \mathcal{S}'_I} \epsilon(\sigma, \varphi) \prod_{j=1}^{h} R_{\varphi_{2j-1}\varphi_{2j}\sigma_{2j-1}\sigma_{2j}}.$$

Using the skew-symmetry $R_{ijk\ell} = -R_{ij\ell k}$ we deduce

$$\sum_{\sigma,\varphi \in \mathcal{S}_I} \epsilon(\sigma,\varphi) Q_{I,\sigma,\varphi} = \sum_{\sigma \in \mathcal{S}_I} \sum_{\varphi \in \mathcal{S}'_I} \epsilon(\sigma,\varphi) \prod_{j=1}^{h} R_{\varphi_{2j-1}\varphi_{2j}\sigma_{2j-1}\sigma_{2j}}$$

$$= \sum_{\sigma \in \mathcal{S}'_I, \tau \in G} \epsilon(\sigma\tau,\varphi) \prod_{j=1}^{h} R_{\varphi_{2j-1}\varphi_{2j}\sigma_{2j-1}\sigma_{2j}}$$

$$= 2^h \underbrace{\sum_{\sigma,\varphi \in \mathcal{S}'_I} \epsilon(\sigma,\varphi) \prod_{j=1}^{h} R_{\varphi_{2j-1}\varphi_{2j}\sigma_{2j-1}\sigma_{2j}}}_{=:\mathfrak{Q}_I(R)}$$

We conclude that

$$\overline{P}_\nu(\psi_{ij}) = 2^h \xi_{m,2h,c} \sum_{\#I=2h} \mathfrak{Q}_I(R). \tag{9.3.11}$$

Using (9.3.2) and (9.3.7) we deduce that

$$V(M,r) = \mathrm{vol}\,(\mathbb{T}_r(M)) = \sum_{h=0}^{\lfloor m/2 \rfloor} \boldsymbol{\omega}_{c+2h} r^{c+2h} \frac{2^h \xi_{m,2h,c}}{(c+2h)\boldsymbol{\omega}_{c+2h}} \int_M \mathfrak{Q}_h(R) |dV_M|.$$

Let us observe that the constant

$$\frac{2^h \xi_{m,2h,c}}{(c+2h)\boldsymbol{\omega}_{c+2h}}$$

is independent of the codimension c. It *depends only* on h. Indeed, we have

$$\frac{2^h \xi_{m,2h,c}}{(c+2h)\boldsymbol{\omega}_{c+2h}} = \frac{2^h \xi_{m,2h,c}}{\boldsymbol{\sigma}_{c+2h-1}} = \frac{2^h}{\boldsymbol{\sigma}_{c+2h-1}} \frac{2\Gamma(\frac{2h+1}{2})\Gamma(1/2)^{c-1}}{(2h)!\Gamma(\frac{c+2h}{2})}$$

$$= \frac{2^h \Gamma(h+1/2)}{\Gamma(1/2)^{1+2h}(2h)!} = \frac{\gamma(2h)}{\pi^h(2h)!} = \frac{1}{(2\pi)^h h!}.$$

We have thus obtained the following celebrated result of Hermann Weyl, [99].

Theorem 9.3.11 (Tube formula). *Suppose M is a closed, compact submanifold of $\mathbb{R}^n$, $\dim M = m$, $c = n - m$. Denote by R the Riemann curvature of the induced metric on M. Then for all $r > 0$ sufficiently small we have*

$$V(M,r) = \mathrm{vol}\,(\mathbb{T}_r(M)) = \sum_{h=0}^{\lfloor m/2 \rfloor} \boldsymbol{\omega}_{c+2h} r^{c+2h} \mu_{m-2h}(M),$$

$$\mu_{m-2h}(M) = \frac{1}{(2\pi)^h h!} \int_M \mathfrak{Q}_h(R) |dV_M|,$$

where $\mathfrak{Q}_h$ is a polynomial of degree h in the curvature. By choosing a local, orthonormal frame $(\boldsymbol{e}_1,\ldots,\boldsymbol{e}_m)$ of TM we can express the polynomial $\mathfrak{Q}_h(R)$ as

$$\mathfrak{Q}_h(R) := \sum_{\#I=2h} \mathfrak{Q}_I(R),$$

where for every $I = \{i_1 < i_2 < \cdots < i_{2h}\} \subset \{1,\ldots,m\}$ we define

$$\mathfrak{Q}_I(R) = \sum_{\sigma,\varphi \in \mathcal{S}'_I} \epsilon(\sigma,\varphi) \prod_{j=1}^{h} R_{\varphi(i_{2j-1})\varphi(i_{2j})\sigma(i_{2j-1})\sigma(i_{2j})}. \qquad \square$$

Example 9.3.12. (a) If $h = 0$ then

$$\mu_m(M) = \text{vol}(M).$$

(b) Assume now that m is even, $m = 2h$, and oriented. Then $c + 2h = m + c = n$ and

$$\mu_0(M) = \frac{1}{(2\pi)^h} \int_M \frac{1}{h!} \mathcal{Q}_h(R) dV_M.$$

Comparing the definition of $\mathcal{Q}_h(R)$ with (9.2.7a) we deduce that the top dimensional form

$$\frac{1}{(2\pi)^h} \frac{1}{h!} \mathcal{Q}_h(R) dV_M \in \Omega^{2h}(M)$$

is precisely the Euler form associated with the orientation of M and the induced metric, so that

$$\mu_0(M, g) = \int_M e(M, g). \tag{9.3.12}$$

(c) Suppose now that M is a hypersurface. Consider the second fundamental form

$$S = (S_{ij})_{1 \leq i,j \leq m}, \quad S_{ij} = (D_{e_i} e_j) \bullet e_{m+1},$$

where we recall that e_{m+1} is in fact the oriented unit normal vector field along M. Fix a point $x_0 \in M$ and assume that at this point the frame $(e_1, \ldots, e_m)$ diagonalizes the second fundamental form so that

$$S_{ij} = \kappa_i \delta_{ij}.$$

The eigenvalues $\kappa_1, \ldots, \kappa_m$ are the principal curvatures at the point x_0. We denote by $c_\nu(\kappa)$ the elementary symmetric polynomial of degree ν in the variables κ_i. In this case $c = 1$, and we have

$$\mathbb{E}_M^* dV_{\mathbb{R}^{m+1}} = \det(\mathbb{1} - tS) dt \wedge dV_M = \sum_{\nu=0}^m (-1)^\nu t^\nu c_\nu(\kappa) dt \wedge dV_M$$

$$\pi_* \mathbb{E}_M^* dV_{\mathbb{R}^{m+1}} = \sum_{\nu=0}^m \left(\int_{-r}^r t^\nu dt \right) c_\nu(\kappa) dV_M = 2 \sum_{h=0}^{\lfloor m/2 \rfloor} \frac{r^{2h+1}}{2h+1} c_{2h}(\kappa) dV_M$$

so that

$$\sum_{h=0}^{\lfloor m/2 \rfloor} \omega_{1+2h} r^{1+2h} \mu_{m-2h}(M) = V(M, r) = 2 \sum_{h=0}^{\lfloor m/2 \rfloor} \frac{r^{2h+1}}{2h+1} \int_M c_{2h}(\kappa) dV_M.$$

We conclude that

$$\mu_{m-2h}(M) = \frac{2}{\sigma_{2h}} \int_M c_{2h}(\kappa) dV_M, \quad \sigma_{2h} = (2h+1)\omega_{1+2h}.$$

If $M = S^m \hookrightarrow \mathbb{R}^{m+1}$ is the unit sphere then $\kappa_i = 1$ and we deduce that

$$\mu_{m-2h}(S^m) = 2 \frac{\sigma_m}{\sigma_{2h}} \binom{m}{2h}. \tag{9.3.13}$$

(d) Using the definition of the scalar curvature we deduce that for any m-dimensional submanifold $M^m \hookrightarrow \mathbb{R}^n$ we have

$$\mu_{m-2}(M,g) = \text{const}_m \int_M s_g |dV_g|,$$

where s denotes the scalar curvature of the induced metric g, and const_m is an universal constant, depending only on m. We see that the map $g \to \mu_{m-2}(M,g)$ is precisely the *Hilbert-Einstein functional* we discussed in Subsection 4.2.5, Definition 4.2.22.

To find const_m we compute $\mu_{m-2}(M)$ when $M = S^m$. Using (9.3.13) we deduce

$$\frac{2\sigma_m}{\sigma_2}\binom{m}{2h} = \text{const}_m \int_M s_{round} |dV_{S^m}|,$$

where s_{round} denotes the scalar curvature of the round metric on the unit sphere.

Observe that the Grassmannian of *oriented* codimension one subspaces of $\mathbb{R}^{m+1}$ can be identified with the unit sphere S^m. Hence, the oriented Gauss map of the unit sphere S^m is the identity. In particular, the shape operator of $S^m \hookrightarrow \mathbb{R}^{m+1}$ is the identity operator From Theorema Egregium we deduce that all the sectional curvatures of S^n are equal to one. Using the equalities (4.2.1) we deduce

$$s_{round} = \sum_{i,j} R_{ijij} = \sum_{i,j} 1 = 2\binom{m}{2}.$$

Hence

$$\text{const}_m = \frac{2}{\sigma_2} = \frac{1}{2\pi} \implies \mu_{m-2}(M,g) = \frac{1}{2\pi}\int_M s_g |dV_g|.$$

(e) The polynomial $\mathcal{Q}_2$ still has a "reasonable form"

$$\mathcal{Q}_2(R) = \sum_{\#I=4} \mathcal{Q}_I$$

Then $\#\mathcal{S}'_I = 6$ and

$$\mathcal{Q}_I = \sum_{\sigma,\varphi \in \mathcal{S}'_I} \epsilon(\sigma,\varphi) R_{\sigma_1\sigma_2\varphi_1\varphi_2} R_{\sigma_3\sigma_4\varphi_3\varphi_4}$$

$$= \sum_{\sigma \in \mathcal{S}'_I} R_{\sigma_1\sigma_2\sigma_1\sigma_2} R_{\sigma_3\sigma_4\sigma_3\sigma_4} + \sum_{\sigma \neq \varphi \in \mathcal{S}'_I} \epsilon(\sigma,\varphi) R_{\sigma_1\sigma_2\varphi_1\varphi_2} R_{\sigma_3\sigma_4\varphi_3\varphi_4}.$$

The first sum has only three different monomials, each of them appearing twice is them sum. The second sum has $\binom{6}{2}$ different monomials (corresponding to subsets of cardinality 2 of $\mathcal{S}'_I$) and each of them appears twice. $\qquad \square$

Definition 9.3.13. If (M,g) is a closed, compact, oriented, Riemann manifold, $m = \dim M$, and w is nonnegative integer. If $m - w$ is odd we set

$$\mu_w(M) = 0.$$

If $m - w$ is an even, nonnegative integer, $m - w = 2h$, then we set

$$\mu_w(M, g) = \frac{1}{(2\pi)^h h!} \int_M \mathfrak{Q}_h(R) |dV_M|.$$

We will say that $\mu_w(M, g)$ is the *weight w curvature measure of* (M, g). We set

$$|d\mu_w| := \frac{1}{(2\pi)^h h!} \mathfrak{Q}_h(R) |dV_M|,$$

and we will refer to it as *the (weight w) curvature density.* $\square$

Remark 9.3.14. (a) Let us observe that for any Riemann manifold M, orientable or not, the quantities $|d\mu_w|$ are indeed well defined, i.e. independent of the choice of local frames used in their definition. The fastest way to argue this is by invoking Nash embedding theorem which implies that any compact manifold is can be isometrically embedded in an Euclidean space. For submanifolds of $\mathbb{R}^n$, the proof of the tube formula then implies that these densities are indeed well defined.

We can prove this by more elementary means by observing that, for any finite set I, the relative signature $\epsilon(\sigma, \varphi)$ of two permutations $\varphi, \sigma : I \to I$ is defined by choosing a linear ordering on I, but it is *independent of this choice.*
(b) The weight of the curvature density has a very intuitive meaning. Namely, we should think of $\mu_w(M, g)$ as a quantity measured in *meter^w*. $\square$

Proof of Lemma 9.3.10. Consider the integral

$$I(h_1, \ldots, h_c) = \int_{\mathbb{R}^c} e^{-|\vec{t}|^2} |t^1|^{2h_1} \cdots |t^c|^{2h_c} |dt|.$$

We have $e^{-|\vec{t}|^2} = e^{-|t+1|^2} \cdots e^{-|t^c|^2}$ so that

$$I(h_1, \ldots, h_c) = \prod_{j=1}^{c} \left(\int_{-\infty}^{\infty} e^{-s^2} s^{2h_j} ds \right) = 2^c \prod_{j=1}^{c} \left(\int_0^{\infty} e^{-s^2} s^{2h_j} ds \right)$$

$(u = s^2)$

$$= \prod_{j=1}^{c} \left(\int_0^{\infty} e^{-u} t^{h_j - 1/2} du \right) = \prod_{j=1}^{c} \Gamma\left(\frac{2h_j + 1}{2} \right).$$

On the other hand, using spherical coordinates, $\rho = |\vec{t}|$, $\omega = \frac{1}{|\vec{t}|} \vec{t}$, and recalling that $h = h_1 + \cdots + h_c$, we deduce that

$$I(h_1, \ldots, h_c) = \left(\int_{S^{c-1}} |\omega^1|^{2h_1} \cdots |\omega^c|^{2h_c} d\omega \right) \left(\int_0^{\infty} e^{-\rho^2} \rho^{2h+c-1} d\rho \right)$$

$(u = \rho^2)$

$$= \frac{1}{2} \left(\int_{S^{c-1}} |\omega^1|^{2h_1} \cdots |\omega^c|^{2h_c} d\omega \right) \int_0^{\infty} e^{-u} u^{\frac{c+2h}{2} - 1} du$$

$$= \frac{1}{2} \Gamma\left(\frac{c+2h}{2} \right) \left(\int_{S^{c-1}} |\omega^1|^{2h_1} \cdots |\omega^c|^{2h_c} d\omega \right).$$ $\square$

9.3.4 *Tube formula* $\implies$ *Gauss-Bonnet formula for arbitrary submanifolds*

Suppose $M^m \subset \mathbb{R}^n$ is a closed, compact submanifold of $\mathbb{R}^n$. *We do not assume that M is orientable.* As usual, set $c = n - m$, and we denote by g the induced metric on M. For every sufficiently small positive real number r we set

$$M_r := \{x \in \mathbb{R}^n; \ \text{dist}\,(x, M) = r\} = \partial \mathbb{T}_r(M).$$

The closed set M_r is a compact hypersurface of $\mathbb{R}^n$, and we denote by g_r the induced metric. Observe that for r and ε sufficiently small we have

$$\mathbb{T}_\varepsilon(M_r) = \mathbb{T}_{r+\varepsilon}(M) - \mathbb{T}_{r-\varepsilon}(M)$$

so that,

$$V(M_r, \varepsilon) = V(M, r + \varepsilon) - V(M, r - \varepsilon),$$

which implies that

$$\sum_{h \geq 0} \omega_{1+2h} \varepsilon^{1+2h} \mu_{n-1-2h}(M_r, g_r)$$

$$= \sum_{k \geq 0} \omega_{c+2k} \Big\{ (r + \varepsilon)^{c+2k} - (r - \varepsilon)^{c+2k} \Big\} \mu_{n-c-2k}(M, g).$$

We deduce

$$\mu_{n-1-2h}(M_r, g_r) = \frac{2}{\omega_{1+2h}} \sum_{k \geq 0} \omega_{c+2k} \binom{c + 2k}{1 + 2h} r^{c-1+2k-2h} \mu_{n-c-2k}(M, g).$$

We make a change in variables. We set

$$p := n - 1 - 2h, \ \ w := n - c - 2k = m - 2k.$$

Then $c + 2k = n - w$, $1 + 2h = n - p$, $c + 2k - 1 - 2h = p - w$, so that we can rewrite the above formula as

$$\mu_p(M_r, g_r) = 2 \sum_{w=0}^{m} \binom{n - w}{n - p} \frac{\omega_{n-w}}{\omega_{n-p}} r^{p-w} \mu_w(M, g). \tag{9.3.14}$$

In the above equality it is understood that $\mu_w(M) = 0$ if $m - w$ is odd. In particular, we deduce that *if the codimension of M is odd* then

$$\lim_{r \to 0} \mu_p(M_r, g_r) = 2\mu_p(M, g), \ \ \forall 0 \leq p \leq m = \dim M. \tag{9.3.15}$$

If in the formula (9.3.14) we assume that the manifold M is a point, then we deduce that M_r is the $(n-1)$-dimensional sphere of radius r, $M_r = S_r^{n-1}$, and we conclude that

$$\mu_p(S_r^{n-1}) = 2 \binom{n}{p} \frac{\omega_n}{\omega_{n-p}} r^p = 2\omega_p \begin{bmatrix} n \\ p \end{bmatrix} r^p, \ \ n - p \equiv 1 \mod 2, \tag{9.3.16}$$

where $\begin{bmatrix} n \\ p \end{bmatrix}$ is defined by (9.1.16). The last equality agrees with our previous computation (9.3.13).

If in the formula (9.3.14) we let $p = 0$ we deduce

$$\mu_0(M_r, g_r) = 2\mu_0(M, g), \quad \forall 0 < r \ll 1, \quad \text{if codim } M \text{ is odd.}$$

Observe that the tube $\mathbb{T}_r(M)$ is naturally oriented, even though the manifold M may not be orientable. The Gauss-Bonnet theorem for oriented hypersurfaces implies

$$\chi(M_r) = \mu_0(M_r, g_r)$$

so that

$$\mu_0(M, g) = \frac{1}{2}\chi(M_r), \quad \forall 0 < r \ll 1. \tag{9.3.17}$$

Theorem 9.3.15 (Gauss-Bonnet). *Suppose M is a closed, compact submanifold of an Euclidean space $\mathbb{R}^n$. Denote by g the induced metric. Then*

$$\mu_0(M, g) = \chi(M).$$

Proof. If $m = \dim M$ is odd, then both $\chi(M)$ and $\mu_0(M)$ are equal to zero and the identity is trivial. Assume therefore that m is even. If the dimension n of the ambient space is odd, then the Poincaré duality for the *oriented* n-dimensional manifold with boundary $\mathbb{T}_r(M)$ implies

$$\chi(M_r) = \chi(\partial \mathbb{T}_r(M)) = 2\chi(\mathbb{T}_r(M)) = 2\chi(M),$$

and the theorem follows from (9.3.17).

If n is even, we apply the above argument to the embedding

$$M \hookrightarrow \mathbb{R}^n \hookrightarrow \mathbb{R}^{n+1},$$

where we observe that the metric induced by the embedding $M \hookrightarrow \mathbb{R}^{n+1}$ coincides with the metric induced by the original embedding $M \hookrightarrow \mathbb{R}^n$. $\qquad\square$

Remark 9.3.16. (a) We want emphasize again that in the above theorem *we did not require that M be orientable* which is the traditional assumption in the Gauss-Bonnet theorem.

(b) By invoking the Nash embedding theorem, we deduce that the tube formula implies the Gauss-Bonnet formula for any compact Riemann manifold, orientable or not. $\qquad\square$

Let us record for later use the following corollary of the above proof.

Corollary 9.3.17. *For every closed compact, smooth submanifold M of an Euclidean space V such that $\dim V - \dim M$ is odd we have*

$$\chi(M) = 2\chi(\partial \mathbb{T}_r(M)), \quad \forall 0 < r \ll 1. \qquad\square$$

9.3.5 *Curvature measures of domains of a Euclidean space*

The second fundamental form of a submanifold is in fact a bilinear form with values in the normal bundle. If the submanifold happens to be the *boundary* of a domain, then the normal bundle admits a *canonical trivialization*, and the second fundamental form will be a scalar valued form. The next definition formalizes this observation.

Definition 9.3.18. For any relatively compact open subset D of an Euclidean space V, with smooth boundary ∂D, we define the *co-oriented second fundamental form* of D to be the symmetric bilinear map

$$S_D : \mathrm{Vect}\,(\partial D) \times \mathrm{Vect}\,(\partial D) \to C^\infty(\partial D),$$

$$S_D(X,Y) = (\boldsymbol{D}_X Y) \bullet \boldsymbol{n}, \quad X, Y \in \mathrm{Vect}\,(\partial D),$$

where $\boldsymbol{n} : \partial D \to V$ denotes the *outer* unit normal vector field along ∂D. $\quad\square$

Suppose $D \subset \mathbb{R}^{m+1}$ is an open, relatively compact subset with smooth boundary $M := \partial D$. We denote by $\boldsymbol{n}$ the unit outer normal vector field along $M := \partial D$, and by $S = S_D$ the co-oriented second fundamental form of D. For every symmetric bilinear form B on a Euclidean space V we define $\mathrm{tr}_j(B)$ the j-th elementary symmetric polynomial in the eigenvalues of B, i.e.,

$$\sum_{j\geq 0} z^j \, \mathrm{tr}_j(B) = \det(\mathbb{1}_V + zB).$$

Equivalently,

$$\mathrm{tr}_j B = \mathrm{tr}\left(\Lambda^j B : \Lambda^j V \to \Lambda^j V\right).$$

We define the tube of radius r around D to be

$$\mathbb{T}_r(D) := \left\{x \in \mathbb{R}^{m+1}; \ \mathrm{dist}\,(x, D) \leq r\right\}.$$

We denote by $\mathbb{E}_M$ the exponential map

$$\mathbb{E}_M : (T\mathbb{R}^{m+1})|_M \to \mathbb{R}^{m+1},$$

$$(X,p) \longmapsto \mathbb{E}_M(X,p) = p + X, \quad p \in M, \ X \in T_p\mathbb{R}^{m+1}.$$

For $r > 0$ we denote by $\Delta_r \subset (T\mathbb{R}^{m+1})|_M$ the closed set

$$\Delta_r := \left\{(t\boldsymbol{n}(p), p); \ p \in M, \ t \in [0, r]\right\}.$$

For sufficiently small r the map $\mathbb{E}_M$ defines a diffeomorphism

$$\mathbb{E}_M : \Delta_r \to \mathbb{T}_r(D) \setminus D,$$

so that

$$\mathrm{vol}\,(\mathbb{T}_r(D)) = \mathrm{vol}\,(D) + \int_{\Delta_r} \mathbb{E}_M^* dV_{\mathbb{R}^{m+1}}.$$

Fix $p_0 \in M$ and a local, positively oriented local orthonormal frame

$$(e_1, \ldots, e_m)$$

of TM defined in a neighborhood U of p_0 in M, such that, for every $p \in U$, the collection

$$(n(p), e_1(p), \ldots, e_m(p))$$

is a positively oriented, orthonormal frame of $\mathbb{R}^n$. We obtain a dual coframe $\theta, \theta^1, \ldots, \theta^n$.

As in the previous section, the pullback of $\mathbb{E}_M^* dV_{\mathbb{R}^{m+1}}$ to Δ_r has the description

$$\mathbb{E}_M^* dV_{\mathbb{R}^{m+1}} = \det(\mathbb{1} - tS_M)\, dt \wedge d\theta^1 \wedge \cdots \wedge \theta^m$$

$$= \left(\sum_{j=1}^m \operatorname{tr}_j(-S_M)t^j \right) dt \wedge d\theta^1 \wedge \cdots \wedge \theta^m.$$

We deduce

$$\int_{\Delta_r} \mathbb{E}_M^* dV_{\mathbb{R}^{m+1}} = \sum_{j \geq 0} \frac{r^{j+1}}{j+1} \left(\int_M \operatorname{tr}_j(-S_M) dV_M \right).$$

Define

$$\mu_{m-j}(D) := \frac{1}{\sigma_j} \left(\int_M \operatorname{tr}_j(-S_M) dV_M \right), \quad 0 \leq j \leq m,$$

and

$$\mu_{m+1}(D) := \operatorname{vol}(D)$$

so that using the equality $\sigma_j = (j+1)\omega_j$ we deduce the *tube formula for domains*,

$$\operatorname{vol}\left(\mathbb{T}_r(D) \right) = \sum_{k=0}^{m+1} \omega_{m+1-k} r^{m+1-k} \mu_k(D). \tag{9.3.18}$$

Theorem 9.2.5 shows that, just as in the case of submanifolds, we have $\mu_0(D) = \chi(D)$.

Definition 9.3.19. Suppose D is a relatively compact domain with smooth boundary of an Euclidean space V, $\dim V = n$. Then *the curvature densities of D* are the densities $|d\mu_j|$ on ∂D defined by

$$|d\mu_j| := \frac{1}{\sigma_{n-j}} \operatorname{tr}_{n-j}(-S_D)|dV_{\partial D}|,$$

where $|dV_{\partial D}|$ denotes the volume density on ∂D induced by the Euclidean metric on V, and S_D denotes the co-oriented second fundamental form of ∂D. $\square$

We denote by $\mathbb{D}_r^{m+1}$ the ball of radius r in $\mathbb{R}^{m+1}$. Then,

$$\mathbb{T}_\varepsilon(\mathbb{D}_r^{m+1}) = \mathbb{D}_{r+\varepsilon}^{m+1},$$

so that

$$\omega_{m+1}(r+\varepsilon)^{m+1} = \sum_{k\geq 0} \omega_{m+1-k}\varepsilon^{m+1-k}\mu_k(\mathbb{D}_r^{m+1}).$$

We conclude

$$\mu_k(\mathbb{D}_r^{m+1}) = \frac{\omega_{m+1}}{\omega_{m+1-k}}\binom{m+1}{k}r^k = \omega_k\begin{bmatrix}m+1\\k\end{bmatrix}r^k. \qquad (9.3.19)$$

Suppose $X \hookrightarrow \mathbb{R}^{m+1}$ is a closed, compact smooth submanifold. Then for every sufficiently small $r > 0$, the tube $D_r := \mathbb{T}_r(X)$ is a compact domain with smooth boundary and

$$\mathbb{T}_\varepsilon(D_r) = \mathbb{T}_{r+\varepsilon}(X).$$

The tube formula for X implies that

$$\sum_{j\geq 0}\omega_j\varepsilon^j\mu_{m+1-j}(D_r) = \sum_{k\geq 0}\omega_k(r+\varepsilon)^k\mu_{m+1-k}(X).$$

We deduce that

$$\mu_{m+1-j}(D_r) = \frac{1}{\omega_j}\sum_{k\geq j}\omega_k\binom{k}{j}r^{k-j}\mu_{m+1-k}(X).$$

We set $n := m+1$ and we make the change in variables

$$p := n-j, \quad w := n-k.$$

Then $k - j = p - w$, and we obtain the following generalization of the tube formula

$$\mu_p\big(\mathbb{T}_r(X)\big) = \frac{1}{\omega_{n-p}}\sum_w \omega_{n-w}\binom{n-w}{n-p}r^{p-w}\mu_w(X)$$
$$= \sum_w \omega_{p-w}\begin{bmatrix}n-w\\p-w\end{bmatrix}r^{p-w}\mu_w(X). \qquad (9.3.20)$$

In particular, we have

$$\lim_{r\to 0}\mu_p\big(\mathbb{T}_r(X)\big) = \mu_p(X), \quad \forall 0 \leq p \leq \dim X. \qquad (9.3.21)$$

9.3.6 *Crofton Formulæ for domains of a Euclidean space*

Suppose D is an open, relatively compact subset of the Euclidean space $\mathbb{R}^n$ with smooth boundary $M = \partial D$. We denote by g the induced metric on M, by $\mathbf{Gr}^c$ the Grassmannian of linear subspaces of $\mathbb{R}^n$ of codimension c, and by $\mathbf{Graff}^c$ the affine Grassmannian of codimension c affine subspaces of $\mathbb{R}^n$.

Recall that on $\mathbf{Gr}^c$ we have a natural metric with volume density $|d\gamma_c|$ and total volume

$$V_c := \frac{\prod_{j=0}^{n-1} \sigma_j}{\left(\prod_{i=0}^{c-1} \sigma_i\right) \cdot \left(\prod_{j=0}^{m-1-c} \sigma_j\right)}.$$

We rescale this volume density as in (9.1.17) to obtain a new volume density $|d\nu_c|$ with total volume

$$\int_{\mathbf{Gr}^c} |d\nu_c| = \begin{bmatrix} n \\ c \end{bmatrix}. \tag{9.3.22}$$

As explained in Subsection 9.1.3, these two densities produce two invariant densities $|d\tilde{\gamma}_c|$ and $|d\tilde{\nu}_c|$ on $\mathbf{Graff}^c$ which differ by a multiplicative constant.

Theorem 9.3.20 (Crofton Formula). *Let $1 \leq p \leq n - c$, and consider the function*

$$f : \mathbf{Graff}^c \to \mathbb{R}, \quad f(L) = \mu_p(L \cap D).$$

If the function f is $|d\tilde{\nu}|$-integrable, then

$$\begin{bmatrix} p+c \\ p \end{bmatrix} \mu_{p+c}(D) = \int_{\mathbf{Graff}^c} \mu_p(L \cap D) |d\tilde{\nu}_c(L)|.$$

Proof.   For simplicity, we set $V = \mathbb{R}^n$, $m = n - 1 = \dim M$. We will carry out the proof in several steps.

Step 1. We will prove that there exists a constant $\xi_{m,c,p}$, depending only on m, c, and p such that

$$\xi_{n,c,p}\mu_{p+c}(D) = \int_{\mathbf{Graff}^c} \mu_p(L \cap D) |d\tilde{\nu}_c(L)|.$$

Step 2. We will show that the constant ξ is equal to $\begin{bmatrix} p+c \\ p \end{bmatrix}$ by explicitly computing both sides of the above equality in the special case $D = \mathbb{D}^{m+1}$.

Step 1. We will rely on a basic trick in integral geometry. For every $S \in \mathbf{Graff}^c$ we denote by $[S] \in \mathbf{Gr}^c$ the parallel translate of S containing the origin, $[S] = S - S$. We introduce the incidence relation

$$\mathcal{I} = \left\{ (v, S) \in V \times \mathbf{Graff}^c; \ v \in S \right\} \subset V \times \mathbf{Graff}^c.$$

Observe that we have a diffeomorphism

$$\mathcal{I} \to V \times \mathbf{Gr}^c, \quad \mathcal{I} \ni (v, S) \longmapsto (v, [S]) \in V \times \mathbf{Gr}^c,$$

with inverse

$$V \times \mathbf{Gr}^c(V) \ni (v, L) \longmapsto (v, v + L) \in \mathcal{I}.$$

We obtain a double fibration

$$\begin{array}{ccc} & \mathcal{I} & \\ {}^{\ell}\swarrow & & \searrow^{r} \\ V & & \mathbf{Graff}^c \end{array}$$

and we set

$$\mathcal{I}(M) := \ell^{-1}(M) = \{\, (v, S) \in V \times \mathbf{Graff}^c;\ v \in S \cap M \,\}.$$

Since $\dim \mathcal{I} = \dim V + \dim \mathbf{Gr}^c = n + c(n-c)$ we deduce

$$\dim \mathcal{I}(M) = n + c(n-c) - \operatorname{codim} M = n + c(n-c).$$

Again we have a diagram

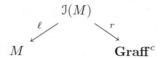

The map r need not be a submersion. Fortunately, Sard's theorem shows that r fails to be a surjection on a rather thin set.

Denote by $\mathbf{Graff}^c(M)$ the set of codimension c affine planes which intersect M transversally. The set $\mathbf{Graff}^c(M)$ is an open subset of $\mathbf{Graff}^c$, and Sard's theorem implies that its complement has *measure zero*. We set

$$\mathcal{I}(M)^* := r^{-1}\big(\mathbf{Graff}^c(M)\big).$$

The set $\mathcal{I}(M)^*$ is an open subset of $\mathcal{I}(M)$, and we obtain a double *fibration*

$$
\begin{array}{ccc}
 & \mathcal{I}(M)^* & \\
{}^{\ell}\swarrow & & \searrow^{r} \\
M & & \mathbf{Graff}^c(M)
\end{array}
\tag{9.3.23}
$$

The fiber of r over $L \in \mathbf{Graff}^c$ is the slice $M_L := L \cap M$ which is the boundary of the domain $D_L := (L \cap D) \subset L$.

The vertical bundle of the fibration $r : \mathcal{I}^*(M) \to \mathbf{Graff}^c(M)$ is equipped with a natural density given along a fiber $L \cap M$ by the curvature density $|d\mu_k|$ of the domain D_L. We will denote this density by $|d\mu_k^L|$. As explained in Subsection 9.1.1, using the pullback $r^*|d\widetilde{\gamma}_c|$ we obtain a density

$$|d\lambda| := |d\mu_p^L| \times r^*|d\widetilde{\gamma}_c|$$

on $\mathcal{I}^*(M)$ satisfying,

$$\int_{\mathcal{I}^*(M)} |d\lambda| = \int_{\mathbf{Graff}^c(M)} \left(\int_{L \cap M} |d\mu_p^L| \right) |d\widetilde{\gamma}_c(L)| = \int_{\mathbf{Graff}^c} \mu_p(L \cap D)|d\widetilde{\gamma}_c(L)|.$$

To complete Step 1 in our strategy it suffices to prove that there exists a constant ξ, depending only on m and c such that

$$\ell_* |d\lambda| = \xi |d\mu_c|,$$

where the curvature density is described in Definition 9.3.13.

Set $h := (m-c)$. The points in $\mathcal{I}(M)$ are pairs (x, L), where $x \in M$, and L is an affine plane of dimension $h + 1$. Suppose $(x_0, L_0) \in \mathcal{I}^*(M)$. Then we can parametrize a small open neighborhood of (x_0, L_0) in $\mathcal{I}^*(M)$ by a family

$$(x, e_0(S), e_1(S), \ldots, e_h(S), e_{h+1}(S), \ldots, e_m(S)),$$

where x runs in a small neighborhood of $x_0 \in M$, S runs in a small neighborhood $\mathcal{U}_0$ of $[L_0]$ in $\mathbf{Gr}^c$ so that the following hold for every S.

- The collection

$$\{e_0(S), e_1(S), \dots, e_h(S), e_{h+1}(S), \dots, e_m(S)\}$$

is an orthonormal frame of $\mathbb{R}^n$.
- The collection $\{e_0, e_1, \dots, e_h\}$ is a basis of S.
- The collection span $\{e_1, \dots, e_h\}$ is a basis of $T_{x_0} M \cap S$.
- The above condition imply that e_0 is a normal vector to the boundary of the domain $D \cap L \subset L$. We require that e_0 is the *outer* normal.

A neighborhood of (x_0, L_0) in $\mathcal{I}$ is parametrized by the family

$$\big(\vec{r}, e_0(S), e_1(S), \dots, e_h(S), e_{h+1}(S), \dots, e_m(S) \big),$$

where $\vec{r}$ runs in a neighborhood of x_0 in the ambient space V.

We denote by S_D, the co-oriented second fundamental form of D, by S_L the co-oriented second fundamental form of $D_L \subset L$, and by $|dV_{L \cap M}|$ the metric volume density on $L \cap M$. Then, if we set $k = \dim L - p = m - c - p$, we deduce

$$|d\mu_p^L| = \frac{1}{\sigma_k} \operatorname{tr}_k(-S_L)|dV_{L \cap M}|.$$

In the sequel we will use the following conventions.

- i, j, k denote indices running in the set $\{0, \dots, h\}$.
- α, β, γ denote indices running in the set $\{h + 1, \dots, m\}$.
- A, B, C denote indices running in the set $\{0, 1, \dots, m\}$.

We denote by (θ^A) the dual coframe of (e_A), and set

$$\theta_{AB} := e_A \bullet (\boldsymbol{D} e_B).$$

Then, the volume density of the natural metric on $\mathbf{Gr}^c$ is

$$|d\gamma_c| = \left| \bigwedge_{\alpha, i} \theta_{\alpha i} \right|.$$

Then

$$|d\widetilde{\gamma}_c| = \left| \bigwedge_\alpha \boldsymbol{D} \vec{r} \bullet e_\alpha \right| \times |d\gamma_c| = \left| \bigwedge_\alpha \theta^\alpha \right| \times |d\gamma_c|,$$

and

$$|d\lambda| = |d\mu_p^L| \times |d\widetilde{\gamma}_c| = \frac{1}{\sigma_k} \det(-S_{L \cap M})|dV_{L \cap M}| \times \left| \bigwedge_\alpha \theta^\alpha \right| \times |d\gamma_c|. \tag{9.3.24}$$

The fiber of $\ell : \mathcal{I}(M) \to M$ over x_0 is described by

$$G_{x_0} := \{ (\vec{r}, e_A(S)) \in \mathcal{I}(M), \ \vec{r} = x_0 \}.$$

We set

$$G_{x_0}^* := G_{x_0} \cap \mathcal{I}^*(M).$$

$G^*_{x_0}(M)$ can be identified with the space of linear subspaces S of codimension c such that $T_{x_0}M + S = V$, i.e., the affine subspace $x_0 + S$ intersects M transversally at x_0.

Denote by $\boldsymbol{n}$ a smooth unit normal vector field defined in a neighborhood of x_0 in M, i.e.,

$$\boldsymbol{n}(x) \perp T_x M, \quad |\boldsymbol{n}(x)| = 1.$$

Lemma 9.3.21. *Suppose $x_0 + S$ intersects M transversally at x_0. We set $\boldsymbol{e}_A := \boldsymbol{e}_A(S)$. Then at the point $x_0 \in M$ we have*

$$|(\boldsymbol{n} \bullet \boldsymbol{e}_0)| \cdot |dV_M| = |\theta^1 \wedge \cdots \wedge \theta^m|,$$

i.e., for any $X_1, \ldots, X_m \in T_{x_0}M$ we have

$$|(\boldsymbol{n} \bullet \boldsymbol{e}_0)| \cdot |dV_M|(X_1, \ldots, X_m) = |\theta^1 \wedge \cdots \wedge \theta^m|(X_1, \ldots, X_m).$$

Proof. It suffices to verify this for one basis $X_1, \ldots, X_m$ of $T_{x_0}M$ which we can choose to consists of the orthogonal projections $\boldsymbol{f}_1, \ldots, \boldsymbol{f}_m$ of $\boldsymbol{e}_1, \ldots, \boldsymbol{e}_m$. These projections form a basis since S intersects $T_{x_0}M$ transversally.

Observe that

$$\boldsymbol{f}_i = \boldsymbol{e}_i, \quad \forall 1 \le i \le h, \quad \boldsymbol{f}_\alpha = \boldsymbol{e}_\alpha - (\boldsymbol{e}_\alpha \bullet \boldsymbol{n})\boldsymbol{n}.$$

Then

$$|dV_M|(\boldsymbol{f}_1, \ldots, \boldsymbol{f}_m)^2 = \det(\boldsymbol{f}_A \bullet \boldsymbol{f}_B)_{1 \le A, B \le m}$$

We observe that

$$\boldsymbol{f}_i \bullet \boldsymbol{f}_j = \delta_{ij}, \quad \boldsymbol{f}_i \bullet \boldsymbol{f}_\alpha = 0, \quad \forall 1 \le i, j \le 2h < \alpha$$

$$\boldsymbol{f}_\alpha \bullet \boldsymbol{f}_\beta = \delta_{\alpha\beta} - n_\alpha n\beta, \quad n_\alpha := \boldsymbol{n} \bullet \boldsymbol{e}_\alpha.$$

We deduce

$$|dV_M|(\boldsymbol{f}_1, \ldots, \boldsymbol{f}_m)^2 = \det(\mathbb{1} - A),$$

where A denotes the $c \times c$ symmetric matrix with entries $n_\alpha n_\beta$, $h + 1 \le \alpha, \beta \le m = h + c$. If we denote by $\vec{u}$ the vector

$$\vec{u} = \begin{bmatrix} n_{h+1} \\ \vdots \\ n_{h+c} \end{bmatrix} \in \mathbb{R}^c,$$

which we also regard as a $c \times 1$ matrix, then we deduce that $A = \vec{u}\vec{u}^\dagger$.

This matrix has a $c - 1$ dimensional kernel corresponding to vectors orthogonal to $\vec{u}$. The vector $\vec{u}$ itself is an eigenvector of A, and the corresponding eigenvalue λ is obtained from the equality

$$\lambda \vec{u} = |\vec{u}|^2 \vec{u} \implies \lambda = |\vec{u}|^2 = \sum_\alpha n_\alpha^2 = |\boldsymbol{n}|^2 - |\boldsymbol{n} \bullet \boldsymbol{e}_0|^2 = 1 - |\boldsymbol{n} \bullet \boldsymbol{e}_0|^2.$$

We conclude that

$$\det(\mathbb{1} - A) = |\boldsymbol{n} \bullet \boldsymbol{e}_0|^2 \implies |dV_M|(\boldsymbol{f}_1, \ldots, \boldsymbol{f}_m) = |\boldsymbol{n} \bullet \boldsymbol{e}_0|.$$

On the other hand,

$$|\theta^1 \wedge \cdots \wedge \theta^m|(\boldsymbol{f}_1, \ldots, \boldsymbol{f}_m) = |\det(\boldsymbol{e}_A \bullet \boldsymbol{f}_B)_{1 \leq A, B \leq m}|.$$

We have again

$$\boldsymbol{e}_i \bullet \boldsymbol{f}_j = \delta_{ij}, \quad \boldsymbol{e}_i \bullet \boldsymbol{f}_\alpha = 0, \quad \forall 1 \leq i, j \leq h < \alpha,$$

$$\boldsymbol{e}_\alpha \bullet \boldsymbol{f}_\beta = \delta_{\alpha\beta} - n_\alpha n_\beta,$$

so that

$$|\theta^1 \wedge \cdots \wedge \theta^m|(\boldsymbol{f}_1, \ldots, \boldsymbol{f}_m) = |\boldsymbol{n} \bullet \boldsymbol{e}_0|^2.$$

The lemma is now proved. $\square$

Lemma 9.3.22 (Euler-Meusnier). *Suppose* $L \in \mathbf{Graff}^c$ *intersects* M *transversally, and* $x_0 \in L \cap M$. *If* $\boldsymbol{n}$ *is a unit vector perpendicular to* $T_{x_0}M$, *then*

$$S_L = (\boldsymbol{n} \bullet \boldsymbol{e}_0) S_D|_{T_{x_0} \cap [L]},$$

that is,

$$S_L(\boldsymbol{e}_i, \boldsymbol{e}_j) = (\boldsymbol{n} \bullet \boldsymbol{e}_0) S_D(\boldsymbol{e}_i, \boldsymbol{e}_j), \quad \forall 1 \leq i, j \leq 2h.$$

Proof.   We have

$$S_L(\boldsymbol{e}_i, \boldsymbol{e}_j) = \boldsymbol{e}_0 \bullet (\boldsymbol{D}_{\boldsymbol{e}_i} \boldsymbol{e}_j).$$

Let us now observe that the vector $(\boldsymbol{D}_{\boldsymbol{e}_i} \boldsymbol{e}_j)$ is parallel with the plane L because the vectors $\boldsymbol{e}_i$ and $\boldsymbol{e}_j$ lie in this plane. Thus, $\boldsymbol{D}_{\boldsymbol{e}_i} \boldsymbol{e}_j$ decomposes into two components, one component parallel to $\boldsymbol{e}_0$, and another component, $(D e_{\boldsymbol{e}_i} \boldsymbol{e}_j)^\tau$, tangent to $L \cap M$. Hence

$$\boldsymbol{D}_{\boldsymbol{e}_i} \boldsymbol{e}_j = S_L(\boldsymbol{e}_i, \boldsymbol{e}_j) \boldsymbol{e}_0 + \sum_{k=1}^h S_{ij}^k \boldsymbol{e}_k.$$

Taking the inner product with $\boldsymbol{n}$ we deduce

$$S_D(\boldsymbol{e}_i, \boldsymbol{e}_j) = (\boldsymbol{D}_{\boldsymbol{e}_i} \boldsymbol{e}_j) \bullet \boldsymbol{n} = S_L(\boldsymbol{e}_i, \boldsymbol{e}_j)(\boldsymbol{e}_0 \bullet \boldsymbol{n}).$$ $\square$

From the above lemma we deduce

$$\mathrm{tr}_k(-S_L|_{x_0}) = |\boldsymbol{n} \bullet \boldsymbol{e}_0|^k \mathrm{tr}_k(-S_D|_{T_{x_0} M \cap [L]}),$$

for any $(x_0, L) \in \mathcal{I}^*(M)$. In a neighborhood of $(x_0, L_0) \in \mathcal{I}^*(M)$ we have

$$|d\lambda|(x, L) = \frac{1}{\sigma_k} \mathrm{tr}_k(-S_L)|dV_{L \cap M}| \times \left| \bigwedge_\alpha \theta^\alpha \right| \times |d\gamma_c|$$

$$= \frac{1}{\sigma_k} |\boldsymbol{n} \bullet \boldsymbol{e}_0|^k \big(\operatorname{tr}_k(-S_D|_{T_x M \cap [L]}) \big) \Big| \bigwedge_{A=1}^m \theta^A \Big| \times |d\gamma_c|$$

(use Lemma 9.3.21)

$$= \frac{1}{\sigma_k} |\boldsymbol{n} \bullet \boldsymbol{e}_0|^{k+1} \big(\operatorname{tr}_k(-S_D|_{T_x M \cap [L]}) \big) |dV_M| \times |d\gamma_c|.$$

This proves that along the fiber $G_{x_0}^*$ we have

$$|d\lambda|/|dV_M| = \frac{1}{\sigma_k} |\boldsymbol{n} \bullet \boldsymbol{e}_0|^{k+1} \big(\operatorname{tr}_k(-S_D|_{T_{x_0} M \cap [L]}) \big) |d\gamma_c|([L]).$$

If we denote by $\theta([L], T_{x_0} M)$ the angle between $[L]$ and the hyperplane $T_{x_0} M$ we deduce

$$|d\lambda|/|dV_M| = \frac{1}{\sigma_k} \cdot |\cos\theta([L], T_{x_0} M)|^{k+1} \big(\operatorname{tr}_k(-S_D|_{T_{x_0} M \cap [L]}) \big) |d\gamma_c|([L]).$$

The map $G_{x_0}^* \ni (x_0, L) \longmapsto [L] \in \mathbf{Gr}^c$ identifies $G_{x_0}^*$ with an open subset of $\mathbf{Gr}^c$ whose complement has measure zero. We now have the following result.

Lemma 9.3.23. *Suppose that V is an Euclidean space, $\dim V = m + 1$, $H \subset V$ is a hyperplane through the origin, and $B : H \times H \to \mathbb{R}$ a symmetric bilinear map. Denote by $O(H)$ the subgroup of orthogonal transformations of V which map H to itself and suppose that*

$$f : \mathbf{Gr}_c \to \mathbb{R}$$

is an $O(H)$-invariant function. Define

$$\mathbf{Gr}_H^c := \big\{ S \in \mathbf{Gr}^c;\ S \text{ intersects } H \text{ transversally} \big\}.$$

Then, for every $0 \le k \le m - c$, there exists a constant $\xi = \xi_{m,c,k}$, depending only on m, c and k, such that

$$I_k(f, B) := \int_{\mathbf{Gr}_H^c} f(S) \operatorname{tr}_k(B|_{H \cap S}) |d\gamma_c|(S) = \xi_{m,c,k} \operatorname{tr}_k(B) \int_{\mathbf{Gr}^c} f |d\gamma_c|(S).$$

Proof. Observe that, for fixed f, the map $B \mapsto I_k(f, B)$ is an $O(H)$-invariant homogeneous polynomial of degree k in the entries of B. We can therefore express it as a polynomial

$$I_k(f, B) = P_f\big(\operatorname{tr}_1(B), \dots, \operatorname{tr}_k(B) \big)$$
$$= \xi_f \operatorname{tr}_k(B) + Q_f\big(\operatorname{tr}_1(B), \dots, \operatorname{tr}_{k-1}(B) \big).$$

Let us prove that $Q_f \equiv 0$. To do this, we apply the above formula to a symmetric bilinear form B such that

$$\dim \ker B > m - k.$$

Thus, at least $m - k + 1$ of the m eigenvalues of B vanish, so that $\operatorname{tr}_k(B) = 0$. For such forms we have

$$I_k(f, B) = Q\big(\operatorname{tr}_1(B), \dots, \operatorname{tr}_{k-1}(B) \big).$$

On the other hand, for almost all $S \in \mathbf{Gr}_H^c$ we have

$$\dim S \cap \ker B > m - c - k.$$

The restriction of B to $S \cap H$ has $m - c$ eigenvalues, and from the above inequality we deduce that at least $m - c - k$ of them are trivial. Hence

$$I_k(f, B) = 0, \quad \forall B, \quad \dim \ker B > m - k \Longrightarrow Q_f = 0.$$

Now choose B to be the bilinear form corresponding to the inner product on H. Then

$$\mathrm{tr}_k(B) = \binom{m}{k} \quad \text{and} \quad \mathrm{tr}_k(B|_{H \cap S}) = \binom{m - c}{k}, \quad \forall S \in \mathbf{Gr}_H^c,$$

and we conclude that

$$\binom{m - c}{k} \int_{\mathbf{Gr}^c} f |d\gamma_c|(S) = \xi_f \binom{m}{k}. \qquad \square$$

Now apply the above lemma in the special case

$$H = T_{x_0} M, \quad B = -S_D, \quad f(S) = \frac{1}{\sigma_k} |\cos \theta(S, H)|^{k+1}$$

to conclude that

$$\ell_* |d\lambda| = \xi \, \mathrm{tr}_k(-S_D) |dV_M| = \xi |d\mu_{p+c}|$$

so that

$$\xi \mu_{p+c}(D) = \int_M \ell_* |d\lambda| = \int_{\mathcal{J}^*(M)} |d\lambda|$$

$$= \int_{\mathbf{Graff}^c(M)} r_* |d\lambda| = \int_{\mathbf{Graff}^c(M)} \mu_p(L \cap D) |d\tilde{\gamma}_c|(L).$$

Thus, rescaling $|d\tilde{\gamma}_c|$ to $|d\tilde{\nu}_c|$, we deduce that there exists a constant ξ depending only on m and c such that

$$\xi \mu_{p+c}(D) = \int_{\mathbf{Graff}^c(M)} \mu_p(L \cap D) |d\tilde{\nu}_c|(L).$$

Step 2. To determine the constant ξ in the above equality we apply it in the special case $D = \mathbb{D}^{m+1}$. Using (9.3.19) we deduce

$$\xi \mu_{p+c}(\mathbb{D}^{m+1}) = \xi \omega_{p+c} \begin{bmatrix} m+1 \\ p+c \end{bmatrix} = \xi \frac{\omega_{m+1}}{\omega_{m+1-c-p}} \binom{m+1}{p+c}.$$

Now observe that for $L \in \mathbf{Graff}^c$ we set $r = r(L) = \mathrm{dist}\,(L, 0)$. Then $L \cap \mathbb{D}^{m+1}$ is empty if $r > 1$, and it is a disk of dimension $(m + 1 - c) = \dim L$ and radius $(1 - r^2)^{1/2}$ if $r < 1$. We conclude that

$$\mu_p(L \cap \mathbb{D}^{m+1}) = \mu_p(\mathbb{D}^{m+1-c}) \times \begin{cases} (1 - r^2)^{p/2} & r < 1 \\ 0 & p > 1. \end{cases}$$

We set

$$\mu_{m,c,p} := \mu_p(\mathbb{D}^{m+1-c}) = \omega_p \begin{bmatrix} m+1-c \\ p \end{bmatrix} = \frac{\omega_{m+1-c}}{\omega_{m+1-c-p}} \binom{m+1-c}{p}.$$

Using Theorem 9.1.14 we deduce

$$\int_{\mathbf{Graff}^c} \mu_p(L \cap \mathbb{D}^{m+1}) |d\tilde{\nu}_c|(L)$$

$$= \int_{\mathbf{Gr}^c} \left(\int_{[L]^\perp} \mu_p \big(\mathbb{D}^{m+1} \cap (x + [L]) \big) |dV_{[L]^\perp}|(x) \right) |d\nu_c|([L])$$

$$= \mu_{m,c,p} \int_{\mathbf{Gr}^c} \underbrace{\left(\int_{x \in [L]^\perp, \ |x| < 1} (1 - |x|^2)^{p/2} |dV_{[L]^\perp}|(x) \right)}_{=:I_{c,p}} |d\nu_c|([L])$$

$$= \mu_{m,c,p} I_{c,p} \int_{\mathbf{Gr}^c} |d\nu_c(S)| \stackrel{(9.3.22)}{=} \mu_{m,c,p} I_{c,p} \begin{bmatrix} m+1 \\ c \end{bmatrix}.$$

Hence

$$\xi \frac{\omega_{m+1}}{\omega_{m+1-c-p}} \binom{m+1}{p+c} = \frac{\omega_{m+1-c}}{\omega_{m+1-c-p}} \binom{m+1-c}{p} I_{c,p} \begin{bmatrix} m+1 \\ c \end{bmatrix}.$$

Using spherical coordinates on $\mathbb{R}^c$ we deduce

$$I_{c,p} = \int_{\mathbb{R}^c} (1 - |x|^2)^{p/2} dV_{\mathbb{R}^c} = \sigma_{c-1} \int_0^1 r^{c-1} (1 - r^2)^{p/2} dr$$

$$\stackrel{s=r^2}{=} \frac{\sigma_{c-1}}{2} \int_0^1 s^{\frac{c-2}{2}} (1 - s)^{p/2} ds$$

$$\stackrel{(9.1.6)}{=} \frac{\sigma_{c-1}}{2} B\left(\frac{c}{2}, \frac{p}{2} + 1\right) \stackrel{(9.1.7)}{=} \frac{\sigma_{c-1}}{2} \frac{\Gamma(\frac{c}{2})\Gamma(1 + \frac{p}{2})}{\Gamma(1 + \frac{c}{2} + \frac{p}{2})}$$

$$\stackrel{(9.1.8)}{=} \Gamma(1/2)^c \frac{\Gamma(1 + \frac{p}{2})}{\Gamma(1 + \frac{c}{2} + \frac{p}{2})} = \frac{\omega_{p+c}}{\omega_p}.$$

Hence

$$\xi \omega_{m+1} \binom{m+1}{p+c} = \frac{\omega_{m+1-c} \omega_{p+c}}{\omega_p} \begin{bmatrix} m+1 \\ c \end{bmatrix} \binom{m+1-c}{p}$$

$$= \frac{\omega_{m+1} \omega_{p+c}}{\omega_p \omega_c} \binom{m+1}{c} \binom{m+1-c}{p}.$$

We deduce

$$\xi = \frac{\omega_{p+c}}{\omega_p \omega_c} \frac{\binom{m+1}{c}}{\binom{m+1}{p+c}\binom{m+1-c}{p}} = \frac{\omega_{p+c}}{\omega_p \omega_c} \binom{p+c}{p} = \begin{bmatrix} p+c \\ p \end{bmatrix}.$$

$\square$

We now describe a simple situation when the function $\mathbf{Graff}^c \ni L \mapsto \mu_0(L \cap M)$ is integrable.

Proposition 9.3.24. *If the bounded domain $D \subset \mathbb{R}^n$ with smooth boundary is tame, then the function*

$$\mathbf{Graff}^c \ni L \mapsto \chi(L \cap D) = \mu_0(L \cap D)$$

is bounded and has compact support. In particular, it is integrable and

$$\mu_c(D) = \int_{\mathbf{Graff}^c} \chi(L \cap D)\, |d\tilde{\nu}|(L).$$

Proof. We know that there exists a tame structure $\mathcal{S}$ such that $D \in \mathcal{S}^n$. In particular, $M = \partial D \in \mathcal{S}^n$. Since the Grassmannian $\mathbf{Graff}^c$ is semialgebraic, we deduce that $\mathbf{Graff}^c$ is also $\mathcal{S}$-definable. Hence, the incidence set

$$\mathcal{I} = \big\{ ((x, L) \in M \times \mathbf{Graff}^c;\; x \in L \big\}$$

is also $\mathcal{S}$-definable, and so is the map

$$\pi : \mathcal{I} \to \mathbf{Graff}^c, \quad (x, L) \mapsto L.$$

The fiber of π over L is the intersection $L \cap M$. From the definability of Euler characteristic we deduce that the map

$$\mathbf{Graff}^c \ni L \mapsto \chi(L \cap M) \in \mathbb{Z}$$

is definable. Its range is a definable subset of $\mathbb{Z}$, i.e., a finite set. To see that it has compact support it suffices to observe that since M is compact, an affine plane which is too far from the origin cannot intersect M. □

Definition 9.3.25. For any compact tame subset $S \subset \mathbb{R}^n$, and for every integer $0 \leq p \leq n$ we define

$$\hat{\boldsymbol{\mu}}_p(S) := \int_{\mathbf{Graff}^p(\mathbb{R}^n)} \chi(L \cap D)\, |d\tilde{\nu}|(L). \qquad\qquad □$$

The proof of Proposition 9.3.24 shows that the integral in the above definition is well defined and finite. This proposition also shows that

$$\hat{\boldsymbol{\mu}}_p(D) = \mu_p(D)$$

if $D \subset \mathbb{R}^n$ is a tame domain with smooth boundary.

The above definition has a "problem": a priori, the quantity $\hat{\boldsymbol{\mu}}_p(S)$ depends on the dimension of the ambient space $\mathbb{R}^n \supset S$. The next exercise asks the reader to prove that this is not the case.

Exercise 9.3.26. Fix tame structure $\mathcal{S}$, and suppose $S \subset \mathbb{R}^n$ is a compact, $\mathcal{S}$-definable set. Let $N > n$ and suppose $i : \mathbb{R}^n \to \mathbb{R}^N$ is an *affine, isometric* embedding.

(a) Prove that

$$\int_{\mathbf{Graff}^p(\mathbb{R}^n)} \chi(L \cap S)\, |d\tilde{\nu}|(L) = \int_{\mathbf{Graff}^p(\mathbb{R}^N)} \chi(U \cap i(S))\, |d\tilde{\nu}|(U),$$

so that $\hat{\boldsymbol{\mu}}_p(S)$ is *independent* of the dimension of the ambient space.

(b) Prove that for every real number λ we have

$$\hat{\boldsymbol{\mu}}_p(\lambda S) = |\lambda|^p \hat{\boldsymbol{\mu}}_p(S).$$

(c) Suppose $S_1, S_2 \subset \mathbb{R}^n$ are two compact $\mathcal{S}$-definable sets. Prove that

$$\hat{\boldsymbol{\mu}}_p(S_1 \cup S_2) = \hat{\boldsymbol{\mu}}_p(S_1) + \hat{\boldsymbol{\mu}}_2(S_2) - \hat{\boldsymbol{\mu}}_p(S_1 \cap S_2).$$

(d) Suppose $T \subset \mathbb{R}^n$ is a triangle, i.e., the convex hull of three non collinear points. Compute $\hat{\boldsymbol{\mu}}_p(T)$, $\forall p \geq 0$. $\qquad\square$

Proposition 9.3.27. *Fix a tame structure $\mathcal{S}$. Suppose $K \subset \mathbb{R}^n$ is a compact, $\mathcal{S}$-definable set, Λ is an arbitrary $\mathcal{S}$-definable set, and $\mathcal{A} \subset K \times \Lambda$ is a closed $\mathcal{S}$-definable subset of $K \times \Lambda$. For every $\lambda \in \Lambda$ we set*

$$A_\lambda := \{ a \in K; \; (a, \lambda) \in \mathcal{A} \}.$$

Then the function $\Lambda \ni \lambda \longmapsto f_p(\lambda) := \hat{\boldsymbol{\mu}}_p(A_\lambda) \in \mathbb{R}$ is bounded and Borel measurable.

Proof. Denote by $\mathbf{Graff}^p(K)$ the set of affine planes of codimension p in $\mathbb{R}^n$ which intersect K. This is a compact, definable subset of $\mathbf{Graff}^p$. Define

$$F : \Lambda \times \mathbf{Graff}^p(K) \to \mathbb{Z}, \quad (\lambda, L) \mapsto \chi(A_\lambda \cap L) \in \mathbb{Z}.$$

Since the family $\{A_\lambda \cap L\}_{\lambda, L}$ is definable, we deduce that the function F is definable. Hence, its range must be a *finite* (!?!) subset of $\mathbb{Z}$. In particular, this also shows that F is measurable and bounded.

Now observe that

$$f_p(\lambda) = \int_{\mathbf{Graff}^p(K)} F(\lambda, L)\, |d\tilde{\nu}|(L).$$

The measurability follows from the classical Fubini theorem. The boundedness is obvious. $\qquad\square$

The above result has the following immediate corollary.

Corollary 9.3.28. *Suppose the bounded domain $D \subset \mathbb{R}^n$ is tame and has smooth boundary. Denote by $\mathbf{Graff}^c_D$ the set of affine planes of codimension c which intersect ∂D transversally. Then for any integer, $0 \leq p \leq n$ the function*

$$\mathbf{Graff}^c_D \ni L \mapsto \mu_p(L \cap D)$$

is bounded, measurable, and has compact support. In particular, it is integrable, and

$$\begin{bmatrix} p+c \\ c \end{bmatrix} \mu_{p+c}(D) = \int_{\mathbf{Graff}^c_D} \mu_p(L \cap D)\, |d\tilde{\nu}(L)| = \int_{\mathbf{Graff}^c} \hat{\boldsymbol{\mu}}_p(L \cap D)|d\tilde{\nu}|(L).$$

Proof.   The set $\mathbf{Graff}^c_D$ is a tame open and dense subset of $\mathbf{Graff}^c$ so that its complement has measure zero. Moreover

$$\mu_p(L \cap D) = \hat{\boldsymbol{\mu}}_p(L \cap D), \;\; \forall L \in \mathbf{Graff}^c_D,$$

while, by Proposition 9.3.27, the function $L \mapsto \hat{\boldsymbol{\mu}}_p(L \cap D)$ is bounded and measurable. This function has compact support because the planes too far from the origin cannot intersect D.   □

9.3.7   *Crofton formulæ for submanifolds of a Euclidean space*

In this last subsection we fix a tame structure $\mathcal{S}$, and we assume that M is a closed, compact, smooth, $\mathcal{S}$-definable submanifold of dimension m of the Euclidean space $\mathbb{R}^n$. We continue to denote by $\mathbf{Graff}^c$ the Grassmannian of affine planes in $\mathbb{R}^n$ of codimension c. We want to prove the following result.

Theorem 9.3.29 (Crofton Formula. Part 1). *Denote by $\mathbf{Graff}^c(M)$ the subset of $\mathbf{Graff}^c$ consisting of affine planes which intersect M transversally. Then*

$$\mu_c(M) = \int_{\mathbf{Graff}^c(M)} \mu_0(L \cap M)|d\tilde{\nu}|(L) = \int_{\mathbf{Graff}^c} \mu_0(L \cap M)|d\tilde{\nu}(L)|,$$

that is,

$$\hat{\boldsymbol{\mu}}_c(M) = \mu_c(M).$$

Proof.   We can assume that $k = \operatorname{codim} M < 1$. For every $x \in \mathbb{R}^n$ set $d(x) := \operatorname{dist}(x, M)$. Fix $R > 0$ such that for any x such that $d(x) \leq R$ there exists a unique point $\bar{x} \in M$ such that

$$|x - \bar{x}| = d(x).$$

For every $r < R$ consider the tube of radius r, around M,

$$D_r := \mathbb{T}_r(M),$$

and set $M_r = \partial \mathbb{T}_r(M)$. For uniformity, we set $D_0 = M$. From (9.3.15) we deduce

$$\mu_c(M) = \lim_{r \to 0} \mu_c(D_r).$$

The tube D_r is a tame domain with smooth boundary, and Theorem 9.3.20 implies

$$\mu_c(D_r) = \int_{\mathbf{Graff}^c} \chi(L \cap D_r)|d\nu|(L).$$

Thus it suffices to show that

$$\lim_{r \to 0} \int_{\mathbf{Graff}^c} \chi(L \cap D_r)|d\nu|(L) = \int_{\mathbf{Graff}^c} \chi(L \cap M).$$

For $r \in [0, R)$ we define

$$f_r : \mathbf{Graff}^c \to \mathbb{R}, \;\; f_r(L) = \chi(L \cap D_r).$$

Lemma 9.3.30. *There exists $C > 0$ such that*

$$|f_r(L)| \leq C, \;\; \forall L \in \mathbf{Graff}^c, \;\; r \in [0, R).$$

Proof. This follows from Proposition 9.3.27. $\qquad\square$

Let
$$\mathbf{Graff}^c(M_r) := \big\{ L \subset \mathbf{Graff}^c;\ \ L \text{ intersects } M_r \text{ transversally}\big\}.$$
Observe that $\mathbf{Graff}^c(M)$ is an open subset of $\mathbf{Graff}^c$ with negligible complement. For every $r > 0$ we set
$$\mathcal{X}_r = \big\{ L \in \mathbf{Graff}^c(M);\ \ L \in \mathbf{Graff}^c(M_s);\ \ \chi(L \cap D_s) = \chi(L \cap M),\ \ \forall s \in (0, r]\big\}.$$
Observe that
$$\mathbf{Graff}^c(M, r_1) \subset \mathbf{Graff}^c(M, r_0),\ \ \forall r_1 \geq r_0.$$
To proceed further we need the following technical result, whose proof will presented at the end of this subsection.

Lemma 9.3.31. *The sets $\mathcal{X}_r$ are measurable in $\mathbf{Graff}^c$ and*
$$\bigcup_{r>0} \mathcal{X}_r = \mathbf{Graff}^c(M).$$

Set
$$\mathbf{Graff}^c_* := \big\{ L \in \mathbf{Graff}^c;\ \ L \cap D_R \neq \emptyset\big\}.$$
The region $\mathbf{Graff}^c_*(M)$ is a relatively compact subset of $\mathbf{Graff}^c$, and thus it has finite measure. Define
$$\mathcal{X}^*_r := \mathcal{X}_r \cap \mathbf{Graff}^c_*,\ \ \mathcal{Y}^*_r := \mathbf{Graff}^c_* \setminus \mathcal{X}^*_r.$$
For $0 < r < R$ we have
$$\mu_c(D_r) = \int_{\mathbf{Graff}^c} f_r(L)|d\tilde{\nu}_c|(L) = \int_{\mathbf{Gr}^c_*} f_r(L)|d\tilde{\nu}_c|(L)$$
$$= \int_{\mathcal{X}^*_r} f_r(L)|d\tilde{\nu}_c| + \int_{\mathcal{Y}^*_r} f_r(L)|d\tilde{\nu}_c| = 2\int_{\mathcal{X}^*_r} f_0(L)|d\tilde{\nu}_c| + \int_{\mathcal{Y}^*_r} f_r(L)|d\tilde{\nu}_c|$$
Hence
$$\left| \mu_c(M_r) - \int_{\mathcal{X}^*_r} f_0(L)|d\tilde{\nu}_c| \right| \leq \int_{\mathcal{Y}^*_r} |f_r(L)||d\tilde{\nu}_c| \leq C\mathrm{vol}\,(\mathcal{Y}^*_r).$$
We now let $r \to 0$, and since $\mathrm{vol}\,(\mathcal{Y}^*_r) \to 0$ we conclude that
$$\mu_c(M) = \lim_{r\to 0} \mu_c(D_r) = \lim_{r\to 0} \int_{\mathcal{X}^*_r} f_0(L)|d\tilde{\nu}_c| = \int_{\mathbf{Graff}^c} f_0(L)|d\tilde{\nu}_c|.$$
This concludes the proof of Theorem 9.3.29. $\qquad\square$

Proof of Lemma 9.3.31. We will prove that for any given $L_0 \in \mathbf{Graff}^c(M)$ there exists a $\rho = \rho(L_0)$ such that
$$L_0 \in \mathcal{X}_\rho.$$

The measurability follows from the fact that $\mathfrak{X}_r$ is described using countably many boolean operations on measurable sets.

Consider the normal bundle

$$N := (TM)^{\perp} \to M.$$

For x in M we denote by N_x the fiber of N over x.

Let $y \in L_0 \cap M$. We denote by N_y^0 the orthogonal complement in L_0 of $T_y(L_0 \cap M)$,

$$N_y^0 = L_0 \cap \left(T_y(L_0 \cap M) \right)^{\perp}.$$

We think of N_y^0 as an affine subspace of $\mathbb{R}^n$ containing y. Because L_0 intersects M transversally we have

$$\dim N_y = \dim N_y^0 = k = \operatorname{codim} M.$$

For every $r > 0$ we set $N_y^0(r) := N_y^0 \cap D_r$; see Figure 9.2.

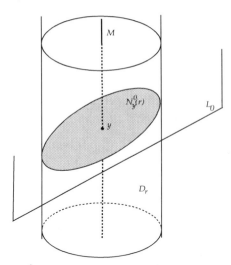

Fig. 9.2   *Slicing the tube D_r around the submanifold M by a plane L_0.*

The collection $(N_y^0)_{y \in L_0 \cap M}$ forms a vector subbundle $N^0 \to L_0 \cap M$ of $(T\mathbb{R}^n)|_{L_0 \cap M}$. The exponential map on $T\mathbb{R}^n$ restricts to a map

$$\mathbb{E}_{L_0 \cap M} : N^0 \to L_0.$$

Denote by δ the pullback to N^0 of the distance function $x \mapsto d(x) = \operatorname{dist}(x, M)$,

$$\delta_{L_0} = d \circ \mathbb{E}_{L_0 \cap M} : N^0 \to \mathbb{R}.$$

For every $y \in L_0 \cap M$, the restriction to N_y^0 of the Hessian of d at y is positive definite. Hence, there exists $\rho = \rho(L_0) > 0$ such that the map

$$\mathbb{E}_{L_0 \cap M} : \left\{ x \in N^0; \ \delta(x) \leq \rho \right\} \to L_0 \cap D_\rho$$

is a diffeomorphism. We deduce that we have a natural projection

$$\pi : L_0 \cap D_\rho \to L_0 \cap M,$$

which is continuous and defines a locally trivial fibration with fibers $N_y^0(\rho)$.

For every $y \in L_0 \cap M$, the fiber $N_y^0(\rho)$ is homeomorphic to a disk of dimension k. Thus, $L_0 \cap D_\rho$ is homeomorphic to a tube in L_0 around $L_0 \cap M \subset L_0$, so that

$$\chi(L_0 \cap D_\rho) = \chi(L_0 \cap M).$$

The downward gradient flow of the restriction to $L_0 \cap D_\rho$ of the distance function $d(x)$ produces diffeomorphisms of manifolds with boundary

$$L_0 \cap D_\rho \cong L_0 \cap D_r, \quad \forall r \in (0, \rho).$$

Hence

$$\chi(L_0 \cap D_r) = \chi(L_0 \cap D_\rho) = \chi(L_0 \cap M), \quad \forall r \in (0, \rho].$$

Since the restriction to $L_0 \cap D_\rho$ of the distance function $d(x)$ has no critical points other than the minima $y \in L_0 \cap M$, we deduce that L_0 is transversal to the level sets

$$\{ d(x) = r \} = M_r, \quad \forall r \in (0, \rho].$$

This proves $L_0 \in \mathfrak{X}_\rho$. $\qquad\qquad\qquad\qquad\qquad\qquad\qquad\qquad\qquad\qquad\qquad\qquad$ □

Corollary 9.3.32. *Suppose $C \subset \mathbb{R}^2$ is a smooth, closed, compact tame curve. For every line $L \in \mathbf{Gr}_1(\mathbb{R}^2) = \mathbf{Gr}^1(\mathbb{R}^2)$ we set*

$$n_C(L) := \#(L \cap C).$$

Then the function $L \mapsto n_C(L)$ belongs to $L^\infty(\mathbf{Gr}_1(\mathbb{R}^2), |d\widetilde{\nu}|_{2,1})$, has compact support and

$$\mathrm{length}\,(C) = \int_{\mathbf{Gr}_1(\mathbb{R}^2)} n_C(L)|d\widetilde{\nu}_{2,1}|(L). \qquad\qquad\qquad □$$

Remark 9.3.33. Theorem 9.3.29 offers a strange interpretation to the Hilbert-Einstein functional . Suppose M is a tame, compact, orientable submanifold of the Euclidean space $\mathbb{R}^n$. Let $m = \dim M$, and denote by h the induced metric on M. Then

$$\mu_{m-2}(M, h) = \frac{1}{2\pi}\mathcal{E}_M(h),$$

where $\mathcal{E}_M(h)$ is the Hilbert-Einstein functional

$$\mathcal{E}_M(h) = \int_M s(h)dV_M(h).$$

Denote by $\mathbf{Graff}^{m-2}(M) = \mathbf{Graff}(M)$ the set of codimension $(m - 2)$ affine planes in $\mathbb{R}^n$ which intersect M along a *nonempty* subset, and by $\mathbf{Graff}_*^{m-2}(M) \subset \mathbf{Graff}^{m-2}(M)$ the subset consisting of those planes intersecting M transversally.

For every $L \in \mathbf{Graff}_*^{m-2}(M)$, the intersection $L \cap M$ is a, possible disconnected, smooth, orientable Riemann surface. By Sard's theorem the complement of $\mathbf{Graff}_*^{m-2}(M)$ in $\mathbf{Graff}^{m-2}(M)$ has zero measure.

For every $R > 0$ we denote by $\mathbf{Graff}^{m-2}(R)$ the set of codimension $(m-2)$ planes in $\mathbb{R}^n$ which intersect the disk $\mathbb{D}_R^n$ of radius R and centered at 0. Note that since M is compact, there exists $R_0 > 0$ such that

$$\mathbf{Graff}^{m-2}(M) \subset \mathbf{Graff}^{m-2}(R), \ \forall R > R_0.$$

We set

$$c(R) = \int_{\mathbf{Graff}^{m-2}(R)} |d\widetilde{\nu}(L)|, \ \ c(M) := \int_{\mathbf{Graff}^{m-2}(M)} |d\widetilde{\nu}(L)|.$$

Using Crofton's formula in Proposition 9.3.24 and the simple observation that

$$\chi(L \cap \mathbb{D}_R^n) = 1, \ \ \forall L \in \mathbf{Graff}^{m-2}(R),$$

we deduce that

$$c(R) = \mu_{m-2}(\mathbb{D}_R^n) = \begin{bmatrix} n \\ m-2 \end{bmatrix} \omega_{m-2} R^{m-2}.$$

Observe now that $\frac{1}{c(R)} |d\widetilde{\nu}(L)|$ is a probability measure on $\mathbf{Graff}^{m-2}(R)$, and we can regard the correspondence

$$\mathbf{Graff}^{m-2}(R) \ni L \longmapsto \chi(L \cap M) \in \mathbb{Z}$$

as a random variable ξ_R whose expectation is

$$\langle \xi_R \rangle := \frac{1}{c(R)} \mu_{m-2}(M) = \frac{1}{2\pi c(R)} \mathcal{E}_M(h).$$

We deduce

$$\mathcal{E}_M(h) = 2\pi \begin{bmatrix} n \\ m-2 \end{bmatrix} \omega_{m-2} R^{m-2} \langle \xi_R \rangle = 2\pi \begin{bmatrix} n \\ m-2 \end{bmatrix} \omega_{m-2} \lim_{R \to \infty} R^{m-2} \langle \xi_R \rangle. \ \ \ \square$$

Theorem 9.3.34 (Crofton Formula. Part 2). *Denote by $\mathbf{Graff}^c(M)$ the subset of $\mathbf{Graff}^c$ consisting of affine planes which intersect M transversally. Then, for every $0 \le p \le m - c$ we have*

$$\begin{bmatrix} p+c \\ p \end{bmatrix} \mu_{p+c}(M) = \int_{\mathbf{Graff}^c(M)} \mu_p(L \cap M) |d\widetilde{\nu}|(L) = \int_{\mathbf{Graff}^c} \hat{\mu}_p(L \cap M) |d\widetilde{\nu}(L)|.$$

Proof. We continue to use the same notations we used in the proof of Theorem 9.3.29. From (9.3.15) we deduce

$$\mu_{p+c}(M) = \lim_{r \to 0} \mu_{p+c}(D_r).$$

The set D_r is a tame domain with smooth boundary, and Corollary 9.3.28 implies

$$\begin{bmatrix} p+c \\ c \end{bmatrix} \mu_{p+c}(D_r) = \int_{\mathbf{Graff}^c} \mu_p(L \cap D_r) |d\widetilde{\nu}|(L) = \int_{\mathbf{Graff}^c} \hat{\mu}_p(L \cap D_r) |d\widetilde{\nu}|(L).$$

Thus it suffices to show that

$$\lim_{r \to 0} \int_{\mathbf{Graff}^c} \hat{\boldsymbol{\mu}}_p(L \cap D_r)|d\widetilde{\nu}(L)| = \int_{\mathbf{Graff}^c} \hat{\boldsymbol{\mu}}_p(L \cap M)|d\widetilde{\nu}(L)|.$$

For $r \in [0, R)$ we define

$$g_r : \mathbf{Graff}^c \to \mathbb{R}, \quad g_r(L) = \hat{\boldsymbol{\mu}}_p(L \cap D_r).$$

From Proposition 9.3.27 we deduce that

$$\exists C > 0 : |g_r(L)| \le C, \quad \forall L \in \mathbf{Graff}^c, \quad r \in [0, R). \tag{9.3.25}$$

Lemma 9.3.35.

$$\lim_{r \searrow 0} g_r(L) = g_0(L),$$

for all $L \in \mathbf{Graff}^c(M)$.

Proof. Let $L \in \mathbf{Graff}^c(M)$ and set $S_r := L \cap D_r$. We have to prove that

$$\lim_{r \searrow 0} \hat{\boldsymbol{\mu}}_p(S_r) = \hat{\boldsymbol{\mu}}_p(S_0).$$

Denote by $\mathbf{Graff}^p(S_0)$ the set of codimension p affine planes in $\mathbb{R}^n$ which intersect S_0 transversally. The set $\mathbf{Graff}^p(S_0)$ is an open subset of $\mathbf{Graff}^p$, and its complement has measure zero. Then,

$$g_r(L) = \hat{\boldsymbol{\mu}}(S_r) = \int_{\mathbf{Graff}^p} \chi(U \cap S_r)|d\widetilde{\nu}^p(U)| = \int_{\mathbf{Graff}^p(S_0)} \chi(U \cap S_r)|d\widetilde{\nu}^p(U)|. \tag{9.3.26}$$

From Proposition 9.3.27 we deduce that the function

$$\mathbf{Graff}^p \times [0, R) \ni (U, r) \mapsto \chi(U \cap S_r) \in \mathbb{Z}$$

is definable, and thus bounded.

Arguing exactly as in the proof of Lemma 9.3.31 we deduce that, for every $U \in \mathbf{Graff}^p(S_0)$, there exists $\rho \in (0, R)$ such that, for every $r \in (0, \rho)$, the set $U \cap S_r$ is homotopy equivalent to $U \cap S_0$. Hence

$$\lim_{r \searrow 0} \chi(U \cap S_r) = \chi(U \cap S_0), \quad \forall U \in \mathbf{Graff}^p(S_0).$$

If we let r go to zero in (9.3.26), we deduce from the dominated convergence theorem that

$$\lim_{r \searrow 0} g_r(L) = \int_{\mathbf{Graff}^p} \chi(U \cap S_0)|d\widetilde{\nu}^p(U)| = \hat{\boldsymbol{\mu}}_p(S_0).$$

Using the fact that S_0 is a smooth, compact tame manifold we deduce from Theorem 9.3.29 that

$$\hat{\boldsymbol{\mu}}_p(S_0) = \mu_p(S_0) = g_0(L). \qquad \square$$

Using (9.3.25) and Lemma 9.3.35, we deduce from the dominated convergence theorem that

$$\lim_{r \searrow 0} \int_{\mathbf{Graff}^c} \hat{\boldsymbol{\mu}}_p(L \cap D_r)|d\widetilde{\nu}(L)| = \int_{\mathbf{Graff}^c} \hat{\boldsymbol{\mu}}_p(L \cap M)|d\widetilde{\nu}(L)|$$

$$= \int_{\mathbf{Graff}^c(M)} \mu_p(L \cap M) |d\widetilde{\nu}(L)|. \qquad\qquad \square$$

Remark 9.3.36. (a) The tameness assumption in the Crofton formulae is not really needed, and they are true in much more general situations; see [33], Sect. 2.10, 3.2.

We have chosen to work in this more restrictive context for two reasons. First, the tameness assumption eliminates the need for sophisticated analytical arguments. Second, we believe that geometers should become more familiar with tame context in which the geometric intuition is not distracted by analytical caveats.

(b) The subject of integral geometry owes a great deal to the remarkable work of S.S. Chern who was the first to have reformulated the main problems from a modern point of view, and push our understanding of this subject to remarkable heights.

It was observed by Chern that the Crofton formulæ that we have proved in this section are only special cases of the so called *kinematic formulæ* which explain how to recover information about a submanifold of a homogeneous space such as $\mathbb{R}^n$, by intersecting it with a large family of simpler shapes. In the Euclidean space, these other simpler shapes could be affine planes, spheres of a given radius, balls of a given radius.

For example, given a compact, m-dimensional submanifold $M \subset \mathbb{R}^n$, and a radius R, we can ask what is the value of the integral

$$\int_{\mathbb{R}^n} \chi(M \cap B_R(x)) |dx|,$$

where $B_R(x)$ denotes the ball $B_R(x) = \{ y \in \mathbb{R}^n; \ |y - x| \leq R \}$. One of the kinematic formulæ states that

$$\int_{\mathbb{R}^n} \chi(M \cap B_R(x)) = \sum_{i=0}^{n} c_{m,n,i} \mu_i(M) R^{n-i},$$

where $c_{m,n,i}$ are universal constants, independent of M.

The above equality is strikingly similar to the tube formula. To see the origin of this similarity note that $B_R(x) \cap M \neq \emptyset \iff x \in \mathbb{T}_R(M)$. If R is sufficiently small, the above formula does indeed specialize to the tube formula, but the kinematic formula extends beyond the range of applicability of the tube formula.

A detailed presentation of these ideas will send us too far, and instead of pursuing further this line of thought, we recommend to the reader to open the classical, yet very much alive monograph of L. Santaló [85], for a more in depth look at this subject. $\qquad\qquad \square$

Chapter 10

Elliptic Equations on Manifolds

Almost all the objects in differential geometry are defined by expressions involving partial derivatives. The curvature of a connection is the most eloquent example.

Many concrete geometric problems lead to studying such objects with specific properties. For example, we inquired whether on a given vector bundle there exist flat connections. This situation can be dealt with topologically, using the Chern-Weil theory of characteristic classes.

Very often, topological considerations alone are not sufficient, and one has look into the microstructure of the problem. This is where analysis comes in, and more specifically, one is led to the study of partial differential equations. Among them, the elliptic ones play a crucial role in modern geometry.

This chapter is an introduction to this vast and dynamic subject which has numerous penetrating applications in geometry and topology.

10.1 Partial differential operators: algebraic aspects

10.1.1 *Basic notions*

We first need to introduce the concept of partial differential operator (p.d.o. for brevity) on a smooth manifold M. To understand the forthcoming formalism, it is best that we begin with the simplest of the situations, $M = \mathbb{R}^N$.

Perhaps the best known partial differential operator is the Laplacian,

$$\Delta : C^\infty(\mathbb{R}^N) \to C^\infty(\mathbb{R}^N), \quad \Delta u := -\sum_i \partial_i{}^2 u,$$

where, as usual, $\partial_i := \frac{\partial}{\partial x_i}$. This is a scalar operator in the sense that it acts on scalar valued functions. Note that our definition of the Laplacian differs from the usual one by a sign. The Laplacian defined as above is sometimes called the *geometers' Laplacian*.

Next in line is the exterior derivative

$$d : \Omega^k(\mathbb{R}^N) \to \Omega^{k+1}(\mathbb{R}^N).$$

This is a vectorial operator in the sense it acts on vector valued functions. A degree k form ω on $\mathbb{R}^N$ can be viewed as a collection of $\binom{N}{k}$ smooth functions or equivalently, as a smooth function $\omega : \mathbb{R}^N \to \mathbb{R}^{\binom{N}{k}}$. Thus d can be viewed as an operator

$$d : C^\infty\left(\mathbb{R}^N, \mathbb{R}^{\binom{N}{k}}\right) \to C^\infty\left(\mathbb{R}^N, \mathbb{R}^{\binom{N}{k+1}}\right).$$

It is convenient to think of $C^\infty(\mathbb{R}^N, \mathbb{R}^\nu)$ as the space of smooth sections of the trivial bundle $\underline{\mathbb{R}}^\nu$ over $\mathbb{R}^N$.

For any smooth $\mathbb{K} = \mathbb{R}, \mathbb{C}$-vector bundles E, F over a smooth manifold M we denote by $\boldsymbol{Op}(E, F)$ the space of $\mathbb{K}$-linear operators

$$C^\infty(E) \to C^\infty(F).$$

The space $\boldsymbol{Op}(E, E)$ is an associative $\mathbb{K}$-algebra.

The spaces of smooth sections $C^\infty(E)$ and $C^\infty(F)$ are more than just $\mathbb{K}$-vector spaces. *They are modules over the ring of smooth functions $C^\infty(M)$.* The partial differential operators are elements of $\boldsymbol{Op}$ which interact in a special way with the above $C^\infty(M)$-module structures. First define

$$\boldsymbol{PDO}^0(E, F) := \mathrm{Hom}(E, F).$$

Given $T \in \boldsymbol{PDO}^0$, $u \in C^\infty(E)$, and $f \in C^\infty(M)$, we have $T(fu) - f(Tu) = 0$ or, in terms of commutators,

$$[T, f]u = T(fu) - f(Tu) = 0. \tag{10.1.1}$$

Each $f \in C^\infty(M)$ defines a map

$$\mathrm{ad}(f) : \boldsymbol{Op}(E, F) \to \boldsymbol{Op}(E, F),$$

by

$$\mathrm{ad}(f)T := T \circ f - f \circ T = [T, f], \ \ \forall T \in \boldsymbol{Op}(E, F).$$

Above, f denotes the $C^\infty(M)$-module multiplication by f. We can rephrase the equality (10.1.1) as

$$\boldsymbol{PDO}^0(E, F) = \{T \in \boldsymbol{Op}(E, F), \ \mathrm{ad}(f)T = 0 \ \forall f \in C^\infty(M)\} =: \ker \mathrm{ad}.$$

Define

$$\boldsymbol{PDO}^{(m)}(E, F) := \ker \mathrm{ad}^{m+1}$$
$$= \big\{ T \in \boldsymbol{Op}(E, F); \ T \in \ker \mathrm{ad}(f_0)\, \mathrm{ad}(f_1) \cdots \mathrm{ad}(f_m), \ \forall f_i \in C^\infty(M) \big\}.$$

The elements of $\boldsymbol{PDO}^{(m)}$ are called *partial differential operators* of order $\leq m$. We set

$$\boldsymbol{PDO}(E, F) := \bigcup_{m \geq 0} \boldsymbol{PDO}^{(m)}(E, F).$$

Remark 10.1.1. Note that we could have defined $\boldsymbol{PDO}^{(m)}$ inductively as

$$\boldsymbol{PDO}^{(m)} = \{T \in \boldsymbol{Op} \ ; \ [T, f] \in \boldsymbol{PDO}^{(m-1)}, \ \forall f \in C^\infty(M)\}.$$

This point of view is especially useful in induction proofs. $\square$

Example 10.1.2. Denote by $\mathbb{R}$ the trivial line bundle over $\mathbb{R}^N$. The sections of $\mathbb{R}$ are precisely the real functions on $\mathbb{R}^N$. We want to analyze $\boldsymbol{PDO}^{(1)} = \boldsymbol{PDO}^{(1)}(\mathbb{R}, \mathbb{R})$.

Let $L \in \boldsymbol{PDO}^{(1)}$, $u, f \in C^\infty(\mathbb{R}^N)$. Then $[L, f]u = \sigma(f) \cdot u$, for some smooth function $\sigma(f) \in C^\infty(\mathbb{R}^N)$. On the other hand, for any $f, g \in C^\infty(\mathbb{R}^N)$

$$\sigma(fg)u = [L, fg]u = [L, f](gu) + f([L, g]u) = \sigma(f)g \cdot u + f\sigma(g) \cdot u.$$

Hence $\sigma(fg) = \sigma(f)g + f\sigma(g)$. In other words, the map $f \mapsto \sigma(f)$ is a derivation of $C^\infty(\mathbb{R}^N)$, and consequently (see Exercise 3.1.9) there exists a smooth vector field X on $\mathbb{R}^N$ such that

$$\sigma(f) = X \cdot f, \quad \forall f \in C^\infty(\mathbb{R}^N).$$

Let $\mu := L(1) \in C^\infty(\mathbb{R}^N)$. Then, for all $u \in C^\infty(\mathbb{R}^N)$, we have

$$Lu = L(u \cdot 1) = [L, u] \cdot 1 + u \cdot L(1) = X \cdot u + \mu \cdot u. \qquad \square$$

Lemma 10.1.3. *Any* $L \in \boldsymbol{PDO}^{(m)}(E, F)$ *is a* local operator, *i.e.,* $\forall u \in C^\infty(E)$,

$$\operatorname{supp} Lu \subset \operatorname{supp} u.$$

Proof. We argue by induction over m. For $m = 0$ the result is obvious. Let $L \in \boldsymbol{PDO}^{(m+1)}$, and $u \in C^\infty(E)$. For every $f \in C^\infty(M)$ we have

$$L(fu) = [L, f]u + fLu.$$

Since $[L, f] \in \boldsymbol{PDO}^{(m)}$ we deduce by induction

$$\operatorname{supp} L(fu) \subset \operatorname{supp} u \cup \operatorname{supp} f, \quad \forall f \in C^\infty(M).$$

For any open set $\mathcal{O}$ such that $\mathcal{O} \supset \operatorname{supp} u$ we can find $f \equiv 1$ on $\operatorname{supp} u$ and $f \equiv 0$ outside $\mathcal{O}$ so that $fu \equiv u$). This concludes the proof of the lemma. $\qquad \square$

The above lemma shows that, in order to analyze the action of a p.d.o., one can work in *local coordinates*. Thus, understanding the structure of an arbitrary p.d.o. boils down to understanding the action of a p.d.o. in $\boldsymbol{PDO}^{(m)}(\mathbb{K}^p, \mathbb{K}^q)$, where $\mathbb{K}^p$, and $\mathbb{K}^q$ are trivial $\mathbb{K}$-vector bundles over $\mathbb{R}^N$. This is done in the exercises at the end of this subsection.

Proposition 10.1.4. *Let* E, F, G *be smooth* $\mathbb{K}$-*vector bundles over the same manifold* M. *If* $P \in \boldsymbol{PDO}^{(m)}(F, G)$ *and* $Q \in \boldsymbol{PDO}^{(n)}(E, F)$ *then* $P \circ Q \in \boldsymbol{PDO}^{(m+n)}(E, G)$.

Proof. We argue by induction over $m + n$. For $m + n = 0$ the result is obvious. In general, if $f \in C^\infty(M)$, then

$$[P \circ Q, f] = [P, f] \circ Q + P \circ [Q, f].$$

By induction, the operators on the right-hand side have orders $\leq m + n - 1$. The proposition is proved. $\square$

Corollary 10.1.5. *The operator*

$$L = \sum_{|\alpha| \leq m} a_\alpha(x)\partial^\alpha : C^\infty(\mathbb{R}^N) \to C^\infty(\mathbb{R}^N),$$

$$\alpha = (\alpha_1, \ldots, \alpha_N) \in \mathbb{Z}_{\geq 0}^N, \ \ |\alpha| = \sum_i \alpha_i, \ \ \partial^\alpha = \partial_1^{\alpha_1} \cdots \partial_N^{\alpha_N},$$

is a p.d.o. of order $\leq m$.

Proof. According to the computation in Example 10.1.2, each partial derivative ∂_i is a 1st order p.d.o. According to the above proposition, multiple compositions of such operators are again p.d.o.'s. $\square$

Lemma 10.1.6. *Let $E, F \to M$ be two smooth vector bundles over the smooth manifold M. Then for any $P \in \boldsymbol{PDO}\,(E, F)$, and any $f, g \in C^\infty(M)$*

$$\mathrm{ad}(f) \cdot (\mathrm{ad}(g)P) = \mathrm{ad}(g) \cdot (\mathrm{ad}(f)P).$$

Proof.

$$\mathrm{ad}(f) \cdot (\mathrm{ad}(g)P) = [[P, g], f] = [[P, f], g] + [P, [f, g] = [[P, f], g] = \mathrm{ad}(g) \cdot (\mathrm{ad}(f)P).$$

$\square$

From the above lemma we deduce that if $P \in \boldsymbol{PDO}^{(m)}$ then, for any $f_1, \ldots, f_m \in C^\infty(M)$, the *bundle morphism*

$$\mathrm{ad}(f_1)\,\mathrm{ad}(f_2) \cdots \mathrm{ad}(f_m)P$$

does not change if we permute the f's.

Proposition 10.1.7. *Let $P \in \boldsymbol{PDO}^{(m)}(E, F)$, $f_i, g_i \in C^\infty(M)$ $(i = 1, \ldots, m)$ such that, at a point $x_0 \in M$,*

$$df_i(x_0) = dg_i(x_0) \in T_{x_0}^* M, \ \ \forall i = 1, \ldots, m.$$

Then

$$\{\mathrm{ad}(f_1)\,\mathrm{ad}(f_2) \cdots \mathrm{ad}(f_m)P\}|_{x_0} = \{\mathrm{ad}(g_1)\,\mathrm{ad}(g_2) \cdots \mathrm{ad}(g_m)P\}|_{x_0}.$$

In the proof we will use the following technical result which we leave to the reader as an exercise.

Lemma 10.1.8. *For each $x_0 \in M$ consider the ideals of $C^\infty(M)$*

$$\mathfrak{m}_{x_0} = \{ f \in C^\infty(M); \ f(x_0) = 0 \},$$

$$\mathfrak{I}_{x_0} = \{ f \in C^\infty(M); \ f(x_0) = 0, \ df(x_0) = 0, \}$$

Then $\mathfrak{I}_{x_0} = \mathfrak{m}_{x_0}^2$, i.e., any function f which vanishes at x_0 together with its derivatives can be written as

$$f = \sum_j g_j h_j \quad g_j, h_j \in \mathfrak{m}_{x_0}. \qquad \square$$

Exercise 10.1.9. Prove the above lemma. $\hspace{3cm} \square$

Proof. Let $P \in \boldsymbol{PDO}^{(m)}$ and $f_i, g_i \in C^\infty(M)$ such that

$$df_i(x_0) = dg_i(x_0) \quad \forall i = 1, \ldots, m.$$

Since $\mathrm{ad}(\mathrm{const}) = 0$, we may assume (eventually altering the f's and the g's by additive constants) that

$$f_i(x_0) = g_i(x_0) \quad \forall i.$$

We will show that

$$\{ \mathrm{ad}(f_1)\,\mathrm{ad}(f_2) \cdots \mathrm{ad}(f_m)P \}|_{x_0} = \{ \mathrm{ad}(g_1)\,\mathrm{ad}(f_2) \cdots \mathrm{ad}(f_m)P \}|_{x_0}.$$

Iterating we get the desired conclusion.

Let $\phi = f_1 - g_1$, and set $Q := \mathrm{ad}(f_2) \cdots \mathrm{ad}(f_m)P \in \boldsymbol{PDO}^{(1)}$. We have to show that

$$\{ \mathrm{ad}(\phi)Q \}|_{x_0} = 0. \qquad (10.1.2)$$

Note that $\phi \in \mathfrak{I}_{x_0}$ so, according to the above lemma, we can write

$$\phi = \sum_j \alpha_j \beta_j, \quad \alpha_j, \beta_j \in \mathfrak{m}_{x_0}.$$

We have

$$\{ \mathrm{ad}(\phi)Q \}|_{x_0} = \left\{ \sum_j \mathrm{ad}(\alpha_j \beta_j)Q \right\}\Big|_{x_0} = \sum_j \{ [Q, \alpha_j]\beta_j \}|_{x_0} + \sum_j \{ \alpha_j[Q, \beta_j] \}|_{x_0} = 0.$$

This proves the equality (10.1.2), and hence the proposition. $\hspace{1cm} \square$

The proposition we have just proved has an interesting consequence. Given $P \in \boldsymbol{PDO}^{(m)}(E, F)$, $x_0 \in M$, and $f_i \in C^\infty(M)$ ($i = 1, \ldots, m$), the linear map

$$\left\{ \frac{1}{m!}\,\mathrm{ad}(f_1) \cdots \mathrm{ad}(f_m)P \right\}\Big|_{x_0} \colon E_{x_0} \to F_{x_0}$$

depends only on the quantities $\xi_i := df_i(x_0) \in T^*_{x_0} M$. Hence, for any $\xi_i \in T^*_{x_0} M$ $(i = 1, \ldots, m)$, the above expression unambiguously induces a linear map

$$\sigma(P)(\xi_1, \ldots, \xi_m)(P) : E_{x_0} \to F_{x_0},$$

that is symmetric in the variables ξ_i. Using the polarization trick of Chapter 8, we see that this map is uniquely determined by the polynomial

$$\sigma(P)(\xi) := \sigma_m(P)(\xi) = \sigma(P)(\underbrace{\xi, \ldots, \xi}_{m}).$$

If we denote by $\pi : T^*M \to M$ the natural projection then, for each $P \in PDO^{(m)}(E, F)$, we have a well defined map

$$\sigma_m(P)(\cdot) \in \mathrm{Hom}\,(\pi^* E, \pi^* F),$$

where $\pi^* E$ and $\pi^* F$ denote the pullbacks of E and F to T^*M via π. Along the fibers of T^*M the map $\xi \mapsto \sigma_m(P)(\xi)$ looks like a degree m homogeneous "polynomial" with coefficients in $\mathrm{Hom}\,(E_{x_0}, F_{x_0})$.

Proposition 10.1.10. *Let* $P \in PDO^{(m)}(E, F)$ *and* $Q \in PDO^{(n)}(F, G)$. *Then*

$$\sigma_{m+n}(Q \circ P) = \sigma_n(Q) \circ \sigma_m(P). \qquad \square$$

Exercise 10.1.11. Prove the above proposition. $\qquad \square$

Definition 10.1.12. A p.d.o. $P \in PDO^{(m)}$ is said to have order m if $\sigma_m(P) \not\equiv 0$. In this case $\sigma_m(P)$ is called the *(principal) symbol* of P. The set of p.d.o.'s of order m will be denoted by PDO^m. $\qquad \square$

Definition 10.1.13. The operator $P \in PDO^m(E, F)$ is said to be *elliptic* if, for any $x \in M$, and any $\xi \in T^*_x M \setminus \{0\}$, the map

$$\sigma_m(P)(\xi) : E_x \to F_x$$

is a linear isomorphism. $\qquad \square$

The following exercises provide a complete explicit description of the p.d.o.'s on $\mathbb{R}^N$.

Exercise 10.1.14. Consider the scalar p.d.o. on $\mathbb{R}^N$ described in Corollary 10.1.5

$$L = \sum_{|\alpha| \le m} a_\alpha(x) \partial^\alpha.$$

Show that

$$\sigma_m(P)(\xi) = \sum_{|\alpha| \le m} a_\alpha(x) \xi^\alpha = \sum_{|\alpha| \le m} a_\alpha(x) \xi_1^{\alpha_1} \cdots \xi_N^{\alpha_N}. \qquad \square$$

Exercise 10.1.15. Let $C : C^\infty(\mathbb{R}^N) \to C^\infty(\mathbb{R}^N)$ be a p.d.o. of order m. Its principal symbol has the form

$$\sigma_m(L)(\xi) = \sum_{|\alpha| \leq m} a_\alpha(x)\xi^\alpha.$$

Show that $L - \sum_{|\alpha|=m} a_\alpha(x)\partial^\alpha$ is a p.d.o. of order $\leq m-1$, and conclude that the only scalar p.d.o.-s on $\mathbb{R}^N$ are those indicated in Corollary 10.1.5. $\quad\square$

Exercise 10.1.16. Let $L \in \boldsymbol{PDO}^{(m)}(E, F)$ and $u \in C^\infty(E)$. Show the operator

$$C^\infty(M) \ni f \longmapsto [L, f]u \in C^\infty(F)$$

belongs to $\boldsymbol{PDO}^{(m)}(\mathbb{R}_M, F)$. $\quad\square$

Exercise 10.1.17. Let $\mathbb{K}^p$ and $\mathbb{K}^q$ denote the trivial $\mathbb{K}$-vector bundles over $\mathbb{R}^N$ of rank p and respectively q. Show that any $L \in \boldsymbol{PDO}^{(m)}(\mathbb{K}^p, \mathbb{K}^q)$ has the form

$$L = \sum_{|\alpha| \leq m} A_\alpha(x)\partial^\alpha,$$

where $A_\alpha \in C^\infty(\mathbb{R}^N, \mathrm{Hom}(\mathbb{K}^p, \mathbb{K}^q))$ for any α .
Hint: Use the previous exercise to reduce the problem to the case $p = 1$. $\quad\square$

10.1.2 *Examples*

At a first glance, the notions introduced so far may look too difficult to "swallow". To help the reader get a friendlier feeling towards them, we included in this subsection a couple of classical examples which hopefully will ease this process. More specifically, we will compute the principal symbols of some p.d.o.'s that we have been extensively using in this book.

In the sequel, we will use the notation "$f\cdot$" to denote the operation of multiplication by the smooth scalar function f.

Example 10.1.18. (The Euclidean Laplacian). This is the second order p.d.o.,

$$\Delta : C^\infty(\mathbb{R}^N) \to C^\infty(\mathbb{R}^N), \quad \Delta = -\sum_i \partial_i^2, \quad \partial_i = \frac{\partial}{\partial x_i}.$$

Let $f \in C^\infty(\mathbb{R}^N)$. Then

$$
\begin{aligned}
\mathrm{ad}(f)(\Delta) &= -\sum_i \mathrm{ad}(f)\,(\partial_i)^2 = -\sum_i \{\mathrm{ad}(f)(\partial_i) \circ \partial_i + \partial_i \circ \mathrm{ad}(f)(\partial_i)\} \\
&= -\sum_i \{f_{x^i} \cdot \partial_i + \partial_i(f_{x^i}\cdot)\} = -\sum_i \{f_{x_i} \cdot \partial_i + f_{x^i x^i}\cdot + f_{x^i} \cdot \partial_i\} \\
&= (\Delta f) \cdot -2\sum_i f_{x^i} \cdot \partial_i.
\end{aligned}
$$

Hence

$$\mathrm{ad}(f)^2(\Delta) = \mathrm{ad}(f)(\Delta f) \cdot -2\sum_i f_{x^i} \cdot \mathrm{ad}(f)\,(\partial_i) = -2\sum_i (f_{x^i})^2 \cdot = -2|df|^2 \cdot \cdot$$

If we set $\xi := df$ in the above equality, we deduce

$$\sigma_2(\Delta)(\xi) = -|\xi|^2 \cdot \cdot$$

In particular, this shows that the Laplacian Δ is an elliptic operator. $\square$

Example 10.1.19. (Covariant derivatives). Consider a vector bundle $E \to M$ over the smooth manifold M, and a connection ∇ on E. We can view ∇ as a p.d.o.

$$\nabla : C^\infty(E) \to C^\infty(T^*M \otimes E).$$

Its symbol can be read from $\mathrm{ad}(f)\nabla$, $f \in C^\infty(M)$. For any $u \in C^\infty(E)$

$$(\mathrm{ad}(f)\nabla)u = \nabla(fu) - f(\nabla u) = df \otimes u.$$

By setting $\xi = df$ we deduce $\sigma_1(\nabla)(\xi) = \xi\otimes$, i.e., the symbol is the tensor multiplication by ξ. $\square$

Example 10.1.20. (The exterior derivative). Let M be a smooth manifold. The exterior derivative

$$d : \Omega^\bullet(M) \to \Omega^{\bullet+1}(M)$$

is a first order p.d.o. To compute its symbol, consider $\omega \in \Omega^k(M)$, and $f \in C^\infty(M)$. Then

$$\big(\,\mathrm{ad}(f)d\,\big)\omega = d(f\omega) - f\,d\omega = df \wedge \omega.$$

If we set $\xi = df$, we deduce $\sigma_1(d) = e(\xi)$ = the left exterior multiplication by ξ. $\square$

Example 10.1.21.   Consider an oriented, n-dimensional Riemann manifold (M, g). As in Chapter 4, we can produce an operator

$$\delta = *d* : \Omega^\bullet(M) \to \Omega^{\bullet-1}(M),$$

where $*$ is the Hodge $*$-operator

$$* : \Omega^\bullet(M) \to \Omega^{n-\bullet}(M).$$

The operator δ is a first order p.d.o., and moreover

$$\sigma_1(\delta) = *\sigma_1(d)* = *e(\xi)*\,.$$

This description can be further simplified.

Fix $\xi \in T_x^*M$, and denote by $\xi^* \in T_x M$ its metric dual. For simplicity we assume $|\xi| = |\xi^*| = 1$. Include ξ in an oriented orthonormal basis $(\xi^1, \ldots, \xi^n)$ of T_x^*M, $\xi^1 = \xi$, and denote by ξ_i the dual basis of $T_x M$.

Consider a monomial $\omega = d\xi^I \in \Lambda^k T_x^* M$, where $I = (i_1, \ldots, i_k)$ denotes as usual an ordered multi-index. Note that if $1 \notin I$, then

$$\sigma_1(\xi^1)\omega = 0. \tag{10.1.3}$$

If $1 \in I$, e.g., $I = (1, \ldots, k)$, then

$$*e(\xi)(*\omega) = *(\xi^1 \wedge \xi^{k+1} \wedge \cdots \wedge \xi^n) = (-1)^{(n-k)(k-1)}\xi^2 \wedge \cdots \wedge \xi^k$$

$$= -(-1)^{\nu_{n,k}} i(\xi^*)\omega, \tag{10.1.4}$$

where $\nu_{n,k} = nk + n + 1$ is the exponent introduced in Subsection 4.1.5 while $i(\xi^*)$ denotes the interior derivative along ξ^*. Putting together (10.1.3) and (10.1.4) we deduce

$$\sigma_1(\delta)(\xi) = -(-1)^{\nu_{n,k}} i(\xi^*). \tag{10.1.5}$$

$\square$

Example 10.1.22. (The Hodge-DeRham operator). Let (M, g) be as in the above example. The Hodge-DeRham operator is

$$d + d^* : \Omega^\bullet(M) \to \Omega^\bullet(M),$$

where $d^* = (-1)^{\nu_{n,k}}\delta$. Hence $d + d^*$ is a first order p.d.o., and moreover,

$$\sigma(d + d^*)(\xi) = \sigma(d)(\xi) + \sigma(d^*)(\xi) = e(\xi) - i(\xi^*).$$

The Hodge Laplacian is the operator $(d + d^*)^2$. We call it Laplacian since

$$\sigma\big((d + d^*)^2\big)(\xi) = \big\{\sigma(d + d^*)(\xi)\big\}^2 = \big(e(\xi) - i(\xi^*)\big)^2,$$

while Exercise 2.2.55 shows

$$\big(e(\xi) - i(\xi^*)\big)^2 = -\big(e(\xi)i(\xi^*) + i(\xi^*)e(\xi)\big) = -|\xi|_g^2.$$

Notice that $d + d^*$ is elliptic since the square of its symbol is invertible. $\square$

Definition 10.1.23. Let $E \to M$ be a smooth vector bundle over the Riemann manifold (M, g). A second order p.d.o.

$$L : C^\infty(E) \to C^\infty(E)$$

is called a *generalized Laplacian* if $\sigma_2(L)(\xi) = -|\xi|_g^2$. $\square$

☞ *Observe that all the generalized Laplacians are elliptic operators.*

10.1.3 *Formal adjoints*

For simplicity, all the vector bundles in this subsection will be assumed complex, unless otherwise indicated.

Let $E_1, E_2 \to M$ be vector bundles over a smooth *oriented* manifold. Fix a Riemann metric g on M and Hermitian metrics $\langle \cdot, \cdot \rangle_i$ on E_i, $i = 1, 2$. We denote by $dV_g = *1$ the volume form on M defined by the metric g. Finally, $C_0^\infty(E_i)$ denotes the space of smooth, *compactly supported* sections of E_i.

Definition 10.1.24. Let $P \in \boldsymbol{PDO}(E_1, E_2)$. The operator $Q \in \boldsymbol{PDO}(E_2, E_1)$ is said to be a *formal adjoint* of P if, $\forall u \in C_0^\infty(E_1)$, and $\forall v \in C_0^\infty(E_2)$ we have

$$\int_M \langle Pu, v \rangle_2 dV_g = \int_M \langle u, Qv \rangle_1 dV_g. \qquad \square$$

Lemma 10.1.25. *Any $P \in \boldsymbol{PDO}(E_1, E_2)$ admits at most one formal adjoint.*

Proof. Let Q_1, Q_2 be two formal adjoints of P. Then, $\forall v \in C_0^\infty(E_2)$, we have

$$\int_M \langle u, (Q_1 - Q_2)v \rangle_1 dV_g = 0 \quad \forall u \in C_0^\infty(E_1).$$

This implies $(Q_1 - Q_2)v = 0 \ \forall v \in C_0^\infty(E_2)$. If now $v \in C^\infty(E_2)$ is not necessarily compactly supported then, choosing a partition of unity $(\alpha) \subset C_0^\infty(M)$, we conclude using the locality of $Q = Q_1 - Q_2$ that

$$Qv = \sum \alpha Q(\alpha v) = 0. \qquad \square$$

The formal adjoint of a p.d.o. $P \in \boldsymbol{PDO}(E_1, E_2)$, whose existence is not yet guaranteed, is denoted by P^*. It is worth emphasizing that P^* depends on the choices of g and $\langle \cdot, \cdot \rangle_i$.

Proposition 10.1.26. *(a) Let $L_0 \in \boldsymbol{PDO}(E_0, E_1)$ and $L_1 \in \boldsymbol{PDO}(E_1, E_2)$ admit formal adjoints $L_i^* \in \boldsymbol{PDO}(E_{i+1}, E_i)$ $(i = 0, 1)$, with respect to a metric g on the base, and metrics $\langle \cdot, \cdot \rangle_j$ on E_j, $j = 0, 1, 2$. Then $L_1 L_0$ admits a formal adjoint, and*

$$(L_1 L_0)^* = L_0^* L_1^*.$$

(b) If $L \in \boldsymbol{PDO}^{(m)}(E_0, E_1)$, then $L^* \in \boldsymbol{PDO}^{(m)}(E_1, E_0)$.

Proof. (a) For any $u_i \in C_0^\infty(E_i)$ we have

$$\int_M \langle L_1 L_0 u_0, u_2 \rangle_2 dV_g = \int_M \langle L_0 u_0, L_1^* u_0 \rangle_1 dV_g = \int_M \langle u_0, L_0^* L_1^* u_2 \rangle_0 dV_g.$$

(b) Let $f \in C^\infty(M)$. Then

$$(\mathrm{ad}(f)L)^* = (L \circ f - f \circ L)^* = -[L^*, f] = -\mathrm{ad}(f)L^*.$$

Thus

$$\mathrm{ad}(f_0)\,\mathrm{ad}(f_1)\cdots\mathrm{ad}(f_m)L^* = (-1)^{m+1}(\mathrm{ad}(f_0)ad(f_1)\cdots\mathrm{ad}(f_m)L)^* = 0. \qquad \square$$

The above computation yields the following result.

Corollary 10.1.27. *If $P \in \mathbf{PDO}^{(m)}$ admits a formal adjoint, then*

$$\sigma_m(P^*) = (-1)^m \sigma_m(P)^*,$$

where the $*$ on the right-hand side denotes the conjugate transpose of a linear map.☐

Let E be a Hermitian vector bundle over the oriented Riemann manifold (M, g).

Definition 10.1.28. A p.d.o. $L \in \mathbf{PDO}(E, E)$ is said to be *formally selfadjoint* if $L = L^*$.
☐

The above notion depends clearly on the various metrics

Example 10.1.29. Using the integration by parts formula of Subsection 4.1.5, we deduce that the Hodge-DeRham operator

$$d + d^* : \Omega^\bullet(M) \to \Omega^\bullet(M),$$

on an oriented Riemann manifold (M, g) is formally selfadjoint with respect with the metrics induced by g in the various intervening bundles. In fact, d^* is the formal adjoint of d.
☐

Proposition 10.1.30. *Let $(E_i, \langle \cdot, \cdot \rangle_i)$, $i = 1, 2$, be two arbitrary Hermitian vector bundle over the oriented Riemann manifold (M, g). Then any $L \in \mathbf{PDO}(E_1, E_2)$ admits at least (and hence exactly) one formal adjoint L^*.*

Sketch of proof We prove this result only in the case when E_1 and E_2 are trivial vector bundles over $\mathbb{R}^N$. However, *we do not assume that the Riemann metric over $\mathbb{R}^N$ is the Euclidean one.* The general case can be reduced to this via partitions of unity and we leave the reader to check it for him/her-self.

Let $E_1 = \mathbb{C}^p$ and $E_2 = \mathbb{C}^q$. By choosing orthonormal moving frames, we can assume that the metrics on E_i are the Euclidean ones. According to the exercises at the end of Subsection 10.1.1, any $L \in \mathbf{PDO}^{(m)}(E_1, E_2)$ has the form

$$L = \sum_{|\alpha| \leq m} A_\alpha(x) \partial^\alpha,$$

where $A_\alpha \in C^\infty(\mathbb{R}^N, M_{q \times p}(\mathbb{C}))$. Clearly, the formal adjoint of A_α is the conjugate transpose

$$A_\alpha^* = \overline{A}_\alpha^t.$$

To prove the proposition it suffices to show that each of the operators $\partial_i \in \mathbf{PDO}^1(E_1, E_1)$ admits a formal adjoint. It is convenient to consider the slightly more general situation.

Lemma 10.1.31. *Let $X = X^i \partial_i \in \mathrm{Vect}(\mathbb{R}^N)$, and denote by ∇_X the first order p.d.o.*

$$\nabla_X u = X^i \partial_i u \quad u \in C_0^\infty(\mathbb{C}^p).$$

Then

$$\nabla_X = -\nabla_X - \boldsymbol{div}_g(X),$$

where $\boldsymbol{div}_g(X)$ denotes the divergence of X with respect to the metric g, i.e., the scalar defined by the equality

$$L_X(dV_g) = \boldsymbol{div}_g(X) \cdot dV_g.$$

Proof.   Let $u, v \in C_0^\infty(\mathbb{R}^N, \mathbb{C}^p)$. Choose $R \gg 0$ such that the Euclidean ball B_R of radius R centered at the origin contains the supports of both u and v. From the equality

$$X \cdot \langle u, v \rangle = \langle \nabla_X u, v \rangle + \langle u, \nabla_X v \rangle,$$

we deduce

$$\int_{\mathbb{R}^N} \langle \nabla_X u, v \rangle dV_g = \int_{B_R} \langle \nabla_X u, v \rangle dV_g$$

$$= \int_{B_R} X \cdot \langle u, v \rangle dV_g - \int_{B_R} \langle u, \nabla_X v \rangle dV_g. \qquad (10.1.6)$$

Set $f := \langle u, v \rangle \in C_0^\infty(\mathbb{R}^N, \mathbb{C})$, and denote by $\alpha \in \Omega^1(\mathbb{R}^N)$ the 1-form dual to X with respect to the Riemann metric g, i.e.,

$$(\alpha, \beta)_g = \beta(X) \quad \forall \beta \in \Omega^1(\mathbb{R}^N).$$

Equivalently,

$$\alpha = g_{ij} X^i dx^j.$$

The equality (10.1.6) can be rewritten

$$\int_{B_R} \langle \nabla_X u, v \rangle dV_g = \int_{B_R} df(X) dV_g - \int_{\mathbb{R}^N} \langle u, \nabla_X v \rangle dV_g$$

$$= \int_{B_R} (df, \alpha)_g dV_g - \int_{\mathbb{R}^N} \langle u, \nabla_X v \rangle dV_g.$$

The integration by parts formula of Subsection 4.1.5 yields

$$\int_{B_R} (df, \alpha)_g dV_g = \int_{\partial B_R} (df \wedge *_g \alpha)|_{\partial B_R} + \int_{B_R} f d^* \alpha dV_g.$$

Since $f \equiv 0$ on a neighborhood of ∂B_R, we get

$$\int_{B_R} (df, \alpha)_g dv_g = \int_{B_R} \langle u, v \rangle d^* \alpha \, dV_g.$$

Since $d^* \alpha = -\boldsymbol{div}_g(X)$, (see Subsection 4.1.5) we deduce

$$\int_{B_R} X \cdot \langle u, v \rangle dV_g = -\int_{B_R} \langle u, v \rangle \boldsymbol{div}_g(X) dV_g = \int_{B_R} \langle u, -\boldsymbol{div}_g(X) v \rangle dV_g.$$

Putting together all of the above we get

$$\int_{\mathbb{R}^N} \langle \nabla_X u, v \rangle dV_g = \int_{\mathbb{R}^N} \langle u, (-\nabla_X - \boldsymbol{div}_g(X))v \rangle dV_g,$$

i.e.,

$$\nabla_X^* = -\nabla_X - \boldsymbol{div}_g(X) = -\nabla_X - \frac{1}{\sqrt{|g|}} \sum_i \partial_i(\sqrt{|g|}X^i). \qquad (10.1.7)$$

The lemma and consequently the proposition is proved. $\qquad\square$

Example 10.1.32. Let E be a rank r smooth vector bundle over the *oriented* Riemann manifold (M, g), $\dim M = m$. Let $\langle \cdot, \cdot \rangle$ denote a Hermitian metric on E, and consider a connection ∇ on E compatible with this metric. The connection ∇ defines a first order p.d.o.

$$\nabla : C^\infty(E) \to C^\infty(T^*M \otimes E).$$

The metrics g and $\langle \cdot, \cdot \rangle$ induce a metric on $T^*M \otimes E$. We want to describe the formal adjoint of ∇ with respect to these choices of metrics.

As we have mentioned in the proof of the previous proposition, this is an entirely local issue. We fix $x_0 \in M$, and we denote by $(\boldsymbol{x}^1, \dots, \boldsymbol{x}^m)$ a collection of g-normal coordinates on a neighborhood U of x_0. We set

$$\nabla_i := \nabla_{\partial_{\boldsymbol{x}^i}}.$$

Next, we pick a local *synchronous* frame of E near x_0, i.e., a local orthonormal frame (e_α) such that, *at x_0*, we have

$$\nabla_i e_\alpha = 0 \quad \forall i = 1, \dots, m.$$

The adjoint of the operator

$$d\boldsymbol{x}^k \otimes : C^\infty(E|_U) \to C^\infty(T^*U \otimes E|_U),$$

is the interior derivative (contraction) along the vector field g-dual to the 1-form $d\boldsymbol{x}^k$, i.e.,

$$(d\boldsymbol{x}^k \otimes)^* = C^k := g^{jk} \cdot i_{\partial_{\boldsymbol{x}^j}} 0.$$

Since ∇ is a metric connection, we deduce as in the proof of Lemma 10.1.31 that

$$\nabla_k{}^* = -\nabla_k - \boldsymbol{div}_g(\partial_{\boldsymbol{x}^k}).$$

Hence,

$$\nabla^* = \sum_k \nabla_k{}^* \circ C^k = \sum_k \left(-\nabla_k - \boldsymbol{div}_g(\partial_{\boldsymbol{x}^k}) \right) \circ C^k$$

$$= -\sum_k (\nabla_k + \partial_{\boldsymbol{x}^k}(\log(\sqrt{|g|}))) \circ C^k. \qquad (10.1.8)$$

In particular, since at x_0 we have $\partial_{x^k} g = 0$, we get

$$\nabla^*|_{x_0} = -\sum_k \nabla_k \circ C^k.$$

The *covariant Laplacian* is the second order p.d.o.

$$\Delta = \Delta_\nabla : C^\infty(E) \to C^\infty(E), \quad \Delta = \nabla^*\nabla.$$

To justify the attribute *Laplacian* we will show that Δ is indeed a generalized Laplacian. Using (10.1.8) we deduce that over U (chosen as above) we have

$$
\begin{aligned}
\Delta &= -\left\{ \sum_k (\nabla_k + \partial_{x^k} \log \sqrt{|g|}) \circ C^k \right\} \circ \left\{ \sum_j d\boldsymbol{x}^j \otimes \nabla_j \right\} \\
&= -\left\{ \sum_k (\nabla_k + \partial_{x^k} \log \sqrt{|g|}) \right\} \circ (g^{kj} \cdot \nabla_j) \\
&= -\sum_{k,j} \{ g^{kj} \nabla_k + \partial_{x^k} g^{kj} + g^{kj} \partial_{x^k} (\log \sqrt{|g|}) \} \circ \nabla_j \\
&= -\sum_{k,j} \{ g^{kj} \nabla_k \nabla_j + \frac{1}{\sqrt{|g|}} \partial_{x^k} (\sqrt{|g|} g^{kj}) \cdot \nabla_j \}.
\end{aligned}
$$

The symbol of Δ can be read easily from the last equality. More precisely,

$$\sigma_2(\Delta)(\xi) = -g^{jk}\xi_j\xi_k = -|\xi|_g^2.$$

Hence Δ is indeed a generalized Laplacian. $\qquad\qquad\square$

In the following exercise we use the notations in the previous example.

Exercise 10.1.33. (a) Show that

$$g^{k\ell}\Gamma_{k\ell}^i = -\frac{1}{\sqrt{|g|}} \partial_{x^k}(\sqrt{|g|} g^{ik}),$$

where $\Gamma_{k\ell}^i$ denote the Christoffel symbols of the Levi-Civita connection associated to the metric g.
(b) Show that

$$\Delta_\nabla = -\operatorname{tr}_g^2(\nabla^{T^*M \otimes E}\nabla^E),$$

where $\nabla^{T^*M \otimes E}$ is the connection on $T^*M \otimes E$ obtained by tensoring the Levi-Civita connection on T^*M and the connection ∇^E on E, while

$$\operatorname{tr}_g^2 : C^\infty(T^*M^{\otimes 2} \otimes E) \to C^\infty(E)$$

denotes the double contraction by g,

$$\operatorname{tr}_g^2(S_{ij} \otimes u) = g^{ij}S_{ij}u. \qquad\qquad\square$$

Proposition 10.1.34. *Let E and M as above, and suppose that $L \in \mathbf{PDO}^2(E)$ is a generalized Laplacian. Then, there exists a unique metric connection ∇ on E and $\mathcal{R} = \mathcal{R}(L) \in \mathrm{End}\,(E)$ such that*

$$L = \nabla^* \nabla + \mathcal{R}.$$

The endomorphism $\mathcal{R}$ is known as the Weitzenböck remainder *of the Laplacian L.*

Exercise 10.1.35. Prove the above proposition.
Hint Try ∇ defined by

$$\nabla_{f\mathbf{grad}(h)} u = \frac{f}{2}\{(\Delta_g h)u - (ad(h)L)u\} \quad f, h \in C^\infty(M), \ u \in C^\infty(E). \qquad \Box$$

Exercise 10.1.36. (General Green formula). Consider a compact Riemannian manifold (M, g) with boundary ∂M. Denote by $\vec{n}$ the unit outer normal along ∂M. Let $E, F \to M$ be Hermitian vector bundles over M and suppose $L \in \mathbf{PDO}^k(E, F)$. Set $g_0 = g|_{\partial M}$, $E_0 = E|_{\partial M}$ and $F_0 = F|_{\partial M}$. The Green formula states that there exists a sesquilinear map

$$B_L : C^\infty(E) \times C^\infty(F) \to C^\infty(\partial M),$$

such that

$$\int_M \langle Lu, v\rangle dv(g) = \int_{\partial M} B_L(u, v) dv(g_0) + \int_M \langle u, L^* v\rangle dv(g).$$

Prove the following.
(a) If L is a zeroth order operator, then $B_L = 0$.
(b) If $L_1 \in \mathbf{PDO}(F, G)$, and $L_2 \in \mathbf{PDO}(E, F)$, then

$$B_{L_1 L_2}(u, v) = B_{L_1}(L_2 u, v) + B_{L_2}(u, L_1^* v).$$

(c)

$$B_{L^*}(v, u) = -\overline{B_L(u, v)}.$$

(d) Suppose ∇ is a Hermitian connection on E, and $X \in \mathrm{Vect}\,(M)$. Then

$$B_{\nabla_X}(u, v) = \langle u, v\rangle g(X, \vec{n}), \quad B_\nabla(u, v) = \langle u, i_{\vec{n}} v\rangle_E,$$

where $i_{\vec{n}}$ denotes the contraction by $\vec{n}$.
(e) Denote by $\vec{\nu}$ the section of $T^* M|_{\partial M}$ g-dual to $\vec{n}$. Suppose L is a first order p.d.o., and set $J := \sigma_L(\vec{\nu})$. Then

$$B_L(u, v) = \langle Ju, v\rangle_F.$$

(f) Using (a)-(e) show that, for all $u \in C^\infty(E)$, $v \in C^\infty(F)$, and any $x_0 \in \partial M$, the quantity $B_L(u, v)(x_0)$ depends only on the jets of u, v at x_0 of order at most $k - 1$. In other words, if all the partial derivatives of u up to order $(k-1)$, with respect to some connection, vanish along the boundary, then $B_L(u, v) = 0$. $\qquad \Box$

10.2 Functional framework

The partial differential operators are linear operators in infinite dimensional spaces, and this feature requires special care in dealing with them. Linear algebra alone is not sufficient. This is where functional analysis comes in.

In this section we introduce a whole range of functional spaces which are extremely useful for most geometric applications.

The presentation assumes the reader is familiar with some basic principles of functional analysis. As a reference for these facts we recommend the excellent monograph [18], or the very comprehensive [30, 103].

10.2.1 *Sobolev spaces in* $\mathbb{R}^N$

Let D denote an open subset of $\mathbb{R}^N$. We denote by $L^1_{loc}(D)$ the space of locally integrable real functions on D, i.e., Lebesgue measurable functions $f : \mathbb{R}^N \to \mathbb{K}$ such that, $\forall \alpha \in C_0^\infty(\mathbb{R}^N)$, the function αf is Lebesgue integrable, $\alpha f \in L^1(\mathbb{R}^N)$.

Definition 10.2.1. Let $f \in L^1_{loc}(D)$, and $1 \leq k \leq N$. A function $g \in L^1_{loc}(D)$ is said to be the *weak k-th partial derivative* of f, and we write this $g = \partial_k f$ weakly, if

$$\int_D g\varphi dx = -\int_D f \partial_k \varphi dx, \ \ \forall \varphi \in C_0^\infty(D). \qquad \square$$

Lemma 10.2.2. *Any $f \in L^1_{loc}(D)$ admits* at most *one weak partial derivative.* $\square$

The proof of this lemma is left to the reader.

☞ *Not all locally integrable functions admit weak derivatives.*

Exercise 10.2.3. Let $f \in C^\infty(D)$, and $1 \leq k \leq N$. Prove that the classical partial derivative $\partial_k f$ is also its weak k-th derivative. $\square$

Exercise 10.2.4. Let $H \in L^1_{loc}(\mathbb{R})$ denote the *Heaviside function*, $H(t) \equiv 1$, for $t \geq 0$, $H(t) \equiv 0$ for $t < 0$. Prove that H is not weakly differentiable. $\square$

Exercise 10.2.5. Let $f_1, f_2 \in L^1_{loc}(D)$. If $\partial_k f_i = g_i \in L^1_{loc}$ weakly, then $\partial_k(f_1 + f_2) = g_1 + g_2$ weakly. $\square$

The definition of weak derivative can be generalized to higher order derivatives as follows. Consider a scalar p.d.o. $L : C^\infty(D) \to C^\infty(D)$, and $f, g \in L^1_{loc}(D)$. Then we say that $Lf = g$ weakly if

$$\int_D g\varphi dx = \int_D f L^* \varphi dx \ \ \forall \varphi \in C_0^\infty(D).$$

Above, L^* denotes the formal adjoint of L with respect to the Euclidean metric on $\mathbb{R}^N$.

Exercise 10.2.6. Let $f, g \in C^\infty(D)$. Prove that

$$Lf = g \text{ classically} \iff Lf = g \text{ weakly.} \qquad \square$$

Definition 10.2.7. Let $k \in \mathbb{Z}_+$, and $p \in [1, \infty]$. The *Sobolev space* $L^{k,p}(D)$ consists of all the functions $f \in L^p(D)$ such that, for any multi-index α satisfying $|\alpha| \leq k$, the mixed partial derivative $\partial^\alpha f$ exists weakly, and moreover, $\partial^\alpha f \in L^p(D)$. For every $f \in L^{k,p}(D)$ we set

$$\|f\|_{k,p} := \|f\|_{k,p,D} = \left(\sum_{|\alpha| \leq k} \int_D |\partial^\alpha f|^p dx \right)^{1/p},$$

if $p < \infty$, while if $p = \infty$

$$\|f\|_{k,\infty} = \sum_{|\alpha| \leq k} \operatorname{ess\,sup} |\partial^\alpha f|.$$

When $k = 0$ we write $\|f\|_p$ instead of $\|f\|_{0,p}$. $\qquad \square$

Definition 10.2.8. Let $k \in \mathbb{Z}_+$ and $p \in [1, \infty]$. Set

$$L^{k,p}_{loc}(D) := \left\{ f \in L^1_{loc}(D) \; ; \; \varphi f \in L^{k,p}(D) \; \forall \varphi \in C^\infty_0(D) \right\}. \qquad \square$$

Exercise 10.2.9. Let $f(t) = |t|^\alpha$, $t \in \mathbb{R}$, $\alpha > 0$. Show that for every $p > 1$ such that $\alpha > 1 - \frac{1}{p}$, we have

$$f(t) \in L^{1,p}_{loc}(\mathbb{R}). \qquad \square$$

Theorem 10.2.10. *Let $k \in \mathbb{Z}_+$ and $1 \leq p \leq \infty$. Then*
(a) $(L^{k,p}(\mathbb{R}^N), \| \cdot \|_{k,p})$ is a Banach space.
(b) If $1 \leq p < \infty$ the subspace $C^\infty_0(\mathbb{R}^N)$ is dense in $L^{k,p}(\mathbb{R}^N)$.
(c) If $1 < p < \infty$ the Sobolev space $L^{k,p}(\mathbb{R}^N)$ is reflexive.

The proof of this theorem relies on a collection of basic techniques frequently used in the study of partial differential equations. This is why we choose to cover the proof of this theorem in some detail. We will consider only the case $p < \infty$, leaving the $p = \infty$ situation to the reader.

Proof. Using Exercise 10.2.5, we deduce that $L^{k,p}$ is a vector space. From the classical Minkowski inequality,

$$\left(\sum_{i=1}^\nu |x_i + y_i|^p \right)^{1/p} \leq \left(\sum_{i=1}^\nu |x_i|^p \right)^{1/p} + \left(\sum_{i=1}^\nu |y_i|^p \right)^{1/p},$$

we deduce that $\| \cdot \|_{k,p}$ is a norm. To prove that $L^{k,p}$ is complete, we will use the well established fact that L^p is complete.

☝ *To simplify the notations, throughout this chapter all the extracted subsequences will be denoted by the same symbols as the sequences they originate from.*

Let $(f_n) \subset L^{k,p}(\mathbb{R}^N)$ be a Cauchy sequence, i.e.,

$$\lim_{m,n\to\infty} \|f_m - f_n\|_{k,p} = 0.$$

In particular, for each multi-index $|\alpha| \leq k$, the sequence $(\partial^\alpha f_n)$ is Cauchy in $L^p(\mathbb{R}^N)$, and thus,

$$\partial^\alpha f_n \xrightarrow{L^p} g_\alpha, \ \ \forall |\alpha| \leq k.$$

Set $f = \lim_n f_n$. We claim that $\partial^\alpha f = g_\alpha$ weakly.

Indeed, for any $\varphi \in C_0^\infty(\mathbb{R}^N)$ we have

$$\int_{\mathbb{R}^N} \partial^\alpha f_n \varphi dx = (-1)^{|\alpha|} \int_{\mathbb{R}^N} f_n \partial^\alpha \varphi.$$

Since $\partial^\alpha f_n \to g_\alpha$, $f_n \to f$ in L^p, and $\varphi \in L^q(\mathbb{R}^N)$, where $1/q = 1 - 1/p$, we conclude

$$\int_{\mathbb{R}^N} g_\alpha \cdot \varphi \, dx = \lim_n \int_{\mathbb{R}^N} \partial^\alpha f_n \cdot \varphi \, dx$$

$$= \lim_n (-1)^{|\alpha|} \int_{\mathbb{R}^N} f_n \cdot \partial^\alpha \varphi \, dx = (-1)^{|\alpha|} \int_{\mathbb{R}^N} f \cdot \partial^\alpha \varphi \, dx.$$

Part (a) is proved.

To prove that $C_0^\infty(\mathbb{R}^N)$ is dense we will use *mollifiers*. Their definition uses the operation of *convolution*. Given $f, g \in L^1(\mathbb{R}^N)$, we define

$$(f * g)(x) := \int_{\mathbb{R}^N} f(x - y)g(y)dy.$$

We let the reader check that $f * g$ is well defined, i.e., $y \mapsto f(x - y)g(y) \in L^1$, for almost all x.

Exercise 10.2.11. (Young's Inequality). If $f \in L^p(\mathbb{R}^N)$, and $g \in L^q(\mathbb{R}^N)$, $1 \leq p, q \leq \infty$, then $f * g$ is well defined, $f * g \in L^r(\mathbb{R}^N)$, $\frac{1}{r} = \frac{1}{p} + \frac{1}{q} - 1$, and

$$\|f * g\|_r \leq \|f\|_p \cdot \|g\|_q. \tag{10.2.1}$$

$\square$

To define the mollifiers one usually starts with a function $\rho \in C_0^\infty(\mathbb{R}^N)$, such that

$$\rho \geq 0, \ \ \mathrm{supp}\, \rho \subset \{|x| < 1\}, \ \ \text{and} \ \int_{\mathbb{R}^N} \rho \, dx = 1.$$

Next, for each $\delta > 0$, we define

$$\rho_\delta(x) := \delta^{-N} \rho(x/\delta).$$

Note that

$$\mathrm{supp}\, \rho_\delta \subset \{|x| < \delta\} \ \ \text{and} \ \int_{\mathbb{R}^N} \rho_\delta dx = 1.$$

The sequence (ρ_δ) is called a *mollifying sequence*. The next result describes the main use of this construction.

Lemma 10.2.12. *(a) For any $f \in L^1_{loc}(\mathbb{R}^N)$ the convolution $\rho_\delta * f$ is a smooth[1] function!*
(b) If $f \in L^p(\mathbb{R}^N)$ $(1 \le p < \infty)$ then

$$\rho_\delta * f \xrightarrow{L^p} f \text{ as } \delta \to 0.$$

Proof. Part (a) is left to the reader as an exercise in the differentiability of integrals with parameters.

To establish part (b), we will use the fact that $C_0^\infty(\mathbb{R}^N)$ is dense in $L^p(\mathbb{R}^N)$. Fix $\varepsilon > 0$, and choose $g \in C_0^\infty(\mathbb{R}^N)$ such that,

$$\|f - g\|_p \le \varepsilon/3.$$

We have

$$\|\rho_\delta * f - f\|_p \le \|\rho_\delta * (f - g)\|_p + \|\rho_\delta * g - g\|_p + \|g - f\|_p.$$

Using the inequality (10.2.1) we deduce

$$\|\rho_\delta * f - f\|_p \le 2\|f - g\|_p + \|\rho_\delta * g - g\|_p. \tag{10.2.2}$$

We need to estimate $\|\rho_\delta * g - g\|_p$. Note that

$$\rho_\delta * g\,(x) = \int_{\mathbb{R}^N} \rho(z)g(x - \delta z)\,dz \text{ and } g(x) = \int_{\mathbb{R}^N} \rho(z)g(x)dz.$$

Hence

$$|\rho_\delta * g\,(x) - g(x)| \le \int_{\mathbb{R}^N} \rho(z)|g(x - \delta z) - g(x)|dz \le \delta \sup |dg|.$$

Since

$$\operatorname{supp} \rho_\delta * g \subset \{\, x + z \;;\; x \in \operatorname{supp} g \,;\, z \in \operatorname{supp} \rho_\delta \,\},$$

there exists a compact set $K \subset \mathbb{R}^N$ such that,

$$\operatorname{supp}(\rho_\delta * g - g) \subset K \quad \forall \delta \in (0, 1).$$

We conclude that

$$\|\rho_\delta * g - g\|_p \le \left(\int_K \delta^p (\sup |dg|)^p dx \right)^{1/p} = \operatorname{vol}(K)^{1/p}\delta \sup |dg|.$$

If now we pick δ such that

$$\operatorname{vol}(K)^{1/p}\delta \sup |dg| \le \varepsilon/3,$$

we conclude from (10.2.2) that

$$\|\rho_\delta * f - f\|_p \le \varepsilon.$$

---

[1]This explains the term *mollifier*: ρ_δ smoothes out the asperities.

The lemma is proved.   □

The next auxiliary result describes another useful feature of the mollification technique, especially versatile in as far as the study of partial differential equations is concerned.

Lemma 10.2.13. *Let* $f, g \in L^1_{loc}(\mathbb{R}^N)$ *such that* $\partial_k f = g$ *weakly. Then* $\partial_k(\rho_\delta * f) = \rho_\delta * g$. *More generally, if*

$$L = \sum_{|\alpha| \le m} a_\alpha \partial^\alpha$$

is a p.d.o with constant coefficients $a_\alpha \in \mathbb{R}$, *and* $Lf = g$ *weakly, then*

$$L(\rho_\delta * f) = \rho_\delta * g \ \text{classically.}$$

Proof.   It suffices to prove only the second part. We will write L_x to emphasize that L acts via derivatives with respect to the variables $x = (x^1, \ldots, x^N)$. Note that

$$L(\rho_\delta * f) = \int_{\mathbb{R}^N} (L_x \rho_\delta(x - y)) f(y) dy. \tag{10.2.3}$$

Since

$$\frac{\partial}{\partial x_i} \rho_\delta(x - y) = -\frac{\partial}{\partial y^i} \rho_\delta(x - y),$$

and $\partial_i^* = -\partial_i$, we deduce from (10.2.3) that

$$L(\rho_\delta * f) = \int_{\mathbb{R}^N} (L_y^* \rho_\delta(x - y)) f(y) dy = \int_{\mathbb{R}^N} \rho_\delta(x - y) L_y f(y) \, dy$$

$$= \int_{\mathbb{R}^N} \rho_\delta(x - y) g(y) dy = \rho_\delta * g. \qquad \square$$

Remark 10.2.14.   The above lemma is a commutativity result. It shows that if L is a p.d.o. with constant coefficients, then

$$[L, \rho_\delta *] f = L(\rho_\delta * f) - \rho_\delta * (Lf) = 0.$$

This fact has a fundamental importance in establishing regularity results for elliptic operators.   □

After this rather long detour, we return to the proof of Theorem 10.2.10. Let $f \in L^{k,p}(\mathbb{R}^N)$. We will construct $f_n \in C_0^\infty(\mathbb{R}^N)$ such that $f_n \to f$ in $L^{k,p}$ using two basic techniques: *truncation* and *mollification* (or *smoothing*).

Truncation. The essentials of this technique are contained in the following result.

Lemma 10.2.15.   *Let* $f \in L^{k,p}(\mathbb{R}^N)$. *Consider for each* $R > 0$ *a smooth function* $\eta_R \in C_0^\infty(\mathbb{R}^N)$ *such that* $\eta(x) \equiv 1$ *for* $|x| \le R$, $\eta_R(x) \equiv 0$ *for* $|x| \ge R + 1$, *and* $|d\eta_R(x)| \le 2 \ \forall x$. *Then* $\eta_R \cdot f \in L^{k,p}(\mathbb{R}^N)$, $\forall R \ge 0$, *and moreover*

$$\eta_R \cdot f \xrightarrow{L^{k,p}} f \ \text{as} \ R \to \infty.$$

Proof. We consider only the case $k = 1$. The general situation can be proved by induction. We first prove that $\partial_i(\eta_R f)$ exists weakly and as expected

$$\partial_i(\eta_R \cdot f) = (\partial_i \eta_R) \cdot f + \eta_R \cdot \partial_i f.$$

Let $\varphi \in C_0^\infty(\mathbb{R}^N)$. Since $\eta_R \varphi \in C_0^\infty(\mathbb{R}^N)$, we have

$$\int_{\mathbb{R}^N} (\partial_i \eta_R)\varphi + \eta_R \cdot \partial_i \varphi dx = \int_{\mathbb{R}^N} \partial_i(\eta_R \varphi)f = -\int_{\mathbb{R}^N} \eta_R \varphi \partial_i f dx.$$

This confirms our claim. Clearly $(\partial_i \eta_R)f + \eta_R \partial_i f \in L^p(\mathbb{R}^N)$, so that $\eta_R f \in L^{1,p}(\mathbb{R}^N)$.

Note that $\partial_i \eta_R \equiv 0$ for $|x| \leq R$ and $|x| \geq R+1$. In particular, we deduce

$$(\partial_i \eta_R)f \to 0 \text{ a.e.}$$

Clearly $|(\partial_i \eta_R)f| \leq 2|f(x)|$, so that by the dominated convergence theorem we conclude

$$(\partial_i \eta_R) \cdot f \xrightarrow{L^p} 0 \text{ as } R \to \infty.$$

Similarly

$$\eta_R \partial_i f \to \partial_i f \text{ a.e.},$$

and $|\eta_R \partial_i f| \leq |\partial_i f|$, which implies $\eta_R \partial_i f \to \partial_i f$ in L^p. The lemma is proved. $\square$

According to the above lemma, the space of compactly supported $L^{k,p}$-functions is dense in $L^{k,p}(\mathbb{R}^N)$. Hence, it suffices to show that any such function can be arbitrarily well approximated in the $L^{k,p}$-norm by smooth, compactly supported functions.

Let $f \in L^{k,p}(\mathbb{R}^N)$, and assume

$$\operatorname{ess\,supp} f \subset \{|x| \leq R\}.$$

Mollification. The sequence $\rho_\delta * f$ converges to f in the norm of $L^{k,p}$ as $\delta \searrow 0$.

Note that each $\rho_\delta * f$ is a smooth function, supported in $\{|x| \leq R+\delta\}$. According to Lemma 10.2.13, we have

$$\partial^\alpha(\rho_\delta * f) = \rho_\delta * (\partial^\alpha f) \ \forall |\alpha| \leq k.$$

The desired conclusion now follows using Lemma 10.2.12.

To conclude the proof of Theorem 10.2.10, we need to show that $L^{k,p}$ is reflexive if $1 < p < \infty$. We will use the fact that L^p is reflexive for p in this range.

Note first that $L^{k,p}(\mathbb{R}^N)$ can be viewed as a closed subspace of the direct product

$$\prod_{|\alpha| \leq k} L^p(\mathbb{R}^N),$$

via the map,

$$T : L^{k,p}(\mathbb{R}^N) \to \prod_{|\alpha| \leq k} L^p(\mathbb{R}^N), \ f \mapsto (\partial^\alpha f)|_{|\alpha| \leq k},$$

which is continuous, one-to-one, and has closed range. Indeed, if $\partial^\alpha f_n \xrightarrow{L^p} f_\alpha$, then, arguing as in the proof of completeness, we deduce that

$$f_\alpha = \partial^\alpha f_0 \text{ weakly,}$$

where $f_0 = \lim f_n$. Hence $(f_\alpha) = T f_0$. We now conclude that $L^{k,p}(\mathbb{R}^N)$ is reflexive as a closed subspace of a reflexive space. Theorem 10.2.10 is proved. $\square$

Remark 10.2.16. (a) For $p = 2$ the spaces $L^{k,2}(\mathbb{R}^N)$ are in fact Hilbert spaces. The inner product is given by

$$\langle u, v \rangle_k = \int_{\mathbb{R}^N} \left(\sum_{|\alpha| \leq k} \partial^\alpha u \cdot \partial^\alpha v dx \right) dx.$$

(b) If $D \subset \mathbb{R}^N$ is open, then $L^{k,p}(D)$ is a Banach space, reflexive if $1 < p < \infty$. However $C_0^\infty(D)$, is no longer dense in $L^{k,p}(D)$. The closure of $C_0^\infty(D)$ in $L^{k,p}(D)$ is denoted by $L_0^{k,p}(D)$. Intuitively, $L_0^{k,p}(D)$ consists of the functions $u \in L^{k,p}(D)$ such that

$$\frac{\partial^j u}{\partial \nu^j} = 0 \text{ on } \partial D, \ \forall j = 0, 1, \ldots, k - 1,$$

where $\partial/\partial\nu$ denotes the normal derivative along the boundary. The above statement should be taken with a grain of salt since at this point it is not clear how one can define $u|_{\partial D}$ when u is defined only almost everywhere. We refer to [3] for a way around this issue.

The larger space $C^\infty(D) \cap L^{k,p}(D)$ is dense in $L^{k,p}(D)$, provided that the boundary of D is sufficiently regular. We refer again to [3] for details. $\square$

Exercise 10.2.17. Prove that the following statements are equivalent.
(a) $u \in L^{1,p}(\mathbb{R}^N)$.
(b) There exists a constant $C > 0$ such that, for all $\varphi \in C^\infty(\mathbb{R}^N)$, we have

$$\left| \int_{\mathbb{R}^N} u \frac{\partial\varphi}{\partial x^i} \right| \leq C \|\varphi\|_{L^{p'}}, \ \forall i = 1, \ldots, N,$$

where $p' = p/(p-1)$.
(c) There exists $C > 0$ such that, for all $h \in \mathbb{R}^N$, we have

$$\|\Delta_h u\|_{L^p} \leq C|h|,$$

where $\Delta_h u(x) = u(x + h) - u(x)$. $\square$

Exercise 10.2.18. Let $f \in L^{1,p}(\mathbb{R}^N)$, and $\phi \in C^\infty(\mathbb{R})$, such that

$$|d\phi| \leq \text{const,}$$

and $\phi(f) \in L^p$. Then $\phi(f) \in L^{1,p}(\mathbb{R}^N)$, and

$$\partial_i \phi(f) = \phi'(f) \cdot \partial_i f.$$ $\square$

Exercise 10.2.19. Let $f \in L^{1,p}(\mathbb{R}^N)$. Show that $|f| \in L^{1,p}(\mathbb{R}^N)$, and

$$\partial_i |f| = \begin{cases} \partial_i f & \text{a.e. on } \{f \geq 0\} \\ -\partial_i f & \text{a.e. on } \{f < 0\} \end{cases}$$

Hint: Show that $f_\varepsilon := (\varepsilon^2 + f^2)^{1/2}$ converges to f in $L^{1,p}$ as $\varepsilon \to 0$. $\square$

10.2.2 *Embedding theorems: integrability properties*

The embedding theorems describe various inclusions between the Sobolev spaces $L^{k,p}(\mathbb{R}^N)$.

Define the "strength" of the Sobolev space $L^{k,p}(\mathbb{R}^N)$ as the quantity

$$\sigma(k,p) := \sigma_N(k,p) = k - N/p.$$

The "strength" is a measure of the size of a Sobolev size. Loosely speaking, the bigger the strength, the more regular are the functions in that space, and thus it consists of "fewer" functions.

Remark 10.2.20. The origin of the quantities $\sigma_N(k,p)$ can be explained by using the notion of *conformal weight*. A function $u : \mathbb{R}^N \to \mathbb{R}$ can be thought of as a dimensionless physical quantity. Its conformal weight is 0. Its partial derivatives $\partial_i u$ are physical quantities measured in $meter^{-1}=$ variation per unit of distance, and they have conformal weight -1. More generally, a mixed partial $\partial^\alpha u$ has conformal weight $-|\alpha|$. The quantities $|\partial^\alpha|^p$ have conformal weight $-p|\alpha|$. The volume form dx is assigned conformal weight N: the volume is measured in $meter^N$. The integral of a quantity of conformal weight w is a quantity of conformal weight $w + N$. For example, the quantity

$$\int_{\mathbb{R}^N} |\partial^\alpha u|^p \, dx$$

has conformal weight $N - p|\alpha|$. In particular the quantity

$$\left(\int \Big\{ \sum_{|\alpha|=k} |\partial^\alpha u|^p \Big\} dx \right)^{1/p}$$

has conformal weight $(N-kp)/p = -\sigma_N(k,p)$. Geometrically, the conformal weight is captured by the behavior under the rescalings $x = \lambda y$.

If we replace the Euclidean metric g on $\mathbb{R}^n$ with the metric $g_\lambda = \lambda^2 g$, $\lambda > 0$, then,

$$dV_{g_\lambda} = \lambda^n dV_g, \;\; |du|_{g_\lambda} = \lambda^{-1}|du|_g, \;\; \forall u \in C^\infty(\mathbb{R}^N).$$

Equivalently, if we introduce the new linear variables y^i related to the canonical Euclidean variables x^i by $x^i = \lambda y^i$, then

$$\partial_{x^i} = \lambda^{-1}\partial_{y^i}, \;\; dx^1 \wedge \cdots \wedge dx^N = \lambda^N dy^1 \wedge \cdots \wedge dy^N. \qquad \square$$

Theorem 10.2.21 (Sobolev). *If*

$$\sigma_N(k,p) = \sigma_N(m,q) < 0 \text{ and } k > m$$

then,

$$L^{k,p}(\mathbb{R}^N) \hookrightarrow L^{m,p}(\mathbb{R}^N),$$

and the natural inclusion is continuous, i.e., there exists $C = C(N,k,m,p,q) > 0$ such that

$$\|f\|_{m,q} \le C\|f\|_{k,p} \;\; \forall f \in L^{k,p}(\mathbb{R}^N).$$

Proof.   We follow the approach of [81] which relies on the following elementary, but ingenious lemma.

Lemma 10.2.22 (Gagliardo-Nirenberg). *Let $N \geq 2$ and $f_1, \ldots, f_N \in L^{N-1}(\mathbb{R}^{N-1})$. For each $x \in \mathbb{R}^N$, and $1 \leq i \leq N$ define*

$$\xi_i = (x^1, \ldots, \hat{x}^i, \ldots, x^N) \in \mathbb{R}^{N-1}.$$

Then

$$f(x) = f_1(\xi_1) f_2(\xi_2) \cdots f_N(\xi_N) \in L^1(\mathbb{R}^N),$$

and moreover,

$$\|f\|_1 \leq \prod_{i=1}^{N} \|f_i\|_{N-1}. \qquad \Box$$

Exercise 10.2.23.   Prove Lemma 10.2.22. $\Box$

We first prove the theorem in the case $k = 1$, $p = 1$, which means $m = 0$, and $q = N/(N-1)$. We will show that

$$\exists C > 0 : \|u\|_{N/(N-1)} \leq C \| |du| \|_1 \quad \forall u \in C_0^\infty(\mathbb{R}^N),$$

where $|du|^2 = |\partial_1 u|^2 + \cdots + |\partial_N u|^2$. This result then extends by density to any $u \in L^{1,1}(\mathbb{R}^N)$.

We have

$$|u(x^1, \ldots, x^N)| \leq \int_{-\infty}^{x^1} |\partial_i u(x^1, \ldots, x^{i-1}, t, x^{i+1}, \ldots, x^N)| dt \stackrel{def}{=} g_i(\xi_i).$$

Note that $g_i \in L^1(\mathbb{R}^{N-1})$, so that

$$f_i(\xi_i) = g_i(\xi_i)^{1/(N-1)} \in L^{N-1}(\mathbb{R}^{N-1}).$$

Since

$$|u(x)|^{N/(N-1)} \leq f_1(\xi_1) \cdots f_N(\xi_N),$$

we conclude from Lemma 10.2.22 that $u(x) \in L^{N/(N-1)}(\mathbb{R}^N)$, and

$$\|u\|_{N/(N-1)} \leq \left(\prod_{1}^{N} \|g_i(\xi_i)\|_1 \right)^{1/N} = \left(\prod_{1}^{N} \|\partial_i u\|_1 \right)^{1/N}.$$

Using the classical arithmetic-geometric means inequality, we conclude

$$\|u\|_{N/(N-1)} \leq \frac{1}{N} \sum_{1}^{N} \|\partial_i u\|_1 \leq \frac{const.}{N} \sum_{1}^{N} \| |du| \|_1.$$

We have thus proved that $L^{1,1}(\mathbb{R}^N)$ embeds continuously in $L^{N/(N-1)}(\mathbb{R}^N)$.

Now let $1 < p < \infty$ such that $\sigma_N(1, p) = 1 - N/p < 0$, i.e., $p < N$. We have to show that $L^{1,p}(\mathbb{R}^N)$ embeds continuously in $L^{p^*}(\mathbb{R}^N)$, where

$$p^* = \frac{Np}{N - p}.$$

Let $u \in C_0^\infty(\mathbb{R}^N)$. Set $v = |v|^{r-1}v$, where $r > 1$ will be specified later. The inequality

$$\|v\|_{N/(N-1)} \leq \left(\prod_1^N \|\partial_i v\|_1 \right)^{1/N}$$

implies

$$\|u\|_{rN/(N-1)}^r \leq r \left(\prod_1^N \| \, |u|^{r-1}\partial_i u\|_1 \right)^{1/N}.$$

If $q = p/(p-1)$ is the conjugate exponent of p, then using the Hölder inequality we get

$$\| \, |u|^{r-1}\partial_i u\| \leq \|u\|_{q(r-1)}^{r-1} \|\partial_i u\|_p.$$

Consequently,

$$\|u\|_{rN/(N-1)}^r \leq r\|u\|_{q(r-1)}^{r-1} \left(\prod_1^N \|\partial_i u\|_p \right)^{1/N}.$$

Now choose r such that $rN/(N-1) = q(r-1)$. This gives

$$r = p^* \frac{N-1}{N},$$

and we get

$$\|u\|_{p^*} \leq r \left(\prod_1^N \|\partial_i u\|_p \right)^{1/N} \leq C(N, p)\| \, |du| \, \|_p.$$

This shows $L^{1,p} \hookrightarrow L^{p^*}$ if $1 \leq p < N$. The general case

$$L^{k,p} \hookrightarrow L^{m,q} \quad \text{if } \sigma_N(k, p) = \sigma_N(m, q) < 0 \;\; k > m$$

follows easily by induction over k. We let the reader fill in the details. $\qquad\square$

Theorem 10.2.24 (Rellich-Kondrachov). *Let $(k, p), (m, q) \in \mathbb{Z}_+ \times [1, \infty)$ such that*

$$k > m \;\; and \;\; 0 > \sigma_N(k, p) > \sigma_N(m, q).$$

Then any bounded sequence $(u_n) \subset L^{k,p}(\mathbb{R}^N)$ supported in a ball $B_R(0)$, $R > 0$ has a subsequence strongly convergent in $L^{m,q}(\mathbb{R}^N)$.

Proof.   We discuss only the case $k = 1$, so that the condition $\sigma_N(1, p) < 0$ imposes $1 \le p < N$. Let (u_n) be a bounded sequence in $L^{1,p}(\mathbb{R}^N)$ such that

$$\text{ess supp}\, u_n \subset \{|x| \le R\} \;\; \forall n.$$

We have to show that, for every $1 \le q < p^* = Np/(N - p)$, the sequence (u_n) contains a subsequence convergent in L^q. The proof will be carried out in several steps.

Step 1. We will prove that, for every $0 < \delta < 1$, the mollified sequence $u_{n,\delta} = \rho_\delta * u_n$ admits a subsequence which is *uniformly convergent on* the ball $\{|x| \le R + 1\}$.

To prove this, we will use the Arzéla-Ascoli theorem, and we will show there exists $C = C(\delta) > 0$, such that

$$|u_{n,\delta}(x)| < C \;\; \forall n, \; \forall |x| \le R + 1,$$

$$|u_{n,\delta}(x_1) - u_{n,\delta}(x_2)| \le C|x_1 - x_2| \;\; \forall n, \; \forall |x_1|, |x_2| \le R + 1.$$

Indeed,

$$|\rho_\delta * u\, (x)| \le \delta^{-N} \int_{|y-x| \le \delta} \rho\left(\frac{x - y}{\delta}\right) |u_n(y)| dy \le \delta^{-N} \int_{B_\delta(x)} |u_n(y)| dy$$

$$\le C(N, p)\delta^{-N} \|u_n\|_{p^*} \cdot \text{vol}\, (B_\delta)^{(p^* - 1)/p^*}$$

$$\le C(\delta)\|u_n\|_{1,p} \;\; \text{(by Sobolev embedding theorem).}$$

Similarly,

$$|u_{n,\delta}(x_1) - u_{n,\delta}(x_2)| \le \int_{B_{R+1}} |\rho_\delta(x_1 - y) - \rho_\delta(x_2 - y)| \cdot |u_n(y)| dy$$

$$\le C(\delta) \cdot |x_1 - x_2| \int_{B_{R+1}} |u_n(y)| dy \le C(\delta) \cdot |x_1 - x_2| \cdot \|u_n\|_{1,p}.$$

Step 1 is completed.

Step 2: Conclusion. Using the diagonal procedure, we can extract a subsequence of (u_n), still denoted by (u_n), and a subsequence $\delta_n \searrow 0$ such that, if we set $g_n := u_{n,\delta_n}$, the following hold.

(i) The sequence g_n is uniformly convergent on B_R.
(ii) $\lim_n \|g_n - u_n\|_{1,p,\mathbb{R}^N} = 0$.

We claim the subsequence (u_n) as above is convergent in $L^q(B_R)$, for all $1 \le q < p$. Indeed ,for all n, m

$$\|u_n - u_m\|_{q,B_R} \le \|u_n - g_n\|_{q,B_R} + \|g_n - g_m\|_{q,B_R} + \|g_m - u_m\|_{q,B_R}.$$

We now examine separately each of the three terms on the right-hand side.

A. The first and the third term.

$$\|u_n - g_n\|_{q,B_R}^q = \int_{B_R} |u_n(x) - g_n(x)| dx$$

$$\text{(Hölder inequality)} \leq \left(\int_{B_R} |u_n(x) - g_n(x)|^q dx \right)^{1/r} \cdot \text{vol}\,(B_R)^{(r-1)/r} \quad (r = p^*/q)$$

$$\leq C(R)\|u_n - g_n\|_{p^*,\mathbb{R}^N}^q \overset{\text{(Sobolev)}}{\leq} C(R)\|u_n - g_n\|_{1,p,\mathbb{R}^N} \to 0.$$

B. The middle term.

$$\|g_n - g_m\|_{q,B_R}^q \leq \{\sup_{B_R} |g_n(x) - g_m(x)|\}^q \cdot \text{vol}\,(B_R) \to 0.$$

Hence the (sub)sequence (u_n) is Cauchy in $L^q(B_R)$, and thus it converges. The compactness theorem is proved. $\qquad\qquad\square$

10.2.3   *Embedding theorems: differentiability properties*

A priori, the functions in the Sobolev spaces $L^{k,p}$ are only measurable, and are defined only almost everywhere. However, if the strength $\sigma_N(k,p)$ is sufficiently large, then the functions of $L^{k,p}$ have a built-in regularity: each can be modified on a negligible set to become continuous, and even differentiable.

To formulate our next results we must introduce another important family of Banach spaces, namely the spaces of Hölder continuous functions.

Let $\alpha \in (0,1)$. A function $u : D \subset \mathbb{R}^N \to \mathbb{R}$ is said to be α-*Hölder continuous* if

$$[u]_\alpha := \sup_{0 < R < 1, z \in D} R^{-\alpha} \text{osc}\,(u;\, B_R(z) \cap D) < \infty,$$

where for any set $S \subset D$, we denoted by $\text{osc}\,(u;\, S)$ the oscillation of u on S, i.e.,

$$\text{osc}\,(u;\, S) := \sup\{|u(x) - u(y)|\,;\, x, y \in S\}.$$

Set

$$\|u\|_{\infty, D} := \sup_{x \in D} |u(x)|,$$

and define

$$C^{0,\alpha}(D) := \{\, u : D \to \mathbb{R}\,;\, \|u\|_{0,\alpha,D} := \|u\|_\infty + [u]_\alpha < \infty\,\}.$$

More generally, for every integer $k \geq 0$, define

$$C^{k,\alpha}(D) := \{\, u \in C^m(D);\, \partial^\beta u \in C^{0,\alpha}(D),\, \forall |\beta| \leq k\,\}.$$

The space $C^{k,\alpha}(D)$ is a Banach space with respect to the norm,

$$\|u\|_{k,\alpha} := \sum_{|\beta| \leq k} \|\partial^\beta u\|_{\infty, D} + \sum_{|\beta| = k} [\partial^\beta u]_{\alpha, D}.$$

Define the strength of the Hölder space $C^{k,\alpha}$ as the quantity

$$\sigma(k, \alpha) := k + \alpha.$$

Theorem 10.2.25 (Morrey). *Consider* $(m, p) \in \mathbb{Z}_+ \times [1, \infty]$ *and* $(k, \alpha) \in \mathbb{Z}_+ \times (0, 1)$ *such that* $m > k$ *and* $\sigma_N(m, p) = \sigma(k, \alpha) > 0$. *Then* $L^{m,p}(\mathbb{R}^N)$ *embeds continuously in* $C^{k,\alpha}(\mathbb{R}^N)$.

Proof. We consider only the case $k = 1$, and (necessarily) $m = 0$. The proof relies on the following elementary observation.

Lemma 10.2.26. *Let* $u \in C^\infty(B_R) \cap L^{1,1}(B_R)$ *and set*

$$\overline{u} := \frac{1}{\text{vol}\,(B_R)} \int_{B_R} u(x)dx.$$

Then

$$|u(x) - \overline{u}| \leq \frac{2^N}{\sigma_{N-1}} \int_{B_R} \frac{|du(y)|}{|x - y|^{N-1}}dy. \tag{10.2.4}$$

In the above inequality $\boldsymbol{\sigma}_{N-1}$ *denotes the "area" of the unit* $(N-1)$-*dimensional round sphere* $S^{N-1} \subset \mathbb{R}^N$.

Proof.

$$u(x) - u(y) = -\int_0^{|x-y|} \frac{\partial}{\partial r} u(x + r\omega)dr \quad (\omega = -\frac{x - y}{|x - y|}).$$

Integrating the above equality with respect to y we get,

$$\text{vol}\,(B_R)(u(x) - \overline{u}) = -\int_{B_R} dy \int_0^{|x-y|} \frac{\partial}{\partial r} u(x + r\omega)dr.$$

If we set $|\partial_r u(x + r\omega)| = 0$ for $|x + r\omega| > R$, then

$$\text{vol}\,(B_R)|u(x) - \overline{u}| \leq \int_{|x-y| \leq 2R} dy \int_0^\infty |\partial_r u(x + r\omega)|dr.$$

If we use polar coordinates (ρ, ω) centered at x, then $dy = \rho^{N-1}d\rho d\omega$, where $\rho = |x - y|$, and $d\omega$ denotes the Euclidean "area" form on the unit round sphere. We deduce

$$\text{vol}\,(B_R)|u(x) - \overline{u}| \leq \int_0^\infty dr \int_{S^{N-1}} d\omega \int_0^{2R} |\partial_r u(x + r\omega)|\rho^{N-1}d\rho$$

$$= \frac{(2R)^N}{N} \int_0^\infty dr \int_{S^{N-1}} |\partial_r (x + r\omega)|d\omega dr$$

$$= \frac{(2R)^N}{N} \int_0^\infty r^{N-1}dr \int_{S^{N-1}} \frac{1}{r^{N-1}} |\partial_r u(x + r\omega)|d\omega$$

$$(z = x + r\omega) \quad = \frac{(2R)^N}{N} \int_{B_R} \frac{|\partial_r u(z)|}{|x - z|^{N-1}} \leq \frac{(2R)^N}{N} \int_{B_R} \frac{|du(z)|}{|x - z|^{N-1}}dy. \qquad \square$$

We want to make two simple observations.

1. In the above lemma, we can replace the round ball B_R centered at origin by any other ball, centered at any other point. In the sequel for any $R > 0$, and any $x_0 \in \mathbb{R}^N$ we set

$$\overline{u}_{x_0,R} := \frac{1}{\mathrm{vol}\,(B_R(x_0))} \int_{B_R(x_0)} u(y)dy.$$

2. The inequality (10.2.4) can be extended by density to any $u \in L^{1,1}(B_R)$.

We will complete the proof of Morrey's theorem in three steps.

Step 1. L^∞-*estimates.* We will show there exists $C > 0$ such that, $\forall u \in L^{1,p}(\mathbb{R}^N) \cap C^1(\mathbb{R}^N)$

$$\|u\|_\infty \leq C\|u\|_{1,p}.$$

For each $x \in \mathbb{R}^N$, denote by $B(x)$ the unit ball centered at x, and set $\overline{u}_x := \overline{u}_{x,1}$. Using (10.2.4) we deduce

$$|u(x)| \leq |\overline{u}_x| + C_N \int_{B(x)} \frac{|du(y)|}{|x-y|^{N-1}} dy \leq C\left(\|u\|_p + \int_{B(x)} \frac{|du(y)|}{|x-y|^{N-1}} dy \right).$$
$$(10.2.5)$$

Since $\sigma_N(1,p) > 0$, we deduce that $p > N$, so that its conjugate exponent q satisfies

$$q = \frac{p}{p-1} < \frac{N}{N-1}.$$

In particular, the function $y \mapsto |x-y|^{-(N-1)}$ lies in $L^q(B(x))$, and

$$\int_{B(x)} \frac{1}{|x-y|^{q(N-1)}} dy \leq C(N,q),$$

where $C(N,q)$ is a universal constant depending only on N, and q. Using the Hölder inequality in (10.2.5) we conclude

$$|u(x)| \leq C\|u\|_{1,p} \quad \forall x.$$

Step 2. *Oscillation estimates.* We will show there exists $C > 0$ such that for all $u \in L^{1,p}(\mathbb{R}^N) \cap C^1(\mathbb{R}^N)$

$$[u]_\alpha \leq C\|u\|_{1,p}.$$

Indeed, from the inequality (10.2.4) we deduce that, for any ball $B_R(x_0)$, and any $x \in B_R(x_0)$,

$$|u(x) - \overline{u}_{x_0,R}| \leq C \int_{B_R(x_0)} \frac{|du(y)|}{|x-y|^{N-1}}$$

$(q = \frac{p}{p-1})$

$$\overset{\text{(Hölder)}}{\leq} C\|\,|du|\,\|_p \cdot \left(\int_{B_R(x_0)} \frac{1}{|x-y|^{q(N-1)}} dy \right)^{1/q} \leq C\|\,|du|\,\|_p R^\nu,$$

where

$$\nu = \frac{1}{q}\big(N - q(N - 1)\big) = 1 - \frac{N}{p} = \alpha.$$

Hence

$$|u(x) - \overline{u}_{x_0,R}| \leq CR^\alpha \quad \forall x \in B_R(x_0),$$

and consequently,

$$\mathrm{osc}\,(u; B_R(x_0)) \leq CR^\alpha.$$

Step 2 is completed.

Step 3:. *Conclusion.* Given $u \in L^{1,p}(\mathbb{R}^N)$ we can find $(u_n) \in C_0^\infty(\mathbb{R}^N)$ such that

$$u_n \to u \text{ in } L^{1,p} \text{ and almost everywhere.}$$

The estimates established at Step 1 and 2 are preserved as $n \to \infty$. In fact, these estimates actually show that the sequence (u_n) converges in the $C^{0,\alpha}$-norm to a function $v \in C^{0,\alpha}(\mathbb{R}^N)$ which agrees almost everywhere with u. $\square$

Exercise 10.2.27. Let $u \in L^{1,1}(\mathbb{R}^N)$ satisfy a (q, ν)-*energy estimate*, i.e.

$$\exists C > 0: \quad \frac{1}{r^N} \int_{B_r(x)} |du(y)|^q dy \leq C_1 r^{-\nu} \quad \forall x \in \mathbb{R}^N,\, 0 < r < 2,$$

where $0 \leq \nu < q$ and $q > 1$. Show that (up to a change on a negligible set) the function u is α-Hölder continuous, $\alpha = 1 - \nu/q$, and moreover

$$[u]_\alpha \leq C_2,$$

where the constant C_2 depends only on N, q, ν and C_1.

Hint: Prove that

$$\int_{B_r(x)} \frac{|du(y)|}{|x - y|^{N-1}} \leq Cr^\alpha,$$

and then use the inequality (10.2.4). $\square$

Remark 10.2.28. The result in the above exercise has a suggestive interpretation. If u satisfies the (q, ν)-energy estimate then although $|du|$ may not be bounded, on average, it "explodes" no worse that $r^{\alpha-1}$ as $r \to 0$. Thus

$$|u(x) - u(0)| \approx C \int_0^{|x|} t^{\alpha-1} dt \approx C|x|^\alpha.$$

The energy estimate is a very useful tool in the study of nonlinear elliptic equations. $\square$

The Morrey embedding theorem can be complemented by a compactness result. Let $(k, p) \in \mathbb{Z}_+ \times [1, \infty]$ and $(m, \alpha) \in \mathbb{Z}_+ \times (0, 1)$ such that

$$\sigma_N(k, p) > \sigma(m, \alpha) \quad k > m.$$

Then a simple application of the Arzela-Ascoli theorem shows that any bounded sequence in $L^{k,p}(\mathbb{R}^N)$ admits a subsequence which converges in the $C^{m,\alpha}$ norm on any bounded open subset of $\mathbb{R}^N$.

The last results we want to discuss are the *interpolation inequalities*. They play an important part in applications, but we chose not to include their long but elementary proofs since they do not use any concept we will need later. The interested reader may consult [3] or [12] for details.

Theorem 10.2.29 (Interpolation inequalities). *For each $R > 0$ choose a smooth, cutoff function $\eta_R \in C_0^\infty(\mathbb{R}^N)$ such that*

$$\eta_R \equiv 1 \text{ if } |x| \leq R,$$

$$\eta_R \equiv 0 \text{ if } |x| \geq R + 1,$$

and

$$|d\eta_R(x)| \leq 2 \ \forall x \in \mathbb{R}^N.$$

Fix $(m, p) \in \mathbb{Z}_+ \times [1, \infty)$, and $(k, \alpha) \in \mathbb{Z}_+ \times (0, 1)$.
(a) For every $0 < r \leq R + 1$, there exists $C = C(r, R, m, p)$ such that, for every $0 \leq j < m$, $\varepsilon > 0$ and for all $u \in L^{m,p}(\mathbb{R}^N)$, we have

$$\|\eta_R u\|_{j,p,\mathbb{R}^N} \leq C\varepsilon \|\eta_R u\|_{m,p,\mathbb{R}^N} + C\varepsilon^{-j(m-j)} \|\eta_R u\|_{p, B_r}.$$

(b) For every $0 < r \leq R + 1$, there exists $C = C(r, R, k, \alpha)$ such that, for every $0 \leq j < k$, $\varepsilon > 0$, and for all $u \in C^{k,\alpha}(\mathbb{R}^N)$, we have

$$\|\eta_R u\|_{j,\alpha,\mathbb{R}^N} \leq C\varepsilon \|\eta_R u\|_{k,\alpha,\mathbb{R}^N} + C\varepsilon^{-j(m-j)} \|\eta_R u\|_{0,\alpha, B_r}. \qquad \square$$

The results in this, and the previous section extend verbatim to slightly more general situations, namely to functions $f : \mathbb{R}^N \to H$, where H is a finite dimensional complex Hermitian space.

10.2.4 *Functional spaces on manifolds*

The Sobolev and the Hölder spaces can be defined over manifolds as well. To define these spaces we need three things: an oriented Riemann manifold (M, g), a $\mathbb{K}$-vector bundle $\pi : E \to M$ endowed with a metric $h = \langle \bullet, \bullet \rangle$, and a connection $\nabla = \nabla^E$ compatible with h. The metric $g = (\bullet, \bullet)$ defines two important objects.

(i) The Levi-Civita connection ∇^g.
(ii) A volume form $dV_g = *1$. In particular, dV_g defines a Borel measure on M. We denote by $L^p(M, \mathbb{K})$ the space of $\mathbb{K}$-valued p-integrable functions on (M, dV_g) (modulo the equivalence relation of equality almost everywhere).

Definition 10.2.30. Let $p \in [1, \infty]$. An L^p-section of E is a Borel measurable map $\psi : M \to E$, i.e., $\psi^{-1}(U)$ is Borel measurable for any open subset $U \subset E$ such that the following hold.

(a) $\pi \circ \psi(x) = x$, for almost all $x \in M$ except possibly a negligible set.
(b) The function $x \mapsto |\psi(x)|_h^p$ is integrable with respect to the measure defined by dV_g. $\qquad \square$

The space of L^p-sections of E (modulo equality almost everywhere) is denoted by $L^p(E)$. We let the reader check the following fact.

Proposition 10.2.31. *$L^p(E)$ is a Banach space with respect to the norm*

$$\|\psi\|_{p,E} = \begin{cases} \left(\int_M |\psi(x)|^p dv_g(x) \right)^{1/p} & \text{if } p < \infty \\ \operatorname{ess\,sup}_x |\psi(x)| & \text{if } p = \infty \end{cases} . \qquad \square$$

Note that if $p, q \in [1, \infty]$ are conjugate, $1/p + 1/q = 1$, then the metric $h : E \times E \to \mathbb{K}_M$ defines a continuous pairing

$$\langle \bullet, \bullet \rangle : L^p(E) \times L^q(E) \to L^1(M, \mathbb{K}),$$

i.e.,

$$\left| \int_M \langle \psi, \phi \rangle dV_g \right| \leq \|\psi\|_{p,E} \cdot \|\phi\|_{q,E}$$

This follows immediately from the Cauchy inequality

$$|h(\psi(x), \phi(x)| \leq |\psi(x)| \cdot |\phi(x)| \quad \text{a.e. on } M,$$

and the usual Hölder inequality.

Exercise 10.2.32. Let $E_i \to M$ $(i = 1, \dots, k)$ be vector bundles with metrics and consider a multilinear bundle map

$$\Xi : E_1 \times \cdots \times E_k \to \mathbb{K}_M.$$

We regard Ξ as a section of $E_1^* \otimes \cdots \otimes E_k^*$. If

$$\Xi \in L^{p_0}(E_1^* \otimes \cdots \otimes E_k^*),$$

then for every $p_1, \dots, p_k \in [1, \infty]$, such that

$$1 - 1/p_0 = 1/p_1 + \cdots + 1/p_k,$$

and $\forall \psi_j \in L^{p_j}(E_j)$, $j = 1, \dots, k$

$$\int_M \left| \Xi(\psi_1, \dots, \psi_k) \right| dV_g \leq \|\Xi\|_{p_0} \cdot \|\psi_1\|_{p_1} \cdots \|\psi_k\|_{p_k}. \qquad \square$$

For each $m = 1, 2, \ldots$, define ∇^m as the composition

$$\nabla^m : \; C^\infty(E) \xrightarrow{\nabla^E} C^\infty(T^*M \otimes E) \xrightarrow{\nabla^{T^*M \otimes E}} \cdots \xrightarrow{\nabla} C^\infty(T^*M^{\otimes m} \otimes E),$$

where we used the symbol ∇ to generically denote the connections in the tensor products $T^*M^{\otimes j} \otimes E$ induced by ∇^g and ∇^E.

The metrics g and h induce metrics in each of the tensor bundles $T^*M^{\otimes m} \otimes E$, and in particular, we can define the spaces $L^p(T^*M^{\otimes m} \otimes E)$.

Definition 10.2.33. (a) Let $u \in L^1_{loc}(E)$ and $v \in L^1_{loc}(T^*M^{\otimes m} \otimes E)$. We say that $\nabla^m u = v$ *weakly* if

$$\int_M \langle v, \phi \rangle dV_g = \int \langle u, (\nabla^m)^* \phi \rangle dV_g \quad \forall u \in C^\infty_0(T^*M^{\otimes m} \otimes E).$$

(b) Define $L^{m,p}(E)$ as the space of sections $u \in L^p(E)$ such that, $\forall j = 1, \ldots, m$, there exist $v_j \in L^p(T^*M^{\otimes j} \otimes E)$, such that $\nabla^j u = v_j$ weakly. We set

$$\|u\|_{m,p} := \|u\|_{m,p,E} = \sum_{j=1}^p \|\nabla^j u\|_p. \qquad \square$$

☞ **A word of warning.** The Sobolev space $L^{m,p}(E)$ introduced above depends on several choices: the metrics on M and E and the connection on E. *When M is non-compact this dependence is very dramatic and has to be seriously taken into consideration.*

Example 10.2.34. Let (M, g) be the space $\mathbb{R}^N$ endowed with the Euclidean metric. The trivial line bundle $E = \underline{\mathbb{R}}_M$ is naturally equipped with the trivial metric and connection. Then, $L^p(\underline{\mathbb{R}}_M) = L^p(M, \mathbb{R})$.

Denote by D the Levi-Civita connection. Then, for every $u \in C^\infty(M)$, and $m \in \mathbb{Z}_+$, we have

$$D^m u = \sum_{|\alpha|=m} dx^{\otimes \alpha} \otimes \partial^\alpha u,$$

where for every multi-index α we denoted by $dx^{\otimes \alpha}$ the monomial

$$dx^{\alpha_1} \otimes \cdots \otimes dx^{\alpha_N}.$$

The length of $D^m u(x)$ is

$$\left(\sum_{|\alpha|=m} |\partial^\alpha u(x)|^2 \right)^{1/2}.$$

The space $L^{m,p}(\underline{\mathbb{R}}_M)$ coincides as a set with the Sobolev space $L^{k,p}(\mathbb{R}^N)$. The norm $\|\bullet\|_{m,p,\underline{\mathbb{R}}_M}$ is equivalent with the norm $\|\bullet\|_{m,p,\mathbb{R}^N}$ introduced in the previous sections. $\qquad \square$

Proposition 10.2.35. $(L^{k,p}(E), \|\cdot\|_{k,p,E})$ *is a Banach space which is reflexive if* $1 < p < \infty$. $\qquad \square$

The proof of this result is left to the reader as an exercise.

The Hölder spaces can be defined on manifolds as well. If (M, g) is a Riemann manifold, then g canonically defines a metric space structure on M, (see Chapter 4) and in particular, we can talk about the oscillation of a function $u : M \to \mathbb{K}$. On the other hand, defining the oscillation of a section of some bundle over M requires a little more work.

Let (E, h, ∇) as before. We assume that the *injectivity radius* ρ_M of M is positive. Set $\rho_0 := \min\{1, \rho_M\}$. If $x, y \in M$ are two points such that $\text{dist}_g(x, y) \leq \rho_0$, then they can be joined by a unique minimal geodesic $\gamma_{x,y}$ starting at x, and ending at y. We denote by $T_{x,y} : E_y \to E_x$ the ∇^E-parallel transport along $\gamma_{x,y}$. For each $\xi \in E_x$, and $\eta \in E_y$ we set

$$|\xi - \eta| := |\xi - T_{x,y}\eta|_x = |\eta - T_{y,x}\xi|_y.$$

If $u : M \to E$ is a section of E, and $S \subset M$ has the diameter $< \rho_0$, we define

$$\text{osc}\,(u;\ S) := \sup\{\ |u(x) - u(y)|;\ \ x, y \in S\ \}.$$

Finally set

$$[u]_{\alpha, E} := \sup\{\ r^{-\alpha}\text{osc}\,(u;\ B_r(x));\ \ 0 < r < \rho_0,\ x \in M\ \}.$$

For any $k \geq 0$, we define

$$\|u\|_{k, \alpha, E} := \sum_{j=0}^{k} \|\nabla^j u\|_{\infty, E} + [\nabla^m u]_{\alpha, T^*M^{\otimes m} \otimes E},$$

and we set

$$C^{k, \alpha}(E) := \{\ u \in C^k(E);\ \|u\|_{k, \alpha} < \infty\ \}.$$

Theorem 10.2.36. *Let (M, g) be a compact, N-dimensional, oriented Riemann manifold, and E a vector bundle over M equipped with a metric h, and compatible connection ∇. Then the following are true.*
(a) The Sobolev space $L^{m,p}(E)$, and the Hölder spaces $C^{k, \alpha}(E)$ do not depend on the metrics g, h and on the connection ∇. More precisely, if g_1 is a different metric on M, and ∇^1 is another connection on E compatible with some metric h_1 then

$$L^{m,p}(E, g, h, \nabla) = L^{m,p}(E, g_1, h_1, \nabla^1) \quad \text{as sets of sections,}$$

and the identity map between these two spaces is a Banach space is continuous. A similar statement is true for the Hölder spaces.
(b) If $1 \leq p < \infty$, then $C^\infty(E)$ is dense in $L^{k,p}(E)$.
(c) If $(k_i, p_i) \in \mathbb{Z}_+ \times [1, \infty)$ $(i = 0, 1)$ are such that

$$k_0 \geq k_1 \quad \text{and} \quad \sigma_N(k_0, p_0) = k_0 - N/p_0 \geq k_1 - N/p_1 = \sigma_N(k_1, p_1),$$

then $L^{k_0, p_0}(E)$ embeds continuously in $L^{k_1, p_1}(E)$. If moreover,

$$k_0 > k_1 \quad \text{and} \quad k_0 - N/p_0 > k_1 - N/p_1,$$

then the embedding $L^{k_0,p_0}(E) \hookrightarrow L^{k_1,p_1}(E)$ is compact, i.e., any bounded sequence of $L^{k_0,p_0}(E)$ admits a subsequence convergent in the L^{k_1,p_1}-norm.

(d) If $(m,p) \in \mathbb{Z}_+ \times [1,\infty)$, $(k,\alpha) \in \mathbb{Z}_+ \times (0,1)$ and

$$m - N/p \geq k + \alpha,$$

then $L^{m,p}(E)$ embeds continuously in $C^{k,\alpha}(E)$. If moreover

$$m - N/p > k + \alpha,$$

then the embedding is also compact. □

We developed all the tools needed to prove this theorem, and we leave this task to the reader. The method can be briefly characterized by two key phrases: partition of unity and interpolation inequalities. We will see them at work in the next section.

10.3 Elliptic partial differential operators: analytic aspects

This section represents the analytical heart of this chapter. We discuss two notions that play a pivotal role in the study of elliptic partial differential equations. More precisely, we will introduce the notion of *weak solution*, and *a priori estimates*.

Consider the following simple example. Suppose we want to solve the partial differential equation

$$\Delta u + u = f \in L^2(S^2, \mathbb{R}), \tag{10.3.1}$$

where Δ denotes the Laplace-Beltrami operator on the round sphere, $\Delta = d^* d$. Riemann suggested that one should look for the minima of the energy functional

$$u \longmapsto E(u) := \int_{S^2} \left\{ \frac{1}{2}(|du|^2 + u^2) - fu \right\} dV_g.$$

If u_0 is a minimum of E, i.e.,

$$E(u_0) \leq E(u), \quad \forall u,$$

then

$$0 = \frac{d}{dt}|_{t=0} E(u_0 + tv) = \int_{S^2} \left\{ (du_0, dv)_g + u_0 \cdot v - f \cdot v \right\} dV_g \quad \forall v. \tag{10.3.2}$$

Integrating by parts we get

$$\int_{S^2} (d^* du_0 + u_0 - f) \cdot v \, dV_g = 0 \quad \forall v,$$

so that necessarily,

$$\Delta u_0 + u_0 = f.$$

There are a few grey areas in this approach, and Weierstrass was quick to point them out: what is the domain of E, u_0 may not exist, and if it does, it may not be C^2, so that the integration by parts is illegal etc. This avenue was abandoned until the dawns of the twentieth century, when Hilbert reintroduced them into the spotlight, and emphasized the need to deal with these issues.

His important new point of view was that the approach suggested by Riemann does indeed produce a solution of (10.3.1) *"provided if need be that the notion of solution be suitable extended"*. The suitable notion of solution is precisely described in (10.3.2). Naturally, one asks when this extended notion of solution coincides with the classical one. Clearly, it suffices that u of (10.3.2) be at least C^2, so that everything boils down to a question of *regularity*.

Riemann's idea was first rehabilitated in Weyl's famous treatise [98] on Riemann surfaces. It took the effort of many talented people to materialize Hilbert's program formulated as his 19th and 20th problem in the famous list of 27 problems he presented at the Paris conference at the beginning of the 20th century. We refer the reader to [1] for more details.

This section takes up the issues raised in the above simple example. The key fact which will allow us to legitimize Riemann's argument is the ellipticity of the partial differential operator involved in this equation.

10.3.1 *Elliptic estimates in* $\mathbb{R}^N$

Let $E = \mathbb{C}^r$ denote the trivial vector bundle over $\mathbb{R}^N$. Denote by $\langle \bullet, \bullet \rangle$ the natural Hermitian metric on E, and by ∂, the trivial connection. The norm of $L^{k,p}(E)$ (defined using the Euclidean volume) will be denoted by $\| \bullet \|_{k,p}$. Consider an elliptic operator of order m

$$L = \sum_{|\alpha| \leq m} A_\alpha(x)\partial_x^\alpha : C^\infty(E) \to C^\infty(E).$$

For any $k \in \mathbb{Z}_+$, and any $R > 0$ define

$$\|L\|_{k,R} = \sum_{|\alpha| \leq m, |\beta| \leq k+m-|\alpha|} \sup_{B_R(0)} \|\partial_x^\beta(A_\alpha(x))\|.$$

In this subsection we will establish the following fundamental result.

Theorem 10.3.1 (Interior elliptic estimates). *(a) Let $(k,p) \in \mathbb{Z}_+ \times (1, \infty)$, and $R > 0$. Then, there exists $C = C(\|L\|_{k+1,R}, k, p, N, R) > 0$ such that, $\forall u \in C_0^\infty(E|_{B_R(0)})$, we have*

$$\|u\|_{k+m,p} \leq C\big(\|Lu\|_{k,p} + \|u\|_p \big). \tag{10.3.3}$$

(b) Let $(k,\alpha) \in \mathbb{Z}_+ \times (0,1)$, and $R > 0$. Then, there exists $C = C(\|L\|_{k+1,R}, k, \alpha, N, R) > 0$ such that, $\forall u \in C_0^\infty(E|_{B_R(0)})$, we have

$$\|u\|_{k+m,\alpha} \leq C\big(\|Lu\|_{k,\alpha} + \|u\|_{0,\alpha} \big). \tag{10.3.4}$$

Proof. The proof consists of two conceptually distinct parts. In the first part we establish the result under the supplementary assumption that L has *constant coefficients*. In the second part, the general result is deduced from the special case using perturbation techniques in which the interpolation inequalities play an important role. Throughout the proof we will use the same letter C to denote various constants $C = C(\|L\|_{k+1,R}, k, p, N, R) > 0$.

Step 1. *Constant coefficients case.* We assume L has the form

$$L = \sum_{|\alpha|=m} A_\alpha \partial_x^\alpha,$$

where A_α are $r \times r$ complex matrices, independent of $x \in \mathbb{R}^N$. We set

$$\|L\| := \sum \|A_\alpha\|.$$

We will prove the conclusions of the theorem hold in this special case. We will rely on a very deep analytical result whose proof goes beyond the scope of this book.

For each $f \in L^1(\mathbb{R}^N, \mathbb{C})$ denote by $\hat{f}(\xi)$ its Fourier transform

$$\hat{f}(\xi) := \frac{1}{(2\pi)^{N/2}} \int_{\mathbb{R}^N} \exp(-ix \cdot \xi) dx.$$

Theorem 10.3.2 (Calderon-Zygmund). *Let $\overline{m} : S^{N-1} \to \mathbb{C}$ be a smooth function, and define*

$$m(\xi) : \mathbb{R}^N \setminus \{0\} \to \mathbb{C}$$

by

$$m(\xi) = \overline{m} \left(\frac{\xi}{|\xi|} \right).$$

Then the following hold.
(a) There exist $\Omega \in C^\infty(S^{N-1}, \mathbb{C})$, and $c \in \mathbb{C}$, such that
  (a_1) $\int_{S^{N-1}} \Omega dv_{S^{N-1}} = 0$.
  (a_2) For any $u \in C_0^\infty(\mathbb{R}^N)$, the limit

$$(Tu)(x) = \lim_{\varepsilon \searrow 0} \int_{|y| \geq \varepsilon} \frac{\Omega(y)}{|y|^N} u(x - y) dy$$

exists for almost every $x \in \mathbb{R}^N$, and moreover

$$cu(x) + Tu(x) = \frac{1}{(2\pi)^{N/2}} \int_{\mathbb{R}^N} \exp(ix \cdot \xi) m(\xi) \hat{u}(\xi) d\xi.$$

(b) For every $1 < p < \infty$ there exists $C = C(p, \|\overline{m}\|_\infty) > 0$ such that,

$$\|Tu\|_p \leq C\|u\|_p \quad \forall u \in C_0^\infty(\mathbb{R}^N).$$

*(c)(**Korn-Lichtenstein**). For every $0 < \alpha < 1$, and any $R > 0$, there exists $C = C(\alpha, \|\overline{m}\|_{0,\alpha}, R) > 0$ such that, $\forall u \in C_0^{0,\alpha}(B_R)$*

$$[Tu]_\alpha \leq C\|u\|_{0,\alpha,B_R}. \qquad \square$$

For a proof of part (a) and (b) we refer to [94], Chap II §4.4. Part (c) is "elementary" and we suggest the reader to try and prove it. In any case a proof, of this inequality can be found in [12], Part II.5.

Let us now return to our problem. Assuming L has the above special form we will prove (10.3.3). The proof of (10.3.4) is entirely similar and is left to the reader. We discuss first the case $k = 0$.

Let $u \in C_0^\infty(E|_{B_R})$. The function u can be viewed as a collection

$$u(x) = (u^1(x), \dots, u^r(x)),$$

of smooth functions compactly supported in B_R. Define

$$\hat{u}(\xi) := (\hat{u}^1(\xi), \dots, \hat{u}^r(\xi)).$$

If we set $v = Lu$, then for any multi-index β such that $|\beta| = m$, we have

$$L\partial^\beta u = \partial^\beta L u = \partial^\beta v,$$

because L has constant coefficients. We Fourier transform the above equality, and we get

$$(-i)^m \sum_{|\alpha|=m} A_\alpha \xi^\alpha \widehat{\partial^\beta u}(\xi) = (-i)^m \xi^\beta \hat{v}(\xi). \tag{10.3.5}$$

Note that since L is *elliptic*, the operator

$$\sigma(L)(\xi) = A(\xi) = \sum_{|\alpha|=m} A_\alpha \xi^\alpha : \mathbb{C}^r \to \mathbb{C}^r$$

is invertible for any $\xi \neq 0$. From (10.3.5) we deduce

$$\widehat{\partial^\beta u}(\xi) = \xi^\beta B(\xi) \hat{v}(\xi) \quad \forall \xi \neq 0,$$

where $B(\xi) = A(\xi)^{-1}$. Note that $B(\xi)$ is homogeneous of degree $-m$, so that $M(\xi) = \xi^\beta B(\xi)$ is homogeneous of degree 0. Thus, we can find functions $m_{ij}(\xi) \in C^\infty(\mathbb{R}^n \setminus \{0\})$, which are homogeneous of degree 0, and such that

$$\widehat{\partial^\beta u^i}(\xi) = \sum_j m_{ij}(\xi) \hat{v}^j(\xi).$$

Using Theorem 10.3.2 (a) and (b) we deduce

$$\|\partial^\beta u\|_p \leq C\|v\|_p = C\|Lu\|_p.$$

This proves (10.3.3) when L has this special form.

Step 2. *The general case.* Suppose now that L is an arbitrary elliptic operator of order m. Let $r > 0$ sufficiently small (to be specified later). Cover B_R by finitely many balls $B_r(x_\nu)$, and consider $\eta_\nu \in C_0^\infty(B_r(x_\nu))$, such that each point in B_R is covered by at most 10^N of these balls, and

$$\eta_\nu \geq 0, \quad \sum_\nu \eta_\nu = 1,$$

$$\|\partial^\beta \eta_\nu\| \le Cr^{-|\beta|} \quad \forall |\beta| \le m.$$

If $u \in C_0^\infty(B_R)$, then

$$v = Lu = L\left(\sum_\nu \eta_\nu u\right) = \sum_\nu L(\eta_\nu u).$$

Set $u_\nu := \eta_\nu u$, and

$$L_\nu := \sum_{|\alpha|=m} A_\alpha(x_\nu)\partial^\alpha.$$

We rewrite the equality $v_\nu \overset{def}{=} Lu_\nu$ as

$$L_\nu u_\nu = (L_\nu - L)u_\nu + v_\nu = \sum_{|\alpha|=m} \varepsilon_{\alpha,\nu}(x)\partial^\alpha u_\nu - \sum_{|\beta|<m} A_\beta(x)\partial^\beta u_\nu + v_\nu,$$

where $\varepsilon_{\alpha,\nu}(x) = A_\alpha(x_\nu) - A_\alpha(x)$. Using (10.3.3) we deduce

$$\|u_\nu\|_{m,p} \le C(\|u_\nu\|_{p,B_r(x_\nu)} + \|v_\nu\|_{p,B_r(x_\nu)} + \sum_{|\alpha|=m} \|\varepsilon_{\alpha,\nu}(x)\partial^\alpha u\|_{p,B_r(x_\nu)} + \|u\|_{m-1,p,B_r(x_\nu)}).$$

Since $|\varepsilon_{\alpha,\nu}(x)| \le Cr$ on $B_r(x_\nu)$, where $C = C(\|L\|_{1,R})$, we deduce

$$\|u_\nu\|_{m,p,B_r(x_\nu)} \le C(r\|u_\nu\|_{p,B_r(x_\nu)} + \|u_\nu\|_{p,B_r(x_\nu)} + \|v_\nu\|_{p,B_r(x_\nu)}).$$

We can now specify $r > 0$ such that $Cr < 1/2$ in the above inequality. Hence

$$\|u_\nu\|_{m,p,B_r(x_\nu)} \le C(\|u_\nu\|_{p,B_r(x_\nu)} + \|v_\nu\|_{p,B_r(x_\nu)}).$$

We need to estimate $\|v_\nu\|_p$. We use the equality

$$v_\nu = L(\eta_\nu u) = \eta_\nu Lu + [L, \eta_\nu]u,$$

in which $[L, \eta_\nu] = ad(\eta_\nu)L$ is a p.d.o. of order $m - 1$, so that

$$\|[L, \eta_\nu]u\|_{p,B_r(x_\nu)} \le Cr^{-(m-1)}\|u\|_{m-1,p,B_r(x_\nu)}.$$

Hence

$$\|v_\nu\|_{p,B_r(x_\nu)} \le C(\|\eta_\nu Lu\|_{p,B_r(x_\nu)} + r^{-(m-1)}\|u\|_{m-1,p,B_r(x_\nu)}),$$

so that

$$\|u_\nu\|_{m,p,B_r(x_\nu)} \le C(\|Lu\|_{p,B_R} + r^{-(m-1)}\|u\|_{m-1,p,B_R}).$$

We sum over ν, and taking into account that the number of spheres $B_r(x_\nu)$ is $O((R/r)^N)$ we deduce

$$\|u\|_{m,p,B_R} \le \sum_\nu \|u_\nu\|_{m,p} \le CR^N(r^{-N}\|Lu\|_{p,B_R} + r^{-(m+N-1)}\|u\|_{m-1,p,B_R}).$$

Note that r depends only on R, p, $\|L\|_{1,R}$, so that

$$\|u\|_{m,p,B_R} \le C(\|Lu\|_{p,B_R} + \|u\|_{m-1,p,B_R}),$$

where C is as in the statement of Theorem 10.3.1.

We still need to deal with the term $\|u\|_{m-1,p,B_R}$ in the above inequality. It is precisely at this point that the interpolation inequalities enter crucially.

View u as a section of $C_0^\infty(E\,|_{B_{2R}})$. If we pick $\eta \in C_0^\infty(B_{2R})$ such that $\eta \equiv 1$ on B_R, we deduce from the interpolation inequalities that there exists $C > 0$ such that

$$\|u\|_{m-1,B_R} \le \varepsilon\|u\|_{m,p,R} + C\varepsilon^{-(m-1)}\|u\|_{p,B_R}.$$

Hence

$$\|u\|_{m,p,B_R} \le C(\|Lu\|_{p,B_R} + \varepsilon\|u\|_{m,p,B_R} + \varepsilon^{-(m-1)}\|u\|_{p,B_R}).$$

If now we choose $\varepsilon > 0$ sufficiently small we deduce the case $k = 0$ of the inequality (10.3.3).

To establish this inequality for arbitrary k we argue by induction. Consider a multi-index $|\beta| = k$. If $u \in C_0^\infty(E\,|_{B_R})$, and $Lu = v$ then

$$L(\partial^\beta u) = \partial^\beta Lu + [L, \partial^\beta]u = \partial^\beta v + [L, \partial^\beta]u.$$

The crucial observation is that $[L, \partial^\beta]$ is a p.d.o. of order $\le m + k - 1$. Indeed,

$$\sigma_{m+k}([L, \partial^\beta]) = [\sigma_m(L), \sigma_k(\partial^\beta)] = 0.$$

Using (10.3.3) with $k = 0$, we deduce

$$\|\partial^\beta u\|_{m,p,B_R} \le C(\|\partial^\beta v\|_{p,B_R} + \|[L, \partial^\beta]u\|_{p,B_R} + \|\partial^\beta u\|_{p,B_R})$$

$$\le C(\|v\|_{k,p,B_R} + \|u\|_{m+k-1,p,B_R} + \|u\|_{k,p,B_R}).$$

The term $\|u\|_{m+k-1,p,B_R} + \|u\|_{k,p,B_R}$ can be estimated from above by

$$\varepsilon\|u\|_{m+k,p,B_R} + C\|u\|_{p,B_R},$$

using the interpolation inequalities as before. The inequality (10.3.3) is completely proved. The Hölder case is entirely similar. It is left to the reader as an exercise. Theorem 10.3.1 is proved. $\qquad\square$

Using the truncation technique and the interpolation inequalities we deduce the following consequence.

Corollary 10.3.3. *Let L as in Theorem 10.3.1, and fix $0 < r < R$. Then, for every $k \in \mathbb{Z}_+$, $1 < p < \infty$, and $\alpha \in (0,1)$, there exists $C = C(k, p, \alpha, N, \|L\|_{k+1,R}, R, r) > 0$ such that, $\forall u \in C^\infty(E)$*

$$\|u\|_{k+m,p,B_r} \le C\big(\|Lu\|_{k,p,B_R} + \|u\|_{p,B_R}\big),$$

and

$$\|u\|_{k+m,\alpha,B_r} \le C\big(\|Lu\|_{k,\alpha,B_R} + \|u\|_{0,\alpha,B_R}\big). \qquad\square$$

Exercise 10.3.4. Prove the above corollary. $\qquad\square$

10.3.2 Elliptic regularity

In this subsection we continue to use the notations of the previous subsection.

Definition 10.3.5. Let $u, v : \mathbb{R}^N \to \mathbb{C}^r$ be measurable functions.
(a) The function u is a *classical solution* of the partial differential equation

$$Lu = \sum_{|\beta| \le m} A_\alpha(x) \partial^\alpha u(x) = v(x) \tag{10.3.6}$$

if there exists $\alpha \in (0,1)$ such that, $v \in C^{0,\alpha}_{loc}(\mathbb{R}^N)$, $u \in C^{m,\alpha}_{loc}(\mathbb{R}^N)$, and (10.3.6) holds everywhere.
(b) The function u is said to be an L^p-*strong solution* of (10.3.6) if $u \in L^{m,p}_{loc}(\mathbb{R}^N)$, $v \in L^p_{loc}(\mathbb{R}^N)$, and (10.3.6) hold almost everywhere. (The partial derivatives of u should be understood in generalized sense.)
(c) The function u is said to be an L^p-*weak solution* if $u, v \in L^p_{loc}(\mathbb{R}^N)$ and

$$\int_{\mathbb{R}^N} \langle u, L^* \phi \rangle dx = \int_{\mathbb{R}^N} \langle v, \phi \rangle \, dx, \quad \forall \phi \in C^\infty_0(E). \qquad \square$$

Note the following obvious inclusion

$$\{L^p \text{ weak solutions}\} \supset \{L^p \text{ strong solutions}\}.$$

The principal result of this subsection will show that, when L is elliptic, then the above inclusion is an equality.

Theorem 10.3.6. *Let $p \in (1, \infty)$ and suppose that $L : C^\infty(E) \to C^\infty(E)$ is an elliptic operator of order m, with smooth coefficients. Then any L^p-weak solution u of*

$$Lu = v \in L^p_{loc}(E)$$

is an L^p strong solution, i.e., $u \in L^{m,p}_{loc}(E)$.

Remark 10.3.7. Loosely speaking the above theorem says that if a "clever" (i.e. elliptic) combination of mixed partial derivatives can be defined weakly, then any mixed partial derivative (up to a certain order) can be weakly defined as well. $\square$

The essential ingredient in the proof is the technique of mollification. For each $\delta > 0$ set

$$u_\delta = \rho_\delta * u \in C^\infty(E) \quad v_\delta = \rho_\delta * v \in C^\infty(E).$$

The decisive result in establishing the regularity of u is the following.

Lemma 10.3.8. *Let $w_\delta = Lu_\delta - v_\delta \in C^\infty(E)$. Then for every $\phi \in C^\infty_0(E)$*

$$\lim_{\delta \to 0} \int_{\mathbb{R}^N} \langle w_\delta, \phi \rangle dx = 0,$$

i.e., w_δ converges weakly to 0 in L^p_{loc}.

Remark 10.3.9. Roughly speaking, Lemma 10.3.8 says that

$$[L, \rho_\delta*] \to 0 \quad \text{as } \delta \to 0. \qquad \square$$

We first show how one can use Lemma 10.3.8 to prove $u \in L^{m,p}_{loc}(E)$. Fix $0 < r < R$. Note that u_δ is a classical solution of

$$Lu_\delta = v_\delta + w_\delta.$$

Using the elliptic estimates of Corollary 10.3.3, we deduce

$$\|u_\delta\|_{m,p,B_{2r}} \leq C(\|u_\delta\|_{p,B_{3r}} + \|v_\delta\|_{p,B_{3r}} + \|w_\delta\|_{p,B_{3r}}).$$

Since $u_\delta \to u$, and $v_\delta \to v$ in $L^p(E|_{B_{3r}})$, we deduce

$$\|u_\delta\|_{p,B_{3r}}, \ \|v_\delta\|_{p,B_{3r}} \leq C.$$

On the other hand, since w_δ is weakly convergent in $L^p(E|_{B_{3r}})$, we deduce that it must be bounded in this norm. Hence

$$\|u\|_{m,p,B_r} \leq C.$$

In particular, because $L^{m,p}(E|_{B_{2r}})$ is *reflexive*, we deduce that a subsequence of (u_δ) converges weakly to some $\bar{u} \in L^{m,p}(E|_{B_{2r}})$. Moreover, using the Rellich-Kondrachov compactness theorem, we deduce that, on a subsequence,

$$u_\delta \to \bar{u} \quad \text{strongly in } L^p(E|_{B_r}).$$

Since $u_\delta \to u$ in L^p_{loc} (as mollifiers), we deduce $u|_{B_r} = \bar{u} \in L^{m,p}(E|_{B_r})$. This shows $u \in L^{m,p}_{loc}$, because r is arbitrary. Theorem 10.3.6 is proved. $\qquad \square$

Proof of Lemma 10.3.8 Pick $\phi \in C_0^\infty(E)$. Assume $\mathrm{supp}\,\phi \subset B = B_R$. We have to show

$$\lim_{\delta \to 0} \left(\int_B \langle Lu_\delta, \phi \rangle dx - \int_B \langle v_\delta, \phi \rangle dx \right) = 0.$$

We analyze each of the above terms separately. Assume the formal adjoint of L has the form

$$L^* = \sum_{|\beta| \leq m} B_\beta(x) \partial_x^\beta.$$

We have

$$\int_B \langle Lu_\delta, \phi \rangle dx = \int_B \langle u_\delta(x), L_x^* \phi(x) \rangle dx = \int_B \left(\int_{\mathbb{R}^N} \langle \rho_\delta(x-y)u(y), L_x^* \phi(x) \rangle dy \right) dx$$

$$= \sum_\beta \int_B \left(\int_{\mathbb{R}^N} \langle \rho_\delta(x-y)u(y), B_\beta(x)\partial_x^\beta \phi(x) \rangle dy \right) dx$$

$$= \sum_\beta \int_B \left(\int_{\mathbb{R}^N} \langle u(y), \rho_\delta(x-y)B_\beta(x)\partial_x^\beta \phi(x) \rangle dy \right) dx.$$

Similarly

$$\int_B \langle v_\delta(x), \phi(x) \rangle dx = \int_B \int_{\mathbb{R}^N} \langle v(y), \rho_\delta(x-y)\phi(x) \rangle dy dx$$

$$= \int_B \int_{\mathbb{R}^N} \langle u(y), L_y^* \rho_\delta(x-y)\phi(x) \rangle dy dx$$

$$= \sum_\beta \int_B \int_{\mathbb{R}^N} \langle u(y), B_\beta(y)\partial_y^\beta(\rho_\delta(x-y)\phi(x)) \rangle dy dx$$

(switch the order of integration)

$$= \sum_\beta \int_{\mathbb{R}^N} \int_B \langle u(y), B_\beta(y)\partial_y^\beta(\rho_\delta(x-y)\phi(x)) \rangle dx dy$$

$(\partial_x \rho_\delta(x-y) = -\partial_y \rho_\delta(x-y))$

$$= \sum_\beta (-1)^\beta \left(\int_{\mathbb{R}^N} \int_B \langle u(y), B_\beta(y)(\partial_x^\beta \rho_\delta(x-y))\phi(x) \rangle dx dy \right)$$

(integrate by parts in the interior integral)

$$= \sum_\beta \int_{\mathbb{R}^N} \int_B \langle u(y), B_\beta(y)\rho_\delta(x-y)\partial_x^\beta \phi(x) \rangle dx dy$$

(switch back the order of integration)

$$= \sum_\beta \int_B \int_{\mathbb{R}^N} \langle u(y), B_\beta(y)\rho_\delta(x-y)\partial_x^\beta \phi(x) \rangle dy dx.$$

Hence

$$\int_B \langle Lu_\delta, \phi \rangle dx - \int_B \langle v_\delta, \phi \rangle dx$$

$$= \sum_\beta \int_B \int_{\mathbb{R}^N} \langle u(y), \rho_\delta(x-y) (B_\beta(x) - B_\beta(y)) \partial_x^\beta \phi(x) \rangle dy dx.$$

We will examine separately each term in the above sum.

$$\int_B \int_{\mathbb{R}^N} \left\langle u(y), \rho_\delta(x-y) (B_\beta(x) - B_\beta(y)) \partial_x^\beta \phi(x) \right\rangle dy dx$$

$$= \int_B \int_{\mathbb{R}^N} \rho_\delta(x-y) \left\langle u(y), (B_\beta(x) - B_\beta(y))\partial_x^\beta \phi(x) \right\rangle dy dx$$

$$= \int_B \left\langle \left(\int_{\mathbb{R}^N} \rho_\delta(x-y)u(y)dy \right), B_\beta(x)\partial_x^\beta \phi(x) \right\rangle dx$$

$$- \int_B \left\langle \left(\int_{\mathbb{R}^N} \rho_\delta(x-y)B_\beta^*(y)u(y)dy \right), \partial_x^\beta \phi(x) \right\rangle dx$$

$$= \int_B \langle\, u_\delta(x), B_\beta(x)\partial_x^\beta \phi(x)\,\rangle dx - \int_B \langle\, (B_\beta^* u)_\delta(x), \partial_x^\beta \phi(x)\,\rangle dx,$$

where $(B_\beta^* u)_\delta$ denotes the mollification of $B_\beta^* u$. As $\delta \to 0$

$$u_\delta \to u \text{ and } (B_\beta^* u)_\delta \to B_\beta^* u \text{ in } L_{loc}^p.$$

Hence

$$\lim_{\delta\to 0} \int_B \int_{\mathbb{R}^N} \langle\, u(y)\,, \rho_\delta(x-y)\,(B_\beta(x) - B_\beta(y))\,\partial_x^\beta\phi(x)\,\rangle dydx$$

$$= \int_B \langle\, u(x), B_\beta(x)\partial_x^\beta\phi(x)\,\rangle dx - \int_B \langle\, B_\beta^*(x)u(x), \partial_x^\beta\phi(x)\,\rangle dx = 0.$$

Lemma 10.3.8 is proved. $\square$

Corollary 10.3.10. *If $u \in L_{loc}^p(E)$ is weak L^p-solution of $Lu = v$, $1 < p < \infty$, and $v \in L_{loc}^{k,p}(E)$, then $u \in L_{loc}^{k+m,p}(E)$, and for every $0 < r < R$ we have*

$$\|u\|_{m+k,p,B_r} \le C(\|v\|_{k,p,B_R} + \|u\|_{p,B_R}),$$

where as usual $C = C(\|L\|_{k+m+1,R}, R, r, \ldots)$.

Proof. We already know that $u \in L_{loc}^{k+m,p}(E)$. Pick a sequence $u_n \in C_0^\infty(E)$ such that

$$u_n \to u \text{ strongly in } L_{loc}^{k+m,p}(E).$$

Then

$$Lu_n \to Lu \text{ strongly in } L_{loc}^{k+m,p}(E),$$

and

$$\|u_n\|_{m+k,p,B_r} \le C(\|u_n\|_{0,p,B_R} + \|Lu_n\|_{k,p,B_R}).$$

The desired estimate is obtained by letting $n \to \infty$ in the above inequality. $\square$

Corollary 10.3.11 (Weyl Lemma). *If $u \in L_{loc}^p(E)$ is an L^p-weak solution of $Lu = v$, and v is smooth, then u must be smooth.*

Proof. Since v is smooth we deduce $v \in L_{loc}^{k,p} \,\forall k \in \mathbb{Z}_+$. Hence $u \in L_{loc}^{k+m,p} \,\forall k$. Using Morrey embedding theorem we deduce that $u \in C_{loc}^{m,\alpha} \,\forall m \ge 0$. $\square$

The results in this, and the previous subsection are local, and so extend to the more general case of p.d.o. on manifolds. They take a particularly nice form for operators on compact manifolds.

Let (M, g) be a compact, oriented Riemann manifold and E, $F \to M$ be two metric vector bundles with compatible connections. Denote by $L \in \boldsymbol{PDO}^m(E, F)$ an elliptic operator of order m.

Theorem 10.3.12. *(a) Let $u \in L^p(E)$ and $v \in L^{k,p}(F)$, $1 < p < \infty$, such that*

$$\int_M \langle v, \phi \rangle_F dv_g = \int_M \langle u, L^*\phi \rangle_E dV_g \quad \forall \phi \in C_0^\infty(F).$$

Then $u \in L^{k+m,p}(E)$, and

$$\|u\|_{k+mp,E} \le C(\|v\|_{k,p,F} + \|u\|_{p,E}),$$

where $C = C(L, k, p)$.
(b) If $u \in C^{m,\alpha}(E)$ and $v \in C^{k,\alpha}(F)$ $(0 < \alpha < 1)$ are such that $Lu = v$, then $u \in C^{m+k,\alpha}(E)$, and

$$\|u\|_{k+m,\alpha,E} \le C(\|u\|_{0,\alpha,E} + \|v\|_{k,\alpha,F}),$$

where $C = C(L, k, \alpha)$.

Remark 10.3.13. The regularity results and the a priori estimates we have established so far represent only the minimal information one needs to become an user of the elliptic theory. Regrettably, we have mentioned nothing about two important topics: equations with non-smooth coefficients, and boundary value problems. For generalized Laplacians these topics are discussed in great detail in [36] and [79]. The boundary value problems for first order elliptic operators require a more delicate treatment. We refer to [16] for a very nice presentation of this subject. □

Exercise 10.3.14. (Kato's inequalities). Let (M, g) denote a compact oriented Riemann manifold without boundary. Consider a metric vector bundle $E \to M$ equipped with a compatible connection ∇.
(a) Show that for every $u \in L^{1,2}(E)$, the function $x \mapsto |u(x)|$ is in $L^{1,2}(M)$, and moreover

$$|d\,|u(x)|\,| \le |\nabla u(x)|,$$

for almost all $x \in M$.
(b) Set $\Delta_E = \nabla^*\nabla$, and denote by Δ_M the scalar Laplacian. Show that, for all $u \in L^{2,2}(E)$, we have

$$\Delta_M(|u|^2) = 2\langle \Delta_E u, u \rangle_E - 2|\nabla u|^2.$$

Conclude that $\forall \phi \in C^\infty(M)$ such that $\phi \ge 0$, we have

$$\int_M (d|u|, d(\phi|u|))_g \, dV_g \le \int_M \langle \Delta_E u, u \rangle_E \phi \, dV_g,$$

i.e.,

$$|u(x)|\Delta_M(|u(x)|) \le \langle \Delta_E u(x), u(x) \rangle_E \quad \text{weakly.} \qquad □$$

Exercise 10.3.15. (Local estimates). Suppose (M, g) is a connected, oriented, Riemann manifolds of dimension m, *not necessarily*, $E_0, E_1 \to M$ are two smooth hermitian vector bundles on M, and ∇^i, $i = 0, 1$ are Hermitian connection on E_i.

Suppose that $L : C^\infty(E_0) \to C^\infty(E_1)$ is an elliptic operator of order ν with smooth coefficients. Prove that for any compact subset $K \subset M$, any relatively compact open neighborhood U of K, and any $p \in (1, \infty)$, $\ell \in \mathbb{Z}_{\geq 0}$, there exists a positive constat C depending only on $K, U, p, \ell, \nabla^0, \nabla^1$, and the coefficients of L such that, $\forall u \in C^\infty(E_0)$, we have

$$\|u\|_{L^{\ell+\nu,p}(K, E_0)} \leq C \big(\|Lu\|_{L^{\ell,p}(U, E_1)} + \|u\|_{L^p(U, E_0)} \big),$$

where the above Sobolev norms are defined in terms of the connections ∇^i. $\square$

10.3.3 *An application: prescribing the curvature of surfaces*

In this subsection we will illustrate the power of the results we proved so far by showing how they can be successfully used to prove an important part of the celebrated uniformization theorem. In the process we will have the occasion to introduce the reader to some tricks frequently used in the study of nonlinear elliptic equations. We will consider a slightly more general situation than the one required by the uniformization theorem.

Let (M, g) be a compact, connected, oriented Riemann manifold of dimension N. Denote by $\Delta = d^*d : C^\infty(M) \to C^\infty(M)$ the scalar Laplacian. We assume for simplicity that

$$\mathrm{vol}_g(M) = \int_M dV_g = 1,$$

so that the average of any integrable function φ is defined by

$$\overline{\varphi} = \int_M \varphi(x) dV_g(x).$$

We will study the following partial differential equation.

$$\Delta u + f(u) = s(x), \tag{10.3.7}$$

where $f : \mathbb{R} \to \mathbb{R}$, and $s \in C^\infty(M)$ satisfy the following conditions.
(C_1) The function f is smooth and strictly increasing.
(C_2) There exist $a > 0$ and $b \in \mathbb{R}$, such that

$$f(t) \geq at + b \quad \forall t \in \mathbb{R}.$$

Set

$$F(u) = \int_0^u f(t) dt.$$

We assume
(C_3) $\lim_{|t| \to \infty} (F(t) - \overline{s}t) = \infty$, where $\overline{s}$ denotes the average of s,

$$\overline{s} = \int_M s(x) dV_g.$$

Theorem 10.3.16. *Let f and $s(x)$ satisfy the conditions $(C_1), (C_2), (C_3)$. Then there exists a unique $u \in C^\infty(M)$ such that*

$$\Delta u(x) + f(u(x)) = s(x) \quad \forall x \in M.$$

Proof. The proof of this theorem will be carried out in two steps.

Step 1. *Existence of a weak solution.* A weak solution of (10.3.7) is a function $u \in L^{1,2}(M)$ such that $f(u(x)) \in L^2(M)$, and

$$\int_M \{(du, d\phi) + f(u)\phi\}dV_g = \int_M s(x)\phi(x)dV_g(x), \quad \forall \phi \in L^{1,2}(M).$$

Step 2. *Regularity.* We show that a weak solution is in fact a classical solution.

Proof of Step 1. We will use *the direct method of the calculus of variations* outlined at the beginning of the current section. Consider the energy functional

$$I : L^{1,2}(M) \to \mathbb{R}, \quad I(u) = \int_M \left\{ \frac{1}{2}|du(x)|^2 + F(u(x)) - g(x)u(x) \right\}dV_g(x).$$

This functional is not quite well defined, since there is no guarantee that $F(u) \in L^1(M)$ for all $u \in L^{1,2}(M)$.

Leaving this issue aside for a moment, we can perform a formal computation à la Riemann. Assume u is a minimizer of I, i.e.,

$$I(u) \leq I(v), \quad \forall v \in L^{1,2}(M).$$

Thus, for all $\phi \in L^{1,2}(M)$

$$I(u) \leq I(u + t\phi) \quad \forall t \in \mathbb{R}.$$

Hence $t = 0$ is a minimum of $h_\phi(t) = I(u + t\phi)$, so that $h'_\phi(0) = 0 \; \forall \phi$. A simple computation shows that

$$h'_\phi(0) = \int_M \{(du, d\phi) + f(u)\phi - s(x)\phi(x)\}dV_g = 0,$$

so that a minimizer of I is a weak solution provided that we deal with the integrability issue raised at the beginning of this discussion. Anyway, the lesson we learn from this formal computation is that minimizers of I are strong candidates for solutions of (10.3.7).

We will circumvent the trouble with the possible non-integrability of $F(u)$ by using a famous trick in elliptic partial differential equations called *the maximum principle.*

Lemma 10.3.17. *Let $h : \mathbb{R} \to \mathbb{R}$ be a continuous, strictly increasing function and $u, v \in L^{1,2}(M)$ such that*
(i) $h(u), h(v) \in L^2(M)$.
(ii) $\Delta u + h(u) \geq \Delta v + h(v)$ weakly i.e.

$$\int_M \{(du, d\phi) + h(u)\phi\}dV_g \geq \int_M \{(dv, d\phi) + h(v)\phi\}dV_g \tag{10.3.8}$$

$\forall \phi \in L^{1,2}(M)$ such that $\phi \geq 0$ a.e. M. Then $u \geq v$ a.e. on M.

Proof.   Let

$$(u-v)^- = \min\{u-v, 0\} = \frac{1}{2}\{(u-v) - |u-v|\}.$$

According to the Exercise 10.2.19 we have $(u-v)^- \in L^{1,2}(M)$, and

$$d(u-v)^- = \begin{cases} d(u-v) & \text{a.e. on } \{u < v\} \\ 0 & \text{a.e. on } \{u \geq v\} \end{cases}$$

Using $\phi = -(u-v)$ in (10.3.8) we deduce

$$-\int_M \left\{ \left(d(u-v), d(u-v)^- \right) + \left(h(u) - h(v) \right)\left(u-v\right)^- \right\} dV_g \geq 0.$$

Clearly

$$\left(d(u-v), d(u-v)^- \right) = |d(u-v)^-|^2,$$

and since h is nondecreasing,

$$(h(u) - h(v))(u-v)^- \geq 0.$$

Hence,

$$\int_M |d(u-v)^-|^2 \, dV_g \leq -\int_M \left(h(u) - h(v) \right)(u-v)^- \, dV_g \leq 0,$$

so that

$$|d(u-v)^-| \equiv 0.$$

Since M is connected, this means $(u-v)^- \equiv c \leq 0$. If $c < 0$, then

$$(u-v) \equiv (u-v)^- \equiv c,$$

so that $u = v + c < v$. Since h is strictly increasing we conclude

$$\int_M h(u) \, dV_g < \int_M h(v) dV_g.$$

On the other hand, using $\phi \equiv 1$ in (10.3.8), we deduce

$$\int_M h(u) dV_g \geq \int_M h(v) dV_g!$$

Hence c cannot be negative, so that $(u-v)^- \equiv 0$ which is another way of saying $u \geq v$. The maximum principle is proved. $\square$

Exercise 10.3.18.   Assume h in the above lemma is only non-decreasing, but u and v satisfy the supplementary condition

$$\overline{u} \geq \overline{v}.$$

Show the conclusion of Lemma 10.3.17 continues to hold. $\square$

We now return to the equation (10.3.7). Note first that if u and v are two weak solutions of this equation, then

$$\Delta u + f(u) \geq (\leq) \Delta v + h(v) \quad \text{weakly},$$

so that by the maximum principle $u \geq (\leq) v$. This shows equation (10.3.7) has *at most one weak solution.*

To proceed further we need the following a priori estimate.

Lemma 10.3.19. *Let u be a weak solution of (10.3.7). If C is a positive constant such that*

$$f(C) \geq \sup_{M} s(x),$$

then $u(x) \leq C$ a.e. on M.

Proof. The equality $f(C) \geq \sup s(x)$ implies

$$\Delta C + f(C) \geq s(x) = \Delta u + f(u) \quad \text{weakly}$$

so the conclusion follows from the maximum principle. $\qquad\square$

Fix $C_0 > 0$ such that $f(C_0) \geq \sup s(x)$. Consider a strictly increasing C^2-function $\tilde{f} : \mathbb{R} \to \mathbb{R}$ such that

$$\tilde{f}(u) = f(u) \quad \text{for } u \leq C_0,$$

$$\tilde{f}(u) \text{ is linear for } u \geq C_0 + 1.$$

The conditions (C_1) and (C_2) imply that there exist A, $B > 0$ such that

$$|\tilde{f}(u)| \leq A|u| + B. \tag{10.3.9}$$

Lemma 10.3.20. *If u is a weak solution of*

$$\Delta u + \tilde{f}(u) = s(x) \tag{10.3.10}$$

then u is also a weak solution of (10.3.7).

Proof. We deduce as in the proof of Lemma 10.3.19 that $u \leq C_0$, which is precisely the range where f coincides with $\tilde{f}$. $\qquad\square$

The above lemma shows that instead of looking for a weak solution of (10.3.7), we should try to find a weak solution of (10.3.10). We will use the direct method of the calculus of variations on a new functional

$$\tilde{I}(u) = \int_{M} \left\{ \frac{1}{2}|du(x)|^2 + \tilde{F}(u(x)) - s(x)u(x) \right\} dV_g(x),$$

where

$$\tilde{F}(u) := \int_{0}^{u} \tilde{f}(t)dt.$$

The advantage we gain by using this new functional is clear. The inequality (10.3.9) shows $\tilde{F}$ has at most quadratic growth, so that $\tilde{F}(u) \in L^1(M)$, for all $u \in L^2(M)$. The existence of a minimizer is a consequence of the following fundamental principle of the calculus of variations.

Proposition 10.3.21. *Let X be a reflexive Banach space and $J : X \to \mathbb{R}$ a convex, weakly lower semi-continuous, coercive functional, i.e., the sublevel sets*

$$J^c = \{x \; ; \; J(x) \leq c\}$$

are respectively convex, weakly closed and bounded in X. Then J admits a minimizer, i.e., there exists $x_0 \in X$ such that

$$J(x_0) \leq J(x) \quad \forall x \in X.$$

Proof. Note that

$$\inf_X J(x) = \inf_{J^c} J(x).$$

Consider $x_n \in J^c$ such that

$$\lim_n J(x_n) = \inf J.$$

Since X is reflexive, and J^c is convex, weakly closed and bounded in the norm of X, we deduce that J^c is weakly compact. Hence a (generalized) subsequence (x_ν) of x_n converges weakly to some $x_0 \in J^c$. Using the lower semi-continuity of J we deduce

$$J(x_0) \leq \liminf_\nu J(x_\nu) = \inf J.$$

Hence x_0 is a minimizer of J. The proposition is proved.   □

The next result will conclude the proof of Step 1.

Lemma 10.3.22. *$\tilde{I}$ is convex, weakly lower semi-continuous and coercive (with respect to the $L^{1,2}$-norm).*

Proof. The convexity is clear since $\tilde{F}$ is convex on account that $\tilde{f}$ is strictly increasing. The functionals

$$u \mapsto \frac{1}{2} \int_M |du|^2 dV_g \text{ and } u \mapsto -\int_M s(x)u(x)dV_g(x)$$

are clearly weakly lower semi-continuous. We need to show the functional

$$u \mapsto \int_M \tilde{F}(u)dV_g$$

is also weakly lower semi-continuous.

Since $\tilde{F}$ is convex, we can find $\alpha > 0, \beta \in \mathbb{R}$ such that

$$\tilde{F}(u) - \alpha u - \beta \geq 0.$$

If $u_n \to u$ strongly in $L^{1,2}(M)$ we deduce using the Fatou lemma that

$$\int_M \{\tilde{F}(u) - \alpha u - \beta\} dV_g \leq \liminf_{n \to \infty} \int_M \{\tilde{F}(u_n) - \alpha u_n - \beta\} dV_g.$$

On the other hand,

$$\lim_n \int_M \alpha u_n + \beta dV_g = \int_M \alpha u + \beta dV_g,$$

which shows that

$$\int_M \tilde{F}(u) dV_g \leq \liminf_n \int \tilde{F}(u_n) dV_g.$$

This means that the functional

$$L^{1,2}(M) \ni u \mapsto \int_M \tilde{F}(u) dV_g$$

is strongly lower semi-continuous. Thus, the sublevel sets

$$\left\{ u; \int_M \tilde{F}(u) dV_g \leq c \right\},$$

are both convex and *strongly closed*. The Hahn-Banach separation principle can now be invoked to conclude these level sets are also *weakly* closed. We have thus established that $\tilde{I}$ is convex and weakly lower semi-continuous.

Remark 10.3.23. We see that the lower semi-continuity, and the convexity conditions are very closely related. In some sense they are almost equivalent. We refer to [24] for a presentation of the direct method of the calculus of variations were the lower semi-continuity issue is studied in great detail. □

The coercivity will require a little more work. The key ingredient will be a *Poincaré inequality*. We first need to introduce some more terminology.

For any $u \in L^2(M)$ we denoted by $\bar{u}$ its average. Now set

$$u^\perp(x) = u(x) - \bar{u}.$$

Note the average of $u^\perp$ is 0. This choice of notation is motivated by the fact that $u^\perp$ is perpendicular (with respect to the $L^2(M)$-inner product) to the kernel of the Laplacian Δ which is the 1-dimensional space spanned by the constant functions.

Lemma 10.3.24 (Poincaré inequality). *There exists $C > 0$ such that*

$$\int_M |du|^2 dV_g \geq C \int_M |u^\perp|^2 dV_g, \quad \forall u \in L^{1,2}(M).$$

Proof. We argue by contradiction. Assume that for any $\varepsilon > 0$ there exists $u_\varepsilon \in L^{1,2}(M)$, such that

$$\int_M |u_\varepsilon^\perp|^2 dV_g = 1, \quad \text{and} \quad \int_M |du_\varepsilon|^2 dV_g \leq \varepsilon.$$

The above two conditions imply that the family $(u_\varepsilon^\perp)$ is bounded in $L^{1,2}(M)$. Since $L^{1,2}(M)$ is reflexive, we deduce that, on a subsequence

$$u_\varepsilon^\perp \rightharpoonup v, \quad \text{weakly in } L^{1,2}(M).$$

The inclusion $L^{1,2}(M) \hookrightarrow L^2(M)$ is compact, (Rellich-Kondrachov) so that, on a subsequence

$$u_\varepsilon^\perp \to v \quad \text{strongly in } L^2(M).$$

This implies $v \neq 0$, since

$$\int_M |v|^2 dV_g = \lim_\varepsilon \int_M |u_\varepsilon^\perp|^2 dV_g = 1.$$

On the other hand,

$$du_\varepsilon = du_\varepsilon^\perp \to 0, \quad \text{strongly in } L^2.$$

We conclude

$$\int_M (dv, d\phi) \, dV_g = \lim_\varepsilon \int_M (du_\varepsilon, d\phi) \, dV_g = 0 \ \forall \phi \in L^{1,2}(M).$$

In particular,

$$\int_M (dv, dv) \, dV_g = 0,$$

so that $dv \equiv 0$. Since M is connected, we deduce $v \equiv c = \text{const}$. Moreover,

$$c = \int_M v(x)dV_g(s) = \lim_\varepsilon \int_M u_\varepsilon^\perp dV_g = 0.$$

This contradicts the fact that $v \neq 0$. The Poincaré inequality is proved.   □

We can now establish the coercivity of $\tilde{I}(u)$. Let $\kappa > 0$, and $u \in L^{1,2}(M)$ such that

$$\int_M \left\{ \frac{1}{2}|du(x)|^2 + \tilde{F}(u(x)) - s(x)u(x) \right\} dV_g(x) \leq \kappa. \tag{10.3.11}$$

Since

$$\int_M |du|^2 \, dV_g = \int_M |du^\perp|^2 dV_g \quad \text{and} \quad \int_M |u|^2 = |\overline{u}|^2 + \int_M |u^\perp|^2 dV_g,$$

it suffices to show that the quantities

$$\overline{u}, \ \int_M |u^\perp|^2 dV_g, \ \int_M |du^\perp| dV_g$$

are bounded. In view of the Poincaré inequality, the boundedness of

$$\int_M |du^\perp|^2 dV_g$$

implies the boundedness of $\|u^\perp\|_{2,M}$, so that we should concentrate only on $\overline{u}$ and $\|du^\perp\|_2$.

The inequality (10.3.11) can be rewritten as

$$\int_M \left\{ \frac{1}{2}|du^\perp|^2 + \tilde{F}(u) - s^\perp u^\perp \right\} dV_g - \bar{s} \cdot \bar{u} \leq c.$$

The Poincaré and Cauchy inequalities imply

$$C\|u^\perp\|_2^2 - \|s^\perp\|_2 \cdot \|u^\perp\|_2 + \int_M \tilde{F}(u) dV_g - \bar{s} \cdot \bar{u} \leq \kappa.$$

Since $\mathrm{vol}_g(M) = 1$, and $\tilde{F}$ is convex, we have a Jensen-type inequality

$$\tilde{F}\left(\int_M u \, dV_g \right) \leq \int_M \tilde{F}(u) dV_g,$$

so that

$$C\|u^\perp\|_2^2 - \|s^\perp\|_2 \cdot \|u^\perp\|_2 + \tilde{F}(\bar{u}) - \bar{s} \cdot \bar{u} \leq \kappa. \qquad (10.3.12)$$

Set $P(t) := Ct^2 - \|s^\perp\|_2 t$, and let $m = \inf P(t)$. From the inequality (10.3.12) we deduce

$$\tilde{F}(\bar{u}) - \bar{s} \cdot \bar{u} \leq \kappa - m.$$

Using condition (C_3) we deduce that $|\bar{u}|$ must be bounded. Feed this information back in (10.3.12). We conclude that $P(\|u^\perp\|_2)$ must be bounded. This forces $\|u^\perp\|_2$ to be bounded. Thus $\tilde{I}$ is coercive, and Lemma 10.3.22 is proved. $\qquad \square$

Step 2. *The regularity of the minimizer.* We will use a technique called *bootstrapping*, which blends the elliptic regularity theory, and the Sobolev embedding theorems, to gradually improve the regularity of the weak solution.

Let u be the weak solution of (10.3.10). Then $u(x)$ is a weak L^2 solution of

$$\Delta u = h(x) \quad \text{on} \quad M,$$

where $h(x) = -\tilde{f}(u(x)) - s(x)$. Note that since the growth of $\tilde{f}$ is at most linear, we have $\tilde{f}(u(x)) \in L^2(M)$. The elliptic regularity theory implies that $u \in L^{2,2}(M)$. Using Sobolev (or Morrey) embedding theorem we can considerably improve the integrability of u. We deduce that
(i) either $u \in L^q(M)$ if $-N/q \leq 2 - N/2 \leq 0$ ($\dim M = N$),
(ii) or u is Hölder continuous if $2 - N/2 > 0$.

In any case, this shows $u(x) \in L^{q_1}(M)$, for some $q_1 > 2$, which implies $h(x) \in L^{q_1}(M)$. Using again elliptic regularity we deduce $u \in L^{2,q_1}$ and Sobolev inequality implies that $u \in L^{q_2}(M)$ for some $q_2 > q_1$.

After a finite number of steps we conclude that $h(x) \in L^q(M)$, for all $q > 1$. Elliptic regularity implies $u \in L^{2,q}(M)$, for all $q > 1$. This implies $h(x) \in L^{2,q}(M)$, for all $q > 1$. Invoking elliptic regularity again we deduce that $u \in L^{4,q}(M)$, for any $q > 1$. (At this point it is convenient to work with f, rather than with $\tilde{f}$ which was only C^2). Feed this back in $h(x)$, and regularity theory improves the regularity of u two orders at a time. In view of Morrey embedding theorem the conclusion is clear: $u \in C^\infty(M)$. The proof of Theorem 10.3.16 is complete. $\qquad \square$

From the theorem we have just proved we deduce immediately the following consequence.

Corollary 10.3.25. *Let (M, g) be a compact, connected, oriented Riemann manifold and $s(x) \in C^\infty(M)$. Assume $\mathrm{vol}_g(M) = 1$. Then the following two conditions are equivalent.*

(a) $\bar{s} = \int_M s(x) dV_g > 0$

(b) For every $\lambda > 0$ there exists a unique $u = u_\lambda \in C^\infty(M)$ such that

$$\Delta u + \lambda e^u = s(x). \tag{10.3.13}$$

Proof. (a) $\Rightarrow$ (b) follows from Theorem 10.3.16.
(b) $\Rightarrow$ (a) follows by multiplying (10.3.13) with $v(x) \equiv 1$, and then integrating by parts so that

$$\bar{s} = \lambda \int_M e^{u(x)} dV_g(x) > 0. \qquad \Box$$

Although the above corollary may look like a purely academic result, it has a very nice geometrical application. We will use it to prove a special case of the celebrated *uniformization theorem*.

Definition 10.3.26. Let M be a smooth manifold. Two Riemann metrics g_1 and g_2 are said to be *conformal* if there exists $f \in C^\infty(M)$, such that $g_2 = e^f g_1$. $\qquad \Box$

Exercise 10.3.27. Let (M, g) be an oriented Riemann manifold of dimension N and $f \in C^\infty(M)$. Denote by $\tilde{g}$ the conformal metric $\tilde{g} = e^f g$. If $s(x)$ is the scalar curvature of g and $\tilde{s}$ is the scalar curvature of $\tilde{g}$ show that

$$\tilde{s}(x) = e^{-f}\{s(x) + (N - 1)\Delta_g f - \frac{(N - 1)(N - 2)}{4}|df(x)|_g^2\}$$

where Δ_g denotes the scalar Laplacian of the metric g while $|\cdot|_g$ denotes the length measured in the metric g. $\qquad \Box$

Remark 10.3.28. A long standing problem in differential geometry, which was only relatively recently solved, is the *Yamabe problem*:

"If (M, g) is a compact oriented Riemann manifold, does there exist a metric conformal to g whose scalar curvature is constant?"

In dimension 2 this problem is related to the uniformization problem of complex analysis.

The solution of the Yamabe problem in its complete generality is due to the combined efforts of T. Aubin, [8, 9] and R. Schoen [86]. For a very beautiful account of its proof we recommend the excellent survey of J. Lee and T. Parker, [63]. One can formulate a more general question than the Yamabe problem.

Given a compact oriented Riemann manifold (M, g) decide whether a smooth function $s(x)$ on M is the scalar curvature of some metric on M conformal to g.

This problem is known as the Kazdan-Warner problem. The case dim $M = 2$ is completely solved in [56]. The higher dimensional situation dim $M > 2$ is far more complicated, both topologically and analytically. $\qquad\square$

Theorem 10.3.29 (Uniformization Theorem). *Let (Σ, g) be a compact, oriented Riemann manifold of dimension 2. Assume* $\mathrm{vol}_g(\Sigma) = 1$. *If $\chi(\Sigma) < 0$ (or equivalently if its genus is ≥ 2) then there exists a unique metric $\tilde{g}$ conformal to g such that*

$$s(\tilde{g}) \equiv -1.$$

Proof. We look for $\tilde{g}$ of the form $\tilde{g} = e^u g$. Using Exercise 10.3.27 we deduce that u should satisfy

$$-1 = e^{-u}\{s(x) + \Delta u\},$$

i.e.

$$\Delta u + e^u = -s(x),$$

where $s(x)$ is the scalar curvature of the metric g. The Gauss-Bonnet theorem implies that

$$\overline{s} = 4\pi\chi(\Sigma) < 0$$

so that the existence of u is guaranteed by Corollary 10.3.25. The uniformization theorem is proved. $\qquad\square$

On a manifold of dimension 2 the scalar curvature coincides up to a positive factor with the sectional curvature. The uniformization theorem implies that the compact oriented surfaces of negative Euler characteristic admit metrics of constant negative sectional curvature. Now, using the Cartan-Hadamard theorem we deduce the following topological consequence.

Corollary 10.3.30. *The universal cover of a compact, oriented surface Σ of negative Euler characteristic is diffeomorphic to $\mathbb{C}$.* $\qquad\square$

Remark 10.3.31. The "inverse" of the covering projection $\pi : \mathbb{C} \to \Sigma$ is classically called a *uniformizing parameter*. It is very similar to the angular coordinate θ on the circle S^1. This is a uniformizing parameter on the circle, which is an "inverse" of the universal covering map $\mathbb{R} \ni \theta \mapsto e^{i\theta} \in S^1$. $\qquad\square$

In the following exercises (M, g) denotes a compact, oriented Riemann manifold without boundary.

Exercise 10.3.32. Fix $c > 0$. Show that for every $f \in L^2(M)$ the equation

$$\Delta u + cu = f$$

has a unique solution $u \in L^{2,2}(M)$. $\qquad\square$

Exercise 10.3.33. Consider a smooth function $f : M \times \mathbb{R} \to \mathbb{R}$ such that for every $x \in M$ the function $u \mapsto f(x, u)$ is increasing. Assume that the equation

$$\Delta_g u = f(x, u) \tag{10.3.14}$$

admits a pair of *comparable* sub/super-solutions, i.e., there exist functions $u_0, U_0 \in L^{1,2}(M) \cap L^\infty(M)$ such that

$$U_0(x) \geq u_0(x), \quad \text{a.e. on } M,$$

and

$$\Delta_g U_0 \geq f(x, U_0(x)) \geq f(x, u_0(x)) \geq \Delta_g u_0, \quad \text{weakly in } L^{1,2}(M).$$

Fix $c > 0$, and define $(u_n)_{n \geq 1} \subset L^{2,2}(M)$ inductively as the unique solution of the equation

$$\Delta u_n(x) + c u_n(x) = c u_{n-1}(x) + f(x, u_{n-1}(x)).$$

(a) Show that

$$u_0(x) \leq u_1(x) \leq u_2(x) \leq \cdots \leq u_n(x) \leq \cdots \leq U_0(x) \quad \forall x \in M.$$

(b) Show that u_n converges uniformly on M to a solution $u \in C^\infty(M)$ of (10.3.14) satisfying

$$u_0 \leq u \leq U_0.$$

(c) Prove that the above conclusions continue to hold, even if the monotonicity assumption on f is dropped. □

10.4 Elliptic operators on compact manifolds

The elliptic operators on compact manifolds behave in many respects as finite dimensional operators. It is the goal of this last section to present the reader some fundamental analytic facts which will transform the manipulation with such p.d.o. into a less painful task.

The main reason why these operators are so "friendly" is the existence of a priori estimates. These estimates, coupled with the Rellich-Kondratchov compactness theorem, are the keys which will open many doors.

10.4.1 *Fredholm theory*

Throughout this section we assume that the reader is familiar with some fundamental facts about unbounded linear operators. We refer to [18] Ch.II for a very concise presentation of these notions. An exhaustive presentation of this subject can be found in [55].

Definition 10.4.1. (a) Let X, Y be two Hilbert spaces over $\mathbb{K} = \mathbb{R}, \mathbb{C}$, and

$$T : D(T) \subset X \to Y$$

be a linear operator, *not necessarily continuous*, defined on the linear subspace $D(T) \subset X$. The operator T is said to be *densely defined* if its domain $D(T)$ is dense in X.

The operator T is said to be *closed* if its graph,

$$\Gamma_T := \big\{ (x, Tx) \in D(T) \times Y \subset X \times Y \big\},$$

is a closed subspace in $X \times Y$.

(b) Let $T : D(T) \subset X \to Y$ be a closed, densely defined linear operator. The *djoint* of T is the operator $T^* : D(T^*) \subset Y \to X$ defined by its graph

$$\Gamma_{T^*} = \{(y^*, x^*) \in Y \times X \; ; \; \langle x^*, x \rangle = \langle y^*, Tx \rangle \; \forall x \in D(T),$$

where $\langle \cdot, \cdot \rangle : Z \times Z \to \mathbb{K}$ denotes the inner product[2] in a Hilbert space Z.

(c) A closed, densely defined operator $T : D(T) \subset X \to X$ is said to be *selfadjoint* if $T = T^*$.   □

Remark 10.4.2.   (a) In more concrete terms, the operator $T : D(T) \subset X \to Y$ is closed if, for any sequence $(x_n) \subset D(T)$, such that $(x_n, Tx_n) \to (x, y)$, it follows that (i) $x \in D(T)$, and (ii) $y = Tx$.

(b) If $T : X \to Y$ is a closed operator, then T is bounded (closed graph theorem). Also note that if $T : D(T) \subset X \to Y$ is a closed, densely defined operator, then $\ker T$ is a *closed* subspace of X.

(c) One can show that the adjoint of any closed, densely defined operator, is a *closed, densely defined* operator.

(d) The closed, densely defined operator $T : D(T) \subset X \to X$ is selfadjoint if the following two conditions hold.

- $\langle Tx, y \rangle = \langle x, Ty \rangle$ for all $x, y \in D(T)$ and
- $D(T) = \big\{ y \in X; \; \exists C > 0, \; |\langle Tx, y \rangle| \leq C|x|, \; \forall x \in D(T) \big\}.$

If only the first condition is satisfied the operator T is called *symmetric*.   □

Let (M, g) be a compact, oriented Riemann manifold, E, F two metric vector bundles with compatible connections, and $L \in \boldsymbol{PDO}^k = \boldsymbol{PDO}^k(E, F)$, a k-th order *elliptic* p.d.o. We will denote the various L^2 norms by $\| \cdot \|$, and the $L^{k,2}$- norms by $\| \cdot \|_k$.

Definition 10.4.3.   The *analytical realization* of L is the linear operator

$$L_a : D(L_a) \subset L^2(E) \to L^2(E), \quad D(L_a) = L^{k,2}(E),$$

given by $u \mapsto Lu$ for all $u \in L^{k,2}(E)$.   □

---

[2]When $\mathbb{K} = \mathbb{C}$, we use the convention that the inner product $\langle \cdot, \cdot \rangle$ is conjugate linear in the second variable, i.e., $\langle z_1, \lambda z_2 \rangle = \bar{\lambda} \langle z_1, z_2 \rangle$, $\forall \lambda \in \mathbb{C}$, $z_1, z_2 \in Z$.

Proposition 10.4.4. *(a) The analytical realization L_a of L is a closed, densely defined linear operator.*
(b) If $L^* : C^\infty(F) \to C^\infty(E)$ is the formal adjoint of L then

$$(L^*)_a = (L_a)^*.$$

Proof. (a) Since $C^\infty(E) \subset D(L_a) = L^{k,2}(E)$ is dense in $L^2(E)$, we deduce that L_a is densely defined. To prove that L_a is also closed, consider a sequence $(u_n) \subset L^{k,2}(E)$ such that

$$u_n \to u \text{ strongly in } L^2(E), \text{ and } Lu_n \to v \text{ strongly in } L^2(F).$$

From the elliptic estimates we deduce

$$\|u_n - u_m\|_k \leq C\big(\|Lu_n - Lu_m\| + \|u_n - u_m\| \big) \to 0 \text{ as } m, n \to \infty.$$

Hence (u_n) is a Cauchy sequence in $L^{k,2}(E)$, so that $u_n \to u$ in $L^{k,2}(E)$. It is now clear that $v = Lu$.
(b) From the equality

$$\int_M \langle Lu, v \rangle dV_g = \int_M \langle u, L^*v \rangle dV_g, \quad \forall u \in L^{k,2}(E), \ v \in L^{k,2}(F),$$

we deduce

$$D\big((L_a)^* \big) \supset D\big((L^*)_a \big) = L^{k,2}(F),$$

and $(L_a)^* = (L^*)_a$ on $L^{k,2}(F)$. To prove that

$$D\big((L_a)^* \big) \subset D\big((L^*)_a \big) = L^{k,2}(F)$$

we need to show that if $v \in L^2(F)$ is such that $\exists C > 0$ with the property that

$$\left| \int_M \langle Lu, v \rangle dV_g \right| \leq C\|u\|, \quad \forall u \in L^{k,2}(E),$$

then $v \in L^{k,2}(F)$. Indeed, the above inequality shows that the functional

$$u \mapsto \int_M \langle Lu, v \rangle dV_g$$

extends to a continuous linear functional on $L^2(E)$. Hence, there exists $\phi \in L^2(E)$ such that

$$\int_M \langle (L^*)^* u, v \rangle dV_g = \int_M \langle u, \phi \rangle dV_g \ \forall u \in L^{k,2}(E).$$

In other words, v is a L^2-weak solution of the elliptic equation $L^* v = \phi$. Using elliptic regularity theory we deduce $v \in L^{k,2}(M)$. The proposition is proved. $\square$

Following the above result we will not make any notational distinction between an elliptic operator (on a compact manifold) and its analytical realization.

Definition 10.4.5. (a) Let X and Y be two Hilbert spaces over $\mathbb{K} = \mathbb{R}, \mathbb{C}$ and suppose that $T : D(T) \subset X \to Y$ is a closed, densely defined linear operator. The operator T is said to be *semi-Fredholm* if the following hold.

(i) $\dim \ker T < \infty$ and

(ii) The range $\mathrm{R}\,(T)$ of T is closed.

(b) The operator T is called *Fredholm* if both T and T^* are semi-Fredholm. In this case, the integer

$$\mathrm{ind}\,T := \dim_{\mathbb{K}} \ker T - \dim_{\mathbb{K}} \ker T^*$$

is called the *Fredholm index* of T. □

Remark 10.4.6. The above terminology has its origin in the work of Ivar Fredholm at the twentieth century. His result, later considerably generalized by F. Riesz, states that if $K : H \to H$ is a compact operator from a Hilbert space to itself, then $\mathbb{1}_H + K$ is a Fredholm operator of index 0. □

Consider again the elliptic operator L of Proposition 10.4.4.

Theorem 10.4.7. *The operator* $L_a : D(L_a) \subset L^2(E) \to L^2(F)$ *is Fredholm.*

Proof. The Fredholm property is a consequence of the following compactness result.

Lemma 10.4.8. *Any sequence* $(u_n)_{n\geq 0} \subset L^{k,2}(E)$ *such that* $\big\{ \|u_n\| + \|Lu_n\| \big\}_{n\geq 0}$ *is bounded contains a subsequence strongly convergent in* $L^2(E)$.

Proof. Using elliptic estimates we deduce that

$$\|u_n\|_k \leq C(\|Lu_n\| + \|u_n\|) \leq const.$$

Hence (u_n) is also bounded in $L^{k,2}(E)$. On the other hand, since M is compact, the space $L^{k,2}(E)$ embeds compactly in $L^2(E)$. The lemma is proved. □

We will first show that $\dim \ker L < \infty$. In the proof we will rely on the classical result of F. Riesz which states that a Banach space is finite dimensional if and only if its bounded subsets are precompact (see [18], Chap. VI).

Note first that, according to Weyl's lemma $\ker L \subset C^\infty(E)$. Next, notice that $\ker L$ is a Banach space with respect to the L^2-norm since according to Remark 10.4.2 (a) $\ker L$ is closed in $L^2(E)$. We will show that any sequence $(u_n) \subset \ker L$ which is also bounded in the L^2-norm contains a subsequence convergent in L^2. This follows immediately from Lemma 10.4.8 since $\|u_n\| + \|Lu_n\| = \|u_n\|$ is bounded.

To prove that the range $\mathrm{R}\,(T)$ is closed, we will rely on the following very useful inequality, a special case of which we have seen at work in Subsection 9.3.3.

Lemma 10.4.9 (Poincaré inequality). *There exists* $C > 0$ *such that*

$$\|u\| \leq C\|Lu\|,$$

for all $u \in L^{k,2}(E)$ *which are* L^2-*orthogonal to* $\ker L$, *i.e.,*

$$\int_M \langle u, \phi \rangle dV_g = 0 \quad \forall \phi \in \ker L.$$

Proof.   We will argue by contradiction. Denote by $X \subset L^{k,2}(E)$ the subspace consisting of sections L^2-orthogonal to $\ker L$. Assume that for any $n > 0$ there exists $u_n \in X$ such that

$$\|u_n\| = 1 \text{ and } \|Lu_n\| \leq 1/n.$$

Thus, $\|Lu_n\| \to 0$, and in particular, $\|u_n\| + \|Lu_n\|$ is bounded. Using Lemma 10.4.8 we deduce that a subsequence of (u_n) is convergent in $L^2(E)$ to some u. Note that $\|u\| = 1$. It is not difficult to see that in fact $u \in X$. We get a sequence

$$(u_n, Lu_n) \subset \Gamma_L = \text{ the graph of } L,$$

such that $(u_n, Lu_n) \to (u, 0)$. Since L is closed, we deduce $u \in D(L)$ and $Lu = 0$. Hence $u \in \ker L \cap X = \{0\}$. This contradicts the condition $\|u\| = 1$.   □

We can now conclude the proof of Theorem 10.4.7. Consider a sequence $(v_n) \subset \mathrm{R}\,(L)$ such that $v_n \to v$ in $L^2(F)$. We want to show $v \in \mathrm{R}\,(L)$.

For each v_n we can find a unique $u_n \in X = (\ker L)^\perp$ such that

$$Lu_n = v_n.$$

Using the Poincaré inequality we deduce

$$\|u_n - u_m\| \leq C\|v_n - v_m\|.$$

When we couple this inequality with the elliptic estimates we get

$$\|u_n - u_m\|_k \leq C(\|Lu_n - Lu_m\| + \|u_n - u_m\|) \leq C\|v_n - v_m\| \to 0 \text{ as } m, n \to \infty.$$

Hence (u_n) is a Cauchy sequence in $L^{k,2}(E)$ so that $u_n \to u$ in $L^{k,2}(E)$. Clearly $Lu = v$, so that $v \in \mathrm{R}\,(L)$.

We have so far proved that $\ker L$ is finite dimensional, and $\mathrm{R}\,(L)$ is closed, i.e. L_a is semi-Fredholm. Since $(L_a)^* = (L^*)_a$, and L^* is also an elliptic operator, we deduce $(L_a)^*$ is also semi-Fredholm. This completes the proof of Theorem 10.4.7.
    □

Using the closed range theorem of functional analysis we deduce the following important consequence.

Corollary 10.4.10 (Abstract Hodge decomposition). *Any k-th order elliptic operator $L : C^\infty(E) \to C^\infty(F)$ over the compact manifold M defines natural orthogonal decompositions of $L^2(E)$ and $L^2(F)$,*

$$L^2(E) = \ker L \oplus \mathrm{R}\,(L^*) \text{ and } L^2(F) = \ker L^* \oplus \mathrm{R}\,(L).$$ □

Corollary 10.4.11.   *If $\ker L^* = 0$ then for every $v \in L^2(F)$ the partial differential equation $Lu = v$ admits at least one weak L^2-solution $u \in L^2(E)$.*   □

The last corollary is really unusual. It states the equation $Lu = v$ has a solution provided the dual equation $L^*v = 0$ has no nontrivial solution. A nonexistence hypothesis implies an existence result! This partially explains the importance of the vanishing results in geometry, i.e., the results to the effect that $\ker L^* = 0$. With an existence result in our hands presumably we are more capable of producing geometric objects. In the next chapter we will describe one powerful technique of producing vanishing theorems based on the so called *Weitzenböck identities*.

Corollary 10.4.12. *Over a* compact *manifold*

$$\ker L = \ker L^*L \quad and \quad \ker L^* = \ker LL^*.$$

Proof. Clearly $\ker L \subset \ker L^*L$. Conversely, let $\psi \in C^\infty(E)$ such that $L^*L\psi = 0$. Then

$$\|L\psi\|^2 = \int_M \langle L\psi L\psi \rangle dV_g = \int_M \langle L^*L\psi, \psi \rangle dV_g = 0. \qquad \square$$

The Fredholm property of an elliptic operator has very deep topological ramifications culminating with one of the most beautiful results in mathematics: the Atiyah-Singer index theorems. Unfortunately, this would require a lot more extra work to include it here. However, in the remaining part of this subsection we will try to unveil some of the natural beauty of elliptic operators. We will show that the index of an elliptic operator has many of the attributes of a topological invariant.

We stick to the notations used so far. Denote by $\boldsymbol{Ell}_k(E, F)$ the space of elliptic operators $C^\infty(E) \to C^\infty(F)$ of order k. By using the attribute *space* when referring to $\boldsymbol{Ell}_k$ we implicitly suggested that it carries some structure. It is not a vector space, it is not an affine space, it is not even a convex set. It is only a cone in the linear space $\boldsymbol{PDO}^{(m)}$, but it carries a natural topology which we now proceed to describe.

Let $L_1, L_2 \in \boldsymbol{Ell}_k(E, F)$. We set

$$\delta(L_1, L_2) = \sup \{ \|L_1u - L_2u\|; \ \|u\|_k = 1 \}.$$

Define

$$d(L_1, L_2) = \max \{ \delta(L_1, L_2), \delta(L_1^*, L_2^*) \}.$$

We let the reader check that $(\boldsymbol{Ell}_k, d)$ is indeed a metric space. A continuous family of elliptic operators $(L_\lambda)_{\lambda \in \Lambda}$, where Λ is a topological space, is then a continuous map

$$\Lambda \ni \lambda \mapsto L_\lambda \in \boldsymbol{Ell}_k.$$

Roughly speaking, this means that the coefficients of L_λ, and their derivatives up to order k depend continuously upon λ.

Theorem 10.4.13. *The index map*

$$\mathrm{ind} : \boldsymbol{Ell}_k(E, F) \to \mathbb{Z}, \quad L \mapsto \mathrm{ind}\,(L)$$

is continuous.

The proof relies on a very simple algebraic trick which requires some analytical foundation.

Let X, Y be two Hilbert spaces. For any Fredholm operator $L : D(L) \subset X \to Y$ denote by $\imath_L : \ker L \to X$ (respectively by $P_L : X \to \ker L$) the natural inclusion $\ker L \hookrightarrow X$ (respectively the orthogonal projection $X \to \ker L$). If $L_i : D(L_i) \subset X \to Y$ ($i = 0, 1$) are two Fredholm operators define

$$\mathcal{R}_{L_0}(L_1) : D(L_1) \oplus \ker L_0^* \subset X \oplus \ker L_0^* \to Y \oplus \ker L_0$$

by

$$\mathcal{R}_{L_0}(L_1)(u, \phi) = (L_1 u + \imath_{L_0^*}\phi, P_{L_0} u), \quad u \in D(L_1), \ \phi \in \ker L_0^*.$$

In other words, the operator $\mathcal{R}_{L_0}(L_1)$ is given by the block decomposition

$$\mathcal{R}_{L_0}(L_1) = \begin{bmatrix} L_1 & \imath_{L_0^*} \\ P_{L_0} & 0 \end{bmatrix}.$$

We will call $\mathcal{R}_{L_0}(L_1)$ is the *regularization of L_1 at L_0*. The operator L_0 is called the *center* of the regularization. For simplicity, when $L_0 = L_1 = L$, we write

$$\mathcal{R}_L = \mathcal{R}_L(L).$$

The result below lists the main properties of the regularization.

Lemma 10.4.14. *(a) $\mathcal{R}_{L_0}(L_1)$ is a Fredholm operator.*
(b) $\mathcal{R}_{L_0}^*(L_1) = \mathcal{R}_{L_0^*}(L_1^*)$.
(c) $\mathcal{R}_{L_0}$ is invertible (with bounded inverse).   □

Exercise 10.4.15. Prove the above lemma.   □

We strongly recommend the reader who feels less comfortable with basic arguments of functional analysis to try to provide the no-surprise proof of the above result. It is a very good "routine booster".

Proof of Theorem 10.4.13   Let $L_0 \in \mathbf{Ell}_k(E, F)$. We have to find $r > 0$ such that, $\forall L \in \mathbf{Ell}_k(E, F)$ satisfying $d(L_0, L) \leq r$, we have

$$\mathrm{ind}\,(L) = \mathrm{ind}\,(L_0).$$

We will achieve this in two steps.

Step 1. We will find $r > 0$ such that, for any L satisfying $d(L_0, L) < r$, the regularization of L at L_0 is invertible, with bounded inverse.

Step 2. We will conclude that if $d(L, L_0) < r$, where $r > 0$ is determined at Step 1, then $\mathrm{ind}\,(L) = \mathrm{ind}\,(L_0)$.

Step 1. Since $\mathcal{R}_{L_0}(L)$ is Fredholm, it suffices to show that both $\mathcal{R}_{L_0}(L)$, and $\mathcal{R}_{L_0^*}(L^*)$ are injective, if L is sufficiently close to L_0. We will do this only for $\mathcal{R}_{L_0}(L)$, since the remaining case is entirely similar.

We argue by contradiction. Assume that there exists a sequence $(u_n, \phi_n) \subset L^{k,2}(E) \times \ker L_0^*$, and a sequence $(L_n) \subset \boldsymbol{Ell}_k(E, F)$ such that

$$\|u_n\|_k + \|\phi_n\| = 1, \qquad (10.4.1)$$

$$\mathcal{R}_{L_0}(L_n)(u_n, \phi_n) = (0, 0), \qquad (10.4.2)$$

and

$$d(L_0, L_n) \le 1/n. \qquad (10.4.3)$$

From (10.4.1) we deduce that (ϕ_n) is a bounded sequence in the *finite dimensional space* $\ker L_0^*$. Hence it contains a subsequence *strongly convergent* in L^2, and in fact in any Sobolev norm. Set $\phi := \lim \phi_n$. Note that $\|\phi\| = \lim_n \|\phi_n\|$. Using (10.4.2) we deduce

$$L_n u_n = -\phi_n,$$

i.e., the sequence $(L_n u_n)$ is strongly convergent to $-\phi$ in $L^2(F)$. The condition (10.4.3) now gives

$$\|L_0 u_n - L_n u_n\| \le 1/n,$$

i.e.,

$$\lim_n L_n u_n = \lim_n L_0 u_n = -\phi \in L^2(F).$$

Since $u_n \perp \ker L_0$ (by (10.4.2)) we deduce from the Poincaré inequality combined with the elliptic estimates that

$$\|u_n - u_m\|_k \le C\|L_0 u_n - L_0 u_m\| \to 0 \text{ as } m, n \to \infty.$$

Hence the sequence u_n strongly converges in $L^{k,2}$ to some u. Moreover,

$$\lim_n \|u_n\|_k = \|u\|_k \text{ and } \|u\|_k + \|\phi\| = 1.$$

Putting all the above together we conclude that there exists a pair $(u, \phi) \in L^{k,2}(E) \times \ker L_0^*$, such that

$$\|u\|_k + \|\phi\| = 1,$$

$$L_0 u = -\phi \text{ and } u \perp \ker L. \qquad (10.4.4)$$

This contradicts the abstract Hodge decomposition which coupled with (10.4.4) implies $u = 0$ and $\phi = 0$. Step 1 is completed.

Step 2. Let $r > 0$ as determined at Step 1, and $L \in \boldsymbol{Ell}_k(E, F)$. Hence

$$\mathcal{R}_{L_0}(L) = \begin{bmatrix} L & \imath_{L_0^*} \\ P_{L_0} & 0 \end{bmatrix}$$

is invertible. We will use the invertibility of this operator to produce an injective operator

$$\ker L^* \oplus \ker L_0 \hookrightarrow \ker L \oplus \ker L_0^*.$$

This implies $\dim \ker L^* + \dim \ker L_0 \leq \dim \ker L + \dim \ker L_0^*$, i.e.,

$$\mathrm{ind}\,(L_0) \leq \mathrm{ind}(L).$$

A dual argument, with L replaced by L^*, and L_0 replaced by L_0^*, will produce the opposite inequality, and thus finish the proof of Theorem 10.4.13. Now let us provide the details.

First, we orthogonally decompose

$$L^2(E) = (\ker L)^\perp \oplus \ker L \quad \text{and} \quad L^2(F) = (\ker L^*)^\perp \oplus \ker L^*.$$

Set $U := \ker L \oplus \ker L_0^*$, and $V := \ker L^* \oplus \ker L_0$. We will regard $\mathcal{R}_{L_0}(L)$ as an operator

$$\mathcal{R}_{L_0}(L) : (\ker L)^\perp \oplus U \to (\ker L^*)^\perp \oplus V.$$

As such, it has a block decomposition

$$\mathcal{R}_{L_0}(L) = \begin{bmatrix} T & A \\ B & C \end{bmatrix},$$

where

$$T : L^{k,2}(E) \cap (\ker L)^\perp \subset (\ker L)^\perp \to (\ker L^*)^\perp = \mathrm{Range}\,(L)$$

denotes the restriction of L to $(\ker L)^\perp$. The operator T *is invertible* and *its inverse is bounded.*

Since $\mathcal{R}_{L_0}(L)$ is invertible, for any $v \in V$ we can find a *unique* pair $(\phi, u) \in (\ker L)^\perp \oplus U$, such that

$$\mathcal{R}_{L_0}(L) \begin{bmatrix} \phi \\ u \end{bmatrix} = \begin{bmatrix} 0 \\ v \end{bmatrix}.$$

This means

$$T\phi + Au = 0 \quad \text{and} \quad B\phi + Cu = v.$$

We can view both ϕ and u as (linear) functions of v, $\phi = \phi(v)$ and $u = u(v)$. We claim the map $v \mapsto u = u(v)$ is injective.

Indeed, if $u(v) = 0$ for some v, then $T\phi = 0$, and since T is injective ϕ must be zero. From the equality $v = B\phi + Cu$ we deduce $v = 0$. We have thus produced the promised injective map $V \hookrightarrow U$. Theorem 10.4.13 is proved. $\square$

The theorem we have just proved has many topological consequences. We mention only one of them.

Corollary 10.4.16. *Let $L_0, L_1 \in \boldsymbol{Ell}_k(E, F)$ if $\sigma_k(L_0) = \sigma_k(L_1)$ then $\mathrm{ind}\,(L_0) = \mathrm{ind}\,(L_1)$.*

Proof. For every $t \in [0,1]$ $L_t = (1-t)L_0 + tL_1$ is a k-th order elliptic operator depending continuously on t. (Look at the symbols). Thus $\mathrm{ind}\,(L_t)$ is an integer depending continuously on t so it must be independent of t. $\qquad\square$

This corollary allows us to interpret the index as as a continuous map from the elliptic symbols to the integers. The analysis has vanished! This is (almost) a purely algebraic-topologic object. There is one (major) difficulty. These symbols are "polynomials with coefficients in some spaces of endomorphism".

The deformation invariance of the index provides a very powerful method for computing it by deforming a "complicated" situation to a "simpler" one. Unfortunately, our deformation freedom is severely limited by the "polynomial" character of the symbols. There aren't that many polynomials around. Two polynomial-like elliptic symbols may be homotopic in a larger classes of symbols which are only positively homogeneous along the fibers of the cotangent bundle).

At this point one should return to analysis, and try to conceive some operators that behave very much like elliptic p.d.o. and have more general symbols. Such objects exist, and are called *pseudo-differential operators*. We refer to [62, 90] for a very efficient presentation of this subject. We will not follow this path, but we believe the reader who reached this point can complete this journey alone.

Exercise 10.4.17. Let $L \in \mathbf{Ell}_k(E, F)$. A finite dimensional subspace $V \subset L^2(F)$ is called a *stabilizer* of L if the operator

$$S_{L,V} : L^{k,2}(E) \oplus V \to L^2(F) \quad S_{L,V}(u \oplus v) = Lu + v$$

is surjective.

(a) Show that any subspace $V \subset L^2(F)$ containing $\ker L^*$ is a stabilizer of L. More generally, any finite dimensional subspace of $L^2(F)$ containing a stabilizer is itself a stabilizer. Conclude that if V is a stabilizer, then

$$\mathrm{ind}\,L = \dim \ker S_{L,V} - \dim V. \qquad\square$$

Exercise 10.4.18. Consider a compact manifold Λ and $L : \Lambda \to \mathbf{Ell}_k(E, F)$ a continuous family of elliptic operators.

(a) Show that the family L admits an *uniform stabilizer*, i.e., there exists a finite dimensional subspace $V \subset L^2(F)$, such that V is a stabilizer of each operator L_λ in the family L.

(b) Show that if V is an uniform stabilizer of the family L, then the family of subspaces $\ker S_{L_\lambda, V}$ defines a vector bundle over Λ.

(c) Show that if V_1 and V_2 are two uniform stabilizers of the family L, then we have a natural isomorphism vector bundles

$$\ker S_{L,V_1} \oplus \underline{V}_2 \cong \ker S_{L,V_2} \oplus \underline{V}_1.$$

In particular, we have an isomorphism of line bundles

$$\det \ker S_{L,V_1} \otimes \det V_1^* \cong \det \ker S_{L,V_2} \otimes \det V_2^*.$$

Thus the line bundle $\det \ker S_{L,V} \otimes \det V^* \to \Lambda$ is independent of the uniform stabilizer V. It is called the *determinant line bundle of the family L* and is denoted by $\det \mathbf{ind}(L)$. $\qquad\qquad\qquad\qquad\qquad\qquad\qquad\qquad\qquad\qquad\qquad\qquad$ □

10.4.2 *Spectral theory*

We mentioned at the beginning of this section that the elliptic operators on compact manifold behave very much like matrices. Perhaps nothing illustrates this feature better than their remarkable spectral properties. This is the subject we want to address in this subsection.

Consider as usual, a compact, oriented Riemann manifold (M, g), and a complex vector bundle $E \to M$ endowed with a Hermitian metric $\langle \bullet, \bullet \rangle$ and compatible connection. Throughout this subsection L will denote a k-th order, formally selfadjoint elliptic operator $L : C^\infty(E) \to C^\infty(E)$. Its analytical realization

$$L_a : L^{k,2}(E) \subset L^2(E) \to L^2(E),$$

is a selfadjoint, elliptic operator, so its spectrum $\mathrm{spec}(L)$ is an unbounded closed subset of $\mathbb{R}$. Note that for any $\lambda \in \mathbb{R}$ the operator $\lambda - L_a$ is the analytical realization of the elliptic p.d.o. $\lambda \mathbb{1}_E - L$, and in particular, the operator $\lambda - L_a$ is a Fredholm, so that

$$\lambda \in \mathrm{spec}(L) \Longleftrightarrow \ker(\lambda - L) \neq 0.$$

Thus, the spectrum of L consists only of eigenvalues of finite multiplicities. The main result of this subsection states that one can find an orthonormal basis of $L^2(E)$ which diagonalizes L_a.

Theorem 10.4.19. *Let $L \in \mathbf{Ell}_k(E)$ be a formally selfadjoint elliptic operator. Then the following are true.*
(a) The spectrum $\mathrm{spec}(L)$ is real, $\mathrm{spec}(L) \subset \mathbb{R}$, and for each $\lambda \in \mathrm{spec}(L)$, the subspace $\ker(\lambda - L)$ is finite dimensional and consists of smooth sections.
(b) The spectrum $\mathrm{spec}(L)$ is a closed, countable, discrete, unbounded set.
(c) There exists an orthogonal decomposition

$$L^2(E) = \bigoplus_{\lambda \in \mathrm{spec}(L)} \ker(\lambda - L).$$

(d) Denote by P_λ the orthogonal projection onto $\ker(\lambda - L)$. Then

$$L^{k,2}(E) = D(L_a) = \left\{ \psi \in L^2(E) \; ; \; \sum_\lambda \lambda^2 \|P_\lambda \psi\|^2 < \infty \right\}.$$

Part (c) of this theorem allows one to write

$$\mathbb{1} = \sum_\lambda P_\lambda \text{ and } L = \sum_\lambda \lambda P_\lambda.$$

The first identity is true over the entire $L^2(E)$ while part (d) of the theorem shows the domain of validity of the second equality is precisely the domain of L.

Proof. (a) We only need to show that $\ker(\lambda - L)$ consists of smooth sections. In view of Weyl's lemma this is certainly the case since $\lambda - L$ is an elliptic operator. (b)&(c) We first show $\mathrm{spec}(L)$ is discrete.

More precisely, given $\lambda_0 \in \mathrm{spec}(L)$, we will find $\varepsilon > 0$ such that $\ker(\lambda - L) = 0$, $\forall |\lambda - \lambda_0|, \varepsilon, \lambda \neq \lambda_0$. Assume for simplicity $\lambda_0 = 0$.

We will argue by contradiction. Thus there exist $\lambda_n \to 0$ and $u_n \in C^\infty(E)$, such that

$$L u_n = \lambda_n u_n, \quad \|u_n\| = 1.$$

Clearly $u_n \in \mathrm{R}(L) = (\ker L^*)^\perp = (\ker L)^\perp$, so that the Poincaré inequality implies

$$1 = \|u_n\| \leq C\|L u_n\| = C\lambda_n \to 0.$$

Thus $\mathrm{spec}(L)$ must be a discrete set.

Now consider $t_0 \in \mathbb{R} \setminus \sigma(L)$. Thus $t_0 - L$ has a bounded inverse

$$T = (t_0 - L)^{-1}.$$

Obviously T is a selfadjoint operator. We claim that T is also a compact operator.

Assume (v_n) is a bounded sequence in $L^2(E)$. We have to show $u_n = T v_n$ admits a subsequence which converges in $L^2(E)$. Note that u_n is a solution of the partial differential equation

$$(t_0 - L)u_n = t_0 u_n - L u_n = v_n.$$

Using the elliptic estimates we deduce

$$\|u_n\|_k \leq C(\|u_n\| + \|v_n\|).$$

Obviously $u_n = T v_n$ is bounded in $L^2(E)$, so the above inequality implies $\|u_n\|_k$ is also bounded. The desired conclusion follows from the compactness of the embedding $L^{k,2}(E) \to L^2(E)$.

Thus T is a compact, selfadjoint operator. We can now use the spectral theory of such well behaved operators as described for example in [18], Chap. 6. The spectrum of T is a closed, bounded, countable set with one accumulation point, $\mu = 0$. Any $\mu \in \mathrm{spec}(T) \setminus \{0\}$ is an eigenvalue of T with finite multiplicity, and since $\ker T = 0$,

$$L^2(E) = \bigoplus_{\mu \in \mathrm{spec}(T) \setminus \{0\}} \ker(\mu - T).$$

Using the equality $L = t_0 - T^{-1}$ we deduce

$$\sigma(L) = \{t_0 - \mu^{-1} \; ; \; \mu \in \mathrm{spec}(T) \setminus \{0\}\}.$$

This proves (b) and (c).

To prove (d), note that if $\psi \in L^{k,2}(E)$, then $L\psi \in L^2(E)$, i.e.,

$$\Big\| \sum_\lambda \lambda P_\lambda \psi \Big\|^2 = \sum_\lambda \lambda^2 \|P_\lambda \psi\|^2 < \infty.$$

Conversely, if

$$\sum_\lambda \lambda^2 \|P_\lambda \psi\|^2 < \infty,$$

consider the sequence of smooth sections

$$\phi_n = \sum_{|\lambda| \le n} \lambda P_\lambda \psi$$

which converges in $L^2(E)$ to

$$\phi = \sum_\lambda \lambda P_\lambda \psi.$$

On the other hand, $\phi_n = L\psi_n$, where

$$\psi_n = \sum_{|\lambda| \le n} P_\lambda \psi$$

converges in $L^2(E)$ to ψ. Using the elliptic estimates we deduce

$$\|\psi_n - \psi_m\|_k \le C(\|\psi_n - \psi_m\| + \|\phi_n - \phi_m\|) \to 0 \ \text{ as } n, m \to \infty.$$

Hence $\psi \in L^{k,2}(E)$ as a $L^{k,2}$-limit of smooth sections. The theorem is proved. □

Example 10.4.20. Let $M = S^1$, $E = \mathbb{C}_M$ and

$$L = -i\frac{\partial}{\partial\theta} : C^\infty(S^1, \mathbb{C}) \to C^\infty(S^1, \mathbb{C}).$$

The operator L is clearly a formally selfadjoint elliptic p.d.o. The eigenvalues and the eigenvectors of L are determined from the periodic boundary value problem

$$-i\frac{\partial u}{\partial\theta} = \lambda u, \ \ u(0) = u(2\pi),$$

which implies

$$u(\theta) = C\exp(i\lambda\theta) \ \text{ and } \ \exp(2\pi\lambda i) = 1.$$

Hence

$$\mathrm{spec}(L) = \mathbb{Z} \ \text{ and } \ \ker(n - L) = \mathrm{span}_\mathbb{C}\{\exp(in\theta)\}.$$

The orthogonal decomposition

$$L^2(S^1) = \bigoplus_n \ker(n - L)$$

is the usual Fourier decomposition of periodic functions. Note that

$$u(\theta) = \sum_n u_n \exp(in\theta) \in L^{1,2}(S^1)$$

if and only if

$$\sum_{n \in \mathbb{Z}} (1 + n^2)|u_n|^2 < \infty.$$ □

The following exercises provide a variational description of the eigenvalues of a formally selfadjoint elliptic operator $L \in \mathbf{Ell}_k(E)$ which is bounded from below i.e.

$$\inf\left\{ \int_M \langle Lu, u \rangle dv_g; \ u \in L^{k,2}(E), \ \|u\| = 1 \right\} > -\infty.$$

Exercise 10.4.21. Let $V \in L^{k,2}(E)$ be a finite dimensional invariant subspace of L, i.e. $L(V) \subset V$. Show that
(a) The space V consists only of smooth sections.
(b) The quantity

$$\lambda(V^\perp) = \inf\left\{ \int_M \langle Lu, u \rangle dv_g; \ u \in L^{k,2}(E) \cap V^\perp, \ \|u\| = 1 \right\}$$

is an eigenvalue of L. ($V^\perp$ denotes the orthogonal complement of V in $L^2(E)$). □

Exercise 10.4.22. Set $V_0 = 0$, and denote $\lambda_0 = \lambda(V_0^\perp)$, According to the previous exercise λ_0 is an eigenvalue of L. Pick ϕ_0 an eigenvector corresponding to λ_0 such that $\|\phi_0\| = 1$ and form $V_1 = V_0 \oplus \text{span}\{\phi_0\}$. Set $\lambda_1 = \lambda(V_1^\perp)$ and iterate the procedure. After m steps we have produced $m+1$ vectors $\phi_0, \phi_1, \ldots, \phi_m$ corresponding to $m + 1$ eigenvalues $\lambda_0, \lambda_1 \leq \cdots \lambda_m$ of L. Set $V_{m+1} = \text{span}_\mathbb{C}\{\phi_0, \phi_1, \ldots, \phi_m\}$ and $\lambda_{m+1} = \lambda(V_{m+1}^\perp)$ etc.
(a) Prove that the set

$$\{\phi_1, \ldots, \phi_m, \ldots\}$$

is a Hilbert basis of $L^2(E)$, and

$$\text{spec}(L) = \{\lambda_1 \leq \cdots \leq \lambda_m \cdots\}.$$

(b) Denote by $\mathbf{Gr}_m$ the Grassmannian of m-dimensional subspaces of $L^{k,2}(E)$. Show that

$$\lambda_m = \inf_{V \in \mathbf{Gr}_m} \max\left\{ \int_M \langle Lu, u \rangle dv_g; \ u \in V \ \|u\| = 1 \right\}. \qquad □$$

Exercise 10.4.23. Use the results in the above exercises to show that if L is a bounded from below, k-th order formally selfadjoint elliptic p.do. over an N-dimensional manifold then

$$\lambda_m(L) = O(m^{k/N}) \ \text{as} \ m \to \infty,$$

and

$$d(\Lambda) = \dim \oplus_{\lambda \leq \Lambda} \ker(\lambda - L) = O(\Lambda^{N/k}) \ \text{as} \ \Lambda \to \infty. \qquad □$$

Remark 10.4.24. (a) When L is a formally selfadjoint generalized Laplacian then the result in the above exercise can be considerably sharpened. More precisely H.Weyl showed that

$$\lim_{\Lambda \to \infty} \Lambda^{-N/2} d(\Lambda) = \frac{\text{rank}(E) \cdot \text{vol}_g(M)}{(4\pi)^{N/2}\Gamma(N/2 + 1)}. \tag{10.4.5}$$

The very ingenious proof of this result relies on another famous p.d.o., namely the heat operator $\partial_t + L$ on $\mathbb{R} \times M$. For details we refer to [11], or [90].

(b) Assume L is the scalar Laplacian Δ on a compact Riemann manifold (M, g) of dimension M. Weyl's formula shows that the asymptotic behavior of the spectrum of Δ contains several geometric informations about M: we can read the dimension and the volume of M from it. If we think of M as the elastic membrane of a drum then the eigenvalues of Δ describe all the frequencies of the sounds the "drum" M can produce. Thus "we can hear" the dimension and the volume of a drum. This is a special case of a famous question raised by V.Kac in [51]: can one hear the shape of a drum? In more rigorous terms this question asks how much of the geometry of a Riemann manifold can be recovered from the spectrum of its Laplacian. This is what spectral geometry is all about.

It has been established recently that the answer to Kac's original question is negative. We refer to [38] and the references therein for more details. $\qquad \square$

Exercise 10.4.25. Compute the spectrum of the scalar Laplacian on the torus T^2 equipped with the flat metric, and then use this information to prove the above Weyl asymptotic formula in this special case. $\qquad \square$

Exercise 10.4.26. Denote by $\Delta_{S^{n-1}}$ the scalar Laplacian on the unit sphere $S^{n-1} \subset \mathbb{R}^n$ equipped with the induced metric. We assume $n \geq 2$. A polynomial $p = p(x_1, \ldots, x_n)$ is called *harmonic* if $\Delta_{\mathbb{R}^n} p = 0$.

(a) Show that if p is a homogenous harmonic polynomial of degree k, and $\bar{p}$ denotes its restriction to the unit sphere S^{n-1} centered at the origin, then

$$\Delta_{S^{n-1}} \bar{p} = n(n + k - 2)\bar{p}.$$

(b) Show that a function $\varphi : S^n \to \mathbb{R}$ is an eigenfunction of $\Delta_{S^{n-1}}$ if and only if there exists a homogeneous harmonic polynomial p such that $\varphi = \bar{p}$.

(c) Set $\lambda_k := n(n + k - 2)$. Show that

$$\dim \ker \left(\lambda_k - \Delta_{S^{n-1}} \right) = \binom{k + n - 1}{n} - \binom{k + n - 3}{n - 2}.$$

(d) Prove Weyl's asymptotic estimate (10.4.5) in the special case of the operator $\Delta_{S^{n-1}}$. $\qquad \square$

10.4.3 *Hodge theory*

We now have enough theoretical background to discuss the celebrated Hodge theorem. It is convenient to work in a slightly more general context than Hodge's original theorem.

Definition 10.4.27. Let (M, g) be an oriented Riemann manifold. An *elliptic complex* is a sequence of first order p.d.o.'s

$$0 \to C^\infty(E_0) \xrightarrow{D_0} C^\infty(E_1) \xrightarrow{D_1} \cdots \xrightarrow{D_{m-1}} C^\infty(E_m) \to 0$$

satisfying the following conditions.

(i) $(C^\infty(E_i), D_i)$ is a cochain complex, i.e., $D_i D_{i-1} = 0$, $\forall 1 \le i \le m$.
(ii) For each $(x, \xi) \in T^*M \setminus \{0\}$ the sequence of principal symbols

$$0 \to (E_0)_x \xrightarrow{\sigma(D_0)(x,\xi)} (E_1)_x \to \cdots \xrightarrow{\sigma(D_{m-1})(x,\xi)} (E_m)_x \to 0$$

is exact.

$\square$

Example 10.4.28. The DeRham complex $(\Omega^*(M), d)$ is an elliptic complex. In this case, the associated sequence of principal symbols is ($e(\xi)$ = exterior multiplication by ξ)

$$0 \to \mathbb{R} \xrightarrow{e(\xi)} T^*_x M \xrightarrow{e(\xi)} \cdots \xrightarrow{e(\xi)} \det(T^*_x M) \to 0$$

is the Koszul complex of Exercise 7.1.23 of Subsection 7.1.3 where it is shown to be exact. Hence, the DeRham complex is elliptic. We will have the occasion to discuss another famous elliptic complex in the next chapter.

$\square$

Consider an elliptic complex $(C^\infty(E_\cdot, D_\cdot))$ over a *compact* oriented Riemann manifold (M, g). Denote its cohomology by $H^\bullet(E_\cdot, D_\cdot)$. A priori these may be infinite dimensional spaces. We will see that the combination ellipticity + compactness prevents this from happening. Endow each E_i with a metric and compatible connection. We can now talk about Sobolev spaces and formal adjoints D_i^*. Form the operators

$$\Delta_i = D_i^* D_i + D_{i-1} D_{i-1}^* : C^\infty(E_i) \to C^\infty(E_i).$$

We can now state and prove the celebrated Hodge theorem.

Theorem 10.4.29 (Hodge). *Assume that M is* compact *and* oriented. *Then the following are true.*
(a) $H^i(E_\cdot, D_\cdot) \cong \ker \Delta_i \subset C^\infty(E_i)$, $\forall i$.
(b) $\dim H^i(E_\cdot, D_\cdot) < \infty$ $\forall i$.
(c) **(Hodge decomposition).** *There exists an orthogonal decomposition*
$$L^2(E_i) = \ker \Delta_i \oplus \mathrm{R}(D_{i-1}) \oplus \mathrm{R}(D_i^*),$$
where we view both D_{i-1}, and D_i as bounded operators $L^{1,2} \to L^2$.

Proof. Set
$$E = \oplus E_i, \quad D = \oplus D_i, \quad D^* = \oplus D_i^*, \quad \Delta = \oplus \Delta_i, \quad \hat{D} = D + D^*.$$
Thus $D, D^*, \Delta \in \mathbf{PDO}(E, E)$. Since $D_i D_{i-1} = 0$, we deduce $D^2 = (D^*)^2 = 0$ which implies
$$\Delta = D^* D + D D^* = (D + D^*)^2 = \hat{D}^2.$$
We now invoke the following elementary algebraic fact which is a consequence of the exactness of the symbol sequence.

Exercise 10.4.30. The operators $\hat{D}$ and Δ are *elliptic*, formally selfadjoint p.d.o. (**Hint:** Use Exercise 7.1.22 in Subsection 7.1.3.)

$\square$

Note that according to Corollary 10.4.12 we have $\ker \Delta = \ker \hat{D}$, so that we have an orthogonal decomposition

$$L^2(E) = \ker \Delta \oplus \mathrm{R}\,(\hat{D}). \tag{10.4.6}$$

This is precisely part (c) of Hodge's theorem.

For each i denote by P_i the orthogonal projection $L^2(E_i) \to \ker \Delta_i$. Set

$$Z^i = \left\{ u \in C^\infty(E_i);\ D_i u = 0 \right\} \ \text{and}\ B^i = D_{i-1}(C^\infty(E_{i-1})),$$

so that

$$H^i(E_\cdot, D_\cdot) = Z^i / B^i.$$

We claim that the map $P_i : Z^i \to \ker \Delta_i$ descends to an isomorphism $H^i(E_\cdot, D_\cdot) \to \ker \Delta_i$. This will complete the proof of Hodge theorem. The above claim is a consequence of several simple facts.

Fact 1. $\ker \Delta_i \subset Z^i$. This follows from the equality $\ker \Delta = \ker \hat{D}$.

Fact 2. If $u \in Z_i$ then $u - P_i u \in B_i$. Indeed, using the decomposition (10.4.6) we have

$$u = P_i u + \hat{D}\psi \ \ \psi \in L^{1,2}(E).$$

Since $u - P_i u \in C^\infty(E_i)$, we deduce from Weyl's lemma that $\psi \in C^\infty(E)$. Thus, there exist $v \in C^\infty(E_{i-1})$ and $w \in C^\infty(E_i)$ such that $\psi = v \oplus w$ and

$$u = P_i u + D_{i-1} v + D_i^* w.$$

Applying D_i on both sides of this equality we get

$$0 = D_i u = D_i P_i u + D_i D_{i-1} u + D_i D_i^* w = D_i D_i^* w.$$

Since $\ker D_i^* = \ker D_i D_i^*$, the above equalities imply

$$u - P_i u = D_{i-1} v \in B^i.$$

We conclude that P_i descends to a linear map $\ker \Delta_i \to H^i(E_\cdot, D_\cdot)$. Thus

$$B^i \subset \mathrm{R}\,(\hat{D}) = (\ker \Delta)^\perp,$$

and we deduce that no two distinct elements in $\ker \Delta_i$ are cohomologous since otherwise their difference would have been orthogonal to $\ker \Delta_i$. Hence, the induced linear map $P_i : \ker \Delta_i \to H^i(E_\cdot, D_\cdot)$ is injective. **Fact 1** shows it is also surjective so that

$$\ker \Delta_i \cong H^i(E_\cdot, D_\cdot).$$

Hodge theorem is proved.   □

Let us apply Hodge theorem to the DeRham complex on a compact oriented Riemann manifold

$$0 \to \Omega^0(M) \xrightarrow{d} \Omega^1(M) \xrightarrow{d} \cdots \xrightarrow{d} \Omega^n(M) \to 0, \ \ n = \dim M.$$

We know that the formal adjoint of $d : \Omega^k(M) \to \Omega^{k+1}(M)$ is

$$d^* = (-1)^{\nu_{n,k}} * d*,$$

where $\nu_{n,k} = nk + n + 1$, and $*$ denotes the Hodge $*$-operator defined by the Riemann metric g and the fixed orientation on M. Set

$$\Delta = dd^* + d^*d.$$

Corollary 10.4.31 (Hodge). *Any smooth k-form $\omega \in \Omega^k(M)$ decomposes uniquely as*

$$\omega = [\omega]_g + d\eta + d^*\zeta, \quad \eta \in \Omega^{k-1}(M), \quad \zeta \in \Omega^{k+1}(M),$$

and $[\omega]_g \in \Omega^k(M)$ is g-harmonic, i.e.,

$$\Delta[\omega]_g = 0 \iff d[\omega]_g = 0 \text{ and } d^*[\omega]_g = 0.$$

The form $[\omega]_g$ is called the g-harmonic part of ω.

If moreover ω is closed, then $d^*\zeta = 0$, and this means any cohomology class $[z] \in H^k(M)$ is represented by a unique *harmonic k-form.* □

Denote by $\mathbf{H}^k(M, g)$ the space of g-harmonic k-forms on M. The above corollary shows

$$\mathbf{H}^k(M, g) \cong H^k(M)$$

for any metric g.

Corollary 10.4.32. *The Hodge $*$-operator defines a bijection*

$$* : \mathbf{H}^k(M, g) \to \mathbf{H}^{n-k}(M, g).$$

Proof. If ω is g-harmonic, then so is $*\omega$ since

$$d * \omega = \pm * d * \omega = 0,$$

and

$$*d * (*\omega) = \pm * (d\omega) = 0.$$

The Hodge $*$-operator is bijective since $*^2 = (-1)^{k(n-k)}$. □

Using the L^2-inner product on $\Omega^\bullet(M)$ we can identify $\mathbf{H}^{n-k}(M, g)$ with its dual, and thus we can view $*$ as an isomorphism

$$\mathbf{H}^k \xrightarrow{*} (\mathbf{H}^{n-k})^*.$$

On the other hand, the Poincaré duality described in Chapter 7 induces another isomorphism

$$\mathbf{H}^k \xrightarrow{PD} (\mathbf{H}^{n-k})^*,$$

defined by

$$\langle PD(\omega), \eta \rangle_0 = \int_M \omega \wedge \eta,$$

where $\langle \cdot, \cdot \rangle_0$ denotes the natural pairing between a vector space and its dual.

Proposition 10.4.33. $PD = *$, *i.e.*

$$\int_M \langle *\omega, \eta \rangle_g dv_g = \int_M \omega \wedge \eta \ \forall \omega \in \mathbf{H}^k \ \eta \in \mathbf{H}^{n-k}.$$

Proof.   We have

$$\langle *\omega, \eta\rangle_g dv_g = \langle \eta, *\omega\rangle_g dv_g = \eta \wedge *^2\omega = (-1)^{k(n-k)}\eta \wedge \omega = \omega \wedge \eta. \qquad \square$$

Exercise 10.4.34.   Let $\omega_0 \in \Omega^k(M)$ be a harmonic k form and denote by $\mathcal{C}_{\omega_0}$ its cohomology class. Show that

$$\int_M |\omega_0|_g^2 dv_g \leq \int_M |\omega|_g^2 dv_g \quad \forall \omega \in \mathcal{C}_{\omega_0},$$

with equality if and only if $\omega = \omega_0$. $\square$

Exercise 10.4.35.   Let G denote a compact connected Lie group equipped with a bi-invariant Riemann metric h. Prove that a differential form on G is h-harmonic if and only if it is bi-invariant. $\square$

Chapter 11

Dirac Operators

We devote this last chapter to a presentation of a very important class of first order elliptic operators which have numerous applications in modern geometry. We will first describe their general features, and then we will spend the remaining part discussing some frequently encountered examples.

11.1 The structure of Dirac operators

11.1.1 *Basic definitions and examples*

Consider a Riemann manifold (M, g), and a smooth vector bundle $E \to M$.

Definition 11.1.1. A *Dirac operator* is a first order p.d.o.

$$D : C^\infty(E) \to C^\infty(E),$$

such that D^2 is a generalized Laplacian, i.e.,

$$\sigma(D^2)(x, \xi) = -|\xi|^2_g \mathbb{1}_{E_x} \quad \forall (x, \xi) \in T^*M.$$

The Dirac operator is said to be *graded* if E splits as $E = E_0 \oplus E_1$, and $D(C^\infty(E_i)) \subset C^\infty(E_{(i+1)\bmod 2})$. In other words, D has a block decomposition

$$D = \begin{bmatrix} 0 & A \\ B & 0 \end{bmatrix}.$$ □

Note that the Dirac operators are first order *elliptic* p.d.o.-s.

Example 11.1.2. (Hamilton-Floer). Denote by E the trivial vector bundle $\mathbb{R}^{2n}$ over the circle S^1. Thus $C^\infty(E)$ can be identified with the space of smooth functions

$$u : S^1 \to \mathbb{R}^{2n}.$$

Let $J : C^\infty(E) \to C^\infty(E)$ denote the endomorphism of E which has the block decomposition

$$J = \begin{bmatrix} 0 & -\mathbb{1}_{\mathbb{R}^n} \\ \mathbb{1}_{\mathbb{R}^n} & 0 \end{bmatrix}$$

with respect to the natural splitting $\mathbb{R}^{2n} = \mathbb{R}^n \oplus \mathbb{R}^n$. We define the *Hamilton-Floer operator*

$$\mathcal{F} : C^\infty(E) \to C^\infty(E), \quad \mathcal{F}u = J\frac{du}{d\theta} \quad \forall u \in C^\infty(E).$$

Clearly, $\mathcal{F}^2 = -\frac{d^2}{d\theta^2}$ is a generalized Laplacian. □

Example 11.1.3. (Cauchy-Riemann). Consider the trivial bundle $\mathbb{C}^2$ over the complex plane $\mathbb{C}$ equipped with the standard Euclidean metric. The *Cauchy-Riemann operator* is the first order p.d.o. $D : \mathbb{C}^2 \to \mathbb{C}^2$ defined by

$$\begin{bmatrix} u \\ v \end{bmatrix} \longmapsto 2\begin{bmatrix} 0 & -\partial_z \\ \partial_{\bar{z}} & 0 \end{bmatrix} \cdot \begin{bmatrix} u \\ v \end{bmatrix},$$

where $z = x + y\boldsymbol{i}$, and

$$\partial_z = \frac{1}{2}(\partial_x - \boldsymbol{i}\partial_y), \quad \partial_{\bar{z}} = \frac{1}{2}(\partial_x + \boldsymbol{i}\partial_y).$$

A simple computation shows that D is a Dirac operator. □

Example 11.1.4. Suppose E is a vector bundle over the Riemann manifold (M, g), and $D : C^\infty(M) \to C^\infty(M)$ is a Dirac operator. Denote by $\hat{M}$ the cylinder $\mathbb{R} \times M$, and by $\hat{g}$ the cylindrical metric $\hat{g} = dt^2 + g$ on $\hat{M}$, where t denotes the coordinate along the factor $\mathbb{R}$.

We have a natural projection $\pi : \hat{M} \to M$, $(t, x) \mapsto x$, and we set $\hat{E} := \pi^*E$, the pullback of E to $\hat{M}$ via π. A section $\hat{u}$ of $\hat{E}$ is then a smooth 1-parameter family of sections $u(t)$ of E, $t \in \mathbb{R}$. In particular, we can define unambiguously

$$\partial_t \hat{u}(t_0, x_0) = \lim_{h \to 0} \frac{1}{h}\big(u(t_0 + h, x_0) - u(t_0, x_0) \big),$$

where the above limit is in the t-independent vector space E_{x_0}. Now define

$$\hat{D} : C^\infty(\hat{E} \oplus \hat{E}) \to C^\infty(\hat{E} \oplus \hat{E}),$$

by

$$\begin{bmatrix} u(t) \\ v(t) \end{bmatrix} = \begin{bmatrix} 0 & -\partial_t + D \\ \partial_t + D & 0 \end{bmatrix}\begin{bmatrix} u(t) \\ v(t) \end{bmatrix}.$$

Then $\hat{D}$ is a Dirac operator called the *suspension* of D. Note that the Cauchy-Riemann operator is the suspension of the Hamilton-Floer operator. □

Example 11.1.5. (Hodge-DeRham). Let (M, g) be an oriented Riemann manifold. Then, according to the computations in the previous chapter, the Hodge-DeRham operator

$$d + d^* : \Omega^\bullet(M) \to \Omega^\bullet(M)$$

is a Dirac operator. □

Let $D : C^\infty(E) \to C^\infty(E)$ be a Dirac operator over the oriented Riemann manifold (M, g). Its symbol is an endomorphism $\sigma(D) : \pi^* E \to \pi^* E$, where $\pi : T^* M \to M$ denotes the natural projection. Thus, for any $x \in M$, and any $\xi \in T_x^* M$, the operator

$$c(\xi) := \sigma(D)(\xi, x)$$

is an endomorphism of E_x depending linearly upon ξ. Since D^2 is a generalized Laplacian we deduce that

$$c(\xi)^2 = \sigma(D^2)(x, \xi) = -|\xi|_g^2 \mathbb{1}_{E_x}.$$

To summarize, we see that each Dirac operator induces a bundle morphism

$$c : T^* M \otimes E \to E \quad (\xi, e) \mapsto c(\xi)e,$$

such that $c(\xi)^2 = -|\xi|^2$. From the equality

$$c(\xi + \eta) = -|\xi + \eta|^2 \quad \forall \xi, \eta \in T_x^* M, \ x \in M$$

we conclude that

$$\{c(\xi), c(\eta)\} = -2g(\xi, \eta) \mathbb{1}_{E_x},$$

where for any linear operators A, B we denoted by $\{A, B\}$ their anticommutator

$$\{A, B\} := AB + BA.$$

Definition 11.1.6. (a) A *Clifford structure* on a vector bundle E over a Riemann manifold (M, g) is a smooth bundle morphism $\boldsymbol{c} : T^* M \otimes E \to E$, such that

$$\{\boldsymbol{c}(\xi), \boldsymbol{c}(\eta)\} = -2g(\xi, \eta) \mathbb{1}_E, \quad \forall \xi, \eta \in \Omega^1(M),$$

where for every 1-form α we denoted by $\boldsymbol{c}(\alpha)$ the bundle morphism $\boldsymbol{c}(\alpha) : E \to E$ given by

$$\boldsymbol{c}(\alpha)u = \boldsymbol{c}(\alpha, u), \quad \forall u \in C^\infty(E).$$

The morphism $\boldsymbol{c}$ is usually called the *Clifford multiplication* of the (Clifford) structure. A pair (vector bundle, Clifford structure) is called a *Clifford bundle*.
(b) A $\mathbb{Z}_2$-grading of a Clifford bundle $E \to M$ is a splitting $E = E^+ \oplus E^-$ such that, $\forall \alpha \in \Omega^1(M)$, the Clifford multiplication by α is an odd endomorphism of the superspace $C^\infty(E_+) \oplus C^\infty(E_-)$, i.e., $\boldsymbol{c}(\alpha)C^\infty(E^\pm) \subset C^\infty(E^\mp)$. $\qquad\square$

Proposition 11.1.7. *Let $E \to M$ be a smooth vector bundle over the Riemann manifold (M, g). Then the following conditions are equivalent.*
(a) There exists a Dirac operator $D : C^\infty(E) \to C^\infty(E)$.
(b) The bundle E admits a Clifford structure.

Proof.   We have just seen that (a)⇒(b). To prove the reverse implication let

$$c : T^*M \otimes E \to E$$

be a Clifford multiplication. Then, for every connection

$$\nabla : C^\infty(E) \to C^\infty(T^*M \otimes E),$$

the composition

$$D = c \circ \nabla : C^\infty(E) \xrightarrow{\nabla} C^\infty(T^*M \otimes E) \xrightarrow{c} C^\infty(E)$$

is a first order p.d.o. with symbol c. Clearly D is a Dirac operator.   □

Example 11.1.8.   Let (M, g) be a Riemann manifold. For each $x \in M$, and $\xi \in T_x^*M$ define

$$c(\xi) : \Lambda^\bullet T_x^*M \to \Lambda^\bullet T_x^*M$$

by

$$c(\xi)\omega = (e_\xi - i_\xi)\omega,$$

where e_ξ denotes the (left) exterior multiplication by ξ, while i_ξ denotes the interior differentiation along $\xi^* \in T_xM$ - the metric dual of ξ. The Exercise 2.2.55 of Section 2.2.4 shows that c defines a Clifford multiplication on $\Lambda^\bullet T^*M$. If ∇ denotes the Levi-Civita connection on $\Lambda^\bullet T^*M$, then the Dirac operator $c \circ \nabla$ is none other than the Hodge-DeRham operator.   □

Exercise 11.1.9.   Prove the last assertion in the above example.   □

The above proposition reduces the problem of describing which vector bundles admit Dirac operators to an algebraic-topological one: find the bundles admitting a Clifford structure. In the following subsections we will address precisely this issue.

11.1.2   *Clifford algebras*

The first thing we want to understand is the object called Clifford multiplication.

Consider a real finite dimensional, Euclidean space (V, g). A *Clifford multiplication associated with* (V, g) is then a pair (E, ρ), where E is a $\mathbb{K}$-vector space, and $\rho : V \to \text{End}(E)$ is an $\mathbb{R}$-linear map such that

$$\{\rho(u), \rho(v)\} = -2g(u, v)\mathbb{1}_E, \;\; \forall u, v \in V.$$

If (e_i) is an orthonormal basis of V, then ρ is completely determined by the linear operators $\rho_i = \rho(e_i)$ which satisfy the anti-commutation rules

$$\{\rho_i, \rho_j\} = -2\delta_{ij}\mathbb{1}_E.$$

The collection (ρ_i) generates an associative subalgebra in $\text{End}(E)$, and it is natural to try to understand its structure. We will look at the following universal situation.

Definition 11.1.10. Let V be a real, finite dimensional vector space, and
$$q : V \times V \to \mathbb{R}$$
a symmetric bilinear form. The Clifford algebra $\mathrm{Cl}(V, q)$ is the associative $\mathbb{R}$-algebra with unit, generated by V, and subject to the relations
$$\{u, v\} = uv + vu = -2q(u, v) \cdot \mathbb{1} \quad \forall u, v \in V. \qquad \square$$

Proposition 11.1.11. *The Clifford algebra $\mathrm{Cl}(V, q)$ exists, and is uniquely defined by its universality property: for every linear map $\jmath : V \to \mathcal{A}$ such that $\mathcal{A}$ is an associative $\mathbb{R}$-algebra with unit, and $\{\jmath(u), \jmath(v)\} = -2q(u, v) \cdot \mathbb{1}$, there exists an unique morphism of algebras $\Phi : \mathrm{Cl}(V, q) \to \mathcal{A}$ such that the diagram below is commutative.*

$\imath$ denotes the natural inclusion $V \hookrightarrow \mathrm{Cl}(V, q)$.

Sketch of proof Let $\mathcal{A} = \oplus_{k \geq 0} V^{\otimes k}$ ($V^{\otimes 0} = \mathbb{R}$) denote the free associative $\mathbb{R}$-algebra with unit generated by V. Set
$$\mathrm{Cl}(V, q) = \mathcal{A}/\mathcal{I},$$
where $\mathcal{I}$ is the ideal generated by
$$\{ u \otimes v + v \otimes u + 2q(u, v) \otimes 1; \ u, v \in V \}.$$
The map $\imath$ is the composition $V \hookrightarrow \mathcal{A} \to \mathrm{Cl}(V, q)$ where the second arrow is the natural projection. We let the reader check the universality property. $\qquad \square$

Exercise 11.1.12. Prove the universality property. $\qquad \square$

Remark 11.1.13. (a) When $q \equiv 0$ then $\mathrm{Cl}(V, 0)$ is the exterior algebra $\Lambda^\bullet V$.
(b) In the sequel the inclusion $V \hookrightarrow \mathrm{Cl}(V, q)$ will be thought of as being part of the definition of a Clifford algebra. This makes a Clifford algebra a structure richer than merely an abstract $\mathbb{R}$-algebra: it is an algebra with a distinguished real subspace. Thus when thinking of automorphisms of this structure one should really concentrate only on those automorphisms of $\mathbb{R}$-algebras preserving the distinguished subspace. $\qquad \square$

Corollary 11.1.14. *Let (V_i, q_i) $(i = 1, 2)$ be two real, finite dimensional vector spaces endowed with quadratic forms $q_i : V \to \mathbb{R}$. Then, any linear map $T : V_1 \to V_2$ such that $q_2(Tv) = q_1(v)$, $\forall v \in V_1$ induces a unique morphism of algebras $T_\# : \mathrm{Cl}(V_1, q_1) \to \mathrm{Cl}(V_2, q_2)$ such that $T_\#(V_1) \subset V_2$, where we view V_i as a linear subspace in $\mathrm{Cl}(V_i, q_i)$.*

The correspondence $T \mapsto T_\#$ constructed above is functorial, i.e., $(\mathbb{1}_{V_i})_\# = \mathbb{1}_{\mathrm{Cl}(V_i, q_i)}$, and $(S \circ T)_\# = S_\# \circ T_\#$, for all admissible S, and T.

The above corollary shows that the algebra $\mathrm{Cl}(V, q)$ depends only on the isomorphism class of the pair $(V, q)=$ vector space $+$ quadratic form. It is known from linear algebra that the isomorphism classes of such pairs are classified by some simple invariants:

$$(\dim V, \operatorname{rank} q, \operatorname{sign} q).$$

We will be interested in the special case when $\dim V = \operatorname{rank} q = \operatorname{sign} q = n$, i.e., when q is an Euclidean metric on the n-dimensional space V. In this case, the Clifford algebra $\mathrm{Cl}(V, q)$ is usually denoted by $\mathrm{Cl}(V)$, or Cl_n. If (e_i) is an orthonormal basis of V, then we can alternatively describe Cl_n as the associative $\mathbb{R}$-algebra with 1 generated by (e_i), and subject to the relations

$$e_i e_j + e_j e_i = -2\delta_{ij}.$$

Using the universality property of Cl_n, we deduce that the map

$$V \to \mathrm{Cl}(V), \quad v \mapsto -v \in \mathrm{Cl}(V)$$

extends to an automorphism of algebras $\alpha : \mathrm{Cl}(V) \to \mathrm{Cl}(V)$. Note that α is involutive, i.e., $\alpha^2 = \mathbb{1}$. Set

$$\mathrm{Cl}^0(V) := \ker(\alpha - \mathbb{1}), \quad \mathrm{Cl}^1(V) = \ker(\alpha + \mathbb{1}).$$

Note that $\mathrm{Cl}(V) = \mathrm{Cl}^0(V) \oplus \mathrm{Cl}^1(V)$, and moreover

$$\mathrm{Cl}^\varepsilon(V) \cdot \mathrm{Cl}^\eta(V) \subset \mathrm{Cl}^{(\varepsilon + \eta) \bmod 2}(V),$$

i.e., the automorphism α naturally defines a $\mathbb{Z}_2$-grading of $\mathrm{Cl}(V)$. In other words, the Clifford algebra $\mathrm{Cl}(V)$ is naturally a super-algebra.

Let $(\tilde{\mathrm{Cl}}(V), +, *)$ denote the opposite algebra of $\mathrm{Cl}(V)$. Then $\tilde{\mathrm{Cl}}(V)$ coincides with $\mathrm{Cl}(V)$ as a vector space, but its multiplication $*$ is defined by

$$x * y := y \cdot x \quad \forall x, y \in \tilde{\mathrm{Cl}}(V),$$

where "$\cdot$" denotes the usual multiplication in $\mathrm{Cl}(V)$. Note that for any $u, v \in V$

$$u \cdot v + v \cdot u = u * v + v * u,$$

so that, using the universality property of Clifford algebras, we conclude that the natural injection $V \hookrightarrow \tilde{\mathrm{Cl}}(V)$ extends to a morphism of algebras $\mathrm{Cl}(V) \to \tilde{\mathrm{Cl}}(V)$. This may as well be regarded as an antimorphism $\mathrm{Cl}(V) \to \mathrm{Cl}(V)$ which we call the *transposition* map, $x \mapsto x^\flat$. Note that

$$(u_1 \cdot u_2 \cdots u_r)^\flat = u_r \cdots u_1, \quad \forall u_i \in V.$$

For $x \in \mathrm{Cl}(V)$ we set

$$x^\dagger := (\alpha(x))^\flat = \alpha(x^\flat).$$

The element $x^\dagger$ is called the *adjoint* of x.

For each $v \in V$ define $c(v) \in \mathrm{End}\,(\Lambda^\bullet V)$ by

$$c(v)\omega := (e_v - i_v)\omega \quad \forall \omega \in \Lambda^\bullet V,$$

where as usual e_v denotes the (left) exterior multiplication by v, while i_v denotes the interior derivative along the metric dual of v. Invoking again Exercise 2.2.55 we deduce

$$\boldsymbol{c}(v)^2 = -|v|_g^2,$$

so that, by the universality property of the Clifford algebras, the map c extends to a morphism of algebras $\boldsymbol{c} : \mathrm{Cl}(V) \to \mathrm{End}\,(\Lambda^\bullet V)$.

Exercise 11.1.15. Prove that $\forall x \in \mathrm{Cl}(V)$ we have

$$\boldsymbol{c}(x^\dagger) = \boldsymbol{c}(x)^*,$$

where the $*$ on the right-hand side denotes the adjoint of $\boldsymbol{c}(x)$ viewed as a linear operator on the linear space $\Lambda^\bullet V$ endowed with the metric induced by the metric on V. $\qquad\square$

For each $x \in \mathrm{Cl}(V)$, we set $\sigma(x) := \boldsymbol{c}(x)1 \in \Lambda^\bullet V$. This is an element of $\Lambda^\bullet V$, called the *symbol* of x. The resulting linear map

$$\mathrm{Cl}(V) \ni x \mapsto \sigma(x) \in \Lambda^\bullet V,$$

is called the *symbol* map. If (e_i) is an orthonormal basis, then

$$\sigma(e_{i_1} \cdots e_{i_k}) = e_{i_1} \wedge \cdots \wedge e_{i_k} \quad \forall e_{i_j}.$$

This shows that the symbol map is bijective since the ordered monomials

$$\{e_{i_1} \cdots e_{i_k} \; ; \; 1 \le i_1, \cdots i_k \le \dim V\}$$

form a basis of $\mathrm{Cl}(V)$. The inverse of the symbol map is called the *quantization* map and is denoted by $\mathfrak{q} : \Lambda^\bullet V \to \mathrm{Cl}(V)$.

Exercise 11.1.16. Show that $\mathfrak{q}(\Lambda^{even/odd}V) = \mathrm{Cl}^{even/odd}(V)$. $\qquad\square$

Definition 11.1.17. (a) A $\mathbb{K}(=\mathbb{R}\text{-},\mathbb{C})$-vector space E is said to be a $\mathbb{K}$-*Clifford module* if there exists a morphism of $\mathbb{R}$-algebras

$$\rho : \mathrm{Cl}(V) \to \mathrm{End}_{\mathbb{K}}(E).$$

(b) A $\mathbb{K}$-superspace E is said to be a $\mathbb{K}$-Clifford s-module if there exists a morphism of s-algebras

$$\rho : \mathrm{Cl}(V) \to \widehat{\mathrm{End}}_{\mathbb{K}}(E).$$

(c) Let (E, ρ) be a $\mathbb{K}$-Clifford module, $\rho : \mathrm{Cl}(V) \to \mathrm{End}_{\mathbb{K}}(E)$. The module (E, ρ) is said to be selfadjoint if there exists a metric on E (Euclidean if $\mathbb{K} = \mathbb{R}$, Hermitian if $\mathbb{K} = \mathbb{C}$) such that

$$\rho(x^\dagger) = \rho(x)^* \quad \forall x \in \mathrm{Cl}(V). \qquad\square$$

We now see that what we originally called a Clifford structure is precisely a Clifford module.

Example 11.1.18. $\Lambda^\bullet V$ is a selfadjoint, real $\mathrm{Cl}(V)$ super-module. $\qquad\square$

In the following two subsections we intend to describe the complex Clifford modules. The real theory is far more elaborate. For more information we refer the reader to the excellent monograph [62].

11.1.3 *Clifford modules: the even case*

In studying complex Clifford modules it is convenient to work with the complexified Clifford algebras

$$\mathbb{Cl}_n = \mathrm{Cl}_n \otimes_{\mathbb{R}} \mathbb{C}.$$

The (complex) representation theory of $\mathbb{Cl}_n$ depends on the parity of n so that we will discuss each case separately. The reader may want to refresh his/her memory of the considerations in Subsection 2.2.5.

Let $n = 2k$, and consider an n-dimensional Euclidean space (V, g). The decisive step in describing the complex $\mathrm{Cl}(V)$-modules is the following.

Proposition 11.1.19. *There exists a complex* $\mathrm{Cl}(V)$-*module* $\mathbb{S} = \mathbb{S}(V)$ *such that*

$$\mathrm{Cl}(V) \cong \mathrm{End}_{\mathbb{C}}(\mathbb{S}) \quad \text{as } \mathbb{C}\text{-algebras.}$$

(The above isomorphism is not natural; it depends on several auxiliary choices.)

Proof. Consider a complex structure on V, i.e., a skew-symmetric operator $J : V \to V$ such that $J^2 = -\mathbb{1}_V$. Such a J exists since V is even dimensional. Let

$$\{e_1, f_1, \ldots, e_k, f_k\}$$

be an orthonormal basis of V such that $Je_i = f_i$, $\forall i$.

Extend J by complex linearity to $V \otimes_{\mathbb{R}} \mathbb{C}$. We can now decompose $V \otimes \mathbb{C}$ into the eigenspaces of J

$$V = V^{1,0} \oplus V^{0,1},$$

where $V^{1,0} = \ker(\boldsymbol{i} - J)$ and $V^{0,1} = \ker(\boldsymbol{i} + J)$. Alternatively,

$$V^{1,0} = \mathrm{span}_{\mathbb{C}}(e_j - \boldsymbol{i}f_j), \quad V^{0,1} = \mathrm{span}_{\mathbb{C}}(e_j + \boldsymbol{i}f_j).$$

The metric on V defines a Hermitian metric on the *complex* vector space (V, J)

$$h(u, v) = g(u, v) + \boldsymbol{i}g(u, Jv),$$

(see (2.2.15) of Subsection 2.2.5) which allows us to identify

$$V^{0,1} \cong_{\mathbb{C}} \overline{(V, J)} \cong_{\mathbb{C}} V_c^* \cong_{\mathbb{C}} (V^{1,0})^*.$$

(Above, V_c^* denotes the complex dual of the complex space (V, J)). With respect to this Hermitian metric the collection

$$\left\{ \varepsilon_j := \frac{1}{\sqrt{2}}(e_j - \boldsymbol{i}f_j); \ 1 \leq j \leq k \right\}$$

is an orthonormal basis of $V^{1,0}$, while the collection

$$\left\{ \bar{\varepsilon}_j = \frac{1}{\sqrt{2}}(e_j + \boldsymbol{i}f_j); \ 1 \leq j \leq k \right\}$$

is an orthonormal basis of $V^{0,1}$. Set

$$\mathbb{S}_{2k} := \Lambda^{\bullet} V^{1,0} = \Lambda^{\bullet,0} V.$$

A morphism of algebras $c : \mathbb{Cl}(V) \to \operatorname{End}(\mathbb{S}_{2k})$ is uniquely defined by its restriction to

$$V \otimes \mathbb{C} = V^{1,0} \oplus V^{0,1}.$$

Hence, we have to specify the action of each of the components $V^{1,0}$ and $V^{0,1}$.

The elements $w \in V^{1,0}$ will act by exterior multiplication

$$c(w)\omega = \sqrt{2}e(w)\omega = \sqrt{2}w \wedge \omega, \quad \forall \omega \in \Lambda^{\bullet,0}V.$$

The elements $\overline{w} \in V^{0,1}$ can be identified with complex linear functionals on $V^{1,0}$, and as such, they will act by interior differentiation

$$c(\overline{w})w_1 \wedge \cdots \wedge w_\ell = -\sqrt{2}i(\overline{w})(w_1 \wedge \cdots \wedge w_\ell)$$

$$= \sqrt{2}\sum_{j=1}^{\ell}(-1)^j g_{\mathbb{C}}(w_j, \overline{w})w_1 \wedge \cdots \wedge \hat{w}_j \wedge \cdots \wedge w_\ell,$$

where $g_{\mathbb{C}}$ denotes the extension of g to $(V \otimes \mathbb{C}) \times (V \otimes \mathbb{C})$ by complex linearity.

To check that the above constructions do indeed define an action of $\mathbb{Cl}(V)$ we need to check that, $\forall v \in V$, we have

$$c(v)^2 = -\mathbb{1}_{\mathbb{S}_{2k}}.$$

This boils down to verifying the anticommutation rules

$$\{c(e_i), c(f_j)\} = 0, \quad \{c(e_i), c(e_j)\} = -2\delta_{ij} = \{c(f_i), c(f_j)\}.$$

We have

$$e_i = \frac{1}{\sqrt{2}}(\varepsilon_i + \overline{\varepsilon}_i), \quad f_j = \frac{i}{\sqrt{2}}(\varepsilon_j - \overline{\varepsilon}_j),$$

so that,

$$c(e_i) = e(\varepsilon_i) - i(\overline{\varepsilon}_i), \quad c(f_j) = i(e(\varepsilon_j) + i(\overline{\varepsilon}_j)).$$

The anti-commutation rules follow as in the Exercise 2.2.55 using the equalities

$$g_{\mathbb{C}}(\varepsilon_i, \overline{\varepsilon}_j) = \delta_{ij}.$$

This shows $\mathbb{S}_{2k}$ is naturally a $\mathbb{Cl}(V)$-module. Note that $\dim_{\mathbb{C}} \mathbb{S}_{2k} = 2^k$ so that

$$\dim_{\mathbb{C}} \operatorname{End}_{\mathbb{C}}(\mathbb{S}) = 2^n = \dim_{\mathbb{C}} \mathbb{Cl}(V).$$

A little work (left to the reader) shows the morphism c is injective. This completes the proof of the proposition. $\square$

Definition 11.1.20. The $\mathbb{Cl}(V)$-module $\mathbb{S}(V)$ is known as the (even) *complex spinor* module. $\square$

Using basic algebraic results about the representation theory of the algebra of endomorphisms of a vector space we can draw several useful consequences. (See [89] for a very nice presentation of these facts.)

Corollary 11.1.21. *There exists an unique (up to isomorphism) irreducible complex $\mathbb{C}l_{2k}$-module and this is the complex spinor module $\mathbb{S}_{2k}$.* □

Corollary 11.1.22. *Any complex $\mathbb{C}l_{2k}$-module has the form $\mathbb{S}_{2k} \otimes W$, where W is an arbitrary complex vector space. The action of $\mathbb{C}l_{2k}$ on $\mathbb{S}_{2k} \otimes W$ is defined by*

$$v \cdot (s \otimes w) = c(v)s \otimes w.$$

The vector space W is called the twisting *space of the given Clifford module.* □

Remark 11.1.23. Given a complex $\mathbb{C}l_{2k}$-module E, its twisting space can be recovered as the space of morphisms of Clifford modules

$$W = \operatorname{Hom}_{\mathbb{C}l_{2k}}(\mathbb{S}_{2k}, E).$$

□

Assume now that (V, g) is an *oriented*, $2k$-dimensional Euclidean space. For any positively oriented orthonormal basis $e_1, \ldots, e_{2k}$ we can form the element

$$\Gamma = i^k e_1 \cdots e_{2k} \in \mathbb{C}l(V).$$

One can check easily this element is independent of the *oriented* orthonormal basis, and thus it is an element intrinsically induced by the orientation. It is called the *chirality operator* defined by the orientation. Note that

$$\Gamma^2 = 1 \text{ and } \Gamma x = (-1)^{\deg x} \Gamma \quad \forall x \in \mathbb{C}l^0(V) \cup \mathbb{C}l^1(V).$$

Let $\mathbb{S} = \mathbb{S}(V)$ denote the spinor module of $\mathbb{C}l(V)$. The chirality operator defines an involutive endomorphism of $\mathbb{S}$, and thus defines a $\mathbb{Z}_2$-grading on $\mathbb{S}$

$$\mathbb{S} = \mathbb{S}^+ \oplus \mathbb{S}^- \quad (\mathbb{S}^\pm = \ker(\pm 1 - \Gamma)),$$

and hence a $\mathbb{Z}_2$-grading of $\operatorname{End}(\mathbb{S})$. Since $\{v, \Gamma\} = 0$, $\forall v \in V$, we deduce the Clifford multiplication by v is an odd endomorphism of $\mathbb{S}$. This means that any algebra isomorphism $\mathbb{C}l(V) \to \operatorname{End}(\mathbb{S}(V))$ is an isomorphism of $\mathbb{Z}_2$-graded algebras.

Exercise 11.1.24. Let J be a complex structure on V. This produces two things: it defines an orientation on V and identifies $\mathbb{S} = \mathbb{S}(V) \cong \Lambda^{*,0}V$. Prove that with respect to these data the chiral grading of $\mathbb{S}$ is

$$\mathbb{S}^{+/-} \cong \Lambda^{even/odd,0}V.$$

□

The above considerations extend to arbitrary Clifford modules. The chirality operator introduces a $\mathbb{Z}_2$-grading in any complex Clifford module which we call the *chiral grading.* However this does not exhaust the family of Clifford s-modules. The family of s-modules can be completely described as

$$\{ \, \mathbb{S} \hat{\otimes} W \; ; \; W \text{ complex s-space} \, \},$$

where $\hat{\otimes}$ denotes the s-tensor product. The modules endowed with the chiral grading form the subfamily in which the twisting s-space W is purely even.

Example 11.1.25. Let (V, g) be a $2k$-dimensional, *oriented, Euclidean space*. Then $\Lambda_{\mathbb{R}}^{\bullet} V \otimes \mathbb{C}$ is naturally a Clifford module. Thus, it has the form

$$\Lambda_{\mathbb{R}}^{\bullet} V \otimes \mathbb{C} \cong \mathbb{S} \otimes W.$$

To find the twisting space, we pick a complex structure on V whose induced orientation agrees with the given orientation of V. This complex structure produces an isomorphism

$$\Lambda_{\mathbb{R}}^{\bullet} V \otimes_{\mathbb{R}} \mathbb{C} \cong \Lambda^{\bullet, 0} V \otimes \Lambda^{0, \bullet} V \cong \mathbb{S} \otimes (\Lambda^{\bullet, 0} V)^* \cong \mathbb{S} \otimes_{\mathbb{R}} \mathbb{S}^*.$$

This shows the twisting space is $\mathbb{S}^*$.

On the other hand, the chirality operator defines $\mathbb{Z}_2$ gradings on both $\mathbb{S}$, and $\mathbb{S}^*$, so that $\Lambda_{\mathbb{R}}^{\bullet} V \otimes \mathbb{C}$ can be given two different s-structures: the chiral superstructure, in which the grading of $\mathbb{S}^*$ is forgotten, and the grading as a super-tensor product $\mathbb{S} \hat{\otimes} \mathbb{S}^*$. Using Exercise 11.1.24 we deduce that the second grading is precisely the degree grading

$$\Lambda_{\mathbb{R}}^{\bullet} V \otimes \mathbb{C} = \Lambda^{even} V \otimes \mathbb{C} \oplus \Lambda^{odd} V \otimes \mathbb{C}.$$

To understand the chiral grading we need to describe the action of the chiral operator on $\Lambda_{\mathbb{R}}^{\bullet} V \otimes \mathbb{C}$. This can be done via the Hodge $*$-operator. More precisely we have

$$\Gamma \cdot \omega = \boldsymbol{i}^{k + p(p-1)} * \omega \quad \forall \omega \in \Lambda^p V \otimes \mathbb{C}. \tag{11.1.1}$$

$\square$

Exercise 11.1.26. Prove the equality 11.1.1. $\square$

In order to formulate the final result of this subsection we need to extend the automorphism α, and the anti-automorphism $^{\flat}$ to the complexified Clifford algebra $\mathbb{C}l(V)$. The automorphism α can be extended by complex linearity

$$\alpha(x \otimes z) = \alpha(x) \otimes z, \quad \forall x \in \text{Cl}(V),$$

while $^{\flat}$ extends according to the rule

$$(x \otimes z)^{\flat} = x^{\flat} \otimes \overline{z}.$$

As in the real case, we set $y^{\dagger} := \alpha(y^{\flat}) = \alpha(y)^{\flat}, \, \forall y \in \mathbb{C}l(V)$.

Proposition 11.1.27. *Let $\mathbb{S}(V)$ denote the spinor module of the $2k$-dimensional Euclidean space (V, g). Then for every isomorphism of algebras*

$$\rho : \mathbb{C}l(V) \to \text{End}_{\mathbb{C}}(\mathbb{S}(V))$$

there exists a Hermitian metric on $\mathbb{S}(V)$ such that

$$\rho(y^{\dagger}) = \rho(y)^*, \quad \forall y \in \mathbb{C}l(V).$$

Moreover, this metric is unique up to a multiplicative constant.

Sketch of proof Choose an orthonormal basis $\{e_1, \ldots, e_{2k}\}$ of V, and denote by G the finite subgroup of $\mathrm{Cl}(V)$ generated by the basic vectors e_i.

Pick a Hermitian metric h on $\mathbb{S}(V)$ and, for each $g \in G$, denote by h_g the pulled-back metric

$$h_g(s_1, s_2) := h(\rho(g)s_1, \rho(g)s_2), \quad \forall s_1, s_2 \in \mathbb{S}.$$

We can now form the averaged metric,

$$h_G := \frac{1}{|G|} \sum_{g \in G} h_g.$$

Each $\rho(g)$ is an unitary operator with respect to this metric. We leave the reader to check that this is the metric we are after. The uniqueness follows from the irreducibility of $\mathbb{S}(V)$ using Schur's lemma. $\square$

Corollary 11.1.28. *Let (V, g) as above, and $\rho : \mathrm{Cl}(V) \to \mathrm{End}_{\mathbb{C}}(E)$ be a complex Clifford module. Then E admits at least one Hermitian metric with respect to which ρ is selfadjoint.*

Proof. Decompose E as $\mathbb{S} \otimes W$, and ρ as $\Delta \otimes \mathbb{1}_W$, for some isomorphism of algebras $\Delta : \mathrm{Cl}(V) \to \mathrm{End}(\mathbb{S})$. The sought for metric is a tensor product of the canonical metric on $\mathbb{S}$, and some metric on W. $\square$

11.1.4 *Clifford modules: the odd case*

The odd dimensional situation can be deduced using the facts we have just established concerning the algebras Cl_{2k}. The bridge between these two situations is provided by the following general result.

Lemma 11.1.29. $\mathrm{Cl}_m \cong \mathrm{Cl}_{m+1}^{even}$.

Proof. Pick an orthonormal basis $\{e_0, e_1, \ldots, e_m\}$ of the standard Euclidean space $\mathbb{R}^{m+1}$. These generate the algebra Cl_{m+1}. We view Cl_m as the Clifford algebra generated by $\{e_1, \ldots, e_m\}$. Now define

$$\Psi : \mathrm{Cl}_m \to \mathrm{Cl}_{m+1}^{even}, \quad \Psi(x^0 + x^1) = x^0 + e_0 \cdot x^1,$$

where $x^0 \in \mathrm{Cl}_m^{even}$, and $x^1 \in \mathrm{Cl}_m^{odd}$. We leave the reader to check that this is indeed an isomorphism of algebras. $\square$

Proposition 11.1.30. *Let (V, g) be a $(2k+1)$-dimensional Euclidean space. Then there exist two complex, irreducible $\mathrm{Cl}(V)$-modules $\mathbb{S}^+(V)$ and $\mathbb{S}^-(V)$ such that*

$$\mathrm{Cl}(V) \cong \mathrm{End}_{\mathbb{C}}(\mathbb{S}^+) \oplus \mathrm{End}_{\mathbb{C}}(\mathbb{S}^-) \quad \text{as ungraded algebras.}$$

The direct sum $\mathbb{S}(V) = \mathbb{S}^+(V) \oplus \mathbb{S}^-(V)$ is called the (odd) spinor module.

Proof. Fix an orientation on V and a positively oriented orthonormal basis $e_1, e_2, \ldots, e_{2k+1}$. Denote by $\mathbb{S}_{2k+2}$ the spinor module of $\mathrm{Cl}(V \oplus \mathbb{R})$, where $V \oplus \mathbb{R}$ is given the direct sum Euclidean metric and the orientation

$$or(V \oplus \mathbb{R}) = or \wedge or(\mathbb{R}).$$

Choose an isomorphism

$$\rho : \mathrm{Cl}(V \oplus \mathbb{R}) \to \mathrm{End}_{\mathbb{C}}(\mathbb{S}_{2k+2}).$$

Then $\mathbb{S}_{2k+2}$ becomes naturally a super Cl_{2k+2}-module

$$\mathbb{S}_{2k+2} = \mathbb{S}_{2k+2}^+ \oplus \mathbb{S}_{2k+2}^-,$$

and we thus we get the isomorphisms of algebras

$$\mathbb{Cl}(V) \cong \mathbb{Cl}(V \oplus \mathbb{R})^{even} \cong \mathrm{End}^{even}(\mathbb{S}_{2k+2}) \cong \mathrm{End}\,(\mathbb{S}_{2k+2}^+) \oplus \mathrm{End}\,(\mathbb{S}_{2k+2}^-). \qquad \square$$

As in the previous section, we can choose Hermitian metrics on $\mathbb{S}^{\pm}(V)$ such that the morphism

$$\rho : \mathbb{Cl}(V) \to \mathrm{End}_{\mathbb{C}}\big(\mathbb{S}(V)\big)$$

is selfadjoint, i.e.,

$$\rho(u^\dagger) = \rho(u)^*, \quad \forall u \in \mathbb{Cl}(V).$$

The above result can be used to describe the complex (super)modules of $\mathbb{Cl}_{2k+1}$. We will not present the details since the applications we have in mind do not require these facts. For more details we refer to [62].

11.1.5 *A look ahead*

In this heuristic section we interrupt a little bit the flow of arguments to provide the reader a sense of direction. The next step in our story is to glue all the pointwise data presented so far into smooth families (i.e. bundles). To produce a Dirac operator on an n-dimensional Riemann manifold (M, g) one needs several things.

(a) A bundle of Clifford algebras $\mathcal{C} \to M$ such that $\mathcal{C}_x \cong \mathrm{Cl}_n$ or $\mathbb{Cl}_n$, $\forall x \in M$.

(b) A fiberwise injective morphism of vector bundles $\imath : T^*M \hookrightarrow \mathcal{C}$ such that, $\forall x \in M$

$$\{\imath(u), \imath(v)\}_{\mathcal{C}} = -2g(u, v), \quad \forall u, v \in T_x^*M.$$

(c) A bundle of Clifford modules, i.e. a vector bundle $\mathcal{E} \to M$, together with a morphism $c : \mathcal{C} \to \mathrm{End}\,(\mathcal{E})$ whose restrictions to the fibers are morphisms of algebras.

(d) A connection on $\mathcal{E}$.

The above collection of data can be constructed from bundles associated to a common principal bundle. The symmetry group of this principal bundle has to be a Lie group with several additional features which we now proceed to describe.

Let (V, g) denote the standard fiber of T^*M, and denote by Aut_V the group of automorphisms ϕ of $\mathrm{Cl}(V)$ such that $\phi(V) \subset V$. The group Aut_V^c is defined similarly, using the complexified algebra $\mathbb{C}l(V)$ instead of $\mathrm{Cl}(V)$. For brevity, we discuss only the real case.

We need a Lie group G which admits a smooth morphism $\rho : G \to \mathrm{Aut}_V$. Tautologically, ρ defines a representation $\rho : G \to \mathrm{GL}(V)$, which we assume is orthogonal.

We also need a Clifford module $\boldsymbol{c} : \mathrm{Cl}(V) \to \mathrm{End}\,(E)$, and a representation $\mu : G \to \mathrm{GL}(E)$, such that, for every $v \in V$, and any $g \in G$, the diagram below is commutative.

$$
\begin{array}{ccc}
E & \xrightarrow{\ \boldsymbol{c}(v)\ } & E \\
{\scriptstyle g}\downarrow & & \downarrow{\scriptstyle g} \\
E & \xrightarrow[\ \boldsymbol{c}(g\cdot v)\ ]{} & E
\end{array}
\qquad (11.1.2)
$$

This commutativity can be given an invariant theoretic interpretation as follows.

View the Clifford multiplication $\boldsymbol{c} : V \to \mathrm{End}\,(E)$ as an element $\boldsymbol{c} \in V^* \otimes E^* \otimes E$. The group G acts on this tensor product, and the above commutativity simply means that $\boldsymbol{c}$ is invariant under this action.

In concrete applications E comes with a metric, and we also need to require that μ is an orthogonal/unitary representation.

To produce all the data (a)-(d) all we now need is a principal G-bundle $P \to M$ such that the associated bundle $P \times_\rho V$ is isomorphic with T^*M. (This may not be always feasible due to possible topological obstructions). Any connection ∇ on P induces by association metric connections ∇^M on[1] T^*M and ∇^E on the bundle of Clifford modules $\mathcal{E} = P \times_\mu E$. With respect to these connections, the Clifford multiplication is covariant constant, i.e.,

$$
\nabla^E(\boldsymbol{c}(\alpha)u) = \boldsymbol{c}(\nabla^M \alpha) + \boldsymbol{c}(\alpha)\nabla^E u, \quad \forall \alpha \in \Omega^1(M), \ \ u \in C^\infty(E).
$$

This follows from the following elementary invariant theoretic result.

Lemma 11.1.31. *Let G be a Lie group, and $\rho : G \to \mathrm{Aut}\,(E)$ be a linear representation of G. Assume that there exists $e_0 \in E$ such that $\rho(g)e_0 = e_0$, $\forall g \in G$. Consider an arbitrary principal G-bundle $P \to X$, and an arbitrary connection ∇ on P. Then e_0 canonically determines a section u_0 on $P \times_\rho E$ which is covariant constant with respect to the induced connection $\nabla^E = \rho_*(\nabla)$, i.e.,*

$$
\nabla^E u_0 = 0. \qquad \qquad \square
$$

---

[1] In practice one requires a little more namely that ∇^M is precisely the Levi-Civita connection on T^*M. This leads to significant simplifications in many instances.

Exercise 11.1.32. Prove the above lemma. □

Apparently, the chances that a Lie group G with the above properties exists are very slim. The very pleasant surprise is that all these, and even more, happen in many geometrically interesting situations.

Example 11.1.33. Let (V, g) be an oriented Euclidean space. Using the universality property of Clifford algebras we deduce that each $g \in SO(V)$ induces an automorphism of $\mathrm{Cl}(V)$ preserving $V \hookrightarrow \mathrm{Cl}(V)$. Moreover, it defines an orthogonal representation on the canonical Clifford module

$$\boldsymbol{c} : \mathrm{Cl}(V) \to \mathrm{End}\,(\Lambda^\bullet V),$$

such that

$$\boldsymbol{c}(g \cdot v)(\omega) = g \cdot (\boldsymbol{c}(v)(g^{-1} \cdot \omega)), \quad \forall g \in SO(V), v \in V, \omega \in \Lambda^\bullet V,$$

i.e., $SO(V)$ satisfies the equivariance property (11.1.2).

If (M, g) is an oriented Riemann manifold, we can now build our bundle of Clifford modules starting from the principal SO bundle of its oriented orthonormal coframes. As connections we can now pick the Levi-Civita connection and its associates. The corresponding Dirac operator is the Hodge-DeRham operator. □

The next two sections discuss two important examples of Lie groups with the above properties. These are the spin groups $\mathrm{Spin}(n)$ and its "complexification" $\mathrm{Spin}^c(n)$. It turns out that all the groups one needs to build Dirac operators are these three classes: SO, Spin, and Spin^c.

11.1.6 *Spin*

Let (V, g) be a finite dimensional Euclidean space. The group of automorphisms of the Clifford algebra $\mathrm{Cl}(V)$ contains a very rich subgroup consisting of the interior ones. These have the form

$$\varphi_x : \mathrm{Cl}(V) \to \mathrm{Cl}(V) \quad u \mapsto \varphi_x(u) = x \cdot u \cdot x^{-1}, \quad \forall u \in \mathrm{Cl}(V),$$

where x is some invertible element in $\mathrm{Cl}(V)$.

The candidates for the Lie groups with the properties outlined in the previous subsection will be sought for amongst subgroups of interior automorphisms. It is thus natural to determine the subgroup

$$\{x \in \mathrm{Cl}(V)^\star; \ x \cdot V \cdot x^{-1} \subset V\},$$

where $\mathrm{Cl}(V)^\star$ denotes the group of invertible elements. We will instead try to understand the *Clifford group*

$$\Gamma(V) = \{x \in \mathrm{Cl}^\star(V) \ \alpha(x) \cdot V \cdot x^{-1} \subset V\},$$

where $\alpha : \mathrm{Cl}(V) \to \mathrm{Cl}(V)$ denotes the involutive automorphism of $\mathrm{Cl}(V)$ defining its $\mathbb{Z}_2$-grading.

In general, the map

$$\mathrm{Cl}(V) \ni u \xmapsto{\rho_x} \alpha(x)ux^{-1} \in \mathrm{Cl}(V)$$

is not an automorphism of algebras, but as we will see by the end of this subsection, if $x \in \Gamma(V)$, then $\rho_x = \pm\varphi_x$, and hence *a posteriori* this alteration has no impact. Its impact is mainly on the æsthetics of the presentation which we borrowed from the elegant paper [7].

By construction, the Clifford group $\Gamma(V)$ comes equipped with a tautological representation

$$\rho : \Gamma(V) \to \mathrm{GL}(V) \quad \rho(x) : v \mapsto \alpha(x) \cdot v \cdot x^{-1}.$$

Proposition 11.1.34. $\ker \rho = (\mathbb{R}^*, \cdot) \subset \mathrm{Cl}(V)^*.$

Proof. Clearly $\mathbb{R}^* \subset \ker \rho$. To establish the opposite inclusion choose an orthonormal basis (e_i) of V, and let $x \in \ker \rho$. The element x decomposes into even/odd components

$$x = x_0 + x_1,$$

and the condition $\alpha(x)e_i x^{-1} = e_i$ translates into

$$(x_0 - x_1)e_i = e_i(x_0 + x_1), \quad \forall i.$$

This is equivalent with the following two conditions

$$[x_0, e_i] = x_0 e_i - e_i x_0 = x_1 e_i + e_i x_1 = \{x_1, e_i\} = 0, \ \forall i.$$

In terms of the s-commutator, the above two equalities can be written as one

$$[e_i, x]_s = 0 \ \forall i.$$

Since $[\cdot, x]$ is a superderivation of $\mathrm{Cl}(V)$ we conclude that

$$[y, x]_s = 0, \quad \forall y \in \mathrm{Cl}(V).$$

In particular, the even part x_0 lies in the center of $\mathrm{Cl}(V)$. We let the reader check the following elementary fact.

Lemma 11.1.35. *The center of the Clifford algebra is the field of scalars* $\mathbb{R} \subset \mathrm{Cl}(V).$ $\square$

Note that since $\{x_1, e_i\} = 0$, then x_1 should be a linear combination of elementary monomials $e_{j_1} \cdots e_{j_s}$ none of which containing e_i as a factor. Since this should happen for every i this means $x_1 = 0$, and this concludes the proof of the proposition. $\square$

Definition 11.1.36. The *spinorial norm* is the map

$$N : \mathrm{Cl}(V) \to \mathrm{Cl}(V) \quad N(x) = x^\flat x.$$ $\square$

Proposition 11.1.37. *(a)* $N(\Gamma(V)) \subset \mathbb{R}^*$.
(b) The map $N : \Gamma(V) \to \mathbb{R}^*$ *is a group morphism.*

Proof. Let $x \in \Gamma(V)$. We first prove that $x^b \in \Gamma(V)$. Since $\alpha(x)vx^{-1} \in V$, $\forall v \in V$, we deduce that

$$\alpha(\{\alpha(x) \cdot v \cdot x^{-1}\}^b) = -\alpha(x) \cdot v \cdot x^{-1} \in V.$$

Using the fact that $x \mapsto x^b$ is an anti-automorphism, we deduce

$$\alpha((x^b)^{-1} \cdot v \cdot \alpha(x^b)) \in V,$$

so that,

$$\alpha((x^b)^{-1}) \cdot v \cdot x^b \in V,$$

that is, $(x^b)^{-1} \in \Gamma(V)$. Hence $x^\dagger \in \Gamma(V)$, $\forall x \in \Gamma(V)$. In particular, since $\alpha(\Gamma(V)) \subset \Gamma(V)$, we deduce $N(\Gamma(V)) \subset \Gamma(V)$.

For any $v \in V$ we have

$$\alpha(N(x)) \cdot v \cdot (N(x))^{-1} = \alpha(x^b x) \cdot v \cdot (x^b x)^{-1} = \alpha(x^b) \cdot \{\alpha(x) \cdot v \cdot x^{-1}\} \cdot (x^b)^{-1}.$$

On the other hand, $y = \alpha(x) \cdot v \cdot x^{-1}$ is an element in V, which implies $y^\dagger = \alpha(y^b) = -y$. Hence

$$\alpha(N(x)) \cdot v \cdot (N(x))^{-1} = -\alpha(x^b) \cdot y^\dagger \cdot (x^b)^{-1}$$

$$= -\alpha(x^b) \cdot \alpha(x^b)^{-1} \cdot v^\dagger \cdot x^b \cdot (x^b)^{-1} = -v^\dagger = v.$$

This means $N(x) \in \ker \rho = \mathbb{R}^*$.
(b) If $x, y \in \Gamma(V)$, then

$$N(x \cdot y) = (xy)^b(xy) = y^b x^b xy = y^b N(x)y = N(x)y^b y = N(x)N(y). \qquad \square$$

Theorem 11.1.38. *(a) For every* $x \in \Gamma(V)$, *the transformation* $\rho(x)$ *of* V *is orthogonal.*
(b) There exists a short exact sequence of groups

$$1 \to \mathbb{R}^* \hookrightarrow \Gamma(V) \xrightarrow{\rho} O(V) \to 1.$$

(c) Every $x \in \Gamma(V)$ *can be written (in a non-unique way) as a product* $x = v_1 \cdots v_k$, $v_j \in V$. *In particular, every element of* $\Gamma(V)$ *is* $\mathbb{Z}_2$-*homogeneous, i.e., it is either purely even, or purely odd.*

Proof. (a) Note that, $\forall v \in V$, we have $N(v) = -|v|_g^2$. For every $x \in \Gamma(V)$ we get

$$N(\rho(x)(v)) = N(\alpha(x)vx^{-1}) = N(\alpha(x))N(v)N(x^{-1}) = N(\alpha(x))N(x)^{-1}N(v).$$

On the other hand, $x^2 = \alpha(x^2) = \alpha(x)^2$ so we conclude that

$$N(x)^2 = N(\alpha(x))^2.$$

Hence, $N(\rho(x)(v)) = \pm N(v)$. Since both $N(v)$ and $N(\rho_x(v))$ are *negative numbers*, we deduce that the only possible choice of signs in the above equality is $+$. This shows that $\rho(x)$ is an orthogonal transformation.

(b) & (c) We only need to show $\rho(\Gamma(V)) = O(V)$. For $x \in V$ with $|x|_g = 1$ we have

$$\alpha(x) = -x = x^{-1}.$$

If we decompose $v \in V$ as $\lambda x + u$, where $\lambda \in \mathbb{R}$ and $u \perp x$, then we deduce

$$\rho(x)v = -\lambda x + u.$$

In other words, $\rho(x)$ is the orthogonal reflection in the hyperplane through origin which is perpendicular to x. Since any orthogonal transformation of V is a composition of such reflections (*Exercise*), we deduce that for each $T \in O(V)$ we can find $v_1, \ldots, v_k \in V$ such that

$$T = \rho(v_1) \cdots \rho(v_k).$$

Incidentally, this also establishes (c). $\qquad\qquad \square$

Exercise 11.1.39. Suppose V is a finite dimensional Euclidean space. Prove that any orthogonal operator $T \in O(V)$ is a product of at most $\dim V$ reflections. $\quad \square$

Set

$$\Gamma^0(V) := \Gamma(V) \cap \mathrm{Cl}^{even}.$$

Note that

$$\rho(\Gamma^0(V)) \subset SO(V) = \big\{ T \in O(V); \ \det T = 1 \big\}.$$

Hence, we have a short exact sequence

$$1 \to \mathbb{R}^* \hookrightarrow \Gamma^0(V) \xrightarrow{\rho} SO(V) \to 1.$$

Definition 11.1.40. Set

$$\mathrm{Pin}(V) := \big\{ x \in \Gamma(V); \ |N(x)| = 1 \big\},$$

and

$$\mathrm{Spin}(V) := \big\{ x \in \Gamma(V); \ N(x) = 1 \big\} = \mathrm{Pin}(V) \cap \Gamma^0(V). \qquad \square$$

The results we proved so far show that $\mathrm{Spin}(V)$ can be alternatively described by the following "friendlier" equality

$$\mathrm{Spin}(V) = \{v_1 \cdots v_{2k} \ ; \ k \geq 0, v_i \in V, \ |v_i| = 1, \ \forall i = 1. \ldots 2k\}.$$

In particular, this shows that $\mathrm{Spin}(V)$ is a compact topological group. Observe that $\mathrm{Spin}(V)$ is a *closed* subgroup of the Lie group $\mathrm{GL}(\mathrm{Cl}(V))$. This implies (see Remark 1.2.29 and [44, 95]) that $\mathrm{Spin}(V)$ is in fact a Lie group, and the map $\mathrm{Spin}(V) \hookrightarrow \mathrm{Cl}(V)$ is a smooth embedding. The proof of the following result is left to the reader.

Proposition 11.1.41. *There exist short exact sequences*

$$1 \to \mathbb{Z}_2 \to \mathrm{Pin}(V) \to O(V) \to 1$$

$$1 \to \mathbb{Z}_2 \to \mathrm{Spin}(V) \to SO(V) \to 1. \qquad\qquad \square$$

Proposition 11.1.42. *The morphism $\rho : \text{Spin}(V) \to SO(V)$ is a covering map. Moreover, the group $\text{Spin}(V)$ is connected if $\dim V \geq 2$, and simply connected if $\dim V \geq 3$. In particular, $\text{Spin}(V)$ is the universal cover of $SO(V)$, when $\dim V \geq 3$.*

Proof. The fact that $\rho : \text{Spin}(V) \to SO(V)$ is a covering map is an elementary consequence of the following simple observations.

(i) The map ρ is a group morphism;
(ii) The map ρ is continuous, and proper;
(iii) The subgroup $\ker \rho$ is discrete.

Since $SO(V)$ is connected if $\dim V \geq 2$, the fact that $\text{Spin}(V)$ is connected would follow if we showed that any points in the same fiber of ρ can be connected by arcs. It suffices to look at the fiber $\rho^{-1}(1) = \{-1, 1\}$.

If $u, v \in V$ are such that $|u| = |v| = 1$, $u \perp v$, then the path

$$\gamma(t) = (u \cos t + v \sin t)(u \cos t - v \sin t), \ \ 0 \leq t \leq \pi/2$$

lies inside $\text{Spin}(V)$, because $|u \cos t \pm v \sin t| = 1$, and moreover

$$\gamma(0) = -1, \ \ \gamma(\pi/2) = 1.$$

To prove that $\text{Spin}(V)$ is simply connected, we argue by induction on $\dim V$.

If $\dim V = 3$, then $\text{Spin}(V)$ is isomorphic to the group of unit quaternions (see Example 11.1.55), and in particular, it is homeomorphic to the sphere S^3 which is simply connected.

Note that if V is an Euclidean space, and U is a subspace, the natural inclusion $U \hookrightarrow V$ induces morphisms

$$\text{Cl}(U) \hookrightarrow \text{Cl}(V), \ \ SO(U) \hookrightarrow SO(V), \ \ \text{Spin}(U) \hookrightarrow \text{Spin}(V),$$

such that the diagrams below are commutative

$$
\begin{array}{ccc}
\text{Spin}(V) & \xrightarrow{\rho} & SO(V) \\
i \uparrow & & \uparrow \\
\text{Spin}(U) & \xrightarrow[\rho]{} & SO(U)
\end{array}
, \qquad
\begin{array}{ccc}
\text{Spin}(V) & \hookrightarrow & \text{Cl}(V) \\
\uparrow & & \uparrow \\
\text{Spin}(U) & \longrightarrow & \text{Cl}(U)
\end{array}
$$

Hence, it suffices to show that if $\dim V > 3$, then for every smooth, closed path

$$\hat{u} : [0, 1] \to \text{Spin}(V), \ \ \hat{u}(0) = \hat{u}(1) = 1,$$

there exists a codimension one subspace $U \hookrightarrow V$ such that $\hat{u}$ is homotopic to a loop in $\text{Spin}(U)$. Fix a unit vector $e \in V$, set $u(t) := \rho_{\hat{u}(t)} \in SO(V)$, and define $v(t) := u(t)e \in V$.

The correspondence $t \mapsto v(t)$ is a smooth, closed path on the unit sphere $S^{\dim V - 1} \subset V$. Using Sard's theorem we deduce that there exists a vector $\boldsymbol{x} \in S^{\dim V - 1}$, such that

$$v(t) \neq \pm \boldsymbol{x}, \ \ \forall t \in [0, 1].$$

In particular, the vectors $v(t)$ and $\boldsymbol{x}$ are linearly independent, for any t. Denote by $\boldsymbol{f}(t)$ the vector obtained by applying the Gramm-Schmidt orthonormalization process to the ordered linearly independent set $\{\boldsymbol{x}, v(t)\}$. In other words,

$$\boldsymbol{f}(t) := \frac{1}{|f_0(t)|} f_0(t), \quad f_0(t) := v(t) - (v(t) \bullet \boldsymbol{x})\boldsymbol{x},$$

where we denoted the inner product in V by $\bullet$. We can find a smooth map

$$\theta : [0, 1] \to [0, 2\pi),$$

such that,

$$\theta(0) = \theta(1) = 0, \quad v(t) = \boldsymbol{x} \cos 2\theta(t) + \boldsymbol{f}(t) \sin 2\theta(t).$$

Set

$$\sigma(t) := \cos\theta + \boldsymbol{x}\boldsymbol{f} \sin\theta \in \mathrm{Cl}(V).$$

Note that

$$\sigma(t) = \left(\boldsymbol{x} \cos\frac{\theta}{2} - \boldsymbol{f} \sin\frac{\theta}{2} \right)\left(-\boldsymbol{x} \cos\frac{\theta}{2} - \boldsymbol{f} \sin\frac{\theta}{2} \right),$$

so that $\sigma(t) \in \mathrm{Spin}(V)$.

On the other hand,

$$\sigma(t)^{-1}(t) = \cos\theta - \boldsymbol{x}\boldsymbol{f} \sin\theta,$$

and

$$\sigma(t)\boldsymbol{x}\sigma(t)^{-1} = (\boldsymbol{x} \cos\theta + \boldsymbol{f} \sin\theta)(\cos\theta - \boldsymbol{x}\boldsymbol{f} \sin\theta) = \boldsymbol{x} \cos 2\theta + \boldsymbol{f} \sin 2\theta = e_0.$$

In other words,

$$\rho_{\sigma(t)}\boldsymbol{x} = v(t).$$

For every $s \in [0, 1]$ define

$$\sigma_s(t) := \hat{u}_s(t) = \cos s\theta + \boldsymbol{x}\boldsymbol{f} \sin s\theta, \quad \hat{u}_s(t) := \sigma_s(t)^{-1}\hat{u}(t)\sigma_s(t),$$

and

$$u_s(t) := \rho_{\sigma_s(t)}^{-1} u(t) \rho_{\sigma_s(t)} \in SO(V).$$

Observe that for $s = 0$ we have $\hat{u}_0(t) = \hat{u}(t)$. On the other hand, for $s = 1$ we have

$$u_1(t)\boldsymbol{x} = \rho_{\sigma_s(t)}^{-1} u(t)\rho_{\sigma_s(t)}\boldsymbol{x} = \rho_{\sigma_s(t)^{-1}}v(t) = \boldsymbol{x}.$$

If we denote by U the orthogonal complement of $\boldsymbol{x}$ in V, we deduce that $u_1(t) \in \mathrm{Spin}(U)$, $\forall t \in [0, 1]$. $\qquad\square$

We can view $\mathrm{Spin}(V)$ as a submanifold of $\mathrm{Cl}(V)$, and as such we can identify its Lie algebra $\underline{\mathrm{spin}}(V)$ with a linear subspace of $\mathrm{Cl}(V)$. The next result offers a more precise description.

Proposition 11.1.43. *Consider the quantization map* $\mathfrak{q} : \Lambda^* V \to \mathrm{Cl}(V)$. *Then*

$$\underline{\mathrm{spin}}(V) = \mathfrak{q}(\Lambda^2 V).$$

The Lie bracket is given by the commutator in $\mathrm{Cl}(V)$.

Proof. The group $\Gamma(V)$ is a Lie group as a closed subgroup of the group of linear transformations of $\mathrm{Cl}(V)$. Since the elements of $\Gamma(V)$ are either purely even, or purely odd, we deduce that the tangent space at $1 \in \Gamma(V)$ is a subspace of

$$E = \{ x \in \mathrm{Cl}^{even}(V); \ xv - vx \in V, \ \forall v \in V \}.$$

Fix $x \in E$, and let $e_1, \ldots, e_n$ be an orthonormal basis of V. We can decompose x as

$$x = x_0 + e_1 x_1,$$

where $x_0 \in \mathrm{Cl}(V)^{even}$, and $x_1 \in \mathrm{Cl}(V)^{odd}$ are linear combinations of monomials involving only the vectors $e_2, \ldots, e_n$. Since $[x_0, e_1] = 0$, and $\{x_1, e_1\} = 0$, we deduce

$$e_1 x_1 = \frac{1}{2}[e_1, x_1] = \frac{1}{2}[e_1, x] \in V.$$

In particular this means $x_1 \in \mathbb{R} \oplus V \in \mathrm{Cl}(V)$.

Repeating the same argument with every vector e_i we deduce that

$$E \subset \mathbb{R} \oplus \mathrm{span}\{e_i \cdot e_j \ ; \ 1 \leq i < j \leq \dim V\} = \mathbb{R} \oplus \mathfrak{q}(\Lambda^2 V).$$

Thus,

$$T_1 \Gamma^0(V) \subset \mathbb{R} \oplus \mathfrak{q}(\Lambda^2 V).$$

The tangent space to $\mathrm{Spin}(V)$ satisfies a further restriction obtained by differentiating the condition $N(x) = 1$. This gives

$$\underline{\mathrm{spin}}(V) \subset \{x \in \mathbb{R} \oplus \mathfrak{q}(\Lambda^2 V) \ ; \ x^\flat + x = 0\} = \mathfrak{q}(\Lambda^2 V).$$

Since $\dim \underline{\mathrm{spin}}(V) = \dim \underline{\mathrm{so}}(V) = \dim \Lambda^2 V$ we conclude that the above inclusion is in fact an equality of vector spaces.

Now consider two smooth paths $x, y : (-\varepsilon, \varepsilon) \to \mathrm{Spin}(V)$ such that $x(0) = y(0) = 1$. The Lie bracket of $\dot{x}(0)$ and $\dot{y}(0)$ is then found (using Exercise 3.1.21) from the equality

$$x(t)y(t)x(t)^{-1}y(t)^{-1} = 1 + [\dot{x}(0), \dot{y}(0)]t^2 + O(t^3) \ \text{(as } t \to 0),$$

where the above bracket is the commutator of $\dot{x}(0)$ and $\dot{y}(0)$ viewed as elements in the associative algebra $\mathrm{Cl}(V)$. $\qquad \square$

To get a more explicit picture of the induced morphism of Lie algebras

$$\rho_* : \underline{\mathrm{spin}}(V) \to \underline{\mathrm{so}}(V),$$

we fix an orientation on V, and then choose a positively oriented orthonormal basis $\{e_1, \ldots, e_n\}$ of V, ($n = \dim V$). For every $x \in \underline{\mathrm{spin}}(V)$, the element $\rho_*(x) \in \underline{\mathrm{spin}}(V)$ acts on V according to

$$\rho_*(x)v = x \cdot v - v \cdot x.$$

If

$$x = \sum_{i<j} x_{ij} e_i e_j,$$

then

$$\rho_*(x)e_j = -2\sum_i x_{ij}e_i, \quad (x_{ij} = -x_{ji}).$$

Note the following often confusing fact. If we identify as usual $\underline{so}(V) \cong \Lambda^2 V$ by

$$\underline{so}(n) \ni A \mapsto \omega_A = \sum_{i<j} g(Ae_i, e_j)e_i \wedge e_j = -\sum_{i<j} g(e_i, Ae_j)e_i \wedge e_j,$$

then the Lie algebra morphism ρ_* takes the form

$$\omega_{\rho_*(x)} = -\sum_{i<j} g(e_i, \rho_*(x)e_j)e_i \wedge e_j = 2\sum_{i<j} x_{ij}e_i \wedge e_j = 2\sigma(x) \in \Lambda^2 V,$$

where $\sigma : \mathrm{Cl}(V) \to \Lambda^\bullet V$ is the symbol map, $e_i e_j \mapsto e_i \wedge e_j$. In particular, this shows ρ_* is an isomorphism, and that the map $\rho : \mathrm{Spin}(V) \to SO(V)$ is also a submersion.

☞ **A word of warning.** If $A \in \underline{so}(V)$ has the matrix description

$$Ae_j = \sum_i a_{ij}e_i$$

with respect to an oriented orthonormal basis $\{e_1, \ldots, e_n\}$, $(n = \dim V)$, then the 2-form associated to A has the form

$$\omega_A = -\sum_{i<j} a_{ij}e_i \wedge e_j,$$

so that

$$\rho_*^{-1}(A) = -\frac{1}{2}\sum_{i<j} a_{ij}e_i e_j = -\frac{1}{4}\sum_{i,j} a_{ij}e_i e_j. \tag{11.1.3}$$

The above negative sign is essential, and in many concrete problems it makes a world of difference.   □

Any Clifford module $\phi : \mathrm{Cl}(V) \to \mathrm{End}_{\mathbb{K}}(E)$ defines by restriction a representation

$$\phi : \mathrm{Spin}(V) \to GL_{\mathbb{K}}(E).$$

The (complex) representation theory described in the previous sections can be used to determine the representations of $\mathrm{Spin}(V)$.

Example 11.1.44. (The complex spinor representations). Consider a finite dimensional oriented Euclidean space (V, g). Assume first that $\dim V$ is even. The orientation on V induces a $\mathbb{Z}_2$-grading on the spinor module $\mathbb{S} = \mathbb{S}^+ \oplus \mathbb{S}^-$. Since $\mathrm{Spin}(V) \subset \mathrm{Cl}^{even}(V)$ we deduce that each of the spinor spaces $\mathbb{S}^\pm$ is a representation space for Spin(V). They are in fact irreducible, nonisomorphic complex Spin(V)-modules. They are called the positive/negative complex spin representations.

Assume next that dim V is odd. The spinor module $\mathbb{S}(V)$

$$\mathbb{S}(V) = \mathbb{S}^+(V) \oplus \mathbb{S}^-(V).$$

is not irreducible as a $\mathbb{Cl}(V)$ modules. Each of the modules $\mathbb{S}^\pm(V)$ is a representation space for Spin(V). They are irreducible but also *isomorphic* as Spin(V)-modules. If we pick an oriented orthonormal basis $e_1, \cdots, e_{2n+1}$ of V then the Clifford multiplication by $\omega = e_1 \cdots e_{2n+1}$ intertwines the $\pm$ components. This is a Spin(V) isomorphism since ω lies in the center of $\mathrm{Cl}(V)$. $\qquad\square$

♙ **Convention.** For each positive integer n we will denote by $\mathbb{S}_n$ the Spin(n)-module defined by

$$\mathbb{S}_{2k} \cong \mathbb{S}^+(\mathbb{R}^{2k}) \oplus \mathbb{S}^-(\mathbb{R}^{2k}) \text{ if } n = 2k,$$

and

$$\mathbb{S}_{2k+1} \cong \mathbb{S}^+(\mathbb{R}^{2k+1}) \cong_{\mathrm{Spin}(2k+1)} \mathbb{S}^-(\mathbb{R}^{2k+1}) \text{ if } n = 2k+1.$$

The module $\mathbb{S}_n$ will be called the *fundamental complex spinor module of* Spin(n).$\square$

Exercise 11.1.45. Let (V, g) be a $2k$-dimensional Euclidean space and $J : V \to V$ a complex structure compatible with the metric g. Thus we have an explicit isomorphism

$$\Delta : \mathbb{Cl}(V) \to \mathrm{End}\,(\Lambda^{\bullet,0} V).$$

Choose $u \in V$ such that $|u|_g = 1$ and then for each $t \in \mathbb{R}$ set

$$q = q(t) = \cos t + u \cdot v \sin t, \quad v = Jv.$$

Note that $q \in \mathrm{Spin}(V)$ so that $\Delta(q)$ preserves the parities when acting on $\Lambda^{*,0}(V)$. Hence $\Delta(q) = \Delta_+(q) \oplus \Delta_-(q)$ where $\Delta_{+/-}(q)$ acts on $\Lambda^{even/odd,0} V$. Compute $\mathrm{tr}\,(\Delta_{+/-}(q))$ and then conclude that $\Lambda^{even,/odd,0} V$ are non-isomorphic Spin(V)-modules. $\qquad\square$

Proposition 11.1.46. *Let $\phi : \mathrm{Cl}(V) \to \mathrm{End}\,(E)$ be a selfadjoint Clifford module. Then the induced representation of* Spin(V) *is orthogonal (unitary).*

Exercise 11.1.47. Prove the above proposition. $\qquad\square$

Exercise 11.1.48. Prove that the group Spin(V) satisfies all the conditions discussed in Subsection 10.1.5. $\qquad\square$

11.1.7 *Spin^c*

The considerations in the previous case have a natural extension to the complexified Clifford algebra $\mathbb{Cl}_n$. The canonical involutive automorphism

$$\alpha : \mathrm{Cl}_n \to \mathrm{Cl}_n$$

extends by complex linearity to an automorphism of $\mathbb{Cl}_n$, while the anti-automorphism $^\flat : \mathrm{Cl}_n \to \mathrm{Cl}_n$ extends to $\mathbb{Cl}_n$ according to the rule

$$(v \otimes z)^\flat = v \otimes \bar{z}.$$

As in the real case set $x^\dagger := \alpha(x)^\flat$, and $N(x) = x^\flat \cdot x$.

Let (V, g) be an Euclidean space. The *complex Clifford group* $\Gamma^c(V)$ is defined by

$$\Gamma^c(V) = \{x \in \mathbb{Cl}(V)^\star; \alpha(x) \cdot v \cdot x^{-1} \in V \ \ \forall v \in V\}.$$

We denote by ρ^c the tautological representation $\rho^c : \Gamma^c(V) \to GL(V, \mathbb{R})$. As in the real case one can check that

$$\rho^c(\Gamma^c(V)) = O(V), \ \ \ker \rho^c = \mathbb{C}^*.$$

The spinorial norm $N(x)$ determines a group morphism $N : \Gamma^c(V) \to \mathbb{C}^*$. Define

$$\mathrm{Pin}^c(V) = \{x \in \Gamma^c(V) \ ; \ |N(x)| = 1\}.$$

We let the reader check the following result.

Proposition 11.1.49. *There exists a short exact sequence*

$$1 \to S^1 \to \mathrm{Pin}^c(V) \to O(V) \to 1. \qquad \qquad \square$$

Corollary 11.1.50. *There exists a natural isomorphism*

$$\mathrm{Pin}^c(V) \cong (\mathrm{Pin}(V) \times S^1)/\sim,$$

where "$\sim$" is the equivalence relation

$$(x, z) \sim (-x, -z) \ \ \ \forall (x, z) \in \mathrm{Pin}(V) \times S^1.$$

Proof. The inclusions $\mathrm{Pin}(V) \subset \mathrm{Cl}(V)$, $S^1 \subset \mathbb{C}$ induce an inclusion

$$(\mathrm{Pin}(V) \times S^1)/\sim \rightarrow \mathbb{Cl}(V).$$

The image of this morphism lies obviously in $\Gamma^c(V) \cap \{|N| = 1\}$ so that $(\mathrm{Pin}(V) \times S^1)/\sim$ can be viewed as a subgroup of $\mathrm{Pin}^c(V)$. The sought for isomorphism now follows from the exact sequence

$$1 \to S^1 \to (\mathrm{Pin}(V) \times S^1)/\sim \rightarrow O(V) \to 1. \qquad \qquad \square$$

We define $\mathrm{Spin}^c(V)$ as the inverse image of $SO(V)$ via the morphism

$$\rho^c : \mathrm{Pin}^c(V) \to O(V).$$

Arguing as in the above corollary we deduce

$$\mathrm{Spin}^c(V) \cong (\mathrm{Spin}(V) \times S^1)/\sim \cong (\mathrm{Spin}(V) \times S^1)/\mathbb{Z}_2.$$

Exercise 11.1.51. Prove $\mathrm{Spin}^c(V)$ satisfies all the conditions outlined in Subsection 10.1.5. $\qquad \qquad \square$

Assume now dim V is even. Then, any any isomorphism $m\mathbb{C}l(V) \cong \mathrm{End}_{\mathbb{C}}(\mathbb{S}(V))$ induces a complex unitary representation

$$\mathrm{Spin}^c(V) \to \mathrm{Aut}(\mathbb{S}(V))$$

called the complex spinorial representation of Spin^c. It is not irreducible since, once we fix an orientation on V, the space $\mathrm{End}(\mathbb{S}(V))$ has a natural superstructure and, by definition, Spin^c acts through even automorphism. As in the real case, $\mathbb{S}(V)$ splits into a direct sum of irreducible representations $\mathbb{S}^{\pm}(V)$.

Fix a complex structure J on V. This complex structure determines two things.

(i) A canonical orientation on V.

(ii) A natural subgroup

$$U(V, J) = \{T \in SO(V) \; ; \; [T, J] = 0\} \subset SO(V).$$

Denote by $\imath_J : U(V, J) \to SO(V)$ the inclusion map.

Proposition 11.1.52. *There exists a natural group morphism* $U(V, J) \xrightarrow{\xi_J} \mathrm{Spin}^c(V)$ *such that the diagram below is commutative.*

$$
\begin{array}{ccc}
U(V, J) & \xrightarrow{\xi_J} & \mathrm{Spin}^c(V) \\
& \searrow{\scriptstyle \imath_J} & \downarrow{\scriptstyle \rho^c} \\
& & SO(V)
\end{array}
$$

Proof. Let $\omega \in U(V)$, and consider a path $\gamma : [0, 1] \to U(V)$ connecting $\mathbb{1}$ to ω. Via the inclusion $U(V) \hookrightarrow SO(V)$ we may regard γ as a path in $SO(V)$. As such, it admits a unique lift $\tilde{\gamma} : [0, 1] \to \mathrm{Spin}(V)$ such that $\tilde{\gamma}(0) = \mathbb{1}$.

Using the double cover $S^1 \to S^1$, $z \mapsto z^2$, we can find a path $t \mapsto \delta(t) \in S^1$ such that

$$\delta(0) = 1 \text{ and } \delta^2(t) = \det \gamma(t).$$

Define $\xi(\omega)$ to be the image of $(\tilde{\gamma}(1), \delta(1))$ in $\mathrm{Spin}^c(V)$. We have to check two things.

(i) The map ξ is well defined.

(ii) The map σ is a smooth group morphism.

To prove (i), we need to show that if $\eta : [0, 1] \to U(V)$ is a different path connecting $\mathbb{1}$ to ω, and $\lambda : [0, 1] \to S^1$ is such that $\lambda(0) = 1$ and $\lambda(t) = \det \eta(t)^2$, then

$$(\tilde{\eta}(1), \lambda(1)) = (\tilde{\gamma}(1), \delta(1)) \text{ in } \mathrm{Spin}^c(V).$$

The elements $\tilde{\gamma}(1)$ and $\tilde{\eta}(1)$ lie in the same fiber of the covering $\mathrm{Spin}(V) \xrightarrow{\rho} SO(V)$ so that they differ by an element in $\ker \rho$. Hence

$$\tilde{\gamma}(1) = \epsilon \tilde{\eta}(1), \quad \epsilon = \pm \mathbb{1}.$$

We can identify ϵ as the holonomy of the covering $\mathrm{Spin}(V) \to SO(V)$ along the loop $\gamma * \eta^-$ which goes from $\mathbb{1}$ to ω along γ, and then back to $\mathbb{1}$ along $\eta^-(t) = \eta(1-t)$.

The map $\det : U(V) \to S^1$ induces an isomorphism between the fundamental groups (see Exercise 6.2.35 of Subsection 6.2.5). Hence, in describing the holonomy ϵ it suffices to replace the loop $\gamma * \eta^- \subset U(V)$ by any loop $\nu(t)$ such that

$$\det \nu(t) = \det(\gamma * \eta^-) = \Delta(t) \in S^1.$$

Such a loop will be homotopic to $\gamma * \eta^{-1}$ in $U(V)$, and thus in $SO(V)$ as well. Select $\nu(t)$ of the form

$$\nu(t)e_1 = \Delta(t)e_1, \quad \nu(t)e_i = e_i \; \forall i \geq 2,$$

where (e_i) is a complex, orthonormal basis of (V, J). Set $f_i = Je_i$. With respect to the real basis $(e_1, f_1, e_2, f_2, \dots)$ the operator $\nu(t)$ (viewed as an element of $SO(V)$) has the matrix description

$$\begin{bmatrix} \cos\theta(t) & -\sin\theta(t) & \cdots \\ \sin\theta(t) & \cos\theta(t) & \cdots \\ \vdots & & \vdots & \mathbb{1} \end{bmatrix},$$

where $\theta : [0,1] \to \mathbb{R}$ is a continuous map such that $\Delta(t) = e^{i\theta(t)}$.

The lift of $\nu(t)$ to $\mathrm{Spin}(V)$ has the form

$$\tilde{\nu}(t) = \left(\cos \frac{\theta(t)}{2} - e_1 f_1 \sin \frac{\theta(t)}{2} \right).$$

We see that the holonomy defined by $\tilde{\nu}(t)$ is nontrivial if and only if the holonomy of the loop $t \mapsto \delta(t)$ in the double cover $S^1 \xrightarrow{z^2} S^1$ is nontrivial. This means that $\delta(1)$ and $\lambda(1)$ differ by the same element of $\mathbb{Z}_2$ as $\tilde{\gamma}(1)$ and $\tilde{\eta}(1)$. This proves ξ is well defined. We leave the reader to check that ξ is indeed a smooth morphism of groups. $\square$

11.1.8 Low dimensional examples

In low dimensions the objects discussed in the previous subsections can be given more suggestive interpretations. In this subsection we will describe some of these interpretations.

Example 11.1.53. (The case $n = 1$). The Clifford algebra Cl_1 is isomorphic with the field of complex numbers $\mathbb{C}$. The $\mathbb{Z}_2$-grading is $\boldsymbol{Re}\,\mathbb{C} \oplus \boldsymbol{Im}\,\mathbb{C}$. The group $\mathrm{Spin}(1)$ is isomorphic with $\mathbb{Z}_2$. $\square$

Example 11.1.54. (The case $n = 2$). The Clifford algebra Cl_2 is isomorphic with the algebra of quaternions $\mathbb{H}$. This can be seen by choosing an orthonormal basis $\{e_1, e_2\}$ in $\mathbb{R}^2$. The isomorphism is given by

$$1 \mapsto 1, \quad e_1 \mapsto \boldsymbol{i}, \quad e_2 \mapsto \boldsymbol{j}, \quad e_1 e_2 \mapsto \boldsymbol{k},$$

where i, j, and k are the imaginary units in $\mathbb{H}$. Note that

$$\text{Spin}(2) = \{a + bk \; ; \; a, b \in \mathbb{R}, \; a^2 + b^2 = 1\} \cong S^1.$$

The natural map $\text{Spin}(1) \to SO(2) \cong S^1$ takes the form $e^{i\theta} \mapsto e^{2i\theta}$. $\qquad\square$

Example 11.1.55. (The case $n = 3$). The Clifford algebra Cl_3 is isomorphic, as an ungraded algebra, to the direct sum $\mathbb{H} \oplus \mathbb{H}$. More relevant is the isomorphism $\text{Cl}_3^{even} \cong \text{Cl}_2 \cong \mathbb{H}$ given by

$$1 \mapsto 1, \;\; e_1 e_2 \mapsto i, \;\; e_2 e_3 \mapsto j, \;\; e_3 e_1 \mapsto k,$$

where $\{e_1, e_2, e_3\}$ is an orthonormal basis in $\mathbb{R}^3$. Under this identification the operation $x \mapsto x^b$ coincides with the conjugation in $\mathbb{H}$

$$x = a + bi + cj + dk \mapsto \bar{x} = a - bi - cj - dk.$$

In particular, the spinorial norm coincides with the usual norm on $\mathbb{H}$

$$N(a + bi + cj + dk) = a^2 + b^2 + c^2 + d^2.$$

Thus, any $x \in \text{Cl}_3^{even} \setminus \{0\}$ is invertible, and

$$x^{-1} = \frac{1}{N(x)} x^b.$$

Moreover, a simple computation shows that $x\mathbb{R}^3 x^{-1} \subset \mathbb{R}^3$, $\forall x \in \text{Cl}_3^{even} \setminus \{0\}$, so that

$$\Gamma^0(\mathbb{R}^3) \cong \mathbb{H} \setminus \{0\}.$$

Hence

$$\text{Spin}(3) \cong \{x \in \mathbb{H}; \; |x| = 1\} \cong SU(2).$$

The natural map $\text{Spin}(3) \to SO(3)$ is precisely the map described in the Exercise 6.2.8 of Subsection 6.2.1.

The isomorphism $\text{Spin}(3) \cong SU(2)$ can be visualized by writing each $q = a + bi + cj + dk$ as

$$q = u + jv, \;\; u = a + bi, \;\; v = (c - di) \in \mathbb{C}.$$

To a quaternion $q = u + jv$ one associates the 2×2 complex matrix

$$S_q = \begin{bmatrix} u & -\bar{v} \\ v & \bar{u} \end{bmatrix} \in SU(2).$$

Note that $S_{\bar{q}} = S_q^*$, $\forall q \in \mathbb{H}$. For each quaternion $q \in \mathbb{H}$ we denote by L_q (respectively R_q) the left (respectively right) multiplication. The right multiplication by i defines a complex structure on $\mathbb{H}$. Define $T : \mathbb{H} \to \mathbb{C}^2$ by

$$q = u + jv \mapsto Tq = \begin{bmatrix} u \\ v \end{bmatrix}.$$

A simple computation shows that $T(R_i q) = iTq$, i.e., T is a complex linear map. Moreover, $\forall q \in \mathrm{Spin}(3) \cong S^3$, the matrix S_q is in $SU(2)$, and the diagram below is commutative.

$$
\begin{array}{ccc}
\mathbb{H} & \xrightarrow{\ T\ } & \mathbb{C}^2 \\
L_q \downarrow & & S_q \downarrow \\
\mathbb{H} & \xrightarrow{\ T\ } & \mathbb{C}^2
\end{array}
$$

In other words, the representation

$$\mathrm{Spin}(3) \ni q \mapsto L_q \in GL_{\mathbb{C}}(\mathbb{H})$$

of $\mathrm{Spin}(3)$ is isomorphic with the tautological representation of $SU(2)$ on $\mathbb{C}^2$.

On the other hand, the correspondences

$$e_1 e_2 \mapsto S_{\boldsymbol{i}} \oplus S_{\boldsymbol{i}} \in \mathrm{End}\,(\mathbb{C}^2) \oplus \mathrm{End}\,(\mathbb{C}^2)$$

$$e_2 e_3 \mapsto S_{\boldsymbol{j}} \oplus S_{\boldsymbol{j}} \in \mathrm{End}\,(\mathbb{C}^2) \oplus \mathrm{End}\,(\mathbb{C}^2)$$

$$e_3 e_1 \mapsto S_{\boldsymbol{k}} \oplus S_{\boldsymbol{k}} \in \mathrm{End}\,(\mathbb{C}^2) \oplus \mathrm{End}\,(\mathbb{C}^2)$$

$$e_1 e_2 e_3 \mapsto R = \mathbb{1}_{\mathbb{C}^2} \oplus (-\mathbb{1}_{\mathbb{C}^2}) \in \mathrm{End}\,(\mathbb{C}^2) \oplus \mathrm{End}\,(\mathbb{C}^2)$$

extend to an isomorphism of algebras $\mathbb{C}\mathrm{l}_3 \to \mathrm{End}\,(\mathbb{C}^2) \oplus \mathrm{End}\,(\mathbb{C}^2)$. This proves that the tautological representation of $SU(2)$ is precisely the complex spinorial representation $\mathbb{S}_3$.

From the equalities

$$[R_j, L_q] = \{R_j, R_i\} = 0,$$

we deduce that R_j defines an isomorphism of $\mathrm{Spin}(3)$-modules

$$R_j : \mathbb{S}_3 \to \overline{\mathbb{S}}_3.$$

This implies there exists a $\mathrm{Spin}(3)$-invariant bilinear map

$$\beta : \mathbb{S}_3 \times \mathbb{S}_3 \to \mathbb{C}.$$

This plays an important part in the formulation of the recently introduced Seiberg-Witten equations (see [102]).   □

Exercise 11.1.56. The left multiplication by $\boldsymbol{i}$ introduces a different complex structure on $\mathbb{H}$. Prove the representation $\mathrm{Spin}(3) \ni q \mapsto (R_{\bar{q}} : \mathbb{H} \to \mathbb{H})$ is

(i) complex with respect to the above introduced complex structure on $\mathbb{H}$, and
(ii) it is isomorphic with complex spinorial representation described by the left multiplication.

□

Example 11.1.57. (The case $n = 4$). The Clifford algebra Cl_4 can be realized as the algebra of 2×2 matrices with entries in $\mathbb{H}$. To describe this isomorphism we have to start from a natural embedding $\mathbb{R}^4 \hookrightarrow M_2(\mathbb{H})$ given by the correspondence

$$\mathbb{H} \cong \mathbb{R}^4 \ni x \mapsto \begin{bmatrix} 0 & -x \\ \overline{x} & 0 \end{bmatrix}$$

A simple computation shows that the conditions in the universality property of a Clifford algebra are satisfied, and this correspondence extends to a bona-fide morphism of algebras $\text{Cl}_4 \to M_2(\mathbb{H})$. We let the reader check this morphism is also injective. A dimension count concludes it must also be surjective. $\square$

Proposition 11.1.58. $\text{Spin}(4) \cong SU(2) \times SU(2)$.

Proof. We will use the description of $\text{Spin}(4)$ as the universal (double-cover) of $SO(4)$ so we will explicitly produce a smooth $2 : 1$ group morphism $SU(2) \times SU(2) \to SO(4)$.

Again we think of $SU(2)$ as the group of unit quaternions. Thus each pair $(q_1, q_2) \in SU(2) \times SU(2)$ defines a real linear map

$$T_{q_1, q_2} : \mathbb{H} \to \mathbb{H}, \quad x \mapsto T_{q_1, q_2} x = q_1 x \overline{q}_2.$$

Clearly $|x| = |q_1| \cdot |x| \cdot |\overline{q_2}| = |T_{q_1, q_2} x| \ \forall x \in \mathbb{H}$, so that each T_{q_1, q_2} is an orthogonal transformation of $\mathbb{H}$. Since $SU(2) \times SU(2)$ is connected, all the operators T_{q_1, q_2} belong to the component of $O(4)$ containing $\mathbb{1}$, i.e,. T defines an (obviously smooth) group morphism

$$T : SU(2) \times SU(2) \to SO(4).$$

Note that $\ker T = \{1, -1\}$, so that T is $2 : 1$. In order to prove T is a double cover it suffices to show it is onto. This follows easily by noticing T is a submersion (*verify this!*), so that its range must contain an entire neighborhood of $\mathbb{1} \in SO(4)$. Since the range of T is closed (*verify this!*) we conclude that T must be onto because the closure of the subgroup (algebraically) generated by an open set in a connected Lie group coincides with the group itself (see Subsection 1.2.3). $\square$

The above result shows that

$$\underline{\text{so}}(4) \cong \underline{\text{spin}}(4) \cong \underline{\text{su}}(2) \oplus \underline{\text{su}}(2) \cong \underline{\text{so}}(3) \oplus \underline{\text{so}}(3).$$

Exercise 11.1.59. Using the identification $\text{Cl}_4 \cong M_2(\mathbb{H})$ show that $\text{Spin}(4)$ corresponds to the subgroup

$$\big\{ \text{diag}\,(p, q); \ \ p, q \in \mathbb{H}, |p| = |q| = 1 \big\} \subset M_2(\mathbb{H}). \qquad \square$$

Exercise 11.1.60. Let $\{e_1, e_2, e_3, e_4\}$ be an oriented orthonormal basis of $\mathbb{R}^4$. Let $*$ denote the Hodge operator defined by the canonical metric and the above chosen orientation. Note that

$$* : \Lambda^2 \mathbb{R}^4 \to \Lambda^2 \mathbb{R}^4$$

is involutive, $*^2 = 1$, so that we can split Λ^2 into the ± 1 eigenspaces of $*$
$$\Lambda^2 \mathbb{R}^4 = \Lambda^2_+ \mathbb{R}^4 \oplus \Lambda^2_- \mathbb{R}^4.$$

(a) Show that
$$\Lambda^2_\pm = \mathrm{span}_{\mathbb{R}} \{\, \eta_1^\pm, \; \eta_2^\pm, \; \eta_3^\pm \,\},$$

where
$$\eta_1^\pm = \frac{1}{\sqrt{2}}(e_1 \wedge e_2 \pm e_3 \wedge e_4), \quad \eta_2^\pm = \frac{1}{\sqrt{2}}(e_1 \wedge e_3 \pm e_4 \wedge e_2),$$

$$\eta_3^\pm = \frac{1}{\sqrt{2}}(e_1 \wedge e_4 \pm e_2 \wedge e_3).$$

(b) Show that the above splitting of $\Lambda^2 \mathbb{R}^4$ corresponds to the splitting $\underline{so}(4) = \underline{so}(3) \oplus \underline{so}(3)$ under the natural identification $\Lambda^2 \mathbb{R}^4 \cong \underline{so}(4)$. $\square$

To obtain an explicit realization of the complex spinorial representations $\mathbb{S}_4^\pm$ we need to describe a concrete realization of the complexification $\mathbb{C}l_4$. We start from the morphism of $\mathbb{R}$-algebras
$$\mathbb{H} \ni x = u + \boldsymbol{j} v \mapsto S_x = \begin{bmatrix} u & -\bar{v} \\ v & \bar{u} \end{bmatrix}$$

This extends by complexification to an isomorphism of $\mathbb{C}$-algebras
$$\mathbb{H} \otimes_{\mathbb{R}} \mathbb{C} \cong M_2(\mathbb{C}).$$

(*Verify this!*) We now use this isomorphism to achieve the identification.
$$\mathrm{End}_{\mathbb{H}}(\mathbb{H} \oplus \mathbb{H}) \otimes_{\mathbb{R}} \mathbb{C} \cong M_2(\mathbb{H}) \otimes_{\mathbb{R}} \mathbb{C} \cong \mathrm{End}_{\mathbb{C}}(\mathbb{C}^2 \oplus \mathbb{C}^2).$$

The embedding $\mathbb{R}^4 \to \mathbb{C}l_4$ now takes the form
$$\mathbb{H} \cong \mathbb{R}^4 \ni x \mapsto T_x = \begin{bmatrix} 0 & -S_{\bar{x}} \\ S_x & 0 \end{bmatrix} \in \mathrm{End}(\mathbb{C}^2 \oplus \mathbb{C}^2). \qquad (11.1.4)$$

Note that the chirality operator $\Gamma = -e_1 e_2 e_3 e_4$ is represented by the canonical involution
$$\Gamma \mapsto \mathbb{1}_{\mathbb{C}^2} \oplus (-\mathbb{1}_{\mathbb{C}^2}).$$

We deduce the $\mathbb{S}_4^\pm$ representations of $\mathrm{Spin}(4) = SU(2) \times SU(2)$ are given by
$$\mathbb{S}_4^+ : SU(2) \times SU(2) \ni (p, q) \mapsto p \in GL(2; \mathbb{C}),$$
$$\mathbb{S}_4^- : SU(2) \times SU(2) \ni (p, q) \mapsto q \in GL(2; \mathbb{C}).$$

Exactly as in the case of $\mathrm{Spin}(3)$, these representations can be given quaternionic descriptions.

Exercise 11.1.61. The space $\mathbb{R}^4 \cong \mathbb{H}$ has a canonical complex structure defined by R_i which defines (following the prescriptions in §10.1.3) an isomorphism
$$\boldsymbol{c} : \mathbb{C}l_4 \to \mathrm{End}(\Lambda^\bullet \mathbb{C}^2).$$

Identify $\mathbb{C}^2$ in the obvious way with $\Lambda^1 \mathbb{C}^2 \Lambda^{odd} \mathbb{C}^2$, and with $\Lambda^{even} \mathbb{C}^2$ via the map
$$e_1 \mapsto 1 \in \Lambda^0 \mathbb{C}^2, \; e_2 \mapsto e_1 \wedge e_2.$$

Show that under these identifications we have
$$c(x) = T_x \; \forall x \in \mathbb{R}^4 \cong (\mathbb{H}, R_i) \cong \mathbb{C}^2,$$

where T_x is the odd endomorphism of $\mathbb{C}^2 \oplus \mathbb{C}^2$ defined in (11.1.4). $\square$

Exercise 11.1.62. Let V be a 4-dimensional, oriented Euclidean space. Denote by

$$\mathfrak{q} : \Lambda^{\bullet}V \to Cl(V)$$

the quantization map, and fix an isomorphism $\Delta : Cl(V) \to \mathrm{End}\,(\mathbb{S}(V))$ of $\mathbb{Z}_2$-graded algebras. Show that for any $\eta \in \Lambda_+^2(V)$, the image $\Delta \circ \mathfrak{q}(\eta) \in \mathrm{End}\,(\mathbb{S}(V))$ is an endomorphism of the form $T \oplus 0 \in \mathrm{End}\,(\mathbb{S}^+(V)) \oplus \mathrm{End}\,(\mathbb{S}^-(V))$. □

Exercise 11.1.63. Denote by V a 4-dimensional oriented Euclidean space.
(a) Show that the representation $\mathbb{S}_4^+ \otimes \mathbb{S}_4^+$ of Spin(4) descends to a representation of $SO(4)$ and moreover

$$\mathbb{S}^+(V) \otimes_{\mathbb{C}} \mathbb{S}^+(V) \cong (\Lambda^0(V) \oplus \Lambda_+^2(V)) \otimes_{\mathbb{R}} \mathbb{C}$$

as $SO(4)$ representations.
(b) Show that $\mathbb{S}^+(V) \cong \bar{\mathbb{S}}^+(V)$ as Spin(4) modules.
(c) The above isomorphism defines an element $\phi \in \mathbb{S}^+(V) \otimes_{\mathbb{C}} \mathbb{S}^+(V)$. Show that via the correspondence at (a) the isomorphism ϕ spans $\Lambda^0(V)$. □

11.1.9 *Dirac bundles*

In this subsection we discuss a distinguished class of Clifford bundles which is both frequently encountered in applications and it is rich in geometric information. We will touch only the general aspects. The special characteristics of the most important concrete examples are studied in some detail in the following section. In the sequel all Clifford bundles will be assumed to be complex.

Definition 11.1.64. Let $E \to M$ be a Clifford bundle over the oriented Riemann manifold (M, g). We denote by $c : \Omega^1(M) \to \mathrm{End}(E)$ the Clifford multiplication.
(a) A *Dirac structure* on E is a pair (h, ∇) consisting of a Hermitian metric h on E, and a *Clifford connection*, i.e., a connection ∇ on E compatible with h and satisfying the following conditions.

(a1) For any $\alpha \in \Omega^1(M)$, the Clifford multiplication by α is a skew-Hermitian endomorphism of E.
(a2) For any $\alpha \in \Omega^1(M)$, $X \in \mathrm{Vect}\,(M)$, $u \in C^\infty(E)$

$$\nabla_X(c(\alpha)u) = c(\nabla_X^M \alpha)u + c(\alpha)(\nabla_X u),$$

where ∇^M denotes the Levi-Civita connection on T^*M. (This condition means that the Clifford multiplication is covariant constant).

A pair (Clifford bundle, Dirac structure) will be called a *Dirac bundle*.

(b) A $\mathbb{Z}_2$-grading on a Dirac bundle (E, h, ∇) is a $\mathbb{Z}_2$ grading of the underlying Clifford structure $E = E_0 \oplus E_1$, such that $h = h_0 \oplus h_1$, and $\nabla = \nabla^0 \oplus \nabla^1$, where h_i (respectively ∇^i) is a metric (respectively a metric connection) on E_i. □

The next result addresses the fundamental consistency question: do there exist Dirac bundles?

Proposition 11.1.65. *Let $E \to M$ be a Clifford bundle over the oriented Riemann manifold (M, g). Then there exist Dirac structures on E.*

Proof.   Denote by $\mathcal{D}$ the (possible empty) family of Dirac structure on E. Note that if $(h_i, \nabla^i) \in \mathcal{D}$, $i = 1, 2$, and $f \in C^\infty(M)$, then

$$(fh_1 + (1-f)h_2, f\nabla^1 + (1-f)\nabla^2) \in \mathcal{D}.$$

This elementary fact shows that the existence of Dirac structures is essentially a local issue: local Dirac structures can be patched-up via partitions of unity.

Thus, it suffices to consider only the case when M is an open subset of $\mathbb{R}^n$, and E is a trivial vector bundle. On the other hand, we cannot assume that the metric g is also trivial (Euclidean) since the local obstructions given by the Riemann curvature cannot be removed. We will distinguish two cases.

A. $n = \dim M$ is even. The proof will be completed in three steps.

Step 1. *A special example.* Consider the (complex) spinor module (representation)

$$c : \mathbb{Cl}_n \to \operatorname{End}(\mathbb{S}_n),$$

We can assume that c is selfadjoint, i.e.,

$$c(u^\dagger) = c(u)^*, \quad \forall u \in \mathbb{Cl}_n.$$

We will continue to denote by c the restriction of the Clifford multiplication to $\operatorname{Spin}(n) \hookrightarrow \mathbb{Cl}_n$.

Fix a global, oriented, orthonormal frame (e_i) of TM, and denote by (e^j) its dual coframe. Denote by $\omega = (\omega_{ij})$ the connection 1-form of the Levi-Civita connection on T^*M with respect to this moving frame, i.e.,

$$\nabla e^j = \omega e^j = \sum_i \omega_{ij} \otimes e^i, \quad \omega \in \Omega^1(M) \otimes \underline{\mathfrak{so}}(n).$$

Using the canonical isomorphism $\rho_* : \underline{\mathfrak{spin}}(n) \to \underline{\mathfrak{so}}(n)$ we define

$$\tilde{\omega} := \rho_*^{-1}(\omega) = -\frac{1}{2} \sum_{i<j} \omega^{ij} e^i \cdot e^j \in \Omega^1(M) \otimes \underline{\mathfrak{spin}}(n).$$

This defines a connection $\nabla^{\mathbb{S}}$ on the trivial vector bundle $\underline{\mathbb{S}}_M = \mathbb{S}_n \times M$ given by

$$\nabla^{\mathbb{S}} u = du - \frac{1}{2} \sum_{i<j} \omega_{ij} \otimes c(e^i) \cdot c(e^j) u, \quad \forall u \in C^\infty(\underline{\mathbb{S}}_M).$$

The considerations in §10.1.5 show that ∇^S is indeed a Clifford connection so that $\underline{\mathbb{S}}_M$ is a Dirac bundle.

Step 2. *Constructing general Dirac bundles.* Fix a Dirac bundle (E, h_E, ∇^E) over M. For any Hermitian vector bundle (W, h_W) equipped with a Hermitian connection ∇^W we can construct the tensor product $E \otimes W$ equipped with the metric $h_E \otimes h_W$ and the product connection $\nabla^{E \otimes W}$. These two data define a Dirac structure on $E \otimes W$ (Exercise 11.1.68 at the end of this subsection).

The representation theory of the Clifford algebra $\mathbb{C}l_n$ with n even shows that any Clifford bundle over M must be a twisting $\mathbb{S}_M \otimes W$ of the spinor bundle $\mathbb{S}_M$. This completes the proof of the proposition when $\dim M$ is even.

Step 3 *Conclusion.* The odd case is dealt with similarly using the different representation theory of the $\mathbb{C}l_{2k+1}$. Now instead of one generating model $\mathbb{S}$ there are two, but the proof is conceptually identical. The straightforward details are left to the reader. □

Denote by (E, h, ∇) a Dirac bundle over the oriented Riemann manifold (M, g). There exists a Dirac operator on E canonically associated to this structure given by

$$D = c \circ \nabla : C^\infty(E) \xrightarrow{\nabla} C^\infty(T^*M \otimes E) \xrightarrow{c} C^\infty(E).$$

A Dirac operator associated to a Dirac structure is said to be a *geometric Dirac operator*.

Proposition 11.1.66. *Any geometric Dirac operator is formally selfadjoint.*

Proof. The assertion in the above proposition is local so we can work with local orthonormal moving frames. Fix $x_0 \in M$, and denote by (x^i) a collection of normal coordinates near x_0. Set $e_i = \frac{\partial}{\partial x_i}$. Denote by (e^i) the dual coframe of (e_i). If (h, ∇) is a Dirac structure on the Clifford bundle E, then *at x_0* the associated Dirac operator can be described as

$$D = \sum_i c(e^i) \nabla_i \quad (\nabla_i = \nabla_{e_i}).$$

We deduce

$$D^* = (\nabla_i)^* c(e^i)^*.$$

Since the connection ∇ is compatible with h and $\boldsymbol{div}_g(e_i)|_{x_0} = 0$, we deduce

$$(\nabla_i)^* = -\nabla_i.$$

Since the Clifford multiplication is skew-Hermitian, $c(e^i)^* = -c(e^i)$, we deduce that at x_0 we have

$$D^* = \sum_i \nabla_i \circ c(e^i) = \sum_i [\nabla_i, c(e^i)] + D.$$

On the other hand, the Clifford multiplication is covariant constant, and $(\nabla_i^M e^i)|_{x_0} = 0$ so we conclude

$$[\nabla_i, c(e^i)]|_{x_0} = 0 \quad \forall i. \qquad \square$$

Let D be a geometric operator associated to the Dirac bundle (E, h, ∇). By definition, D^2 is a generalized Laplacian, and according to Proposition 10.1.34 we must have an equality of the form

$$D^2 = \tilde{\nabla}^* \tilde{\nabla} + \mathcal{R},$$

where $\tilde{\nabla}$ is a connection on E, and $\mathcal{R}$ is the Weitzenböck remainder which is an endomorphism of E. For geometric Dirac operators, $\tilde{\nabla} = \nabla$, and this remainder can be given a very explicit description with remarkable geometric consequences. To formulate it we need a little bit of foundational work.

Let (E, h, ∇) be a Dirac bundle over the oriented Riemann manifold (M, g). Denote by $\mathrm{Cl}(T^*MM) \to M$ the bundle of Clifford algebras generated by (T^*M, g).

The curvature $F(\nabla)$ of ∇ is a section of $\Lambda^2 T^*M \otimes \mathrm{End}\,(E)$. Using the quantization map $\mathfrak{q} : \Lambda^\bullet T^*M \to \mathrm{Cl}(T^*M)$ we get a section

$$\mathfrak{q}(F) \in \mathrm{Cl}(T^*M) \otimes \mathrm{End}\,(E).$$

On the other hand, the Clifford multiplication $\boldsymbol{c} : \mathrm{Cl}(T^*M) \to \mathrm{End}\,(E)$ defines a linear map

$$\mathrm{Cl}(T^*M) \otimes \mathrm{End}\,(E) \to \mathrm{End}(E), \quad \omega \otimes T \mapsto \boldsymbol{c}(\omega) \circ T.$$

This map associates to the element $\mathfrak{q}(F)$ an endomorphism of E which we denote by $\boldsymbol{c}(F)$. If (e^i) is a local, oriented, orthonormal moving frame for T^*M, then we can write

$$F(\nabla) = \sum_{i<j} e^i \wedge e^j \otimes F_{ij},$$

and

$$\boldsymbol{c}(F) = \sum_{i<j} \boldsymbol{c}(e^i)\boldsymbol{c}(e^j) F_{ij} = \frac{1}{2} \sum_{i,j} \boldsymbol{c}(e^i)\boldsymbol{c}(e^j) F_{ij}.$$

Theorem 11.1.67 (Bochner-Weitzenböck). *Let D be the geometric Dirac operator associated to the Dirac bundle (E, h, ∇) over the oriented Riemann manifold (M, g). Then*

$$D^2 = \nabla^* \nabla + \boldsymbol{c}(F(\nabla)).$$

Proof. Fix $x \in M$, and then choose an oriented, local orthonormal moving frame (e_i) of TM near x such that

$$[e_i, e_j]\,|_x = (\nabla^M e_i)\,|_x = 0,$$

where ∇^M denotes the Levi-Civita connection. Such a choice is always possible because the torsion of the Levi-Civita connection is zero. Finally, denote by (e^i) the dual coframe of (e_i). Then

$$D^2\,|_x = \sum_i \boldsymbol{c}(e^i)\nabla_i \left(\sum_j \boldsymbol{c}(e^j)\nabla_j \right).$$

Since $[\nabla_i, \boldsymbol{c}(e_j)]|_x = 0$ we deduce

$$D^2|_x = \sum_{i,j} \boldsymbol{c}(e^i)\boldsymbol{c}(e^j)\nabla_i\nabla_j = -\sum_i \nabla_i^2 + \sum_{i \neq j} \boldsymbol{c}(e^i)\boldsymbol{c}(e^j)\nabla_i\nabla_j$$

$$= -\sum_i \nabla_i^2 + \sum_{i<j} \boldsymbol{c}(e^i)\boldsymbol{c}(e^j)[\nabla_i, \nabla_j] = -\sum_i \nabla_i^2 + \sum_{i<j} \boldsymbol{c}(e^i)\boldsymbol{c}(e^j)F_{ij}(\nabla).$$

We want to emphasize again *the above equalities hold only at* x. The theorem now follows by observing that

$$(\nabla^*\nabla)|_x = -\left(\sum_i \nabla_i^2\right)\Big|_x. \qquad \qquad \square$$

Exercise 11.1.68. Let (E, h, ∇) be a Dirac bundle over the oriented Riemann manifold (M, g) with associated Dirac operator D. Consider a Hermitian bundle $W \to M$ and a connection ∇^W compatible with the Hermitian metric h_W.
(a) Show that $(E \otimes W, h \otimes h_W, \hat{\nabla} = \nabla \otimes \mathbf{1}_W + \mathbf{1}_E \otimes \nabla^W)$ defines a Dirac structure on $E \otimes W$ in which the Clifford multiplication by $\alpha \in \Omega^1(M)$ is defined by

$$\boldsymbol{c}(\alpha)(e \otimes w) = (\boldsymbol{c}(\alpha)e) \otimes w, \quad e \in C^\infty(E),\ w \in C^\infty(W).$$

We denote by D_W the corresponding geometric Dirac operator.
(b) Denote by $\boldsymbol{c}_W(F(\nabla^W))$ the endomorphism of $E \otimes W$ defined by the sequence

$$F(\nabla^W) \in C^\infty(\Lambda^2 T^*M \otimes \mathrm{End}\,(W)) \xrightarrow{q} C^\infty(\mathrm{Cl}(T^*M) \otimes \mathrm{End}\,(W)) \xrightarrow{c} C^\infty(\mathrm{End}\,(E \otimes W)).$$

Show that

$$D_W^2 = \hat{\nabla}^*\hat{\nabla} + \boldsymbol{c}(F(\nabla)) + \boldsymbol{c}_W(F(\nabla^W)). \qquad \qquad \square$$

11.2 Fundamental examples

This section is entirely devoted to the presentation of some fundamental examples of Dirac operators. More specifically we will discuss the Hodge-DeRham operator, the Dolbeault operator the *spin* and *spin^c* Dirac. We will provide more concrete descriptions of the Weitzenböck remainder presented in Subsection 10.1.9 and show some of its uses in establishing vanishing theorems.

11.2.1 *The Hodge-DeRham operator*

Let (M, g) be an oriented Riemann manifold and set

$$\Lambda_{\mathbb{C}}^\bullet T^*M := \Lambda^\bullet T^*M \otimes \mathbb{C}.$$

For simplicity we continue to denote by $\Omega^\bullet(M)$ the space of smooth differential forms on M with complex coefficients. We have already seen that the Hodge-DeRham operator

$$d + d^* : \Omega^\bullet(M) \to \Omega^\bullet(M)$$

is a Dirac operator. In fact, we will prove this operator is a *geometric* Dirac operator.

Continue to denote by g the Hermitian metric induced by the metric g on the complexification $\Lambda_{\mathbb{C}}^{\bullet}T^*M$. We will denote by ∇^g all the Levi-Civita connection on the tensor bundles of M. When we want to be more specific about which Levi-Civita connection we are using at a given moment we will indicate the bundle it acts on as a superscript. E.g., ∇^{T^*M} is the Levi-Civita connection on T^*M.

Proposition 11.2.1. *The pair (g, ∇^g) defines a Dirac structure on the Clifford bundle $\Lambda_{\mathbb{C}}T^*M$ and $d + d^*$ is the associated Dirac operator.*

Proof.   In Subsection 4.1.5 we have proved that d can be alternatively described as the composition

$$C^{\infty}(\Lambda^{\bullet}T^*M) \xrightarrow{\nabla} C^{\infty}(T^*M \otimes \Lambda^{\bullet}T^*M) \xrightarrow{\varepsilon} C^{\infty}(\Lambda^{\bullet}T^*M),$$

where ε denotes the exterior multiplication map. Thus

$$d + d^* = \varepsilon \circ \nabla + \nabla^* \circ \varepsilon^*.$$

If $X_1, \cdots, X_n$ is a local orthonormal frame of TM, and $\theta^1, \cdots, \theta^n$ is its dual coframe, then for any ordered multi-index I we have

$$\varepsilon^*(\theta^I) = \sum_j i_{X_j}\theta^I.$$

Thus, for any $\omega \in C^{\infty}(\Lambda^{\bullet}T^*M)$ we have

$$(\nabla^* \circ \varepsilon^*)\omega = -\left(\sum_k \nabla_{X_k} i_{X_k} + \boldsymbol{div}_g(X_k) i_{X_k}\right)\omega.$$

Fortunately, we have the freedom to choose the frame (X_k) in any manner we find convenient.

Fix an *arbitrary* point $x_0 \in M$ and choose (X_k) such that, at x_0, we have $X_k = \frac{\partial}{\partial x_k}$, where (x^k) denotes a collection of normal coordinates near x_0. With such a choice $\boldsymbol{div}_g(X_k) = 0$ at x_0, and thus

$$d^*\omega|_{x_0} = -\sum_k i_{\partial_k}\nabla_{\partial_k}\omega.$$

This shows $d + d^*$ can be written as the composition

$$C^{\infty}(\Lambda^{\bullet}T^*M) \xrightarrow{\nabla} C^{\infty}(T^*M \otimes \Lambda^{\bullet}T^*M) \xrightarrow{c} C^{\infty}(\Lambda^{\bullet}T^*M),$$

where c denotes the usual Clifford multiplication on an exterior algebra. We leave the reader to verify that the Levi-Civita connection on $\Lambda^{\bullet}T^*M$ is indeed a Clifford connection, i.e., the Clifford multiplication is covariant constant. Passing to the complexification $\Lambda_{\mathbb{C}}^{\bullet}T^*M$ we deduce that $d + d^*$ is a geometric Dirac operator. $\square$

We want to spend some time elucidating the structure of the Weitzenböck remainder. First of all, we need a better description of the curvature of ∇^g viewed as a connection on $\Lambda^{\bullet}T^*M$.

Denote by R the Riemann curvature tensor, i.e., the curvature of the Levi-Civita connection

$$\nabla^g : C^\infty(TM) \to C^\infty(T^*M \otimes TM).$$

Thus R is a bundle morphism

$$R : C^\infty(TM) \to C^\infty(\Lambda^2 T^*M \otimes TM).$$

We have a dual morphism

$$\tilde{R} : C^\infty(T^*M) \to C^\infty(\Lambda^2 T^*M \otimes T^*M),$$

uniquely determined by the equality

$$(\tilde{R}(Y,Z)\alpha)(X) = -\alpha(R(Y,Z)X) \quad \forall \alpha \in \Omega^1(M) \ X, Y, Z \in \text{Vect}(M).$$

Lemma 11.2.2. *$\tilde{R}$ is the curvature of the Levi-Civita connection ∇^{T^*M}.*

Proof. The Levi-Civita connection on T^*M is determined by the equalities

$$(\nabla_Z^{T^*M}\alpha)(X) = Z \cdot \alpha(X) - \alpha(\nabla_Z^{TM}X) \quad \forall \alpha \in \Omega^1(M), \ X, Z \in \text{Vect}(M).$$

Derivating along $Y \in \text{Vect}(M)$ we get

$$(\nabla_Y\nabla_Z\alpha)(X) = Y \cdot (\nabla_Z\alpha)(X) - (\nabla_Z\alpha)(\nabla_Y X)$$

$$= Y \cdot Z \cdot \alpha(X) - Y \cdot \alpha(\nabla_Z X) - Z \cdot \alpha(\nabla_Y X) - \alpha(\nabla_Z\nabla_Y X).$$

Similar computations give $\nabla_Z\nabla_Y\alpha$ and $\nabla_{[Y,Z]}$ and we get

$$(R^{T^*M}(Y,Z)\alpha)(X) = -\alpha(R^{TM}(Y,Z)X). \qquad \square$$

The Levi-Civita connection $\nabla^g = \nabla^{T^*M}$ extends as an even derivation to a connection on $\Lambda^\bullet T^*M$. More precisely, for every $X \in \text{Vect}(M)$, and any $\alpha_1, \ldots, \alpha_k \in \Omega^1(M)$, we define

$$\nabla_X^g(\alpha_1 \wedge \cdots \wedge \alpha_k) = (\nabla_X^g\alpha_1) \wedge \cdots \wedge \alpha_k + \cdots + \alpha_1 \wedge \cdots \wedge (\nabla_X^g\alpha_k).$$

Lemma 11.2.3. *The curvature of the Levi-Civita connection on $\Lambda^\bullet T^*M$ is defined by*

$$R^{\Lambda^\bullet T^*M}(X,Y)(\alpha_1 \wedge \cdots \wedge \alpha_k)$$

$$= (\tilde{R}(X,Y)\alpha_1) \wedge \cdots \wedge \alpha_k + \cdots + \alpha_1 \wedge \cdots \wedge (\tilde{R}(X,Y)\alpha_k),$$

for any vector fields X, Y, and any 1-forms $\alpha_1, \ldots, \alpha_k$. $\qquad \square$

Exercise 11.2.4. Prove the above lemma. $\qquad \square$

The Weitzenböck remainder of the Dirac operator $d + d^*$ is $\boldsymbol{c}(R^{\Lambda^\bullet T^* M})$. To better understand its action we need to pick a local, oriented, orthonormal moving frame (e_i) of TM. We denote by (e^i) its dual coframe. The Riemann curvature tensor can be expressed as

$$R = \sum_{i<j} e^i \wedge e^j R_{ij},$$

where R_{ij} is the skew-symmetric endomorphism $R_{ij} = R(e_i, e_j) : TM \to TM$. Thus

$$\boldsymbol{c}(R^{\Lambda^\bullet T^* M}) = \sum_{i<j} \boldsymbol{c}(e^i)\boldsymbol{c}(e^j) R^{\Lambda^\bullet T^* M}(e_i, e_j) = \frac{1}{2} \sum_{i,j} \boldsymbol{c}(e^i)\boldsymbol{c}(e^j) R^{\Lambda^\bullet T^* M}(e_i, e_j),$$

where $\boldsymbol{c}(e^j) = e(e^j) - i(e_j)$.

Note that since both $\nabla^* \nabla$, and $(d + d^*)^2$ preserve the $\mathbb{Z}$-grading of $\Lambda_\mathbb{C}^\bullet T^* M$, so does the Weitzenböck remainder, and consequently it must split as

$$\boldsymbol{c}(R^{\Lambda^\bullet T^* M}) = \oplus_{k \geq 0} \mathcal{R}^k.$$

Since $\mathcal{R}^0 \equiv 0$, the first interesting case is $\mathcal{R}^1$. To understand its form, pick normal coordinates (x^i) near $\boldsymbol{x}_0 \in M$. Set $e_i = \frac{\partial}{\partial x_i}|_{\boldsymbol{x}_0} \in T_{\boldsymbol{x}_0} M$ and $e^i = dx^i|_{\boldsymbol{x}_0} \in T_{\boldsymbol{x}_0}^* M$.

At $\boldsymbol{x}_0$ the Riemann curvature tensor R has the form

$$R = \sum_{k<\ell} e^k \wedge e^\ell R(e_k, e_\ell)$$

where $R(e_k, e_\ell)e_j = R_{jk\ell}^i e_i = R_{ijk\ell} e_i$. Using Lemma 11.2.2 we get

$$\tilde{R}(e_k, e_\ell)e^j = R_{ijk\ell} e^i.$$

Using this in the expression of $\mathcal{R}^1$ at $\boldsymbol{x}_0$ we get

$$\mathcal{R}^1(\sum_j \alpha_j e^j) = \frac{1}{2} \sum_{k,\ell} \sum_{i,j} \alpha_j R_{ijk\ell} \boldsymbol{c}(e^k)\boldsymbol{c}(e^\ell)e^i.$$

We need to evaluate the Clifford actions in the above equality. We have

$$\boldsymbol{c}(e^\ell)e^i = e^\ell \wedge e^i - \delta_{i\ell},$$

and

$$\boldsymbol{c}(e^k)\boldsymbol{c}(e^\ell)e^i = e^k \wedge e^\ell \wedge e^i - \delta_{i\ell}e^k - \delta_{k\ell}e^i + \delta_{ik}e^\ell.$$

Hence

$$\mathcal{R}^1(\sum_j \alpha_j e^j) = \frac{1}{2} \sum_{i,j,k,\ell} \alpha_j R_{ijk\ell}(e^k \wedge e^\ell \wedge e^i - \delta_{i\ell}e^k - \delta_{k\ell}e^i + \delta_{ik}e^\ell).$$

Using the first Bianchi identity we deduce that

$$\sum_{i,k,\ell} R_{ijk\ell} e^k \wedge e^\ell \wedge e^i = -\sum_{i,k,\ell} R_{jik\ell} e^i \wedge e^k \wedge e^\ell = 0, \quad \forall j.$$

Because of the skew-symmetry $R_{ijk\ell} = -R_{ij\ell k}$ we conclude that

$$\sum_{ijk\ell} \alpha_j R_{ijk\ell} \delta_{k\ell} e^i = 0.$$

Hence

$$\mathcal{R}^1 \left(\sum_j \alpha_j e^j \right) = \frac{1}{2} \sum_{i,j,k,\ell} \alpha_j R_{ijk\ell} (\delta_{ik} e^\ell - \delta_{i\ell} e^k)$$

$$= \frac{1}{2} \sum_{i,j,\ell} \alpha_j R_{iji\ell} e^\ell - \frac{1}{2} \sum_{i,j,k} \alpha_j R_{ijki} e^k = \sum_{i,j,k} \alpha_j R_{ijik} e^k = \sum_{jk} \alpha_j \operatorname{Ric}_{jk} e^k,$$

where $\operatorname{Ric}_{jk} = \sum_i R_{ijik}$ denotes the Ricci tensor at x_0. Hence

$$\mathcal{R}^1 = \operatorname{Ric}. \tag{11.2.1}$$

In the above equality, the Ricci tensor Ric is regarded (via the metric duality) as a selfadjoint endomorphism of T^*M.

The identity (11.2.1) has a beautiful consequence.

Theorem 11.2.5 (Bochner). *Let (M, g) be a compact, connected, oriented Riemann manifold.*
(a) If the Ricci tensor is non-negative definite, then $b_1(M) \leq \dim M$.
(b) If Ricci tensor is non-negative definite, but it is somewhere strictly positive definite then $b_1(M) = 0$.
(Recall that $b_1(M)$ denotes the first Betti number of M.)

The above result is truly remarkable. The condition on the Ricci tensor is purely local, but it has global consequences. We have proved a similar result using geodesics (see Myers Theorem, Subsection 5.2.2, 6.2.4 and 6.2.5). Under more restrictive assumptions on the Ricci tensor (uniformly positive definite) one deduces a stronger conclusion namely that the fundamental group is finite. If the uniformity assumption is dropped then the conclusion of the Myers theorem no longer holds (think of the flat torus).

This result gives yet another explanation for the equality $H^1(G) = 0$, where G is a compact semisimple Lie group. Recall that in this case the Ricci curvature is $\frac{1}{4} \times \{\text{the Killing metric}\}$.

Proof. (a) Let Δ_1 denote the metric Laplacian

$$\Delta_1 = dd^* + d^*d : \Omega^1(M) \to \Omega^1(M).$$

Hodge theory asserts that $b_1(M) = \dim \ker \Delta_1$ so that, to find the first Betti number, we need to estimate the "number" of solutions of the elliptic equation

$$\Delta_1 \eta = 0, \quad \eta \in \Omega^1.$$

Using the Bochner-Weitzenböck theorem and the equality (11.2.1) we deduce that if $\eta \in \ker \Delta_1$, then

$$\nabla^* \nabla \eta + \operatorname{Ric} \eta = 0, \quad \text{on } M.$$

Taking the L^2-inner product by η, and then integrating by parts, we get

$$\int_M |\nabla\eta|^2 dv_g + \int_M (\mathrm{Ric}\ \eta, \eta) dv_g = 0. \tag{11.2.2}$$

Since Ric is non-negative definite we deduce $\nabla\eta = 0$, so that any harmonic 1-form must be covariant constant. In particular, since M is connected, the number of linearly independent harmonic 1-forms is no greater than the rank of T^*M which is $\dim M$.

(b) Using the equality (11.2.2) we deduce that any harmonic 1-form η must satisfy

$$(\mathrm{Ric}(x)\eta_x, \eta_x)_x = 0 \quad \forall x \in M.$$

If the Ricci tensor is positive at some $x_0 \in M$, then $\eta(x_0) = 0$. Since η is also covariant constant, and M is connected, we conclude that $\eta \equiv 0$. □

Remark 11.2.6. For a very nice survey of some beautiful applications of this technique we refer to [10]. □

11.2.2 *The Hodge-Dolbeault operator*

This subsection introduces the reader to the Dolbeault operator which plays a central role in complex geometry. Since we had almost no contact with this beautiful branch of geometry we will present only those aspects concerning the "Dirac nature" of these operators. To define this operator we need a little more differential geometric background.

Definition 11.2.7. (a) Let $E \to M$ be a smooth real vector bundle over the smooth manifold M. An *almost complex* structure on E is an endomorphism $J : E \to E$ such that $J^2 = -\mathbb{1}_E$.

(b) An almost complex structure on a smooth manifold M is an almost complex structure J on the tangent bundle. An *almost complex manifold* is a pair (manifold, almost complex structure). □

Note that any almost complex manifold is necessarily even dimensional and orientable, so from the start we know that not any manifold admits almost complex structures. In fact, the existence of such a structure is determined by topological invariants finer than the dimension and orientability.

Example 11.2.8. (a) A complex manifold M is almost complex. Indeed, the manifold M is locally modelled by $\mathbb{C}^n$, and the transition maps are holomorphic maps $\mathbb{C}^n \to \mathbb{C}^n$. The multiplication by i defines a real endomorphism on $\mathbb{R}^{2n} \cong \mathbb{C}^n$, and the differential of a holomorphic map $\mathbb{C}^n \to \mathbb{C}^n$ commutes with the endomorphism of $T\mathbb{C}^n$ given by multiplication by i in the fibers. Hence, this endomorphism induces the almost complex structure on TM.

(b) For any manifold M, the total space of its tangent bundle TM can be equipped in many different ways with an almost complex structure.

To see this, denote by E the tangent bundle of TM, $E = T(TM)$. The bundle E has a natural subbundle $\mathcal{V}$, the *vertical subbundle*, which is the kernel of the differential of the canonical projection $\pi : TM \to M$.

If $x \in M$, $v \in T_x M$, then a tangent vector in $T_{(v,x)} TM$ is vertical if and only if it is tangent to a smooth path which entirely the fiber $T_x M \subset TM$. Note that we have a canonical identification

$$\mathcal{V}_{(v,x)} \cong T_x M.$$

Fix a Riemann metric h on the manifold TM, and denote by $\mathcal{H} \subset E$, the subbundle of E which is the orthogonal complement of $\mathcal{V}$ in E. For every $(v, x) \in TM$, the differential of π induces an isomorphism

$$\pi_* : \mathcal{H}_{(v,x)} \to T_x M \cong \mathcal{V}_{v,x}.$$

Thus, π_* induces a bundle isomorphism $A : \mathcal{H} \to \mathcal{V}$. Using the direct sum decomposition $E = \mathcal{H} \oplus \mathcal{V}$ we define the endomorphism J of E in the block form

$$J = \begin{bmatrix} 0 & -A^{-1} \\ A & 0 \end{bmatrix}. \qquad \square$$

Exercise 11.2.9. Prove that a smooth manifold M of dimension $2n$ admits an almost complex structure if an only if there exists a 2-form $\omega \in \Omega^2(M)$ such that the top exterior power ω^n is a volume form on M. $\qquad \square$

Let (M, J) be an almost complex manifold. Using the results of Subsection 2.2.5 we deduce that the complexified tangent bundle $TM \otimes \mathbb{C}$ splits as

$$TM \otimes \mathbb{C} = (TM)^{1,0} \otimes (TM)^{0,1}.$$

The complex bundle $TM^{1,0}$ is isomorphic (over $\mathbb{C}$) with (TM, J).

By duality, the operator J induces an almost complex structure in the cotangent bundle T^*M, and we get a similar decomposition

$$T^*M \otimes \mathbb{C} = (T^*M)^{1,0} \oplus (T^*M)^{0,1}.$$

In turn, this defines a decomposition

$$\Lambda^\bullet_\mathbb{C} T^*M = \bigoplus_{p,q} \Lambda^{p,q} T^*M.$$

We set $\Omega^{p,q}(M) := C^\infty(\Lambda^{p,q} T^*M)$.

Example 11.2.10. Let M be a complex manifold. If $(z^j = x^j + iy^j)$ are local holomorphic coordinates on M, then $(TM)^{1,0}$ is generated (locally) by the complex tangent vectors

$$\partial_{z^j} = \frac{1}{2} \left(\partial_{x^j} - i\partial_{y^j} \right),$$

while $(T^*M)^{1,0}$ is locally generated by the complex 1-forms $dz^j = dx^j + iy^j$.

The bundle $(TM)^{1,0}$ is also known as the *holomorphic tangent space*, while $(T^*M)^{1,0}$ is called the *holomorphic cotangent space*. The space $(T^*M)^{0,1}$ is called the anti-holomorphic cotangent space, and it is locally spanned by the complex 1-forms $d\bar{z}^j = dx^j - iy^j$.

Note that any (p,q)-form $\eta \in \Omega^{p,q}(M)$ can be locally described as

$$\eta = \sum_{|A|=p,|B|=q} \eta_{A,B} dz^A \wedge d\bar{z}^B, \quad \eta_{A,B} \in C^\infty(M,\mathbb{C}),$$

where we use capital Latin letters $A, B, C \ldots$ to denote ordered multi-indices, and for each such index A we set

$$dz^A =: dz^{A_1} \wedge \cdots \wedge dz^{A_p}.$$

The differential form $d\bar{z}^B$ is defined similarly.   □

Exercise 11.2.11. Let (M, J) be an arbitrary smooth almost complex manifold.
(a) Prove that

$$d\Omega^{p,q}(M) \subset \Omega^{p+2,q-1}(M) \oplus \Omega^{p+1,q}(M) \oplus \Omega^{p,q+1}(M)\Omega^{p-1,q+2}(M).$$

(b) Show that if M is a *complex* manifold then

$$d\Omega^{p,q}(M) \subset \Omega^{p+1,q}(M) \oplus \Omega^{p,q+1}(M). \tag{11.2.3}$$

(The converse is also true, and it is known as the *Newlander-Nirenberg theorem*. Its proof is far from trivial. For details we refer to the original paper [80].)   □

Definition 11.2.12. An almost complex structure on a smooth manifold is called *integrable* if it can derived from a holomorphic atlas, i.e. an atlas in which the transition maps are holomorphic.   □

Thus, the Newlander-Nirenberg theorem mentioned above states that the condition (11.2.3) is necessary and sufficient for an almost complex structure to be integrable.

Exercise 11.2.13. Assuming the Newlander-Nirenberg theorem prove that an almost complex structure J on the smooth manifold M is integrable if and only if the *Nijenhuis tensor* $N \in C^\infty(T^*M^{\otimes 2} \otimes TM)$ defined by

$$N(X,Y) = [JX, JY] - [X,Y] - J[X, JY] - J[JX, Y] \quad \forall X, Y \in \text{Vect}(M)$$

vanishes identically.   □

In the sequel we will exclusively consider only complex manifolds. Let M be such a manifold. The above exercise shows that the exterior derivative

$$d : \Omega_{\mathbb{C}}^k(M) \to \Omega_{\mathbb{C}}^{k+1}(M)$$

splits as a direct sum $d = \oplus_{p+q=k} d^{p,q}$, where

$$d^{p,q} = \{d : \Omega^{p,q}(M) \to \Omega^{p+1,q}(M) \oplus \Omega^{p,q+1}(M)\}.$$

The component $\Omega^{p,q} \to \Omega^{p+1,q}$ is denoted by $\partial = \partial^{p,q}$, while the component $\Omega^{p,q} \to \Omega^{p,q+1}$ is denoted by $\bar{\partial} = \bar{\partial}^{p,q}$.

Example 11.2.14. In local holomorphic coordinates (z^i) the action of the operator $\bar{\partial}$ is described by

$$\bar{\partial}\left(\sum_{A,B} \eta_{AB} dz^A \wedge d\bar{z}^B\right) = \sum_{j,A,B} (-1)^{|A|} \frac{\partial \eta_{AB}}{\partial \bar{z}^j} dz^A \wedge d\bar{z}^j \wedge d\bar{z}^B. \qquad \square$$

It is not difficult to see that

$$\bar{\partial}^{p,q+1} \circ \bar{\partial}^{p,q} = 0 \quad \forall p, q.$$

In other words, for any $0 \leq p \leq \dim_{\mathbb{C}} M$, the sequence

$$0 \to \Omega^{p,0}(M) \xrightarrow{\bar{\partial}} \Omega^{p,1}(M) \xrightarrow{\bar{\partial}} \cdots$$

is a cochain complex known as the *p*-th *Dolbeault complex* of the complex manifold M. Its cohomology groups are denoted by $H^{p,q}_{\bar{\partial}}(M)$.

Lemma 11.2.15. *The Dolbeault complex is an elliptic complex.*

Proof. The symbol of $\bar{\partial}^{p,q}$ is very similar to the symbol of the exterior derivative. For any $x \in M$ and any $\xi \in T^*X$

$$\sigma(\bar{\partial}^{p,q})(\xi) : \Lambda^{p,q} T^*_x M \to \Lambda^{p,q+1} T^*_x M$$

is (up to a multiplicative constant) the (left) exterior multiplication by $\xi^{0,1}$, where $\xi^{0,1}$ denotes the $(0,1)$ component of ξ viewed as an element of the complexified tangent space. More precisely,

$$\xi^{0,1} = \frac{1}{2}(\xi + iJ_0\xi),$$

where $J_0 : T^*_x M \to T^*_x M$ denotes the canonical complex structure induced on $T^*_x M$ by the holomorphic charts. The sequence of symbols is the cochain complex

$$0 \to \Lambda^{p,0} \otimes \Lambda^{0,0} \xrightarrow{1 \otimes (-1)^p \xi^{0,1} \wedge} \Lambda^{p,0} \otimes \Lambda^{p,0} \otimes \Lambda^{0,1} \xrightarrow{1 \otimes (-1)^p \xi^{0,1} \wedge} \cdots .$$

This complex is the ($\mathbb{Z}$-graded) tensor product of the trivial complex

$$0 \to \Lambda^{p,0} \xrightarrow{1} \Lambda^{p,0} \to 0,$$

with the Koszul complex

$$0 \to \Lambda^{0,0} \xrightarrow{(-1)^p \xi^{0,1} \wedge} \Lambda^{0,1} \xrightarrow{(-1)^p \xi^{0,1} \wedge} \cdots .$$

Since $\xi^{0,1} \neq 0$, for any $\xi \neq 0$, the Koszul complex is exact (see Subsection 7.1.3). This proves that the Dolbeault complex is elliptic. $\qquad \square$

To study the Dirac nature of this complex we need to introduce a Hermitian metric h on TM. Its real part is a Riemann metric g on M, and the canonical almost complex structure on TM is a skew-symmetric endomorphism with respect to this real metric. The associated 2-form $\Omega_h = -\boldsymbol{Im}\, h$ is nondegenerate in the sense that Ω^n ($n = \dim_{\mathbb{C}} M$) is a volume form on M. According to the results of Subsection 2.2.5 the orientation of M defined by Ω^n coincides with the orientation induced by the complex structure.

We form the *Hodge-Dolbeault operator*

$$\bar{\partial} + \bar{\partial}^* : \Omega^{p,\bullet}(M) \to \Omega^{p,\bullet}(M).$$

Proposition 11.2.16. *The Hodge-Dolbeault operator $\sqrt{2}(\bar{\partial}+\bar{\partial}^*)$ is a Dirac operator.*

Proof. We need to show that

$$\left(\sigma(\bar{\partial})(\xi) - \sigma(\bar{\partial})(\xi)^* \right)^2 = -\frac{1}{2}|\xi|^2\mathbb{1}, \ \ \forall \xi \in T^*M.$$

Denote by J the canonical complex structure on T^*M and set $\eta = J\xi$. Note that $\xi \perp \eta$ and $|\xi| = |\eta|$. Then

$$\sigma(\bar{\partial})(\xi) = (-1)^p\frac{1}{2}e(\xi + i\eta) = \frac{1}{2}(e(\xi) + ie(\eta)),$$

where as usual $e(\cdot)$ denotes the (left) exterior multiplication. The adjoint of $\sigma(\bar{\partial})(\xi)$ is

$$(-1)^p\sigma(\bar{\partial})(\xi)^* = \frac{1}{2}\left(\xi^* \lrcorner -i\eta^*\lrcorner\right),$$

where ξ^* (respectively η^*) denotes the metric dual of ξ (respectively η), and $\xi^*\lrcorner$ denotes the contraction by ξ^*. We set

$$\boldsymbol{c}(\xi) := e(\xi) - \xi^*\lrcorner, \ \ \tilde{\boldsymbol{c}}(\eta) := e(\eta) + \eta^*\lrcorner,$$

and we deduce

$$\left(\sigma(\bar{\partial})(\xi) - \sigma(\bar{\partial})(\xi)^* \right)^2 = \frac{1}{4}\left\{ e(\xi) + ie(\eta) - \xi^* \lrcorner +i\eta^*\lrcorner \right\}^2$$

$$= \frac{1}{4}\left\{ \left(e(\xi) - \xi^*\lrcorner \right) + i\left(e(\eta) + \eta^*\lrcorner \right) \right\}^2 = \frac{1}{4}\left\{ \boldsymbol{c}(\xi) + i\tilde{\boldsymbol{c}}(\eta) \right\}^2$$

$$= \frac{1}{4}\left\{ \boldsymbol{c}(\xi)^2 - \tilde{\boldsymbol{c}}(\eta)^2 + i\left(\boldsymbol{c}(\xi)\tilde{\boldsymbol{c}}(\eta) + \boldsymbol{c}(\eta)\boldsymbol{c}(\xi) \right) \right\}.$$

Note that

$$\boldsymbol{c}(\xi)^2 = -\left(e(\xi)\xi^* \lrcorner +\xi^* \lrcorner e(\xi) \right) = -|\xi|^2,$$

and

$$\tilde{\boldsymbol{c}}(\eta)^2 = e(\eta)\eta^* \lrcorner +\eta^* \lrcorner e(\eta) = |\eta|^2.$$

On the other hand, since $\xi \perp \eta$, we deduce as above that

$$c(\xi)\tilde{c}(\eta) + \tilde{c}(\eta)c(\xi) = 0.$$

(Verify this!) Hence

$$\left\{ \sigma(\bar{\partial})(\xi) - \sigma(\bar{\partial})(\xi)^* \right\}^2 = -\frac{1}{4}(|\xi|^2 + |\eta|^2) = -\frac{1}{2}|\eta|^2. \qquad \square$$

A natural question arises as to when the above operator is a *geometric* Dirac operator. Note first that the Clifford multiplication is certainly skew-adjoint since it is the symbol of a formally selfadjoint operator. Thus all we need to inquire is when the Clifford multiplication is covariant constant. Since

$$c(\xi) = \frac{(-1)^p}{\sqrt{2}}\left(e(\xi) + ie(\eta) - \xi^* \lrcorner - i\eta^* \lrcorner \right),$$

where $\eta = J\xi$, we deduce the Clifford multiplication is covariant constant if $\nabla^g J = 0$.

Definition 11.2.17. Let M be a complex manifold, and h a Hermitian metric on TM (viewed as a complex bundle). Then h is said to be a *Kähler metric* if $\nabla J = 0$, where ∇ denotes the Levi-Civita connection associated to the Riemann metric *Re* h, and J is the canonical almost complex structure on TM. A pair (complex manifold, Kähler metric) is called a *Kähler manifold*. $\qquad \square$

Exercise 11.2.18. Let (M, J) be an almost complex manifold and h a Hermitian metric on TM. Let $\Omega = -\textbf{\textit{Im}}\, h$. Using the exercise 11.2.13 show that if $d\Omega_h = 0$, then the almost complex is integrable, and the metric h is Kähler.

Conversely, assuming that J is integrable, and h is Kähler show that $d\Omega = 0$. $\square$

We see that on a Kähler manifold the above Clifford multiplication is covariant constant. In fact, a more precise statement is true.

Proposition 11.2.19. *Let M be a complex manifold and h a Kähler metric on TM. Then the Levi-Civita induced connection on $\Lambda^{0,\bullet}T^*M$ is a Clifford connection with respect to the above Clifford multiplication, and moreover, the Hodge-Dolbeault operator $\sqrt{2}(\bar{\partial} + \bar{\partial}^*)$ is the geometric Dirac operator associated to this connection.* $\square$

Exercise 11.2.20. Prove the above proposition. $\qquad \square$

Example 11.2.21. Let (M, g) be an oriented 2-dimensional Riemann manifold (surface). The Hodge $*$-operator defines an endomorphism

$$* : TM \to TM,$$

satisfying $*^2 = -\mathbb{1}_{TM}$, i.e., the operator $*$ is an almost complex structure on M. Using the Exercise 11.2.18 we deduce this almost complex structure is integrable since, by dimensionality

$$d\Omega = 0,$$

where Ω is the natural 2-form $\Omega(X,Y) = g(*X,Y)$ $X,Y \in \text{Vect}\,(M)$. This complex structure is said to be canonically associated to the metric. $\qquad\square$

Example 11.2.22. Perhaps the favorite example of Kähler manifold is the complex projective space $\mathbb{C}^n$. To describe this structure consider the tautological line bundle $L_1 \to \mathbb{C}\mathbb{P}^n$. It can be naturally viewed as a subbundle of the trivial bundle $\mathbb{C}^{n+1} \to \mathbb{C}\mathbb{P}^n$.

Denote by h_0 the canonical Hermitian metric on $\mathbb{C}^{n+1}$, and by ∇^0 the trivial connection. If we denote by $P : \mathbb{C}^{n+1} \to L_1$ the orthogonal projection, then $\nabla = P \circ \nabla^0|_{L_1}$ defines a connection on L_1 compatible with $h_1 = h|_{L_1}$. Denote by ω the first Chern form associated to this connection

$$\omega = \frac{i}{2\pi}F(\nabla),$$

and set

$$h_{FS}(X,Y)_x := -\omega_x(X, JY) + i\omega(X,Y) \quad \forall x \in M,\ X,Y \in T_xM.$$

Then h_{FS} is a Hermitian metric on $\mathbb{C}\mathbb{P}^n$ (verify!) called the *Fubini-Study* metric. It is clearly a Kähler metric since (see Exercise 11.2.18) $d\Omega_h = -d\omega = -dc_1(\nabla) = 0$. $\qquad\square$

Exercise 11.2.23. Describe h_{FS} in projective coordinates, and then prove that h is indeed a Hermitian metric, i.e. it is positive definite. $\qquad\square$

Remark 11.2.24. Any complex submanifold of a Kähler manifold is obviously Kähler. In particular, any complex submanifold of $\mathbb{C}\mathbb{P}^n$ is automatically Kähler.

A celebrated result of Chow states that any complex submanifold of $\mathbb{C}\mathbb{P}^n$ is automatically algebraic, i.e., it can be defined as the zero set of a family of homogeneous polynomials. Thus, all complex nonsingular algebraic varieties admit a natural Kähler structure. It is thus natural to ask whether there exist Kähler manifolds which are not algebraic.

The answer is positive, and a very thorough resolution of this problem is contained in the famous Kodaira embedding theorem which provides a simple necessary and sufficient condition for a compact complex manifold to be algebraic.

For this work, Kodaira was awarded the Fields medal in 1954. His proofs rely essentially on some vanishing results deduced from the Weitzenböck formulæ for the Dolbeault operator $\bar{\partial} + \bar{\partial}^*$, and its twisted versions. A very clear presentation of this subject can be found in the beautiful monograph [40]. $\qquad\square$

11.2.3 *The spin Dirac operator*

Like the Dolbeault operator, the *spin* Dirac operator exists only on manifolds with a bit of extra structure. We will first describe this new structure.

Let (M^n, g) be an n-dimensional, oriented Riemann manifold. In other words, the tangent bundle TM admits an $SO(n)$ structure so that it can be defined by an open cover (U_α), and transition maps

$$g_{\alpha\beta} : U_{\alpha\beta} \to SO(n)$$

satisfying the cocycle condition.

The manifold is said to *spinnable* if there exist smooth maps

$$\tilde{g}_{\alpha\beta} : U_{\alpha\beta} \to \mathrm{Spin}(n)$$

satisfying the cocycle condition, and such that

$$\rho(\tilde{g}_{\alpha\beta}) = g_{\alpha\beta} \ \forall \alpha, \beta,$$

where $\rho : \mathrm{Spin}(n) \to SO(n)$ denotes the canonical double cover. The collection $\tilde{g}_{\alpha\beta}$ as above is called a *spin structure*. A pair (manifold, spin structure) is called a *spin manifold*.

Not all manifolds are spinnable. To understand what can go wrong, let us start with a trivializing cover $\mathcal{U} = (U_\alpha)$ for TM, with transition maps $g_{\alpha\beta}$, and such that all the multiple intersection $U_{\alpha\beta\cdots\gamma}$ are contractible. In other words, $\mathcal{U}$ is a good cover.

Since each of the overlaps $U_{\alpha\beta}$ is contractible, each map $g_{\alpha\beta} : U_{\alpha\beta} \to SO(n)$ admits at least one lift

$$\tilde{g}_{\alpha\beta} : U_{\alpha\beta} \to \mathrm{Spin}(n).$$

From the equality $\rho(\tilde{g}_{\alpha\beta}\tilde{g}_{\beta\gamma}\tilde{g}_{\gamma\alpha}) = g_{\alpha\beta}g_{\beta\gamma}g_{\gamma\alpha} = \mathbb{1}$ we deduce

$$\epsilon_{\alpha\beta\gamma} = \tilde{g}_{\alpha\beta}\tilde{g}_{\beta\gamma}\tilde{g}_{\gamma\alpha} \in \ker\rho = \mathbb{Z}_2.$$

Thus, any lift of the gluing data $g_{\alpha\beta}$ to $\mathrm{Spin}(n)$ produces a degree 2 Čech cochain of the trivial sheaf $\mathbb{Z}_2$, namely the 2-cochain

$$(\epsilon_\bullet) : \quad U_{\alpha\beta\gamma} \mapsto \epsilon_{\alpha\beta\gamma}.$$

Note that for any $\alpha, \beta, \gamma, \delta$ such that $U_{\alpha\beta\gamma\delta} \neq \emptyset$, we have

$$\epsilon_{\beta\gamma\delta} - \epsilon_{\alpha\gamma\delta} + \epsilon_{\alpha\beta\delta} - \epsilon_{\alpha\beta\gamma} = 0 \in \mathbb{Z}_2.$$

In other words, $\epsilon_\bullet$ defines a Čech 2-cocycle, and thus defines an element in the Čech cohomology group $H^2(M, \mathbb{Z}_2)$.

It is not difficult to see that this element is independent of the various choices: the cover $\mathcal{U}$, the gluing data $g_{\alpha\beta}$, and the lifts $\tilde{g}_{\alpha\beta}$. This element is intrinsic to the tangent bundle TM. It is called the *second Stiefel-Whitney class* of M, and it is denoted by $w_2(M)$. We see that if $w_2(M) \neq 0$ then M cannot admit a spin structure. In fact, the converse is also true.

Proposition 11.2.25. *An oriented Riemann manifold M admits a spin structure if and only if $w_2(M) = 0$.* $\qquad\square$

Exercise 11.2.26. Prove the above result. □

Remark 11.2.27. The usefulness of the above proposition depends strongly on the ability of computing w_2. This is a good news/bad news situation. The good news is that algebraic topology has produced very efficient tools for doing this. The bad news is that we will not mention them since it would lead us far astray. See [62] and [75] for more details. □

Remark 11.2.28. The definition of isomorphism of spin structures is rather subtle (see [72]). More precisely, two spin structures defined by the cocycles $\tilde{g}_{\bullet\bullet}$ and $\tilde{h}_{\bullet\bullet}$ are isomorphic if there exists a collection $\varepsilon_\alpha \in \mathbb{Z}_2 \subset \mathrm{Spin}(n)$ such that the diagram below is commutative for all $x \in U_{\alpha\beta}$

$$\begin{array}{ccc} \mathrm{Spin}(n) & \xrightarrow{\varepsilon_\alpha} & \mathrm{Spin}(n) \\ \downarrow{\scriptstyle \tilde{g}_{\beta\alpha}(x)} & & \downarrow{\scriptstyle \tilde{h}_{\beta\alpha}(x)} \\ \mathrm{Spin}(n) & \xrightarrow{\varepsilon_\beta} & \mathrm{Spin}(n) \end{array}$$

The group $H^1(M, \mathbb{Z}_2)$ acts on $\mathrm{Spin}(M)$ as follows. Take an element $\varepsilon \in H^1(M, \mathbb{Z}_2)$ represented by a Čech cocycle, i.e., a collection of *continuous maps* $\varepsilon_{\alpha\beta} : U_{\alpha\beta} \to \mathbb{Z}_2 \subset \mathrm{Spin}(n)$ satisfying the cocycle condition

$$\varepsilon_{\alpha\beta} \cdot \varepsilon_{\beta\gamma} \cdot \varepsilon_{\gamma\alpha} = 1.$$

Then the collection $\varepsilon_{\bullet\bullet} \cdot \tilde{g}_{\bullet\bullet}$ is a $\mathrm{Spin}(n)$ gluing cocycle defining a spin structure that we denote by $\varepsilon \cdot \sigma$.

It is easy to check that the isomorphism class of $\varepsilon \cdot \sigma$ is independent of the various choice, i.e, the Čech representatives for ε and σ. Clearly, the correspondence

$$H^1(M, \mathbb{Z}_2) \times \mathrm{Spin}(M) \ni (\varepsilon, \sigma) \mapsto \varepsilon \cdot \sigma \in \mathrm{Spin}(M)$$

defines a left action of $H^1(M, \mathbb{Z}_2)$ on $\mathrm{Spin}(M)$. This action is transitive and free. □

Exercise 11.2.29. Prove the statements in the above remark. □

Exercise 11.2.30. Describe the only two spin structures on S^1. □

Example 11.2.31. (a) A simply connected Riemann manifold M of dimension ≥ 5 is spinnable if and only if every compact orientable surface embedded in M has trivial normal bundle.
(b) A simply connected four-manifold M is spinnable if and only if the normal bundle N_Σ of any embedded compact, orientable surface Σ has even Euler class, i.e.,

$$\int_\Sigma e(N_\Sigma)$$

is an even integer.

(c) Any compact oriented surface is spinnable. Any sphere S^n admits a unique spin structure. The product of two spinnable manifolds is canonically a spinnable manifold.

(d) $w_2(\mathbb{RP}^n) = 0$ if and only if $n \equiv 3 \,(\mathrm{mod}\, 4)$, while $\mathbb{CP}^n$ admits spin structures if and only if n is odd. $\qquad\qquad\qquad\qquad\qquad\qquad\qquad\qquad\qquad\qquad\qquad\qquad$ $\square$

Let (M^n, g) be a spin manifold. Assume the tangent bundle TM is defined by the open cover (U_α), and transition maps

$$g_{\alpha\beta} : U_{\alpha\beta} \to SO(n).$$

Moreover assume the spin structure is given by the lifts

$$\tilde{g}_{\alpha\beta} : U_{\alpha\beta} \to \mathrm{Spin}(n).$$

As usual, we regard the collection $g_{\alpha\beta}$ as defining the principal $SO(n)$-bundle of oriented frames of TM. We call this bundle $P_{SO(M)}$.

The collection $\tilde{g}_{\alpha\beta}$ defines a principal $\mathrm{Spin}(n)$-bundle which we denote by $P_{\mathrm{Spin}(M)}$. We can regard $P_{SO(M)}$ as a bundle associated to $P_{\mathrm{Spin}(M)}$ via $\rho : \mathrm{Spin}(n) \to SO(n)$. Using the unitary spinorial representation

$$\Delta_n : \mathrm{Spin}(n) \to \mathrm{Aut}\,(\mathbb{S}_n)$$

we get a Hermitian vector bundle

$$\mathbb{S}(M) = P_{\mathrm{Spin}(M)} \times_{\Delta_n} \mathbb{S}_n$$

called the *complex spinor bundle*. Its sections are called (complex) spinors. Let us pont out that when M is even dimensional, the spinor bundle is equipped with a natural $\mathbb{Z}/2$-grading

$$\mathbb{S}(M) = \mathbb{S}^+(M) \oplus \mathbb{S}^-(M).$$

From Lemma 11.1.31 we deduce that $\mathbb{S}(M)$ is naturally a bundle of $\mathrm{Cl}(TM)$ modules. Then, the natural isomorphism $\mathrm{Cl}(TM) \cong \mathrm{Cl}(T^*M)$ induces on $\mathbb{S}(M)$ a structure of $\mathrm{Cl}(T^*M)$-module,

$$\boldsymbol{c} : \mathrm{Cl}(TM) \to \mathrm{End}_{\mathbb{C}}(\,\mathbb{S}(M)\,).$$

Moreover, when M is even dimensional, the above morphism is compatible with the $\mathbb{Z}/2$-gradings of $\mathrm{Cl}(TM)$ and $\mathrm{End}_{\mathbb{C}}(\,\mathbb{S}(M)\,)$ As it turns out, the bundle $\mathbb{S}(M)$ has a natural Dirac structure whose associated Dirac operator is the spin Dirac operator on M. We will denote it by $\mathfrak{D}$.

Since Δ_n is a unitary representation, we can equip the spinor bundle $\mathbb{S}(M)$ with a natural metric with respect to which the Clifford multiplication

$$\boldsymbol{c} : \mathrm{Cl}(TM) \to \mathrm{End}_{\mathbb{C}}(\,\mathbb{S}(M)\,)$$

is self-adjoint, i.e.,

$$\boldsymbol{c}(u^\dagger) = \boldsymbol{c}(u)^*, \quad \forall u \in C^\infty(\mathrm{Cl}(TM)).$$

Since $\mathrm{Cl}(TM)$ is locally generated as an algebra by $\mathrm{Vect}\,(M)$, we can equivalently rewrite the selfadjointness condition as

$$\boldsymbol{c}(X) = -\boldsymbol{c}(X)^{*}, \quad \forall X \in \mathrm{Vect}\,(M) \subset C^{\infty}(\,\mathrm{Cl}(T)\,).$$

All we now need is to describe a natural connection on $\mathbb{S}(M)$ with respect to which the Clifford multiplication is covariant constant.

We start with the Levi-Civita connection ∇^{g} which we can regard as a connection on the principal bundle $P_{SO(M)}$. Alternatively, ∇^{g} can be defined by a collection of $\underline{so}(n)$-valued 1-forms $\omega_{\alpha} \in \Omega^{1}(U_{\alpha}) \otimes \underline{so}(n)$ such that

$$\omega_{\beta} = g_{\alpha\beta}^{-1} dg_{\alpha\beta} + g_{\alpha\beta}^{-1} \omega_{\alpha} g_{\alpha\beta} \quad \text{on } U_{\alpha\beta}.$$

Consider the canonical isomorphism of Lie algebras

$$\rho_{*} : \underline{\mathrm{spin}}(n) \rightarrow \underline{so}(n).$$

Then the collection $\tilde{\omega}_{\alpha} = \rho_{*}^{-1}(\omega_{\alpha})$ defines a connection $\hat{\nabla}$ on the principal bundle $P_{\mathrm{Spin}(M)}$, and thus via the representation Δ_{n} it defines a connection $\nabla = \nabla^{\mathbb{S}}$ on the spinor bundle $\mathbb{S}(M)$.

The above construction can be better visualized if we work in local coordinates. Choose a local, oriented orthonormal frame (e_{i}) of $TM\,|_{U_{\alpha}}$, and denote by (e^{j}) is dual coframe. The Levi-Civita connection has the form

$$\nabla e_{j} = e^{k} \otimes \omega_{kj}^{i} e_{i},$$

so that $\omega_{\alpha} = e^{k} \otimes (\omega_{ki}^{j})$, where for each k the collection $(\omega_{ki}^{j})_{i,j}$ is a skew-symmetric matrix.

Using the concrete description of ρ_{*}^{-1} given in (11.1.3) we deduce

$$\tilde{\omega}_{\alpha} = -\sum_{k} e^{k} \otimes \left(\frac{1}{2} \sum_{i<j} \omega_{kj}^{i} e_{i} e_{j} \right) = -\frac{1}{4} \sum_{i,j,k} \omega_{kj}^{i} e^{k} \otimes e_{i} e_{j}.$$

Lemma 11.2.32. *The connection $\hat{\nabla}$ is a Clifford connection on $\mathbb{S}(M)$ so that $(\mathbb{S}(M), \hat{\nabla})$ is a Dirac bundle called the bundle of complex spinors.*

Proof. Use Lemma 11.1.31. $\square$

We denote by $\mathfrak{D} = \mathfrak{D}_{M}$ the geometric Dirac operator associated to the bundle of complex spinors $\mathbb{S}$. When the manifold M is even dimensional, this operator is graded. We recall that this means that, with respect to the decomposition $\mathbb{S} = \mathbb{S}^{+} \oplus \mathbb{S}^{-}$, the Dirac operator $\mathfrak{D}$ has the block form

$$\mathfrak{D} = \begin{bmatrix} 0 & \mathfrak{D}^{*} \\ \mathfrak{D} & 0 \end{bmatrix},$$

where $\mathfrak{D} : C^{\infty}(\mathbb{S}^{+}) \rightarrow C^{\infty}(\mathbb{S}^{-})$ is a first order operator whose symbol is given by the Clifford multiplication.

We now want to understand the structure of the Weitzenböck remainder of the geometric Dirac operator $\mathfrak{D}$. If

$$R = \sum_{k<\ell} e^k \wedge e^\ell R_{k\ell} \quad R_{k\ell} = (R^i_{jk\ell}) = (R_{ijk\ell})$$

is the Riemann curvature tensor, (in the above trivializations over U_α) we deduce that the curvature of $\hat{\nabla}$ is

$$\tilde{R} = \sum_{k<\ell} e^k \wedge e^\ell \otimes \rho_*^{-1}(R_{k\ell}) = -\frac{1}{4} \sum_{k<\ell} \sum_{ij} e^k \wedge e^\ell \otimes R_{ijk\ell} e_i e_j.$$

From this we obtain

$$\boldsymbol{c}(F(\hat{\nabla})) = -\frac{1}{8} \sum_{ijk\ell} R_{ijk\ell} \boldsymbol{c}(e^i) \boldsymbol{c}(e^j) \boldsymbol{c}(e^k) \boldsymbol{c}(e^\ell).$$

In the above sum, the terms corresponding to indices (i, j, k, ℓ) such that $i = j$ or $k = \ell$ vanish due to the corresponding skew-symmetry of the Riemann tensor. Thus, we can write

$$\boldsymbol{c}(F(\hat{\nabla})) = -\frac{1}{8} \sum_{i \neq j} \sum_{k \neq \ell} R_{ijk\ell} \boldsymbol{c}(e^i) \boldsymbol{c}(e^j) \boldsymbol{c}(e^k) \boldsymbol{c}(e^\ell)$$

Using the equalities $\boldsymbol{c}(e^i)\boldsymbol{c}(e^j) + \boldsymbol{c}(e^j)\boldsymbol{c}(e^i) = -2\delta_{ij}$ we deduce that the monomial $\boldsymbol{c}(e^i)\boldsymbol{c}(e^j)$ anti-commutes with $\boldsymbol{c}(e^k)\boldsymbol{c}(e^\ell)$ if the two sets $\{i, j\}$ and $\{k, \ell\}$ have a unique element in common. Such pairs of monomials will have no contributions in the above sum due to the curvature symmetry

$$R_{ijk\ell} = R_{k\ell ij}.$$

Thus, we can split the above sum into two parts

$$\boldsymbol{c}(F(\hat{\nabla})) = -\frac{1}{4} \sum_{i,j} R_{ijij} \boldsymbol{c}(e^i) \boldsymbol{c}(e^j) \boldsymbol{c}(e^i) \boldsymbol{c}(e^j) + \sum_{i,j,k,\ell \text{ distinct}} R_{ijk\ell} \boldsymbol{c}(e^i) \boldsymbol{c}(e^j) \boldsymbol{c}(e^k) \boldsymbol{c}(e^\ell).$$

Using the first Bianchi identity we deduce that the second sum vanishes. The first sum is equal to

$$-\frac{1}{4} \sum_{i,j} R_{ijij} (\boldsymbol{c}(e^i) \boldsymbol{c}(e^j))^2 = \frac{1}{4} \sum_{i,j} R_{ijij} = \frac{s}{4},$$

where s denotes the scalar curvature of M. We have thus proved the following result, [64].

Theorem 11.2.33 (Lichnerowicz).

$$\mathfrak{D}^2 = \nabla^* \nabla + \frac{1}{4} s. \qquad \square$$

A section $\psi \in C^\infty(\mathbb{S}(M))$ such that $\mathfrak{D}^2\psi = 0$ is called a *harmonic spinor*. Lichnerowicz theorem shows that a compact spin manifold with positive scalar curvature admits no harmonic spinors.

Exercise 11.2.34. Consider an oriented 4-dimensional Riemann spin manifold (M, g) and $W \to M$ a Hermitian vector bundle equipped with a Hermitian connection ∇. Form the twisted Dirac operator

$$\mathfrak{D}_W : C^\infty \left((\mathbb{S}^+(M) \oplus \mathbb{S}^-(M)) \otimes W \right) \to C^\infty \left((\mathbb{S}^+(M) \oplus \mathbb{S}^-(M)) \otimes W \right)$$

defined in Exercise 11.1.68. The operator $\mathfrak{D}_W$ is $\mathbb{Z}_2$-graded, and hence it has a block decomposition

$$\mathfrak{D}_W = \begin{bmatrix} 0 & \mathcal{D}^*_{W,+} \\ \mathcal{D}_{W,+} & 0 \end{bmatrix},$$

where $\mathcal{D}_{W,+} : C^\infty(\mathbb{S}^+(M) \otimes W) \to C^\infty(\mathbb{S}^-(M) \otimes W)$. Show that

$$\mathcal{D}^*_{W,+}\mathcal{D}_{W,+} = \tilde{\nabla}^*\tilde{\nabla} + \frac{s}{4} + c\big(F_+(\nabla) \big),$$

where $\tilde{\nabla}$ denotes the product connection of $\mathbb{S}(M) \otimes W$, and

$$F_+(\nabla) = \frac{1}{2}\big(F(\nabla) + *F(\nabla) \big),$$

denotes the self-dual part of the curvature of the bundle (W, ∇).   □

Exercise 11.2.35. The torus $T^3 = S^1 \times S^1 \times S^1$ equipped with the product metric (of volume $(2\pi)^3$) has eight spin structures. Since T^3 is also a Lie group, its tangent bundle admits a canonical trivialization. We denote by σ_0 the spin structure determined by this trivialization. Using the free and transitive action of $H^1(T^3, \mathbb{Z}_2)$ on $\mathrm{Spin}(T^3)$ we deduce that we have a canonical bijection

$$H^1(T^3, \mathbb{Z}_2) \to \mathrm{Spin}(T^3), \quad H^1(T^3, \mathbb{Z}_2) \ni \epsilon \mapsto \sigma_\epsilon := \epsilon \cdot \sigma_0.$$

Compute the spectrum of the spin Dirac operator $\mathfrak{D}_\epsilon$ determined by the spin structure σ_ϵ.   □

11.2.4   *The spin^c Dirac operator*

Our last example of Dirac operator generalizes both the spin Dirac operator, and the Hodge-Dolbeault operator. The common ingredient behind both these examples is the notion of spin^c structure. We begin by introducing it to the reader.

Let (M^n, g) be an oriented, n-dimensional Riemann manifold. As in the previous section we can regard the tangent bundle as associated to the principal bundle $P_{SO(M)}$ of oriented orthonormal frames. Assume $P_{SO(M)}$ is defined by a good open cover $\mathcal{U} = (U_\alpha)$ and transition maps

$$g_{\alpha\beta} : U_{\alpha\beta} \to SO(n).$$

The manifold M is said to posses a spin^c structure (or complex spin structure) if there exists a principal Spin$^c(n)$ -bundle P_{Spin^c} such that $P_{SO(M)}$ is associated to P_{Spin^c} via the natural morphism $\rho^c : \text{Spin}^c(n) \to SO(n)$:

$$P_{SO(M)} = P_{\text{Spin}^c} \times_{\rho^c} SO(n).$$

Equivalently, this means there exist smooth maps $\tilde{g}_{\alpha\beta} : U_{\alpha\beta} \to \text{Spin}^c(n)$, satisfying the cocycle condition, such that

$$\rho^c(\tilde{g}_{\alpha\beta}) = g_{\alpha\beta}.$$

As for *spin* structures, there are obstructions to the existence spin^c structures, but they are less restrictive. Let us try to understand what can go wrong. We stick to the assumption that all the overlaps $U_{\alpha\beta\cdots\gamma}$ are contractible.

Since $\text{Spin}^c(n) = (\text{Spin}(n) \times S^1)/\mathbb{Z}_2$, lifting the $SO(n)$ structure $(g_{\alpha\beta})$ reduces to finding smooth maps

$$h_{\alpha\beta} : U_{\alpha\beta} \to \text{Spin}(n) \text{ and } z_{\alpha\beta} : U_{\alpha\beta} \to S^1,$$

such that

$$\rho(h_{\alpha\beta}) = g_{\alpha\beta},$$

and

$$(\epsilon_{\alpha\beta\gamma}, \zeta_{\alpha\beta\gamma}) := \left(h_{\alpha\beta}h_{\beta\gamma}h_{\gamma\alpha}, z_{\alpha\beta}z_{\beta\gamma}z_{\gamma\alpha} \right) \in \left\{ (-1,-1), (1,1) \right\}. \tag{11.2.4}$$

If we set $\lambda_{\alpha\beta} := z_{\alpha\beta}^2 : U_{\alpha\beta} \to S^1$, we deduce from (11.2.4) that the collection $(\lambda_{\alpha\beta})$ should satisfy the cocycle condition. In particular, it defines a principal S^1-bundle over M, or equivalently, a complex Hermitian line bundle[2] $\mathcal{L}$. This line bundle should be considered as part of the data defining a spin^c structure.

The collection $(\epsilon_{\alpha\beta\gamma})$ is an old acquaintance: it is a Čech 2-cocycle representing the second Stieffel-Whitney class.

As in Subsection 8.2.2, we can represent the cocycle $\lambda_{\alpha\beta}$ as

$$\lambda_{\alpha\beta} = \exp(i\theta_{\alpha\beta}).$$

The collection

$$n_{\alpha\beta\gamma} = \frac{1}{2\pi} \left(\theta_{\alpha\beta} + \theta_{\beta\gamma} + \theta_{\gamma\alpha} \right),$$

defines a 2-cocycle of the constant sheaf $\mathbb{Z}$ representing the topological first Chern class of $\mathcal{L}$. The condition (11.2.4) shows that

$$n_{\alpha\beta\gamma} = \epsilon_{\alpha\beta\gamma} \pmod 2.$$

To summarize, we see that the existence of a spin^c structure implies the existence of a complex Hermitian line bundle $\mathcal{L}$ such that

$$c_1^{top}(\mathcal{L}) = w_2(M) \pmod 2.$$

---

[2] In this subsection, by complex Hermitian line bundle we understand a complex line bundle equipped with a $U(1)$-structure.

It is not difficult to prove that the above condition is also sufficient. In fact, one can be more precise.

Denote by $\mathrm{Spin}^c(M)$ the collection of isomorphism classes of spin^c structures on the manifold M. Any $\sigma \in \mathrm{Spin}^c(M)$ is defined by a lift $(h_{\alpha\beta}, z_{\alpha\beta})$ as above. We denote by $\det\sigma$ the complex Hermitian line bundle defined by the gluing data $(z_{\alpha\beta})$. We have seen that

$$c_1^{top}(\det\sigma) \equiv w_2(M) \ (\mathrm{mod}\ 2).$$

Denote by $\mathcal{L}_M \subset H^2(M, \mathbb{Z})$ the "affine" subspace consisting of those cohomology classes satisfying the above congruence modulo 2. We thus have a map

$$\mathrm{Spin}^c(M) \to \mathcal{L}_M, \quad \sigma \mapsto c_1^{top}(\det\sigma).$$

Proposition 11.2.36. *The above map is a surjection.*   □

Exercise 11.2.37. Complete the proof of the above proposition.   □

The smooth Picard group $\mathrm{Pic}^\infty(M)$ of isomorphisms classes of complex line bundles (with group operation given by the tensor product) acts on $\mathrm{Spin}^c(M)$ by

$$\mathrm{Spin}^c(M) \times \mathrm{Pic}^\infty(M) \ni (\sigma, L) \mapsto \sigma \otimes L.$$

More precisely, if $\sigma \in \mathrm{Spin}^c(M)$ is given by the cocycle

$$\sigma = [h_{\alpha\beta}, z_{\alpha\beta}] : U_{\alpha\beta} \to \mathrm{Spin}(n) \times S^1 / \sim,$$

and L is given by the S^1 cocycle $\zeta_{\alpha\beta} : U_{\alpha\beta} \to S^1$, then $\sigma \otimes L$ is given by the cocycle $[h_{\alpha\beta}, z_{\alpha\beta}\zeta_{\alpha\beta}]$. Note that

$$\det(\sigma \otimes L) = \det\sigma \otimes L^2,$$

so that

$$c_1^{top}\big(\det(\sigma \otimes L)\big) = c_1^{top}(\det\sigma) + 2c_1^{top}(L).$$

Proposition 11.2.38. *The above action of $\mathrm{Pic}^\infty(M)$ on $\mathrm{Spin}^c(M)$ is free and transitive.*

Proof. Consider two spin^c structures σ^1 and σ^2 defined by the good cover (U_α), and the gluing cocycles

$$[h_{\alpha\beta}^{(i)}, z_{\alpha\beta}^{(i)}], \quad i = 1, 2.$$

Since $\rho^c(h_{\alpha\beta}^{(1)}) = \rho^c(h_{\alpha\beta}^{(2)}) = g_{\alpha\beta}$, we can assume (eventually modifying the maps $h_{\alpha\beta}^{(2)}$ by a sign) that

$$h_{\alpha\beta}^{(1)} = h_{\alpha\beta}^{(2)}.$$

This implies that the collection

$$\zeta_{\alpha\beta} = z_{\alpha\beta}^{(2)} / z_{\alpha\beta}^{(1)}$$

is an S^1-cocycle defining a complex Hermitian line bundle L. Obviously $\sigma^2 = \sigma^1 \otimes L$. This shows the action of $\mathrm{Pic}^\infty(M)$ is transitive. We leave the reader verify this action is indeed free. The proposition is proved. $\qquad\qquad\square$

Given two spin^c structures σ_1 and σ_2 we can define their "difference" σ_2/σ_1 as the unique complex Hermitian line bundle L such that $\sigma_2 = \sigma_1 \otimes L$. This shows that the collection of spin^c structures is (non-canonically) isomorphic with $H^2(M, \mathbb{Z}) \cong \mathrm{Pic}^\infty(M)$.

It is a sort of affine space modelled on $H^2(M, \mathbb{Z})$ in the sense that the "difference " between two spin^c structures is an element in $H^2(M, \mathbb{Z})$, but there is no distinguished origin of this space. A structure as above is usually called a $H^2(M, \mathbb{Z})$-*torsor*.

The set $\mathrm{Spin}^c(M)$ is equipped with a natural involution

$$\mathrm{Spin}^c(M) \ni \sigma \mapsto \bar{\sigma} \in \mathrm{Spin}^c(M),$$

defined as follows. If σ is defined by the cocycle $[\tilde{g}_{\alpha\beta}, z_{\alpha\beta}]$, then $\bar{\sigma}$ is defined by the cocycle $[\tilde{g}_{\alpha\beta}, z_{\alpha\beta}{}^{-1}]$. Observe that

$$\sigma/\bar{\sigma} = \det \sigma.$$

Without sufficient background in algebraic topology the above results may look of very little help in detecting spin^c structures. This is not the case, and to convince the reader we will list below (without proofs) some examples of spin^c manifolds.

Example 11.2.39. (a) Any *spin* manifold admits a spin^c structure.
(b) Any almost complex manifold has a natural spin^c structure.
(c) (Hirzebruch-Hopf, [47]; see also [79]) Any oriented manifold of dimension ≤ 4 admits a spin^c structure. $\qquad\qquad\square$

Let us analyze the first two examples above. If M is a spin manifold, then the lift

$$\tilde{g}_{\alpha\beta} : U_{\alpha\beta} \to \mathrm{Spin}(n)$$

of the SO-structure to a spin structure canonically defines a spin^c structure via the trivial morphism

$$\mathrm{Spin}(n) \to \mathrm{Spin}^c(n) \times_{\mathbb{Z}_2} S^1, \quad g \mapsto (g, 1) \text{ mod the } \mathbb{Z}_2\text{-action.}$$

We see that in this case the associated complex line bundle is the trivial bundle. This is called the *canonical* spin^c structure of a spin manifold. We thus have a map

$$\mathrm{Spin}(M) \to \mathrm{Spin}^c(M).$$

Suppose we have fixed a spin structure on M given by a Spin-lift $\tilde{g}_{\beta\alpha}$ of an SO-gluing cocycle $g_{\beta\alpha}$.

To any complex Hermitian line bundle L defined by the S^1-cocycle $(z_{\alpha\beta})$ we can associate the spin^c structure $\sigma \otimes L$ defined by the gluing data

$$\left\{ (\tilde{g}_{\alpha\beta}, z_{\alpha\beta}) \right\}.$$

The complex Hermitian line bundle $\det \sigma \otimes L$ associated to this structure is $\det \sigma \otimes L = L^{\otimes 2}$. Since the topological Picard group $\mathrm{Pic}^{\infty}(M)$ acts freely and transitively on $\mathrm{Spin}^c(M)$, we deduce that to any pair $(\epsilon, \sigma) \in \mathrm{Spin}(M) \times \mathrm{Spin}^c(M)$ we can canonically associate a complex line bundle $L = L_{\epsilon, \sigma}$ such that $L^{\otimes 2} \cong \det \sigma$, i.e., $L_{\epsilon, \sigma}$ is a "square root of $\det \sigma$.

Exercise 11.2.40. Show that for any $\sigma \in \mathrm{Spin}^c(M)$ there exists a natural bijection between the set $\mathrm{Spin}(M)$, and the set of isomorphisms of complex line bundles L such that $L^{\otimes 2} \cong \det \sigma$. $\square$

Exercise 11.2.41. Prove that the image of the natural map $\mathrm{Spin}(M) \to \mathrm{Spin}^c(M)$ coincides with the fixed point set of the involution $\sigma \mapsto \bar{\sigma}$. $\square$

Exercise 11.2.42. The torus T^3 is the base of a principal bundle

$$\mathbb{Z}^3 \hookrightarrow \mathbb{R}^3 \to T^3,$$

so that, to any group morphism $\rho : \mathbb{Z}^3 \to U(1)$ we can associate a complex line bundle $L_\rho \to T^3$.
(a) Prove that $L_{\rho_1 \cdot \rho_2} \cong L_{\rho_1} \otimes L_{\rho_2}$, $\forall \rho_1, \rho_1 \in \mathrm{Hom}(\mathbb{Z}^3, S^1)$, and use this fact to describe the first Chern class of L_ρ by explicitly producing a Hermitian connection on L_ρ.
(b) Prove that any complex line bundle on T^3 is isomorphic to a line bundle of the form L_ρ, for some morphism $\rho \in \mathrm{Hom}(\mathbb{Z}^3, S^1)$.
(c) Show that the image of the map $\mathrm{Spin}(T^3) \to \mathrm{Spin}^c(T^3)$ consists of a *single point*. $\square$

Exercise 11.2.43. Prove that $\mathrm{Spin}(\mathbb{RP}^3)$ consists of precisely two isomorphisms classes of spin structures and moreover, the natural map

$$\mathrm{Spin}(\mathbb{RP}^3) \to \mathrm{Spin}^c(\mathbb{RP}^3)$$

is a bijection. $\square$

To understand why an almost complex manifold admits a canonical spin^c structure it suffices to recall the natural morphism $U(k) \to SO(2k)$ factors through a morphism

$$\xi : U(k) \to \mathrm{Spin}^c(2k).$$

The $U(k)$-structure of TM, defined by the gluing data

$$h_{\alpha\beta} : U_{\alpha\beta} \to U(k),$$

induces a spin^c structure defined by the gluing data $\xi(h_{\alpha\beta})$. Its associated line bundle is given by the S^1-cocycle

$$\det_{\mathbb{C}}(h_{\alpha\beta}) : U_{\alpha\beta} \to S^1,$$

and it is precisely the determinant line bundle

$$\det{}_{\mathbb{C}} T^{1,0} M = \Lambda^{k,0} T M.$$

The dual of this line bundle, $\det_{\mathbb{C}} (T^* M)^{1,0} = \Lambda^{k,0} T^* M$ plays a special role in algebraic geometry. It usually denoted by K_M, and it is called the *canonical line bundle*. Thus the line bundle associated to this spin^c structure is $K_M^{-1} \stackrel{def}{=} K_M^*$.

From the considerations in Subsection 11.1.5 and 11.1.7 we see that many (complex) vector bundles associated to the principal Spin^c bundle of a spin^c manifold carry natural Clifford structures, and in particular, one can speak of Dirac operators. We want to discuss in some detail a very important special case.

Assume that (M, g) is an oriented, n-dimensional Riemann manifold. Fix $\sigma \in$ Spin$^c(M)$ (assuming there exist spin^c structures). Denote by $(g_{\alpha\beta})$ a collection of gluing data defining the SO structure $P_{SO(M)}$ on M with respect to some good open cover (U_α). Moreover, we assume σ is defined by the data

$$h_{\alpha\beta} : U_{\alpha\beta} \to \text{Spin}^c(n).$$

Denote by Δ_n^c the fundamental complex spinorial representation defined in Subsection 11.1.6,

$$\Delta_n : \text{Spin}^c(n) \to \text{Aut}(\mathbb{S}_n).$$

We obtain a complex Hermitian vector bundle

$$\mathbb{S}_\sigma(M) = P_{\text{Spin}^c} \times_{\Delta_n} \mathbb{S}_n,$$

which has a natural Clifford structure. This is called the bundle of complex spinors associated to σ.

Example 11.2.44. (a) Assume M is a spin manifold. We denote by σ_0 the spin^c structure corresponding to the fixed spin structure. The corresponding bundle of spinors $\mathbb{S}_0(M)$ coincides with the bundle of pure spinors defined in the previous section. Moreover for any complex line bundle L we have

$$\mathbb{S}_L := \mathbb{S}_\sigma \cong \mathbb{S}_0 \otimes L,$$

where $\sigma = \sigma_0 \otimes L$. Note that in this case $L^2 = \det \sigma$ so one can write

$$\mathbb{S}_\sigma \cong \mathbb{S}_0 \otimes (\det \sigma)^{1/2}.$$

(b) Assume M is an almost complex manifold. The bundle of complex spinors associated to the canonical spin^c structure σ (such that $\det \sigma = K_M^{-1}$) is denoted by $\mathbb{S}_{\mathbb{C}}(M)$. Note that

$$\mathbb{S}_{\mathbb{C}}(M) \cong \Lambda^{0,\bullet} T^* M. \qquad \square$$

We will construct a natural family of Dirac structures on the bundle of complex spinors associated to a spin^c structure.

Consider the warm-up case when TM is trivial. Then we can assume $g_{\alpha\beta} \equiv \mathbb{1}$, and

$$h_{\alpha\beta} = (\mathbb{1}, z_{\alpha\beta}) : U_{\alpha\beta} \to \mathrm{Spin}(n) \times S^1 \to \mathrm{Spin}^c(n).$$

The S^1-cocycle $(z_{\alpha\beta}^2)$ defines the complex Hermitian line bundle $\det \sigma$. In this case something more happens.

The collection $(z_{\alpha\beta})$ is also an S^1-cocycle defining a complex Hermitian line bundle $\hat{L}$, such that $\hat{L}^2 \cong \det \sigma$. Traditionally, $\hat{L}$ is denoted by $(\det \sigma)^{1/2}$, though the square root may not be uniquely defined.

We can now regard $\mathbb{S}_\sigma(M)$ as a bundle associated to the trivial $\mathrm{Spin}(n)$-bundle P_{Spin}, and as such, there exists an isomorphism of complex $\mathrm{Spin}(n)$ vector bundles

$$\mathbb{S}_\sigma(M) \cong \mathbb{S}(M) \otimes (\det \sigma)^{1/2}.$$

As in the exercise 11.1.68 of Subsection 11.1.9, we deduce that twisting the canonical connection on the bundle of pure spinors $\mathbb{S}_0(M)$ with any $U(1)$-connection on $\det \sigma^{1/2}$ we obtain a Clifford connection on $\mathbb{S}_\sigma(M)$. Notice that if the collection

$$\left\{ \omega_\alpha \in \underline{u}(1) \otimes \Omega^1(U_\alpha) \right\}$$

defines a connection on $\det \sigma$, i.e.,

$$\omega_\beta = \frac{dz_{\alpha\beta}^2}{z_{\alpha\beta}^2} + \omega_\alpha \ \text{ over } U_{\alpha\beta},$$

then the collection

$$\hat{\omega}_\alpha = \frac{1}{2}\omega_\alpha$$

defines a Hermitian connection on $\hat{L} = \det \sigma^{1/2}$. Moreover, if F denotes the curvature of $(\omega_\cdot)$, then the curvature of $(\hat{\omega}_\cdot)$ is given by

$$\hat{F} = \frac{1}{2}F. \tag{11.2.5}$$

Hence any connection on $\det \sigma$ defines in a unique way a Clifford connection on $\mathbb{S}_\sigma(M)$.

Assume now that TM is not necessarily trivial. We can however cover M by open sets (U_α) such that each TU_α is trivial. If we pick from the start a Hermitian connection on $\det \sigma$, this induces a Clifford connection on each $\mathbb{S}_\sigma(U_\alpha)$. These can be glued back to a Clifford connection on $\mathbb{S}_\sigma(M)$ using partitions of unity. We let the reader check that the connection obtained in this way is independent of the various choices.

Here is an equivalent way of associating a Clifford connection on $\mathbb{S}_\sigma$ to any Hermitian connection on $\det \sigma$. Fix an open cover (U_α) of M consisting of geodesically convex open sets. The restriction of TM to any U_α is trivializable. Fixing such (orthogonal) trivializations leading to the gluing cocycle

$$g_{\beta\alpha} : U_{\alpha\beta} \to SO(n), \ \ n := \dim M.$$

We can describe the Levi-Civita connection on TM as a collection

$$A_\alpha \in \Omega^1(U_\alpha) \otimes \underline{\mathrm{so}}(n),$$

satisfying the transition rules

$$A_\beta = -(dg_{\beta\alpha})g_{\beta\alpha}^{-1} + g_{\beta\alpha}A_\alpha g_{\beta\alpha}^{-1}.$$

The spin^c structure σ is described by a gluing cocycle

$$h_{\beta\alpha} : U_{\alpha\beta} \to \mathrm{Spin}^c(n), \quad h_{\beta\alpha} = [\hat{g}_{\beta\alpha}, z_{\beta\alpha}] \in \mathrm{Spin}(n) \times U(1)/\{\pm 1\},$$

such that, if $\rho : \mathrm{Spin}(n) \to SO(n)$ denotes the natural $2 : 1$ morphism, then

$$g_{\beta\alpha} = \rho(\hat{g}_{\beta\alpha}).$$

Define $\hat{A}_\alpha \in \Omega^1(U_\alpha) \otimes \underline{\mathrm{spin}}(n)$ by setting

$$\hat{A}_\alpha(x) = \rho_*^{-1}\big(A_\alpha(x)\big),$$

where $\rho_* : \underline{\mathrm{spin}}(n) \to \underline{\mathrm{so}}(n)$ is the natural Lie algebra isomorphism described in (11.1.3).

The collection $(\hat{A}_\alpha)$ satisfies the transition rules

$$\hat{A}_\beta = -(d\hat{g}_{\beta\alpha})\hat{g}_{\beta\alpha}^{-1} + \hat{g}_{\beta\alpha}\hat{A}_\alpha\hat{g}_{\beta\alpha}^{-1}.$$

Observe that although $\hat{g}_{\bullet\bullet}$ is only defined up to a ± 1 ambiguity, this ambiguity is lost in the above equality.

Consider now a connection ω on the Hermitian line bundle $\det \sigma$ defined by the gluing cocycle $(z_{\bullet\bullet}^2)$. It is defined by a collection

$$\omega_\alpha \in \Omega^1(U_\alpha) \otimes \underline{\mathrm{u}}(1) = \Omega^1(U_\alpha) \otimes i\mathbb{R},$$

satisfying the transition rules

$$\omega_\beta = -2\frac{dz_{\beta\alpha}}{z_{\beta\alpha}} + \omega_\alpha.$$

The collection

$$A_\alpha^\omega := \hat{A}_\alpha \oplus \frac{1}{2}\omega_\alpha \in \Omega^1(U_\alpha) \otimes \big(\underline{\mathrm{spin}}(n) \oplus \underline{\mathrm{u}}(1)\big) = \Omega^1(U_\alpha) \otimes \underline{\mathrm{spin}}^c(n)$$

defines a connection A^ω on the principal $\mathrm{Spin}^c(n)$-bundle $P_{\mathrm{Spin}^c}^\sigma \to M$ determined by the spin^c-structure σ. In induces a connection ∇^ω on the associated vector bundle $\mathbb{S}_\sigma$, and as in the previous subsection one can verify that ∇^ω is a Clifford connection.

Example 11.2.45. Assume (M, g) is both complex and spin. Then a choice of a spin structure canonically selects a square root $K_M^{-1/2}$ of the line bundle K_M^{-1} because K_M^{-1} is the line bundle associated to the spin^c structure determined by the complex structure on M. Then

$$\mathbb{S}_\mathbb{C} \cong \mathbb{S}_0 \otimes K_M^{-1/2}.$$

Any Hermitian connection on K_M induces a connection of $\mathbb{S}_\mathbb{C}$. If M happens to be Kähler, then the Levi-Civita connection induces a complex Hermitian connection K_M and thus a Clifford connection on $\mathbb{S}_\mathbb{C}(M) \cong \Lambda^{0,\bullet}T^*M$. $\qquad\square$

Let ω be a connection on $\det \sigma$. Denote by ∇^ω the Clifford connection it induces on $\mathbb{S}_\sigma(M)$ and by $\mathfrak{D}_\omega$ the associated geometric Dirac operator. Since the Weitzenböck remainder of this Dirac operator is a *local object* so to determine its form we may as well assume $\mathbb{S}_\sigma = \mathbb{S} \otimes \det \sigma^{1/2}$. Using the computation of Exercise 11.1.68 and the form of the Weitzenböck remainder for the *spin* operator we deduce

$$\mathfrak{D}_{\sigma,\omega}^2 = (\nabla^\omega)^* \nabla^\omega + \frac{1}{4}s + \frac{1}{2}\mathbf{c}(\, F_\sigma(\omega)\,), \tag{11.2.6}$$

where F_σ denotes the curvature of the connection ω on $\det \sigma$, and $c(F_\sigma)$ denotes the Clifford multiplication by the purely imaginary 2-form F_σ.

Exercise 11.2.46. Consider the torus $T^3 = S^1 \times S^1 \times S^1$ equipped with the product round metric

$$g = (d\theta^1)^2 + (d\theta^2)^2 + (d\theta^3)^2,$$

where for $j = 1, 2, 3$, $\theta^j \in [0, 2\pi)$ denotes the angular coordinate on the j-th circle.

Denote by σ_0 the unique spin^c structure on T^3, such that $\det \sigma_0$ is a trivial line bundle. As in the Exercise 11.2.42, for every morphism $\rho : \mathbb{Z}^3 \to S^1$, we denote by L_ρ the associated complex line bundle. We set $\sigma_\rho := \sigma \otimes L_\rho$, and we fix a Hermitian metric on L_ρ. The spin^c structure σ_ρ, and a Hermitian connection ∇ on L_ρ determines spin^c Dirac operator $\mathfrak{D}_\rho(\nabla)$.

(a) Prove that if is a Hermitian a unitary automorphism of L_ρ then

$$\mathfrak{D}_\rho(\gamma \nabla \gamma^{-1}) = \gamma \mathfrak{D}_\rho(\nabla) \gamma^{-1}.$$

In particular, the operators $\mathfrak{D}_\rho(\gamma \nabla \gamma^{-1})$ and $\mathfrak{D}_\rho(\nabla)$ have the same spectrum.

(b)* We say that two Hermitian connections ∇, ∇' on L_ρ are gauge equivalent if and only if there exists a unitary automorphism γ of L such that $\nabla' = \gamma \nabla \gamma^{-1}$. Prove that two connections ∇ and ∇' are gauge equivalent if and only if

$$\frac{1}{2\pi} \int_{T^3} d\theta^j \wedge \big(F(\nabla') - F(\nabla) \big) \in 2\pi i \mathbb{Z}, \quad \forall j = 1, 2, 3.$$

(c) Fix a Hermitian connection ∇ on L_ρ. Describe the spectrum of $\mathfrak{D}_\rho(\nabla)$ in terms of the real numbers

$$s_j = \frac{1}{2\pi i} \int_{T^3} d\theta^j \wedge F(\nabla), \quad j = 1, 2, 3. \qquad \square$$

Bibliography

[1] *Mathematical Developments Arising from Hilbert Problems*, Proceedings of Sym. in Pure Math., vol . **28**, Amer. Math. Soc. Providence RI, 1976.

[2] J. F. Adams: *Vector fields on spheres*, Ann.of Math **75** (1960), 603-632.

[3] R. A. Adams: *Sobolev Spaces*, Academic Press, New York, 1975.

[4] V. I. Arnold: *Ordinary Differential Equations*, Springer Texts, Springer-Verlag, 1992.

[5] V. I. Arnold: *Mathematical Methods of Classical Mechanics*, Springer Texts, Springer Verlag, 1989.

[6] M. F. Atiyah, R. Bott, V. K. Patodi: *On the heat equation and the index theorem*, Invent. Math., **19** (1973), 279-330.

[7] M. F. Atiyah, R. Bott, A. Shapiro: *Clifford modules*, Topology **3**, Suppl. 1 (1964), 3-38.

[8] T. Aubin: *Problèmes isopérimétriques et espaces de Sobolev*, J. Diff. Geom. **11** (1976), 533-598.

[9] T. Aubin: *Équations différentielles non linéaires et problème de Yamabe concernant la courbure scalaire*, J. Math. Pures et Appl., **55** (1976), 269-296.

[10] Bérard: *From vanishing theorems to estimating theorems: the Bochner technique revisited*, Bull. A.M.S. **19** (1988), 371-406.

[11] N. Berline, E. Getzler, M. Vergne: *Heat Kernels and Dirac Aperators*, Springer Verlag, 1992.

[12] L. Bers, F. John, M. Schechter: *Partial Differential Equations*, Lectures in Applied Mathematics, vol. **3A**, Amer. Math. Soc., Providence RI, 1964.

[13] A. Besse: *Einstein Manifolds*, Ergebnisse der Mathematik, Bd.10, Springer Verlag, 1987.

[14] A. Bjorner: *Topological methods*, Hadbook of Combinatorics, R. Graham, M. Grötschel and L. Lovász, (Eds), North-Holland, Amsterdam, 1995, 1819-1872.

[15] J. Bochnak, M. Coste, M.-F. Roy: *Real Algebraic Geometry*, Springer Verlag, 1998.

[16] B. Boos, K. Wojciechowski: *Elliptic Boundary Problems for Dirac Operators*, Birkhauser, 1993.

[17] R. Bott, L. Tu: *Differential Forms in Algebraic Topology*, Springer-Verlag, 1982.

[18] H. Brezis: *Analyse Fonctionelle. Théorie et Applications*, Masson, 1983.

[19] T. Bröcker, K. Jänich: *Introduction To Differential Topology*, Cambridge University Press, 1982.

[20] E. Cartan: *Sur les invariants integraux des certains espaces homogènes et les propriétés topologiques de ces espaces*, Ann. Soc. Pol. Math, **8** (1929), 181-225.

[21] H. Cartan: *La transgression dans un group de Lie et dans un espace fibré principal*,

1268-1282, vol. III of H. Cartan's *"Collected Works"*, Springer-Verlag, 1979.

[22] S. S. Chern: *A simple intrinsic proof of the Gauss-Bonnet formula for closed Riemann manifolds*, Ann. Math, **45** (1946), 747-752.

[23] M. Coste: *An Introduction to o-minimal geometry*, Real Algebraic and Analytic Geometry Network,
`http://www.ihp-raag.org/publications.php`

[24] B. Dacorogna: *Direct Methods in the Calculus of Variations*, Applied Mathematical Sciences, vol. **78**, Springer-Verlag, 1989.

[25] B. A. Dubrovin, A.T. Fomenko, S.P. Novikov: *Modern Geometry - Methods And Applications*, Vol. 1-3, Springer Verlag, 1984, 1985, 1990.

[26] J. Dieudonne: *Foundations of Modern Analysis*, Academic Press, 1963.

[27] M. P. DoCarmo: *Riemannian Geometry*, Birkhäuser, 1992.

[28] L. van der Dries: *Tame Topology and o-minimal Structures*, London Math. Soc. Series, vol. 248, Cambridge University Press, 1998.

[29] L. van der Dries, C. Miller: *Geometric categories and o-minimal structures* Duke Math. J. **84** (1996), 497-540.

[30] R. E. Edwards: *Functional Analysis. Theory and Applications*, Dover Publications, 1995.

[31] C. Ehresmann: *Sur la topologie de certains espaces homogènes*, Ann. of Math., **35** (1934), 396-443.

[32] S. Eilenberg, N. Steenrod: *Foundations of Algebraic Topology*, Princeton University Press, 1952.

[33] H. Federer: *Geometric Measure Theory*, Springer Verlag, 1969.

[34] W. Fulton, J. Harris: *Representation Theory*, Springer-Verlag, 1991.

[35] R. P. Feynman, R. B. Leighton, M. Sands: *The Feynamn Lectures on Physics*, Addison Wesley, 1964.

[36] D. Gilbarg, N. S. Trudinger: *Elliptic Partial Differential Equations of Second Order*, Springer Verlag, 1983.

[37] R. Godement: *Topologie Algébrique et Théorie des Faisceaux*, Hermann et Cie, Paris, 1964.

[38] C. Gordon. D. L. Webb, S. Wolpert: *One cannot hear the shape of a drum*, Bull. A.M.S., **27** (1992), 134-138.

[39] I. S. Gradshteyn, I. M. Ryzhik: *Tables of Integrals, Series and Products*, Academic Press, 1994.

[40] P. Griffiths, J. Harris: *Principles of Algebraic Geometry*, John Wiley&Sons, 1978.

[41] B. Hall: *Lie Groups, Lie Algebras, and Reprsentations. An Elementary Introduction*, Springer Verlag 2003.

[42] G. H. Hardy, E. M. Wright: *An Introduction to the Theory of Numbers*, Oxford University Press, 1962.

[43] P. Hartman: *Ordinary Differential Equations*, 2nd Edition, Classics in Applied Mathematics, SIAM 2002.

[44] S. Helgason: *Differential Geometry, Lie Groups, and Symmetric Spaces*, Graduate Studies in Math. v. 34, Amer. Math. Soc., 2001.

[45] M. Hirsch: *Differential Topology*, Springer-Verlag, 1976.

[46] F. Hirzebruch: *Topological Methods in Algebraic Geometry*, Springer Verlag, New York, 1966.

[47] F. Hirzebruch, H. Hopf: *Felder von Flachenelementen in 4-dimensionalen Manigfaltigkeiten* Math. Ann. **136**(1958).

[48] B. Iversen: *Cohomology of Sheaves*, Springer Verlag, 1986.

[49] H. Iwamoto: *On integral invariants and Betti numbers of symmetric Riemannian*

manifolds, I, J. of the Math. Soc. of Japan, **1**(1949), 91-110.

[50] J. Jost: *Riemannian Goemtry and Geometric Analysis*, Universitext, Springer Verlag, 1998.

[51] M. Kac: *Can one hear the shape of a drum?*, Amer. Math. Monthly, **73**(1966) Part II, 1-23.

[52] M. Karoubi: *K-Theory*, Springer-Verlag, 1978.

[53] M. Kashiwara, P. Schapira: *Sheaves on Manifolds*, Springer Verlag, 1990.

[54] M. Kashiwara, P. Schapira: *Categories and Sheaves*, Springer Verlag, 2006.

[55] T. Kato: *Perturbation Theory for Linear Operators*, Springer Verlag, 1984.

[56] J. Kazdan, F. Warner: *Curvature functions for compact 2-manifolds*, Ann. Math. **99**(1974), 14-47.

[57] J. L. Kelley: *General Topology*, Van Nostrand, 1955.

[58] A. Khovanskii: *Fewnomials*, Transl. Math. Monogr., **88**, Amer. Math. Soc., 1991.

[59] D. A. Klain, G.-C. Rota: *Introduction to Geometric Probability*, Cambridge University Press, 1997.

[60] S. Kobayashi, K. Nomizu: *Foundations of Differential Geometry*, Interscience Publishers, New York, 1963.

[61] K. Kodaira, D. C. Spencer: *Groups of complex line bundles over compact Kähler varieties. Divisor class group on algebraic varieties*, Proc. Nat. Acad. Sci. USA **39** (1953), 868-877.

[62] H. B. Lawson, M.-L. Michelson: *Spin Geometry*, Princeton University Press, 1989.

[63] J. M. Lee, T. H. Parker: *The Yamabe problem*, Bull. A.M.S. **17** (1987), 37-91.

[64] A. Lichnerowicz: *Laplacien sur une variété riemannienne et spineure*, Atti Accad. Naz. dei Lincei, Rendicotti **33** (1962), 187-191.

[65] D. E. Littlewood: *The Theory of Group Characters and Matrix Representations of Groups*, Clarendon Press, Oxford, 1940.

[66] I. G. Macdonald: *Symmetric Functions and Hall Polynomials*, Clarendon Press, Oxford, 1979.

[67] S. MacLane: *Categories for the Working Mathematician*, Springer Verlag, 1971.

[68] W. S. Massey: *A Basic Course in Algebraic Topology*, Springer Verlag, 1991.

[69] W. S. Massey: *Cross products of vectors in higher dimensional euclidian spaces*, Amer. Math. Monthly, **93** (1983), 697-700.

[70] J. Milnor: *On manifolds homeomorphic to the 7-sphere*, Ann. of Math. **64** (1956), 399-405.

[71] J. Milnor: *Morse Theory*, Princeton University Press, 1963.

[72] J. W. Milnor: *Spin structures on manifolds*, L'Enseignement Math., **9** (1963), 198-203.

[73] J. Milnor: *Curvatures of left invariant metrics on Lie groups*, Adv. in Math., **21** (1976), 293-329.

[74] J. Milnor: *Topology from the Differentiable Viewpoint*, Princeton University Press, 1965.

[75] J. Milnor, J. D. Stasheff: *Characteristic Classes*, Ann. Math. Studies 74, Princeton University Press, Princeton, 1974.

[76] M. Mimura, H. Toda: *Topology of Lie Groups, I and II*, Translations of Mathematical Monographs, vol. 91, Amer. Math. Soc., 1991.

[77] R. Miron, M. Anastasiei: *The Geometry of Lagrange Spaces: Theory and Applications*, Kluwer Academic Publishers, 1994.

[78] J. W. Morgan: *The Seiberg-Witten Equations and Applications to the Topology of Smooth Manifolds*, Mathematical Notes, Princeton University Press, 1996.

[79] C. Morrey: *Multiple Integrals in the Calculus of Variations*, Springer Verlag, 1966.

[80] A. Newlander, L. Nirenberg: *Complex analytic coordinates on almost complex manifolds*, Ann. Math., **54** (1954), 391-404.

[81] L. Nirenberg: *On elliptic partial differential equations*, Ann. Scuola Norm. Sup. Pisa, **13**(1959), 1-48.

[82] L. S. Pontryagin: *Smooth manifolds and their applications in homotopy theory*, Amer. Math. Soc. Trans., II (1959), 1-114.

[83] I. R. Porteous: *Topological Geometry*, Cambridge University Press, 1981.

[84] M. M. Postnikov: *Lectures in Geometry. Semester V: Lie Groups and Lie Algebras*, Mir Publishers, Moscow, 1986.

[85] L. Santaló: *Integral Geometry and Geometric Probability*, Cambridge University Press, 2004.

[86] R. Schoen: *Conformal deformations of a Riemannian metric to constant curvature*, J. Diff. Geom. **20** (1984), 479-495.

[87] J. A. Schouten, D. J. Struik: *On some properties of general manifolds relating to Einstein's theory of gravitation*, Amer. J. Math **43** (1921), 213-216.

[88] J. P. Serre: *Faisceaux algébriques cohérentes*, Ann. Math. **61** (1955), 197-278.

[89] I. R. Shafarevitch: *Algebra I. Basic Notions*, Encyclopedia of Mathematical Sciences, vol. **11**, Springer Verlag, 1990.

[90] M. A. Shubin: *Pseudodifferential Operators and Spectral Theory*, Springer Verlag, 2001.

[91] P. Speissegger: *The Pfaffian closure of an o-minimal structure*, J. Reine. Angew. Math., **508** (1999), 189-211.

[92] M. Spivak: *A Comprehensive Introduction to Differential Geometry*, vol. 1-5, Publish or Perish, 1979.

[93] N. E. Steenrod: *The Topology of Fibre bundles*, Princeton University Press, 1951.

[94] E. M. Stein: *Singular Integrals and Differentiability Properties of Functions*, Princeton University Press, 1970.

[95] F. W. Warner: *Foundations of Differentiable Manifolds and Lie Groups*, Springer Verlag, 1983.

[96] H. Weyl: Nachr. Akad. Wiss Göttingen, 1914, 235-236.

[97] H. Weyl: Math. Ann., **77** (1916), 313-315.

[98] H. Weyl: *Die Idee Der Riemannschen Fläche* B. G. Tuebner, Berlin 1913.

[99] H. Weyl: *On the volume of tubes*, Amer. J. Math., **61** (1939), 461-472.

[100] H. Weyl: *The Classical Groups. Their Invariants and Representations*, Princeton University Press, 1939.

[101] E. T. Whittaker, G. N. Watson: *A Course of Modern Analysis*, Cambridge Univ. Press, 1996.

[102] E. Witten: *Monopoles and four-manifolds*, Math.Res. Letters, **1** (1994), 769-796.

[103] K. Yosida: *Functional Analysis*, Springer Verlag, 1974.

Index